Metals

Metalloids

Nonmetals

10 8B	11 1B	12 2B	13 3A	14 4A	15 5A	16 6A	17 7A	18 8A
								2 He 4.00 helium
			5 B 10.81 boron	6 C 12.01 carbon	7 N 14.01 nitrogen	8 O 16.00 oxygen	9 F 19.00 fluorine	10 Ne 20.18 neon
			13 Al 26.98 aluminum	14 Si 28.09 silicon	15 P 30.97 phosphorus	16 S 32.06 sulfur	17 Cl 35.45 chlorine	18 Ar 39.95 argon
28 Ni 58.69 nickel	29 Cu 63.55 copper	30 Zn 65.39 zinc	31 Ga 69.72 gallium	32 Ge 72.63 germanium	33 As 74.92 arsenic	34 Se 78.97 selenium	35 Br 79.90 bromine	36 Kr 83.80 krypton
46 Pd 106.42 palladium	47 Ag 107.87 silver	48 Cd 112.41 cadmium	49 In 114.82 indium	50 Sn 118.71 tin	51 Sb 121.75 antimony	52 Te 127.60 tellurium	53 I 126.90 iodine	54 Xe 131.29 xenon
78 Pt 195.08 platinum	79 Au 196.97 gold	80 Hg 200.59 mercury	81 Tl 204.38 thallium	82 Pb 207.2 lead	83 Bi 208.98 bismuth	84 Po (209) polonium	85 At (210) astatine	86 Rn (222) radon
110 Ds (281) darmstadtium	111 Rg (280) roentgenium	112 Cn (285) copernicium	113 Nh (284) nihonium	114 Fl (289) flerovium	115 Mc (289) moscovium	116 Lv (293) livermorium	117 Ts (294) tennessine	118 Og (294) oganesson

64 Gd 157.25 gadolinium	65 Tb 158.93 terbium	66 Dy 162.50 dysprosium	67 Ho 164.93 holmium	68 Er 167.26 erbium	69 Tm 168.93 thulium	70 Yb 173.04 ytterbium	71 Lu 174.97 lutetium
96 Cm (247) curium	97 Bk (247) berkelium	98 Cf (251) californium	99 Es (252) einsteinium	100 Fm (257) fermium	101 Md (258) mendelevium	102 No (259) nobelium	103 Lr (260) lawrencium

PEARSON ALWAYS LEARNING

Nivaldo J. Tro

Introductory Chemistry

Fourth Custom Edition for Portland Community College
Rock Creek and Sylvania Campuses

Taken from:
Introductory Chemistry, Sixth Edition
by Nivaldo J. Tro

Cover Art: Courtesy of Pearson Learning Solutions.

Taken from:

Introductory Chemistry, Sixth Edition
by Nivaldo J. Tro
Copyright © 2018, 2015 by Pearson Education, Inc.
Publishing as Benjamin Cummings
San Francisco, California 94111

All rights reserved. No part of this book may be reproduced, in any form or by any means, without permission in writing from the publisher.

This special edition published in cooperation with Pearson Learning Solutions.

All trademarks, service marks, registered trademarks, and registered service marks are the property of their respective owners and are used herein for identification purposes only.

Pearson Education, Inc. 330 Hudson Street, New York, New York 10013
A Pearson Education Company
www.pearsoned.com

Printed in the United States of America

1 2 3 4 5 6 7 8 9 10 XXXX 20 19 18 17

000200010272104055

HG

ISBN 10: 1-323-76510-7
ISBN 13: 978-1-323-76510-4

About the Author

Nivaldo Tro is a professor of chemistry at Westmont College in Santa Barbara, California, where he has been a faculty member since 1990. He received his Ph.D. in chemistry from Stanford University for work on developing and using optical techniques to study the adsorption and desorption of molecules to and from surfaces in ultrahigh vacuum. He then went on to the University of California at Berkeley, where he did postdoctoral research on ultrafast reaction dynamics in solution. Since coming to Westmont, Professor Tro has been awarded grants from the American Chemical Society Petroleum Research Fund, from the Research Corporation, and from the National Science Foundation to study the dynamics of various processes occurring in thin adlayer films adsorbed on dielectric surfaces. He has been honored as Westmont's Outstanding Teacher of the Year three times and has also received the college's Outstanding Researcher of the Year award. Professor Tro lives in Santa Barbara with his wife, Ann, and their four children, Michael, Ali, Kyle, and Kaden. In his leisure time, Professor Tro enjoys mountain biking, surfing, and being outdoors with his family.

To Annie

Brief Contents

Contents

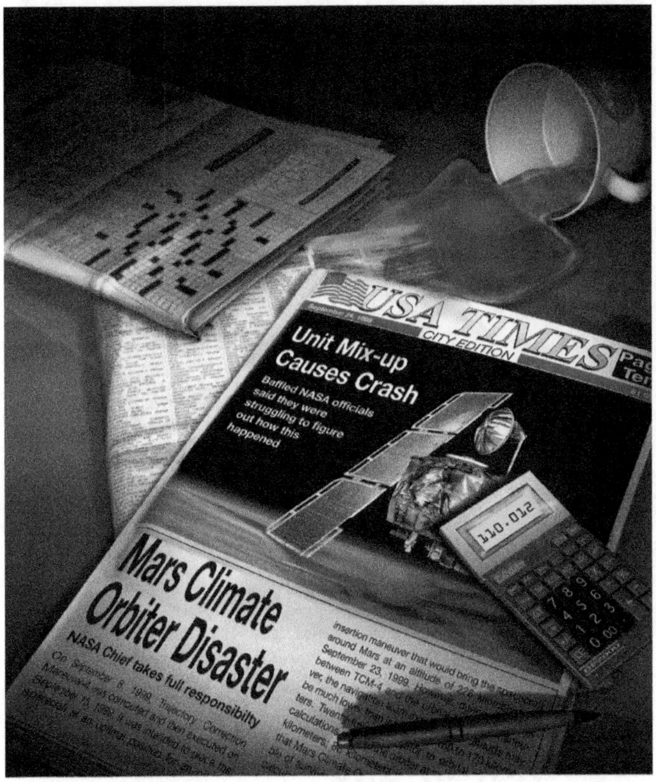

Three-Column Problem Solving Strategies

Interactive Media Contents

Key Concept Videos

The eText 2.0 icon indicates that this feature is embedded and interactive in the eText.

Interactive Worked Examples

To the Student

This book is for *you*, and every text feature is meant to help you learn and succeed in your chemistry course. I wrote this book with two main goals for you in mind: to see chemistry as you never have before and to develop the problem-solving skills you need to succeed in chemistry.

I want you to experience chemistry in a new way. I have written each chapter to show you that chemistry is not just something that happens in a laboratory; chemistry surrounds you at every moment. Several outstanding artists have helped me to develop photographs and art that will help you visualize the molecular world. From the opening example to the closing chapter, you will *see* chemistry. My hope is that when you finish this course, you will think differently about your world because you understand the molecular interactions that underlie everything around you.

My second goal is for you to develop problem-solving skills. No one succeeds in chemistry—or in life, really—without the ability to solve problems. I can't give you a one-size-fits-all formula for problem solving, but I can and do give you strategies that will help you develop the *chemical intuition* you need to understand chemical reasoning.

Look for several recurring features throughout this book designed to help you master problem solving. The most important ones are: (1) a four-step process (Sort, Strategize, Solve, and Check) designed to help you learn how to develop a problem-solving approach; (2) the solution map, a visual aid that helps you navigate your way through problems; (3) two-column Examples, in which the left column explains in clear and simple language the purpose of each step of the solution shown in the right column; and (4) three-column Examples, which describe a problem-solving procedure while demonstrating how it is applied to two different Examples. In addition, the For More Practice feature at the end of each worked Example directs you to the end-of-chapter Problems that provide more opportunity to practice the skill(s) covered in each Example. In addition, Interactive Worked Examples are digital versions of select worked Examples from the text that help you break down problems using the book's "Sort, Strategize, Solve, and Check" technique. Interactive Worked Examples can be found in the eText 2.0 and can be accessed directly at: https://media.pearsoncmg.com/ph/esm/esm_tro_intro_6/media/index.html.

Recent research has demonstrated that you will do better on your exams if you take a multiple-choice pre-exam before your actual exam. At the end of each chapter, you will find a Self-Assessment Quiz to help you check your understanding of the material in that chapter. You can string these together to make a pre-exam. For example, if your exam covers Chapters 5–7, complete the Self-Assessment Quizzes for those chapters as part of your preparation for the exam. The questions you miss on the quiz will reveal the areas you need to spend the most time studying. Studies show that if you do this, you will do better on the actual exam.

Lastly, I hope this book leaves you with the knowledge that chemistry is *not* reserved only for those with some superhuman intelligence level. With the right amount of effort and some clear guidance, anyone can master chemistry, including you.

Sincerely,

Nivaldo J. Tro
tro@westmont.edu

To the Instructor

I thank all of you who have used any of the first five editions of *Introductory Chemistry*—you have made this book the best-selling book in its market, and for that I am extremely grateful. The preparation of the sixth edition has enabled me to continue to refine the book to meet its fundamental purpose: teaching chemical skills in the context of relevance.

Introductory Chemistry is designed for a one-semester, college-level, introductory or preparatory chemistry course. Students taking this course need to develop problem-solving skills—but they also must see *why* these skills are important to them and to their world. *Introductory Chemistry* extends chemistry from the laboratory to the student's world. It motivates students to learn chemistry by demonstrating the role it plays in their daily lives.

This is a visual book. Wherever possible, I use images to help communicate the subject. In developing chemical principles, for example, I worked with several artists to develop multipart images that show the connection between everyday processes visible to the eye and the molecular interactions responsible for those processes. This art has been further refined and improved in the sixth edition, making the visual impact sharper and more targeted to student learning. For example, I have continued to expand and refine a hierarchical system of labeling in many of the images: the white-boxed labels are the most important, the tan boxes are second in importance, and the unboxed labels are the third most important. In many cases, this system allows information to be placed closest to its point of relevance, instead of being lumped together in the caption. In addition, this allows me to treat related labels and annotations within an image in the same way, so that the relationships between them are immediately evident. My intent is to create an art program that teaches and presents complex information clearly and concisely. Many of the illustrations showing molecular depictions of a real-world object or process have three parts: macroscopic (what we can see with our eyes); molecular and atomic (space-filling models that depict what the molecules and atoms are doing); and symbolic (how chemists represent the molecular and atomic world). Students can begin to see the connections between the macroscopic world, the molecular world, and the representation of the molecular world with symbols and formulas.

The problem-solving pedagogy employs four steps as it has done in the previous five editions: Sort, Strategize, Solve, and Check. This four-step procedure guides students as they learn chemical problem solving. Students will also encounter extensive flowcharts throughout the book, allowing them to better visualize the organization of chemical ideas and concepts.

Throughout the worked Examples in this book, I use a *two-* or *three-column* layout in which students learn a general procedure for solving problems of a particular type as they see this procedure applied to one or two worked Examples. In this format, the *explanation* of how to solve a problem is placed directly beside the actual steps in the *solution* of the problem. Many of you have told me that you use a similar technique in lecture and office hours. Since students have specifically asked for connections between worked Examples and end-of-chapter Problems, I include a For More Practice feature at the end of each worked Example that lists the end-of-chapter review Examples and end-of-chapter Problems that provide additional opportunities to practice the skill(s) covered in the example. Also in this edition, we have 39 Interactive Worked Examples, which can be accessed in the eText 2.0.

A successful feature of previous editions is the Conceptual Checkpoints, a series of short questions that students can use to test their mastery of key concepts as they read through a chapter. For this edition, all Conceptual Checkpoints are embedded in eText 2.0. Emphasizing understanding rather than calculation, they are designed to encourage active learning even while reading. Your continued embrace of this feature prompted me to add more of these to the sixth edition.

In this edition, I have also added a new category of End-of-Chapter Questions called *Data Interpretation and Analysis*. These questions present real data in real-life situations and ask students to analyze and interpret that data. They are designed to give students much needed practice in reading graphs, understanding tables, and making data-driven decisions.

In my own teaching, I have been influenced by two studies published in the last few years. The first one is a mega analysis of the effect of active learning on student learning in STEM disciplines.[1] In this study, Freeman and his coworkers convincingly demonstrate that students learn better when they are active in the process. The second study focuses on the effect of multiple-choice pretests on student exam performance.[2] Here, Pyburn and his coworkers show that students who take a multiple-choice pretest do better on exams than those who do not. Even more interesting, the enhancement is greater for lower performing students. In my courses, I have implemented both active learning and multiple-choice pretesting with good results. In my books, I have developed tools to allow you to incorporate these techniques as well.

To help you with active learning, I have added 12 Key Concept Videos to the media package that accompanies this book. These three- to five-minute videos each introduce a key concept from the chapter. They are themselves interactive because every video has an embedded question posed to the student to test understanding. In addition, there are 19 new Interactive Worked Examples adding to a total of 39 new and revised Interactive Worked Example videos in the media package. This means that you now have a library of 31 new interactive videos and a total of 51 new and revised interactive videos to enhance your course.

In my courses, I use these videos in conjunction with the book to implement a *before, during, after* strategy for my students. My goal is simple: *Engage students in active learning before class, during class, and after class.* To that end, I assign a video *before* most class sessions. All Key Concept Videos and Interactive Worked Examples are embedded and interactive in eText 2.0, allowing students to review and test their understanding in real-time. The video introduces students to a concept or problem that I will cover in the lecture. *During* class, I expand on the concept or problem using *Learning Catalytics*™ to question my students. Instead of simply passively listening to a lecture, they are interacting with the concepts through questions that I pose. Sometimes I ask my students to answer individually, other times in pairs or even groups. This approach has changed my classroom. Students engage in the material in new ways. They are actively learning and have to think and process and interact. Finally, *after* class, I give them another assignment, usually a short follow-up question or problem. At this point, they must apply what they have learned to solve a problem.

To help you with multiple-choice pretesting, each chapter contains a Self-Assessment Quiz. Like the Conceptual Checkpoints and the videos, these quizzes are embedded in eText 2.0. These quizzes are designed so that students can test themselves on the core concepts and skills of each chapter. I encourage my students to use these quizzes as they prepare for exams. For example, if my exam covers Chapters 5–8, I assign the quizzes for those chapters for credit (you can do this in MasteringChemistry™). Students then get a pretest on the core material that will be on the exam.

My goal with this new edition is to continue to help you make learning a more active (rather than passive) process for your students. I hope the tools that I have provided here continue to aid you in teaching your students better and more effectively. Please feel free to email me with any questions or comments you might have. I look forward to hearing from you as you use this book in your course.

Sincerely,

Nivaldo J. Tro
tro@westmont.edu

---

[1] Freeman, Scott; Eddy, Sarah L.; McDonough, Miles; Smith, Michelle K.; Okoroafor, Nnadozie; Jordt, Hannah; and Wenderoth, Mary Pat; Active learning increases student performance in science, engineering, and mathematics, 2014, Proc. Natl. Acad. Sci.

[2] Daniel T. Pyburn, Samuel Pazicni, Victor A. Benassi, and Elizabeth M. Tappin *J. Chem. Educ.*, 2014, 91 (12), pp. 2045–2057.

Preface

What's New in This Edition?

The book has been extensively revised and contains more small changes than can be detailed here. The most significant changes to the book and its supplements are listed below.

- I have added a new category of end-of-chapter questions called *Data Interpretation and Analysis*. These questions present real data in real-life situations and ask students to analyze that data. They give students much needed practice in reading graphs, digesting tables, and making data-driven decisions. A new section (Section 1.4), including a new in-chapter worked Example (Example 1.4), introduces these skills.
- There are 12 new Key Concept Videos and 19 new Interactive Worked Examples to accompany the book. That means there are 31 new videos and 51 total new and revised interactive videos to accompany the material in the sixth edition. All Key Concept Videos and Interactive Worked Examples are embedded and interactive in eText 2.0, allowing students to review and test their understanding in real-time. These tools are designed to help professors engage their students in active learning. Recent research has conclusively demonstrated that students learn better when they are active in the learning process, as opposed to passively listening and simply taking in content. The Key Concept Videos are brief (three to five minutes), and each introduces and explains a key concept from a chapter. The student does not just passively listen to the video; the video stops in the middle and poses a question to the student. The student must answer the question before the video continues. Each video also includes a follow-up question that is assignable in MasteringChemistry™. The Interactive Worked Examples are similar in concept, but instead of explaining a key concept, they walk the student through one of the in-chapter worked examples from the book. Like the Key Concept Videos, Interactive Worked Examples stop in the middle and force the student to interact by completing a step in the example. The examples also have a follow-up question that is assignable in MasteringChemistry™. The power of interactivity to make connections in problem solving is immense. I did not quite realize this power until we started making the Interactive Worked Examples and I saw how I could use the animations to make connections that are just not possible on the static page.
- All chapter-ending Self-Assesment Quizzes are embedded in eText 2.0.
- I have added 13 new Conceptual Checkpoint questions throughout the book. For this edition, all Conceptual Checkpoints are embedded in eText 2.0.
- I have updated the data throughout the book to reflect the most recent measurements and developments available. I changed the half-life of carbon-14 to 5715 years in Table 17.2 and throughout Chapter 17 to reflect the current accepted value, and I also added new information about *thermoluminescent dosimeters* (and deleted the information on film badge dosimeters) to Section 17.4. Other updates include changes to Figure 8.2, *Climate change;* Section 10.1, *Bonding Models and AIDS Drugs;* Table 11.5, *Changes in Pollutant Levels for Major U.S. Cities, 1980–2014;* the *Chemistry in the Environment* box in Section 12.8, *Water: A Remarkable Molecule;* and Section 17.8, *Nuclear Power: Using Fission to Generate Electricity.*

- Several chapter-opening sections and (or) the corresponding art, including Sections 1.1, 2.1, 12.1, and 16.1, have been replaced or significantly modified.
- I added a new section (Section 2.8) and new worked example (Example 2.12) as well as new end-of-chapter Problems to address conversions involving quantities with combined units such as mL/kg or km/hr.
- I have extensively modified the art program to move information from the captions and into the art itself. This allows relevant information to be placed right where it is most needed and makes the art a more accessible study and review tool. I have modified 70 figures in this way.
- I have modified end-of-chapter Problems that were showing low levels of student success when assigned in MasteringChemistry™.
- I have added temporary symbols for elements 113, 115, 117, and 118 (Uut, Uup, Uus, and Uuo, respectively) to all periodic tables.
- In all chapters, chapter text was edited for clarity and to limit use of passive voice and extraneous words and phrases.

Teaching Principles

The development of basic chemical principles—such as those of atomic structure, chemical bonding, chemical reactions, and the gas laws—is one of the main goals of this text. Students must acquire a firm grasp of these principles in order to succeed in the general chemistry sequence or the chemistry courses that support the allied health curriculum. To that end, the book integrates qualitative and quantitative material and proceeds from concrete concepts to more abstract ones.

Organization of the Text

The main divergence in topic ordering among instructors teaching introductory and preparatory chemistry courses is the placement of electronic structure and chemical bonding. Should these topics come early, at the point where models for the atom are being discussed? Or should they come later, after the student has been exposed to chemical compounds and chemical reactions? Early placement gives students a theoretical framework within which they can understand compounds and reactions. However, it also presents students with abstract models before they understand why they are necessary. I have chosen a later placement; nonetheless, I know that every course is unique and that each instructor chooses to cover topics in his or her own way. Consequently, I have written each chapter for maximum flexibility in topic ordering. In addition, the book is offered in two formats. The full version, *Introductory Chemistry*, contains 19 chapters, including organic chemistry and biochemistry. The shorter version, *Introductory Chemistry Essentials*, contains 17 chapters and omits these topics.

Print and Media Resources

Instructor and Student Supplements	
0134564057 / 9780134564050	Instructor Resource Manual with Complete Solutions
013455342X / 9780134553429	TestGen Test Bank (Download only)
0134553446 / 9780134553443	Instructor Resource Materials (Download only)
	Instructor's Guide (Download only) for Student's Guided Activity Workbook
0134553411 / 9780134553412	Study Guide
0134564065 / 9780134564067	Student Selected Solutions Manual
0134555554 / 9780134555553	Modified MasteringChemistry™ with Pearson eText–Instant Access
0134555570 / 9780134555577	MasteringChemistry™ with Pearson eText–Instant Access

Acknowledgments

This book has been a group effort, and I am grateful for all of those who helped me. First and foremost, I would like to thank my editor Scott Dustan. I have known Scott for many years and in various roles, and am grateful to have him as my editor. I appreciate his straightforward style, constant support, and commitment to my work. I am also in a continual state of awe and gratitude to Erin Mulligan, my development editor and friend. Thanks, Erin, for all your outstanding help and advice. Thanks also to Jackie Jakob, media editor extraordinaire. Jackie is the force behind the media elements that accompany this book, and I am grateful for her vision, guidance, and friendship. Thanks also to Jennifer Hart, with whom I have now worked for over a decade. Thanks Jennifer for your constant attention, guidance, and wisdom on all of my projects at Pearson. I am also grateful for Jeanne Zalesky, Adam Jaworski, Paul Corey and the rest of Pearson leadership. You have supported my projects and my vision from the beginning, and I am privileged to work with you.

I would also like to thank Elizabeth Ellsworth, my marketing manager, whose creativity in describing and promoting the book is without equal. I am also grateful to Coleen Morrison, whose help with editing and manuscript preparation was invaluable. Thanks also to the MasteringChemistry™ team who continue to provide and promote the best online homework system on the planet. I also appreciate the expertise and professionalism of my copy editor, Betty Pessagno, as well as the skill and diligence of Francesca Monaco and her colleagues at codeMantra. I am a picky author, and they always accommodated my seemingly endless requests. Thank you, Francesca. Thanks as well to my content producer, Chandrika Madhavan and the rest of the Pearson editorial and production team—they are part of a first-class operation. This text has benefited immeasurably from their talents and hard work. I owe a special debt of gratitude to Quade Paul, who continues to make my ideas come alive in his chapter-opener and cover art.

I am grateful for the assistance of my colleagues, Allan Nishimura, David Marten, Stephen Contakes, Kristi Lazar, Carrie Hill, Michael Everest, Amanda Silberstein, and Heidi Henes-Vanbergen, who have supported me in my department while I worked on this book. I owe a special debt of gratitude to Michael Tro. He has been helping me with manuscript preparation, proofreading, organizing art manuscripts, and tracking changes in end-of-chapter material for the past six years. Michael has been reliable, accurate, and invaluable. Thanks Mikee! I also owe a special thanks to my colleagues Michael Everest and Tom Greenbowe, who collaborated with me in creating some of the end of chapter questions.

I am grateful to those who have given so much to me personally while writing this book. First on that list is my wife, Ann. Her patience and love for me are beyond description. I also thank my children, Michael, Ali, Kyle, and Kaden, whose smiling faces and love of life always inspire me. I come from a large Cuban family, whose closeness and support most people would envy. Thanks to my parents, Nivaldo and Sara; my siblings, Sarita, Mary, and Jorge; my siblings-in-law, Jeff, Nachy, Karen, and John; my nephews and nieces, Germain, Danny, Lisette, Sara, and Kenny. These are the people with whom I celebrate life.

Lastly, I am indebted to the many reviewers, listed next, whose ideas are found throughout this book. They have corrected me, inspired me, and sharpened my thinking on how best to teach this subject we call chemistry. I deeply appreciate their commitment to this project.

Reviewers of the 6th Edition

Premilla Arasasingham
El Camino College

Crystal Bendenbaugh
Southeastern University

Charles Carraher
Florida Atlantic University

Cassidy Dobson
St. Cloud University

David Futoma
Roger Williams University

Galen George
Santa Rosa Junior College

Marcia Gillette
Indiana University Kokomo

Ganna Lyubartseva
Southern Arkansas University

Helen Motokane
El Camino College

David Rodgers
North Central Michigan College

Mu Zheng
Tennessee State University

6th Edition Accuracy Reviewers

Kelly Befus
Anoka-Ramsey Community College

Stevenson Flemer
University of Vermont

Lance Lund
Anoka-Ramsey Community College

Tanea Reed
Eastern Kentucky University

Jennifer Zabzydar
Palomar College

Reviewers of the 5th Edition

Scott Bunge
Kent State University

Ebru Buyuktanir
Stark State College

Claire Cohen
University of Toledo

Robert Culp
California State University — Fresno

Rosa Davila
College of Southern Idaho

Alyse Dilts
Harrisburg Area Community College

Sylvia Esjornson
Southwestern Oklahoma State University

Jennifer Firestine
Lindenwood University

Kathy Flynn
College of the Canyons

Sara Harvey
Los Angeles Pierce College

Michael Hauser
St. Louis Community College — Meramec

Edward Lee
Texas Tech University

Craig McClure
University of Alabama — Birmingham

Virginia Miller
Montgomery College

Michael Rodgers
Southeast Missouri State University

Janice Webster
Ivy Tech Community College —Terre Haute

James Zubricky
University of Toledo

5th Edition Accuracy Reviewers

Alyse Dilts
Harrisburg Area Community College

Stevenson Flemer Jr.
University of Vermont

Connie Lee
Montgomery County Community College

Lance Lund
Anoka-Ramsey Community College

Kent McCorkle
Fresno City College

Reviewers of the 4th Edition

Jeffrey Allison
Austin Community College

Mikhail V. Barybin
The University of Kansas

Lara Baxley
California Polytechnic State University

Kelly Befus
Anoka-Ramsey Community College

Joseph Bergman
Illinois Central College

Simon Bott
University of Houston

Carmela Byrnes
MiraCosta College

Carmela Magliocchi Brynes
MiraCosta College

Guy Dadson
Fullerton College

Maria Cecilia D. de Mesa
Baylor University

Brian G. Dixon
Massachusetts Maritime Academy

Timothy Dudley
Villanova University

Jeannine Eddleton
Virginia Tech

Ron Erickson
University of Iowa

Donna Friedman
St. Louis Community College—Florissant Valley

Luther D. Giddings
Salt Lake Community College

Marcus Giotto
Quinsigamond Community College

Melodie Graber
Oakton Community College

Maru Grant
Ohlone College

Jerod Gross
Roanoke Benson High School

Tammy S. Gummersheimer
Schenectady County Community College

Tamara E. Hanna
Texas Tech University

Michael A. Hauser
St. Louis Community College

Bruce E. Hodson
Baylor University

Donald R. Jones
Lincoln Land Community College
Martha R. Kellner
Westminster College
Farkhondeh Khalili
Massachusetts Bay Community College
Margaret Kiminsky
Monroe Community College
Rebecca Krystyniak
Saint Cloud State
Chuck Laland
Black Hawk College
Richard Lavallee
Santa Monica College
Laurie Leblanc
Cuyamaca College
Nancy Lee
MiraCosta College
Vicki MacMurdo
Anoka-Ramsey Community College
Jack F. McKenna
St. Cloud State University
Virginia Miller
Montgomery College
Geoff Mitchell
Washington International School
Meg Osterby
Western Technical College
John Petty
University of South Alabama
Jason Serin
Glendale Community College
Steven Socol
McHenry Community College
Youngju Sohn
Florida Institute of Technology
Jie Song
University of Michigan—Flint
Clarissa Sorenson-Unruh
Central New Mexico Community College
David Vanderlinden
Des Moines Area Community College
Vidyullata C. Waghulde
St. Louis Community—Meramec

Reviewers of the 3rd Edition

Benjamin Arrowood
Ohio University
Joe Bergman
Illinois Central College
Timothy Dudley
Villanova University
Sharlene J. Dzugan
University of Cumberlands
Thomas Dzugan
University of Cumberlands
Donna G. Friedman
St. Louis Community College
Erick Fuoco
Daley College
Melodie A. Graber
Oakton Community College
Michael A. Hauser
St. Louis Community College, Meramec Campus
Martha R. Joseph
Westminster College
Timothy Kreider
University of Medicine & Dentistry of New Jersey
Laurie Leblanc
Grossmont College
Carol A. Martinez
Central New Mexico Community College
Kresimir Rupnik
Louisiana State University
Kathleen Thrush Shaginaw
Particular Solutions, Inc.
Pong (David) Shieh
Wharton College
Mary Sohn
Florida Tech
Kurt Allen Teets
Okaloosa-Walton College
John Thurston
University of Iowa
Anthony P. Toste
Missouri State University
Carrie Woodcock
Eastern Michigan University

Reviewers of the 2nd Edition

David S. Ballantine, Jr.
Northern Illinois University
Colin Bateman
Brevard Community College
Michele Berkey
San Juan College
Steven R. Boone
Central Missouri State University
Morris Bramlett
University of Arkansas—Monticello
Bryan E. Breyfogle
Southwest Missouri State University
Frank Carey
Wharton County Junior College
Robbey C. Culp
Fresno City College
Michelle Driessen
University of Minnesota—Minneapolis
Donna G. Friedman
St. Louis Community College—Florissant Valley
Crystal Gambino
Manatee Community College
Steve Gunther
Albuquerque Technical Vocational Institute
Michael Hauser
St. Louis Community College—Meramec
Newton P. Hillard, Jr.
Eastern New Mexico University
Carl A. Hoeger
University of California—San Diego
Donna K. Howell
Angelo State University
Nichole Jackson
Odessa College
T. G. Jackson
University of South Alabama

Donald R. Jones
Lincoln Land Community College

Kirk Kawagoe
Fresno City College

Roy Kennedy
Massachusetts Bay Community College

Blake Key
Northwestern Michigan College

Rebecca A. Krystyniak
St. Cloud State University

Laurie LeBlanc
Cuyamaca College

Ronald C. Marks
Warner Southern College

Carol A. Martinez
Albuquerque Technical Vocational Institute

Charles Michael McCallum
University of the Pacific

Robin McCann
Shippensburg University

Victor Ryzhov
Northern Illinois University

Theodore Sakano
Rockland Community College

Deborah G. Simon
Santa Fe Community College

Mary Sohn
Florida Institute of Technology

Peter-John Stanskas
San Bernardino Valley College

James G. Tarter
College of Southern Idaho

Ruth M. Topich
Virginia Commonwealth University

Eric L. Trump
Emporia State University

Mary Urban
College of Lake County

Richard Watt
University of New Mexico

Lynne Zeman
Kirkwood Community College

Reviewers of the 1st Edition

Lori Allen
University of Wisconsin—Parkside

Laura Andersson
Big Bend Community College

Danny R. Bedgood
Arizona State University

Christine V. Bilicki
Pasadena City College

Warren Bosch
Elgin Community College

Bryan E. Breyfogle
Southwest Missouri State University

Carl J. Carrano
Southwest Texas State University

Donald C. Davis
College of Lake County

Donna G. Friedman
St. Louis Community College at Florissant Valley

Leslie Wo-Mei Fung
Loyola University of Chicago

Dwayne Gergens
San Diego Mesa College

George Goth
Skyline College

Jan Gryko
Jacksonville State University

Roy Kennedy
Massachusetts Bay Community College

C. Michael McCallum
University of the Pacific

Kathy Mitchell
St. Petersburg Junior College

Bill Nickels
Schoolcraft College

Bob Perkins
Kwantlen University College

Mark Porter
Texas Tech University

Caryn Prudenté
University of Southern Maine

Rill Ann Reuter
Winona State University

Connie M. Roberts
Henderson State University

Jeffery A. Schneider
SUNY—Oswego

Kim D. Summerhays
University of San Francisco

Ronald H. Takata
Honolulu Community College

Calvin D. Tormanen
Central Michigan University

Eric L. Trump
Emporia State University

Help students develop 21st-century skills to succeed in chemistry courses, future careers, and beyond.

Nivaldo Tro's approach introduces students to 21st-century skills, encouraging them to think critically when they encounter complex information and real-world problems.

NEW! Data Interpretation and Analysis Questions at the end of each chapter allow students to work with real data to develop 21st-century problem-solving skills. These questions ask students to sort, analyze and interpret actual data from real-life situations. Students practice reading graphs, digesting tables, and making data-driven decisions.

Data Interpretation and Analysis

124. The graph at right shows U.S. energy consumption by source from 1980 to 2040 (based on projections). The consumption is measured in quadrillion BTUs or quads (1 quad = 1.055×10^{18} J).
    (a) What were the three largest sources of U.S. energy in 2013 in descending order? What total percent of U.S. energy do these three sources provide?
    (b) What percent of total U.S. energy is provided by renewables in 2013?
    (c) Which two sources of U.S. energy decline as a percentage of total energy use between 1989 and 2040 (based on projections)?
    (d) How much U.S. energy (in joules) was produced by nuclear power in 1990?

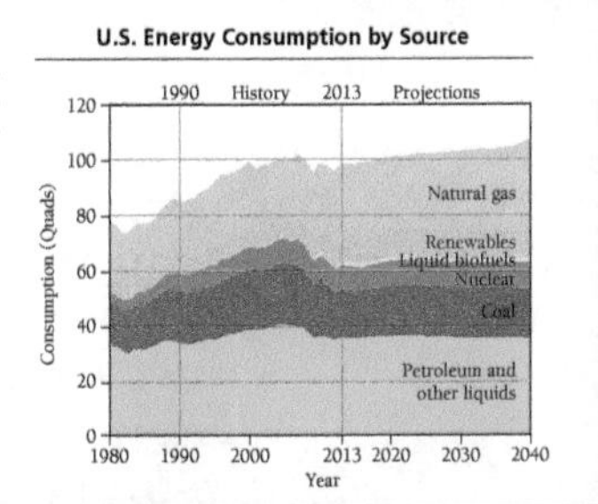

A new section (Section 1.4), which includes a new in-chapter worked example (Example 1.1), introduces data interpretation and analysis skills and emphasizes their importance in student success.

All Data Interpretation and Analysis Questions are assignable in MasteringChemistry™

1.4 Analyzing and Interpreting Data

▶ Identify patterns in data and interpret graphs.

We just learned how early scientists such as Lavoisier and Dalton saw patterns in a series of related measurements. Sets of measurements constitute scientific *data*, and learning to analyze and interpret data is an important scientific skill.

Identifying Patterns in Data

Suppose you are an early chemist trying to understand the composition of water. You know that water is composed of the elements hydrogen and oxygen. You do several experiments in which you decompose different samples of water into hydrogen and oxygen, and you get the following results:

Sample	Mass of Water Sample	Mass of Hydrogen Formed	Mass of Oxygen Formed
A	20.0 g	2.2 g	17.8 g
B	50.0 g	5.6 g	44.4 g
C	100.0 g	11.1 g	88.9 g

Do you notice any patterns in this data? The first and easiest pattern to see is that the sum of the masses of oxygen and hydrogen always sums to the mass of the water sample. For example, for the first water sample, 2.2 g hydrogen + 17.8 g oxygen = 20.0 g water. The same is true for the other samples. Another pattern, which is a bit more difficult to see, is that the ratio of the masses of oxygen and hydrogen is the same for each sample.

Sample	Mass of Hydrogen Formed	Mass of Oxygen Formed	Mass Oxygen / Mass Hydrogen
A	2.2 g	17.8 g	8.1
B	5.6 g	44.4 g	7.9
C	11.1 g	88.9 g	8.01

The ratio is 8—the small variations are due to experimental error, which is common in all measurements and observations.

1.4 Analyzing and Interpreting Data | 9

Atmospheric Carbon Dioxide

▲ **FIGURE 1.5 Atmospheric carbon dioxide levels from 1860 to present.**

Seeing patterns in data is a creative process that requires you to not just merely tabulate laboratory measurements, but to see relationships that may not always be obvious. The best scientists see patterns that others have missed. As you learn to interpret data in this course, be creative and try looking at data in new ways.

Interpreting Graphs

Data is often visualized using graphs or images, and scientists must constantly analyze and interpret graphs. For example, the graph in ◀ FIGURE 1.5 shows the concentration of carbon dioxide in Earth's atmosphere as a function of time. Carbon dioxide is a greenhouse gas that has been rising as result of the burning of fossil fuels (such as gasoline and coal). When you look at a graph such as this one, you should first examine the x and y axes and make sure you understand what each axis represents. You should also examine the numerical range of the axes. In Figure 1.5, the y axis does not begin at zero in order to better display the change that is occurring. How would this graph look different if the y axis began at zero instead of at 290? Notice also that, in this graph, the increase in carbon dioxide has not been constant over time. The rate of increase—represented by the slope of the line—has intensified since about 1960.

EXAMPLE 1.1 Interpreting Graphs

Examine the graph in Figure 1.5 and answer each question.

(a) What was the concentration of carbon dioxide in 1960?
(b) What was the concentration in 2000?
(c) How much did the concentration increase between 1960 and 2000?
(d) What is the average rate of increase over this time?
(e) If the average rate of increase remains constant, estimate the carbon dioxide concentration in 2030.

SOLUTION

a) What was the concentration of carbon dioxide in 1960?

To determine the concentration of carbon dioxide in 1960, draw a vertical line at the year 1960. At the point where the vertical line intersects the carbon dioxide concentration curve, draw a horizontal line. The point where the horizontal line intercepts the y axis represents the concentration in 1960. So, the concentration in 1960 was 318 ppm.

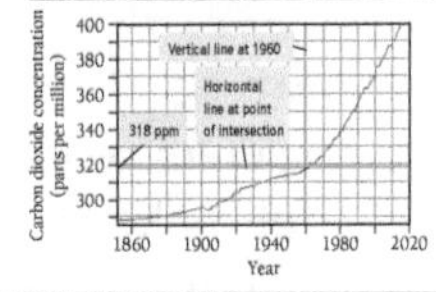

b) What was the concentration in 2000?

Apply the same procedure as in part a, but now shift the vertical line to the year 2000. The concentration in the year 2000 was 370 ppm.

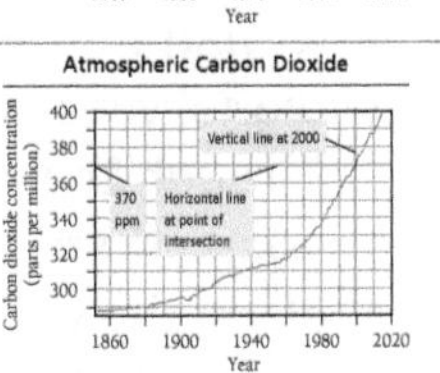

continued on page 10 ▶

Students build a framework for solving problems.

Nivaldo Tro's unique problem-solving technique, "Sort, Strategize, Solve, and Check," teaches students how to successfully approach, set up, and solve the problems they encounter in their introductory chemistry course. Solution maps visually walk students through problems and help them learn how to organize and use given information to successfully solve problems.

Two- and three-column example formats help students break down the steps of each problem and learn and practice problem-solving techniques they can apply in other assignments.

EXAMPLE **3.5** **Conversion of Energy Units**

A candy bar contains 225 Cal of nutritional energy. How many joules does it contain?	
SORT Begin by sorting the information in the problem. Here you are *given* energy in Calories and asked to *find* energy in joules.	**GIVEN:** 225 Cal **FIND:** J
STRATEGIZE Draw a solution map. Begin with Cal, convert to cal, and then convert to J.	**SOLUTION MAP** Cal → cal → J $\frac{1000\ \text{cal}}{1\ \text{Cal}}$ $\frac{4.184\ \text{J}}{1\ \text{cal}}$ **RELATIONSHIPS USED** 1000 calories = 1 Cal (Table 3.2) 4.184 J = 1 cal (Table 3.2)
SOLVE Follow the solution map to solve the problem. Begin with 225 Cal and multiply by the appropriate conversion factors to arrive at J. Round the answer to the correct number of significant figures (in this case, three because of the three significant figures in 225 Cal).	**SOLUTION** $225\ \text{Cal} \times \frac{1000\ \text{cal}}{1\ \text{Cal}} \times \frac{4.184\ \text{J}}{1\ \text{cal}} = 9.41 \times 10^5\ \text{J}$
CHECK Check your answer. Are the units correct? Does the answer make physical sense?	The units of the answer (J) are the desired units. The magnitude of the answer makes sense because the J is a smaller unit than the Cal; therefore, the quantity of energy in J should be greater than the quantity in Cal.

▶ **SKILLBUILDER 3.5 | Conversion of Energy Units**
The complete combustion of a small wooden match produces approximately 512 cal of heat. How many kilojoules are produced?

▶ **SKILLBUILDER PLUS** Convert 2.75×10^4 kJ to calories.

▶ **FOR MORE PRACTICE** Example 3.16; Problems 51, 52, 53, 54, 55, 56, 57, 58.

Problem-Solving Procedure	EXAMPLE **2.8** **UNIT CONVERSION** Convert 7.8 km to miles.	EXAMPLE **2.9** **UNIT CONVERSION** Convert 0.825 m to millimeters.
SORT Begin by sorting the information in the problem into *given* and *find*.	**GIVEN:** 7.8 km **FIND:** mi	**GIVEN:** 0.825 m **FIND:** mm
STRATEGIZE Draw a *solution map* for the problem. Begin with the *given* quantity and symbolize each step with an arrow. Below the arrow, write the conversion factor for that step. The solution map ends at the *find* quantity. (In these examples, the relationships used in the conversions are below the solution map.)	**SOLUTION MAP** km → mi $\frac{0.6214\ \text{mi}}{1\ \text{km}}$ **RELATIONSHIPS USED** 1 km = 0.6214 mi (This conversion factor is from Table 2.3.)	**SOLUTION MAP** m → mm $\frac{1\ \text{mm}}{10^{-3}\ \text{m}}$ **RELATIONSHIPS USED** $1\ \text{mm} = 10^{-3}\ \text{m}$ (This conversion factor is from Table 2.2.)
SOLVE Follow the *solution map* to solve the problem. Begin with the *given* quantity and its units. Multiply by the appropriate conversion factor, canceling units to arrive at the *find* quantity.	**SOLUTION** $7.8\ \text{km} \times \frac{0.6214\ \text{mi}}{1\ \text{km}} = 4.84692\ \text{mi}$ 4.84692 mi = 4.8 mi	**SOLUTION** $0.825\ \text{m} \times \frac{1\ \text{mm}}{10^{-3}\ \text{m}} = 825\ \text{mm}$ 825 mm = 825 mm
Round the answer to the correct number of significant figures. (If possible, obtain conversion factors to enough significant figures so that they do not limit the number of significant figures in the answer.)	Round the answer to two significant figures because the quantity given has two significant figures.	Leave the answer with three significant figures because the quantity given has three significant figures and the conversion factor is a definition and therefore does not limit the number of significant figures in the answer.
CHECK Check your answer. Are the units correct? Does the answer make sense?	The units, mi, are correct. The magnitude of the answer is reasonable. A mile is longer than a kilometer, so the value in miles should be smaller than the value in kilometers.	The units, mm, are correct, and the magnitude is reasonable. A millimeter is shorter than a meter, so the value in millimeters should be larger than the value in meters.
	▶ **SKILLBUILDER 2.8 | Unit Conversion** Convert 56.0 cm to inches.	▶ **SKILLBUILDER 2.9 | Unit Conversion** Convert 5678 m to kilometers.
	▶ **FOR MORE PRACTICE** Example 2.26; Problems 73, 74, 75, 76.	▶ **FOR MORE PRACTICE** Problems 69, 70, 71, 72.

NEW! and UPDATED! Interactive Worked Examples are digital versions of select worked examples from the text that make Nivaldo Tro's unique problem-solving strategies interactive. In these digital versions the author walks students through the problem-solving process, asking them to pause and answer questions along the way. Worked example videos are embedded in eText 2.0 and assignable in MasteringChemistry™.

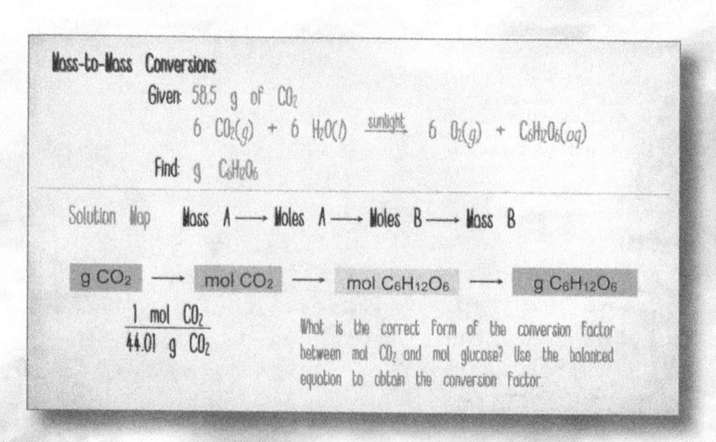

Students learn to think critically about information in the classroom and in everyday life.

NEW! Key Concept Videos combine artwork from the textbook with 2D and 3D animations to create a dynamic on-screen viewing experience and help students understand and apply important concepts throughout the text. Key Concept Videos are embedded in eText 2.0 and are assignable in MasteringChemistry™.

These short (3–5 minutes) videos combine animation and live-action clips of author Nivaldo Tro explaining the key concept of each chapter. Embedded questions in each video increase engagement and test student understanding. Follow-up questions are assignable in MasteringChemistry™.

UPDATED! Chapter-in-Review Exercises and Self-Assessment Quizzes have been revised using MasteringChemistry™ metadata to identify questions that students struggled with in previous editions. In addition to a full complement of end-of-chapter questions, each chapter features a 10–15 multiple-choice question quiz that help students assess their understanding of chapter content, building critical thinking skills and reinforcing key concepts.

Chapter 3 in Review

MasteringChemistry™ provides end-of-chapter exercises, feedback-enriched tutorial problems, animations, and interactive activities to encourage problem solving practice and deeper understanding of key concepts and topics.

Self-Assessment Quiz

Q1. Which substance is a pure compound?
(a) Gold (b) Water
(c) Milk (d) Fruit cake

Q2. Which property of trinitrotoluene (TNT) is most likely a chemical property?
(a) Yellow color
(b) Melting point is 80.1 °C
(c) Explosive
(d) None of the above

Q3. Which change is a chemical change?
(a) The condensation of dew on a cold night
(b) A forest fire
(c) The smoothening of rocks by ocean waves
(d) None of the above

Q4. Which process is endothermic?
(a) The burning of natural gas in a stove
(b) The metabolism of glucose by your body
(c) The melting of ice in a soft drink
(d) None of the above

Q5. A 35-g sample of potassium completely reacts with chlorine to form 67 g of potassium chloride. How many grams of chlorine must have reacted?
(a) 67 g (b) 35 g (c) 32 g (d) 12 g

Q6. A runner burns 2.56×10^3 kJ during a five-mile run. How many nutritional Calories did the runner burn?
(a) 1.07×10^1 Cal (b) 612 Cal
(c) 6.12×10^5 Cal (d) 1.07×10^4 Cal

Q7. Convert the boiling point of water (100.00 °C) to K.
(a) −173.15 K
(b) 0 K
(c) 100.00 K
(d) 373.15 K

Q8. A European doctor reports that you have a fever of 39.2 °C. What is your fever in degrees Fahrenheit?
(a) 102.6 °F (b) 128.26 °F
(c) 71.2 °F (d) 4 °F

Q9. How much heat must be absorbed by 125 g of ethanol to change its temperature from 21.5 °C to 34.8 °C?
(a) 6.95 kJ
(b) 4.02×10^3 kJ
(c) 86.6 kJ
(d) 4.02 kJ

Q10. Substance A has a heat capacity that is much greater than that of substance B. If 10.0 g of substance A initially at 25.0 °C is brought into thermal contact with 10.0 g of B initially at 75.0 °C, what can you conclude about the final temperature of the two substances once the exchange of heat between the substances is complete?
(a) The final temperature will be between 25.0 °C and 50.0 °C.
(b) The final temperature will be between 50.0 °C and 75.0 °C.
(c) The final temperature will be 50.0 °C.
(d) You can conclude nothing about the final temperature without more information.

Answers: 1:b, 2:c, 3:b, 4:c, 5:c, 6:b, 7:d, 8:a, 9:d, 10:a

Multipart macroscopic and molecular images engage students in chemistry.

Multipart images allow students to see the relationship between the formulas they write down on paper (symbolic), the world they see around them (macroscopic), and the atoms and molecules that compose the world (molecular).

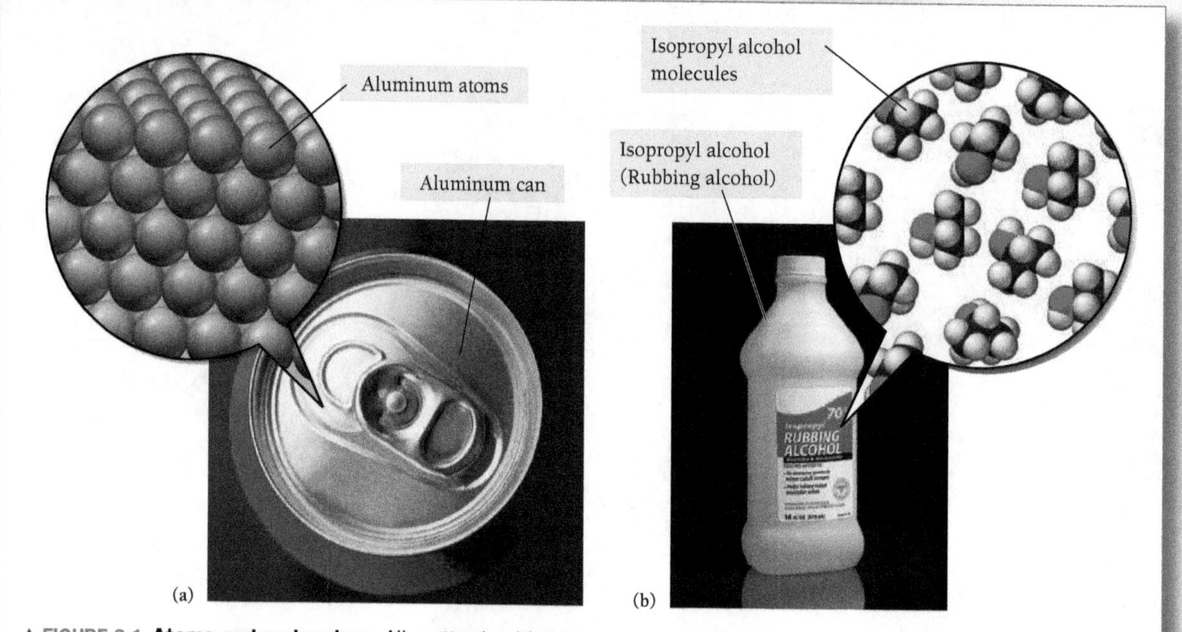

▲ FIGURE 3.1 **Atoms and molecules** All matter is ultimately composed of atoms. **(a)** In some substances, such as aluminum, the atoms exist as independent particles. **(b)** In other substances, such as rubbing alcohol, several atoms bond together in well-defined structures called molecules.

Abundant molecular-level views show students the connection between everyday processes that are visible to the eye and the behavior of atoms and molecules.

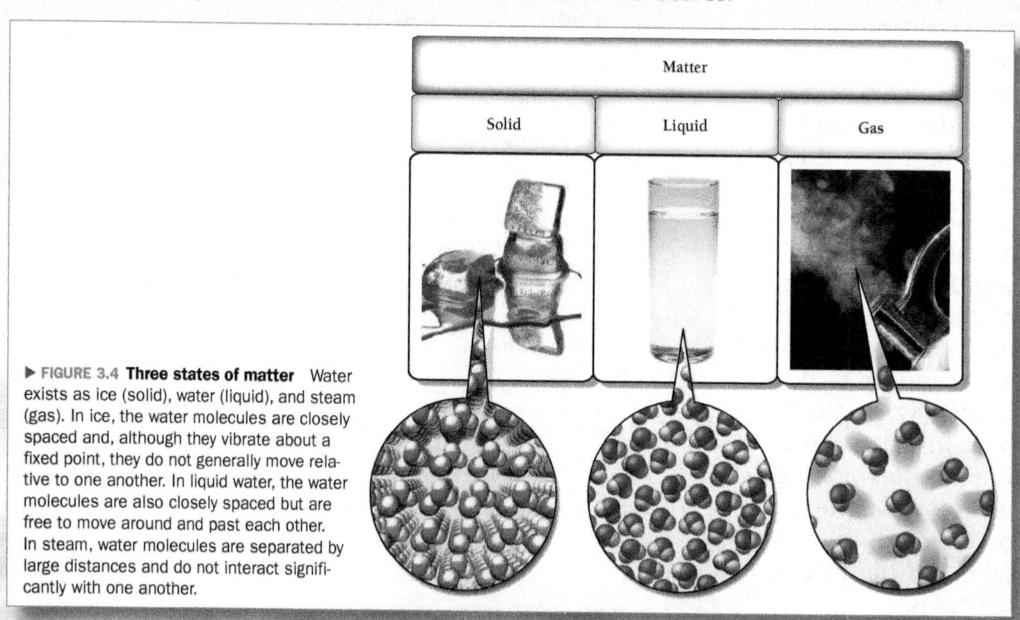

▶ FIGURE 3.4 **Three states of matter** Water exists as ice (solid), water (liquid), and steam (gas). In ice, the water molecules are closely spaced and, although they vibrate about a fixed point, they do not generally move relative to one another. In liquid water, the water molecules are also closely spaced but are free to move around and past each other. In steam, water molecules are separated by large distances and do not interact significantly with one another.

A revised art program helps students make connections and see that chemistry is all around them.

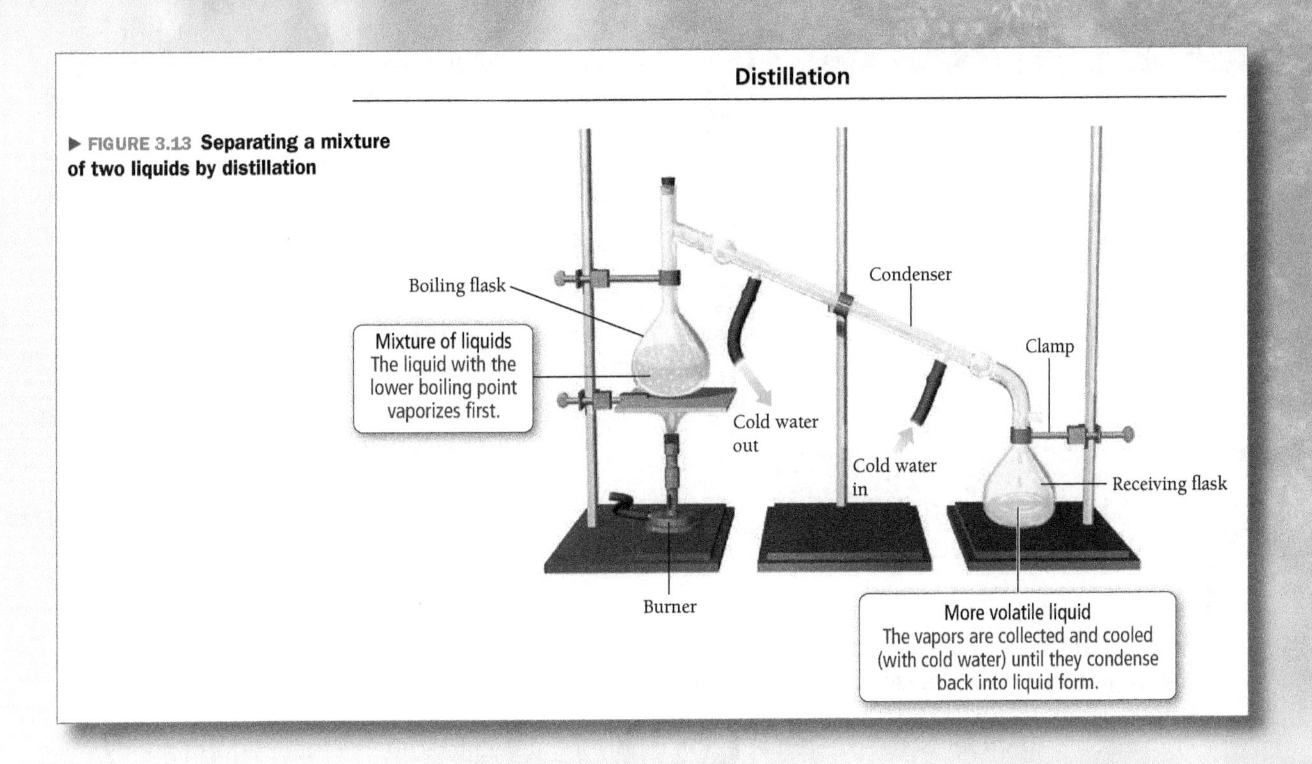

▶ FIGURE 3.13 **Separating a mixture of two liquids by distillation**

▲ FIGURE 3.9 **A physical property** The boiling point of water is a physical property, and boiling is a physical change. When water boils, it turns into a gas, but the water molecules are the same in both the liquid water and the gaseous steam.

NEW and UPDATED! Illustrations include extensive labels and annotations to direct student attention to key elements in the art and promote understanding of the processes depicted. Numerous figures in the sixth edition have updated labels and annotations to focus readers on key concepts. Relevant information is placed where it is most needed and makes the art a vital study and review tool.

Dynamic Study Modules and the Chemistry Primer help students come to class prepared.

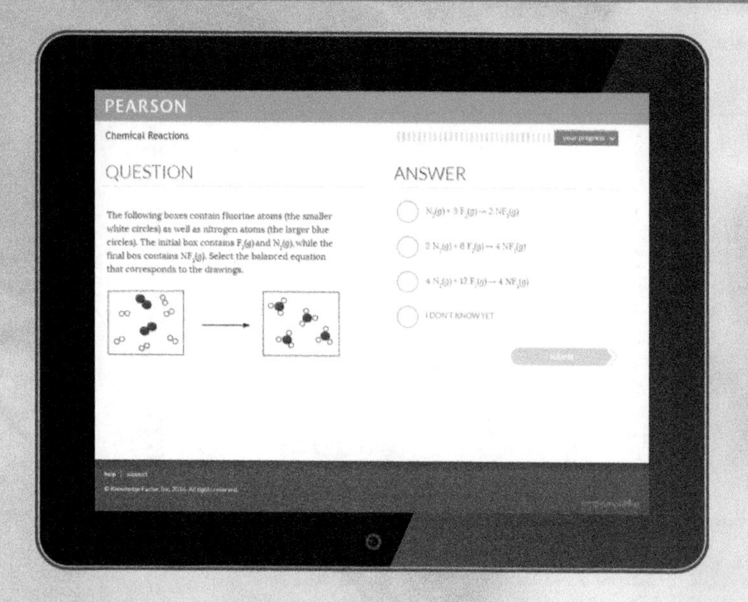

66 Dynamic Study Modules adapt to students' individual levels of understanding and help them study effectively on their own. Dynamic Study Modules continuously assess student activity and performance in real time. These are available as graded assignments prior to class and are accessible on smartphones, tablets, and computers.

Topics include key math skills and general chemistry concepts such as phases of matter, redox reactions, acids and bases, solutions, and chemical equilibrium.

The Chemistry Primer's pre-built diagnostic assignments get students up-to-speed at the beginning of the course, addressing topics such as math in the context of chemistry, basic chemical literacy, balancing chemical equations, mole theory, and stoichiometry. The Chemistry Primer scales to students' needs – remediation is only suggested to students that perform poorly on initial assessment, and involves Tutorials, Wrong-Answer Specific Feedback, Video Instruction, and Step-Wise Scaffolding to build student understanding.

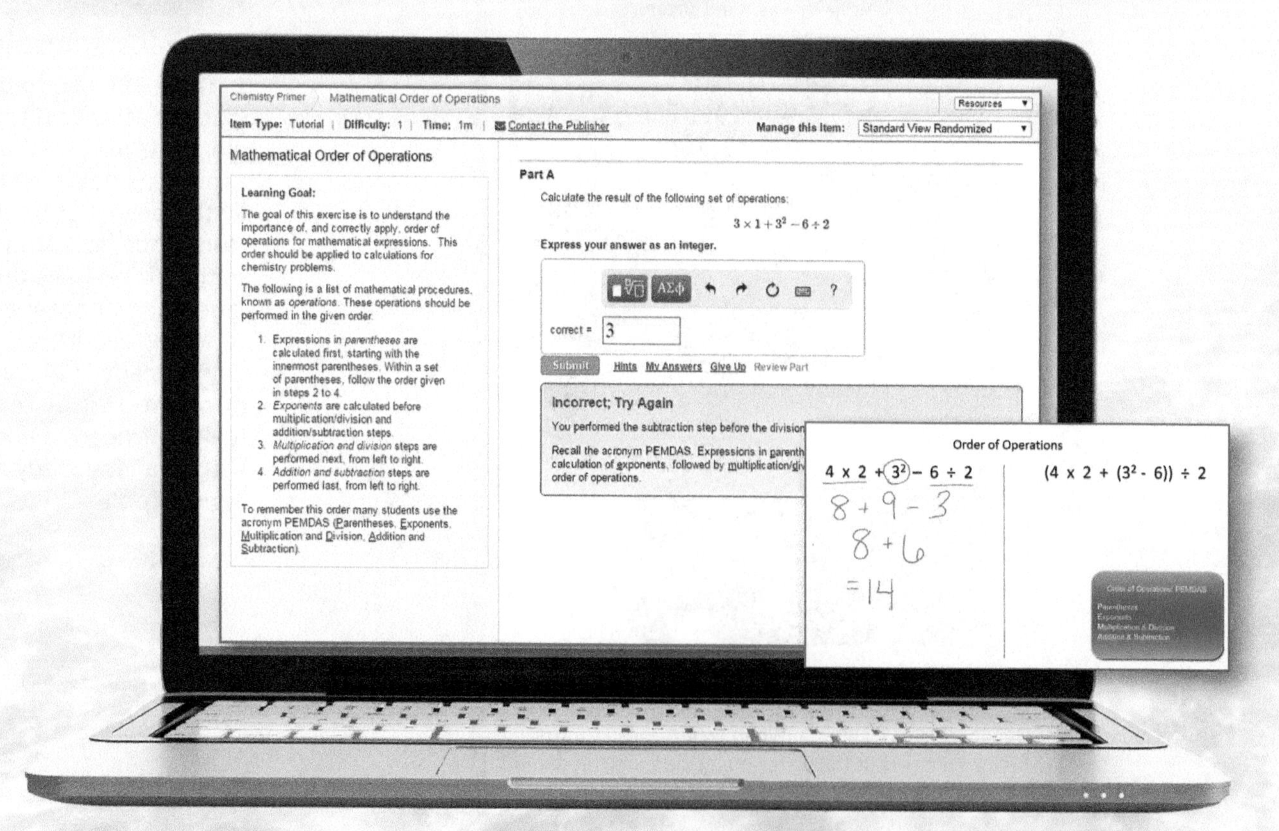

MasteringChemistry™ ensures student engagement before, during, and after class.

With questions specific to Tro's Introductory Chemistry, **Learning Catalytics™** generates class discussion, guides lecture, and promotes peer-to-peer learning with real-time analytics. Instructors can:

- **NEW!** Upload a full PowerPoint® deck for easy creation of slide questions
- Help students develop critical thinking skills
- Monitor responses to find out where students are struggling
- Adjust teaching strategy with real-time data
- Automatically group students for discussion, team-work, and peer-to-peer learning

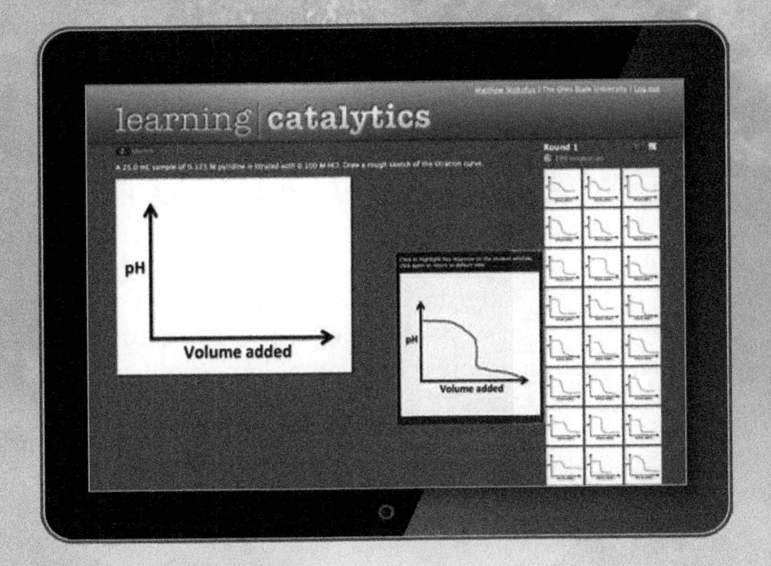

UPDATED! MasteringChemistry™ offers a wide variety of problems, ranging from multi-step tutorials with extensive hints and feedback to multiple-choice End-of-Chapter Problems and Test Bank questions.

To provide additional scaffolding for students moving from Tutorial Problems to End-of-Chapter Problems we created **NEW! Enhanced End-of-Chapter** Problems that now contain specific wrong-answer feedback.

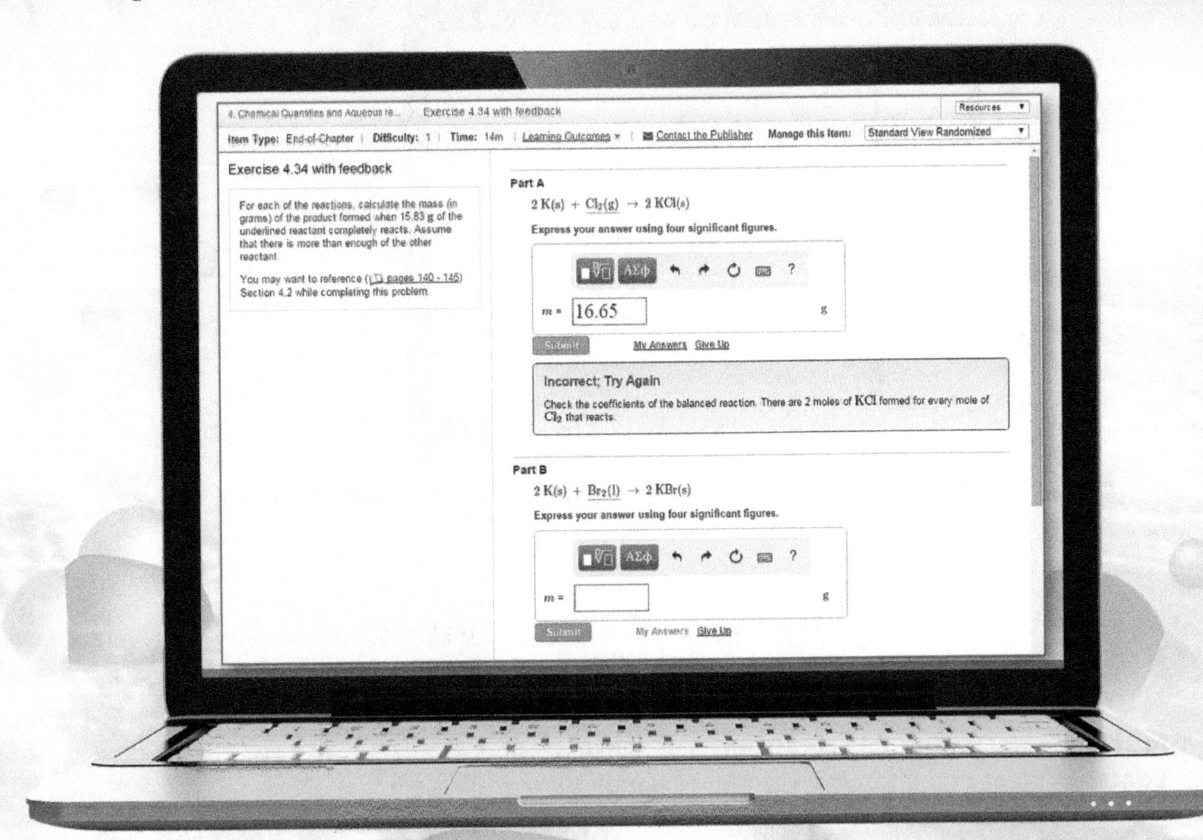

Students can study anywhere with fully interactive and mobile eText 2.0 features.

NEWLY INTERACTIVE! Self-Assessment Quizzes and Conceptual Checkpoints allow students to interact with all Conceptual Checkpoints and Self-Assessment Quizzes within eText 2.0! With one click these activities are brought to life, allowing students to study on their own and test their understanding in real-time. These interactives help students extinguish misconceptions and deepen their understanding of important concepts and topics.

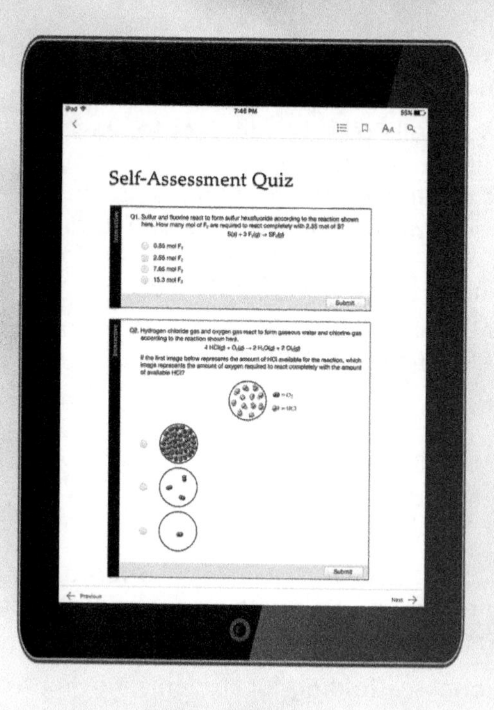

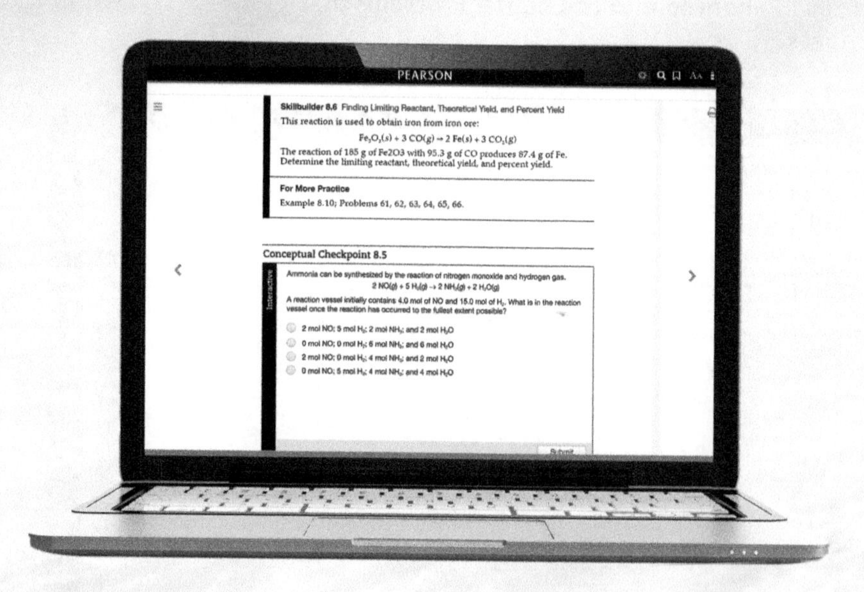

An icon in the text alerts students to interactive eText 2.0 features. The eText is fully optimized for use on mobile devices, allowing students to study anywhere.

In addition to **Conceptual Checkpoints** and **Self-Assessment Quizzes**, all **Key Concept Videos** and **Interactive Worked Examples** are available in the eText.

INTRODUCTORY CHEMISTRY

SiO2

1 The Chemical World

"Imagination is more important than knowledge."
—Albert Einstein (1879–1955)

CHAPTER OUTLINE

1.1 Sand and Water

I love the beach but hate sand. Sand gets everywhere and even comes home with you. Sand is annoying because sand particles are so small. They stick to your hands, to your feet, and to any food you might be trying to eat for lunch. But the smallness of sand particles pales in comparison to the smallness of the particles that compose them. Sand—like all other kinds of ordinary matter—is composed of atoms. Atoms are unimaginably small. A single sand grain contains more atoms than there are sand grains on the largest of beaches.

▲ Richard Feynman (1918–1988), Nobel Prize–winning physicist and popular professor at California Institute of Technology.

The idea that matter is composed of tiny particles is among the greatest discoveries of humankind. Nobel laureate Richard Feynman (1918–1988), in a lecture to first-year physics students at the California Institute of Technology, said that the most important idea in all human knowledge is that *all things are made of atoms.* Why is this idea so important? Because it establishes how we should go about understanding the properties of the things around us. If we want to understand how matter behaves, we must understand how the particles that compose that matter behave.

Atoms, and the molecules they compose, determine how matter behaves—if they were different, matter would be different. The nature of water molecules, for example, determines how water behaves. If water molecules were different—even in a small way—then water would be a different sort of substance. For example, we know that a water molecule is composed of two hydrogen atoms bonded to an oxygen atom with a shape that looks like this:

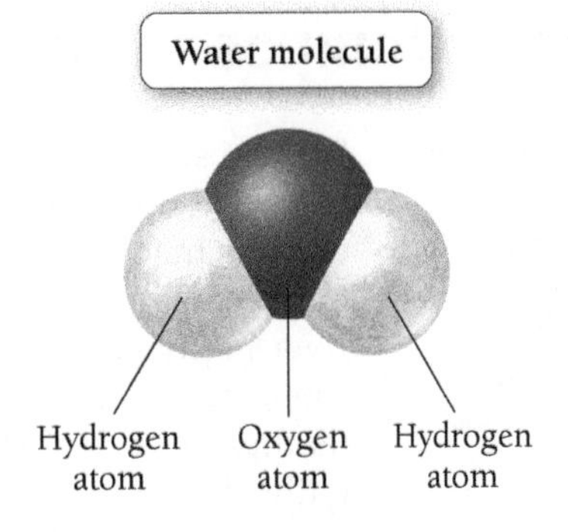

◀ A single grain of sand on a large beach contains more atoms than there are grains of sand on the entire beach.

How would water be different if the shape of the water molecule was different? What if the hydrogen atoms bonded to oxygen to form a linear molecule instead of a bent one?

Hypothetical linear water molecule

The answer to this question is not altogether simple. We don't know exactly how our hypothetical linear water would behave, but we do know it would be much different than ordinary water. For example, linear water would probably have a much lower boiling point than ordinary water. In fact, it may even be a gas (instead of a liquid) at room temperature. Imagine what our world would be like if water was a gas at room temperature. There would be no rivers, no lakes, no oceans, and probably no people (since liquid water is such an important part of what composes us).

There is a direct connection between the world of atoms and molecules and the world we experience every day (▼ FIGURE 1.1). Chemists explore this connection. They seek to understand it. A good, simple definition of **chemistry** is *the science that tries to understand how matter behaves by studying how atoms and molecules behave.*

▲ FIGURE 1.1 **Virtually everything around you is composed of chemicals.**

1.2 Chemicals Compose Ordinary Things

▶ Recognize that chemicals make up virtually everything we come into contact with in our world. *(Note: Most of the sections in the chapters in this book link to a Learning Objective (LO), which is listed at the beginning of the section.)*

We just saw how chemists are interested in substances such as sand and water. But are these substances chemicals? Yes. In fact, everything that we can hold or touch is made of chemicals. When most people think of chemicals, however, they may envision a can of paint thinner in their garage, or recall a headline about a river polluted by industrial waste. But chemicals compose ordinary things, too. Chemicals compose the air we breathe and the water we drink. They compose toothpaste, Tylenol®, and toilet paper. Chemicals make up virtually everything with which we come into contact. Chemistry explains the properties and behavior of chemicals, in the broadest sense, by helping us understand the molecules that compose them.

▲ Chemists are interested in knowing why ordinary things, such as water, are the way they are. When a chemist sees a pitcher of water, she thinks of the molecules that compose the liquid and how those molecules determine its properties.

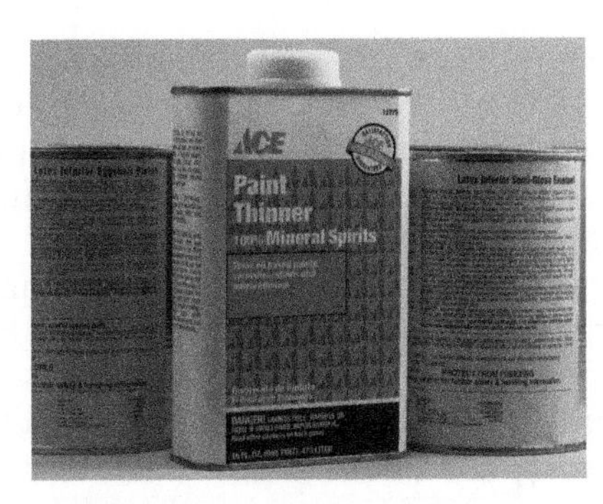

▲ People often have a very narrow view of chemicals, thinking of them only as dangerous poisons or pollutants.

As you experience the world around you, molecules are interacting to create your reality. Imagine watching a sunset. Molecules are involved in every step. Molecules in the air interact with light from the sun, scattering away the blue and green light and leaving the red and orange light to create the color you see. Molecules in your eyes absorb that light and, as a result, are altered in a way that sends a signal to your brain. Molecules in your brain then interpret the signal to produce images and emotions. This whole process—mediated by molecules—creates the evocative experience of seeing a sunset.

Chemists are interested in why ordinary substances are the way they are. Why is water a liquid? Why is salt a solid? Why does soda fizz? Why is a sunset red? Throughout this book, you will learn the answers to these questions and many others. *You will learn the connections between the behavior of matter and the structure of the particles that compose it.*

1.3 The Scientific Method: How Chemists Think

▶ Identify and understand the key characteristics of the scientific method: observation, the formulation of hypotheses, the testing of hypotheses by experiment, and the formulation of laws and theories.

Chemists use the **scientific method**—a way of learning that emphasizes observation and experimentation—to understand the world. The scientific method stands in contrast to ancient Greek philosophies that emphasized *reason* as the way to understand the world. Although the scientific method is not a rigid procedure that automatically leads to a definitive answer, it does have key characteristics that distinguish it from other ways of acquiring knowledge. These key characteristics include observation, the formulation of hypotheses, the testing of hypotheses by experiment, and the formulation of laws and theories.

The first step in acquiring scientific knowledge (▼ FIGURE 1.2) is often the **observation** or measurement of some aspect of nature. Some observations are simple, requiring nothing more than the naked eye. Other observations rely

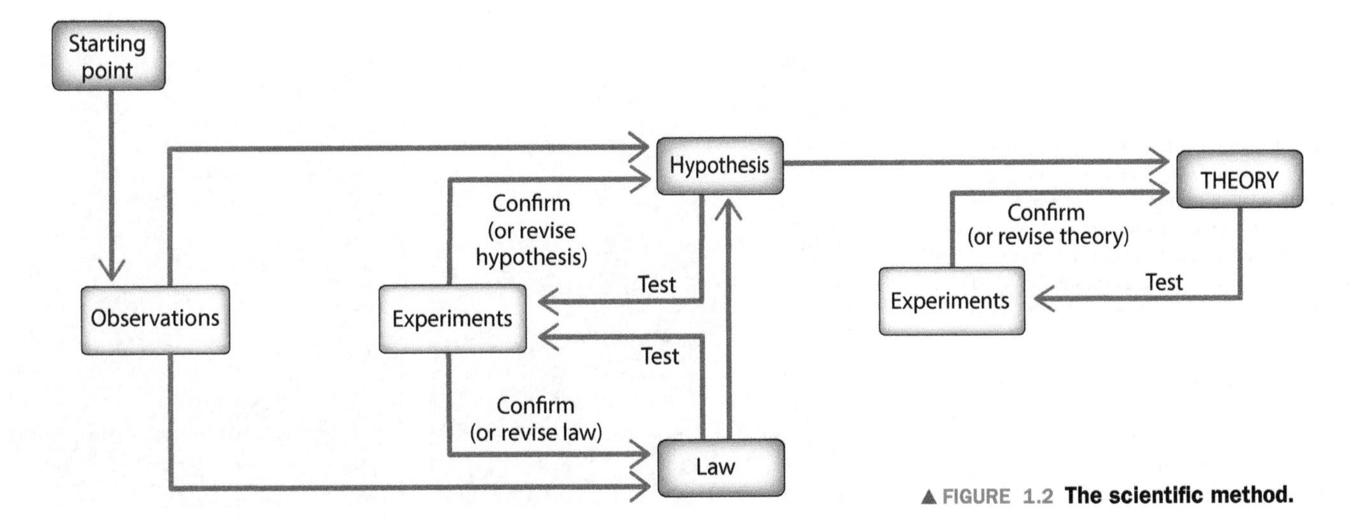

▲ FIGURE 1.2 **The scientific method.**

on the use of sensitive instrumentation. Occasionally, an important observation happens entirely by chance. Alexander Fleming (1881–1955), for example, discovered penicillin when he observed a bacteria-free circle around a certain mold that had accidentally grown on a culture plate. Regardless of how these observations occur, they usually involve the measurement or description of some aspect of the physical world. For example, Antoine Lavoisier (1743–1794), a French chemist who studied *combustion*, burned substances in closed containers. He carefully measured the mass of each container and its contents before and after burning the substance inside, noting that there was no change in the mass during combustion. Lavoisier made an *observation* about the physical world.

Combustion means burning. The mass of an object is a measure of the quantity of matter within it.

Observations often lead scientists to formulate a **hypothesis**, a tentative interpretation or explanation of the observations. Lavoisier explained his observations on combustion by hypothesizing that combustion involved the combination of a substance with a component of air. A good hypothesis is *falsifiable*, which means that further testing has the potential to prove it wrong. Hypotheses are tested by **experiments**, highly controlled observations designed to validate or invalidate hypotheses. The results of an experiment may confirm a hypothesis or show the hypothesis to be mistaken in some way. In the latter case, the hypothesis may have to be modified, or even discarded and replaced by an alternative hypothesis. Either way, the new or revised hypothesis must also be tested through further experimentation.

Sometimes a number of similar observations lead to the development of a **scientific law**, a brief statement that summarizes past observations and predicts future ones. For example, based on his observations of combustion, Lavoisier developed the **law of conservation of mass**, which states, "In a chemical reaction matter is neither created nor destroyed." This statement grew out of Lavoisier's observations, and it predicted the outcome of similar experiments on *any* chemical reaction. Laws are also subject to experiments, which can prove them wrong or validate them.

Scientific theories are also called *models*.

One or more well-established hypotheses may form the basis for a scientific **theory**. Theories provide a broader and deeper explanation for observations and laws. They are models of the way nature is, and they often predict behavior that extends well beyond the observations and laws on which they are founded. A good example of a theory is the **atomic theory** of John Dalton (1766–1844). Dalton explained the law of conservation of mass, as well as other laws and observations, by proposing that all matter was composed of small, indestructible particles called atoms. Dalton's theory was a model of the physical world—it went beyond the laws and observations of the time to explain these laws and observations.

▶ (Right) Painting of the French chemist Antoine Lavoisier and his wife, Marie, who helped him in his work by illustrating his experiments, recording results, and translating scientific articles from English. [*Source:* Jacques Louis David (French, 1748–1825). "Antoine-Laurent Lavoisier (1743–1794) and His Wife (Marie-Anne-Pierrette Paulze, 1758–1836)," 1788, oil on canvas, H. 102-1/4 in. W. 76-5/8 in. (259.7 × 194.6 cm). The Metropolitan Museum of Art, Purchase, Mr. and Mrs. Charles Wrightsman Gift, in honor of Everett Fahy, 1977. (1977.10) Image copyright © The Metropolitan Museum of Art.] (Far right) John Dalton, the English chemist who formulated the atomic theory.

▲ FIGURE 1.3 **Are atoms real?** The atomic theory has 200 years of experimental evidence to support it, including recent images, such as this one, of atoms themselves. This image shows the Kanji (a system of Japanese writing using Chinese characters) for "atom" written with individual iron atoms on top of a copper surface.

Theories are also tested and validated by experiments. Notice that the scientific method begins with observation, and then laws, hypotheses, and theories are developed based on those observations. Experiments, which are carefully controlled observations, determine the validity of laws, hypotheses, or theories. If a law, hypothesis, or theory is inconsistent with the findings of an experiment, it must be revised and new experiments must be conducted to test the revisions. Over time, scientists eliminate poor theories, and good theories—those consistent with experiments—remain. Established theories with strong experimental support are the most powerful pieces of scientific knowledge. People unfamiliar with science sometimes say, "That is just a theory," as if theories were mere speculations. However, well-tested theories are as close to truth as we get in science. For example, the idea that all matter is made of atoms is "just a theory," but it is a theory with 200 years of experimental evidence to support it, including the recent imaging of atoms themselves (◀ FIGURE 1.3). Established theories should not be taken lightly—they are the pinnacle of scientific understanding.

The eText 2.0 icon indicates that this feature is embedded and interactive in the eText.

CONCEPTUAL ✔ CHECKPOINT 1.1

Which statement most resembles a scientific theory?

(a) When the pressure on a sample of oxygen gas is increased 10%, the volume of the gas decreases by 10%.

(b) The volume of a gas is inversely proportional to its pressure.

(c) A gas is composed of small particles in constant motion.

(d) A gas sample has a mass of 15.8 g and a volume of 10.5 L.

Note: The answers to all Conceptual Checkpoints appear at the end of the chapter.

EVERYDAY CHEMISTRY

Combustion and the Scientific Method

Early chemical theories attempted to explain common phenomena such as combustion. Why did things burn? What was happening to a substance when it burned? Could something that was burned be unburned? Early chemists burned different substances and made observations to try to answer these questions. They observed that substances stop burning when placed in a closed container. They found that many metals burn to form a white powder that they called a *calx* (now we know that these white powders are oxides of the metal) and that the metal could be recovered from the calx, or unburned, by combining the calx with charcoal and heating it.

Chemists in the first part of the eighteenth century formed a theory about combustion to explain these observations. In this theory, combustion involved a fundamental substance that they called *phlogiston*. This substance was present in anything that burned and was released during combustion. Flammable objects were flammable because they contained phlogiston. When things burned in a closed container, they didn't burn for very long because the space within the container became saturated with phlogiston. When things burned in the open, they continued to burn until all of the phlogiston within them was gone. This theory also explained how metals that had burned could be unburned. Charcoal was a phlogiston-rich material—they knew this because it burned so well—and when it was combined with a calx, which was a metal that had been emptied of its phlogiston, it transferred some of its phlogiston into the calx, converting the calx back into the unburned form of the metal. The phlogiston theory was consistent with all of the observations of the time and was widely accepted as valid.

Like any theory, the phlogiston theory was tested continually by experiment. One set of experiments, conducted in the mid-eighteenth century by Louis-Bernard Guyton de Morveau (1737–1816), consisted of weighing metals before and after burning them. In every case the metals *gained* weight when they were burned. This observation is inconsistent with the phlogiston theory, which predicted that metals should *lose* weight because phlogiston was supposed to be lost during combustion. Clearly, the phlogiston theory needed modification.

continued on page 8

continued from page 7

The first modification was that phlogiston was a very light substance, so that it actually "buoyed up" the materials that contained it. Thus when phlogiston was released, the material became heavier. Such a modification seemed to fit the observations but also seemed far-fetched. Antoine Lavoisier developed a more likely explanation by devising a completely new theory of combustion. Lavoisier proposed that, when a substance burned, it actually took something *out* of the air, and when it unburned, it released something back into the air. Lavoisier said that burning objects *fixed* (attached or bonded) the air and that the *fixed* air was released during unburning. In a confirming experiment (▶ FIGURE 1.4), Lavoisier roasted a mixture of calx and charcoal with the aid of sunlight focused by a giant burning lens, and found that a huge volume of "fixed air" was released in the process. The scientific method worked. The phlogiston theory was proven wrong, and a new theory of combustion took its place—a theory that, with a few refinements, is still valid today.

B1.1 CAN YOU ANSWER THIS? *What is the difference between a law and a theory? How does the example of the phlogiston theory demonstrate this difference?*

▲ FIGURE 1.4 **Focusing on combustion** The great burning lens belonging to the Academy of Sciences. Lavoisier used a similar lens in 1777 to show that a mixture of *calx* (metal oxide) and charcoal released a large volume of *fixed air* when heated.

1.4 Analyzing and Interpreting Data

▶ Identify patterns in data and interpret graphs.

We just learned how early scientists such as Lavoisier and Dalton saw patterns in a series of related measurements. Sets of measurements constitute scientific *data*, and learning to analyze and interpret data is an important scientific skill.

Identifying Patterns in Data

Suppose you are an early chemist trying to understand the composition of water. You know that water is composed of the elements hydrogen and oxygen. You do several experiments in which you decompose different samples of water into hydrogen and oxygen, and you get the following results:

Sample	Mass of Water Sample	Mass of Hydrogen Formed	Mass of Oxygen Formed
A	20.0 g	2.2 g	17.8 g
B	50.0 g	5.6 g	44.4 g
C	100.0 g	11.1 g	88.9 g

Do you notice any patterns in this data? The first and easiest pattern to see is that the sum of the masses of oxygen and hydrogen always sums to the mass of the water sample. For example, for the first water sample, 2.2 g hydrogen + 17.8 g oxygen = 20.0 g water. The same is true for the other samples. Another pattern, which is a bit more difficult to see, is that the ratio of the masses of oxygen and hydrogen is the same for each sample.

Sample	Mass of Hydrogen Formed	Mass of Oxygen Formed	$\frac{\text{Mass Oxygen}}{\text{Mass Hydrogen}}$
A	2.2 g	17.8 g	8.1
B	5.6 g	44.4 g	7.9
C	11.1 g	88.9 g	8.01

The ratio is 8—the small variations are due to experimental error, which is common in all measurements and observations.

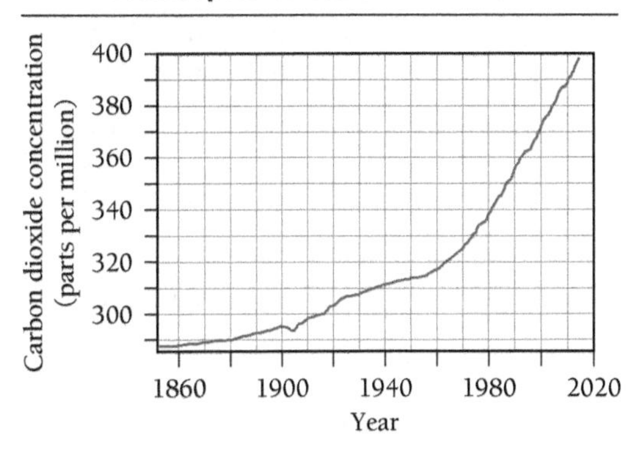

▲ **FIGURE 1.5 Atmospheric carbon dioxide levels from 1860 to present.**

Seeing patterns in data is a creative process that requires you to not just merely tabulate laboratory measurements, but to see relationships that may not always be obvious. The best scientists see patterns that others have missed. As you learn to interpret data in this course, be creative and try looking at data in new ways.

Interpreting Graphs

Data is often visualized using graphs or images, and scientists must constantly analyze and interpret graphs. For example, the graph in ◀ FIGURE 1.5 shows the concentration of carbon dioxide in Earth's atmosphere as a function of time. Carbon dioxide is a greenhouse gas that has been rising as result of the burning of fossil fuels (such as gasoline and coal). When you look at a graph such as this one, you should first examine the x and y axes and make sure you understand what each axis represents. You should also examine the numerical range of the axes. In Figure 1.5, the y axis does not begin at zero in order to better display the change that is occurring. How would this graph look different if the y axis began at zero instead of at 290? Notice also that, in this graph, the increase in carbon dioxide has not been constant over time. The rate of increase—represented by the slope of the line—has intensified since about 1960.

EXAMPLE 1.1 Interpreting Graphs

Examine the graph in Figure 1.5 and answer each question.

(a) What was the concentration of carbon dioxide in 1960?
(b) What was the concentration in 2000?
(c) How much did the concentration increase between 1960 and 2000?
(d) What is the average rate of increase over this time?
(e) If the average rate of increase remains constant, estimate the carbon dioxide concentration in 2030.

SOLUTION

a) What was the concentration of carbon dioxide in 1960?

To determine the concentration of carbon dioxide in 1960, draw a vertical line at the year 1960. At the point where the vertical line intersects the carbon dioxide concentration curve, draw a horizontal line. The point where the horizontal line intercepts the y axis represents the concentration in 1960. So, the concentration in 1960 was 318 ppm.

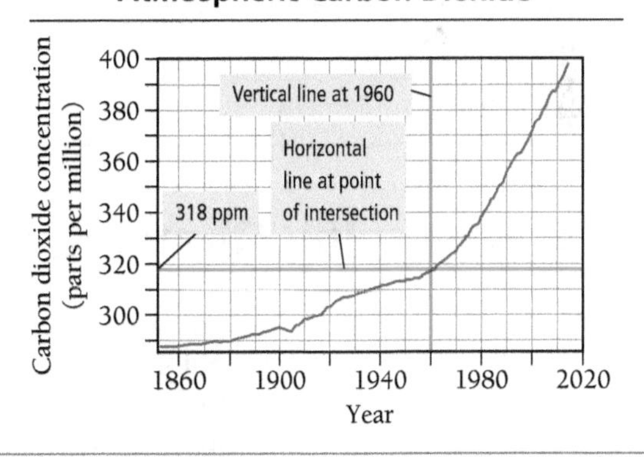

b) What was the concentration in 2000?

Apply the same procedure as in part a, but now shift the vertical line to the year 2000. The concentration in the year 2000 was 370 ppm.

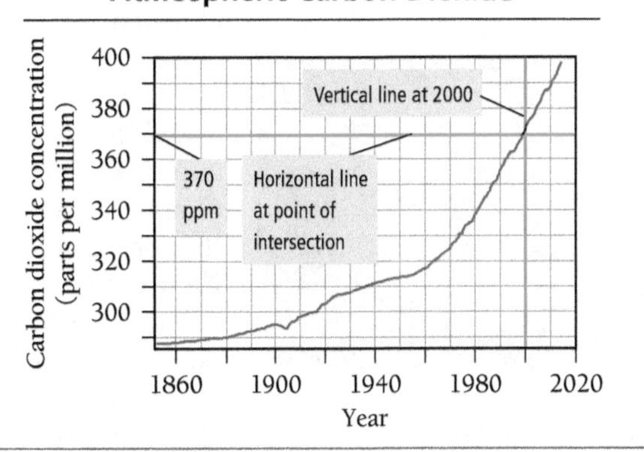

continued on page 10 ▶

continued from page 9

c) How much did the concentration increase between 1960 and 2000? The increase in the carbon dioxide concentration is the difference between the two concentrations. When calculating changes in quantities such as this, take the final quantity minus the initial quantity.	increase in concentration = concentration in 2000 − concentration in 1960 = 390 ppm − 318 ppm = 72 ppm
d) What is the average rate of increase over this time? The average rate of increase over this time is the change in the concentration divided by the number of years that passed. Determine the number of years that have passed by subtracting the initial year from the final year. Determine the average rate of increase by dividing the change in concentration (from part c) by the number of years that you just calculated.	number of years = final year − initial year = 2000 − 1960 = 40 years $\text{average rate} = \dfrac{\text{change in concentration}}{\text{number of years}}$ $= \dfrac{72\text{ ppm}}{40\text{ years}}$ $= \dfrac{1.8\text{ ppm}}{\text{year}}$
e) If the average rate of increase remains constant, estimate the carbon dioxide concentration in 2030. Determine the increase in concentration between 2000 and 2030 by multiplying the number of years that pass in that time interval by the average rate of change (from part d). Lastly, determine the concentration in 2030 by adding the increase between 2000 and 2030 to the concentration in 2000.	$\text{increase} = 30\text{ years} \times \dfrac{1.8\text{ ppm}}{\text{year}}$ = 54 ppm concentration in 2030 = 370 ppm + 54 ppm = 424 ppm

▶ **SKILLBUILDER 1.1** | What was the average rate of increase in carbon dioxide concentration between 1880 and 1920? Why might that rate be different than the rate between 1960 and 2000?

▶ **FOR MORE PRACTICE** Problem 25.

1.5 A Beginning Chemist: How to Succeed

You are a beginning chemist. This may be your first chemistry course, but it may not be your last. To succeed as a beginning chemist, keep the following ideas in mind. First, chemistry requires curiosity and imagination. If you are content knowing that the sky is blue but don't care *why* it is blue, then you may have to rediscover your curiosity. I say "rediscover" because even children—or better, *especially* children—have this kind of curiosity. To succeed as a chemist, you must have the curiosity and imagination of a child—*you must want to know the why of things.*

▲ To succeed as a scientist, you must have the curiosity of a child.

Second, chemistry requires calculation. Throughout this course, you will be asked to calculate answers and quantify information. *Quantification* involves measurement as part of observation—it is one of the most important tools in science. Quantification allows you to go beyond merely saying that this object is hot and that one is cold or that this one is large and that one is small. It allows you to specify the difference precisely. For example, two samples of water may feel equally hot to your hand, but when you measure their temperatures, you may find that one is 40 °C and the other is 44 °C. Even small differences can be important in a calculation or experiment, so assigning numbers to observations and manipulating those numbers become very important in chemistry.

Lastly, chemistry requires commitment. To succeed in this course, you must commit to learning chemistry. Roald Hoffmann (1937–), winner of the 1981 Nobel Prize for chemistry, said,

> *I like the idea that human beings can do anything they want to. They need to be trained sometimes. They need a teacher to awaken the intelligence within them. But to be a chemist requires no special talent, I'm glad to say. Anyone can do it, with hard work.*

Professor Hoffmann is right. The key to success in this course is hard work—that requires commitment. You must do your work regularly and carefully. If you do, you will succeed, and you will be rewarded by seeing a whole new world—the world of molecules and atoms. This world exists beneath the surface of nearly everything you encounter. I welcome you to this world and consider it a privilege, together with your professor, to be your guide.

Chapter 1 in Review

MasteringChemistry™ provides end-of-chapter exercises, feedback-enriched tutorial problems, animations, and interactive activities to encourage problem solving practice and deeper understanding of key concepts and topics.

Self-Assessment Quiz

The eText 2.0 icon indicates that this feature is embedded and interactive in the eText.

Q1. Where can you find chemicals?
(a) In a hardware store
(b) In a chemical stockroom
(c) All around you and even inside of you
(d) All of the above

Q2. Which statement best defines chemistry?
(a) The science that studies solvents, drugs, and insecticides
(b) The science that studies the connections between the properties of matter and the particles that compose that matter
(c) The science that studies air and water pollution
(d) The science that seeks to understand processes that occur only in chemical laboratories

Q3. According to the scientific method, what is a law?
(a) A short statement that summarizes a large number of observations
(b) A fact that can never be refuted
(c) A model that gives insight into how nature is
(d) An initial guess with explanatory power

Q4. Which statement is an example of an observation?
(a) In a chemical reaction, matter is conserved.
(b) All matter is made of atoms.
(c) When a given sample of gasoline is burned in a closed container, the mass of the container and its contents does not change.
(d) Atoms bond to one another by sharing electrons.

Q5. The graph below shows the area of a circle as a function of its radius. What is the radius of a circle that has an area of 155 square inches?

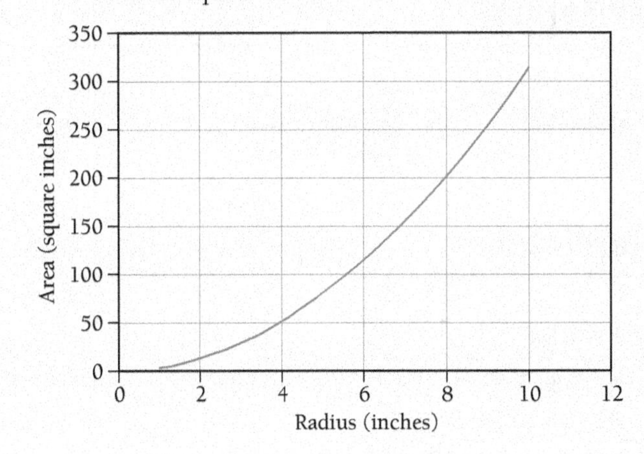

(a) 7.0 inches **(b)** 6.5 inches
(c) 6.8 inches **(d)** 6.2 inches

Q6. Which characteristic is necessary for success in understanding chemistry?
(a) Curiosity **(b)** Calculation
(c) Commitment **(d)** All of the above

Answers: 1:d, 2:b, 3:a, 4:c, 5:a, 6:d

Chemical Principles

Matter and Molecules

Chemists are interested in all matter, even ordinary matter such as water or air. You don't need to go to a chemical storeroom to find chemicals because they are all around you. Chemistry is the science that studies the connections between the properties of matter and the particles that compose that matter.

Relevance

Chemists want to understand matter for several reasons. First, chemists are simply curious—they want to know why. Why are some substances reactive and others not? Why are some substances gases, some liquids, and others solids? Chemists are also practical; they want to understand matter so that they can control it and produce substances that are useful to society and to humankind.

The Scientific Method

Chemists employ the scientific method, which makes use of observations, hypotheses, laws, theories, and experiments. Observations involve measuring or observing some aspect of nature. Hypotheses are tentative interpretations of observations. Laws summarize the results of a large number of observations, and theories are models that explain and give the underlying causes for observations and laws. Hypotheses, laws, and theories must be tested and validated by experiment. If they are not confirmed, they are revised and tested through further experimentation.

The scientific method is a way to understand the world. Since the inception of the scientific method, knowledge about the natural world has grown rapidly. The application of the scientific method has produced technologies that have raised living standards throughout the world with advances such as increased food production, rapid transportation, unparalleled access to information, and longer life spans.

Analyzing and Interpreting Data

A series of measurements are often referred to as data. Scientific data can be graphed to better see relationships between variables.

Virtually all scientists have to analyze and interpret the data they collect. This skill is an important part of understanding chemistry.

Success as a Beginning Chemist

To succeed as a beginning chemist, you must be curious and imaginative, be willing to do calculations, and be committed to learning the material.

To succeed as a beginning chemist, you must be curious and imaginative, be willing to do calculations, and be committed to learning the material.

Key Terms

atomic theory [1.3]
chemistry [1.1]
experiment [1.3]
hypothesis [1.3]
law of conservation of mass [1.3]
observation [1.3]
scientific law [1.3]
scientific method [1.3]
theory [1.3]

Exercises

Questions

Answers to all questions numbered in blue appear in the Answers section at the back of the book.

1. Why does soda fizz?
2. What are chemicals? Give some examples.
3. What do chemists try to do? How do they understand the natural world?
4. What is meant by the statement, "Matter does what molecules do"? Give an example.
5. Define *chemistry*.
6. How is chemistry connected to everyday life? How is chemistry relevant outside the chemistry laboratory?
7. Explain the scientific method.
8. Cite an example from this chapter of the scientific method at work.
9. What is the difference between a law and a theory?
10. What is the difference between a hypothesis and a theory?
11. What is wrong with the statement, "It is just a theory"?
12. What is the law of conservation of mass, and who discovered it?
13. What is the atomic theory, and who formulated it?
14. What are three things you need to do to succeed in this course?

Problems

Note: The exercises in the Problems section are paired, and the answers to the odd-numbered exercises (numbered in blue) appear in the Answers section at the back of the book.

15. Classify each statement as an observation, a law, or a theory.
    - **(a)** When a metal is burned in a closed container, the sum of the masses of the container and its contents does not change.
    - **(b)** Matter is made of atoms.
    - **(c)** Matter is conserved in chemical reactions.
    - **(d)** When wood is burned in a closed container, its mass does not change.
16. Classify each statement as an observation, a law, or a theory.
    - **(a)** The star closest to Earth is moving away from Earth at high speed.
    - **(b)** A body in motion stays in motion unless acted upon by a force.
    - **(c)** The universe began as a cosmic explosion called the Big Bang.
    - **(d)** A stone dropped from an altitude of 450 m falls to the ground in 9.6 s.

17. A student prepares several samples of the same gas and measures their mass and volume. The results are tabulated as follows. Formulate a tentative law from the measurements.

Mass of Gas (in g)	Volume of Gas (in L)
22.5	1.60
35.8	2.55
70.2	5.00
98.5	7.01

18. A student measures the volume of a gas sample at several different temperatures. The results are tabulated as follows. Formulate a tentative law from the measurements.

Temperature of Gas (in K)	Volume of Gas (in L)
298	4.55
315	4.81
325	4.96
335	5.11

19. A chemist in an imaginary universe does an experiment that attempts to correlate the size of an atom with its chemical reactivity. The results are tabulated as follows.

Size of Atom	Chemical Reactivity
small	low
medium	intermediate
large	high

(a) Formulate a law from this data.
(b) Formulate a theory to explain the law.

20. A chemist decomposes several samples of water into hydrogen and oxygen and measures the mass of the hydrogen and the oxygen obtained. The results are tabulated as follows.

Sample Number	Grams of Hydrogen	Grams of Oxygen
1	1.5	12
2	2	16
3	2.5	20

(a) Summarize these observations in a short statement.
Next, the chemist decomposes several samples of carbon dioxide into carbon and oxygen. The results are tabulated as follows:

Sample Number	Grams of Carbon	Grams of Oxygen
1	0.5	1.3
2	1.0	2.7
3	1.5	4.0

(b) Summarize these observations in a short statement.
(c) Formulate a law from the observations in (a) and (b).
(d) Formulate a theory that might explain your law in (c).

Questions for Group Work

Discuss these questions with the group and record your consensus answer.

21. The manufacturer of a particular brand of toothpaste claims that the brand contains "no chemicals." Using a few grammatically correct English sentences, describe what you think the company means by that statement. Would a scientist consider the manufacturer's statement to be correct? Why or why not?
22. Make a list (including up to ten items) of all the atoms or molecules group members can name off the top of their heads. Get at least one contribution from each group member.
23. In your own words, provide a brief definition for each of the following: observation, law, hypothesis, and theory.
24. How curious are you? How good are your quantitative skills? How hard are you willing to work to succeed in chemistry? Answer these questions individually on a scale of 1 (=not at all) to 5 (=very), then share your answers with your group. Report the group average for each question.

Data Interpretation and Analysis

25. The graph displays world population over time. Study the graph and answer the following questions.

(a) What was the world population in 1950?
(b) What was the world population in 2010?
(c) How much did the population increase between 1950 and 2010?
(d) What is the average rate of increase over this time?
(e) If the average rate of increase remains constant, estimate the world population in 2035.

World Population Versus Time

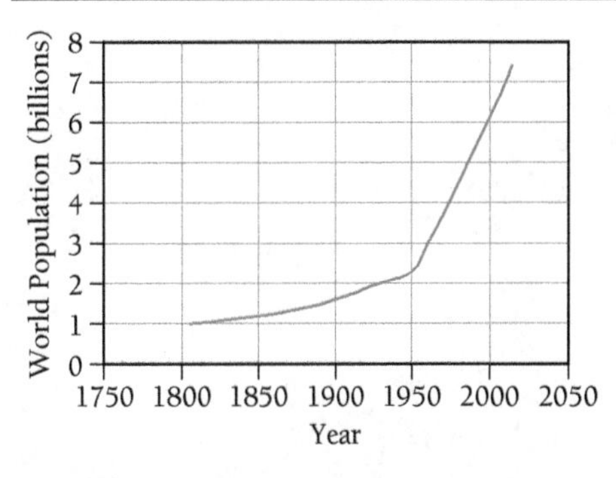

Source: http://www.worldometers.info/world-population/

Answers to Skillbuilder Exercises

Skillbuilder 1.1..........0.33 ppm/yr; The rate is smaller because less fossil fuels were being used.

Answers to Conceptual Checkpoints

1.1 (c) Answers a and d are observations. Answer b is a scientific law. Answer c is the only answer that proposes a *model* for what a gas is like.

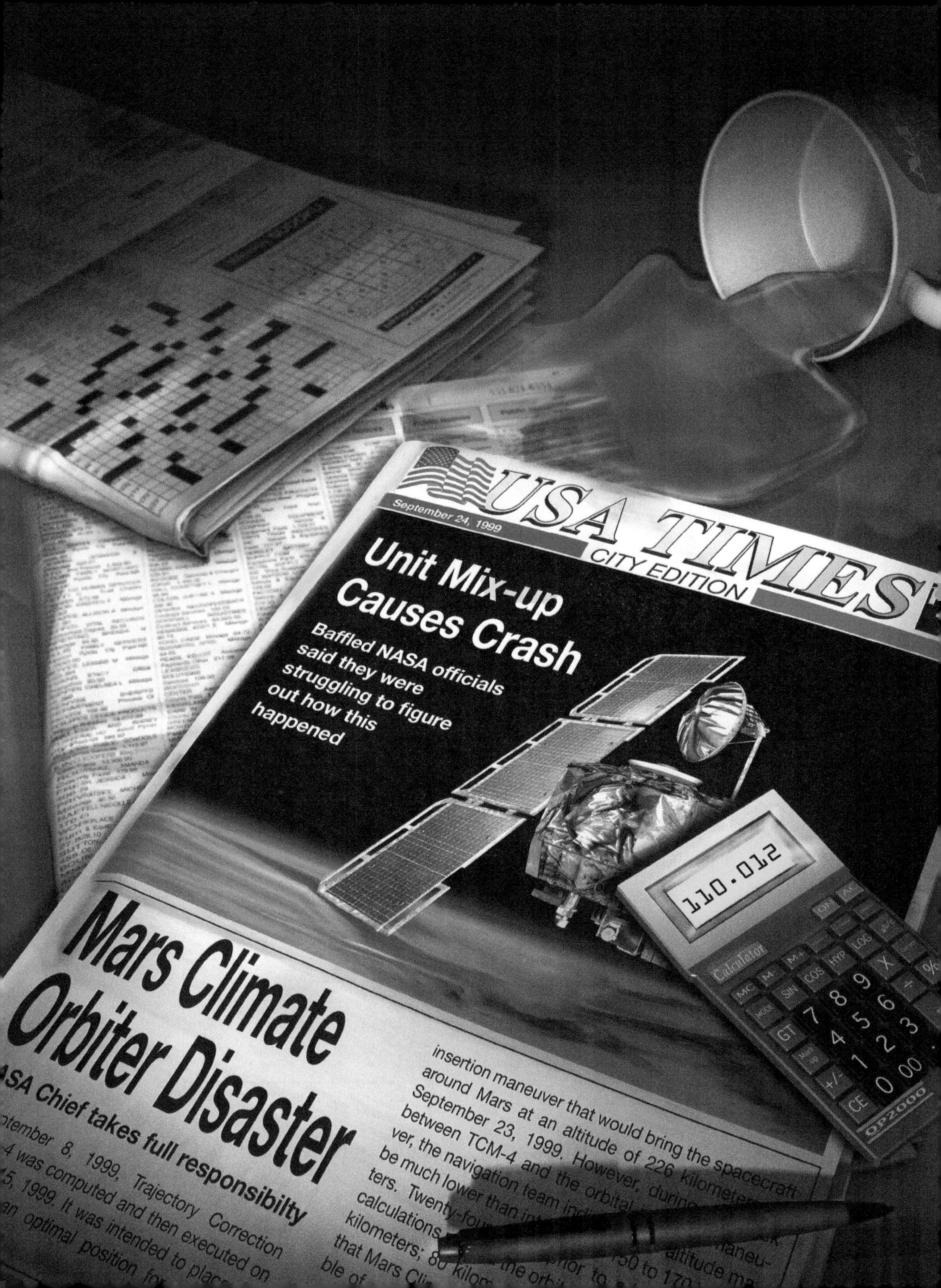

USA TIMES

CITY EDITION

September 24, 1999

Unit Mix-up Causes Crash

Baffled NASA officials said they were struggling to figure out how this happened

Mars Climate Orbiter Disaster

ASA Chief takes full responsibilty

2 Measurement and Problem Solving

"The important thing in science is not so much to obtain new facts as to discover new ways of thinking about them."

—Sir William Lawrence Bragg (1890–1971)

CHAPTER OUTLINE

2.1 The Metric Mix-up: A $125 Million Unit Error

On December 11, 1998, NASA launched the Mars Climate Orbiter, which was to become the first weather satellite for a planet other than Earth. The Orbiter's mission was to monitor the atmosphere on Mars and to serve as a communications relay for the Mars Polar Lander, a probe that was to follow the Orbiter and land on the planet's surface three weeks later. Unfortunately, the mission ended in disaster. A unit mix-up caused the Orbiter to enter the Martian atmosphere at an altitude that was too low. Instead of settling into a stable orbit, the Orbiter likely disintegrated. The cost of the failed mission was estimated at $125 million.

Later investigations showed that the Orbiter had come within 57 km of the planet surface, which was too close. When a spacecraft enters a planet's atmosphere too close to the planet's surface, friction with the atmosphere can cause the spacecraft to burn up. The on-board computers that controlled the trajectory corrections were programmed in metric units (newton · second), but the ground engineers entered the trajectory corrections in English units (pound · second). The English and the metric units are not equivalent (1 pound · second = 4.45 newton · second). The corrections that the ground engineers entered were 4.45 times too small and did not alter the trajectory enough to keep the Orbiter at a sufficiently high altitude. In chemistry as in space exploration, *units* (see Section 2.5) are critical. If we get them wrong, the consequences can be disastrous.

A unit is a standard, agreed on quantity by which other quantities are measured.

2.2 Scientific Notation: Writing Large and Small Numbers

▶ Express very large and very small numbers using scientific notation.

Science constantly pushes the boundaries of the very large and the very small. We can, for example, now measure time periods as short as 0.000000000000001 second and distances as great as 14,000,000,000 light-years. Because the many zeros

▲ Lasers such as this one can measure time periods as short as 1×10^{-15} s.

in these numbers are cumbersome to write, we use **scientific notation** to write them more compactly. In scientific notation, 0.000000000000001 is 1×10^{-15}, and 14,000,000,000 is 1.4×10^{10}. A number written in scientific notation consists of a **decimal part**, a number that is usually between 1 and 10, and an **exponential part**, 10 raised to an **exponent**, n.

$$\underbrace{1.2}_{\text{decimal part}} \times \underbrace{10^{-10 \leftarrow \text{exponent } (n)}}_{\text{exponential part}}$$

A positive exponent (n) means 1 multiplied by 10 n times.

$$10^0 = 1$$
$$10^1 = 1 \times 10 = 10$$
$$10^2 = 1 \times 10 \times 10 = 100$$
$$10^3 = 1 \times 10 \times 10 \times 10 = 1000$$

A negative exponent ($-n$) means 1 divided by 10 n times.

$$10^{-1} = \frac{1}{10} = 0.1$$
$$10^{-2} = \frac{1}{10 \times 10} = 0.01$$
$$10^{-3} = \frac{1}{10 \times 10 \times 10} = 0.001$$

To convert a number to scientific notation, we move the decimal point (either to the left or to the right, as needed) to obtain a number between 1 and 10 and then multiply that number (the decimal part) by 10 raised to the power that reflects the movement of the decimal point. For example, to write 5983 in scientific notation, we move the decimal point to the left three places to get 5.983 (a number between 1 and 10) and then multiply the decimal part by 1000 to compensate for moving the decimal point.

$$5983 = 5.983 \times \underbrace{1000}_{1000 \text{ is } 10^3}$$
$$= 5.983 \times 10^3$$

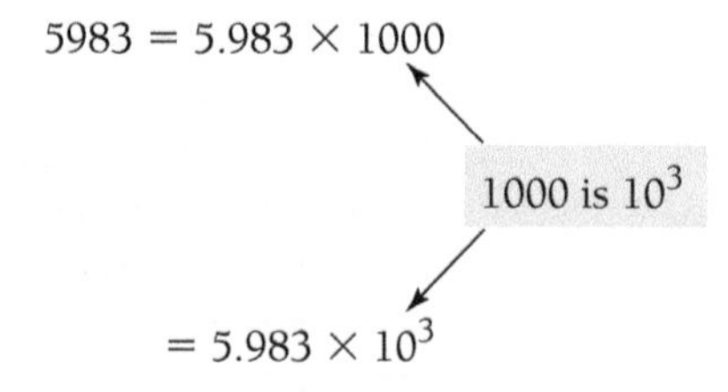

We can do this in one step by counting how many places we move the decimal point to obtain a number between 1 and 10 and then by writing the decimal part multiplied by 10 raised to the number of places we moved the decimal point.

$$\underset{3\,2\,1}{5983} = 5.983 \times 10^3$$

The Mathematics Review Appendix (p. MR-2) includes a review of mathematical operations for numbers written in scientific notation.

If the decimal point is moved to the left, as in the previous example, the exponent is positive. If the decimal is moved to the right, the exponent is negative.

$$\underset{1\,2\,3\,4}{0.00034} = 3.4 \times 10^{-4}$$

To express a number in scientific notation:

Remember, large numbers have positive exponents and small numbers have negative exponents.

1. Move the decimal point to obtain a number between 1 and 10.
2. Write the result from Step 1 multiplied by 10 raised to the number of places you moved the decimal point.
   - The exponent is positive if you moved the decimal point to the left.
   - The exponent is negative if you moved the decimal point to the right.

EXAMPLE 2.1 Scientific Notation

The 2016 U.S. population was estimated to be 323,000,000 people. Express this number in scientific notation.

To obtain a number between 1 and 10, move the decimal point to the left eight decimal places; the exponent is 8. Because you move the decimal point to the left, the sign of the exponent is positive.

SOLUTION

323,000,000 people $= 3.23 \times 10^8$ people

▶ **SKILLBUILDER 2.1 | Scientific Notation**

The total U.S. national debt in 2016 was approximately $18,416,000,000,000. Express this number in scientific notation.

Note: The answers to all Skillbuilders appear at the end of the chapter.

▶ **FOR MORE PRACTICE** Example 2.19; Problems 31, 32.

EXAMPLE 2.2 Scientific Notation

The radius of a carbon atom is approximately 0.000000000070 m. Express this number in scientific notation.

To obtain a number between 1 and 10, move the decimal point to the right 11 decimal places; therefore, the exponent is 11. Because you moved the decimal point to the right, the sign of the exponent is negative.

SOLUTION

0.000000000070 m $= 7.0 \times 10^{-11}$ m

▶ **SKILLBUILDER 2.2 | Scientific Notation**

Express the number 0.000038 in scientific notation.

▶ **FOR MORE PRACTICE** Problems 33, 34.

The eText 2.0 icon indicates that this feature is embedded and interactive in the eText.

CONCEPTUAL ✔ CHECKPOINT 2.1

The radius of a dust speck is 4.5×10^{-3} mm. What is the correct value of this number in decimal notation (i.e., express the number without using scientific notation)?

(a) 4500 mm **(b)** 0.045 mm **(c)** 0.0045 mm **(d)** 0.00045 mm

Note: The answers to all Conceptual Checkpoints appear at the end of the chapter.

2.3 Significant Figures: Writing Numbers to Reflect Precision

- ▶ Report measured quantities to the right number of digits.
- ▶ Determine which digits in a number are significant.

Climate change has become a household term. Global climate affects agriculture, weather, and ocean levels. The report that global temperatures are increasing is based on the continuing work of scientists who analyze records from thousands of temperature-measuring stations around the world. To date, these scientists conclude that the average global temperature has risen by 0.7 °C since 1880.

Notice how the scientists reported their results. What if the scientists had included additional zeros in their results—for example, 0.70 °C or 0.700 °C—or if they had reported the number their computer displayed after averaging many measurements, something like 0.68759824 °C. Would these reported numbers convey the same information? Not really. Scientists adhere to a standard way of reporting measured quantities in which the number of reported digits reflects the precision in the measurement—more digits, more precision; fewer digits, less precision. Numbers are usually written so that the uncertainty is indicated by the last reported digit. For example, by reporting a temperature increase of 0.7 °C, the scientists mean 0.7 ± 0.1 °C (± means plus or minus). The temperature rise could be as much as 0.8 °C or as little as 0.6 °C, but it is not 1.0 °C. The degree of certainty in this particular measurement is critical, influencing political decisions that directly affect people's lives.

We report scientific numbers so that every digit is certain except the last, which we estimate. For example, if a reported measurement is:

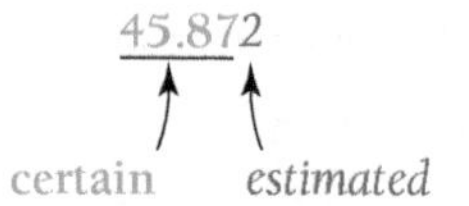

We know that the first four digits are certain; the last digit is estimated. The number of digits reported depends on the precision of the measuring device. For example, consider the measurement shown here:

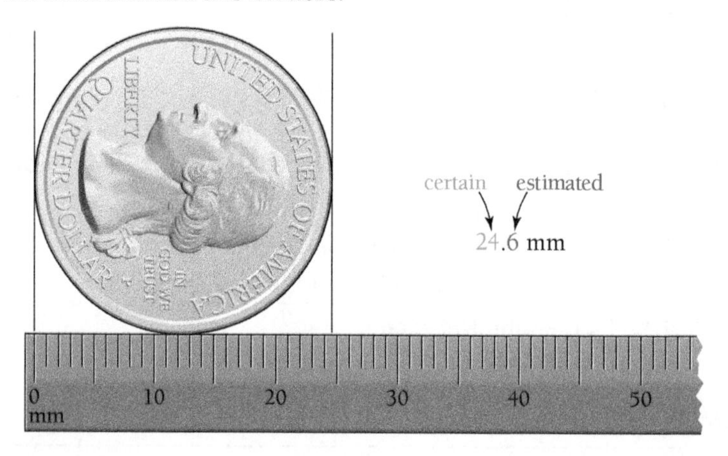

How should we report the width of the quarter as measured with this particular ruler? Since the ruler has markings every 1 mm, we can say for certain that the quarter is between 24 and 25 mm wide. We can then estimate the remaining width to the nearest 0.1 mm. We do this by mentally dividing the space between the 24 and 25 mm marks into 10 equal spaces and estimate the position of the line that indicates the width of the quarter. In this case, the line is a bit beyond the halfway point between the two markings on the ruler, so we might report 24.6 mm. Another person might report 24.7 mm or 24.5 mm—both of which are within the realm of the precision of the ruler. However, it would be incorrect to report 24.60 mm because reporting the measurement with four digits instead of three overstates the precision of the measuring device.

Suppose that we weigh an object on a balance with marks at every 1 g, and the pointer is between the 1-g mark and the 2-g mark (▶ FIGURE 2.1) but much closer to the 1-g mark. To record this measurement, we mentally divide the space between the 1- and 2-g marks into 10 equal spaces and estimate the position of the pointer. In this case, the pointer indicates about 1.2 g. We write the measurement as 1.2 g, indicating that we are sure of the "1" but have estimated the ".2."

If we measure an object using a balance with marks every tenth of a gram, we need to write the result with more digits. For example, suppose that on this more precise balance the pointer is between the 1.2-g mark and the 1.3-g mark (▶ FIGURE 2.2). We again divide the space between the two marks into 10 equal spaces and estimate the third digit. In the case of the nut shown in Figure 2.2, we report 1.26 g. Digital balances usually have readouts that report the mass to the correct number of digits.

1.2 g

Balance has marks every one gram.

▲ FIGURE 2.1 **Estimating tenths of a gram** This balance has markings every 1 g, so we estimate to the tenths place. To estimate between markings, we mentally divide the space into 10 equal spaces and estimate the last digit. This reading is 1.2 g.

▲ FIGURE 2.2 **Estimating hundredths of a gram** Because this scale has markings every 0.1 g, we estimate to the hundredths place. The correct reading is 1.26 g.

EXAMPLE **2.3** **Reporting the Right Number of Digits**

The bathroom scale in ▼ FIGURE 2.3 has markings at every 1 lb. Report the reading to the correct number of digits.

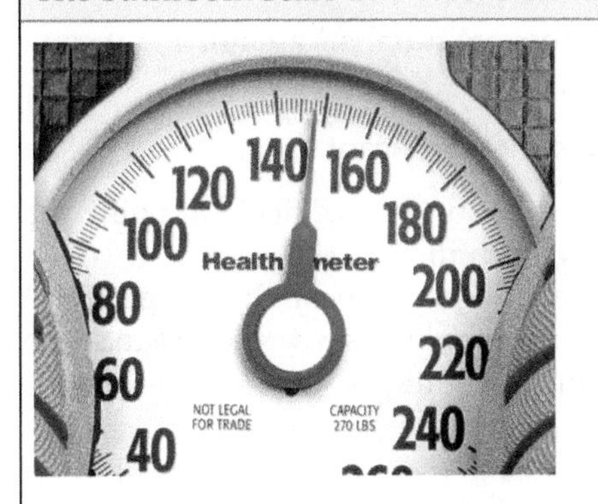

▲ FIGURE 2.3 **Reading a bathroom scale**

SOLUTION

Because the pointer is between the 147- and 148-lb markings, you mentally divide the space between the markings into 10 equal spaces and estimate the next digit. In this case, you should report the result as:

147.7 lb

What if you estimated a little differently and wrote 147.6 lb? In general, one unit of difference in the last digit is acceptable because the last digit is estimated and different people might estimate it slightly differently. However, if you wrote 147.2 lb, you would clearly be wrong.

▶ **SKILLBUILDER 2.3 | Reporting the Right Number of Digits**

You use a thermometer to measure the temperature of a backyard hot tub, and you obtain the reading shown in ▶ FIGURE 2.4. Record the temperature reading to the correct number of digits.

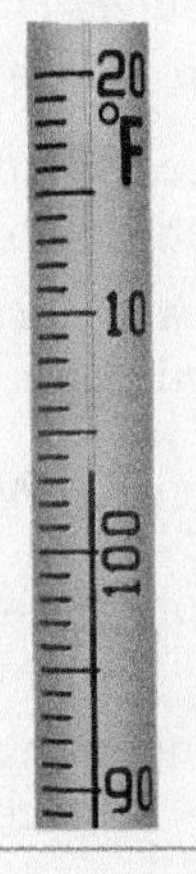

◀ FIGURE 2.4 **Reading a thermometer**

▶ **FOR MORE PRACTICE** Example 2.20; Problems 41, 42.

Counting Significant Figures

The non–place-holding digits in a measurement are **significant figures** (or **significant digits**). As we have seen, these significant figures represent the precision of a measured quantity—the greater the number of significant figures, the greater the precision of the measurement. We can determine the number of significant figures in a written number fairly easily; however, if the number contains zeros, we must distinguish between the zeros that are significant and those that simply mark the decimal place. In the number 0.002, for example, the leading zeros mark the decimal place; they *do not* add to the precision of the measurement. In the number 0.00200, however, the trailing zeros *do* add to the precision of the measurement.

To determine the number of significant figures in a number, we follow these rules:

1. All nonzero digits are significant.

   $\underline{1}.0\underline{5}$ $\quad$ $0.0\underline{11}0$

2. Interior zeros (zeros between two numbers) are significant.

   $4.\underline{0}2\underline{0}8$ $\quad$ $5\underline{0}.1$

3. Trailing zeros (zeros to the right of a nonzero number) that fall after a decimal point are significant.

   $5.1\underline{0}$ $\quad$ $3.\underline{00}$

When a number is expressed in scientific notation, all trailing zeros are significant.

4. Trailing zeros that fall before a decimal point are significant.

   $5\underline{0}.00$ $\quad$ $17\underline{00}.24$

5. Leading zeros (zeros to the left of the first nonzero number) are not significant. They only serve to locate the decimal point.

   For example, the number 0.0005 has only one significant digit.

Some books put a decimal point after one or more trailing zeros if the zeros are to be considered significant. We avoid that practice in this book, but you should be aware of it.

6. Trailing zeros at the end of a number, but before an *implied* decimal point, are ambiguous and should be avoided by using scientific notation.

   For example, it is unclear if the number 350 has two or three significant figures. We can avoid confusion by writing the number as 3.5×10^2 to indicate two significant figures or as 3.50×10^2 to indicate three.

Exact Numbers

Exact numbers have an unlimited number of significant figures. Exact numbers originate from three sources:

- Exact counting of discrete objects. For example, 10 pencils means 10.0000... pencils and 3 atoms means 3.00000...atoms.
- *Defined quantities*, such as the number of centimeters in 1 m. Because 100 cm is defined as 1 m,

  $$100 \text{ cm} = 1 \text{ m means } 100.00000\ldots \text{ cm} = 1.0000000\ldots \text{ m}$$

  Note that some conversion factors (see Section 2.6) are defined quantities while others are not.
- Integral numbers that are part of an equation. For example, in the equation, $radius = \dfrac{diameter}{2}$, the number 2 is exact and therefore has an unlimited number of significant figures.

EXAMPLE 2.4 Determining the Number of Significant Figures in a Number

How many significant figures are in each number?

(a) 0.0035 **(b)** 1.080 **(c)** 2371 **(d)** 2.97×10^5 **(e)** 1 dozen = 12 **(f)** 100.00 **(g)** 100,000

	SOLUTION
The 3 and the 5 are significant (rule 1). The leading zeros only mark the decimal place and are not significant (rule 5).	**(a)** 0.0035 two significant figures
The interior zero is significant (rule 2), and the trailing zero is significant (rule 3). The 1 and the 8 are also significant (rule 1).	**(b)** 1.080 four significant figures
All digits are significant (rule 1).	**(c)** 2371 four significant figures
All digits in the decimal part are significant (rule 1).	**(d)** 2.97×10^5 three significant figures
Defined numbers are exact and therefore have an unlimited number of significant figures.	**(e)** 1 dozen = 12 unlimited significant figures
The 1 is significant (rule 1), and the trailing zeros before the decimal point are significant (rule 4). The trailing zeros after the decimal point are also significant (rule 3).	**(f)** 100.00 five significant figures
This number is ambiguous. Write as 1×10^5 to indicate one significant figure or as 1.00000×10^5 to indicate six significant figures.	**(g)** 100,000 ambiguous

▶ SKILLBUILDER 2.4 | Determining the Number of Significant Figures in a Number

How many significant figures are in each number?

(a) 58.31 **(b)** 0.00250 **(c)** 2.7×10^3 **(d)** 1 cm = 0.01 m **(e)** 0.500 **(f)** 2100

▶ FOR MORE PRACTICE Example 2.21; Problems 43, 44, 45, 46, 47, 48.

CHEMISTRY IN THE MEDIA

The COBE Satellite and Very Precise Measurements That Illuminate Our Cosmic Past

Since the earliest times, humans have wondered about the origins of our planet. Scientists have probed this question and developed theories for how the universe and the Earth began. The most accepted theory today about the origin of the universe is the Big Bang theory. According to this theory, the universe began in a tremendous expansion about 13.7 billion years ago and has been expanding ever since. A measurable prediction of this theory is the presence of a remnant "background radiation" from the expansion of the universe. That remnant radiation is characteristic of the current temperature of the universe. When the Big Bang occurred, the temperature of the universe was very hot and the associated radiation very bright. Today, 13.7 billion years later, the temperature of the universe is very cold and the background radiation is very faint.

In the early 1960s, Robert H. Dicke, P. J. E. Peebles, and their colleagues at Princeton University began to build a device to measure this background radiation and by doing so, they took a direct look into the cosmological past and provided evidence for the Big Bang theory. At about the same time, quite by accident, Arno Penzias and Robert Wilson of Bell Telephone Laboratories measured excess radio noise on one of their communications satellites. As it turned out, this noise was the background radiation that the Princeton scientists were looking for. The two groups published papers together in 1965 reporting their findings along with the corresponding current temperature of the universe, about 3 degrees above absolute zero, or 3 K. We will define temperature measurement scales in Chapter 3. For now, know that 3 K is an extremely low temperature (454 degrees below zero on the Fahrenheit scale).

In 1989, NASA's Goddard Space Flight Center developed the Cosmic Background Explorer (COBE) satellite to measure the background radiation more precisely. The COBE satellite determined that the background radiation corresponded to a universe with a temperature of 2.735 K. (Notice the difference in significant figures from the previous measurement.) It went on to measure tiny fluctuations in the background radiation that amount to temperature differences of 1 part in 100,000. These fluctuations, though small, are an important prediction of the Big Bang theory. Scientists announced that the COBE satellite had produced the strongest evidence yet for the Big Bang theory of the creation of the universe. This is the way that science works. Measurement, and precision in measurement, are important to understanding the world—so important that we dedicate most of this chapter to the concept of measurement.

B2.1 CAN YOU ANSWER THIS? *How many significant figures are there in each of the preceding temperature measurements (3 K, 2.735 K)?*

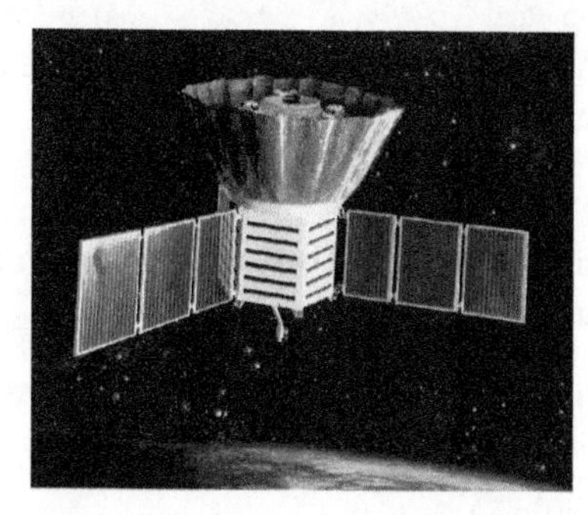

▲ The COBE satellite, launched in 1989 to measure background radiation. Background radiation is a remnant of the Big Bang—the expansion that is believed to have formed the universe.

CONCEPTUAL ✔ CHECKPOINT 2.2

The Curiosity Rover on the surface of Mars recently measured a daily low temperature of −65.19 °C. What is the implied range of the actual temperature?

(a) between −65.190 °C and −65.199 °C **(b)** between −65.18 °C and −65.20 °C
(c) between −65.1 °C and −65.2 °C **(d)** exactly −65.19 °C

2.4 Significant Figures in Calculations

- Round numbers to the correct number of significant figures.
- Determine the correct number of significant figures in the results of multiplication and division calculations.
- Determine the correct number of significant figures in the results of addition and subtraction calculations.
- Determine the correct number of significant figures in the results of calculations involving both addition/subtraction and multiplication/division.

When we use measured quantities in calculations, the results of the calculation must reflect the precision of the measured quantities. We should not lose or gain precision during mathematical operations.

Multiplication and Division

In multiplication or division, the result carries the same number of significant figures as the factor with the fewest significant figures.

For example:

$$\underset{\text{(3 sig. figures)}}{5.02} \times \underset{\text{(5 sig. figures)}}{89.665} \times \underset{\text{(2 sig. figures)}}{0.10} = 45.0118 = \underset{\text{(2 sig. figures)}}{45}$$

We round the intermediate result (in blue) to two significant figures to reflect the least precisely known factor (0.10), which has two significant figures.

In division, we follow the same rule.

$$\underset{\text{(4 sig. figures)}}{5.892} \div \underset{\text{(3 sig. figures)}}{6.10} = 0.96590 = \underset{\text{(3 sig. figures)}}{0.966}$$

We round the intermediate result (in blue) to three significant figures to reflect the least precisely known factor (6.10), which has three significant figures.

Rounding

When we round to the correct number of significant figures:

We round down if the last (or leftmost) digit dropped is 4 or less;
we round up if the last (or leftmost) digit dropped is 5 or more.

For example, when we round each of these numbers to two significant figures:

2.33 rounds to 2.3
2.37 rounds to 2.4

2.34 rounds to 2.3
2.35 rounds to 2.4

We consider only the *last (or leftmost) digit being dropped* when we decide in which direction to round—we ignore all digits to the right of it. For example, to round 2.349 to two significant figures, only the 4 in the hundredths place (2.349) determines which direction to round—the 9 is irrelevant.

2.349 rounds to 2.3

For calculations involving multiple steps, we round only the final answer—we do not round the intermediate results. This prevents small rounding errors from affecting the final answer.

EXAMPLE **2.5** **Significant Figures in Multiplication and Division**

Perform each calculation to the correct number of significant figures.

(a) $1.01 \times 0.12 \times 53.51 \div 96$
(b) $56.55 \times 0.920 \div 34.2585$

	SOLUTION
Round the intermediate result (in blue) to two significant figures to reflect the two significant figures in the least precisely known quantities (0.12 and 96).	**(a)** $1.01 \times 0.12 \times 53.51 \div 96 = 0.067556 = 0.068$
Round the intermediate result (in blue) to three significant figures to reflect the three significant figures in the least precisely known quantity (0.920).	**(b)** $56.55 \times 0.920 \div 34.2585 = 1.51863 = 1.52$

▶ **SKILLBUILDER 2.5 | Significant Figures in Multiplication and Division**
Perform each calculation to the correct number of significant figures.

(a) $1.10 \times 0.512 \times 1.301 \times 0.005 \div 3.4$
(b) $4.562 \times 3.99870 \div 89.5$

▶ **FOR MORE PRACTICE** Examples 2.22, 2.23; Problems 57, 58, 59, 60.

Addition and Subtraction

In addition or subtraction, the result has the same number of decimal places as the quantity with the fewest decimal places.

For example:

$$\begin{array}{r} 5.74 \\ 0.823 \\ +\ 2.651 \\ \hline 9.214 \end{array} = 9.21$$

It is sometimes helpful to draw a vertical line directly to the right of the number with the fewest decimal places. The line shows the number of decimal places that should be in the answer.

We round the intermediate answer (in blue) to two decimal places because the quantity with the fewest decimal places (5.74) has two decimal places.

For subtraction, we follow the same rule. For example:

$$\begin{array}{r} 4.8 \\ -\ 3.965 \\ \hline 0.835 \end{array} = 0.8$$

We round the intermediate answer (in blue) to one decimal place because the quantity with the fewest decimal places (4.8) has one decimal place. Remember: *For multiplication and division, the quantity with the fewest* ***significant figures*** *determines* *the number of significant figures in the answer. For addition and subtraction, the quantity with the fewest* ***decimal places*** *determines the number of decimal places in the answer.* In multiplication and division we focus on significant figures, but in addition and subtraction we focus on decimal places. When a problem involves addition and subtraction, the answer may have a different number of significant figures than the initial quantities. For example:

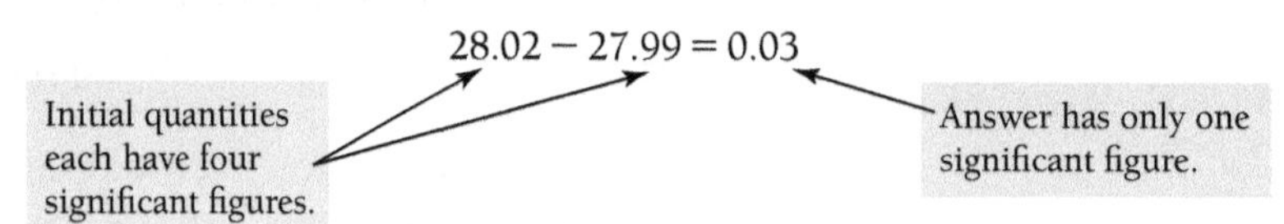

The answer has only one significant figure, even though the initial quantities each had four significant figures.

EXAMPLE 2.6 Significant Figures in Addition and Subtraction

Perform the calculations to the correct number of significant figures.

(a)
$$\begin{array}{r} 0.987 \\ +125.1 \\ -1.22 \\ \hline \end{array}$$

(b)
$$\begin{array}{r} 0.765 \\ -3.449 \\ -5.98 \\ \hline \end{array}$$

SOLUTION

Round the intermediate answer (in blue) to one decimal place to reflect the quantity with the fewest decimal places (125.1). Notice that 125.1 is not the quantity with the fewest significant figures—it has four while the other quantities only have three—but because it has the fewest decimal places, it determines the number of decimal places in the answer.

(a)
$$\begin{array}{r} 0.987 \\ +125.1 \\ -1.22 \\ \hline 124.867 \end{array} = 124.9$$

Round the intermediate answer (in blue) to two decimal places to reflect the quantity with the fewest decimal places (5.98).

(b)
$$\begin{array}{r} 0.765 \\ -3.449 \\ -5.98 \\ \hline -8.664 \end{array} = -8.66$$

▶ **SKILLBUILDER 2.6 | Significant Figures in Addition and Subtraction**

Perform the calculations to the correct number of significant figures.

(a)
$$\begin{array}{r} 2.18 \\ +5.621 \\ +1.5870 \\ -1.8 \\ \hline \end{array}$$

(b)
$$\begin{array}{r} 7.876 \\ -0.56 \\ +123.792 \\ \hline \end{array}$$

▶ **FOR MORE PRACTICE** Example 2.24; Problems 61, 62, 63, 64.

Calculations Involving Both Multiplication/Division and Addition/Subtraction

In calculations involving both multiplication/division and addition/subtraction, we do the steps in parentheses first; determine the correct number of significant figures in the intermediate answer; then complete the remaining steps.

For example:

$$3.489 \times (5.67 - 2.3)$$

We complete the subtraction step first.

$$5.67 - 2.3 = 3.37$$

We use the subtraction rule to determine that the intermediate answer (3.37) has only one significant decimal place. To avoid small errors, we do not round at this point; instead, we underline the least significant figure as a reminder.

$$= 3.489 \times 3.\underline{3}7$$

We then complete the multiplication step.

$$3.489 \times 3.\underline{3}7 = 11.758 = 12$$

The multiplication rule indicates that the intermediate answer (11.758) rounds to two significant figures (12) because it is limited by the two significant figures in $3.\underline{3}7$.

EXAMPLE 2.7 Significant Figures in Calculations Involving Both Multiplication/Division and Addition/Subtraction

Perform the calculations to the correct number of significant figures.

(a) $6.78 \times 5.903 \times (5.489 - 5.01)$
(b) $19.667 - (5.4 \times 0.916)$

SOLUTION

Do the step in parentheses first. Use the subtraction rule to mark 0.479 to two decimal places because 5.01, the number in the parentheses with the least number of decimal places, has two.

(a) $6.78 \times 5.903 \times (5.489 - 5.01)$

$= 6.78 \times 5.903 \times (0.479)$

$= 6.78 \times 5.903 \times 0.4\underline{7}9$

Then perform the multiplication and round the answer to two significant figures because the number with the least number of significant figures has two.

$6.78 \times 5.903 \times 0.4\underline{7}9 = 19.1707$

$= 19$

Do the step in parentheses first. The number with the least number of significant figures within the parentheses (5.4) has two, so mark the answer to two significant figures.

(b) $19.667 - (5.4 \times 0.916)$

$= 19.667 - (4.9464)$

$= 19.667 - 4.\underline{9}464$

Then perform the subtraction and round the answer to one decimal place because the number with the least number of decimal places has one.

$19.667 - 4.\underline{9}464 = 14.7206$

$= 14.7$

▶ **SKILLBUILDER 2.7 | Significant Figures in Calculations Involving Both Multiplication/Division and Addition/Subtraction**

Perform each calculation to the correct number of significant figures.

(a) $3.897 \times (782.3 - 451.88)$ **(b)** $(4.58 \div 1.239) - 0.578$

▶ **FOR MORE PRACTICE** Example 2.25; Problems 65, 66, 67, 68.

PEARSON eText 2.0

CONCEPTUAL ✔ CHECKPOINT 2.3

Which calculation would have its result reported to the *greater* number of significant figures?

(a) $3 + (15/12)$
(b) $(3 + 15)/12$

2.5 The Basic Units of Measurement

▶ Recognize and work with the SI base units of measurement, prefix multipliers, and derived units.

By themselves, numbers have little meaning. Read this sentence: When my son was 7 he walked 3, and when he was 4 he could throw his baseball 8 and tell us that his school was 5 away. The sentence is confusing. We don't know what the numbers mean because the **units** are missing. The meaning becomes clear, however, when we add the missing units to the numbers: When my son was 7 *months old* he walked 3 *steps*, and when he was 4 *years old* he could throw his baseball 8 *feet* and tell us that his school was 5 *minutes* away. Units make all the difference. In chemistry, units are critical. Never write a number by itself; always use its associated units—otherwise your work will be as confusing as the initial sentence.

The two most common unit systems are the **English system**, used in the United States, and the **metric system**, used in most of the rest of the world. The English system uses units such as inches, yards, and pounds, while the metric system uses centimeters, meters, and kilograms. The most convenient system for science measurements is based on the metric system and is called the **International System** of units or **SI units**. SI units are a set of standard units agreed on by scientists throughout the world.

The abbreviation *SI* comes from the French *le Système International*.

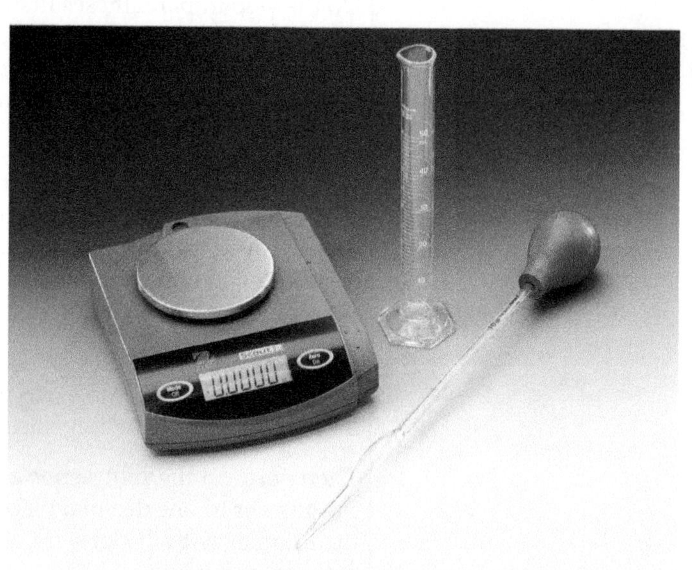

▲ Science uses instruments to make measurements. Every instrument is calibrated in a particular unit without which the measurements would be meaningless.

TABLE 2.1 Important SI Base Units

Quantity	Unit	Symbol
length	meter	m
mass	kilogram	kg
time	second	s
temperature*	kelvin	K

*Temperature units are discussed in Chapter 3.

The Base Units

Table 2.1 lists the common SI base units. They include the **meter (m)** as the base unit of length; the **kilogram (kg)** as the base unit of mass; and the **second (s)** as the base unit of time. Each of these base units is precisely defined. The meter is defined as the distance light travels in a certain period of time: 1/299,792,458 s (▶ FIGURE 2.5). (The speed of light is 3.0×10^8 m/s.) The kilogram is defined as the mass of a block of metal kept at the International Bureau of Weights and Measures at Sèvres, France (▶ FIGURE 2.6). The second is defined using an atomic standard (▶ FIGURE 2.7).

Because the mass of the block of metal used to define a kilogram has changed slightly over the years, scientists are currently working on alternate ways to define the kilogram.

Most people are familiar with the SI base unit of time, the second. However, if you live in the United States, you may be less familiar with the meter and the kilogram. The meter is slightly longer than a yard (a yard is 36 in., while a meter is 39.37 in.). A 100-yd football field measures only 91.4 m.

▲ FIGURE 2.5 **The base unit of length** The definition of a meter, established by international agreement in 1983, is the distance that light travels in a vacuum in 1/299,792,458 s.

Question: Why is such a precise standard necessary?

▲ FIGURE 2.6 **The base unit of mass** A duplicate of the international standard kilogram, called kilogram 20, is kept at the National Institute of Standards and Technology near Washington, D.C.

▲ FIGURE 2.7 **The base unit of time** The second is defined, using an atomic clock, as the duration of 9,192,631,770 periods of the radiation emitted from a certain transition in a cesium-133 atom.

A nickel (5 cents) has a mass of about 5 g.

The kilogram is a measure of mass, which is different from weight. The **mass** of an object is a measure of the quantity of matter within it, while the weight of an object is a measure of the gravitational pull on that matter. Consequently, weight depends on gravity, while mass does not. If you were to weigh yourself on Mars, for example, the lower gravity would pull you toward the scale less than Earth's gravity would, resulting in a lower weight. A 150-lb person on Earth weighs only 57 lb on Mars. However, the person's mass, the quantity of matter in his or her body, remains the same. A kilogram of mass is the equivalent of 2.205 lb of weight on Earth, so if we express mass in kilograms, a 150-lb person on Earth has a mass of approximately 68 kg. A second common unit of mass is the gram (g), defined as follows:

$$1000 \text{ g} = 10^3 \text{ g} = 1 \text{ kg}$$

Prefix Multipliers

The SI system employs **prefix multipliers** (Table 2.2, on the next page) with the base units. These multipliers change the value of the unit by powers of 10. For example, the kilometer (km) features the prefix *kilo-*, meaning 1000 or 10^3. Therefore:

$$1 \text{ km} = 1000 \text{ m} = 10^3 \text{ m}$$

Similarly, the millisecond (ms) has the prefix *milli-*, meaning 0.001 or 10^{-3}.

$$1 \text{ ms} = 0.001 \text{ s} = 10^{-3} \text{ s}$$

The prefix multipliers allow us to express a wide range of measurements in units that are similar in size to the quantity we are measuring. We choose the prefix multiplier that is most convenient for a particular measurement. For example, to measure the diameter of a coin, we might use centimeters or millimeters because coins have diameters between 1–3 cm (or 10–30 mm). A centimeter is a common metric unit and is about equivalent to the width of a pinky finger (2.54 cm = 1 in.). The millimeter could also conveniently express the diameter of a coin, but the kilometer would not work as well because, in that unit, a coin's diameter is 0.000010–0.000030 km. We pick a unit similar in size to (or smaller than) the quantity we are measuring. Consider expressing the length of a short chemical bond, about 1.2×10^{-10} m. Which prefix multiplier should we use? The most convenient one is probably the picometer (pico = 10^{-12}). Chemical bonds measure about 120 pm.

TABLE 2.2 SI Prefix Multipliers

Prefix	Symbol	Meaning	Multiplier	
tera-	T	trillion	1,000,000,000,000	(10^{12})
giga-	G	billion	1,000,000,000	(10^{9})
mega-	M	million	1,000,000	(10^{6})
kilo-	k	thousand	1,000	(10^{3})
hecto-	h	hundred	100	10^{2}
deca-	da	ten	10	10^{1}
deci-	d	tenth	0.1	(10^{-1})
centi-	c	hundredth	0.01	(10^{-2})
milli-	m	thousandth	0.001	(10^{-3})
micro-	μ	millionth	0.000001	(10^{-6})
nano-	n	billionth	0.000000001	(10^{-9})
pico-	p	trillionth	0.000000000001	(10^{-12})
femto-	f	quadrillionth	0.000000000000001	(10^{-15})

TABLE 2.3 Some Common Units and Their Equivalents

Length
1 kilometer (km) = 0.6214 mile (mi)
1 meter (m) = 39.37 inches (in.)
= 1.094 yards (yd)
1 foot (ft) = 30.48 centimeters (cm)
1 inch (in.) = 2.54 centimeters (cm) (exact)

Mass
1 kilogram (kg) = 2.205 pounds (lb)
1 pound (lb) = 453.59 grams (g)
1 ounce (oz) = 28.35 grams (g)

Volume
1 liter (L) = 1000 milliliters (mL)
= 1000 cubic centimeters (cm^3)
1 liter (L) = 1.057 quarts (qt)
1 U.S. gallon (gal) = 3.785 liters (L)

PEARSON eText 2.0

CONCEPTUAL CHECKPOINT 2.4

Which is the most convenient unit to use to express the dimensions of a polio virus, which is about 2.8×10^{-8} m in diameter?

(a) Mm **(b)** mm **(c)** μm **(d)** nm

Derived Units

A derived unit is formed from other units. For example, many units of **volume**, a measure of space, are derived units. Any unit of length, when cubed (raised to the third power), becomes a unit of volume. Therefore, cubic meters (m^3), cubic centimeters (cm^3), and cubic millimeters (mm^3) are all units of volume. A three-bedroom house has a volume of about 630 m^3, a can of soda pop has a volume of about 350 cm^3, and a rice grain has a volume of about 3 mm^3. We also use the **liter (L)** and milliliter (mL) to express volume (although these are not derived units). A gallon is equal to 3.785 L. A milliliter is equivalent to 1 cm^3. Table 2.3 lists some common units and their equivalents.

2.6 Problem Solving and Unit Conversion

▶ Convert between units.

Problem solving is one of the most important skills you will acquire in this course. Not only will this skill help you succeed in chemistry, but it will also help you learn how to think critically, which is important in every area of knowledge. When my daughter was a freshman in high school, she came to me for help on an algebra problem. The problem went something like this:

> Sam and Sara live 11 miles apart. Sam leaves his house traveling at 6 miles per hour toward Sara's house. Sara leaves her house traveling at 3 miles per hour toward Sam's house. How much time elapses until Sam and Sara meet?

The Mathematics Review Appendix (p. MR-1) includes a review of how to solve algebraic problems for a variable.

Solving the problem requires setting up the equation $11 - 6t = 3t$. Although my daughter could solve this equation for t quite easily, getting to the equation from the problem statement was another matter—that process requires *critical thinking*, and that was the skill she needed to learn to successfully solve the problem. You can't succeed in chemistry—or in life, really—without developing critical thinking skills. Learning how to solve chemical problems helps you develop these kinds of skills.

Although no simple formula applies to every problem, you can learn problem-solving strategies and begin to develop some chemical intuition. You can think of many of the problems in this book as *unit conversion problems*, where you are given one or more quantities and asked to convert them into different units. Other problems require the use of *specific equations* to get to the information you are trying to find. In the sections that follow, you will find strategies to help you solve both of these types of problems. Of course, many problems contain both conversions and equations, requiring the combination of these strategies, and some problems may require an altogether different approach, but the basic tools you learn here can be applied to those problems as well.

Converting Between Units

Using units as a guide to solving problems is called dimensional analysis.

Units are critical in calculations. Knowing how to work with and manipulate units in calculations is a crucial part of problem solving. In calculations, units help determine correctness. You should always include units in calculations, and you can think of many calculations as converting from one unit to another. You can multiply, divide, and cancel units like any other algebraic quantity.

Remember:

1. Always write every number with its associated unit. Never ignore units; they are critical.
2. Always include units in your calculations, dividing them and multiplying them as if they were algebraic quantities. Do not let units magically appear or disappear in calculations. Units must flow logically from beginning to end.

Consider converting 17.6 in. to centimeters. You know from Table 2.3 that 1 in. = 2.54 cm. To determine how many centimeters are in 17.6 in., perform the conversion:

$$17.6\ \cancel{\text{in.}} \times \frac{2.54\ \text{cm}}{1\ \cancel{\text{in.}}} = 44.7\ \text{cm}$$

The unit in. cancels and you are left with cm as your final unit. The quantity $\frac{2.54\ \text{cm}}{1\ \text{in.}}$ is a **conversion factor** between in. and cm—it is a quotient with cm on top and in. on bottom.

For most conversion problems, you are given a quantity in some units and asked to convert the quantity to another unit. These calculations take the form:

$$\text{information given} \times \text{conversion factor(s)} = \text{information sought}$$

$$\cancel{\text{given unit}} \times \frac{\text{desired unit}}{\cancel{\text{given unit}}} = \text{desired unit}$$

You can construct conversion factors from any two quantities known to be equivalent. In this example, 2.54 cm = 1 in., so construct the conversion factor by dividing both sides of the equality by 1 in. and canceling the units.

$$2.54\ \text{cm} = 1\ \text{in.}$$

$$\frac{2.54\ \text{cm}}{1\ \text{in.}} = \frac{1\ \cancel{\text{in.}}}{1\ \cancel{\text{in.}}}$$

$$\frac{2.54\ \text{cm}}{1\ \text{in.}} = 1$$

The quantity $\frac{2.54\ \text{cm}}{1\ \text{in.}}$ is equal to 1, and you can use it to convert between inches and centimeters.

What if you want to perform the conversion the other way, from centimeters to inches? If you try to use the same conversion factor, the units do not cancel correctly.

$$44.7 \text{ cm} \times \frac{2.54 \text{ cm}}{1 \text{ in.}} = \frac{114 \text{ cm}^2}{\text{in.}}$$

The units in the answer, as well as the value of the answer, are incorrect. The unit $\text{cm}^2/\text{in.}$ is not correct, and, based on your knowledge that centimeters are smaller than inches, you know that 44.7 cm cannot be equivalent to 114 in. In solving problems, always check if the final units are correct, and consider whether or not the magnitude of the answer makes sense. In this case, the mistake was in how the conversion factor was used. You must invert it.

$$44.7 \cancel{\text{cm}} \times \frac{1 \text{ in.}}{2.54 \cancel{\text{cm}}} = 17.6 \text{ in.}$$

You can invert conversion factors because they are equal to 1 and the inverse of 1 is 1.

$$\frac{1}{1} = 1$$

Therefore,

$$\frac{2.54 \text{ cm}}{1 \text{ in.}} = 1 = \frac{1 \text{ in.}}{2.54 \text{ cm}}$$

You can diagram conversions using a **solution map**. A solution map is a visual outline that shows the strategic route required to solve a problem. For unit conversion, the solution map focuses on units and how to convert from one unit to another. The solution map for converting from inches to centimeters is:

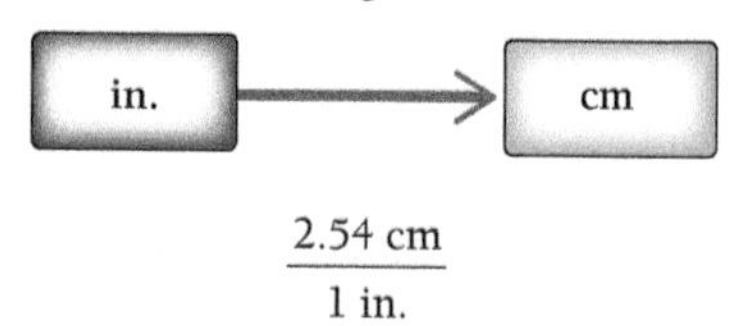

The solution map for converting from centimeters to inches is:

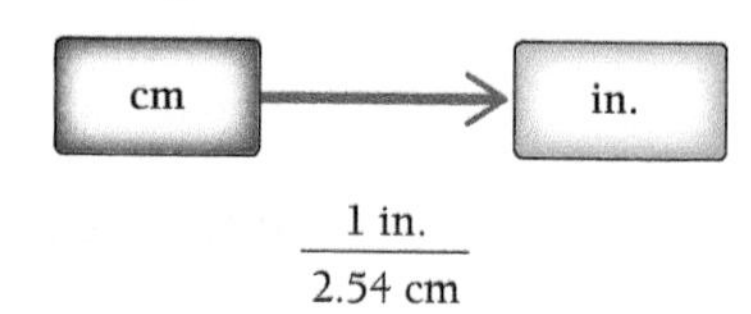

Each arrow in a solution map for a unit conversion has an associated conversion factor, with the units of the previous step in the denominator and the units of the following step in the numerator. For one-step problems such as these, the solution map is only moderately helpful, but for multistep problems, it becomes a powerful way to develop a problem-solving strategy. In the section that follows, you will learn how to incorporate solution maps into an overall problem-solving strategy.

PEARSON eText 2.0

CONCEPTUAL CHECKPOINT 2.5

Which conversion factor should you use to convert 4 ft to inches (12 in. = 1 ft)?

(a) $\frac{12 \text{ in.}}{1 \text{ ft}}$ (b) $\frac{1 \text{ ft}}{12 \text{ in.}}$ (c) $\frac{1 \text{ in.}}{12 \text{ ft}}$ (d) $\frac{12 \text{ ft}}{1 \text{ in.}}$

General Problem-Solving Strategy

In this book, we use a standard problem-solving procedure that you can adapt to many of the problems encountered in chemistry and beyond. Solving any problem essentially requires that you assess the information given in the problem and devise a way to get to the requested information. In other words, you need to:

- Identify the starting point (the ***given*** information).
- Identify the endpoint (what you must ***find***).
- Devise a way to get from the starting point to the endpoint using what is given as well as what you already know or can look up. You can use a *solution map* to diagram the steps required to get from the starting point to the endpoint.

In graphic form, this progression looks like this:

Given $\longrightarrow$ **Solution Map** $\longrightarrow$ **Find**

Beginning students often have trouble knowing how to start solving a chemistry problem. Although no problem-solving procedure is applicable to all problems, the following four-step procedure can be helpful in working through many of the numerical problems you will encounter in chemistry.

1. **Sort.** Begin by sorting the information in the problem. *Given* information is the basic data provided by the problem—often one or more numbers with their associated units. The given information is the starting point for the problem. *Find* indicates what the problem is asking you to find (the endpoint of the problem).
2. **Strategize.** This is usually the hardest part of solving a problem. In this process, you must create a solution map—the series of steps that will get you from the given information to the information you are trying to find. You have already seen solution maps for simple unit conversion problems. Each arrow in a solution map represents a computational step. On the left side of the arrow is the quantity (or quantities) you had before the step; on the right side of the arrow is the quantity (or quantities) you will have after the step; and below the arrow is the information you need to get from one to the other—the relationship between the quantities.

   Often such relationships will take the form of conversion factors or equations. These may be given in the problem, in which case you will have written them down under "Given" in Step 1. Usually, however, you will need other information—such as physical constants, formulas, or conversion factors—to help get you from what you are given to what you must find.You may recall this information from what you have learned, or you can look it up in the chapters or tables within the book.

   In some cases, you may get stuck at the strategize step. If you cannot figure out how to get from the given information to the information you are asked to find, you might try working backwards. For example, you may want to look at the units of the quantity you are trying to find and look for conversion factors to get to the units of the given quantity. You may even try a combination of strategies; work forward, backward, or some of both. If you persist, you will develop a strategy to solve the problem.
3. **Solve.** This is the most straightforward part of solving a problem. Once you set up the problem properly and devise a solution map, you follow the map to solve the problem. Carry out mathematical operations (paying attention to the rules for significant figures in calculations) and cancel units as needed.
4. **Check.** Beginning students often overlook this step. Experienced problem solvers always ask, Does this answer make physical sense? Are the units correct? Is the number of significant figures correct? When solving multistep problems, errors easily creep into the solution. You can catch most of these errors by simply checking the answer. For example, suppose you are calculating the number of atoms in a gold coin and end up with an answer of 1.1×10^{-6} atoms. Could the gold coin really be composed of one-millionth of one atom?

In Examples 2.8 and 2.9, you will find this problem-solving procedure applied to unit conversion problems. The left column summarizes the procedure, and the middle and right columns show two examples of applying the procedure. You will encounter this three-column format in selected examples throughout this text. It allows you to see how a particular procedure can be applied to two different problems. Work through one problem first (from top to bottom) and then examine how the other problem applies the same procedure. Recognizing the commonalities and differences between problems is a key part of problem solving.

Problem-Solving Procedure	EXAMPLE **2.8** **UNIT CONVERSION** Convert 7.8 km to miles.	EXAMPLE **2.9** **UNIT CONVERSION** Convert 0.825 m to millimeters.
SORT Begin by sorting the information in the problem into *given* and *find*.	**GIVEN:** 7.8 km **FIND:** mi	**GIVEN:** 0.825 m **FIND:** mm
STRATEGIZE Draw a *solution map* for the problem. Begin with the *given* quantity and symbolize each step with an arrow. Below the arrow, write the conversion factor for that step. The solution map ends at the *find* quantity. (In these examples, the relationships used in the conversions are below the solution map.)	**SOLUTION MAP** km → mi $\frac{0.6214\text{ mi}}{1\text{ km}}$ **RELATIONSHIPS USED** 1 km = 0.6214 mi (This conversion factor is from Table 2.3.)	**SOLUTION MAP** m → mm $\frac{1\text{ mm}}{10^{-3}\text{ m}}$ **RELATIONSHIPS USED** $1\text{ mm} = 10^{-3}\text{ m}$ (This conversion factor is from Table 2.2.)
SOLVE Follow the *solution map* to solve the problem. Begin with the *given* quantity and its units. Multiply by the appropriate conversion factor, canceling units to arrive at the *find* quantity.	**SOLUTION** $7.8\text{ km} \times \frac{0.6214\text{ mi}}{1\text{ km}} = 4.84692\text{ mi}$ $4.84692\text{ mi} = 4.8\text{ mi}$	**SOLUTION** $0.825\text{ m} \times \frac{1\text{ mm}}{10^{-3}\text{ m}} = 825\text{ mm}$ $825\text{ mm} = 825\text{ mm}$
Round the answer to the correct number of significant figures. (If possible, obtain conversion factors to enough significant figures so that they do not limit the number of significant figures in the answer.)	Round the answer to two significant figures because the quantity given has two significant figures.	Leave the answer with three significant figures because the quantity given has three significant figures and the conversion factor is a definition and therefore does not limit the number of significant figures in the answer.
CHECK Check your answer. Are the units correct? Does the answer make sense?	The units, mi, are correct. The magnitude of the answer is reasonable. A mile is longer than a kilometer, so the value in miles should be smaller than the value in kilometers.	The units, mm, are correct, and the magnitude is reasonable. A millimeter is shorter than a meter, so the value in millimeters should be larger than the value in meters.
	▶ **SKILLBUILDER 2.8 | Unit Conversion** Convert 56.0 cm to inches.	▶ **SKILLBUILDER 2.9 | Unit Conversion** Convert 5678 m to kilometers.
	▶ **FOR MORE PRACTICE** Example 2.26; Problems 73, 74, 75, 76.	▶ **FOR MORE PRACTICE** Problems 69, 70, 71, 72.

CONCEPTUAL CHECKPOINT 2.6

Which conversion factor should you use to convert a distance in meters to kilometers?

(a) $\frac{1\text{ m}}{10^3\text{ km}}$ (b) $\frac{10^3\text{ m}}{1\text{ km}}$ (c) $\frac{1\text{ km}}{10^3\text{ m}}$ (d) $\frac{10^3\text{ km}}{1\text{ m}}$

2.7 Solving Multistep Unit Conversion Problems

▶ Convert between units.

When solving multistep unit conversion problems, follow the preceding procedure, but add more steps to the solution map. Each step in the solution map should have a conversion factor, with the units of the previous step in the denominator and the units of the following step in the numerator. For example, suppose you want to convert 194 cm to ft. The solution map begins with cm, and you use the relationship 2.54 cm = 1 in. to convert to in. Then use the relationship 12 in. = 1 ft to convert to ft.

SOLUTION MAP

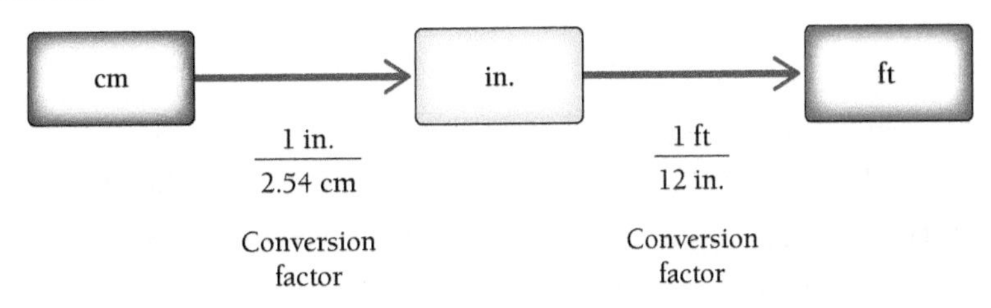

Once the solution map is complete, follow it to solve the problem.

SOLUTION

$$194\text{ cm} \times \frac{1\text{ in.}}{2.54\text{ cm}} \times \frac{1\text{ ft}}{12\text{ in.}} = 6.3648\text{ ft}$$

You then round to the correct number of significant figures—in this case, three (from 194 cm, which has three significant figures).

Because 1 foot is defined as 12 in., it does not limit significant figures.

$$6.3648\text{ ft} = 6.36\text{ ft}$$

Finally, check the answer. The units of the answer, feet, are the correct ones, and the magnitude seems about right. A foot is larger than a centimeter, so it is reasonable that the value in feet is smaller than the value in centimeters.

The eText 2.0 icon indicates that this feature is embedded and interactive in the eText.

Interactive Worked Example Video 2.10

EXAMPLE 2.10 Solving Multistep Unit Conversion Problems

A recipe for making creamy pasta sauce calls for 0.75 L of cream. Your measuring cup measures only in cups. How many cups of cream should you use? (4 cups = 1 quart)

SORT Begin by sorting the information in the problem into given and find.	**GIVEN:** 0.75 L **FIND:** cups

continued on page 34 ▶

continued from page 33

STRATEGIZE Draw a solution map for the problem. Begin with the *given* quantity and symbolize each step with an arrow. Below the arrow, write the conversion factor for that step. The solution map ends at the *find* quantity.	**SOLUTION MAP** 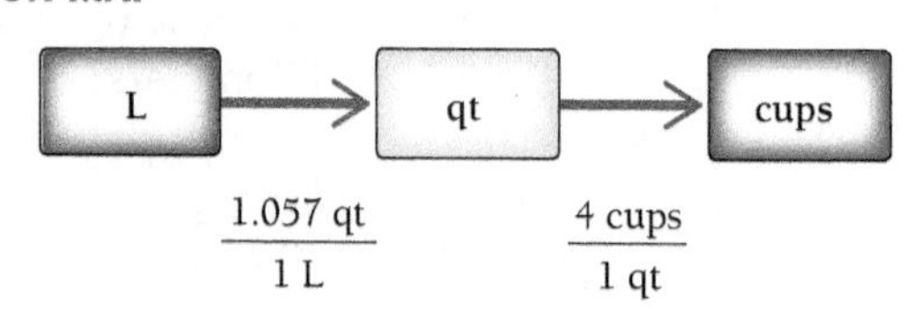 **RELATIONSHIPS USED** 1.057 qt = 1 L (from Table 2.3) 4 cups = 1 qt (given in problem statement)
SOLVE Follow the solution map to solve the problem. Begin with 0.75 L and multiply by the appropriate conversion factor, canceling units to arrive at qt. Then, use the second conversion factor to arrive at cups. Round the answer to the correct number of significant figures. In this case, you round the answer to two significant figures because the quantity given has two significant figures.	**SOLUTION** $0.75\ \cancel{L} \times \frac{1.057\ \cancel{qt}}{1\ \cancel{L}} \times \frac{4\ \text{cups}}{1\ \cancel{qt}} = 3.171\ \text{cups}$ $3.171\ \text{cups} = 3.2\ \text{cups}$
CHECK Check your answer. Are the units correct? Does the answer make physical sense?	The answer has the right units (cups) and seems reasonable. A cup is smaller than a liter, so the value in cups should be larger than the value in liters.

▶ **SKILLBUILDER 2.10 | Solving Multistep Unit Conversion Problems**

A recipe calls for 1.2 cups of oil. How many liters of oil is this?

▶ **FOR MORE PRACTICE** Problems 85, 86.

EXAMPLE 2.11 Solving Multistep Unit Conversion Problems

One lap of a running track measures 255 m. To run 10.0 km, how many laps should you run?

SORT Begin by sorting the information in the problem into given and find. You are given a distance in km and asked to find the distance in laps. You are also given the quantity 255 m per lap, which is a conversion factor between m and laps.	**GIVEN:** 10.0 km 255 m = 1 lap **FIND:** number of laps
STRATEGIZE Build the solution map beginning with km and ending at laps. Focus on the units.	**SOLUTION MAP** 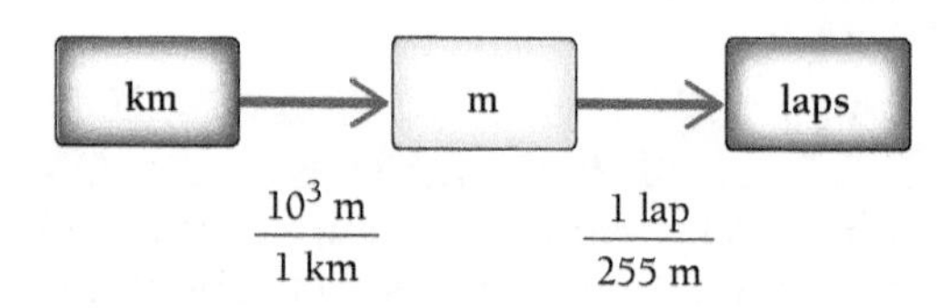 **RELATIONSHIPS USED** 1 km = 10^3 m (from Table 2.2) 1 lap = 255 m (given in problem)

SOLVE Follow the solution map to solve the problem. Begin with 10.0 km and multiply by the appropriate conversion factor, canceling units to arrive at m. Then, use the second conversion factor to arrive at laps. Round the intermediate answer (in blue) to three significant figures because it is limited by the three significant figures in the given quantity, 10.0 km.	**SOLUTION** $10.0\ \text{km} \times \frac{10^3\ \text{m}}{1\ \text{km}} \times \frac{1\ \text{lap}}{255\ \text{m}} = 39.216\ \text{laps} = 39.2\ \text{laps}$
CHECK Check your answer. Are the units correct? Does the answer make physical sense?	The units of the answer are correct, and the value of the answer makes sense. If a lap is 255 m, there are about 4 laps to each km (1000 m), so it seems reasonable that you would have to run about 40 laps to cover 10 km.

▶ **SKILLBUILDER 2.11 | Solving Multistep Unit Conversion Problems**
A running track measures 1056 ft per lap. To run 15.0 km, how many laps should you run? (1 mi = 5280 ft)

▶ **SKILLBUILDER PLUS**
An island is 5.72 nautical mi from the coast. How far away is the island in meters? (1 nautical mi = 1.151 mi)

▶ **FOR MORE PRACTICE** Problems 83, 84.

2.8 Unit Conversion in Both the Numerator and Denominator

▶ Convert units in a quantity that has units in the numerator and the denominator.

Some unit conversion problems require converting the units in both the numerator and denominator of a fraction. For example, the Toyota Prius has an EPA estimated city gas mileage of 48.0 miles per gallon. In Europe, gasoline is sold in liters (L), and distances are measured in kilometers (km). How do we convert the Prius's mileage estimate from miles per gallon to kilometers per liter? The answer is to use two conversion factors: one from miles to kilometers and another from gallons to liters:

$$1\ \text{mi} = 1.609\ \text{km}$$
$$1\ \text{gal} = 3.785\ \text{L}$$

Begin with the quantity 48.0 mi/gal and write the conversion factors so that the units cancel correctly. First, convert the numerator to km and then the denominator to L:

SOLUTION MAP

$$\frac{\text{mi}}{\text{gal}} \xrightarrow{\frac{1.609\ \text{km}}{\text{mi}}} \frac{\text{km}}{\text{gal}} \xrightarrow{\frac{1\ \text{gal}}{3.785\ \text{L}}} \frac{\text{km}}{\text{L}}$$

SOLUTION

$$48.0\frac{\text{mi}}{\text{gal}} \times \frac{1.609\ \text{km}}{\text{mi}} \times \frac{1\ \text{gal}}{3.785\ \text{L}} = 20.4\frac{\text{km}}{\text{L}}$$

Notice that to convert the denominator from gal to L, you write the conversion factor with gal in the numerator and L in the denominator.

EXAMPLE **2.12** **Solving Unit Conversions in the Numerator and Denominator**

A prescription medication requires 11.5 mg per kg of body weight. Convert this quantity to the number of grams required per pound of body weight and determine the correct dose (in g) for a 145-lb patient.

SORT
Begin by sorting the information in the problem into given and find. You are given the dose of the drug in mg/kg and the weight of the patient in lb. You are asked to find the dose in g/lb and the dose in g for the 145-lb patient.

GIVEN: $11.5\,\frac{\text{mg}}{\text{kg}}$

145 lb

FIND: $\frac{\text{g}}{\text{lb}}$; dose in g

STRATEGIZE
The solution map has two parts. In the first part, convert from mg/kg to g/lb. In the second part, use the result from the first part to determine the correct dose for a 145-lb patient.

SOLUTION MAP

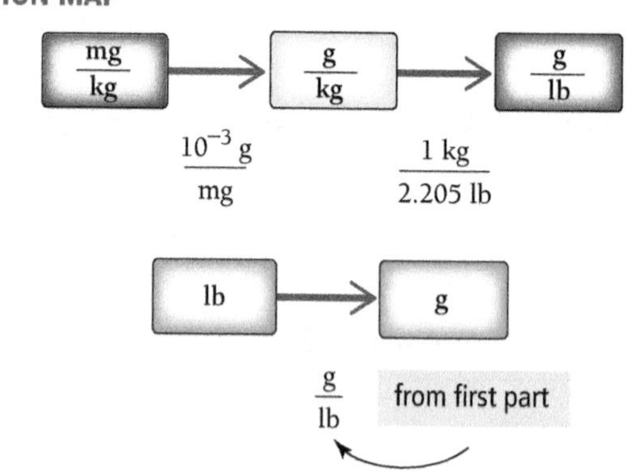

RELATIONSHIPS USED

$1\text{ mg} = 10^{-3}\text{ g}$ (from Table 2.2)

$1\text{ kg} = 2.205\text{ lb}$ (from Table 2.3)

SOLVE
Follow the solution map to solve the problem. For the first part, begin with 11.5 mg/kg and multiply by the two conversion factors to arrive at the dose in g/lb. Mark the answer to three significant figures to reflect the three significant figures in the least precisely known quantity.

SOLUTION

$$11.5\,\frac{\text{mg}}{\text{kg}} \times \frac{10^{-3}\text{ g}}{\text{mg}} \times \frac{1\text{ kg}}{2.205\text{ lb}} = 0.005215\,\frac{\text{g}}{\text{lb}}$$

For the second part, begin with 145 lb and use the dose obtained in the first part to convert to g. Then round the answer to the correct number of significant figures, which is three.

$$145\text{ lb} \times \frac{0.005215\text{ g}}{\text{lb}} = 0.75617\text{ g} = 0.756\text{ g}$$

CHECK
Check your answer. Are the units correct? Does the answer make physical sense?

The units of the answer are correct, and the value of the answer makes sense. Drug doses can vary over some range, but in many cases they are between 0 and 1 gram.

▶ **SKILLBUILDER 2.12 | Solving Unit Conversions in the Numerator and Denominator**
A car is driving at a velocity of 65 km/hr. What is the car's velocity in m/s?

▶ **FOR MORE PRACTICE** Example 2.27; Problems 95–98.

2.9 Units Raised to a Power

▶ Convert units raised to a power.

When converting quantities with units raised to a power, such as cubic centimeters (cm^3)*, you must raise the conversion factor to that power.* For example, suppose you want to convert the size of a motorcycle engine reported as 1255 cm^3 to cubic inches. You know that:

$$2.54\text{ cm} = 1\text{ in.}$$

The unit cm^3 is often abbreviated as cc.

Most tables of conversion factors do not include conversions between cubic units, but you can derive them from the conversion factors for the basic units. You cube both sides of the preceding equality to obtain the proper conversion factor.

$$(2.54 \text{ cm})^3 = (1 \text{ in.})^3$$
$$(2.54)^3 \text{ cm}^3 = 1^3 \text{ in.}^3$$
$$16.387 \text{ cm}^3 = 1 \text{ in.}^3$$

You can do the same thing in fractional form.

$$\frac{1 \text{ in.}}{2.54 \text{ cm}} = \frac{(1 \text{ in.})^3}{(2.54 \text{ cm})^3} = \frac{1 \text{ in.}^3}{16.387 \text{ cm}^3}$$

You then proceed with the conversion in the usual manner.

SOLUTION MAP

$$\text{cm}^3 \rightarrow \text{in.}^3$$

$$\frac{1 \text{ in.}^3}{16.387 \text{ cm}^3}$$

2.54 cm = 1 in. is an exact conversion factor. After cubing, you retain five significant figures so that the conversion factor does not limit the four significant figures of your original quantity (1255 cm^3).

SOLUTION

$$1255 \text{ cm}^3 \times \frac{1 \text{ in.}^3}{16.387 \text{ cm}^3} = 76.5851 \text{ in.}^3 = 76.59 \text{ in.}^3$$

CHEMISTRY AND HEALTH

Drug Dosage

The unit of choice in specifying drug dosage is the milligram (mg). A bottle of aspirin, Tylenol®, or any other common drug lists the number of milligrams of the active ingredient contained in each tablet, as well as the number of tablets to take per dose. The following table shows the mass of the active ingredient per pill in several common pain relievers, all reported in milligrams. The remainder of each tablet is composed of inactive ingredients such as cellulose (or fiber) and starch.

The recommended adult dose for many of these pain relievers is one or two tablets every four to eight hours (depending on the specific pain reliever). Notice that the extra-strength version of each pain reliever just contains a higher dose of the same compound found in the regular-strength version. For the pain relievers listed, three regular-strength tablets are the equivalent of two extra-strength tablets (and probably cost less).

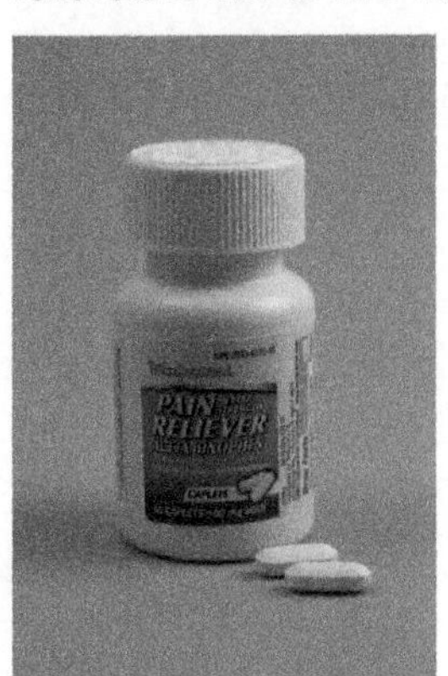

The dosages given in the table are fairly standard for each drug, regardless of the brand. On most drugstore shelves, there are many different brands of regular-strength ibuprofen, some sold under the generic name and others sold under their brand names (such as Advil®). However, if you look closely at the labels, you will find that they all contain the same thing: 200 mg of the compound ibuprofen. There is no difference in the compound or in the amount of the compound. Yet these pain relievers will most likely all have different prices. Choose the least expensive. Why pay more for the same thing?

B2.2 CAN YOU ANSWER THIS? *Convert each of the doses in the table to ounces. Why are drug dosages not listed in ounces?*

Drug Mass per Pill for Common Pain Relievers

Pain Reliever	Mass of Active Ingredient per Pill
aspirin	325 mg
aspirin, extra strength	500 mg
ibuprofen (Advil®)	200 mg
ibuprofen, extra strength	300 mg
acetaminophen (Tylenol®)	325 mg
acetaminophen, extra strength	500 mg

EXAMPLE 2.13 Converting Quantities Involving Units Raised to a Power

A circle has an area of 2659 cm^2. What is its area in square meters?

SORT You are given an area in square centimeters and asked to convert the area to square meters.	**GIVEN:** 2659 cm^2 **FIND:** m^2
STRATEGIZE Build a solution map beginning with cm^2 and ending with m^2. Remember that you must square the conversion factor.	**SOLUTION MAP** $cm^2 \rightarrow m^2$ $\frac{(0.01\ m)^2}{(1\ cm)^2}$ **RELATIONSHIPS USED** 1 cm = 0.01 m (from Table 2.2)
SOLVE Follow the solution map to solve the problem. Square the conversion factor (both the units and the number) as you carry out the calculation. Round the answer to four significant figures to reflect the four significant figures in the given quantity. The conversion factor is exact and therefore does not limit the number of significant figures.	**SOLUTION** $2659\ cm^2 \times \frac{(0.01\ m)^2}{(1\ cm)^2} = 2659\ \cancel{cm^2} \times \frac{10^{-4}\ m^2}{1\ \cancel{cm^2}}$ $= 0.265900\ m^2$ $= 0.2659\ m^2$
CHECK Check your answer. Are the units correct? Does the answer make physical sense?	The units of the answer are correct, and the magnitude makes physical sense. A square meter is much larger than a square centimeter, so the value in square meters should be much smaller than the value in square centimeters.

▶ **SKILLBUILDER 2.13 | Converting Quantities Involving Units Raised to a Power**

An automobile engine has a displacement (a measure of the size of the engine) of 289.7 $in.^3$ What is its displacement in cubic centimeters?

▶ **FOR MORE PRACTICE** Example 2.28; Problems 87, 88, 89, 90, 91, 92.

Interactive Worked Example Video 2.14

EXAMPLE 2.14 Solving Multistep Conversion Problems Involving Units Raised to a Power

The average annual per person crude oil consumption in the United States is 15,615 dm^3. What is this value in cubic inches?

SORT You are given a volume in cubic decimeters and asked to convert it to cubic inches.	**GIVEN:** 15,615 dm^3 **FIND:** $in.^3$
STRATEGIZE Build a solution map beginning with dm^3 and ending with $in.^3$ You must cube each of the conversion factors because the quantities involve cubic units.	**SOLUTION MAP** $dm^3 \rightarrow m^3 \rightarrow cm^3 \rightarrow in.^3$ $\frac{(0.1\ m)^3}{(1\ dm)^3}$ $\frac{(1\ cm)^3}{(0.01\ m)^3}$ $\frac{(1\ in.)^3}{(2.54\ cm)^3}$

	RELATIONSHIPS USED 1 dm = 0.1 m (from Table 2.2) 1 cm = 0.01 m (from Table 2.2) 2.54 cm = 1 in. (from Table 2.3)
SOLVE Follow the solution map to solve the problem. Begin with the given value in dm^3 and multiply by the string of conversion factors to arrive at $in.^3$ Be sure to cube each conversion factor as you carry out the calculation. Round the answer to five significant figures to reflect the five significant figures in the least precisely known quantity (15,615 dm^3). The conversion factors are all exact and therefore do not limit the number of significant figures.	**SOLUTION** $15{,}615\ \cancel{dm^3} \times \frac{(0.1\ \cancel{m})^3}{(1\ \cancel{dm})^3} \times \frac{(1\ \cancel{cm})^3}{(0.01\ \cancel{m})^3} \times \frac{(1\ in.)^3}{(2.54\ \cancel{cm})^3}$ $= 9.5289 \times 10^5\ in.^3$
CHECK Check your answer. Are the units correct? Does the answer make physical sense?	The units of the answer are correct, and the magnitude makes sense. A cubic inch is smaller than a cubic decimeter, so the value in cubic inches should be larger than the value in cubic decimeters.

▶ **SKILLBUILDER 2.14 | Solving Multistep Problems Involving Units Raised to a Power**

How many cubic inches are there in 3.25 yd^3?

▶ **FOR MORE PRACTICE** Problems 93, 94.

CONCEPTUAL ✔ CHECKPOINT 2.7

You know that there are 3 ft in a yard. How many cubic feet are there in a cubic yard?

(a) 3 **(b)** 6 **(c)** 9 **(d)** 27

2.10 Density

- ▶ Calculate the density of a substance.
- ▶ Use density as a conversion factor.

Why do some people pay over $3000 for a bicycle made of titanium? A steel frame would be just as strong for a fraction of the cost. The difference between the two bikes is their masses—the titanium bike is lighter. For a given volume of metal, titanium has less mass than steel. Titanium is said to be *less dense* than steel. The **density** of a substance is the ratio of its mass to its volume.

$$\text{density} = \frac{\text{mass}}{\text{volume}} \quad \text{or} \quad d = \frac{m}{V}$$

Density is a fundamental property of a substance and differs from one substance to another. The units of density are those of mass divided by those of volume, most conveniently expressed in grams per cubic centimeter (g/cm^3) or grams per milliliter (g/mL). Table 2.4 on the next page lists the densities of some common substances. Aluminum is among the least dense structural metals with a density of 2.70 g/cm^3, while platinum is among the densest with a density of 21.4 g/cm^3. Titanium has a density of 4.50 g/cm^3.

▲ Top-end bicycle frames are made of titanium because of titanium's low density and high relative strength. Titanium has a density of 4.50 g/cm^3, while iron, for example, has a density of 7.86 g/cm^3.

TABLE 2.4 Densities of Some Common Substances

Substance	Density (g/cm³)
charcoal, oak	0.57
ethanol	0.789
ice	0.92
water	1.0
glass	2.6
aluminum	2.7
titanium	4.50
iron	7.86
copper	8.96
lead	11.4
gold	19.3
platinum	21.4

Calculating Density

You can calculate the density of a substance by dividing the mass of a given amount of the substance by its volume. For example, a sample of liquid has a volume of 22.5 mL and a mass of 27.2 g. To find its density, use the equation $d = m/V$.

$$d = \frac{m}{V} = \frac{27.2\text{ g}}{22.5\text{ mL}} = 1.21\text{ g/mL}$$

You can use a solution map to solve problems involving equations, but the solution map takes a slightly different form than one for a pure conversion problem. In a problem involving an equation, the solution map shows how the *equation* takes you from the *given* quantities to the *find* quantity. The solution map for this problem is:

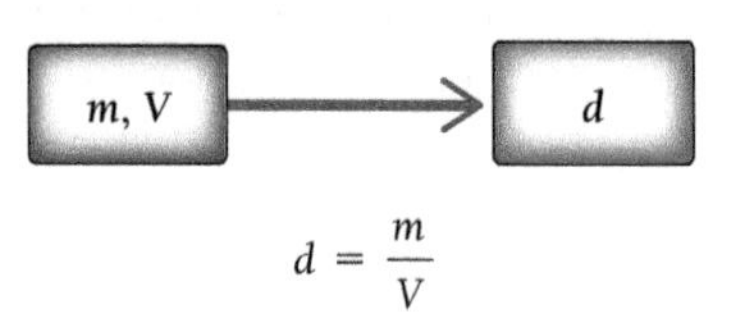

$$d = \frac{m}{V}$$

Remember that cubic centimeters and milliliters are equivalent units.

The solution map illustrates that the values of m and V, when substituted into the equation $d = \frac{m}{V}$, give the desired result, d.

EXAMPLE 2.15 Calculating Density

A jeweler offers to sell a ring to a woman and tells her that it is made of platinum. Noting that the ring feels a little light, the woman decides to perform a test to determine the ring's density. She places the ring on a balance and finds that it has a mass of 5.84 g. She also finds that the ring *displaces* 0.556 cm³ of water. Is the ring made of platinum? The density of platinum is 21.4 g/cm³. (The displacement of water is a common way to measure the volume of irregularly shaped objects. To say that an object *displaces* 0.556 cm³ of water means that when the object is submerged in a container of water filled to the brim, 0.556 cm³ overflows. Therefore, the volume of the object is 0.556 cm³.)

SORT

You are given the mass and volume of the ring and asked to find the density.

GIVEN: $m = 5.84\text{ g}$

$V = 0.556\text{ cm}^3$

FIND: density in g/cm³

STRATEGIZE

If the ring is platinum, its density should match that of platinum. Build a solution map that represents how you get from the given quantities (mass and volume) to the find quantity (density). Unlike in conversion problems, where you write a conversion factor beneath the arrow, here you write the equation for density beneath the arrow.

SOLUTION MAP

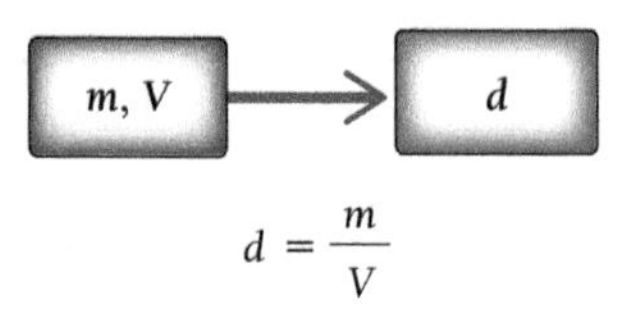

$$d = \frac{m}{V}$$

RELATIONSHIPS USED

$d = \frac{m}{V}$ (equation for density)

SOLVE

Follow the solution map. Substitute the given values into the density equation and calculate the density.

Round the answer to three significant figures to reflect the three significant figures in the given quantities.

SOLUTION

$$d = \frac{m}{V} = \frac{5.84\text{ g}}{0.556\text{ cm}^3} = 10.5\text{ g/cm}^3$$

The density of the ring is much too low to be platinum; therefore, the ring is a fake.

CHECK Check your answer. Are the units correct? Does the answer make physical sense?	The units of the answer are correct, and it seems that the magnitude could be an actual density. As you can see from Table 2.4, the densities of liquids and solids range from below 1 g/cm^3 to just over 20 g/cm^3.

▶ SKILLBUILDER 2.15 | Calculating Density

The woman takes the ring back to the jewelry shop, where she is met with endless apologies. The jeweler had accidentally made the ring out of silver rather than platinum. The jeweler gives her a new ring that she promises is platinum. This time when the customer checks the density, she finds the mass of the ring to be 9.67 g and its volume to be 0.452 cm^3. Is this ring genuine?

▶ FOR MORE PRACTICE Example 2.29; Problems 95, 96, 97, 98, 99, 100.

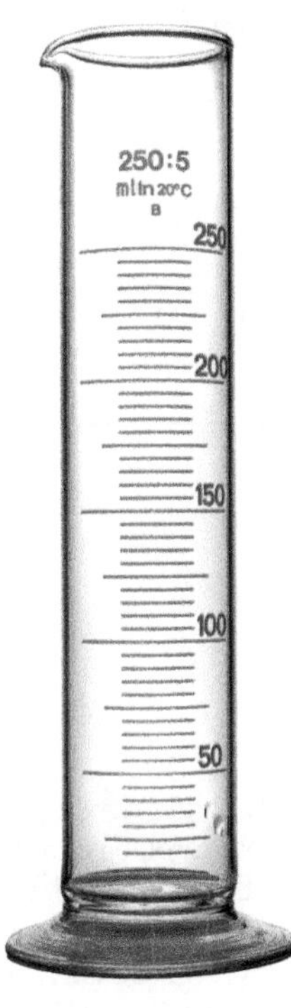

▲ A graduated cylinder is used to measure the volume of a liquid in the laboratory.

Density as a Conversion Factor

You can use the density of a substance as a conversion factor between the mass of the substance and its volume. For example, suppose you need 68.4 g of a liquid with a density of 1.32 g/cm^3 and want to measure the correct volume with a graduated cylinder (a piece of laboratory glassware used to measure volume). How much volume should you measure?

Start with the mass of the liquid and use the density as a conversion factor to convert mass to volume. However, you must use the inverted density expression 1 cm^3/1.32 g because you want g, the unit you are converting from, to be on the bottom (in the denominator) and cm^3, the unit you are converting to, on the top (in the numerator). The solution map takes this form:

SOLUTION MAP

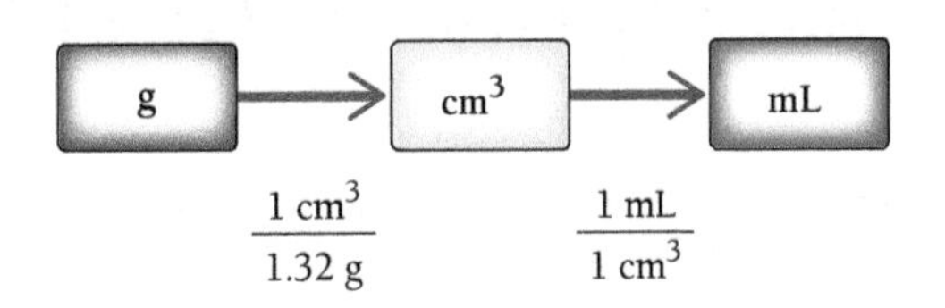

$$\frac{1\ \text{cm}^3}{1.32\ \text{g}} \qquad \frac{1\ \text{mL}}{1\ \text{cm}^3}$$

SOLUTION

$$68.4\ \text{g} \times \frac{1\ \text{cm}^3}{1.32\ \text{g}} \times \frac{1\ \text{mL}}{1\ \text{cm}^3} = 51.8\ \text{mL}$$

You must measure 51.8 mL to obtain 68.4 g of the liquid.

Interactive Worked Example Video 2.16

EXAMPLE 2.16 Density as a Conversion Factor

The gasoline in an automobile gas tank has a mass of 60.0 kg and a density of 0.752 g/cm^3. What is its volume in cm^3?

SORT You are given the mass in kilograms and asked to find the volume in cubic centimeters. Density is the conversion factor between mass and volume.	GIVEN: 60.0 kg density = 0.752 g/cm^3 FIND: volume in cm^3
STRATEGIZE Build the solution map starting with kg and ending with cm^3. Use the density (inverted) to convert from g to cm^3.	SOLUTION MAP 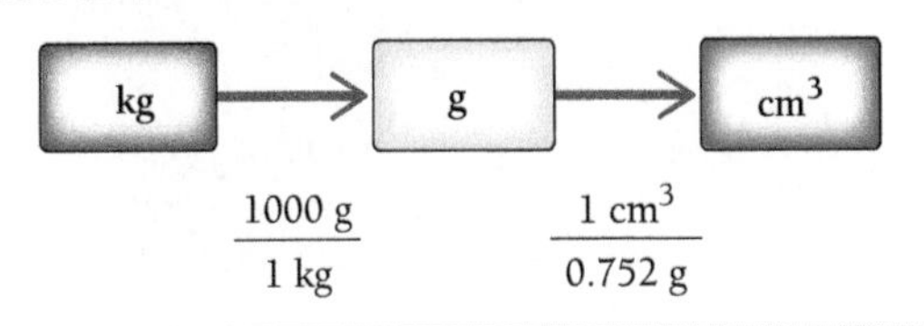 $\frac{1000\ \text{g}}{1\ \text{kg}} \qquad \frac{1\ \text{cm}^3}{0.752\ \text{g}}$

	RELATIONSHIPS USED $0.752\ g/cm^3$ (given in problem) $1000\ g = 1\ kg$ (from Table 2.2)
SOLVE Follow the solution map to solve the problem. Round the answer to three significant figures to reflect the three significant figures in the given quantities.	**SOLUTION** $60.0\ \cancel{kg} \times \frac{1000\ \cancel{g}}{1\ \cancel{kg}} \times \frac{1\ cm^3}{0.752\ \cancel{g}} = 7.98 \times 10^4\ cm^3$
CHECK Check your answer. Are the units correct? Does the answer make physical sense?	The units of the answer are those of volume, so they are correct. The magnitude seems reasonable because the density is somewhat less than $1\ g/cm^3$; therefore, the volume of 60.0 kg should be somewhat more than $60.0 \times 10^3\ cm^3$.

▶ **SKILLBUILDER 2.16 | Density as a Conversion Factor**

A drop of acetone (nail polish remover) has a mass of 35 mg and a density of $0.788\ g/cm^3$. What is its volume in cubic centimeters?

▶ **SKILLBUILDER PLUS**

A steel cylinder has a volume of $246\ cm^3$ and a density of $7.93\ g/cm^3$. What is its mass in kilograms?

▶ **FOR MORE PRACTICE** Example 2.30; Problems 101, 102.

CHEMISTRY AND HEALTH

Density, Cholesterol, and Heart Disease

Cholesterol is the fatty substance found in animal-derived foods such as beef, eggs, fish, poultry, and milk products. The body uses cholesterol for several purposes. However, excessive amounts of cholesterol in the blood—which can be caused by both genetic factors and diet—may result in the deposition of cholesterol in arterial walls, leading to a condition called atherosclerosis, or blocking of the arteries. These blockages are dangerous because they inhibit blood flow to important organs, causing heart attacks and strokes. The risk of stroke and heart attack increases with increasing blood cholesterol levels (Table 2.5). Cholesterol is carried in the bloodstream by a class of substances known as lipoproteins. Lipoproteins are often separated and classified according to their density.

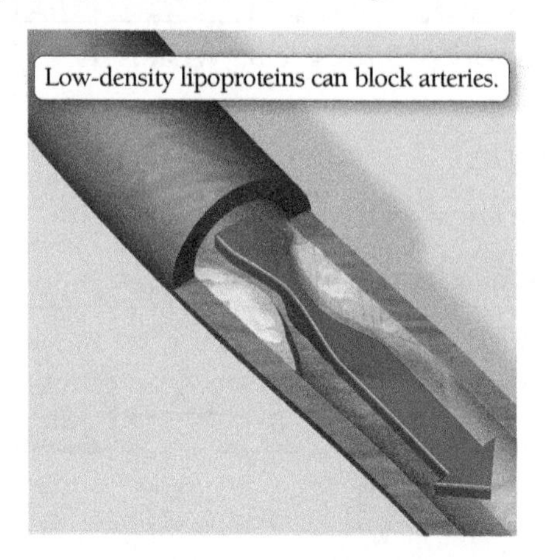

Low-density lipoproteins can block arteries.

The main carriers of blood cholesterol are low-density lipoproteins (LDLs). LDLs, also called bad cholesterol, have a density of $1.04\ g/cm^3$. They are bad because they tend to deposit cholesterol on arterial walls, increasing the risk of stroke and heart attack. Cholesterol is also carried by high-density lipoproteins (HDLs). HDLs, called good cholesterol, have a density of $1.13\ g/cm^3$. HDLs transport cholesterol to the liver for processing and excretion and therefore have a tendency to reduce cholesterol on arterial walls. Too low a level of HDLs (below 35 mg/100 mL) is considered a risk factor for heart disease. Exercise, along with a diet low in saturated fats, is believed to raise HDL levels in the blood while lowering LDL levels.

B2.3 CAN YOU ANSWER THIS? *What mass of low-density lipoprotein is contained in a cylinder that is 1.25 cm long and 0.50 cm in diameter? (The volume of a cylinder, V, is given by $V = \pi r^2 \ell$, where r is the radius of the cylinder and ℓ is its length.)*

TABLE 2.5 Risk of Stroke and Heart Attack versus Blood Cholesterol Level

Risk Level	Total Blood Cholesterol (mg/100 mL)	LDL (mg/100 mL)
low	< 200	< 130
borderline	200–239	130–159
high	> 240	> 160

2.11 Numerical Problem-Solving Strategies and the Solution Map

In this chapter, you have seen a few examples of how to solve numerical problems. In Section 2.6, you encountered a procedure to solve simple unit conversion problems. You then learned how to modify that procedure to work with multistep unit conversion problems and problems involving an equation. This section summarizes and generalizes these procedures and includes two additional examples. Just as was done in Section 2.6, the left column shows the general procedure for solving numerical problems, and the center and right columns illustrate the application of the procedure to the two examples.

	EXAMPLE 2.17	EXAMPLE 2.18
Solving Numerical Problems	**UNIT CONVERSION** A chemist needs a 23.5-kg sample of ethanol for a large-scale reaction. What volume in liters of ethanol should the chemist use? The density of ethanol is 0.789 g/cm^3.	**UNIT CONVERSION WITH EQUATION** A 55.9-kg person displaces 57.2 L of water when submerged in a water tank. What is the density of the person in grams per cubic centimeter?
1. SORT • Scan the problem for one or more numbers and their associated units. This number (or numbers) is (are) the starting point(s) of the calculation. Write them down as given. • Scan the problem to determine what you are asked to find. Sometimes the units of this quantity are implied; other times they are specified. Write down the quantity and/or units you are asked to find.	**GIVEN:** 23.5 kg ethanol density $= 0.789\ g/cm^3$ **FIND:** volume in L	**GIVEN:** $m = 55.9$ kg $V = 57.2$ L **FIND:** density in g/cm^3
2. STRATEGIZE • For problems involving only conversions, focus on units. The solution map shows how to get from the units in the given quantity to the units in the quantity you are asked to find. • For problems involving equations, focus on the equation. The solution map shows how the equation takes you from the given quantity (or quantities) to the quantity you are asked to find. • Some problems may involve both unit conversions and equations, in which case the solution map employs both of the above points.	**SOLUTION MAP** kg → g → cm^3 → mL → L $\frac{1000\ g}{1\ kg}$ $\frac{1\ cm^3}{0.789\ g}$ $\frac{1\ mL}{1\ cm^3}$ $\frac{1\ L}{1000\ mL}$ **RELATIONSHIPS USED** $0.789\ g/cm^3$ (given in problem) 1000 g = 1 kg (Table 2.2) 1000 mL = 1 L (Table 2.2) 1 mL = 1 cm^3 (Table 2.3)	**SOLUTION MAP** m, V → d $d = \frac{m}{V}$ **RELATIONSHIPS USED** $d = \frac{m}{V}$ (definition of density)

3. SOLVE • For problems involving only conversions, begin with the given quantity and its units. Multiply by the appropriate conversion factor(s), canceling units, to arrive at the quantity you are asked to find. • For problems involving equations, solve the equation to arrive at the quantity you are asked to find. (Use algebra to rearrange the equation so that the quantity you are asked to find is isolated on one side.) Gather each of the quantities that must go into the equation in the correct units. (Convert to the correct units using additional solution maps if necessary.) Finally, substitute the numerical values and their units into the equation and calculate the answer. • Round the answer to the correct number of significant figures. Use the significant figure rules from Sections 2.3 and 2.4.	**SOLUTION** $23.5\ \text{kg} \times \frac{1000\ \text{g}}{1\ \text{kg}} \times \frac{1\ \text{cm}^3}{0.789\ \text{g}} \times \frac{1\ \text{mL}}{1\ \text{cm}^3} \times \frac{1\ \text{L}}{1000\ \text{mL}} = 29.7845\ \text{L}$ $29.7845\ \text{L} = 29.8\ \text{L}$	The equation is already solved for the find quantity. Convert mass from kilograms to grams. $m = 55.9\ \text{kg} \times \frac{1000\ \text{g}}{1\ \text{kg}} = 5.59 \times 10^4\ \text{g}$ Convert volume from liters to cubic centimeters. $V = 57.2\ \text{L} \times \frac{1000\ \text{mL}}{1\ \text{L}} \times \frac{1\ \text{cm}^3}{1\ \text{mL}} = 57.2 \times 10^3\ \text{cm}^3$ Calculate density. $d = \frac{m}{V} = \frac{55.9 \times 10^3\ \text{g}}{57.2 \times 10^3\ \text{cm}^3} = 0.9772727\ \frac{\text{g}}{\text{cm}^3} = 0.977\ \frac{\text{g}}{\text{cm}^3}$
4. CHECK • Does the magnitude of the answer make physical sense? Are the units correct?	The units are correct (L) and the magnitude is reasonable. Because the density is less than 1 g/cm^3, the calculated volume (29.8 L) should be greater than the mass (23.5 kg).	The units are correct. Because the mass in kilograms and the volume in liters are very close to each other in magnitude, it makes sense that the density is close to 1 g/cm^3.
	▶ **SKILLBUILDER 2.17 | Unit Conversion** A pure gold metal bar displaces 0.82 L of water. What is its mass in kilograms? (The density of gold is 19.3 g/cm^3.)	▶ **SKILLBUILDER 2.18 | Unit Conversion with Equation** A gold-colored pebble is found in a stream. Its mass is 23.2 mg, and its volume is 1.20 mm^3. What is its density in grams per cubic centimeter? Is it gold? (The density of gold = 19.3 g/cm^3.)
	▶ **FOR MORE PRACTICE** Problems 103, 109, 110, 111, 112.	▶ **FOR MORE PRACTICE** Problems 103, 104, 105, 106.

Chapter 2 in Review

MasteringChemistry™ provides end-of-chapter exercises, feedback-enriched tutorial problems, animations, and interactive activities to encourage problem solving practice and deeper understanding of key concepts and topics.

Self-Assessment Quiz

PEARSON eText 2.0 The eText 2.0 icon indicates that this feature is embedded and interactive in the eText.

Q1. Express the number 0.000042 in scientific notation.
(a) 0.42×10^{-4} (b) 4.2×10^{-5}
(c) 4.2×10^{-4} (d) 4×10^{-5}

Q2. A graduated cylinder has markings every milliliter. Which measurement is accurately reported for this graduated cylinder?
(a) 21 mL (b) 21.2 mL
(c) 21.23 mL (d) 21.232 mL

Q3. How many significant figures are in the number 0.00620?
(a) 2 (b) 3 (c) 4 (d) 5

Q4. Round the number 89.04997 to three significant figures.
(a) 89.03 (b) 89.04 (c) 89.1 (d) 89.0

Q5. Perform this multiplication to the correct number of significant figures: $65.2 \times 0.0015 \times 12.02$
(a) 1.17 (b) 1.18 (c) 1.2 (d) 1.176

Q6. Perform this addition to the correct number of significant figures: $8.32 + 12.148 + 0.02$
(a) 20.488 (b) 20.49 (c) 20.5 (d) 21

Q7. Perform this calculation to the correct number of significant figures: $78.222 \times (12.02 - 11.52)$
(a) 39 (b) 39.1 (c) 39.11 (d) 39.111

Q8. Convert 76.8 cm to m.
(a) 0.0768 m (b) 7.68 m
(c) 0.768 m (d) 7.68×10^{-2} m

Q9. Convert 2855 mg to kg.
(a) 2.855×10^{-3} kg
(b) 2.855 kg
(c) 0.02855 kg
(d) 3.503×10^{-4} kg

Q10. A runner runs 4875 ft in 6.85 minutes. What is the runner's average speed in miles per hour?
(a) 1.34 mi/hr (b) 0.0022 mi/hr
(c) 8.09 mi/hr (d) 8.087 mi/hr

Q11. An automobile travels 97.2 km on 7.88 L of gasoline. What is the gas mileage for the automobile in miles per gallon?
(a) 2.02 mi/gal (b) 7.67 mi/gal
(c) 0.034 mi/gal (d) 29.0 mi/gal

Q12. Convert 876.9 $in.^3$ to m^3.
(a) 0.01437 m^3 (b) 22.27 m^3
(c) 5.351×10^7 m^3 (d) 0.014 m^3

Q13. Convert 27 m/s to km/hr.
(a) 97 km/hr (b) 7.5 km/hr
(c) 1.6 km/hr (d) 0.027 km/hr

Q14. A cube measures 2.5 cm on each edge and has a mass of 66.9 g. Calculate the density of the material that composes the cube. (The volume of a cube is equal to the edge length cubed.)
(a) 10.7 g/cm^3 (b) 4.3 g/cm^3
(c) 0.234 g/cm^3 (d) 26.7 g/cm^3

Q15. What is the mass of 225 mL of a liquid that has a density of 0.880 g/mL?
(a) 198 g (b) 0.0039 g (c) 0.198 g (d) 0.256 g

Q16. What is the edge length of a 155-g iron cube? (The density of iron is 7.86 g/cm^3, and the volume of a cube is equal to the edge length cubed.)
(a) 0.0197 cm (b) 19.7 cm
(c) 1218 cm (d) 2.70 cm

Answers: 1:b, 2:b, 3:b, 4:d, 5:c, 6:b, 7:a, 8:c, 9:a, 10:c, 11:d, 12:a, 13:a, 14:b, 15:a, 16:d

Chemical Principles	*Relevance*
Uncertainty Scientists report measured quantities so that the number of digits reflects the certainty in the measurement. Write measured quantities so that every digit is certain except the last, which is estimated.	Measurement is a hallmark of science, and you must communicate the precision of a measurement so that others know how reliable the measurement is. When you write or manipulate measured quantities, you must show and retain the precision with which the original measurement was made.
Units Measured quantities usually have units associated with them. The SI unit for length is the meter; for mass, the kilogram; and for time, the second. Prefix multipliers such as *kilo-* or *milli-* are often used in combination with these basic units. The SI units of volume are units of length raised to the third power; liters or milliliters are often used as well.	The units in a measured quantity communicate what the quantity actually is. Without an agreed-on system of units, scientists could not communicate their measurements. Units are also important in calculations, and the tracking of units throughout a calculation is essential.

Chemical Principles

Density

The density of a substance is its mass divided by its volume, $d = m/V$, and is usually reported in units of grams per cubic centimeter or grams per milliliter. Density is a fundamental property of all substances and generally differs from one substance to another.

Relevance

The density of substances is an important consideration in choosing materials for manufacturing and production. Airplanes, for example, are made of low-density materials, while bridges are made of higher-density materials. Density is important as a conversion factor between mass and volume and vice versa.

Chemical Skills

LO: Express very large and very small numbers using scientific notation (Section 2.2).

To express a number in scientific notation:

- Move the decimal point to obtain a number between 1 and 10.
- Write the decimal part multiplied by 10 raised to the number of places you moved the decimal point.
- The exponent is positive if you moved the decimal point to the left and negative if you moved the decimal point to the right.

Examples

EXAMPLE **2.19** **Scientific Notation**

Express the number 45,000,000 in scientific notation.

45,000,000

7 6 5 4 3 2 1

4.5×10^7

LO: Report measured quantities to the right number of digits (Section 2.3).

Report measured quantities so that every digit is certain except the last, which is estimated.

EXAMPLE **2.20** **Reporting Measured Quantities to the Right Number of Digits**

Record the volume of liquid in the graduated cylinder to the correct number of digits. Laboratory glassware is calibrated (and should therefore be read) from the bottom of the meniscus, the curved surface at the top of a column of liquid (see figure).

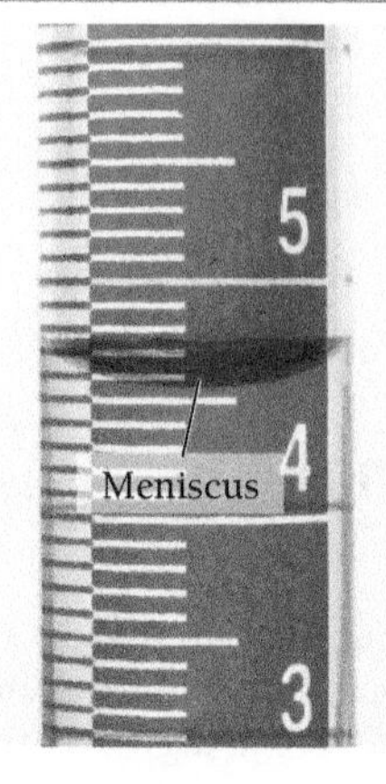

Because the graduated cylinder has markings every 0.1 mL, you should record the measurement to the nearest 0.01 mL. In this case, that is 4.57 mL.

LO: Determine which digits in a number are significant (Section 2.3).

Always count the following as significant:

- nonzero digits
- interior zeros
- trailing zeros after a decimal point
- trailing zeros before a decimal point but after a nonzero number

Never count the following digits as significant:

- zeros to the left of the first nonzero number

The following digits are ambiguous, and you should avoid them by using scientific notation:

- zeros at the end of a number but before a decimal point

EXAMPLE **2.21** **Counting Significant Digits**

How many significant figures are in the following numbers?

1.0050	five significant figures
0.00870	three significant figures
100.085	six significant digits
5400	It is not possible to tell in its current form.

The number must be written as 5.4×10^3, 5.40×10^3, or 5.400×10^3, depending on the number of significant figures intended.

LO: Round numbers to the correct number of significant figures (Section 2.4).

When rounding numbers to the correct number of significant figures, round down if the last digit dropped is 4 or less; round up if the last digit dropped is 5 or more.

EXAMPLE **2.22** **Rounding**

Round 6.442 and 6.456 to two significant figures each.

6.442 rounds to 6.4
6.456 rounds to 6.5

LO: Determine the correct number of significant figures in the results of multiplication and division calculations (Section 2.4).

The result of a multiplication or division should carry the same number of significant figures as the factor with the least number of significant figures.

EXAMPLE **2.23** **Significant Figures in Multiplication and Division**

Perform the calculation and report the answer to the correct number of significant figures.

$$8.54 \times 3.589 \div 4.2$$
$$= 7.2976$$
$$= 7.3$$

Round the final result to two significant figures to reflect the two significant figures in the factor with the least number of significant figures (4.2).

LO: Determine the correct number of significant figures in the results of addition and subtraction calculations (Section 2.4).

The result of an addition or subtraction should carry the same number of decimal places as the quantity carrying the least number of decimal places.

EXAMPLE **2.24** **Significant Figures in Addition and Subtraction**

Perform the operation and report the answer to the correct number of significant figures.

$$\begin{array}{r} 3.098 \\ +0.67 \\ \underline{-0.9452} \\ 2.8228 \end{array} = 2.82$$

Round the final result to two decimal places to reflect the two decimal places in the quantity with the least number of decimal places (0.67).

LO: Determine the correct number of significant figures in the results of calculations involving both addition/subtraction and multiplication/division (Section 2.4).

In calculations involving both addition/subtraction and multiplication/division, do the steps in parentheses first, keeping track of how many significant figures are in the answer by underlining the least significant figure, then proceeding with the remaining steps. Do not round off until the very end.

EXAMPLE 2.25 Significant Figures in Calculations Involving Both Addition/Subtraction and Multiplication/Division

Perform the operation and report the answer to the correct number of significant figures.

$$8.16 \times (5.4323 - 5.411)$$
$$= 8.16 \times 0.02\underline{1}3$$
$$= 0.1738 = 0.1\underline{7}$$

LO: Convert between units (Sections 2.6, 2.7).

Solve unit conversion problems by following these steps.

1. **Sort** Write down the given quantity and its units and the quantity you are asked to find and its units.
2. **Strategize** Draw a solution map showing how to get from the given quantity to the quantity you are asked to find.
3. **Solve** Follow the solution map. Starting with the given quantity and its units, multiply by the appropriate conversion factor(s), canceling units, to arrive at the quantity to find in the desired units. Round the final answer to the correct number of significant figures.
4. **Check** Are the units correct? Does the answer make physical sense?

EXAMPLE 2.26 Unit Conversion

Convert 108 ft to meters.

GIVEN: 108 ft

FIND: m

SOLUTION MAP

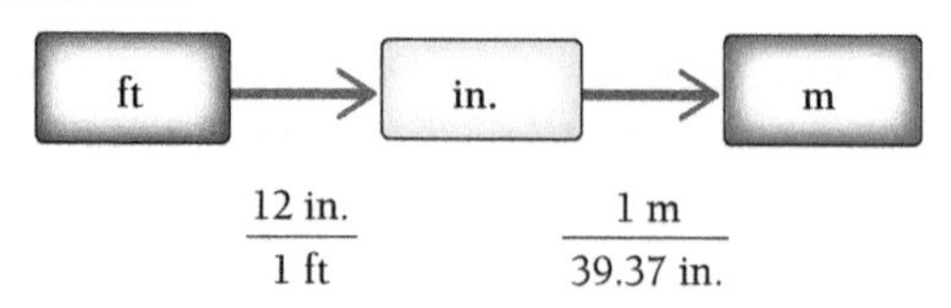

$$\frac{12 \text{ in.}}{1 \text{ ft}} \qquad \frac{1 \text{ m}}{39.37 \text{ in.}}$$

RELATIONSHIPS USED

1 m = 39.37 in. (Table 2.3)
1 ft = 12 in. (by definition)

SOLUTION

$$108 \text{ ft} \times \frac{12 \text{ in.}}{1 \text{ ft}} \times \frac{1 \text{ m}}{39.37 \text{ in.}}$$
$$= 32.918 \text{ m}$$
$$= 32.9 \text{ m}$$

The answer has the right units (meters), and it makes sense; because a meter is longer than a foot, the number of meters should be less than the number of feet.

LO: Convert units in a quantity that has units in the numerator and the denominator (Section 2.8).

1. **Sort** Write down the given quantity and its units and the quantity you are asked to find and its units.
2. **Strategize** Draw a solution map showing how to get from the given quantity to the quantity you are asked to find. Because the units are squared, you must square the conversion factor.

EXAMPLE 2.27 Unit Conversion in the Numerator and Denominator

A drug company lists the dose of a drug as 0.012 g per pound of body weight. What is the dose of the drug in mg per kilogram of body weight?

GIVEN: $0.012 \frac{\text{g}}{\text{lb}}$

FIND: $\frac{\text{mg}}{\text{kg}}$

SOLUTION MAP

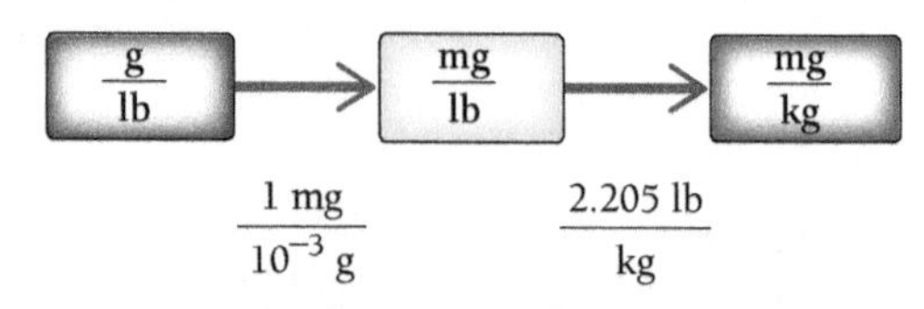

$$\frac{1 \text{ mg}}{10^{-3} \text{ g}} \qquad \frac{2.205 \text{ lb}}{\text{kg}}$$

3. **Solve** Follow the solution map. Starting with the given quantity and its units, multiply by the appropriate conversion factor(s), canceling units, to arrive at the quantity you are asked to find in the desired units. Don't forget to square the conversion factor for squared units.
4. **Check** Are the units correct? Does the answer make physical sense?

RELATIONSHIPS USED

$1\ \text{mg} = 10^{-3}\ \text{g}$ (Table 2.2)
$1\ \text{kg} = 2.205\ \text{lb}$ (Table 2.3)

SOLUTION

$$0.012\,\frac{\text{g}}{\text{lb}} \times \frac{\text{mg}}{10^{-3}\ \text{g}} \times \frac{2.205\ \text{lb}}{1\ \text{kg}} = 26.46\,\frac{\text{mg}}{\text{kg}}$$

The units of the answer are correct, and the value of the answer makes sense. The dose in mg/kg should be larger than the dose in g/lb because the mg is a smaller unit than kg (so it takes more to make the same amount) and the kg is a larger unit than lb (so it takes more of the drug per kg than per lb).

LO: Convert units raised to a power (Section 2.9).

When working problems involving units raised to a power, raise the conversion factors to the same power.

EXAMPLE **2.28** **Unit Conversion Involving Units Raised to a Power**

How many square meters are in $1.0\ \text{km}^2$?

1. **Sort** Begin by sorting the information in the problem into given and find. You are given the dose of the drug in g/lb. You are asked to find the dose in mg/kg.
2. **Strategize** Convert the numerator from g to mg; then convert the denominator from lb to kg.

GIVEN: $1.0\ \text{km}^2$

FIND: m^2

SOLUTION MAP

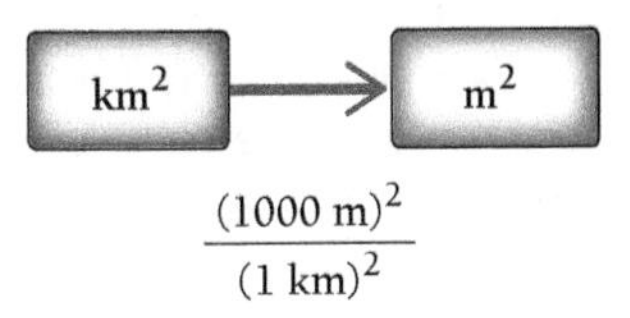

$$\frac{(1000\ \text{m})^2}{(1\ \text{km})^2}$$

RELATIONSHIPS USED

$1\ \text{km} = 1000\ \text{m}$ (Table 2.2)

3. **Solve** Follow the solution map to solve the problem. Begin with 0.012 g/lb and multiply by the two conversion factors to arrive at the dose in mg/kg. Round to two significant figures to reflect the two significant figures in the least precisely known quantity.
4. **Check** Check your answer. Are the units correct? Does the answer make physical sense?

SOLUTION

$$1.0\ \text{km}^2 \times \frac{(1000\ \text{m})^2}{(1\ \text{km})^2}$$

$$= 1.0\ \text{km}^2 \times \frac{1 \times 10^6\ \text{m}^2}{1\ \text{km}^2}$$

$$= 1.0 \times 10^6\ \text{m}^2$$

The units are correct. The answer makes physical sense. A square meter is much smaller than a square kilometer, so the number of square meters should be much larger than the number of square kilometers.

LO: Calculate the density of a substance (Section 2.10).

The density of an object or substance is its mass divided by its volume.

$$d = \frac{m}{V}$$

1. **Sort** Write down the given quantity and its units and the quantity you are asked to find and its units.
2. **Strategize** Draw a solution map showing how to get from the given quantity to the quantity you are asked to find. Use the definition of density as the equation that takes you from the mass and the volume to the density.
3. **Solve** Substitute the correct values into the equation for density.
4. **Check** Are the units correct? Does the answer make physical sense?

EXAMPLE 2.29 Calculating Density

An object has a mass of 23.4 g and displaces 5.7 mL of water. Determine its density in grams per milliliter.

GIVEN:

$m = 23.4\text{ g}$

$V = 5.7\text{ mL}$

FIND: density in g/mL

SOLUTION MAP

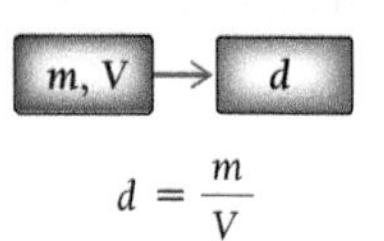

$$d = \frac{m}{V}$$

RELATIONSHIPS USED

$d = \frac{m}{V}$ (definition of density)

SOLUTION

$$d = \frac{m}{V} = \frac{23.4\text{ g}}{5.7\text{ mL}} = 4.11\text{ g/mL} = 4.1\text{ g/mL}$$

The units (g/mL) are units of density. The answer is in the range of values for the densities of liquids and solids (see Table 2.4).

LO: Use density as a conversion factor (Section 2.10).

You can use density as a conversion factor from mass to volume or from volume to mass. To convert between volume and mass, use density directly. To convert between mass and volume, invert the density.

1. **Sort** Write down the given quantity and its units and the quantity you are asked to find and its units.
2. **Strategize** Draw a solution map showing how to get from the given quantity to the quantity you are asked to find. Use the inverse of the density to convert from g to mL.
3. **Solve** Begin with the given quantity and multiply by the appropriate conversion factors to arrive at the quantity you are asked to find. Round to the correct number of significant figures.
4. **Check** Are the units correct? Does the answer make physical sense?

EXAMPLE 2.30 Density as a Conversion Factor

What is the volume in liters of 321 g of a liquid with a density of 0.84 g/mL?

GIVEN: 321 g

FIND: volume in L

SOLUTION MAP

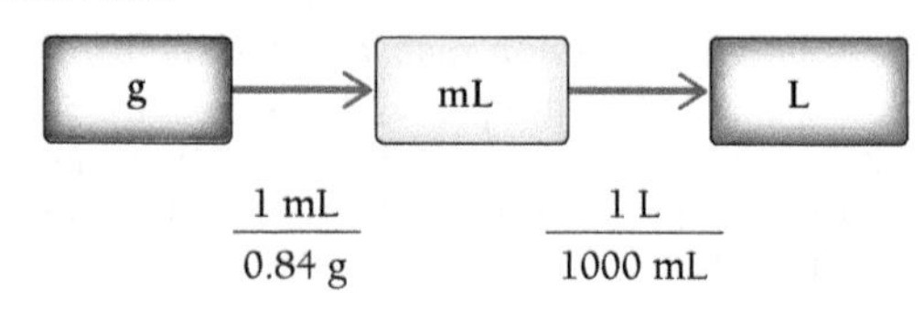

$$\frac{1\text{ mL}}{0.84\text{ g}} \qquad \frac{1\text{ L}}{1000\text{ mL}}$$

RELATIONSHIPS USED

0.84 g/mL (given in the problem)

1 L = 1000 mL (Table 2.2)

SOLUTION

$$321\text{ g} \times \frac{1\text{ mL}}{0.84\text{ g}} \times \frac{1\text{ L}}{1000\text{ mL}}$$

$$= 0.382\text{ L} = 0.38\text{ L}$$

The answer is in the correct units. The magnitude seems right because the density is slightly less than 1; therefore, the volume (382 mL) should be slightly greater than the mass (321 g).

Key Terms

conversion factor [2.6]
decimal part [2.2]
density (*d*) [2.10]
English system [2.5]
exponent [2.2]
exponential part [2.2]
International System [2.5]
kilogram (kg) [2.5]
liter (L) [2.5]
mass [2.5]
meter (m) [2.5]
metric system [2.5]
prefix multipliers [2.5]
scientific notation [2.2]
second (s) [2.5]
SI units [2.5]
significant figures (digits) [2.3]
solution map [2.6]
units [2.5]
volume [2.5]

Exercises

Questions

Answers to all questions numbered in blue appear in the Answers section at the back of the book.

1. Why is it necessary to include units when reporting scientific measurements?
2. Why are the number of digits reported in scientific measurements important?
3. Why is scientific notation useful?
4. If a measured quantity is written correctly, which digits are certain? Which are uncertain?
5. When do zeros count as significant digits, and when don't they count?
6. How many significant digits are there in exact numbers? What kinds of numbers are exact?
7. What limits the number of significant digits in a calculation involving only multiplication and division?
8. What limits the number of significant digits in a calculation involving only addition and subtraction?
9. How do we determine significant figures in calculations involving both addition/subtraction and multiplication/division?
10. What are the rules for rounding numbers?
11. What are the basic SI units of length, mass, and time?
12. List the common units of volume.
13. Suppose you are trying to measure the diameter of a Frisbee. What unit and prefix multiplier should you use?
14. What is the difference between mass and weight?
15. Using a metric ruler, measure these objects to the correct number of significant figures.
    (a) quarter (diameter) **(b)** dime (diameter)
    (c) notebook paper (width) **(d)** this book (width)
16. Using a stopwatch, measure each time to the correct number of significant figures.
    (a) time between your heartbeats
    (b) time it takes you to do the next problem
    (c) time between your breaths
17. Explain why units are important in calculations.
18. How are units treated in a calculation?
19. What is a conversion factor?
20. Why does the fundamental value of a quantity not change when you multiply the quantity by a conversion factor?
21. Write the conversion factor that converts a measurement in inches to feet. How does the conversion factor change for converting a measurement in feet to inches?
22. Write conversion factors for each.
    (a) miles to kilometers
    (b) kilometers to miles
    (c) gallons to liters
    (d) liters to gallons
23. This book outlines a four-step problem-solving strategy. Describe each step and its significance.
    (a) Sort
    (b) Strategize
    (c) Solve
    (d) Check
24. Experienced problem solvers always consider both the value and units of their answer to a problem. Why?
25. Draw a solution map to convert a measurement in grams to pounds.
26. Draw a solution map to convert a measurement in milliliters to gallons.
27. Draw a solution map to convert a measurement in meters to feet.
28. Draw a solution map to convert a measurement in ounces to grams. (1 lb = 16 oz)
29. What is density? Explain why density can work as a conversion factor. Between what quantities does it convert?
30. Explain how you would calculate the density of a substance. Include a solution map in your explanation.

Problems

Note: The exercises in the Problems section are paired, and the answers to the odd-numbered exercises (numbered in blue) appear in the Answers section at the back of the book.

SCIENTIFIC NOTATION

31. Express each number in scientific notation.
    (a) 38,802,000 (population of California)
    (b) 1,419,000 (population of Hawaii)
    (c) 19,746,000 (population of New York)
    (d) 584,000 (population of Wyoming)
32. Express each number in scientific notation.
    (a) 7,376,000,000 (population of the world)
    (b) 1,404000,000 (population of China)
    (c) 11,258,000 (population of Cuba)
    (d) 4,677,000 (population of Ireland)

33. Express each number in scientific notation.
   (a) 0.00000000007461 m (length of a hydrogen–hydrogen chemical bond)
   (b) 0.0000158 mi (number of miles in an inch)
   (c) 0.000000632 m (wavelength of red light)
   (d) 0.000015 m (diameter of a human hair)

34. Express each number in scientific notation.
   (a) 0.000000001 s (time it takes light to travel 1 ft)
   (b) 0.143 s (time it takes light to travel around the world)
   (c) 0.000000000001 s (time it takes a chemical bond to undergo one vibration)
   (d) 0.000001 m (approximate size of a dust particle)

35. Express each number in decimal notation (i.e., express the number without using scientific notation).
   (a) 6.022×10^{23} (number of carbon atoms in 12.01 g of carbon)
   (b) 1.6×10^{-19} C (charge of a proton in coulombs)
   (c) 2.99×10^{8} m/s (speed of light)
   (d) 3.44×10^{2} m/s (speed of sound)

36. Express each number in decimal notation (i.e., express the number without using scientific notation).
   (a) 450×10^{-9} m (wavelength of blue light)
   (b) 13.7×10^{9} years (approximate age of the universe)
   (c) 5×10^{9} years (approximate age of Earth)
   (d) 5.0×10^{1} years (approximate age of this author)

37. Express each number in decimal notation (i.e., express the number without using scientific notation).
   (a) 3.22×10^{7}
   (b) 7.2×10^{-3}
   (c) 1.18×10^{11}
   (d) 9.43×10^{-6}

38. Express each number in decimal notation (i.e., express the number without using scientific notation).
   (a) 1.30×10^{6}
   (b) 1.1×10^{-4}
   (c) 1.9×10^{2}
   (d) 7.41×10^{-10}

39. Complete the table.

Decimal Notation	Scientific Notation
2,000,000,000	______
______	1.211×10^{9}
0.000874	______
______	3.2×10^{11}

40. Complete the table.

Decimal Notation	Scientific Notation
______	4.2×10^{-3}
315,171,000	______
______	1.8×10^{-11}
1,232,000	______

SIGNIFICANT FIGURES

41. Read each instrument to the correct number of significant figures. Laboratory glassware should always be read from the bottom of the *meniscus* (the curved surface at the top of the liquid column).

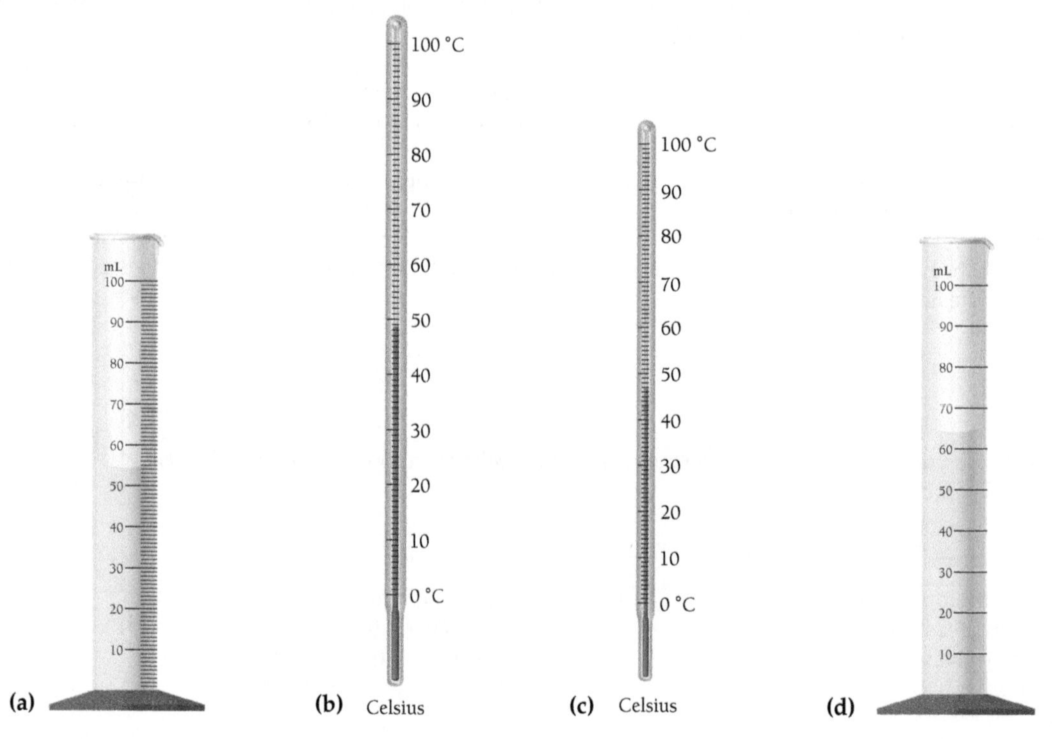

42. Read each instrument to the correct number of significant figures. Laboratory glassware should always be read from the bottom of the meniscus (the curved surface at the top of the liquid column).

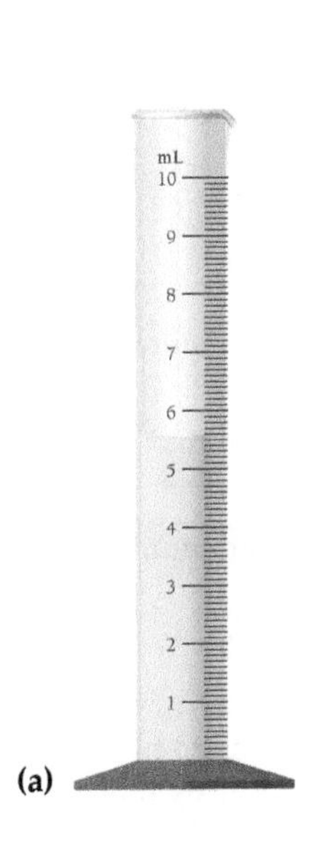

(a)

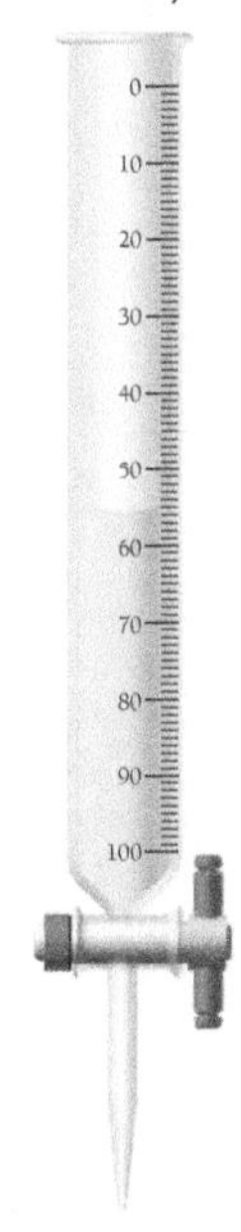

(b)

Note: A burette reads from the top down.

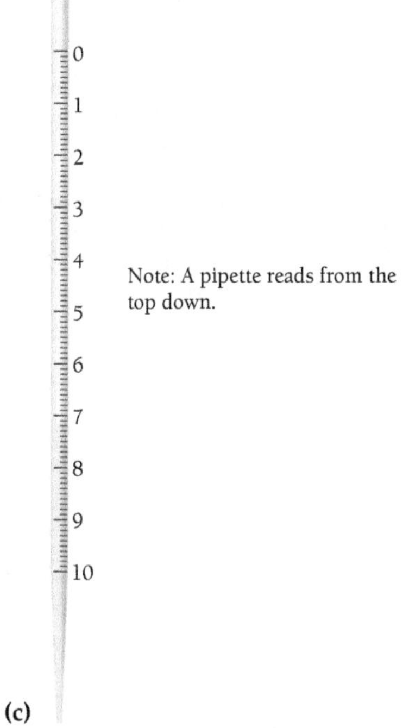

(c)

Note: A pipette reads from the top down.

(d) Note: Digital balances normally display mass to the correct number of significant figures for that particular balance.

43. For each measured quantity, underline the zeros that are significant and draw an X through the zeros that are not.
(a) 0.005050 m
(b) 0.0000000000000060 s
(c) 220,103 kg
(d) 0.00108 in.

44. For each measured quantity, underline the zeros that are significant and draw an X through the zeros that are not.
(a) 0.00010320 s
(b) 1,322,600,324 kg
(c) 0.0001240 in.
(d) 0.02061 m

45. How many significant figures are in each measured quantity?
(a) 0.001125 m
(b) 0.1125 m
(c) 1.12500×10^4 m
(d) 11205 m

46. How many significant figures are in each measured quantity?
(a) 13001 kg
(b) 13111 kg
(c) 1.30×10^4 kg
(d) 0.00013 kg

47. Correct any entries in the table that are wrong.

Quantity	Significant Figures
(a) 895675 m	6
(b) 0.000869 kg	6
(c) 0.5672100 s	5
(d) 6.022×10^{23} atoms	4

48. Correct any entries in the table that are wrong.

Quantity	Significant Figures
(a) 24 days	2
(b) 5.6×10^{-12} s	3
(c) 3.14 m	3
(d) 0.00383 g	5

ROUNDING

49. Round each number to four significant figures.
(a) 255.98612 **(b)** 0.0004893222
(c) 2.900856×10^{-4} **(d)** 2,231,479

50. Round each number to three significant figures.
(a) 10,776.522 **(b)** 4.999902×10^{6}
(c) 1.3499999995 **(d)** 0.0000344988

51. Round each number to two significant figures.
(a) 2.34 **(b)** 2.35
(c) 2.349 **(d)** 2.359

52. Round each number to three significant figures.
(a) 65.74 **(b)** 65.749
(c) 65.75 **(d)** 65.750

53. Each number is supposed to be rounded to three significant figures. Correct the ones that are incorrectly rounded.
(a) 42.3492 to 42.4 **(b)** 56.9971 to 57.0
(c) 231.904 to 232 **(d)** 0.04555 to 0.046

54. Each number is supposed to be rounded to two significant figures. Correct the ones that are incorrectly rounded.
(a) 1.249×10^{3} to 1.3×10^{3} **(b)** 3.999×10^{2} to 40
(c) 56.21 to 56.2 **(d)** 0.009964 to 0.010

55. Round the number on the left to the number of significant figures indicated by the example in the first row. (Use scientific notation as needed to avoid ambiguity.)

Number	Rounded to 4 Significant Figures	Rounded to 2 Significant Figures	Rounded to 1 Significant Figure
1.45815	1.458	1.5	1
8.32466			
84.57225			
132.5512			

56. Round the number on the left to the number of significant figures indicated by the example in the first row. (Use scientific notation as needed to avoid ambiguity.)

Number	Rounded to 4 Significant Figures	Rounded to 2 Significant Figures	Rounded to 1 Significant Figure
94.52118	94.52	95	9×10^{1}
105.4545			
0.455981			
0.009999991			

SIGNIFICANT FIGURES IN CALCULATIONS

57. Perform each calculation to the correct number of significant figures.
(a) $4.5 \times 0.03060 \times 0.391$
(b) $5.55 \div 8.97$
(c) $(7.890 \times 10^{12}) \div (6.7 \times 10^{4})$
(d) $67.8 \times 9.8 \div 100.04$

58. Perform each calculation to the correct number of significant figures.
(a) $89.3 \times 77.0 \times 0.08$
(b) $(5.01 \times 10^{5}) \div (7.8 \times 10^{2})$
(c) $4.005 \times 74 \times 0.007$
(d) $453 \div 2.031$

59. Correct any answers that have the incorrect number of significant figures.
(a) $34.00 \times 567 \div 4.564 = 4.2239 \times 10^{3}$
(b) $79.3 \div 0.004 \times 35.4 = 7 \times 10^{5}$
(c) $89.763 \div 22.4581 = 3.997$
(d) $(4.32 \times 10^{12}) \div (3.1 \times 10^{-4}) = 1.4 \times 10^{16}$

60. Correct any answers that have the incorrect number of significant figures.
a) $45.3254 \times 89.00205 = 4034.05$
b) $0.00740 \times 45.0901 = 0.334$
c) $49857 \div 904875 = 0.05510$
d) $0.009090 \times 6007.2 = 54.605$

61. Perform each calculation to the correct number of significant figures.
(a) $87.6 + 9.888 + 2.3 + 10.77$ **(b)** $43.7 - 2.341$
(c) $89.6 + 98.33 - 4.674$ **(d)** $6.99 - 5.772$

62. Perform each calculation to the correct number of significant figures.
(a) $1459.3 + 9.77 + 4.32$ **(b)** $0.004 + 0.09879$
(c) $432 + 7.3 - 28.523$ **(d)** $2.4 + 1.777$

63. Correct any answers that have the incorrect number of significant figures.
(a) $(3.8 \times 10^5) - (8.45 \times 10^5) = -4.7 \times 10^5$
(b) 0.00456 + 1.0936 = 1.10
(c) 8475.45 − 34.899 = 8440.55
(d) 908.87 − 905.34095 = 3.5291

64. Correct any answers that have the incorrect number of significant figures.
(a) $78.9 + 890.43 - 23 = 9.5 \times 10^2$
(b) 9354 − 3489.56 + 34.3 = 5898.74
(c) 0.00407 + 0.0943 = 0.0984
(d) 0.00896 − 0.007 = 0.00196

65. Perform each calculation to the correct number of significant figures.
(a) (78.4 − 44.889) ÷ 0.0087
(b) (34.6784 × 5.38) + 445.56
(c) $(78.7 \times 10^5 \div 88.529) + 356.99$
(d) (892 ÷ 986.7) + 5.44

66. Perform each calculation to the correct number of significant figures.
(a) $(1.7 \times 10^6 \div 2.63 \times 10^5) + 7.33$
(b) (568.99 − 232.1) ÷ 5.3
(c) $(9443 + 45 - 9.9) \times 8.1 \times 10^6$
(d) (3.14 × 2.4367) − 2.34

67. Correct any answers that have the incorrect number of significant figures.
(a) (78.56 − 9.44) × 45.6 = 3152
(b) $(8.9 \times 10^5 \div 2.348 \times 10^2) + 121 = 3.9 \times 10^3$
(c) (45.8 ÷ 3.2) − 12.3 = 2
(d) $(4.5 \times 10^3 - 1.53 \times 10^3) \div 34.5 = 86$

68. Correct any answers that have the incorrect number of significant figures.
(a) $(908.4 - 3.4) \div 3.52 \times 10^4 = 0.026$
(b) $(1206.7 - 0.904) \times 89 = 1.07 \times 10^5$
(c) (876.90 + 98.1) ÷ 56.998 = 17.11
(d) (4.55 ÷ 407859) + 1.00098 = 1.00210

UNIT CONVERSION

69. Perform each conversion.
(a) 3.55 kg to grams
(b) 8944 mm to meters
(c) 4598 mg to kilograms
(d) 0.0187 L to milliliters

70. Perform each conversion.
(a) 155.5 cm to meters
(b) 2491.6 g to kilograms
(c) 248 cm to millimeters
(d) 6781 mL to liters

71. Perform each conversion.
(a) 5.88 dL to liters
(b) 3.41×10^{-5} g to micrograms
(c) 1.01×10^{-8} s to nanoseconds
(d) 2.19 pm to meters

72. Perform each conversion.
(a) 1.08 Mm to kilometers
(b) 4.88 fs to picoseconds
(c) 7.39×10^{11} m to gigameters
(d) 1.15×10^{-10} m to picometers

73. Perform each conversion.
(a) 22.5 in. to centimeters
(b) 126 ft to meters
(c) 825 yd to kilometers
(d) 2.4 in. to millimeters

74. Perform each conversion.
(a) 78.3 in. to centimeters
(b) 445 yd to meters
(c) 336 ft to centimeters
(d) 45.3 in. to millimeters

75. Perform each conversion.
(a) 40.0 cm to inches
(b) 27.8 m to feet
(c) 10.0 km to miles
(d) 3845 kg to pounds

76. Perform each conversion.
(a) 254 cm to inches
(b) 89 mm to inches
(c) 7.5 L to quarts
(d) 122 kg to pounds

77. Complete the table.

m	km	Mm	Gm	Tm
5.08×10^8 m	____	508 Mm	____	____
____	____	27,976 Mm	____	____
____	____	____	____	1.77 Tm
____	1.5×10^5 km	____	____	____
____	____	____	423 Gm	____

78. Complete the table.

s	ms	μs	ns	ps
1.31×10^{-4} s	____	131 μs	____	____
____	____	____	____	12.6 ps
____	____	____	155 ns	____
____	1.99×10^{-3} ms	____	____	____
____	____	8.66×10^{-5} μs	____	____

79. Convert 2.255×10^{10} g to each unit.
(a) kg
(b) Mg
(c) mg
(d) metric tons (1 metric ton = 1000 kg)

80. Convert 1.88×10^{-6} g to each unit.
(a) mg
(b) cg
(c) ng
(d) μg

81. A student loses 3.3 lb in one month. How many grams did he lose?

82. A student gains 1.9 lb in two weeks. How many grams did he gain?

83. A runner wants to run 10.0 km. She knows that her running pace is 7.5 mi/h. How many minutes must she run?
Hint: Use 7.5 mi/h as a conversion factor between distance and time.

84. A cyclist rides at an average speed of 24 mi/h. If he wants to bike 195 km, how long (in hours) must he ride?

85. A recipe calls for 5.0 qt of milk. What is this quantity in cubic centimeters?

86. A gas can holds 2.0 gal of gasoline. What is this quantity in cubic centimeters?

UNITS RAISED TO A POWER

87. Fill in the blanks.
(a) $1.0 \text{ km}^2 =$ _____ m^2
(b) $1.0 \text{ cm}^3 =$ _____ m^3
(c) $1.0 \text{ mm}^3 =$ _____ m^3

88. Fill in the blanks.
(a) $1.0 \text{ ft}^2 =$ _____ in.^2
(b) $1.0 \text{ yd}^2 =$ _____ ft^2
(c) $1.0 \text{ m}^2 =$ _____ yd^2

89. The hydrogen atom has a volume of approximately $6.2 \times 10^{-31} \text{ m}^3$. What is this volume in each unit?
(a) cubic picometers
(b) cubic nanometers
(c) cubic angstroms (1 angstrom $= 10^{-10}$ m)

90. Earth has a surface area of 197 million square miles. What is its area in each unit?
(a) square kilometers
(b) square megameters
(c) square decimeters

91. A house has an area of 215 m^2. What is its area in each unit?
(a) km^2 **(b)** dm^2 **(c)** cm^2

92. A classroom has a volume of 285 m^3. What is its volume in each unit?
(a) km^3 **(b)** dm^3 **(c)** cm^3

93. Total U.S. farmland occupies 954 million acres. How many square miles is this?
(1 acre $= 43{,}560 \text{ ft}^2$; 1 mi $=$ 5280 ft)

94. The average U.S. farm occupies 435 acres. How many square miles is this?
(1 acre $= 43{,}560 \text{ ft}^2$; 1 mi $=$ 5280 ft)

UNIT CONVERSION IN BOTH THE NUMERATOR AND DENOMINATOR

95. The speed limit on many U.S. highways is 65 mi/hr. Convert this speed into each alternative unit.
(a) km/day **(b)** ft/s **(c)** m/s **(d)** yd/min

96. A form of children's Tylenol is sold as a suspension (in which the drug is suspended in water) that contains 32 mg/mL. Convert this concentration into each alternative unit.
(a) g/L **(b)** g/dL **(c)** kg/hL **(d)** μg/μL

97. A prescription medication requires 7.55 mg per kg of body weight. Convert this quantity to the number of grams required per pound of body weight and determine the correct dose (in g) for a 175-lb patient.

98. A prescription medication requires 0.00225 g per lb of body weight. Convert this quantity to the number of mg required per kg of body weight and determine the correct dose (in mg) for a 105-kg patient.

DENSITY

99. A sample of an unknown metal has a mass of 35.4 g and a volume of 3.11 cm^3. Calculate its density and identify the metal by comparison to Table 2.4.

100. A new penny has a mass of 2.49 g and a volume of 0.349 cm^3. Is the penny pure copper?

101. Glycerol is a syrupy liquid often used in cosmetics and soaps. A 2.50-L sample of pure glycerol has a mass of 3.15×10^3 g. What is the density of glycerol in grams per cubic centimeter?

102. An aluminum engine block has a volume of 4.77 L and a mass of 12.88 kg. What is the density of the aluminum in grams per cubic centimeter?

103. A supposedly gold crown is tested to determine its density. It displaces 10.7 mL of water and has a mass of 206 g. Could the crown be made of gold?

104. A vase is said to be solid platinum. It displaces 18.65 mL of water and has a mass of 157 g. Could the vase be solid platinum?

105. Ethylene glycol (antifreeze) has a density of $1.11\ g/cm^3$.
(a) What is the mass in grams of 387 mL of ethylene glycol?
(b) What is the volume in liters of 3.46 kg of ethylene glycol?

106. Acetone (fingernail-polish remover) has a density of $0.7857\ g/cm^3$.
(a) What is the mass in grams of 17.56 mL of acetone?
(b) What is the volume in milliliters of 7.22 g of acetone?

Cumulative Problems

107. A thief uses a bag of sand to replace a gold statue that sits on a weight-sensitive, alarmed pedestal. The bag of sand and the statue have exactly the same volume, 1.75 L. (Assume that the mass of the bag is negligible.)
(a) Calculate the mass of each object. (density of gold $= 19.3\ g/cm^3$; density of sand $= 3.00\ g/cm^3$)
(b) Did the thief set off the alarm? Explain.

108. One of the particles in an atom is the proton. A proton has a radius of approximately 1.0×10^{-13} cm and a mass of 1.7×10^{-24} g. Determine the density of a proton. (*Hint*: Find the volume of the proton and then divide the mass by the volume to get the density.)

$$\text{(volume of a sphere} = \frac{4}{3}\pi r^3;\ \pi = 3.14)$$

109. A block of metal has a volume of $13.4\ in.^3$ and weighs 5.14 lb. What is its density in grams per cubic centimeter?

110. A log is either oak or pine. It displaces 2.7 gal of water and weighs 19.8 lb. Is the log oak or pine? (density of oak $= 0.9\ g/cm^3$; density of pine $= 0.4\ g/cm^3$)

111. The density of aluminum is $2.7\ g/cm^3$. What is its density in kilograms per cubic meter?

112. The density of platinum is $21.4\ g/cm^3$. What is its density in pounds per cubic inch?

113. A typical backyard swimming pool holds $150\ yd^3$ of water. What is the mass in pounds of the water?

114. An iceberg has a volume of $8975\ ft^3$. What is the mass in kilograms of the iceberg? The density of ice is $0.917\ g/cm^3$.

115. The mass of fuel in an airplane must be determined before takeoff. A jet contains 155,211 L of fuel after it has been filled with fuel. What is the mass of the fuel in kilograms if the fuel's density is $0.768\ g/cm^3$?

116. A backpacker carries 2.5 L of white gas as fuel for her stove. How many pounds does the fuel add to her load? Assume the density of white gas to be $0.79\ g/cm^3$.

117. Honda produces a hybrid electric car called the Honda Insight. The Insight has both a gasoline-powered engine and an electric motor and has an EPA gas mileage rating of 43 mi per gallon on the highway. What is the Insight's rating in kilometers per liter?

118. You rent a car in Germany with a gas mileage rating of 12.8 km/L. What is its rating in miles per gallon?

119. A car has a mileage rating of 38 mi per gallon of gasoline. How many miles can the car travel on 76.5 L of gasoline?

120. A hybrid SUV consumes fuel at a rate of 12.8 km/L. How many miles can the car travel on 22.5 gal of gasoline?

121. Consider these observations on two blocks of different unknown metals:

	Volume
Block A	$125\ cm^3$
Block B	$145\ cm^3$

If block A has a greater mass than block B, what can be said of the relative densities of the two metals? (Assume that both blocks are solid.)

122. Consider these observations on two blocks of different unknown metals:

	Volume
Block A	$125\ cm^3$
Block B	$105\ cm^3$

If block A has a greater mass than block B, what can be said of the relative densities of the two metals? (Assume that both blocks are solid.)

123. You measure the masses and volumes of two cylinders. The mass of cylinder 1 is 1.35 times the mass of cylinder 2. The volume of cylinder 1 is 0.792 times the volume of cylinder 2. If the density of cylinder 1 is 3.85 g/cm^3, what is the density of cylinder 2?

124. A bag contains a mixture of copper and lead BBs. The average density of the BBs is 9.87 g/cm^3. Assuming that the copper and lead are pure, determine the relative amounts of each kind of BB.

125. An aluminum sphere has a mass of 25.8 g. Find the radius of the sphere. (The density of aluminum is 2.7 g/cm^3, and the volume of a sphere is given by the equation $V = \frac{4}{3}\pi r^3$.)

126. A copper cube has a mass of 87.2 g. Find the edge length of the cube. (The density of copper is 8.96 g/cm^3, and the volume of a cube is equal to the edge length cubed.)

Highlight Problems

127. Recall from Section 2.1 that NASA lost the Mars Climate Orbiter because one group of engineers used metric units in their calculations while another group used English units. Consequently, the Orbiter descended too far into the Martian atmosphere and burned up. Suppose that the Orbiter was to have established orbit at 155 km and that one group of engineers specified this distance as 1.55×10^5 m. Suppose further that a second group of engineers programmed the Orbiter to go to 1.55×10^5 ft. What was the difference in kilometers between the two altitudes? How low did the probe go?

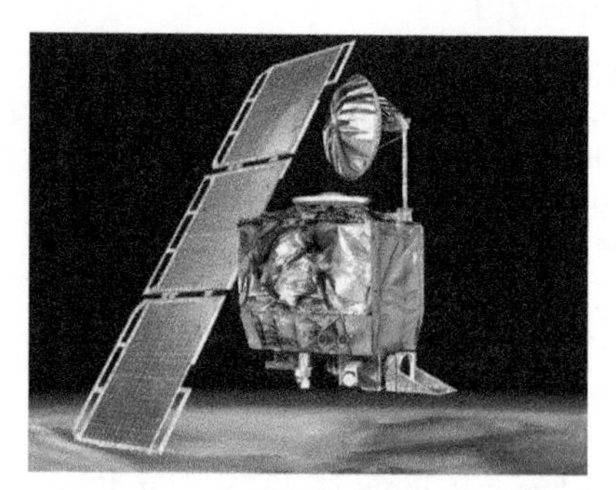

◀ The $94 million Mars Climate Orbiter was lost in the Martian atmosphere in 1999 because two groups of engineers failed to communicate with each other about the units in their calculations.

128. A NASA satellite showed that in 2012 the ozone hole over Antarctica had a maximum surface area of 21.2 million km^2. The largest ozone hole on record occurred in 2006 and had a surface area of 29.6 million km^2. Calculate the difference in diameter (in meters) between the ozone hole in 2012 and in 2006.

129. In 1999, scientists discovered a new class of black holes with masses 100 to 10,000 times the mass of our sun but occupying less space than our moon. Suppose that one of these black holes has a mass of 1×10^3 suns and a radius equal to one-half the radius of our moon. What is its density in grams per cubic centimeter? The mass of the sun is 2.0×10^{30} kg, and the radius of the moon is 2.16×10^3 mi. (Volume of a sphere $= \frac{4}{3}\pi r^3$.)

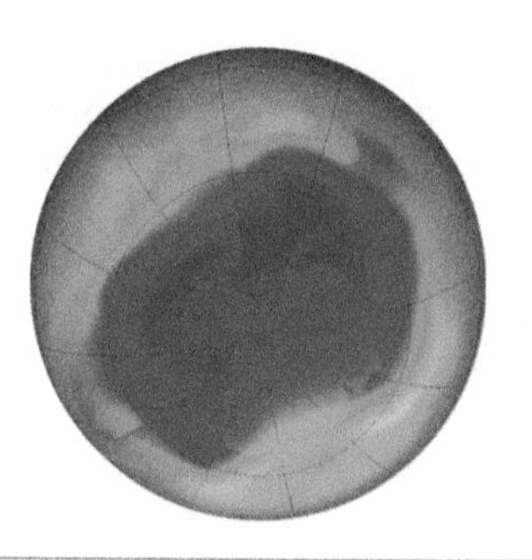

◀ A layer of ozone gas (a form of oxygen) in the upper atmosphere protects Earth from harmful ultraviolet radiation in sunlight. Human-made chemicals react with the ozone and deplete it, especially over the Antarctic at certain times of the year (the so-called ozone hole). The region of low-ozone concentration in 2006 (represented here by the dark purple color) was the largest on record.

130. A titanium bicycle frame contains the same amount of titanium as a titanium cube measuring 6.8 cm on a side. Use the density of titanium to calculate the mass in kilograms of titanium in the frame. What would be the mass of a similar frame composed of iron?

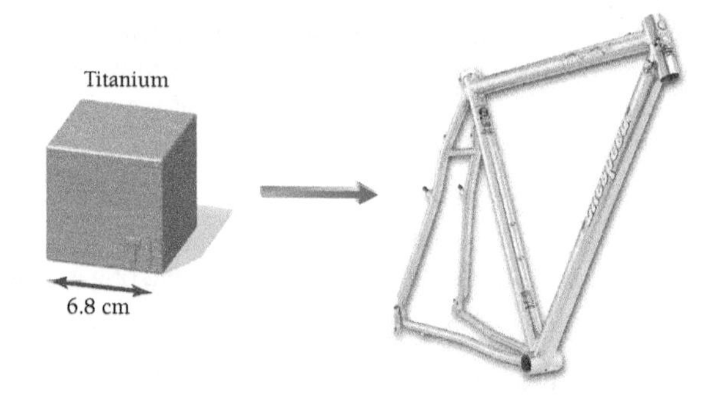

▶ A titanium bicycle frame contains the same amount of titanium as a titanium cube measuring 6.8 cm on a side.

Questions for Group Work

Discuss these questions with the group and record your consensus answer.

131. Look up the thickness of a human hair.

(a) Convert it to an SI standard unit (if it isn't already).

(b) Write it in scientific notation.

(c) Write it without scientific notation (you may need some zeros!).

(d) Write it with an appropriate prefix on a base unit.

Now complete the same exercises with the distance from the Earth to the sun.

132. The following statements are all true.
(a) Jessica's house is 5 km from the grocery store.
(b) Jessica's house is 4.73 km from the grocery store.
(c) Jessica's house is 4.73297 km from the grocery store.
How can they all be true? What does the number of digits communicate? What sort of device would Jessica need to make the claim in each statement?

133. Convert the height of each group member from feet and inches into meters. Once you determine your heights in meters, calculate the sum of all your heights. Use appropriate rules for significant figures at each step.

Data Interpretation and Analysis

134. The density of a substance can change with temperature. ▶ FIGURE A displays the density of water from −150 °C to 100 °C. Examine the graph and answer the questions that follow.
(a) Water undergoes a large change in density at 0 °C as it freezes to form ice. Calculate the percent change in density that occurs at 0 °C.
(b) Calculate the volume (in cm^3) of 54 g of water at 1 °C and the volume of the same mass of ice at −1 °C. What is the change in volume?
(c) Antarctica contains 26.5 million cubic kilometers of ice. Assume the temperature of the ice is −20 °C. If all of this ice were heated to 1 °C and melted to form water, what volume of liquid water would form?
(d) A 1.00-L sample of water is heated from 1 °C to 100 °C, What is the volume of the water after it is heated?

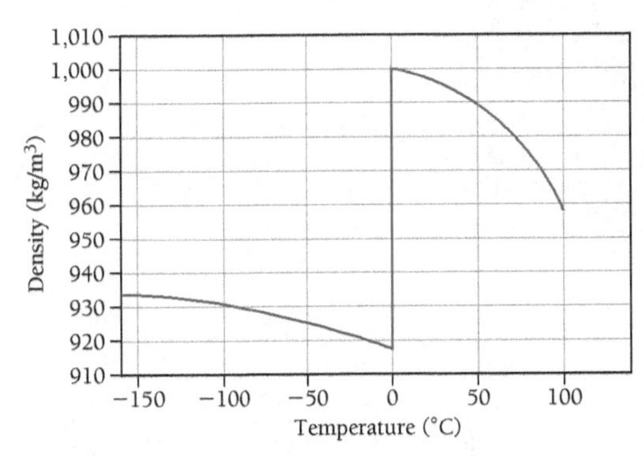

▲ FIGURE A **Density of Water as a Function of Temperature at 1.00 atm**

Answers to Skillbuilder Exercises

Skillbuilder 2.1.......... $\$1.8416 \times 10^{13}$
Skillbuilder 2.2.......... 3.8×10^{-5}
Skillbuilder 2.3.......... 103.4 °F
Skillbuilder 2.4.......... (a) four significant figures
(b) three significant figures
(c) two significant figures
(d) unlimited significant figures
(e) three significant figures
(f) ambiguous
Skillbuilder 2.5.......... (a) 0.001 or 1×10^{-3}
(b) 0.204
Skillbuilder 2.6.......... (a) 7.6
(b) 131.11
Skillbuilder 2.7.......... (a) 1288
(b) 3.12
Skillbuilder 2.8.......... 22.0 in.
Skillbuilder 2.9.......... 5.678 km
Skillbuilder 2.10.......... 0.28 L
Skillbuilder 2.11.......... 46.6 laps
Skillbuilder Plus, p. 35 1.06×10^4 m
Skillbuilder 2.12.......... 18 m/s
Skillbuilder 2.13.......... 4747 cm^3
Skillbuilder 2.14.......... 1.52×10^5 $in.^3$
Skillbuilder 2.15.......... Yes, the density is 21.4 g/cm^3 and matches that of platinum.
Skillbuilder 2.16.......... 4.4×10^{-2} cm^3
Skillbuilder Plus, p. 42 1.95 kg
Skillbuilder 2.17.......... 16 kg
Skillbuilder 2.18.......... $d = 19.3$ g/cm^3; yes, the density is consistent with that of gold.

Answers to Conceptual Checkpoints

2.1 (c) Multiplying by 10^{-3} is equivalent to moving the decimal point three places to the left.
2.2 (b) The last digit is considered to be uncertain by ± 1.
2.3 (b) The result of the calculation in **(a)** would be reported as 4; the result of the calculation in **(b)** would be reported as 1.5.
2.4 (d) The diameter would be expressed as 28 nm.
2.5 (a) The unit of ft should be in the denominator, and the conversion factor in **(c)** is incorrect (because 1 in. ≠ 12 ft).
2.6 (c) Kilometers must appear in the numerator and meters in the denominator, and the conversion factor in **(d)** is incorrect (10^3 km ≠ 1 m).
2.7 (d) $(3\text{ ft}) \times (3\text{ ft}) \times (3\text{ ft}) = 27\text{ ft}^3$

3 Matter and Energy

"Thus, the task is, not so much to see what no one has yet seen; but to think what nobody has yet thought, about that which everybody sees."

—Erwin Schrödinger (1887–1961)

CHAPTER OUTLINE

3.1 In Your Room

Look around the room you are in—what do you see? You might see your desk, your bed, or a glass of water. Maybe you have a window and can see trees, grass, or mountains. You can certainly see this book and possibly the table it sits on. What are these things made of? They are all made of *matter*, which we will define more carefully shortly. For now, know that all you see is matter—your desk, your bed, the glass of water, the trees, the mountains, and this book. Some of what you don't see is matter as well. For example, you are constantly breathing air, which is also matter, into and out of your lungs. You feel the matter in air when you feel wind on your skin. Virtually everything is made of matter.

Think about the differences between different kinds of matter. Air is different from water, and water is different from wood. One of our first tasks as we learn about matter is to identify the similarities and differences among different kinds of matter. How are sugar and salt similar? How are air and water different? Why are they different? Why is a mixture of sugar and water similar to a mixture of salt and water but different from a mixture of sand and water? As students of chemistry, we are particularly interested in the similarities and differences between various kinds of matter and how these reflect the similarities and differences between their component atoms and molecules. We strive to understand the connection between the macroscopic world and the molecular one.

◀ Everything that you can see in this room is made of matter. As students of chemistry, we are interested in how the differences between different kinds of matter are related to the differences between the molecules and atoms that compose the matter. The molecular structures shown here are water molecules on the left and carbon atoms in graphite on the right.

3.2 What Is Matter?

▶ Define matter, atoms, and molecules.

We define **matter** as anything that occupies space and has mass. Some types of matter—such as steel, water, wood, and plastic—are readily visible to our eyes. Other types of matter—such as air or microscopic dust—are not visible to our naked eyes. Matter may sometimes appear smooth and continuous, but actually it is not. Matter is ultimately composed of **atoms**, submicroscopic particles that are the fundamental building blocks of matter (▼ FIGURE 3.1a). In many cases, these atoms are bonded together to form **molecules**, two or more atoms joined to one another in specific geometric arrangements (▼ FIGURE 3.1b). Advances in microscopy have allowed us to image the atoms (▼ FIGURE 3.2) and molecules (▼ FIGURE 3.3) that compose matter, sometimes with stunning clarity.

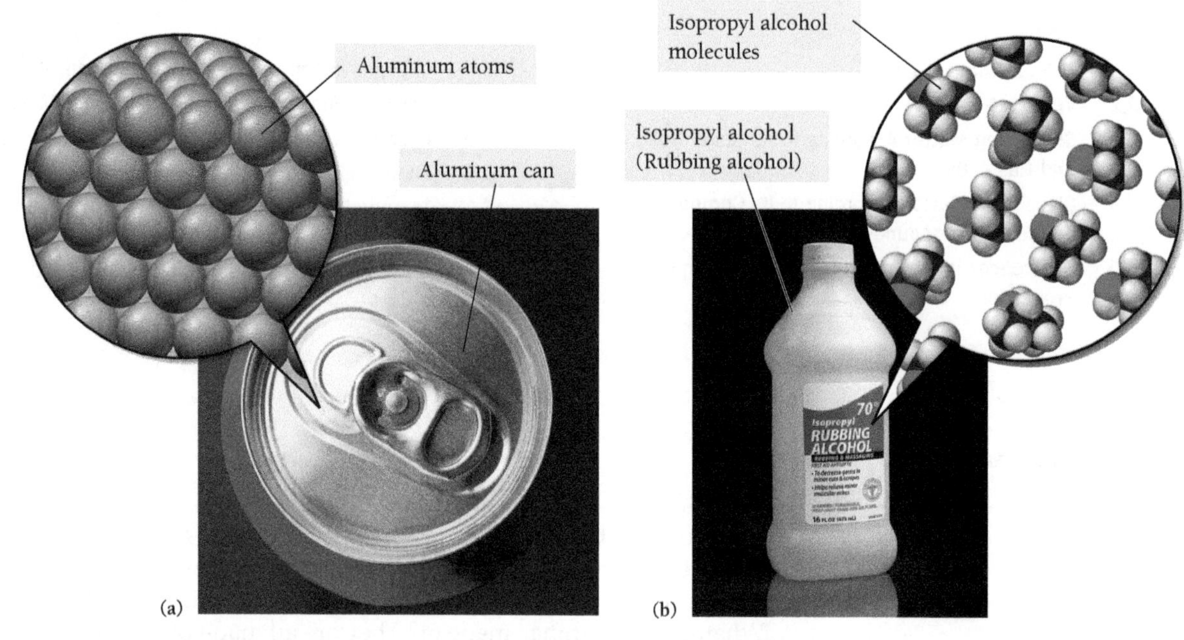

▲ FIGURE 3.1 **Atoms and molecules** All matter is ultimately composed of atoms. (a) In some substances, such as aluminum, the atoms exist as independent particles. (b) In other substances, such as rubbing alcohol, several atoms bond together in well-defined structures called molecules.

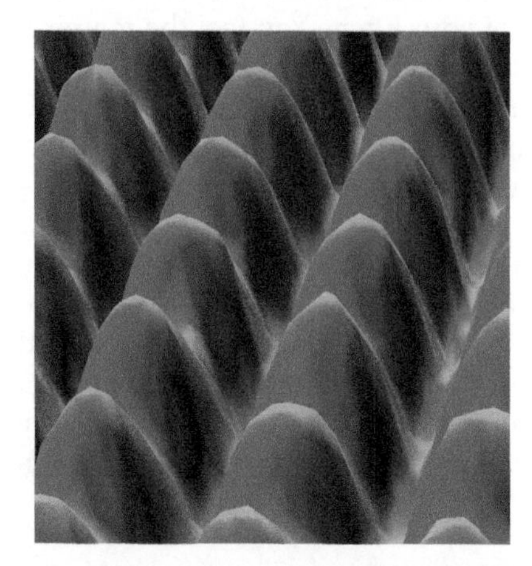

▲ FIGURE 3.2 **Scanning tunneling microscope image of nickel atoms** A scanning tunneling microscope (STM) creates an image by scanning a surface with a tip of atomic dimensions. It can distinguish individual atoms, seen as blue bumps in this image.

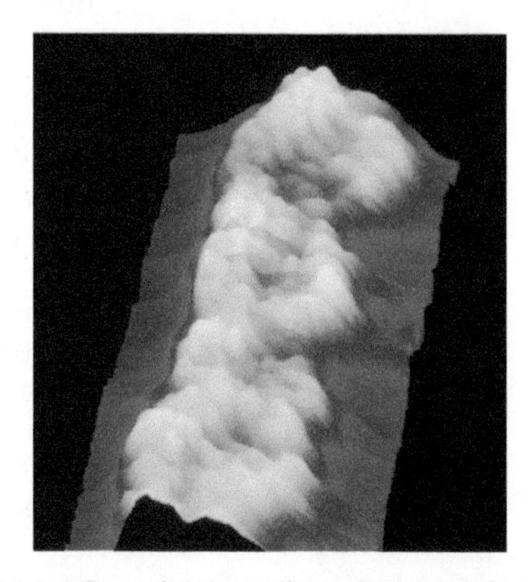

▲ FIGURE 3.3 **Scanning tunneling microscope image of a DNA molecule** DNA is the hereditary material that encodes the operating instructions for most cells in living organisms. In this image, the DNA molecule is yellow, and the double-stranded structure of DNA is discernible.

3.3 Classifying Matter According to Its State: Solid, Liquid, and Gas

- Classify matter as solid, liquid, or gas.

Key Concept Video Classifying Matter

The eText 2.0 icon indicates that this feature is embedded and interactive in the eText.

The common **states of matter** are **solid**, **liquid**, and **gas** (▼ FIGURE 3.4). In solid matter, atoms or molecules pack close to each other in fixed locations. Although neighboring atoms or molecules in a solid may vibrate or oscillate, they do not move around each other, giving solids their familiar fixed volume and rigid shape.

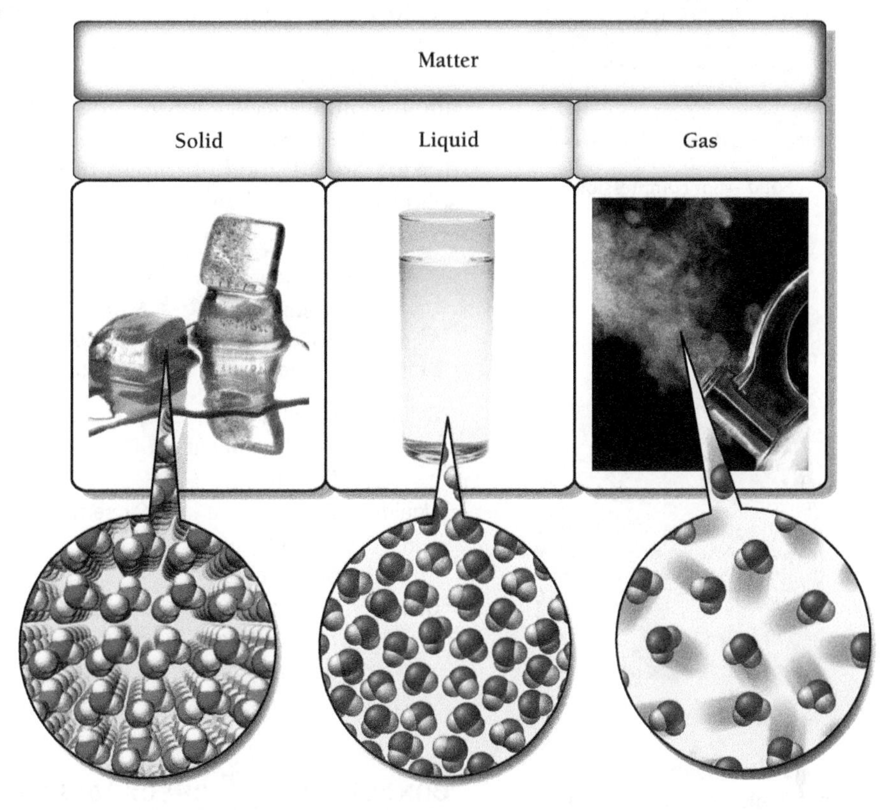

▶ FIGURE 3.4 **Three states of matter** Water exists as ice (solid), water (liquid), and steam (gas). In ice, the water molecules are closely spaced and, although they vibrate about a fixed point, they do not generally move relative to one another. In liquid water, the water molecules are also closely spaced but are free to move around and past each other. In steam, water molecules are separated by large distances and do not interact significantly with one another.

Ice, diamond, quartz, and iron are examples of solid matter. Solid matter may be **crystalline**, in which case its atoms or molecules arrange in geometric patterns with long-range, repeating order (◀ FIGURE 3.5a), or it may be **amorphous**, in which case its atoms or molecules do not have long-range order (◀ FIGURE 3.5b). Examples of *crystalline* solids include salt (▼ FIGURE 3.6) and diamond; the well-ordered, geometric shapes of salt and diamond crystals

▲ FIGURE 3.5 **Types of solid matter** (a) In a crystalline solid, atoms or molecules occupy specific positions to create a well-ordered, three-dimensional structure. (b) In an amorphous solid, atoms do not have any long-range order.

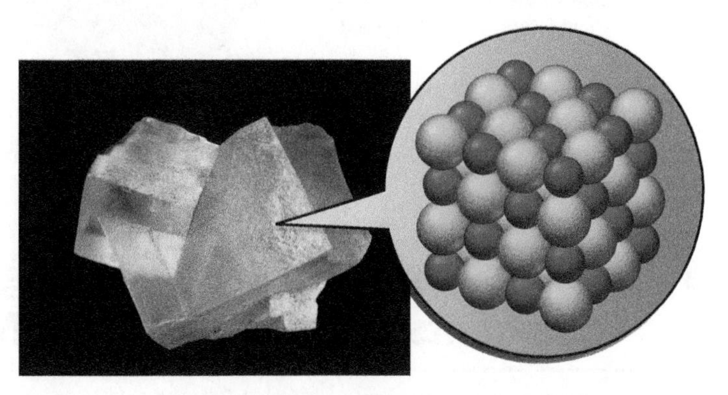

▲ FIGURE 3.6 **Salt: a crystalline solid** Sodium chloride is an example of a crystalline solid. The well-ordered, cubic shape of salt crystals is due to the well-ordered, cubic arrangement of its atoms.

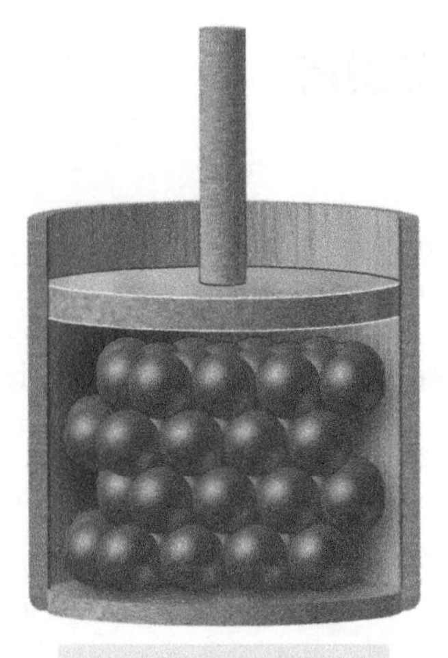

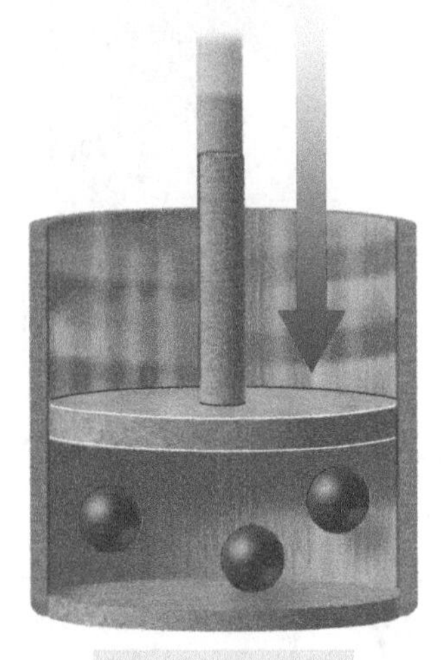

▲ FIGURE 3.7 **Gases are compressible** Because the atoms or molecules that compose gases are not in contact with one another, we can compress gases.

reflect the well-ordered geometric arrangement of their atoms. Examples of *amorphous* solids include glass, rubber, and plastic.

In liquid matter, atoms or molecules are close to each other (about as close as molecules in a solid), but they are free to move around and by each other. Like solids, liquids have a fixed volume because their atoms or molecules are in close contact. Unlike solids, however, liquids assume the shape of their container because the atoms or molecules are free to move relative to one another. Water, gasoline, alcohol, and mercury are examples of liquid matter.

In gaseous matter, atoms or molecules are separated by large distances and are free to move relative to one another. Because the atoms or molecules that compose gases are not in contact with one another, gases are **compressible** (◀ FIGURE 3.7). When we inflate a bicycle tire, for example, we push more atoms and molecules into the same space, compressing them and making the tire harder. Gases always assume the shape and volume of their containers. Oxygen, helium, and carbon dioxide are all good examples of gases. Table 3.1 summarizes the properties of solids, liquids, and gases.

TABLE 3.1 Properties of Solids, Liquids, and Gases

State	Atomic/ Molecular Motion	Atomic/ Molecular Spacing	Shape	Volume	Compressibility
Solid	Oscillation/ vibration about fixed point	Close together	Definite	Definite	Incompressible
Liquid	Free to move relative to one another	Close together	Indefinite	Definite	Incompressible
Gas	Free to move relative to one another	Far apart	Indefinite	Indefinite	Compressible

CONCEPTUAL CHECKPOINT 3.1

Which image best represents matter in the gas state?

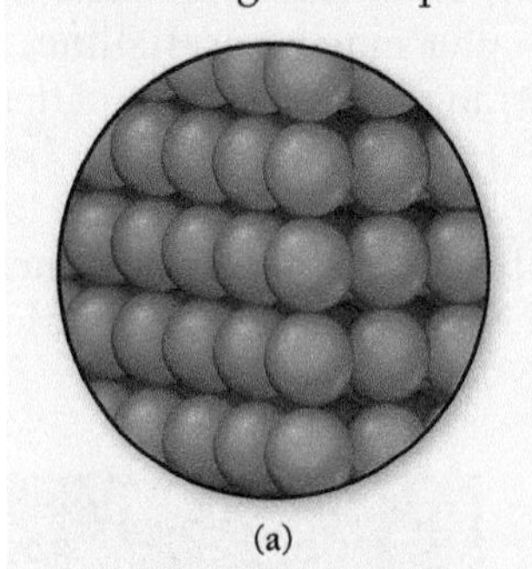
(a)

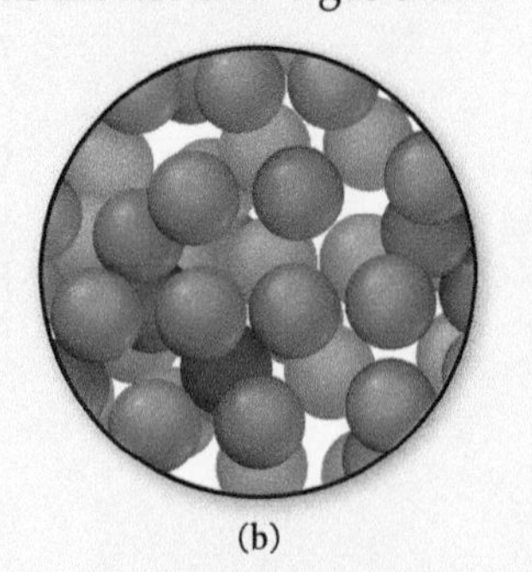
(b)

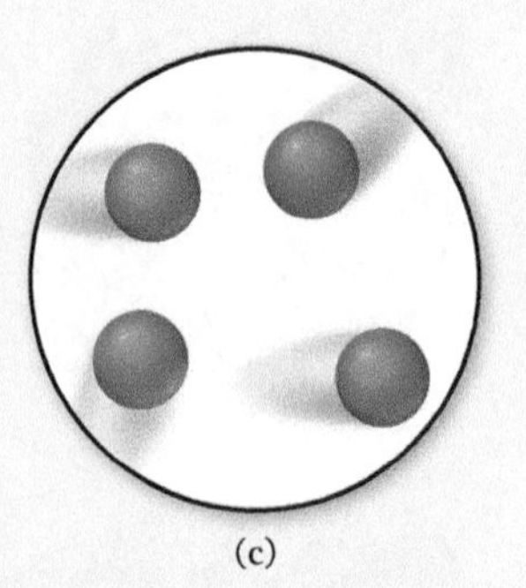
(c)

Note: You can find the answers to all Conceptual Checkpoints at the end of the chapter.

PEARSON eText 2.0

The eText 2.0 icon indicates that this feature is embedded and interactive in the eText.

3.4 Classifying Matter According to Its Composition: Elements, Compounds, and Mixtures

▶ Classify matter as element, compound, or mixture.

In addition to classifying matter according to its state, we can classify it according to its composition (▶ FIGURE 3.8). Matter may be either a **pure substance**, composed of only one type of atom or molecule, or a **mixture**, composed of two or more different types of atoms or molecules combined in variable proportions.

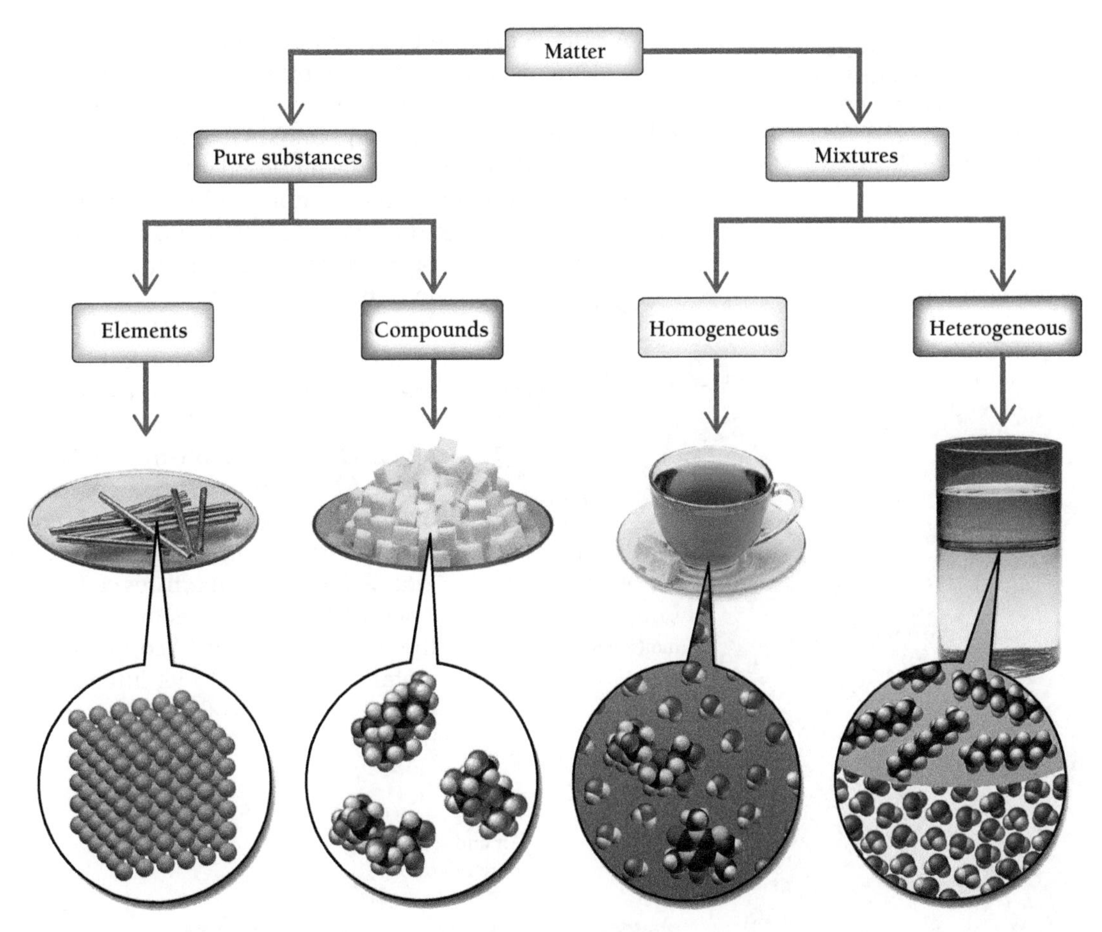

▲ FIGURE 3.8 **Classification of matter** Matter may be a pure substance or a mixture. A pure substance may be either an element (such as copper) or a compound (such as sugar), and a mixture may be either homogeneous (such as sweetened tea) or heterogeneous (such as gasoline and water).

Pure substances are composed of only one type of atom or molecule. Helium and water are both pure substances. The atoms that compose helium are all helium atoms, and the molecules that compose water are all water molecules—no other atoms or molecules are mixed in.

Pure substances can themselves be divided into two types: elements and compounds. Copper is an example of an **element**, a substance that cannot be broken down into simpler substances. The graphite in pencils is also an element—carbon. No chemical transformation can decompose graphite into simpler substances; it is pure carbon. All known elements are listed in the periodic table in the inside front cover of this book and in alphabetical order on the inside back cover of this book.

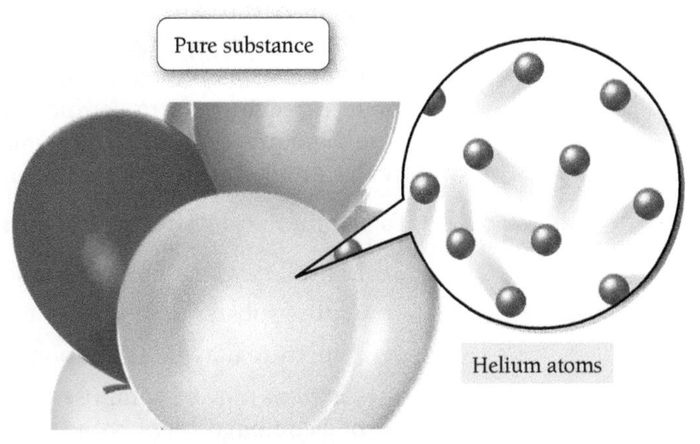

▲ Helium is a pure substance composed only of helium atoms.

A compound is composed of different atoms that are chemically united (bonded). A mixture is composed of different substances that are not chemically united, but simply mixed together.

A pure substance can also be a **compound**, a substance composed of two or more elements in fixed definite proportions. Compounds are more common than pure elements because most elements are chemically reactive and combine with other elements to form compounds. Water, table salt, and sugar are examples of compounds; they can all be decomposed into simpler substances. If you heat sugar in a pan over a flame, you decompose it into several substances including carbon (an element) and gaseous water (a different compound). The black substance left on your pan after burning is the carbon; the water escapes into the air as steam.

Pure substance

Water molecules

▲ Water is a pure substance composed only of water molecules.

The majority of matter that we encounter is in the form of mixtures. Apple juice, a flame, salad dressing, and soil are all examples of mixtures; they each contain several substances with proportions that vary from one sample to another. Other common mixtures include air, seawater, and brass. Air is a mixture composed primarily of nitrogen and oxygen gas; seawater is a mixture composed primarily of salt and water; and brass is a mixture composed of copper and zinc. Each of these mixtures can have different proportions of its constituent components. For example, metallurgists vary the relative amounts of copper and zinc in brass to tailor the metal's properties to its intended use—the higher the zinc content relative to the copper content, the more brittle the brass.

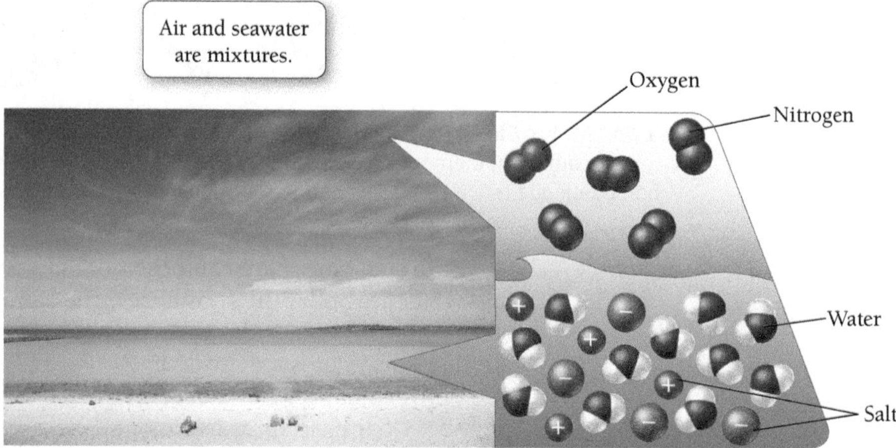

▲ Air and seawater are examples of mixtures. Air contains primarily nitrogen and oxygen. Seawater contains primarily salt and water.

We can also classify mixtures according to how uniformly the substances within them mix. In a **heterogeneous mixture**, such as oil and water, the composition varies from one region to another. In a **homogeneous mixture**, such as salt water or sweetened tea, the composition is the same throughout. Homogeneous mixtures have uniform compositions because the atoms or molecules that compose them mix uniformly. Remember that the properties of matter are determined by the atoms or molecules that compose it.

To summarize, as shown in Figure 3.8 (on p. 65):

- Matter may be a pure substance, or it may be a mixture.
- A pure substance may be either an element or a compound.
- A mixture may be either homogeneous or heterogeneous.
- Mixtures may be composed of two or more elements, two or more compounds, or a combination of both.

EXAMPLE **3.1** **Classifying Matter**

Classify each type of matter as a pure substance or a mixture. If it is a pure substance, classify it as an element or a compound; if it is a mixture, classify it as homogeneous or heterogeneous.

(a) a lead weight **(b)** seawater
(c) distilled water **(d)** Italian salad dressing

SOLUTION

Begin by examining the alphabetical listing of pure elements inside the back cover of this text. If the substance appears in that table, it is a pure substance and an element. If it is not in the table but is a pure substance, then it is a compound. If the substance is not a pure substance, then it is a mixture. Think about your everyday experience with each mixture to determine if it is homogeneous or heterogeneous.

(a) Lead is listed in the table of elements. It is a pure substance and an element.

(b) Seawater is composed of several substances, including salt and water; it is a mixture. It has a uniform composition, so it is a homogeneous mixture.

(c) Distilled water is not listed in the table of elements, but it is a pure substance (water); therefore, it is a compound.

(d) Italian salad dressing contains a number of substances and is therefore a mixture. It usually separates into at least two distinct regions with different composition and is therefore a heterogeneous mixture.

▶ SKILLBUILDER 3.1 | Classifying Matter

Classify each type of matter as a pure substance or a mixture. If it is a pure substance, classify it as an element or a compound. If it is a mixture, classify it as homogeneous or heterogeneous.

(a) mercury in a thermometer
(b) exhaled air
(c) chicken noodle soup
(d) sugar

▶ FOR MORE PRACTICE Example 3.12; Problems 31, 32, 33, 34, 35, 36.

Note: The answers to all Skillbuilder exercises appear at the end of the chapter.

CONCEPTUAL CHECKPOINT 3.2

Which of the substances depicted here is a pure substance?

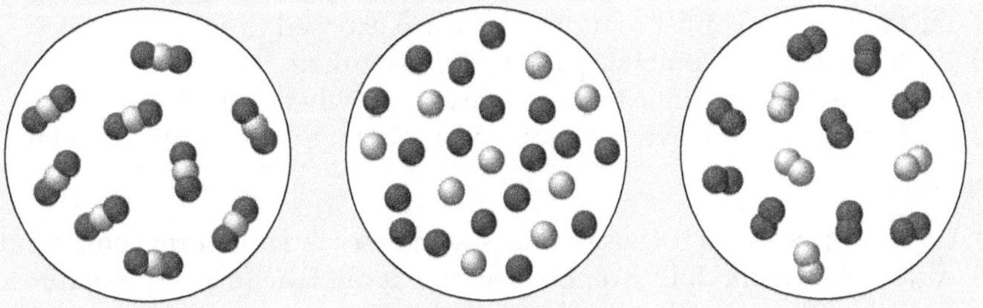

(a) (b) (c)

3.5 Differences in Matter: Physical and Chemical Properties

▶ Distinguish between physical and chemical properties.

The characteristics that distinguish one substance from another are called **properties**. Different substances have unique properties that characterize them and distinguish them from other substances. For example, we can distinguish water from alcohol based on their different smells, or we can distinguish gold from silver based on their different colors.

In chemistry, we categorize properties into two different types: physical and chemical. A **physical property** is one that a substance displays without changing

▲ FIGURE 3.9 **A physical property** The boiling point of water is a physical property, and boiling is a physical change. When water boils, it turns into a gas, but the water molecules are the same in both the liquid water and the gaseous steam.

its composition. A **chemical property** is one that a substance displays only through changing its composition. For example, the characteristic odor of gasoline is a physical property—gasoline does not change its composition when it exhibits its odor. In contrast, the flammability of gasoline is a chemical property—gasoline does change its composition when it burns.

The atomic or molecular composition of a substance does not change when the substance displays its physical properties. For example, the boiling point of water—a physical property—is 100 °C. When water boils, it changes from a liquid to a gas, but the gas is still water (◀ FIGURE 3.9). On the contrary, the susceptibility of iron to rust is a chemical property—iron must change into iron(III) oxide to display this property (▼ FIGURE 3.10). Physical properties include odor, taste, color, appearance, melting point, boiling point, and density. Chemical properties include corrosiveness, flammability, acidity, and toxicity.

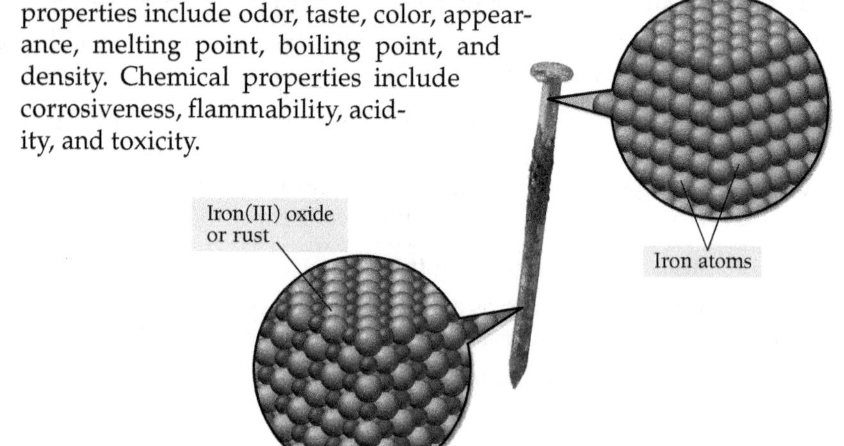

▲ FIGURE 3.10 **A chemical property** The susceptibility of iron to rusting is a chemical property, and rusting is a chemical change. When iron rusts, it turns from iron to iron(III) oxide.

EXAMPLE 3.2 Physical and Chemical Properties

Classify each property as physical or chemical.

(a) the tendency of copper to turn green when exposed to air
(b) the tendency of automobile paint to dull over time
(c) the tendency of gasoline to evaporate quickly when spilled
(d) the low mass (for a given volume) of aluminum relative to other metals

SOLUTION

(a) Copper turns green because it reacts with gases in air to form compounds; this is a chemical property.
(b) Automobile paint dulls over time because it can fade (decompose) due to sunlight or it can react with oxygen in air. In either case, this is a chemical property.
(c) Gasoline evaporates quickly because it has a low boiling point; this is a physical property.
(d) Aluminum's low mass (for a given volume) relative to other metals is due to its low density; this is a physical property.

▶ **SKILLBUILDER 3.2 | Physical and Chemical Properties**

Classify each property as physical or chemical.

(a) the explosiveness of hydrogen gas
(b) the bronze color of copper
(c) the shiny appearance of silver
(d) the ability of dry ice to sublime (change from solid directly to vapor)

▶ **FOR MORE PRACTICE** Example 3.13; Problems 37, 38, 39, 40.

3.6 Changes in Matter: Physical and Chemical Changes

▶ Distinguish between physical and chemical changes.

Every day, we witness changes in matter: Ice melts, iron rusts, and fruit ripens. What happens to the atoms and molecules that make up these substances during these changes? The answer depends on the kind of change. In a **physical change**, matter changes its appearance but not its composition. For example, when ice melts, it looks different—water looks different from ice—but its composition is the same. Solid ice and liquid water are both composed of water molecules, so melting is a physical change. Similarly, when glass shatters, it looks different, but its composition remains the same—it is still glass. Again, this is a physical change. In a **chemical change**, however, matter *does* change its composition. For example, copper turns green upon continued exposure to air because it reacts with gases in air to form new compounds. This is a chemical change. Matter undergoes a chemical change when it undergoes a **chemical reaction**. In a chemical reaction, the substances present before the chemical change are called **reactants**, and the substances present after the change are called **products**:

$$\text{Reactants} \xrightarrow[\text{Chemical Change}]{} \text{Products}$$

We will cover chemical reactions in much more detail in Chapter 7.

State changes—transformations from one state of matter (such as solid or liquid) to another—are always physical changes.

The differences between physical and chemical changes are not always apparent. Only chemical examination of the substances before and after the change can verify whether the change is physical or chemical. For many cases, however, we can identify chemical and physical changes based on what we know about the changes. Changes in state, such as melting or boiling, or changes that involve merely appearance, such as those produced by cutting or crushing, are always physical changes. Changes involving chemical reactions—often evidenced by heat exchange or color changes—are always chemical changes.

The main difference between chemical and physical changes is related to the changes at the molecular and atomic level. In physical changes, the atoms that compose the matter *do not* change their fundamental associations, even though the matter may change its appearance. In chemical changes, atoms do change their fundamental associations, resulting in matter with a new identity. *A physical change results in a different form of the same substance, while a chemical change results in a completely new substance.*

Consider physical and chemical changes in liquid butane, the substance used to fuel butane lighters. In many lighters, you can see the liquid butane through the plastic case of the lighter. If you push the fuel button on the lighter without turning the flint, some of the liquid butane *vaporizes* (changes from liquid to gas). You cannot see the gaseous butane, but if you listen carefully you can usually hear hissing as it leaks out (◀ FIGURE 3.11). Since the liquid butane and the gaseous butane are both composed of butane molecules, the change is physical. On the other hand, if you push the button *and* turn the flint to create a spark, a chemical change occurs. The butane molecules react with oxygen molecules in air to form new molecules, carbon dioxide, and water (◀ FIGURE 3.12). The change is chemical because the molecular composition changes upon burning.

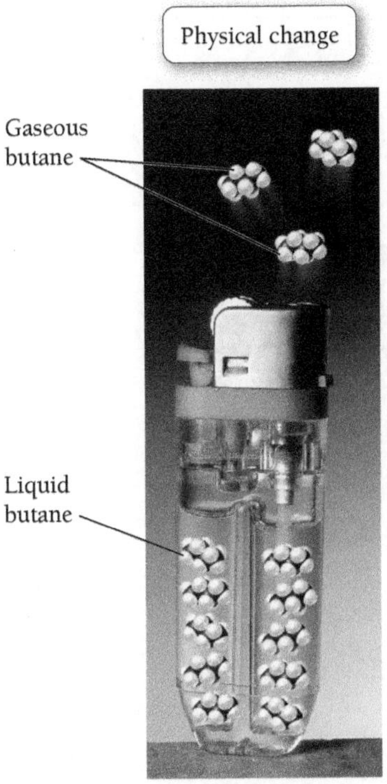

▲ **FIGURE 3.11 Vaporization: a physical change** If you push the button on a lighter without turning the flint, some of the liquid butane vaporizes to gaseous butane. Since the liquid butane and the gaseous butane are both composed of butane molecules, this is a physical change.

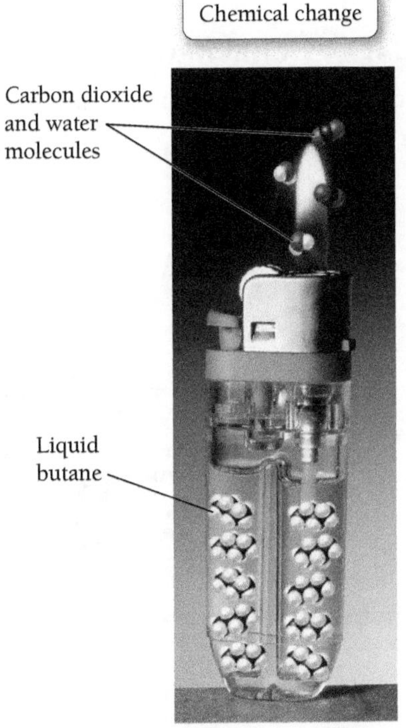

▲ **FIGURE 3.12 Burning: a chemical change** If you push the button *and* turn the flint to create a spark, you produce a flame. The butane molecules react with oxygen molecules in air to form new molecules, carbon dioxide, and water. This is a chemical change.

EXAMPLE 3.3 Physical and Chemical Changes

Classify each change as physical or chemical.

(a) the rusting of iron
(b) the evaporation of fingernail-polish remover (acetone) from the skin
(c) the burning of coal
(d) the fading of a carpet upon repeated exposure to sunlight

SOLUTION

(a) Iron rusts because it reacts with oxygen in air to form iron(III) oxide; therefore, this is a chemical change.
(b) When fingernail-polish remover (acetone) evaporates, it changes from liquid to gas, but it remains acetone; therefore, this is a physical change.
(c) Coal burns because it reacts with oxygen in air to form carbon dioxide; this is a chemical change.
(d) A carpet fades on repeated exposure to sunlight because the molecules that give the carpet its color are decomposed by sunlight; this is a chemical change.

▶ SKILLBUILDER 3.3 | Physical and Chemical Changes

Classify each change as physical or chemical.

(a) copper metal forming a blue solution when it is dropped into colorless nitric acid
(b) a train flattening a penny placed on a railroad track
(c) ice melting into liquid water
(d) a match igniting a firework

▶ FOR MORE PRACTICE Example 3.14; Problems 41, 42, 43, 44.

CONCEPTUAL ✔ CHECKPOINT 3.3

In this figure, liquid water is vaporizing into steam.

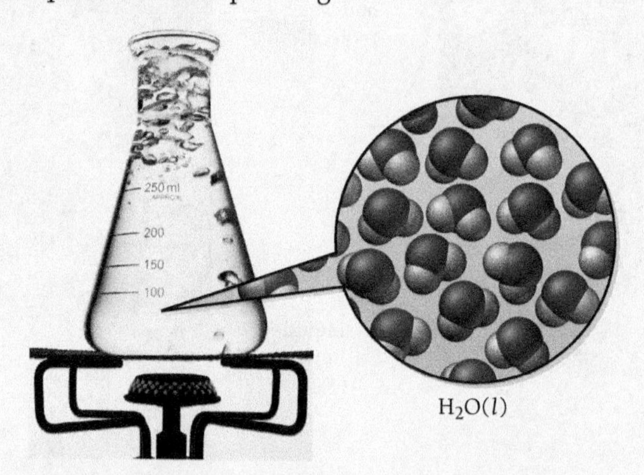

Which diagram best represents the molecules in the steam?

(a)

(b)

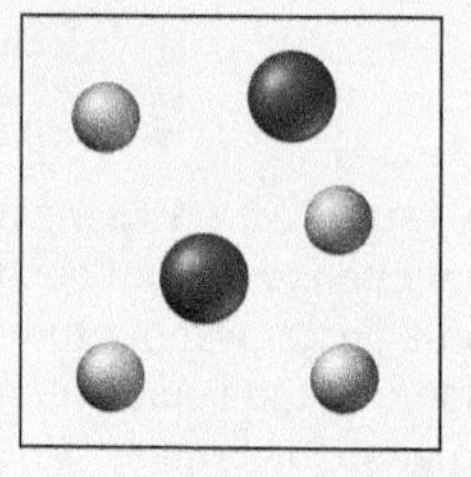

(c)

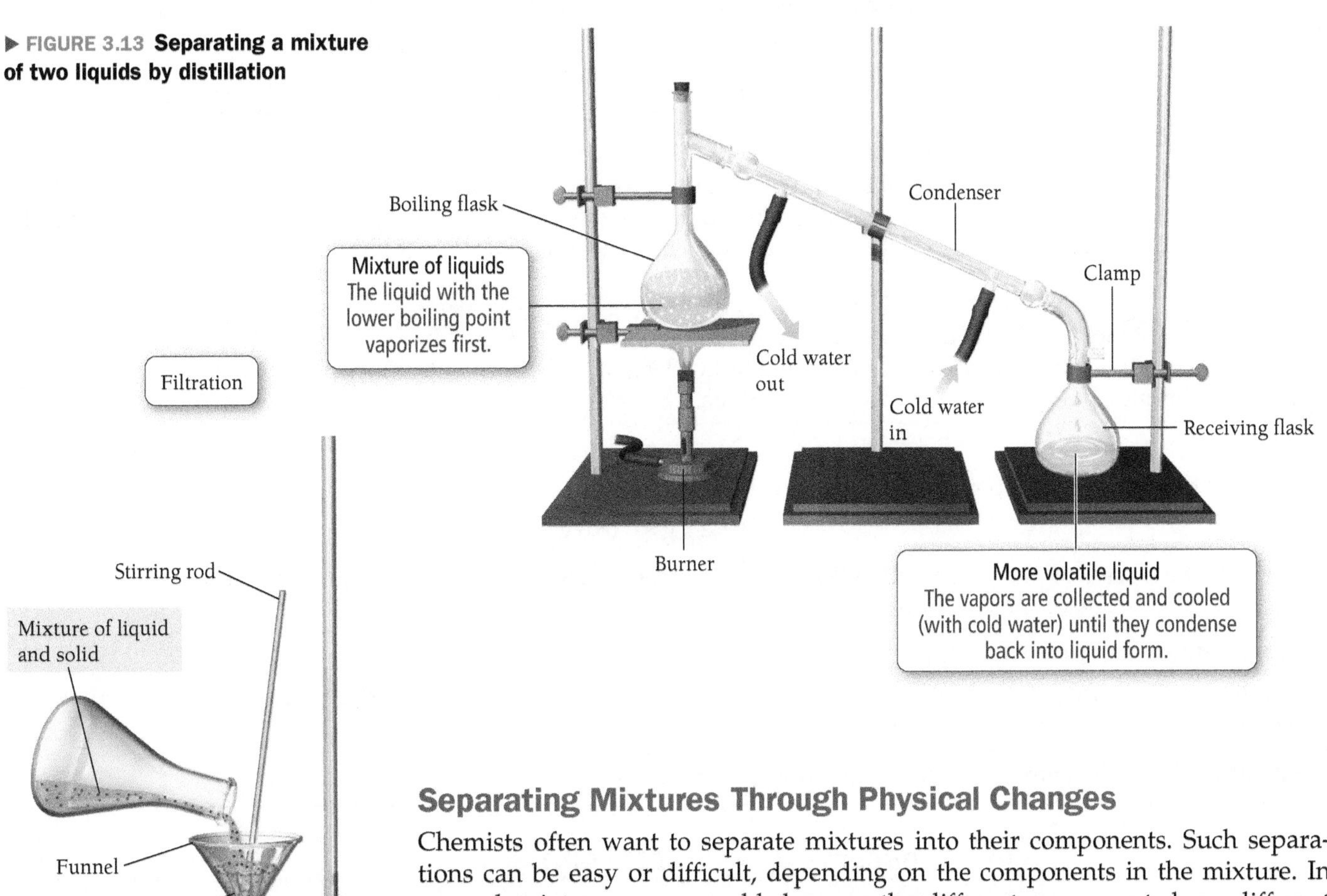

▶ FIGURE 3.13 **Separating a mixture of two liquids by distillation**

▲ FIGURE 3.14 **Separating a solid from a liquid by filtration**

Separating Mixtures Through Physical Changes

Chemists often want to separate mixtures into their components. Such separations can be easy or difficult, depending on the components in the mixture. In general, mixtures are separable because the different components have different properties. We can use various techniques that exploit these differences to achieve separation. For example, oil and water are immiscible (do not mix) and have different densities. For this reason, oil floats on top of water, and we can separate it from water by **decanting**—carefully pouring off—the oil into another container. We can separate mixtures of miscible liquids by **distillation**, in which we heat the mixture to boil off the more **volatile**—the more easily vaporizable—liquid. We then recondense the volatile liquid in a condenser and collect it in a separate flask (▲ FIGURE 3.13). If a mixture is composed of a solid and a liquid, we can separate the two by **filtration**, in which we pour the mixture through filter paper usually held in a funnel (◀ FIGURE 3.14).

3.7 Conservation of Mass: There Is No New Matter

▶ Apply the law of conservation of mass.

This law is a slight oversimplification. In nuclear reactions, covered in Chapter 17, significant changes in mass can occur. In chemical reactions, however, the changes are so minute that they can be ignored.

We examine the quantitative relationships in chemical reactions in Chapter 8.

As we have seen, our planet, our air, and even our own bodies are composed of matter. Physical and chemical changes do not destroy matter, nor do they create new matter. Recall from Chapter 1 that Antoine Lavoisier, by studying combustion, established the law of conservation of mass, which states:

Matter is neither created nor destroyed in a chemical reaction.

During physical and chemical changes, the total amount of matter remains constant even though it may not initially appear that it has. When we burn butane in a lighter, for example, the butane slowly disappears. Where does it go? It combines with oxygen to form carbon dioxide and water that travel into the surrounding air. The mass of the carbon dioxide and water that forms, however, exactly equals the mass of the butane and oxygen that combined.

Suppose that we burn 58 g of butane in a lighter. It will react with 208 g of oxygen to form 176 g of carbon dioxide and 90 g of water.

$$\underbrace{\underset{58\text{ g} + 208\text{ g}}{\text{Butane} + \text{Oxygen}}}_{266\text{ g}} \longrightarrow \underbrace{\underset{176\text{ g} + 90\text{ g}}{\text{Carbon Dioxide} + \text{Water}}}_{266\text{ g}}$$

The sum of the masses of the butane and oxygen, 266 g, is equal to the sum of the masses of the carbon dioxide and water, which is also 266 g. In this chemical reaction, as in all chemical reactions, matter is conserved.

EXAMPLE 3.4 Conservation of Mass

A chemist forms 16.6 g of potassium iodide by combining 3.9 g of potassium with 12.7 g of iodine. Show that these results are consistent with the law of conservation of mass.

SOLUTION

The sum of the masses of the potassium and iodine is:

$$3.9\text{ g} + 12.7\text{ g} = 16.6\text{ g}$$

The sum of the masses of potassium and iodine equals the mass of the product, potassium iodide. The results are consistent with the law of conservation of mass.

▶ **SKILLBUILDER 3.4 | Conservation of Mass**

Suppose 12 g of natural gas combines with 48 g of oxygen in a flame. The chemical change produces 33 g of carbon dioxide. How many grams of water form?

▶ **FOR MORE PRACTICE** Example 3.15; Problems 45, 46, 47, 48, 49, 50.

CONCEPTUAL CHECKPOINT 3.4

Consider a drop of water that is put into a flask, sealed with a cap, and heated until the droplet vaporizes. Is the mass of the container and water different after heating?

3.8 Energy

- ▶ Recognize the different forms of energy.
- ▶ Identify and convert between energy units.

Matter is one of the two major components of our universe. The other major component is **energy**, *the capacity to do work.* **Work** is defined as the result of a force acting on a distance. For example, if you push this book across your desk, you have done work. You may at first think that in chemistry we are only concerned with matter, but the behavior of matter is driven in large part by energy, so understanding energy is critical to understanding chemistry. Like matter, energy is conserved. The **law of conservation of energy** states that *energy is neither created nor destroyed.* The total amount of energy is constant; energy can be changed from one form to another or transferred from one object to another, but it cannot be created out of nothing, and it does not vanish into nothing.

Virtually all samples of matter have energy. The total energy of a sample of matter is the sum of its **kinetic energy**, the energy associated with its motion, and its **potential energy**, the energy associated with its position or composition. For example, a moving billiard ball contains *kinetic energy* because it is *moving* at some speed across the billiard table. Water behind a dam contains *potential energy* because it is held at a high *position* in the Earth's gravitational field by the dam. When the water flows through the dam from a higher position to a lower position, it can turn a turbine and produce electrical

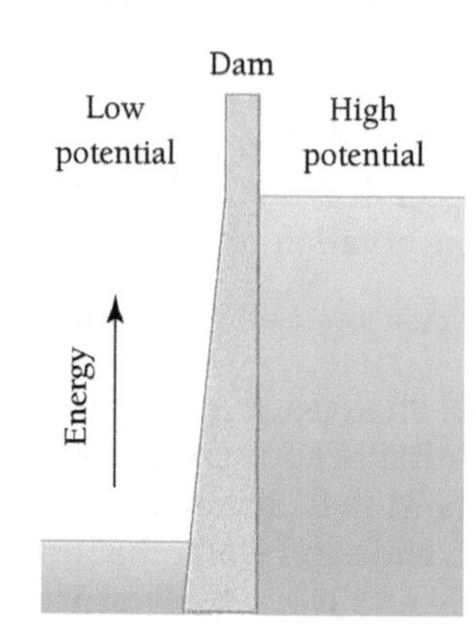

▲ Water behind a dam contains potential energy.

CHEMISTRY IN THE ENVIRONMENT

Getting Energy out of Nothing?

The law of conservation of energy has significant implications for energy use. The best we can do with energy is break even (and even that is not really possible); we can't continually draw energy from a device without putting energy into it. A device that supposedly produces energy without the need for energy input is sometimes called a *perpetual motion machine* (▶ FIGURE 3.15). According to the law of conservation of energy, such a machine cannot exist. Occasionally, the media report or speculate on the discovery of a system that appears to produce more energy than it consumes. For example, I once heard a radio talk show on the subject of energy and gasoline costs. The reporter suggested that we simply design an electric car that recharges itself while being driven. The battery in the electric car would charge during operation in the same way that the battery in a conventional car recharges, except the electric car would run with energy from the battery. Although people have dreamed of machines such as this for decades, such ideas violate the law of conservation of energy because they produce energy without any energy input. In the case of the perpetually moving electric car, the fault lies in the idea that driving the electric car can recharge the battery—it can't.

The battery in a conventional car recharges because energy from gasoline combustion is converted into electrical energy that then charges the battery. The electric car needs energy to move forward, and the battery eventually discharges as it provides that energy. Hybrid cars (electric and gasoline-powered) such as the Toyota Prius can capture some limited energy from braking and use that energy to recharge the battery. However, they could never run indefinitely without the addition of fuel. Our society has a continual need for energy, and as our current energy resources dwindle, new energy sources are required. Unfortunately, those sources must also follow the law of conservation of energy—energy must be conserved.

◀ FIGURE 3.15 **A proposed perpetual motion machine** The rolling balls supposedly keep the wheel perpetually spinning. Question: Can you explain why this would not work?

B3.1 CAN YOU ANSWER THIS? *A friend asks you to invest in a new flashlight he invented that never needs batteries. What questions should you ask before writing a check?*

energy. **Electrical energy** is the energy associated with the flow of electrical charge. **Thermal energy** is the energy associated with the random motions of atoms and molecules in matter. The hotter an object, the more thermal energy it contains.

Chemical systems contain **chemical energy**, a form of potential energy associated with the positions of the particles that compose the chemical system. For example, the molecules that compose gasoline contain a substantial amount of chemical energy. They are a bit like the water behind a dam. Burning the gasoline is analogous to releasing the water from the dam. The chemical energy present in the gasoline is released upon burning. When we drive a car, we use that chemical energy to move the car forward. When we heat a home, we use chemical energy stored in natural gas to produce heat and warm the air in the house.

Units of Energy

We commonly use several different energy units. The SI unit of energy is the joule (J), named after the English scientist James Joule (1818–1889), who demonstrated that energy could be converted from one type to another as long as the total energy was conserved. A second unit of energy is the **calorie (cal)**, the amount of energy required to raise the temperature of 1 g of water by 1 °C. A calorie is a larger unit than a joule: 1 cal = 4.184 J. A related energy unit is the nutritional or *capital C* **Calorie (Cal)**, equivalent to 1000 *little c* calories. Electricity bills usually come in yet another energy unit, the **kilowatt-hour (kWh)**. The average cost of residential electricity in the United States is about $0.12 per kilowatt-hour. Table 3.2 lists various energy units and their conversion factors. Table 3.3 shows the amount of energy required for various processes in each of these units.

TABLE 3.2 Energy Conversion Factors

1 calorie (cal)	= 4.184 joules (J)
1 Calorie (Cal)	= 1000 calories (cal)
1 kilowatt-hour (kWh)	= 3.60×10^6 joules (J)

TABLE 3.3 Energy Use in Various Units

Unit	Energy Required to Raise Temperature of 1 g of Water by 1 °C	Energy Required to Light 100-W Bulb for 1 Hour	Total Energy Used by Average U.S. Citizen in 1 Day
joule (J)	4.18	3.6×10^5	9.0×10^8
calorie (cal)	1.00	8.60×10^4	2.2×10^8
Calorie (Cal)	0.00100	86.0	2.2×10^5
kilowatt-hour (kWh)	1.16×10^{-6}	0.100	2.50×10^2

PEARSON eText 2.0 **Interactive Worked Example Video 3.5**

EXAMPLE 3.5 Conversion of Energy Units

A candy bar contains 225 Cal of nutritional energy. How many joules does it contain?

SORT

Begin by sorting the information in the problem. Here you are *given* energy in Calories and asked to *find* energy in joules.

GIVEN: 225 Cal

FIND: J

STRATEGIZE

Draw a solution map. Begin with Cal, convert to cal, and then convert to J.

SOLUTION MAP

Cal → cal → J

$$\frac{1000 \text{ cal}}{1 \text{ Cal}} \qquad \frac{4.184 \text{ J}}{1 \text{ cal}}$$

RELATIONSHIPS USED

1000 calories = 1 Cal (Table 3.2)
4.184 J = 1 cal (Table 3.2)

SOLVE

Follow the solution map to solve the problem. Begin with 225 Cal and multiply by the appropriate conversion factors to arrive at J. Round the answer to the correct number of significant figures (in this case, three because of the three significant figures in 225 Cal).

SOLUTION

$$225 \text{ Cal} \times \frac{1000 \text{ cal}}{1 \text{ Cal}} \times \frac{4.184 \text{ J}}{1 \text{ cal}} = 9.41 \times 10^5 \text{ J}$$

CHECK

Check your answer. Are the units correct? Does the answer make physical sense?

The units of the answer (J) are the desired units. The magnitude of the answer makes sense because the J is a smaller unit than the Cal; therefore, the quantity of energy in J should be greater than the quantity in Cal.

▶ **SKILLBUILDER 3.5 | Conversion of Energy Units**

The complete combustion of a small wooden match produces approximately 512 cal of heat. How many kilojoules are produced?

▶ **SKILLBUILDER PLUS** Convert 2.75×10^4 kJ to calories.

▶ **FOR MORE PRACTICE** Example 3.16; Problems 51, 52, 53, 54, 55, 56, 57, 58.

CONCEPTUAL CHECKPOINT 3.5

Suppose a salesperson wants to make an appliance seem as efficient as possible. In which units does the yearly energy consumption of the appliance have the lowest numerical value and therefore seem most efficient?

(a) J **(b)** cal **(c)** Cal **(d)** kWh

3.9 Energy and Chemical and Physical Change

▶ Distinguish between exothermic and endothermic reactions.

When discussing energy transfer, we often define the object of our study (such as a flask in which a chemical reaction is occurring) as the *system*. The system then exchanges energy with its *surroundings*. In other words, we view energy changes as an exchange of energy between the system and the surroundings.

The physical and chemical changes that we discussed in Section 3.6 are usually accompanied by energy changes. For example, when water evaporates from skin (a physical change), the water molecules absorb energy, cooling the skin. When we burn natural gas on a stove (a chemical change), energy is released, heating the food we are cooking.

The release of energy during a chemical reaction is analogous to the release of energy that occurs when a weight falls to the ground. When you lift a weight, you raise its potential energy; when you drop it, you release the potential energy (◀ FIGURE 3.16). *Systems with high potential energy—like the raised weight—have a tendency to change in a way that lowers their potential energy*. For this reason, objects or systems with high potential energy tend to be *unstable*. A weight lifted several meters from the ground is unstable because it contains a significant amount of localized potential energy. Unless restrained, the weight will fall, lowering its potential energy.

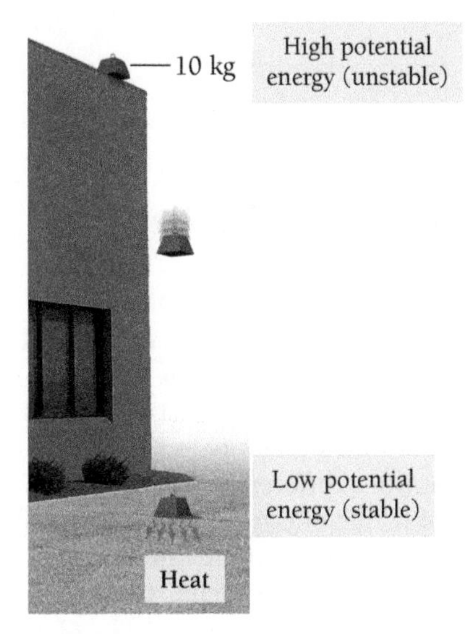

▲ FIGURE 3.16 **Potential energy of raised weight** A weight lifted off the ground has a high potential and will tend to fall toward the ground to lower its potential energy.

Some chemical substances are like a raised weight. For example, the molecules that compose TNT (trinitrotoluene) have a relatively high potential energy—energy is concentrated in them just as energy is concentrated in the raised weight. TNT molecules therefore tend to undergo rapid chemical changes that lower their potential energy, which is why TNT is explosive. Chemical reactions that *release* energy, like the explosion of TNT, are **exothermic**.

Some chemical reactions behave in just the opposite way—they *absorb* energy from their surroundings as they occur. Such reactions are **endothermic**. The reaction that occurs in a chemical cold pack is a good example of an endothermic reaction. When you break the barrier separating the reactants in the chemical cold pack, the substances mix, react, and absorb heat from the surroundings. The surroundings—possibly including your bruised ankle—get colder.

We can represent the energy changes that occur during a chemical reaction with an energy diagram, as shown in Figure 3.17. In an exothermic reaction (▼ FIGURE 3.17a), the reactants have greater energy than the products, and energy is released as the reaction occurs. In an endothermic reaction (▼ FIGURE 3.17b), the products have more energy than the reactants, and energy is absorbed as the reaction occurs.

If a particular reaction or process is exothermic, then the reverse process must be endothermic. For example, the evaporation of water from skin is endothermic (and therefore cools you off), but the condensation of water onto skin is exothermic (which is why steam burns can be so painful).

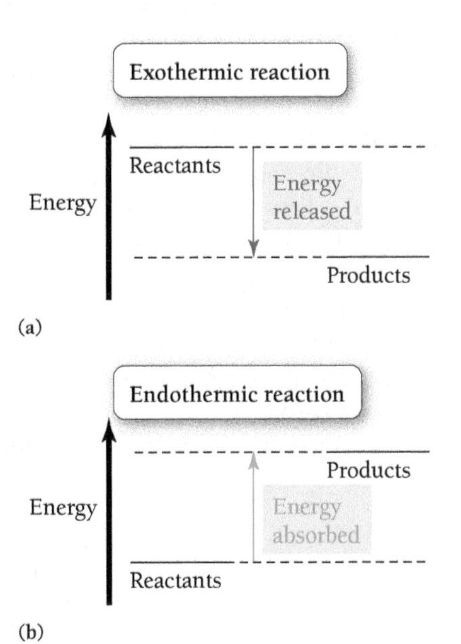

◀ FIGURE 3.17 **Exothermic and endothermic reactions** (a) In an exothermic reaction, energy is released. (b) In an endothermic reaction, energy is absorbed.

EXAMPLE **3.6** **Exothermic and Endothermic Processes**

Classify each change as exothermic or endothermic.

(a) wood burning in a fire
(b) ice melting

SOLUTION

(a) When wood burns, it emits heat into the surroundings. Therefore, the process is exothermic.
(b) When ice melts, it absorbs heat from the surroundings. For example, when ice melts in a glass of water, it cools the water as the melting ice absorbs heat from the water. Therefore, the process is endothermic.

▶ **SKILLBUILDER 3.6 | Exothermic and Endothermic Processes**

Classify each change as exothermic or endothermic.

(a) water freezing into ice
(b) natural gas burning

▶ **FOR MORE PRACTICE** Problems 61, 62, 63, 64.

3.10 Temperature: Random Motion of Molecules and Atoms

▶ Convert between Fahrenheit, Celsius, and Kelvin temperature scales.

The atoms and molecules that compose matter are in constant random motion—they contain *thermal energy*. The **temperature** of a substance is a measure of its thermal energy. The hotter an object, the greater the random motion of the atoms and molecules that compose it, and the higher its temperature. We must be careful to not confuse *temperature* with *heat*. **Heat**, which has units of energy, is the *transfer* or *exchange* of thermal energy caused by a temperature difference. For example, when a cold ice cube is dropped into a warm cup of water, energy is transferred as heat from the water to the ice, resulting in the cooling of the water. Temperature, by contrast, is a *measure* of the thermal energy of matter (not the exchange of thermal energy).

The most familiar temperature scale in the United States is the **Fahrenheit (°F) scale**. On the Fahrenheit scale, water freezes at 32 °F and boils at 212 °F. Room temperature is approximately 72 °F. The Fahrenheit scale was initially set up by assigning 0 °F to the freezing point of a concentrated saltwater solution and 96 °F to normal body temperature (although body temperature is now known to be 98.6 °F).

The scale scientists use is the **Celsius (°C) scale**. On this scale, water freezes at 0 °C and boils at 100 °C. Room temperature is approximately 22 °C.

The Fahrenheit and Celsius scales differ in both the size of their respective degrees and the temperature each calls "zero" (▶ FIGURE 3.18). Both the Fahrenheit and Celsius scales contain negative temperatures. A third temperature scale, called the **Kelvin (K) scale**, avoids negative temperatures by assigning 0 K to the coldest temperature possible, absolute zero. Absolute zero (−273.15 °C or −459.7 °F) is the temperature at which molecular motion virtually stops. There is no lower temperature. The kelvin degree, or kelvin (K), is the same size as the Celsius degree; the only difference is the temperature that each scale designates as zero.

We can convert between these temperature scales using the following formulas.

The degree symbol is used with the Celsius and Fahrenheit scales but not with the Kelvin scale.

$$K = °C + 273.15$$

$$°C = \frac{°F - 32}{1.8}$$

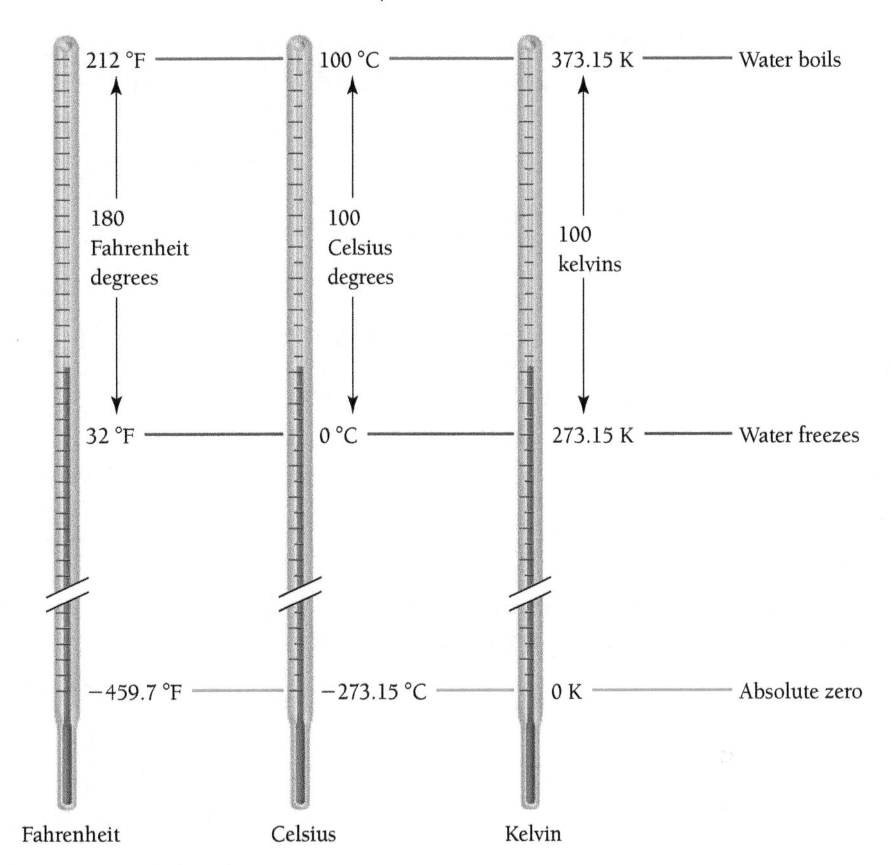

▶ FIGURE 3.18 **Comparison of the Fahrenheit, Celsius, and Kelvin temperature scales** The Fahrenheit degree is five-ninths the size of a Celsius degree. The Celsius degree and the kelvin degree are the same size.

For example, suppose we want to convert 212 K to Celsius. Following the procedure for solving numerical problems (Section 2.6), we first sort the information in the problem statement:

GIVEN: 212 K

FIND: °C

We then strategize by building a solution map.

SOLUTION MAP

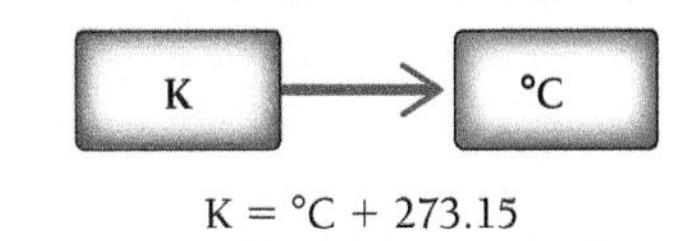

$$K = °C + 273.15$$

In a solution map involving a formula, the formula establishes the relationship between the variables. However, the formula under the arrow is not necessarily solved for the correct variable until later, as is the case here.

RELATIONSHIPS USED

K = °C + 273.15 (This equation relates the *given* quantity (K) to the *find* quantity (°C) and is given in this section.)

SOLUTION

Finally, we follow the solution map to solve the problem. The equation below the arrow shows the relationship between K and °C, but it is not solved for the correct variable. Before using the equation, we must solve it for °C.

$$K = °C + 273.15$$
$$°C = K - 273.15$$

We can now substitute the given value for K and calculate the answer to the correct number of significant figures.

$$°C = 212 - 273.15$$
$$= -61\ °C$$

EXAMPLE 3.7 Converting between Celsius and Kelvin Temperature Scales

Convert −25 °C to kelvins.

SORT You are given a temperature in degrees Celsius and asked to find the value of the temperature in kelvins.	**GIVEN:** −25 °C **FIND:** K
STRATEGIZE Draw a solution map. Use the equation that relates the temperature in kelvins to the temperature in Celsius to convert from the given quantity to the quantity you want to find.	**SOLUTION MAP** °C → K $K = °C + 273.15$ **RELATIONSHIPS USED** $K = °C + 273.15$ (presented in this section)
SOLVE Follow the solution map by substituting the correct value for °C and calculating the answer to the correct number of significant figures.	**SOLUTION** $K = °C + 273.15$ $K = -25\ °C + 273.15 = 248\ K$ The significant figures are limited to the ones place because the given temperature (−25 °C) is only given to the ones place. Remember that for addition and subtraction, the number of *decimal places* in the least precisely known quantity determines the number of *decimal places* in the result.
CHECK Check your answer. Are the units correct? Does the answer make physical sense?	The units (K) are correct. The answer makes sense because the value in kelvins should be a more positive number than the value in degrees Celsius.

▶ **SKILLBUILDER 3.7 | Converting between Celsius and Kelvin Temperature Scales**

Convert 358 K to Celsius.

▶ **FOR MORE PRACTICE** Example 3.17; Problems 65c, 66d.

EXAMPLE 3.8 Converting between Fahrenheit and Celsius Temperature Scales

Convert 55 °F to Celsius.

SORT You are given a temperature in degrees Fahrenheit and asked to find the value of the temperature in degrees Celsius.	**GIVEN:** 55 °F **FIND:** °C
STRATEGIZE Draw the solution map. Use the equation that shows the relationship between the *given* quantity (°F) and the *find* quantity (°C).	**SOLUTION MAP** °F → °C $°C = \frac{(°F - 32)}{1.8}$ **RELATIONSHIPS USED** $°C = \frac{(°F - 32)}{1.8}$ (presented in this section)
SOLVE Substitute the given value into the equation and calculate the answer to the correct number of significant figures.	**SOLUTION** $°C = \frac{(°F - 32)}{1.8}$ $°C = \frac{(55 - 32)}{1.8} = 12.778\ °C = 13\ °C$

CHECK Check your answer. Are the units correct? Does the answer make physical sense?	The units (°C) are correct. The value of the answer (13 °C) is smaller than the value in degrees Fahrenheit. For positive temperatures, the value of a temperature in degrees Celsius will always be smaller than the value in degrees Fahrenheit because the Fahrenheit degree is smaller than the Celsius degree and the Fahrenheit scale is offset by 32 degrees (see Figure 3.18).

▶ **SKILLBUILDER 3.8 | Converting between Fahrenheit and Celsius Temperature Scales**

Convert 139 °C to Fahrenheit.

▶ **FOR MORE PRACTICE** Example 3.18; Problems 65a, 66a, c.

EXAMPLE 3.9 Converting between Fahrenheit and Kelvin Temperature Scales

Convert 310 K to Fahrenheit.

SORT You are given a temperature in kelvins and asked to find the value of the temperature in degrees Fahrenheit.	**GIVEN:** 310 K **FIND:** °F
STRATEGIZE Build the solution map, which requires two steps: one to convert kelvins to degrees Celsius and one to convert degrees Celsius to degrees Fahrenheit.	**SOLUTION MAP** 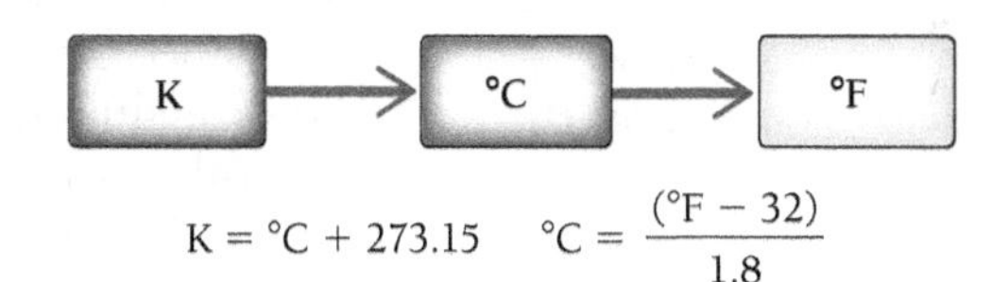 $K = °C + 273.15 \quad °C = \frac{(°F - 32)}{1.8}$ **RELATIONSHIPS USED** $K = °C + 273.15$ (presented in this section) $°C = \frac{(°F - 32)}{1.8}$ (presented in this section)
SOLVE Solve the first equation for °C and substitute the *given* quantity in K to convert it to °C. Solve the second equation for °F. Substitute the value of the temperature in °C (from the previous step) to convert it to °F and round the answer to the correct number of significant figures.	**SOLUTION** $K = °C + 273.15$ $°C = K - 273.15$ $°C = 310 - 273.15 = 3\underline{6}.85\ °C$ $°C = \frac{(°F - 32)}{1.8}$ $1.8\,(°C) = (°F - 32)$ $°F = 1.8\,(°C) + 32$ $°F = 1.8\,(3\underline{6}.85) + 32 = 98.33\ °F = 98\ °F$
CHECK Check your answer. Are the units correct? Does the answer make physical sense?	The units (°F) are correct. The magnitude of the answer is a bit trickier to judge. In this temperature range, a temperature in Fahrenheit should indeed be smaller than a temperature in kelvins. However, because the Fahrenheit degree is smaller, temperatures in Fahrenheit become larger than temperatures in kelvins above 575 °F.

▶ **SKILLBUILDER 3.9 | Converting between Fahrenheit and Kelvin Temperature Scales**

Convert −321 °F to kelvins.

▶ **FOR MORE PRACTICE** Problems 65b, d, 66b.

CONCEPTUAL CHECKPOINT 3.6

Which temperature is identical on both the Celsius and the Fahrenheit scales?

(a) 100°

(b) 32°

(c) 0°

(d) −40°

3.11 Temperature Changes: Heat Capacity

▶ Relate energy, temperature change, and heat capacity.

All substances change temperature when they are heated, but how much they change for a given amount of heat varies significantly from one substance to another. For example, if you put a steel skillet on a flame, its temperature rises rapidly. However, if you put some water in the skillet, the temperature increases more slowly. Why? One reason is that when you add water, the same amount of heat energy must warm more matter, so the temperature rise is slower. The second, and more interesting reason, is that water is more resistant to temperature change than steel because water has a higher *specific heat capacity*. The **specific heat capacity** (or **specific heat**) of a substance is the quantity of heat (usually in joules) required to change the temperature of 1 g of the substance by 1 °C. Specific heat capacity has units of joules per gram per degree Celsius (J/g °C). Table 3.4 lists the values of the specific heat capacity for several substances.

TABLE 3.4 Specific Heat Capacities of Some Common Substances

Substance	Specific Heat Capacity (J/g °C)
Lead	0.128
Gold	0.128
Silver	0.235
Copper	0.385
Iron	0.449
Aluminum	0.903
Ethanol	2.42
Water	4.184

Notice that water has the highest specific heat capacity on the list—changing the temperature of water requires a lot of heat. If you have traveled from an inland geographical region to a coastal one and have felt a drop in temperature, you have experienced the effects of water's high specific heat capacity. On a summer day in California, for example, the temperature difference between Sacramento (an inland city) and San Francisco (a coastal city) can be 30 °F (17 °C); San Francisco enjoys a cool 68 °F (20 °C), while Sacramento bakes at near 100 °F (37 °C), yet the intensity of sunlight falling on these two cities is the same. Why the large temperature difference? The difference is due to the presence of the Pacific Ocean, which practically surrounds San Francisco. On the one hand, water, with its high heat capacity, absorbs much of the sun's heat without undergoing a large increase in temperature, keeping San Francisco cool. The land surrounding Sacramento, on the other hand, with its lower heat capacity, cannot absorb a lot of heat without a large increase in temperature—it has a lower *capacity* to absorb heat without a large temperature increase.

▲ San Francisco enjoys cool weather even in summer months because of the high heat capacity of the surrounding ocean.

Similarly, only two U.S. states have never recorded a temperature above 100 °F (37 °C). One of them is obvious: Alaska. It is too far north to get that hot. The other one, however, may come as a surprise. It is Hawaii. The water that surrounds America's only island state moderates the temperature, preventing Hawaii from ever getting too hot.

EVERYDAY CHEMISTRY

Coolers, Camping, and the Heat Capacity of Water

Have you ever loaded a cooler with ice and then added room-temperature drinks? If you have, you know that the ice quickly melts. In contrast, if you load your cooler with chilled drinks, the ice lasts for hours. Why the difference? The answer is related to the high heat capacity of the water within the drinks. As you just learned, water must absorb a lot of heat to raise its temperature, and it must also release a lot of heat to lower its temperature. When the warm drinks are placed into the ice, they release heat, which then melts the ice. The chilled drinks, however, are already cold, so they do not release much heat. It is always better to load your cooler with chilled drinks—that way, the ice will last the rest of the day.

B3.2 CAN YOU ANSWER THIS? *Suppose you are cold-weather camping and decide to heat some objects to bring into your sleeping bag for added warmth. You place a large water jug and a rock of equal mass close to the fire. Over time, both the rock and the water jug warm to about 38 °C (100 °F). If you could bring only one into your sleeping bag, which one would keep you warmer? Why?*

▲ The ice in a cooler loaded with cold drinks lasts much longer than the ice in a cooler loaded with warm drinks.
QUESTION: Can you explain why?

CONCEPTUAL CHECKPOINT 3.7

If you want to heat a metal plate to as high a temperature as possible for a given energy input, what metal should you use? (Assume all the plates have the same mass.)

(a) copper

(b) iron

(c) aluminum

(d) it would make no difference

3.12 Energy and Heat Capacity Calculations

▶ Perform calculations involving transfer of heat and changes in temperature.

When a substance absorbs heat (which we represent with the symbol q), its temperature change (which we represent as ΔT) is in direct proportion to the amount of heat absorbed.

$$\xrightarrow[q]{} \text{System} \quad \Delta T$$

In other words, the more heat absorbed, the greater the temperature change. We can use the specific heat capacity of the substance to *quantify* the relationship between the amount of heat added to a given amount of the substance and the corresponding temperature increase. The equation that relates these quantities is:

$$\text{heat} = \text{mass} \times \text{specific heat capacity} \times \text{temperature change}$$
$$q = m \times C \times \Delta T$$

ΔT in °C is equal to ΔT in K but is not equal to ΔT in °F.

where q is the amount of heat in joules, m is the mass of the substance in grams, C is the specific heat capacity in joules per gram per degree Celsius, and ΔT is the temperature change in Celsius. The symbol Δ means *the change in*, so ΔT means *the change in temperature*. For example, suppose we are making a cup of tea and want to know how much heat energy will warm 235 g of water (about 8 oz) from 25 °C to 100.0 °C (boiling).

We begin by sorting the information in the problem statement.

GIVEN: 235 g water (m)
25 °C initial temperature (T_i)
100.0 °C final temperature (T_f)

FIND: amount of heat needed (q)

Then we strategize by building a solution map.

SOLUTION MAP

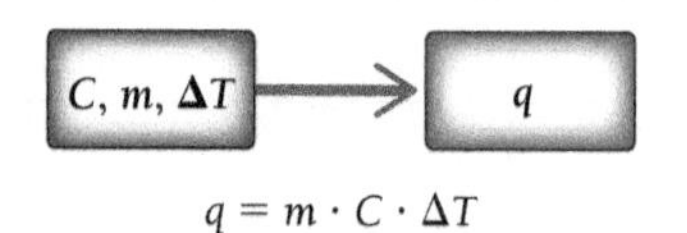

$$q = m \cdot C \cdot \Delta T$$

In addition to m and ΔT, the equation requires C, the specific heat capacity of water. The next step is to gather all of the required quantities for the equation (C, m, and ΔT) in the correct units. These are:

$$C = 4.18\ \text{J/g °C}$$
$$m = 235\ \text{g}$$

The other quantity we require is ΔT. The change in temperature is the difference between the final temperature (T_f) and the initial temperature (T_i).

$$\Delta T = T_f - T_i$$
$$= 100.0\ °\text{C} - 25\ °\text{C} = 75\ °\text{C}$$

SOLUTION

Finally, we solve the problem. We substitute the correct values into the equation and calculate the answer to the correct number of significant figures.

$$q = m \cdot C \cdot \Delta T$$
$$= 235\ \cancel{\text{g}} \times 4.18\ \frac{\text{J}}{\cancel{\text{g}}\ \cancel{°\text{C}}} \times 75\ \cancel{°\text{C}}$$
$$= 7.367 \times 10^4\ \text{J} = 7.4 \times 10^4\ \text{J}$$

It is critical that we substitute each of the correct variables into the equation in the correct units and cancel units as we calculate the answer. If, during this process, we learn that one of our variables is not in the correct units, we convert it to the correct units using the skills we learned in Chapter 2. Notice that the sign of q is positive (+) if the substance is increasing in temperature (heat entering the substance) and negative (−) if the substance is decreasing in temperature (heat leaving the substance).

Interactive Worked Example Video 3.10

EXAMPLE 3.10 Relating Heat Energy to Temperature Changes

Gallium is a solid metal at room temperature but melts at 29.9 °C. If you hold gallium in your hand, it melts from your body heat. How much heat must 2.5 g of gallium absorb from your hand to raise the temperature of the gallium from 25.0 °C to 29.9 °C? The specific heat capacity of gallium is 0.372 J/g °C.

SORT

You are given the mass of gallium, its initial and final temperatures, and its specific heat capacity, and are asked to find the amount of heat absorbed by the gallium.

GIVEN: 2.5 g gallium (m)

$T_{\text{i}} = 25.0\ ^\circ\text{C}$

$T_{\text{f}} = 29.9\ ^\circ\text{C}$

$C = 0.372\ \text{J/g}\ ^\circ\text{C}$

FIND: q

STRATEGIZE

The equation that relates the *given* and *find* quantities is the specific heat capacity equation. The solution map indicates that this equation takes you from the *given* quantities to the quantity you are asked to *find*.

SOLUTION MAP

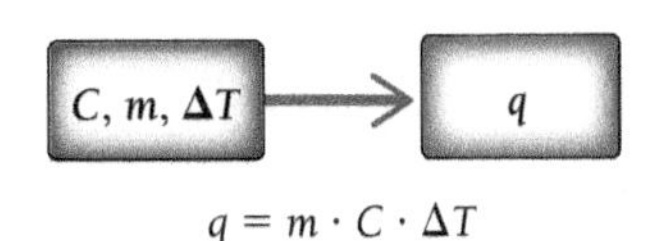

$q = m \cdot C \cdot \Delta T$

RELATIONSHIPS USED

$q = m \cdot C \cdot \Delta T$ (presented in this section)

SOLVE

Before solving the problem, you must gather the necessary quantities—C, m, and ΔT—in the correct units.

SOLUTION

$$C = 0.372\ \text{J/g}\ ^\circ\text{C}$$

$$m = 2.5\ \text{g}$$

$$\Delta T = T_{\text{f}} - T_{\text{i}}$$

$$= 29.9\ ^\circ\text{C} - 25.0\ ^\circ\text{C}$$

$$= 4.9\ ^\circ\text{C}$$

Substitute C, m, and ΔT into the equation, canceling units, and calculate the answer to the correct number of significant figures.

$$q = m \cdot C \cdot \Delta T$$

$$= 2.5\ \cancel{\text{g}} \times 0.372 \frac{\text{J}}{\cancel{\text{g}}\ \cancel{^\circ\text{C}}} \times 4.9\ \cancel{^\circ\text{C}} = 4.557\ \text{J} = 4.6\ \text{J}^*$$

CHECK

Check your answer. Are the units correct? Does the answer make physical sense?

The units (J) are correct. The magnitude of the answer makes sense because it takes almost 1 J to heat the 2.5 g sample of the metal by 1 °C; therefore, it should take about 5 J to heat the sample by 5 °C.

▶ **SKILLBUILDER 3.10 | Relating Heat Energy to Temperature Changes**

You find a 1979 copper penny (pre-1982 pennies are nearly pure copper) in the snow and pick it up. How much heat does the penny absorb as it warms from the temperature of the snow, −5.0 °C, to the temperature of your body, 37.0 °C? Assume the penny is pure copper and has a mass of 3.10 g. You can find the heat capacity of copper in Table 3.4.

▶ **SKILLBUILDER PLUS** The temperature of a lead fishing weight rises from 26 °C to 38 °C as it absorbs 11.3 J of heat. What is the mass of the fishing weight in grams?

▶ **FOR MORE PRACTICE** Example 3.19; Problems 75, 76, 77, 78.

* This is the amount of heat required to raise the temperature to the melting point. Actually, melting the gallium requires additional heat.

EXAMPLE 3.11 Relating Heat Capacity to Temperature Changes

A chemistry student finds a shiny rock that she suspects is gold. She weighs the rock on a balance and determines that its mass is 14.3 g. She then finds that the temperature of the rock rises from 25 °C to 52 °C upon absorption of 174 J of heat. Find the heat capacity of the rock and determine whether the value is consistent with the heat capacity of gold (which is listed in Table 3.4).

SORT

You are given the mass of the "gold" rock, the amount of heat absorbed, and the initial and final temperatures. You are asked to find the heat capacity of the rock.

GIVEN: 14.3 g
174 J of heat absorbed
$T_i = 25\ °C$
$T_f = 52\ °C$

FIND: C

STRATEGIZE

The solution map shows how the heat capacity equation relates the *given* and *find* quantities.

SOLUTION MAP

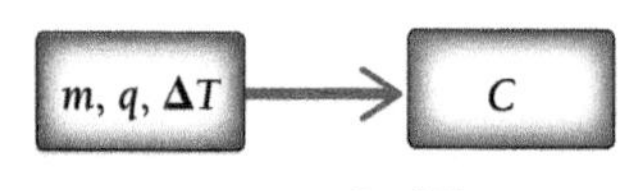

$q = m \cdot C \cdot \Delta T$

RELATIONSHIPS USED

$q = m \cdot C \cdot \Delta T$ (presented in this section)

SOLVE

First, gather the necessary quantities—m, q, and ΔT—in the correct units.

SOLUTION

$$m = 14.3\ g$$
$$q = 174\ J$$
$$\Delta T = 52\ °C - 25\ °C = 27\ °C$$

Then solve the equation for C and substitute the correct variables into the equation. Finally, calculate the answer to the right number of significant figures.

$$q = m \cdot C \cdot \Delta T$$
$$C = \frac{q}{m \cdot \Delta T}$$
$$C = \frac{174\ J}{14.3\ g \times 27\ °C}$$
$$= 0.4507 \frac{J}{g\ °C} = 0.45 \frac{J}{g\ °C}$$

By comparing the calculated value of the specific heat capacity (0.45 J/g °C) with the specific heat capacity of gold from Table 3.4 (0.128 J/g °C), you conclude that the rock is not pure gold.

CHECK

Check your answer. Are the units correct? Does the answer make physical sense?

The units of the answer are those of specific heat capacity, so they are correct. The magnitude of the answer falls in the range of specific heat capacities given in Table 3.4. A value of heat capacity that falls far outside this range would immediately be suspect.

▶ SKILLBUILDER 3.11 | Relating Heat Capacity to Temperature Changes

A 328-g sample of water absorbs 5.78×10^3 J of heat. Calculate the change in temperature for the water. If the water is initially at 25.0 °C, what is its final temperature?

▶ FOR MORE PRACTICE Problems 85, 86, 87, 88.

CONCEPTUAL CHECKPOINT 3.8

The heat capacity of substance A is twice that of substance B. If samples of equal mass of the two substances absorb the same amount of heat, which substance undergoes the larger change in temperature?

Chapter 3 in Review

MasteringChemistry™ provides end-of-chapter exercises, feedback-enriched tutorial problems, animations, and interactive activities to encourage problem solving practice and deeper understanding of key concepts and topics.

Self-Assessment Quiz

PEARSON eText 2.0

Q1. Which substance is a pure compound?
(a) Gold (b) Water
(c) Milk (d) Fruit cake

Q2. Which property of trinitrotoluene (TNT) is most likely a chemical property?
(a) Yellow color
(b) Melting point is 80.1 °C
(c) Explosive
(d) None of the above

Q3. Which change is a chemical change?
(a) The condensation of dew on a cold night
(b) A forest fire
(c) The smoothening of rocks by ocean waves
(d) None of the above

Q4. Which process is endothermic?
(a) The burning of natural gas in a stove
(b) The metabolism of glucose by your body
(c) The melting of ice in a soft drink
(d) None of the above

Q5. A 35-g sample of potassium completely reacts with chlorine to form 67 g of potassium chloride. How many grams of chlorine must have reacted?
(a) 67 g (b) 35 g (c) 32 g (d) 12 g

Q6. A runner burns 2.56×10^3 kJ during a five-mile run. How many nutritional Calories did the runner burn?
(a) 1.07×10^1 Cal (b) 612 Cal
(c) 6.12×10^5 Cal (d) 1.07×10^4 Cal

Q7. Convert the boiling point of water (100.00 °C) to K.
(a) −173.15 K
(b) 0 K
(c) 100.00 K
(d) 373.15 K

Q8. A European doctor reports that you have a fever of 39.2 °C. What is your fever in degrees Fahrenheit?
(a) 102.6 °F (b) 128.26 °F
(c) 71.2 °F (d) 4 °F

Q9. How much heat must be absorbed by 125 g of ethanol to change its temperature from 21.5 °C to 34.8 °C?
(a) 6.95 kJ
(b) 4.02×10^3 kJ
(c) 86.6 kJ
(d) 4.02 kJ

Q10. Substance A has a heat capacity that is much greater than that of substance B. If 10.0 g of substance A initially at 25.0 °C is brought into thermal contact with 10.0 g of B initially at 75.0 °C, what can you conclude about the final temperature of the two substances once the exchange of heat between the substances is complete?
(a) The final temperature will be between 25.0 °C and 50.0 °C.
(b) The final temperature will be between 50.0 °C and 75.0 °C.
(c) The final temperature will be 50.0 °C.
(d) You can conclude nothing about the final temperature without more information.

Answers: 1:b, 2:c, 3:b, 4:c, 5:c, 6:b, 7:d, 8:a, 9:d, 10:a

Chemical Principles

Matter

Matter is anything that occupies space and has mass. It is composed of atoms, which are often bonded together as molecules. Matter can exist as a solid, a liquid, or a gas. Solid matter can be either amorphous or crystalline.

Classification of Matter

We can classify matter according to its composition. Pure matter is composed of only one type of substance; that substance may be an element (a substance that cannot be decomposed into simpler substances), or it may be a compound (a substance composed of two or more elements in fixed definite proportions). Mixtures are composed of two or more different substances, the proportions of which may vary from one sample to the next. Mixtures can be either homogeneous, having the same composition throughout, or heterogeneous, having a composition that varies from region to region.

Relevance

Everything is made of matter—you, me, the chair you sit on, and the air we breathe. The physical universe basically contains only two things: matter and energy. We begin our study of chemistry by defining and classifying these two building blocks of the universe.

Since ancient times, humans have tried to understand matter and harness it for their purposes. The earliest humans shaped matter into tools and used the transformation of matter—especially fire—to keep warm and to cook food. To manipulate matter, we must understand it. Fundamental to this understanding is the connection between the properties of matter and the molecules and atoms that compose it.

Properties and Changes of Matter

We can divide the properties of matter into two types: physical and chemical. The physical properties of matter do not involve a change in composition. The chemical properties of matter involve a change in composition. We can divide changes in matter into physical and chemical. In a physical change, the appearance of matter may change, but its composition does not. In a chemical change, the composition of matter changes.

The physical and chemical properties of matter make the world around us the way it is. For example, a physical property of water is its boiling point at sea level—100 °C. The physical properties of water—and all matter—are determined by the atoms and molecules that compose it. If water molecules were different—even slightly different—water would boil at a different temperature. Imagine a world where water boiled at room temperature.

Conservation of Mass

Whether the changes in matter are chemical or physical, matter is always conserved. In a chemical change, the masses of the matter undergoing the chemical change must equal the sum of the masses of matter resulting from the chemical change.

The conservation of matter is relevant to, for example, pollution. We often think that humans create pollution, but, actually, we are powerless to create anything. Matter cannot be created. So, pollution is simply misplaced matter—matter that we have put into places where it does not belong.

Energy

Besides matter, energy is the other major component of our universe. Like matter, energy is conserved—it can be neither created nor destroyed. Energy exists in various different types, and these can be converted from one to another. Some common units of energy are the joule (J), the calorie (cal), the nutritional Calorie (Cal), and the kilowatt-hour (kWh). Chemical reactions that emit energy are exothermic; those that absorb energy are endothermic.

Our society's energy sources will not last forever because as we burn fossil fuels—our primary energy source—we convert chemical energy, stored in molecules, to kinetic and thermal energy. The kinetic and thermal energy is not readily available to be used again. Consequently, our energy resources are dwindling, and the conservation of energy implies that we will not be able simply to create new energy—it must come from somewhere. All of the chemical reactions that we use for energy are exothermic.

Temperature

The temperature of matter is related to the random motions of the molecules and atoms that compose it—the greater the motion, the higher the temperature. To measure temperature we use three scales: Fahrenheit (°F), Celsius (°C), and Kelvin (K).

The temperature of matter and its measurement are relevant to many everyday phenomena. Humans are understandably interested in the weather, and air temperature is a fundamental part of weather. We use body temperature as one measure of human health and global temperature as one measure of the planet's health.

Heat Capacity

The temperature change that a sample of matter undergoes upon absorption of a given amount of heat relates to the heat capacity of the substance composing the matter. Water has one of the highest heat capacities, meaning that it is most resistant to rapid temperature changes.

The heat capacity of water explains why it is cooler in coastal areas, which are near large bodies of high-heat-capacity water, than in inland areas, which are surrounded by low-heat-capacity land. It also explains why it takes longer to cool a refrigerator filled with liquids than an empty one.

Chemical Skills

LO: Classify matter as element, compound, or mixture (Section 3.4).

Begin by examining the alphabetical listing of elements in the back of this book. If the substance is listed in that table, it is a pure substance and an element.

If the substance is not listed in that table, refer to your everyday experience with the substance to determine whether it is a pure substance. If it is a pure substance not listed in the table, then it is a compound.

If it is not a pure substance, then it is a mixture. Refer to your everyday experience with the mixture to determine whether it has uniform composition (homogeneous) or nonuniform composition (heterogeneous).

Examples

EXAMPLE **3.12** **Classifying Matter**

Classify each type of matter as a pure substance or a mixture. If it is a pure substance, classify it as an element or a compound. If it is a mixture, classify it as homogeneous or heterogeneous.

(a) pure silver
(b) swimming-pool water
(c) dry ice (solid carbon dioxide)
(d) blueberry muffin

SOLUTION

(a) Pure element; silver appears in the element table.
(b) Homogeneous mixture; pool water contains at least water and chlorine, and it is uniform throughout.
(c) Compound; dry ice is a pure substance (carbon dioxide), but it is not listed in the table.
(d) Heterogeneous mixture; a blueberry muffin is a mixture of several things and has nonuniform composition.

LO: Distinguish between physical and chemical properties (Section 3.5).

To distinguish between physical and chemical properties, consider whether the substance changes composition while displaying the property. If it *does not* change composition, the property is physical; if it *does*, the property is chemical.

EXAMPLE **3.13** **Distinguishing between Physical and Chemical Properties**

Classify each property as physical or chemical.

(a) the tendency for platinum jewelry to scratch easily
(b) the ability of sulfuric acid to burn the skin
(c) the ability of hydrogen peroxide to bleach hair
(d) the density of lead relative to other metals

SOLUTION

(b) Physical; scratched platinum is still platinum.
(c) Chemical; the acid chemically reacts with the skin to produce the burn.
(d) Chemical; the hydrogen peroxide chemically reacts with hair to change the hair.
(e) Physical; you can determine the density of lead by measuring the volume and mass of a lead sample.

LO: Distinguish between physical and chemical changes (Section 3.6).

To distinguish between physical and chemical changes, consider whether the substance changes composition during the change. If it *does not* change composition, the change is physical; if it *does*, the change is chemical.

EXAMPLE **3.14** **Distinguishing between Physical and Chemical Changes**

Classify each change as physical or chemical.

(a) the explosion of gunpowder in the barrel of a gun
(b) the melting of gold in a furnace
(c) the bubbling that occurs when you mix baking soda and vinegar
(d) the bubbling that occurs when water boils

SOLUTION

(a) Chemical; the gunpowder reacts with oxygen during the explosion.
(b) Physical; the liquid gold is still gold.
(c) Chemical; the bubbling is a result of a chemical reaction between the two substances to form new substances, one of which is carbon dioxide released as bubbles.
(d) Physical; the bubbling is due to liquid water turning into gaseous water, but it is still water.

LO: Apply the law of conservation of mass (Section 3.7).

The sum of the masses of the substances involved in a chemical change must be the same before and after the change.

EXAMPLE **3.15** **Applying the Law of Conservation of Mass**

An automobile runs for 10 minutes and burns 47 g of gasoline. The gasoline combines with oxygen from air and forms 132 g of carbon dioxide and 34 g of water. How much oxygen is consumed in the process?

SOLUTION

The total mass after the chemical change is:

$$132\text{ g} + 34\text{ g} = 166\text{ g}$$

The total mass before the change must also be 166 g.

$$47\text{ g} + \text{oxygen} = 166\text{ g}$$

So, the mass of oxygen consumed is the total mass (166 g) minus the mass of gasoline (47 g).

$$\text{grams of oxygen} = 166\text{ g} - 47\text{ g} = 119\text{ g}$$

LO: Identify and convert among energy units (Section 3.8).

Solve unit conversion problems using the problem-solving strategies outlined in Section 2.6.

EXAMPLE 3.16 Converting Energy Units

Convert 1.7×10^3 kWh (the amount of energy the average U.S. citizen uses in one week) into calories.

SORT
You are given an amount of energy in kilowatt-hours and asked to find the amount in calories.

GIVEN: 1.7×10^3 kWh

FIND: cal

STRATEGIZE
Draw a solution map. Begin with kilowatt-hours and determine the conversion factors to get to calories.

SOLUTION MAP

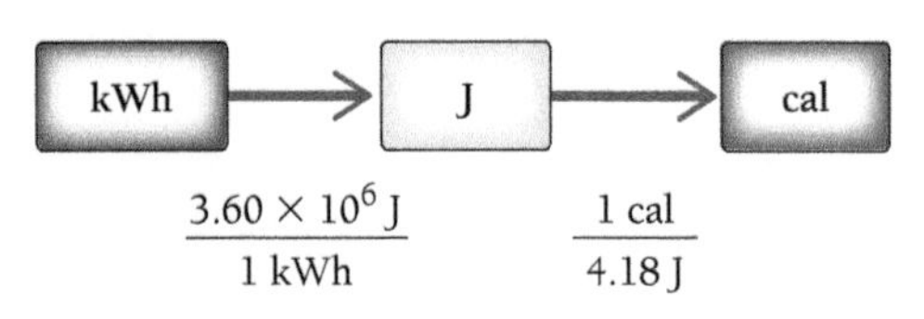

$$\frac{3.60 \times 10^6 \text{ J}}{1 \text{ kWh}} \qquad \frac{1 \text{ cal}}{4.18 \text{ J}}$$

RELATIONSHIPS USED

$1 \text{ kWh} = 3.60 \times 10^6 \text{ J}$ (Table 3.2)

$1 \text{ cal} = 4.18 \text{ J}$ (Table 3.2)

SOLVE
Follow the solution map to solve the problem. Begin with the *given* quantity and multiply by the conversion factors to arrive at calories. Round the answer to the correct number of significant figures.

SOLUTION

$$1.7 \times 10^3 \text{ kWh} \times \frac{3.60 \times 10^6 \text{ J}}{1 \text{ kWh}} \times \frac{1 \text{ cal}}{4.18 \text{ J}} = 1.464 \times 10^9 \text{ cal}$$

$$1.464 \times 10^9 \text{ cal} = 1.5 \times 10^9 \text{ cal}$$

CHECK
Are the units correct? Does the answer make physical sense?

The unit of the answer, cal, is correct. The magnitude of the answer makes sense because cal is a smaller unit than kWh; therefore, the value in cal should be larger than the value in kWh.

LO: Convert between Fahrenheit, Celsius, and Kelvin temperature scales (Section 3.10).

Solve temperature conversion problems using the problem-solving procedure in Sections 2.6 and 2.10. Take the steps appropriate for equations.

EXAMPLE 3.17 Converting between Celsius and Kelvin Temperature Scales

Convert 257 K to Celsius.

SORT
You are given the temperature in kelvins and asked to convert it to degrees Celsius.

GIVEN: 257 K

FIND: °C

STRATEGIZE
Draw a solution map. Use the equation that relates the *given* quantity to the *find* quantity.

SOLUTION MAP

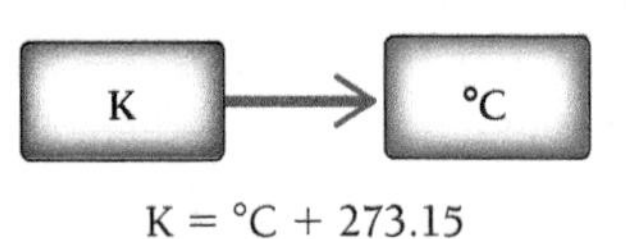

$$K = °C + 273.15$$

RELATIONSHIPS USED

$K = °C + 273.15$ (Section 3.10)

SOLVE
Solve the equation for the *find* quantity (°C) and substitute the temperature in K into the equation. Calculate the answer to the correct number of significant figures.

SOLUTION

$$K = °C + 273.15$$
$$°C = K - 273.15$$
$$°C = 257 - 273.15 = -16 \text{ °C}$$

CHECK
Are the units correct? Does the answer make physical sense?

The answer has the correct unit, and its magnitude seems correct (see Figure 3.18).

LO: Convert between Fahrenheit, Celsius, and Kelvin temperature scales (Section 3.10).

Solve temperature conversion problems using the problem-solving procedure in Sections 2.6 and 2.10. Take the steps appropriate for equations.

EXAMPLE **3.18** **Converting between Fahrenheit and Celsius Temperature Scales**

Convert 62.0 °C to Fahrenheit.

SORT
You are given the temperature in degrees Celsius and asked to convert it to degrees Fahrenheit.

GIVEN: 62.0 °C

FIND: °F

STRATEGIZE
Draw a solution map. Use the equation that relates the *given* quantity to the *find* quantity.

SOLUTION MAP

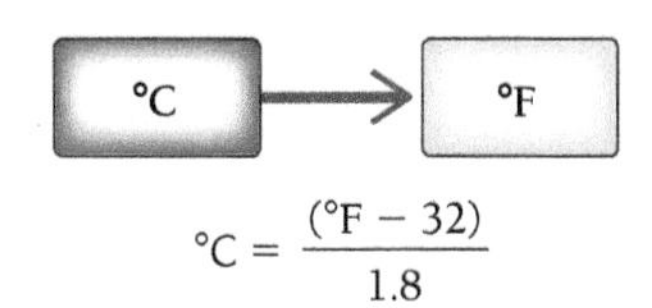

$$°C = \frac{(°F - 32)}{1.8}$$

RELATIONSHIPS USED

$$°C = \frac{(°F - 32)}{1.8}$$

SOLVE
Solve the equation for the *find* quantity (°F) and substitute the temperature in °C into the equation. Calculate the answer to the correct number of significant figures.

SOLUTION

$$°C = \frac{(°F - 32)}{1.8}$$

$$1.8\,(°C) = °F - 32$$

$$°F = 1.8\,(°C) + 32$$

$$°F = 1.8\,(62.0) + 32 = 143.60\,°F = 144\,°F$$

CHECK
Are the units correct? Does the answer make physical sense?

The answer is in the correct units, and its magnitude seems correct (see Figure 3.18).

LO: Perform calculations involving transfer of heat and changes in temperature (Section 3.12).

Solve heat capacity problems using the problem-solving procedure in Sections 2.6 and 2.10. Take the steps appropriate for equations.

EXAMPLE **3.19** **Relating Energy, Temperature Change, and Heat Capacity**

What is the temperature change in 355 mL of water upon absorption of 34 kJ of heat?

SORT
You are given the volume of water and the amount of heat absorbed. You are asked to find the change in temperature.

GIVEN: 355 mL water
34 kJ of heat

FIND: ΔT

STRATEGIZE
The solution map shows how the heat-capacity equation relates the *given* and *find* quantities.

SOLUTION MAP

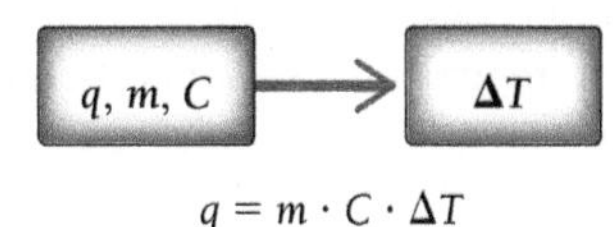

$$q = m \cdot C \cdot \Delta T$$

RELATIONSHIPS USED

$$q = m \cdot C \cdot \Delta T$$

SOLVE
First, gather the necessary quantities—m, q, and ΔT—in the correct units. You must convert the value for q from kJ to J.

SOLUTION

$$q = 34\text{ kj} \times \frac{1000\text{ J}}{1\text{ kJ}} = 3.4 \times 10^4\text{ J}$$

You also must convert the value for m from milliliters to grams; use the density of water, 1.0 g/mL, to convert milliliters to grams.

$$m = 355\text{ mL} \times \frac{1.0\text{ g}}{1\text{ mL}} = 355\text{ g}$$

Look up the heat capacity for water in Table 3.4.

$$C = 4.18\text{ J/g °C}$$

$$q = m \cdot C \cdot \Delta T$$

Then solve the equation for ΔT and substitute the correct variables into the equation. Finally, calculate the answer to the right number of significant figures.

$$\Delta T = \frac{q}{mC}$$

$$\Delta T = \frac{3.4 \times 10^4 \text{ J}}{355 \text{ g} \times 4.18 \text{ J/g °C}}$$

$$= 22.91 \text{ °C} = 23 \text{ °C}$$

CHECK

Check your answer. Are the units correct? Does the answer make physical sense?

The answer has the correct units, and the magnitude seems correct. If the magnitude of the answer were a huge number—3×10^6, for example—you would need to go back and look for a mistake. Above 100 °C, water boils, so such a large answer would be unlikely.

Key Terms

amorphous **[3.3]**
atom **[3.2]**
calorie (cal) **[3.8]**
Calorie (Cal) **[3.8]**
Celsius (°C) scale **[3.10]**
chemical change **[3.6]**
chemical energy **[3.8]**
chemical property **[3.5]**
chemical reaction **[3.6]**
compound **[3.4]**
compressible **[3.3]**
crystalline **[3.3]**
decanting **[3.6]**
distillation **[3.6]**
electrical energy **[3.8]**
element **[3.4]**
endothermic **[3.9]**
energy **[3.8]**
exothermic **[3.9]**
Fahrenheit (°F) scale **[3.10]**
filtration **[3.6]**
gas **[3.3]**
heat **[3.10]**
heterogeneous mixture **[3.4]**
homogeneous mixture **[3.4]**
Kelvin (K) scale **[3.10]**
kilowatt-hour (kWh) **[3.8]**
kinetic energy **[3.8]**
law of conservation of energy **[3.8]**
liquid **[3.3]**
matter **[3.2]**
mixture **[3.4]**
molecule **[3.2]**
physical change **[3.6]**
physical property **[3.5]**
potential energy **[3.8]**
product **[3.6]**
property **[3.5]**
pure substance **[3.4]**
reactants **[3.6]**
solid **[3.3]**
specific heat capacity (specific heat) **[3.11]**
state of matter **[3.3]**
temperature **[3.10]**
thermal energy **[3.8]**
volatile **[3.6]**
work **[3.8]**

Exercises

Questions

Answers to all odd-numbered questions (numbered in blue) appear in the Answers section at the back of the book.

1. Define matter and list some examples.
2. What is matter composed of?
3. What are the three states of matter?
4. What are the properties of a solid?
5. What is the difference between a crystalline solid and an amorphous solid?
6. What are the properties of a liquid?
7. What are the properties of a gas?
8. Why are gases compressible?
9. What is a mixture?
10. What is the difference between a homogeneous mixture and a heterogeneous mixture?
11. What is a pure substance?
12. What is an element? A compound?
13. What is the difference between a mixture and a compound?
14. What is the definition of a physical property? What is the definition of a chemical property?
15. What is the difference between a physical change and a chemical change?
16. What is the law of conservation of mass?
17. What is the definition of energy?
18. What is the law of conservation of energy?
19. Explain the difference between kinetic energy and potential energy.
20. What is chemical energy? List some examples of common substances that contain chemical energy.
21. List three common units for energy.
22. What is an exothermic reaction? Which has greater energy in an exothermic reaction, the reactants or the products?
23. What is an endothermic reaction? Which has greater energy in an endothermic reaction, the reactants or the products?
24. List three common units for measuring temperature.
25. Explain the difference between heat and temperature.
26. How do the three temperature scales differ?
27. What is heat capacity?
28. Why are coastal geographic regions normally cooler in the summer than inland geographic regions?

29. The following equation can be used to convert Fahrenheit temperature to Celsius temperature.

$$°C = \frac{(°F - 32)}{1.8}$$

Use algebra to change the equation to convert Celsius temperature to Fahrenheit temperature.

30. The following equation can be used to convert Celsius temperature to Kelvin temperature.

$$K = °C + 273.15$$

Use algebra to change the equation to convert Kelvin temperature to Celsius temperature.

Problems

Note: The exercises in the Problems section are paired, and the answers to the odd-numbered exercises (numbered in blue) appear in the Answers section at the back of the book.

CLASSIFYING MATTER

31. Classify each pure substance as an element or a compound.
    (a) aluminum
    (b) sulfur
    (c) methane
    (d) acetone

32. Classify each pure substance as an element or a compound.
    (a) carbon
    (b) baking soda (sodium bicarbonate)
    (c) nickel
    (d) gold

33. Classify each mixture as homogeneous or heterogeneous.
    (a) coffee
    (b) chocolate sundae
    (c) apple juice
    (d) gasoline

34. Classify each mixture as homogeneous or heterogeneous.
    (a) baby oil
    (b) chocolate chip cookie
    (c) water and gasoline
    (d) wine

35. Classify each substance as a pure substance or a mixture. If it is a pure substance, classify it as an element or a compound. If it is a mixture, classify it as homogeneous or heterogeneous.
    (a) helium gas
    (b) clean air
    (c) rocky road ice cream
    (d) concrete

36. Classify each substance as a pure substance or a mixture. If it is a pure substance, classify it as an element or a compound. If it is a mixture, classify it as homogeneous or heterogeneous.
    (a) urine
    (b) pure water
    (c) Snickers™ bar
    (d) soil

PHYSICAL AND CHEMICAL PROPERTIES AND PHYSICAL AND CHEMICAL CHANGES

37. Classify each property as physical or chemical.
    (a) the tendency of silver to tarnish
    (b) the shine of chrome
    (c) the color of gold
    (d) the flammability of propane gas

38. Classify each property as physical or chemical.
    (a) the boiling point of ethyl alcohol
    (b) the temperature at which dry ice sublimes (turns from a solid into a gas)
    (c) the flammability of ethyl alcohol
    (d) the smell of perfume

39. Which of the following properties of ethylene (a ripening agent for bananas) are physical properties, and which are chemical?
    - colorless
    - odorless
    - flammable
    - gas at room temperature
    - 1 L has a mass of 1.260 g under standard conditions
    - mixes with acetone
    - polymerizes to form polyethylene

40. Which of the following properties of ozone (a pollutant in the lower atmosphere but part of a protective shield against UV light in the upper atmosphere) are physical, and which are chemical?
    - bluish color
    - pungent odor
    - very reactive
    - decomposes on exposure to ultraviolet light
    - gas at room temperature

41. Classify each change as physical or chemical.
   (a) A balloon filled with hydrogen gas explodes upon contact with a spark.
   (b) The liquid propane in a barbecue evaporates away because someone left the valve open.
   (c) The liquid propane in a barbecue ignites upon contact with a spark.
   (d) Copper metal turns green on exposure to air and water.

42. Classify each change as physical or chemical.
   (a) Sugar dissolves in hot water.
   (b) Sugar burns in a pot.
   (c) A metal surface becomes dull because of continued abrasion.
   (d) A metal surface becomes dull on exposure to air.

43. A block of aluminum is **(a)** ground into aluminum powder and then **(b)** ignited. It then emits flames and smoke. Classify **(a)** and **(b)** as chemical or physical changes.

44. Several pieces of graphite from a mechanical pencil are **(a)** broken into tiny pieces. Then the pile of graphite is **(b)** ignited with a hot flame. Classify **(a)** and **(b)** as chemical or physical changes.

THE CONSERVATION OF MASS

45. An automobile gasoline tank holds 42 kg of gasoline. When the gasoline burns, 168 kg of oxygen are consumed and carbon dioxide and water are produced. What total combined mass of carbon dioxide and water is produced?

46. In the explosion of a hydrogen-filled balloon, 0.50 g of hydrogen reacts with 4.0 g of oxygen. How many grams of water vapor are formed? (Water vapor is the only product.)

47. Are these data sets on chemical changes consistent with the law of conservation of mass?
   (a) A 7.5-g sample of hydrogen gas completely reacts with 60.0 g of oxygen gas to form 67.5 g of water.
   (b) A 60.5-g sample of gasoline completely reacts with 243 g of oxygen to form 206 g of carbon dioxide and 88 g of water.

48. Are these data sets on chemical changes consistent with the law of conservation of mass?
   (a) A 12.8-g sample of sodium completely reacts with 19.6 g of chlorine to form 32.4 g of sodium chloride.
   (b) An 8-g sample of natural gas completely reacts with 32 g of oxygen gas to form 17 g of carbon dioxide and 16 g of water.

49. In a butane lighter, 9.7 g of butane combine with 34.7 g of oxygen to form 29.3 g carbon dioxide and how many grams of water?

50. A 56-g sample of iron reacts with 24 g of oxygen to form how many grams of iron oxide?

CONVERSION OF ENERGY UNITS

51. Perform each conversion.
   (a) 588 cal to joules
   (b) 17.4 J to Calories
   (c) 134 kJ to Calories
   (d) 56.2 Cal to joules

52. Perform each conversion.
   (a) 45.6 J to calories
   (b) 355 cal to joules
   (c) 43.8 kJ to calories
   (d) 215 cal to kilojoules

53. Perform each conversion.
   (a) 25 kWh to joules
   (b) 249 cal to Calories
   (c) 113 cal to kilowatt-hours
   (d) 44 kJ to calories

54. Perform each conversion.
   (a) 345 Cal to kilowatt-hours
   (b) 23 J to calories
   (c) 5.7×10^3 J to kilojoules
   (d) 326 kJ to joules

55. Complete the table:

J	cal	Cal	kWh
225 J	______	5.38×10^{-2} Cal	______
______	8.21×10^5 cal	______	______
______	______	______	295 kWh
______	______	155 Cal	______

56. Complete the table:

J	cal	Cal	kWh
7.88×10^6 J	1.88×10^6 cal	______	______
______	______	1154 Cal	______
______	88.4 cal	______	______
______	______	______	125 kWh

57. An energy bill indicates that a customer used 1027 kWh in July. How many joules did the customer use?

58. A television uses 32 kWh of energy per year. How many joules does it use?

59. An adult eats food whose nutritional energy totals approximately 2.2×10^3 Cal per day. The adult burns 2.0×10^3 Cal per day. How much excess nutritional energy, in kilojoules, does the adult consume per day? If 1 lb of fat is stored by the body for each 14.6×10^3 kJ of excess nutritional energy consumed, how long will it take this person to gain 1 lb?

60. How many joules of nutritional energy are in a bag of chips with a label that lists 245 Cal? If 1 lb of fat is stored by the body for each 14.6×10^3 kJ of excess nutritional energy consumed, how many bags of chips contain enough nutritional energy to result in 1 lb of body fat?

ENERGY AND CHEMICAL AND PHYSICAL CHANGE

61. A common type of handwarmer contains iron powder that reacts with oxygen to form an oxide of iron. As soon as the handwarmer is exposed to air, the reaction begins and heat is emitted. Is the reaction between the iron and oxygen exothermic or endothermic? Draw an energy diagram showing the relative energies of the reactants and products in the reaction.

62. In a chemical cold pack, two substances are kept separate by a divider. When the divider is broken, the substances mix and absorb heat from the surroundings. The chemical cold pack feels cold. Is the reaction exothermic or endothermic? Draw an energy diagram showing the relative energies of the reactants and products in the reaction.

63. Classify each process as exothermic or endothermic.
(a) gasoline burning in a car
(b) isopropyl alcohol evaporating from skin
(c) water condensing as dew during the night

64. Classify each process as exothermic or endothermic.
(a) dry ice subliming (changing from a solid directly to a gas)
(b) the wax in a candle burning
(c) a match burning

CONVERTING BETWEEN TEMPERATURE SCALES

65. Perform each temperature conversion.
(a) 212 °F to Celsius (temperature of boiling water)
(b) 77 K to Fahrenheit (temperature of liquid nitrogen)
(c) 25 °C to kelvins (room temperature)
(d) 98.6 °F to kelvins (body temperature)

66. Perform each temperature conversion.
(a) 102 °F to Celsius
(b) 0 K to Fahrenheit
(c) −48 °C to Fahrenheit
(d) 273 K to Celsius

67. The coldest temperature ever measured in the United States was −80 °F on January 23, 1971, in Prospect Creek, Alaska. Convert that temperature to degrees Celsius and Kelvin. (Assume that −80 °F is accurate to two significant figures.)

68. The warmest temperature ever measured in the United States was 134 °F on July 10, 1913, in Death Valley, California. Convert that temperature to degrees Celsius and Kelvin.

69. Vodka does not freeze in the freezer because it contains a high percentage of ethanol. The freezing point of pure ethanol is −114 °C. Convert that temperature to degrees Fahrenheit and Kelvin.

70. Liquid helium boils at 4.2 K. Convert this temperature to degrees Fahrenheit and Celsius.

71. The temperature in the South Pole during the Antarctic winter is so cold that planes cannot land or take off, effectively leaving the inhabitants of the South Pole isolated for the winter. The average daily temperature at the South Pole in July is −59.7 °C. Convert this temperature to degrees Fahrenheit.

72. The coldest temperature ever recorded in Iowa was −47 °F on February 3, 1998. Convert this temperature to kelvins and degrees Celsius.

73. Complete the table.

Kelvin	Fahrenheit	Celsius
0.0 K	______	−273.0 °C
______	82.5 °F	______
______	______	8.5 °C

74. Complete the table.

Kelvin	Fahrenheit	Celsius
273.0 K	______	0.0 °C
______	−40.0 °F	______
385 K	______	______

ENERGY, HEAT CAPACITY, AND TEMPERATURE CHANGES

75. Calculate the amount of heat required to raise the temperature of a 65-g sample of water from 32 °C to 65 °C.

76. Calculate the amount of heat required to raise the temperature of a 22-g sample of water from 7 °C to 18 °C.

77. Calculate the amount of heat required to heat a 45-kg sample of ethanol from 11.0 °C to 19.0 °C.

78. Calculate the amount of heat required to heat a 3.5-kg gold bar from 21 °C to 67 °C.

79. If 89 J of heat are added to a pure gold coin with a mass of 12 g, what is its temperature change?

80. If 57 J of heat are added to an aluminum can with a mass of 17.1 g, what is its temperature change?

81. An iron nail with a mass of 12 g absorbs 15 J of heat. If the nail was initially at 28 °C, what is its final temperature?

82. A 45-kg sample of water absorbs 345 kJ of heat. If the water was initially at 22.1 °C, what is its final temperature?

83. Calculate the temperature change that occurs when 248 cal of heat are added to 24 g of water.

84. A lead fishing weight with a mass of 57 g absorbs 146 cal of heat. If its initial temperature is 47 °C, what is its final temperature?

85. An unknown metal with a mass of 28 g absorbs 58 J of heat. Its temperature rises from 31.1 °C to 39.9 °C. Calculate the heat capacity of the metal and identify it, referring to Table 3.4.

86. When 2.8 J of heat are added to 5.6 g of an unknown metal, its temperature rises by 3.9 °C. Are these data consistent with the metal being gold?

87. When 56 J of heat are added to 11 g of a liquid, its temperature rises from 10.4 °C to 12.7 °C. What is the heat capacity of the liquid?

88. When 47.5 J of heat are added to 13.2 g of a liquid, its temperature rises by 1.72 °C. What is the heat capacity of the liquid?

89. Two identical coolers are packed for a picnic. Each cooler is packed with eighteen 12-oz soft drinks and 3 lb of ice. However, the drinks that went into cooler A were refrigerated for several hours before they were packed in the cooler, while the drinks that went into cooler B were at room temperature. When the two coolers are opened three hours later, most of the ice in cooler A is still ice, while nearly all of the ice in cooler B has melted. Explain.

90. A 100-g block of iron metal and 100 g of water are each warmed to 75 °C and placed into two identical insulated containers. Two hours later, the two containers are opened and the temperature of each substance is measured. The iron metal has cooled to 38 °C while the water has cooled only to 69 °C. Explain.

91. How much energy (in J) is lost when a sample of iron with a mass of 25.7 g cools from 75.0 °C to 22.0 °C?

92. A sample of aluminum with mass of 53.2 g is initially at 155 °C. What is the temperature of the aluminum after it loses 2.87×10^3 J?

Cumulative Problems

93. 245 mL of water with an initial temperature of 32 °C absorbs 17 kJ of heat. Find the final temperature of the water. (density of water = 1.0 g/mL)

94. 32 mL of ethanol with an initial temperature of 11 °C absorbs 562 J of heat. Find the final temperature of the ethanol. (density of ethanol = 0.789 g/mL)

95. A pure gold ring with a volume of 1.57 cm^3 is initially at 11.4 °C. When it is put on, it warms to final temperature of 29.5 °C. How much heat (in J) does the ring absorb? (density of gold = 19.3 g/cm^3)

96. A block of aluminum with a volume of 98.5 cm^3 absorbs 67.4 J of heat. If its initial temperature is 32.5 °C, what is its final temperature? (density of aluminum = 2.70 g/cm^3)

97. How much heat in kilojoules is required to heat 56 L of water from an initial temperature of 85 °F to a final temperature of 212 °F? (The density of water is 1.00 g/mL.)

98. How much heat in joules is required to heat a 43-g sample of aluminum from an initial temperature of 72 °F to a final temperature of 145 °F? (The density of water is 1.00 g/mL.)

99. What is the temperature change (ΔT) in Celsius when 29.5 L of water absorbs 2.3 kWh of heat?

100. If 1.45 L of water has an initial temperature of 25.0 °C, what is its final temperature after absorption of 9.4×10^{22} kWh of heat?

101. A water heater contains 55 gal of water. How many kilowatt-hours of energy are necessary to heat the water in the water heater by 25 °C? (*Hint*: $\Delta T = 25$ °C)

102. A room contains 48 kg of air. How many kilowatt-hours of energy are necessary to heat the air in the room from an initial temperature of 7 °C to a final temperature of 28 °C? The heat capacity of air is 1.03 J/g °C.

103. A backpacker wants to carry enough fuel to heat 2.5 kg of water from 25 °C to 100.0 °C. If the fuel she carries produces 36 kJ of heat per gram when it burns, how much fuel should she carry? (For the sake of simplicity, assume that the transfer of heat is 100% efficient.)

104. A cook wants to heat 1.35 kg of water from 32.0 °C to 100.0 °C. If he uses the combustion of natural gas (which is exothermic) to heat the water, how much natural gas will he need to burn? Natural gas produces 49.3 kJ of heat per gram. (For the sake of simplicity, assume that the transfer of heat is 100% efficient.)

105. Evaporating sweat cools the body because evaporation is endothermic and absorbs 2.44 kJ per gram of water evaporated. Estimate the mass of water that must evaporate from the skin to cool a body by 0.50 °C, if the mass of the body is 95 kg and its heat capacity is 4.0 J/g °C. (Assume that the heat transfer is 100% efficient.)

106. When ice melts, it absorbs 0.33 kJ per gram. How much ice is required to cool a 12.0-oz drink from 75 °F to 35 °F, if the heat capacity of the drink is 4.18 J/g °C? (Assume that the heat transfer is 100% efficient.)

107. A 15.7-g aluminum block is warmed to 53.2 °C and plunged into an insulated beaker containing 32.5 g of water initially at 24.5 °C. The aluminum and the water are allowed to come to thermal equilibrium. Assuming that no heat is lost, what is the final temperature of the water and aluminum?

108. A 25.0-mL sample of ethanol (density = 0.789 g/mL) initially at 7.0 °C is mixed with 35.0 mL of water (density = 1.0 g/mL) initially at 25.3 °C in an insulated beaker. Assuming that no heat is lost, what is the final temperature of the mixture?

109. The wattage of an appliance indicates its average power consumption in watts (W), where 1 W = 1 J/s. What is the difference in the number of kJ of energy consumed per month between a refrigeration unit that consumes 625 W and one that consumes 855 W? If electricity costs $0.15 per kWh, what is the monthly cost difference to operate the two refrigerators? (Assume 30.0 days in one month and 24.0 hours per day.)

110. A portable electric water heater transfers 255 watts (W) of power to 5.5 L of water, where 1 W = 1 J/s. How much time (in minutes) does it take for the water heater to heat the 5.5 L of water from 25 °C to 42 °C? (Assume that the water has a density of 1.0 g/mL.)

111. What temperature is the same whether it is expressed on the Celsius or Fahrenheit scale?

112. What temperature on the Celsius scale is equal to twice its value when expressed on the Fahrenheit scale?

Highlight Problems

113. Classify each as a pure substance or a mixture.

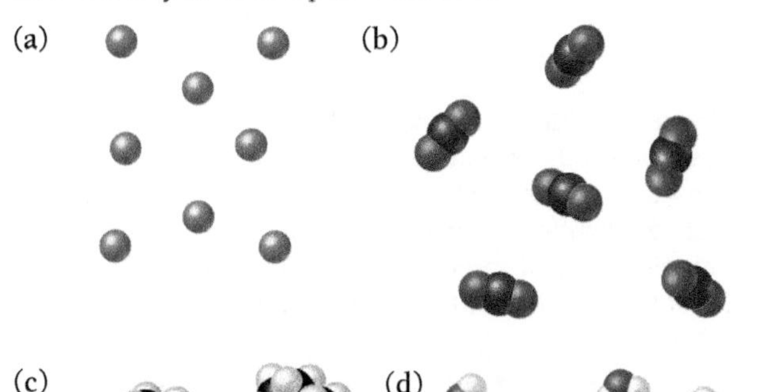

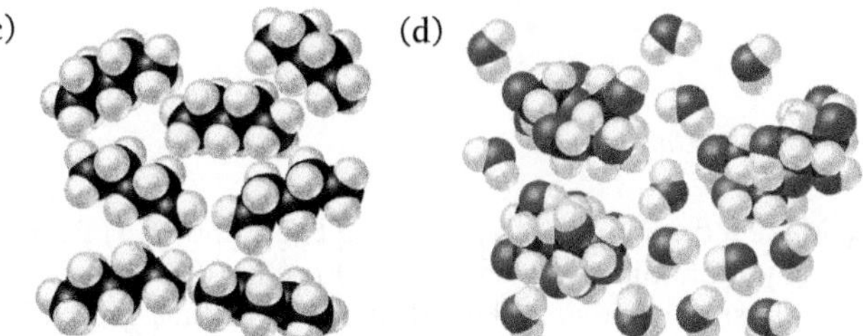

114. Classify each as a pure substance or a mixture. If it is a pure substance, classify it as an element or a compound. If it is a mixture, classify it as homogeneous or heterogeneous.

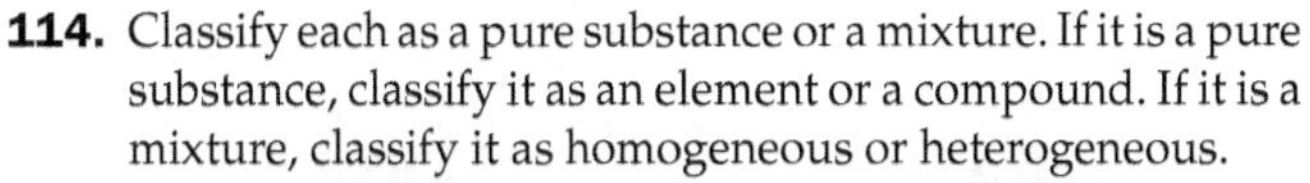

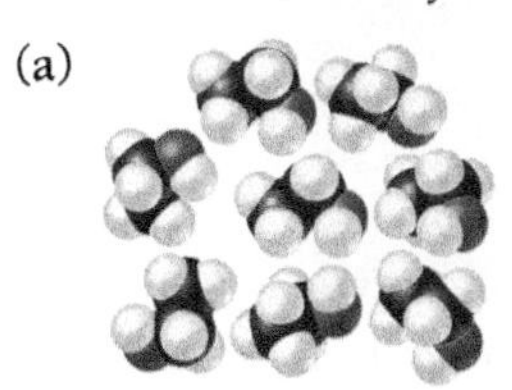

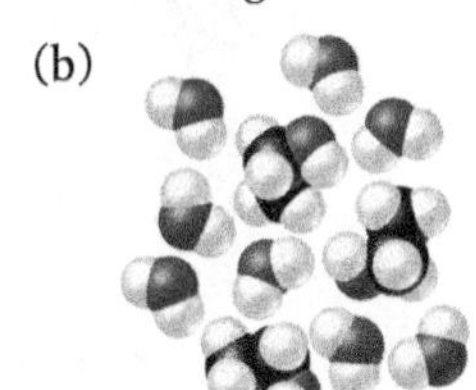

115. This molecular drawing shows images of acetone molecules before and after a change. Was the change chemical or physical?

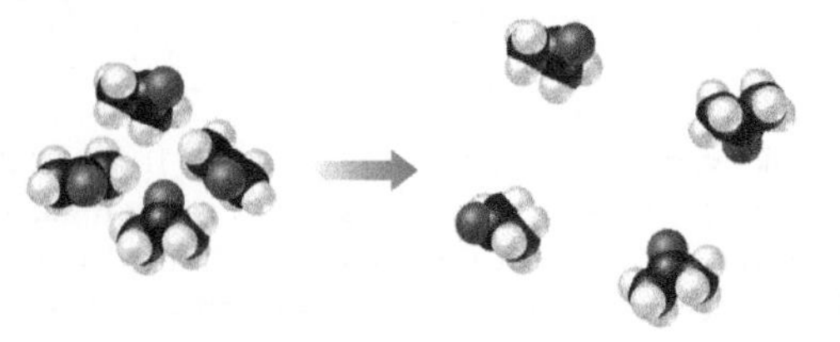

116. This molecular drawing shows images of methane molecules and oxygen molecules before and after a change. Was the change chemical or physical?

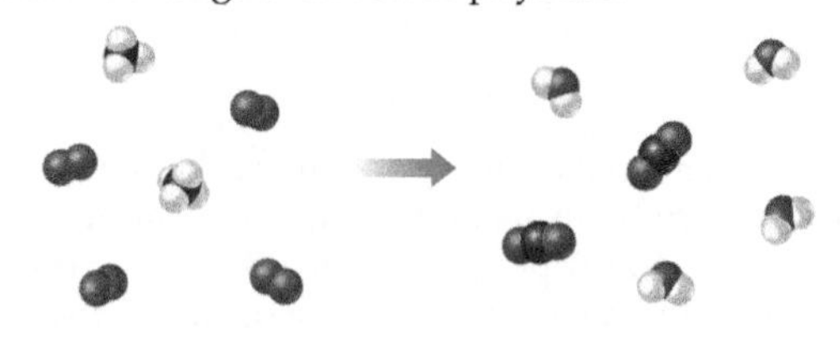

117. A major event affecting global climate is the El Niño/La Niña cycle. In this cycle, equatorial Pacific Ocean waters warm by several degrees Celsius above normal (El Niño) and then cool by several degrees Celsius below normal (La Niña). This cycle affects weather not only in North and South America, but also in places as far away as Africa. Why does a seemingly small change in ocean temperature have such a large impact on weather?

March 1, 2016

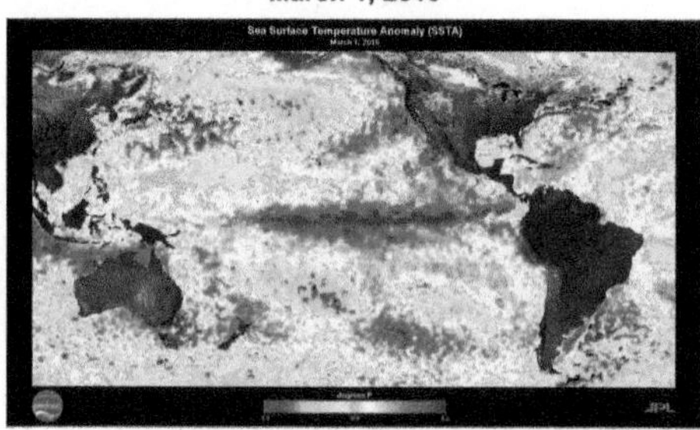

◀ Temperature anomaly plot of the world's oceans. The large red-orange section in the middle of the map indicates the El Niño effect, a warming of the Pacific Ocean along the equator.

118. Global warming refers to the rise in average global temperature due to the increased concentration of certain gases, called greenhouse gases, in our atmosphere. Earth's oceans, because of their high heat capacity, absorb heat and therefore act to slow down global warming. How much heat would be required to warm Earth's oceans by 1.0 °C? Assume that the volume of water in Earth's oceans is 137×10^7 km^3 and that the density of seawater is 1.03 g/cm^3. Also assume that the heat capacity of seawater is the same as that of water.

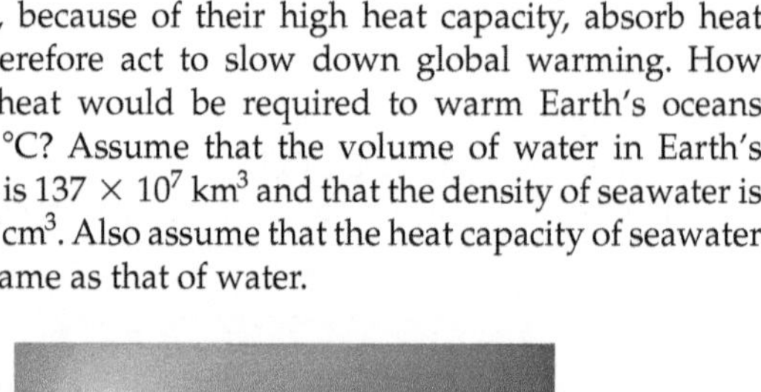

▲ Earth's oceans moderate temperatures by absorbing heat during warm periods.

119. Examine the data for the maximum and minimum average temperatures of San Francisco and Sacramento in the summer and in the winter.

San Francisco (Coastal City)

January		August	
High	Low	High	Low
57.4 °F	43.8 °F	64.4 °F	54.5 °F

Sacramento (Inland City)

January		August	
High	Low	High	Low
53.2 °F	37.7 °F	91.5 °F	57.7 °F

(a) Notice the difference between the August high in San Francisco and Sacramento. Why is it much hotter in the summer in Sacramento?

(b) Notice the difference between the January low in San Francisco and Sacramento. How might the heat capacity of the ocean contribute to this difference?

Questions for Group Work

Discuss the following questions with your group. Record the answer that your group agrees on.

120. Using white and black circles to represent different kinds of atoms, make a drawing that accurately represents each of the following: a solid element, a liquid compound, a homogeneous mixture, and a heterogeneous mixture.

121. Make a drawing (clearly showing *before* and *after*) depicting your liquid compound from Question 120 undergoing a physical change. Make a drawing depicting your solid element undergoing a chemical change.

122. A friend asks you to invest in a new motorbike that he invented that never needs gasoline and does not use batteries. What questions should you ask before investing?

123. In a grammatically correct sentence or two (and in your own words), describe what heat and temperature have in common and how they are different. Make sure you are using the words correctly with their precise scientific meanings.

Data Interpretation and Analysis

124. The graph at right shows U.S. energy consumption by source from 1980 to 2040 (based on projections). The consumption is measured in quadrillion BTUs or quads (1 quad = 1.055×10^{18} J).

(a) What were the three largest sources of U.S. energy in 2013 in descending order? What total percent of U.S. energy do these three sources provide?

(b) What percent of total U.S. energy is provided by renewables in 2013?

(c) Which two sources of U.S. energy decline as a percentage of total energy use between 1989 and 2040 (based on projections)?

(d) How much U.S. energy (in joules) was produced by nuclear power in 1990?

U.S. Energy Consumption by Source

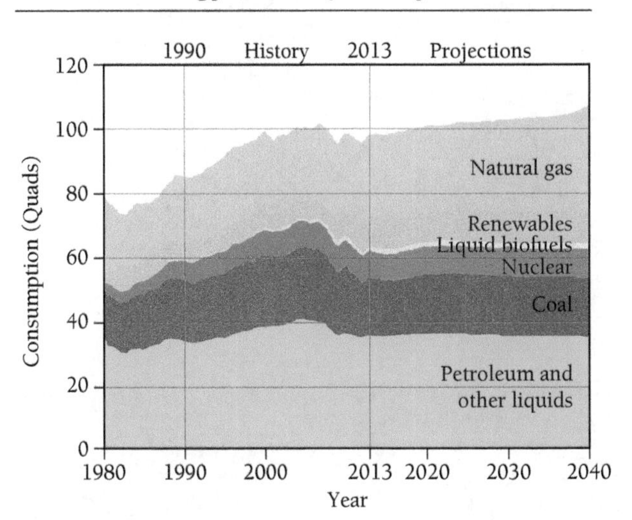

Answers to Skillbuilder Exercises

Skillbuilder 3.1.................. **(a)** pure substance, element
(b) mixture, homogeneous
(c) mixture, heterogeneous
(d) pure substance, compound

Skillbuilder 3.2.................. **(a)** chemical
(b) physical
(c) physical
(d) physical

Skillbuilder 3.3.................. **(a)** chemical
(b) physical
(c) physical
(d) chemical

Skillbuilder 3.4.................. 27 g
Skillbuilder 3.5.................. 2.14 kJ
Skillbuilder Plus, p. 74 6.57×10^6 cal
Skillbuilder 3.6.................. **(a)** exothermic
(b) exothermic
Skillbuilder 3.7.................. 85 °C
Skillbuilder 3.8.................. 282 °F
Skillbuilder 3.9.................. 77 K
Skillbuilder 3.10................ 50.1 J
Skillbuilder Plus, p. 83 7.4 g
Skillbuilder 3.11................ $\Delta T = 4.21$ °C; $T_f = 29.2$ °C

Answers to Conceptual Checkpoints

3.1 (c) The particles are far apart and moving relative to one another.

3.2 (a) The substance is composed of only one type of particle (even though each particle is composed of two different type of atoms), so it is a pure substance.

3.3 (a) Vaporization is a physical change, so the water molecules are the same before and after the boiling.

3.4 No In the vaporization, the liquid water becomes gaseous, but its mass does not change. Like chemical changes, physical changes also follow the law of conservation of mass.

3.5 (d) kWh is the largest of the four units listed, so the numerical value of the yearly energy consumption is lowest if expressed in kWh.

3.6 (d) You can confirm this by substituting each of the Fahrenheit temperatures into the equation in Section 3.10 and solving for the Celsius temperature.

3.7 (a) Because copper has the lowest specific heat capacity of the three metals, it experiences the greatest temperature change for a given energy input.

3.8 Substance B will undergo a greater change in temperature because it has the lower heat capacity. A substance with a lower heat capacity is less resistant to temperature changes.

4 Atoms and Elements

"Nothing exists except atoms and empty space; everything else is opinion."
—Democritus (460–370 B.C.)

CHAPTER OUTLINE

4.1 Experiencing Atoms at Tiburon

My wife and I recently enjoyed a visit to the northern California seaside town of Tiburon. Tiburon sits next to San Francisco Bay and enjoys views of the water, the city of San Francisco, and the surrounding mountains. As we walked along a waterside path, I could feel the wind as it blew over the bay. I could hear the water splashing on the shore, and I could smell the sea air. What was the cause of these sensations? The answer is simple—atoms.

As we learned in Chapter 3, many atoms exist not as free particles but as groups of atoms bound together to form molecules. Nevertheless, all matter is ultimately made of atoms.

Because all matter is made of atoms, atoms are at the foundation of our sensations. The atom is the fundamental building block of everything you hear, feel, see, and experience. When you feel wind on your skin, you are feeling atoms. When you hear sounds, you are in a sense hearing atoms. When you touch a shoreside rock, you are touching atoms, and when you smell sea air, you are smelling atoms. You eat atoms, you breathe atoms, and you excrete atoms. Atoms are the building blocks of matter; they are the basic units from which nature builds. They are all around us and compose everything, including our own bodies.

Atoms are incredibly small. A single pebble from the shoreline contains more atoms than you could ever count. The number of atoms in a single pebble far exceeds the number of pebbles on the bottom of San Francisco Bay. To get an idea of how small atoms are, imagine this: if every atom within a small pebble were the size of the pebble itself, the pebble would be larger than Mount Everest (▶ FIGURE 4.1 on the next page). Atoms are small—yet they compose everything.

The exact number of naturally occurring elements is controversial because some elements previously considered only synthetic may actually occur in nature in very small quantities.

The key to connecting the submicroscopic world with the macroscopic world is the atom. Atoms compose matter; the properties of atoms determine the properties of matter. An **atom** is the smallest identifiable unit of an element. Recall from Section 3.4 that an *element* is a substance that cannot be broken down into simpler substances. There are about 91 different elements in nature and consequently about 91 different

◀ Seaside rocks are typically composed of silicates, compounds of silicon and oxygen atoms. Seaside air, like all air, contains nitrogen and oxygen molecules, and it often also contains substances called amines. The amine shown here is triethylamine, which is emitted by decaying fish. Triethylamine is one of the compounds responsible for the fishy smell of the seaside.

▲ FIGURE 4.1 **The size of the atom** If every atom within a pebble were the size of the pebble itself, then the pebble would be larger than Mount Everest.

kinds of atoms. In addition, scientists have succeeded in making over 20 synthetic elements (not found in nature). In this chapter, we examine atoms: what they are made of, how they differ from one another, and how they are structured. We also examine the elements that atoms compose and some of the properties of those elements.

4.2 Indivisible: The Atomic Theory

▶ Recognize that all matter is composed of atoms.

If we look at matter, even under a microscope, it is not obvious that matter is composed of tiny particles. In fact, it appears to be just the opposite. If we divide a sample of matter into smaller and smaller pieces, it seems that we could divide it forever. From our perspective, matter seems continuous. The first people recorded as thinking otherwise were Leucippus (fifth century B.C.E., exact dates unknown) and Democritus (460–370 B.C.E.). These Greek philosophers theorized that matter was ultimately composed of small, indivisible particles. Democritus suggested that if you divided matter into smaller and smaller pieces, you would eventually end up with tiny, indestructible particles called *atomos*, or "atoms." The word *atomos* means "indivisible."

▲ Diogenes and Democritus, as imagined by a medieval artist. Democritus is the first person on record to have postulated that matter was composed of atoms.

The ideas of Leucippus and Democritus were not widely accepted. It was not until 1808—over 2000 years later—that John Dalton formalized a theory of atoms that gained broad acceptance. Dalton's atomic theory has three parts:

1. Each element is composed of tiny indestructible particles called atoms.
2. All atoms of a given element have the same mass and other properties that distinguish them from the atoms of other elements.
3. Atoms combine in simple, whole-number ratios to form compounds.

Today, the evidence for the atomic theory is overwhelming. Recent advances in microscopy have allowed scientists not only to image individual atoms but also to pick them up and move them (▼ FIGURE 4.2). Matter is indeed composed of atoms.

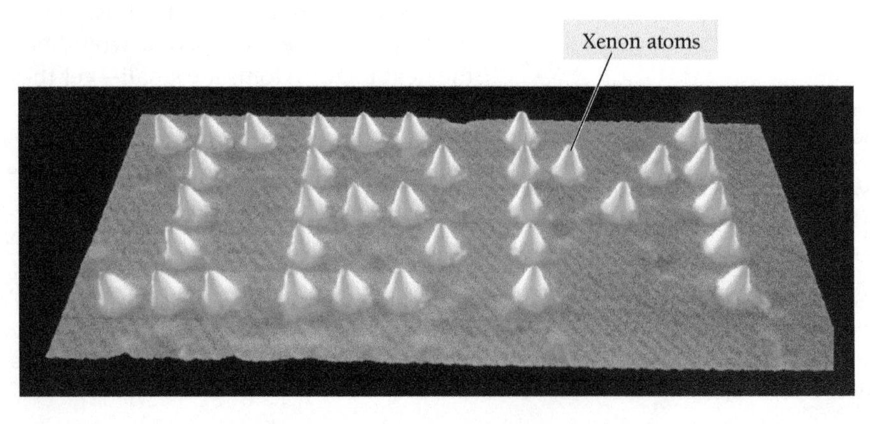

▶ FIGURE 4.2 **Writing with atoms** Scientists at IBM used a special microscope, called a scanning tunneling microscope (STM), to move xenon atoms to form the letters I, B, and M. The cone shape of these atoms is due to the peculiarities of the instrumentation. Atoms are, in general, spherical in shape.

4.3 The Nuclear Atom

▶ Explain how the experiments of Thomson and Rutherford led to the development of the nuclear theory of the atom.

Electric charge is more fully defined in Section 4.4. For now, think of it as an inherent property of electrons that causes them to interact with other charged particles.

By the end of the nineteenth century, scientists were convinced that matter was composed of atoms, the permanent, indestructible building blocks from which all substances are constructed. However, an English physicist named J. J. Thomson (1856–1940) complicated the picture by discovering an even smaller and more fundamental particle called the **electron**. Thomson discovered that electrons are negatively charged, that they are much smaller and lighter than atoms, and that they are uniformly present in many different kinds of substances. His experiments proved that the indestructible building block called the atom could apparently be "chipped."

The discovery of negatively charged particles within atoms raised the question of a balancing positive charge. Atoms were known to be charge-neutral, so it was believed that they must contain positive charge that balanced the negative charge of electrons. But how do the positive and negative charges within the atom fit together? Are atoms just a jumble of even more fundamental particles? Are they solid spheres, or do they have some internal structure? Thomson proposed that the negatively charged electrons were small particles held within a positively charged sphere. This model, the most popular of the time, became known as the plum-pudding model (plum pudding is an English dessert) (◀ FIGURE 4.3). The picture suggested by Thomson was—to those of us not familiar with plum pudding—like a blueberry muffin, where the blueberries are the electrons and the muffin is the positively charged sphere.

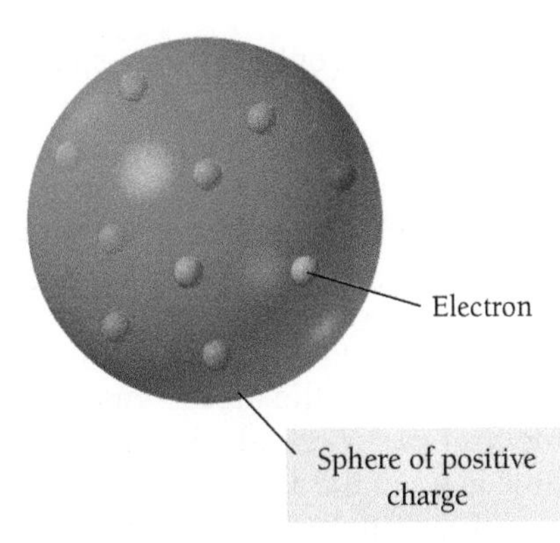

▲ FIGURE 4.3 **Plum-pudding model of the atom** In the model suggested by J. J. Thomson, negatively charged electrons (yellow) are held in a sphere of positive charge (red).

In 1909, Ernest Rutherford (1871–1937), who had worked under Thomson and adhered to his plum-pudding model, performed an experiment in an attempt to confirm it. His experiment instead proved the plum-pudding model wrong. In his experiment, Rutherford directed tiny, positively charged particles—called alpha particles—at an ultrathin sheet of gold foil (▼ FIGURE 4.4). Alpha particles are about 7000 times more massive than electrons and carry a positive charge. These particles were to act as probes of the gold atoms' structure. If the gold atoms were indeed like blueberry muffins or plum pudding—with their mass

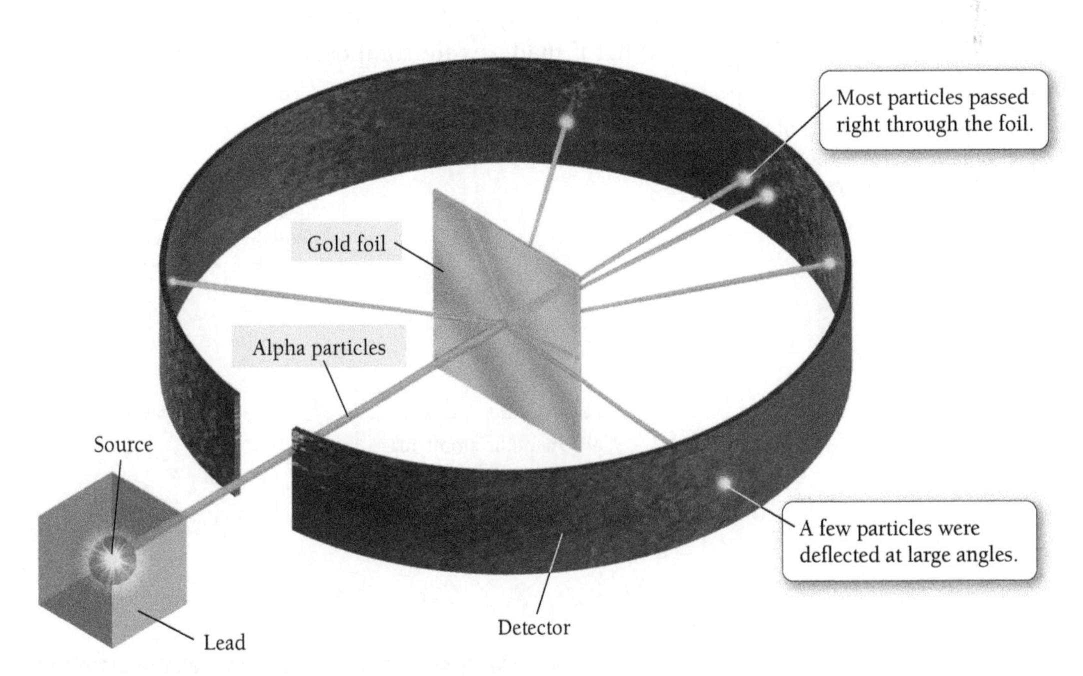

▲ FIGURE 4.4 **Rutherford's gold foil experiment** Rutherford directed tiny particles called alpha particles at a thin sheet of gold foil. Most of the particles passed directly through the foil. A few, however, were deflected—some of them at sharp angles.

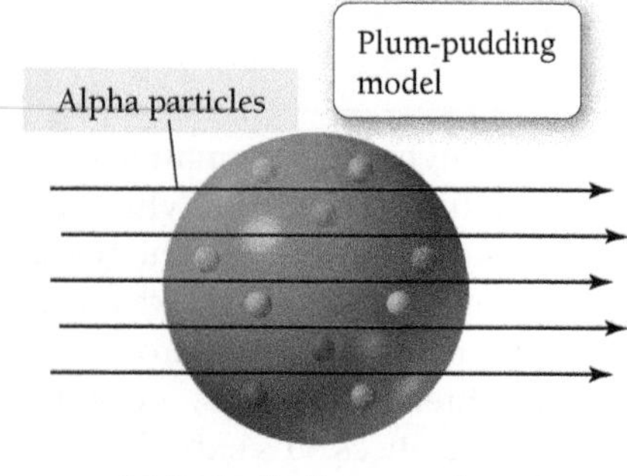

(a) Rutherford's expected result

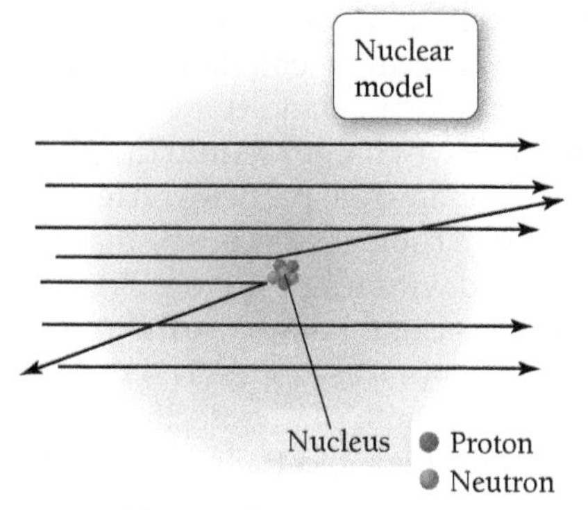

(b) Rutherford's actual result

▲ FIGURE 4.5 **Discovery of the atomic nucleus** **(a)** If the plum-pudding model were correct, the alpha particles would pass right through the gold foil with minimal deflection. **(b)** A small number of alpha particles were deflected or bounced back. The only way to explain the deflections was to suggest that most of the mass and all of the positive charge of an atom must be concentrated in a space much smaller than the size of the atom itself—the nucleus. The nucleus itself is composed of positively charged particles (protons) and neutral particles (neutrons).

and charge spread throughout the entire volume of the atom—these speeding probes should pass right through the gold foil with minimum defection. Rutherford's results were not as he expected. A majority of the particles did pass directly through the foil, but some particles were deflected, and some (1 in 20,000) even bounced back. The results puzzled Rutherford, who found them "about as credible as if you had fired a 15-inch shell at a piece of tissue paper and it came back and hit you." What must the structure of the atom be in order to explain this odd behavior?

Rutherford created a new model to explain his results (◀ FIGURE 4.5). He concluded that matter must not be as uniform as it appears. It contains large regions of empty space dotted with small regions of very dense matter. In order to explain the deflections he observed, the mass and positive charge of an atom must all be concentrated in a space much smaller than the size of the atom itself. Based on this idea, he developed the **nuclear theory of the atom**, which has three basic parts:

1. Most of the atom's mass and all of its positive charge are contained in a small core called the **nucleus**.
2. Most of the volume of the atom is empty space through which the tiny, negatively charged electrons are dispersed.
3. There are as many negatively charged electrons outside the nucleus as there are positively charged particles (*protons*) inside the nucleus, so that the atom is electrically neutral.

Later work by Rutherford and others demonstrated that the atom's nucleus contains both positively charged **protons** and neutral particles called **neutrons**. The dense nucleus makes up more than 99.9% of the mass of the atom but occupies only a small fraction of its volume. The electrons are distributed through a much larger region but don't have much mass (▼ FIGURE 4.6). For now, you can think of these electrons as akin to the water droplets that make up a cloud—they are dispersed throughout a large volume but weigh almost nothing.

Rutherford's nuclear theory is still valid today. The revolutionary part of this theory is the idea that matter—at its core—is much less uniform than it appears. If the nucleus of the atom were the size of this dot ·, the average electron would be about 10 m away. Yet the dot would contain almost the entire mass of the atom. What if matter were composed of atomic nuclei piled on top of each other like

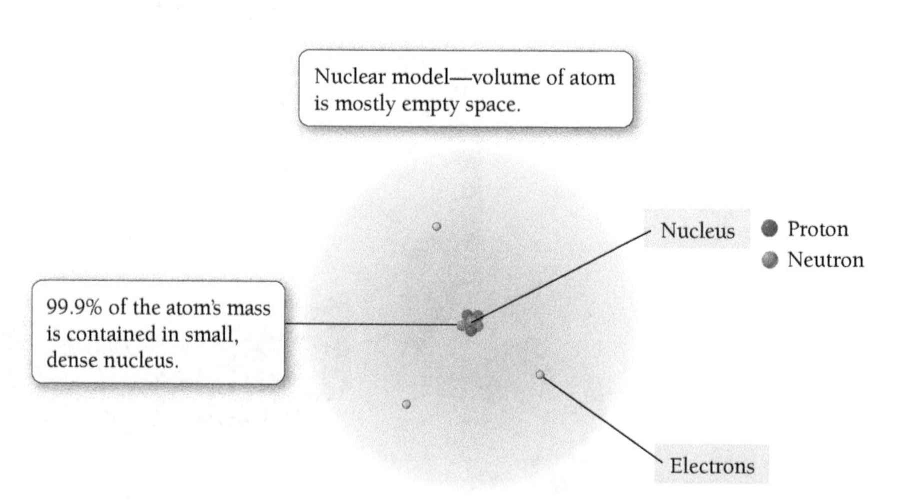

▲ FIGURE 4.6 **The nuclear atom** In this model, the atom's mass is concentrated in a nucleus that contains protons and neutrons. The number of electrons outside the nucleus is equal to the number of protons inside the nucleus. In this image, the nucleus is greatly enlarged and the electrons are portrayed as particles.

marbles? Such matter would be incredibly dense; a single grain of sand composed of solid atomic nuclei would have a mass of 5 million kg (or a weight of about 10 million lb). Astronomers believe that black holes and neutron stars are composed of this kind of incredibly dense matter.

4.4 The Properties of Protons, Neutrons, and Electrons

▶ Describe the respective properties and charges of electrons, neutrons, and protons.

Key Concept Video Subatomic Particles and Isotope Symbols

Protons and neutrons have similar masses. In SI units, the mass of the proton is 1.67262×10^{-27} kg, and the mass of the neutron is a close 1.67493×10^{-27} kg. A more common unit to express these masses, however, is the **atomic mass unit (amu)**, defined as one-twelfth of the mass of a carbon atom containing six protons and six neutrons. A proton has a mass of 1.0073 amu, and a neutron has a mass of 1.0087 amu. Electrons, by contrast, have an almost negligible mass of 0.00091×10^{-27} kg, or approximately 0.00055 amu.

The proton and the electron both have electrical **charge**. The proton's charge is 1+ and the electron's charge is 1−. The charges of the proton and the electron are equal in magnitude but opposite in sign, so that when paired, the two particles' charges exactly cancel. The neutron has no charge.

What is electrical charge? Electrical charge is a fundamental property of protons and electrons, just as mass is a fundamental property of matter. Most matter is charge-neutral because protons and electrons occur together and their charges cancel. However, you may have experienced excess electrical charge when brushing your hair on a dry day. The brushing action results in the accumulation of electrical charge on the hair strands, which then repel each other, causing your hair to stand on end.

◀ If a proton had the mass of a baseball, an electron would have the mass of a rice grain. The proton is nearly 2000 times as massive as an electron.

EVERYDAY CHEMISTRY

Solid Matter?

If matter really is mostly empty space as Rutherford suggested, then why does it appear so solid? Why can you tap your knuckles on the table and feel a solid thump? Matter appears solid because the variation in the density is on such a small scale that our eyes can't see it. Imagine a jungle gym 100 stories high and the size of a football field. It is mostly empty space. Yet if you were to view it from an airplane, it would appear as a solid mass. Matter is similar. When you tap your knuckles on the table, it is much like one giant jungle gym (your finger) crashing into another (the table). Even though they are both primarily empty space, one does not fall into the other.

B4.1 CAN YOU ANSWER THIS? *Use the jungle gym analogy to explain why most of Rutherford's alpha particles went right through the gold foil and why a few bounced back. Remember that his gold foil was extremely thin.*

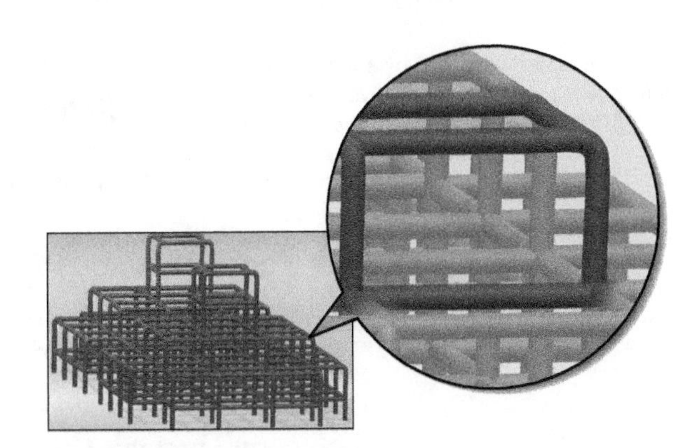

▲ Matter appears solid and uniform because the variation in density is on a scale too small for our eyes to see. Just as this scaffolding appears solid at a distance, so matter appears solid to us.

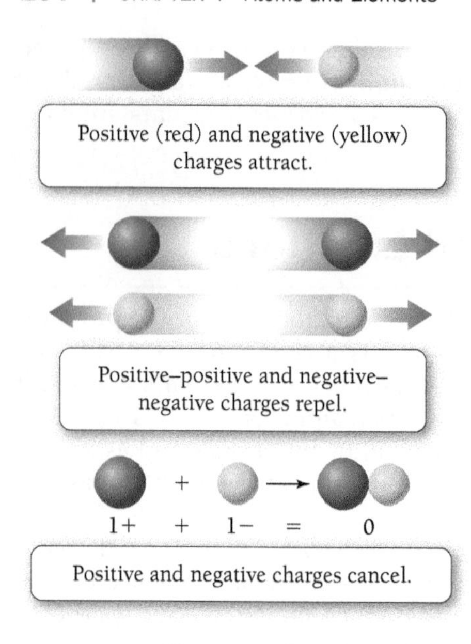

▲ FIGURE 4.7 **The properties of electrical charge**

We can summarize the nature of electrical charge as follows (◀ FIGURE 4.7).

- Electrical charge is a fundamental property of protons and electrons.
- Positive and negative electrical charges attract each other.
- Positive–positive and negative–negative charges repel each other.
- Positive and negative charges cancel each other so that a proton and an electron, when paired, are charge-neutral.

Note that matter is usually charge-neutral due to the canceling effect of protons and electrons. When matter does acquire charge imbalances, these imbalances usually equalize quickly, often in dramatic ways. For example, the shock you receive when touching a doorknob during dry weather is the equalization of a charge imbalance that developed as you walked across the carpet. Lightning is an equalization of charge imbalances that develop during electrical storms.

If you had a sample of matter—even a tiny sample, such as a sand grain—that was composed of only protons or only electrons, the forces around that matter would be extraordinary, and the matter would be unstable. Fortunately, matter is not that way—protons and electrons exist together, canceling each other's charge and making matter charge-neutral. Table 4.1 summarizes the properties of protons, neutrons, and electrons.

TABLE 4.1 Subatomic Particles

	Mass (kg)	Mass (amu)	Charge
proton	1.67262×10^{-27}	1.0073	1+
neutron	1.67493×10^{-27}	1.0087	0
electron	0.00091×10^{-27}	0.00055	1−

▶ Matter is normally charge-neutral, having equal numbers of positive and negative charges that exactly cancel. When the charge balance of matter is disturbed, as in an electrical storm, it quickly rebalances, often in dramatic ways such as lightning.

CONCEPTUAL CHECKPOINT 4.1

An atom composed of which of these particles would have a mass of approximately 12 amu and be charge-neutral?

(a) 6 protons and 6 electrons

(b) 3 protons, 3 neutrons, and 6 electrons

(c) 6 protons, 6 neutrons, and 6 electrons

(d) 12 neutrons and 12 electrons

4.5 Elements: Defined by Their Numbers of Protons

▶ Determine an element's atomic symbol and atomic number using the periodic table.

We have seen that atoms are composed of protons, neutrons, and electrons. However, it is the number of protons in the nucleus of an atom that identifies it as a particular element. For example, atoms with 2 protons in their nucleus are helium atoms, atoms with 13 protons in their nucleus are aluminum atoms, and atoms with 92 protons in their nucleus are uranium atoms. The number of protons in an atom's nucleus defines the element (▼ FIGURE 4.8). Every aluminum atom has 13 protons in its nucleus; if it had a different number of protons, it would be a different element. The number of protons in the nucleus of an atom is its **atomic number** and is represented with the symbol Z.

The periodic table of the elements (▼ FIGURE 4.9) lists all known elements according to their atomic numbers. Each element is represented by a unique

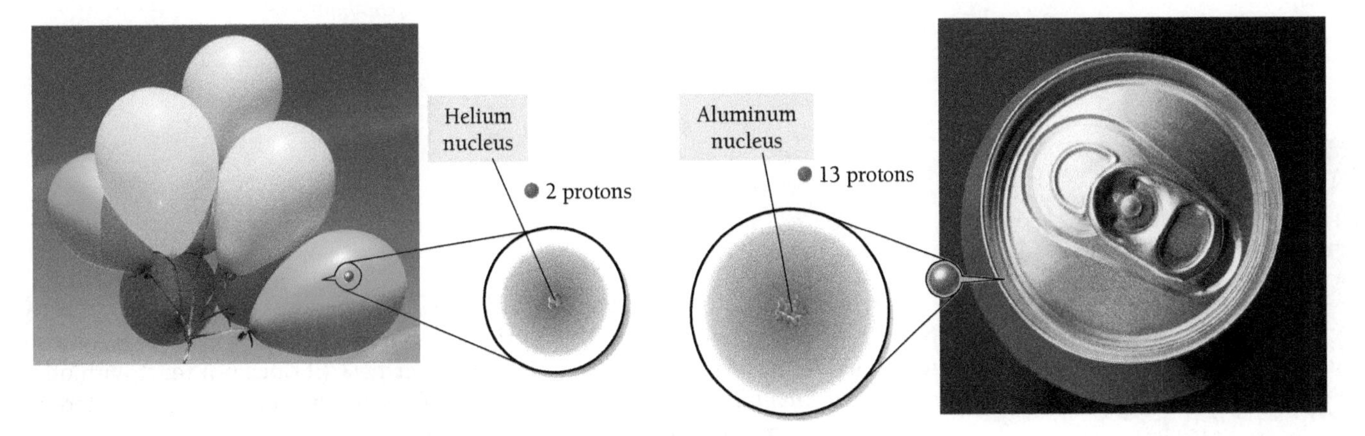

▲ FIGURE 4.8 **The number of protons in the nucleus defines the element**

Key: 4 — Atomic number (Z); Be — Chemical symbol; 9.01 — Atomic mass (defined in Section 4.9); beryllium — Name

Periods	1A 1	2A 2	3B 3	4B 4	5B 5	6B 6	7B 7	8B 8	8B 9	8B 10	1B 11	2B 12	3A 13	4A 14	5A 15	6A 16	7A 17	8A 18
1	1 H 1.01 hydrogen																	2 He 4.00 helium
2	3 Li 6.94 lithium	4 Be 9.01 beryllium											5 B 10.81 boron	6 C 12.01 carbon	7 N 14.01 nitrogen	8 O 16.00 oxygen	9 F 19.00 fluorine	10 Ne 20.18 neon
3	11 Na 22.99 sodium	12 Mg 24.31 magnesium											13 Al 26.98 aluminum	14 Si 28.09 silicon	15 P 30.97 phosphorus	16 S 32.06 sulfur	17 Cl 35.45 chlorine	18 Ar 39.95 argon
4	19 K 39.10 potassium	20 Ca 40.08 calcium	21 Sc 44.96 scandium	22 Ti 47.88 titanium	23 V 50.94 vanadium	24 Cr 52.00 chromium	25 Mn 54.94 manganese	26 Fe 55.85 iron	27 Co 58.93 cobalt	28 Ni 58.69 nickel	29 Cu 63.55 copper	30 Zn 65.39 zinc	31 Ga 69.72 gallium	32 Ge 72.63 germanium	33 As 74.92 arsenic	34 Se 78.96 selenium	35 Br 79.90 bromine	36 Kr 83.80 krypton
5	37 Rb 85.47 rubidium	38 Sr 87.62 strontium	39 Y 88.91 yttrium	40 Zr 91.22 zirconium	41 Nb 92.91 niobium	42 Mo 95.94 molybdenum	43 Tc (99) technetium	44 Ru 101.07 ruthenium	45 Rh 102.91 rhodium	46 Pd 106.42 palladium	47 Ag 107.87 silver	48 Cd 112.41 cadmium	49 In 114.82 indium	50 Sn 118.71 tin	51 Sb 121.75 antimony	52 Te 127.60 tellurium	53 I 126.90 iodine	54 Xe 131.29 xenon
6	55 Cs 132.91 cesium	56 Ba 137.33 barium	57 La 138.91 lanthanum	72 Hf 178.49 hafnium	73 Ta 180.95 tantalum	74 W 183.85 tungsten	75 Re 186.21 rhenium	76 Os 190.2 osmium	77 Ir 192.22 iridium	78 Pt 195.08 platinum	79 Au 196.97 gold	80 Hg 200.59 mercury	81 Tl 204.38 thallium	82 Pb 207.2 lead	83 Bi 208.98 bismuth	84 Po (209) polonium	85 At (210) astatine	86 Rn (222) radon
7	87 Fr (223) francium	88 Ra (226) radium	89 Ac (227) actinium	104 Rf (261) rutherfordium	105 Db (262) dubnium	106 Sg (263) seaborgium	107 Bh (262) bohrium	108 Hs (265) hassium	109 Mt (266) meitnerium	110 Ds (281) darmstadtium	111 Rg (280) roentgenium	112 Cn (285) copernicium	113 Nh (284) nihonium	114 Fl (289) flerovium	115 Mc (289) moscovium	116 Lv (292) livermorium	117 Ts (294) tennessine	118 Og (294) oganesson

Lanthanides	58 Ce 140.12 cerium	59 Pr 140.91 praseodymium	60 Nd 144.24 neodymium	61 Pm (147) promethium	62 Sm 150.36 samarium	63 Eu 151.97 europium	64 Gd 157.25 gadolinium	65 Tb 158.93 terbium	66 Dy 162.50 dysprosium	67 Ho 164.93 holmium	68 Er 167.26 erbium	69 Tm 168.93 thulium	70 Yb 173.04 ytterbium	71 Lu 174.97 lutetium
Actinides	90 Th (232) thorium	91 Pa (231) protactinium	92 U (238) uranium	93 Np (237) neptunium	94 Pu (244) plutonium	95 Am (243) americium	96 Cm (247) curium	97 Bk (247) berkelium	98 Cf (251) californium	99 Es (252) einsteinium	100 Fm (257) fermium	101 Md (258) mendelevium	102 No (259) nobelium	103 Lr (260) lawrencium

▲ FIGURE 4.9 **The periodic table of the elements**

chemical symbol, a one- or two-letter abbreviation for the element that appears directly below its atomic number on the periodic table. The chemical symbol for helium is He; for aluminum, Al; and for uranium, U. The chemical symbol and the atomic number always go together. If the atomic number is 13, the chemical symbol must be Al. If the atomic number is 92, the chemical symbol must be U. This is just another way of saying that the number of protons defines the element.

Most chemical symbols are based on the English name of the element. For example, the symbol for carbon is C; for silicon, Si; and for bromine, Br. Some elements, however, have symbols based on their Latin names. For example, the symbol for potassium is K, from the Latin *kalium*, and the symbol for sodium is Na, from the Latin *natrium*. Additional elements with symbols based on their Greek or Latin names include:

lead	Pb	*plumbum*
mercury	Hg	*hydrargyrum*
iron	Fe	*ferrum*
silver	Ag	*argentum*
tin	Sn	*stannum*
copper	Cu	*cuprum*

▲ The name *bromine* originates from the Greek word *bromos*, meaning "stench." Bromine vapor, seen as the red-brown gas in this photograph, has a strong odor.

Early scientists often gave newly discovered elements names that reflected their properties. For example, *argon* originates from the Greek word *argos*, meaning "inactive," referring to argon's chemical inertness (it does not react with other elements). *Bromine* originates from the Greek word *bromos*, meaning "stench," referring to bromine's strong odor. Scientists named other elements after countries. For example, polonium was named after Poland, francium after France, and americium after the United States of America. Still other elements were named after scientists. Curium was named after Marie Curie, and mendelevium after Dmitri Mendeleev. You can find every element's name, symbol, and atomic number in the periodic table (inside front cover) and in an alphabetical listing (inside back cover) in this book.

▶ Curium is named after Marie Curie (1867–1934), a chemist who helped discover radioactivity and also discovered two new elements. Curie won two Nobel Prizes for her work.

Curium
96
Cm
(247)

EXAMPLE 4.1 Atomic Number, Atomic Symbol, and Element Name

List the atomic symbol and atomic number for each element.

(a) silicon
(b) potassium
(c) gold
(d) antimony

SOLUTION

As you become familiar with the periodic table, you will be able to quickly locate elements on it. At first you may find it easier to locate them in the alphabetical listing on the inside back cover of this book, but you should become familiar with their positions in the periodic table.

Element	Symbol	Atomic Number
silicon	Si	14
potassium	K	19
gold	Au	79
antimony	Sb	51

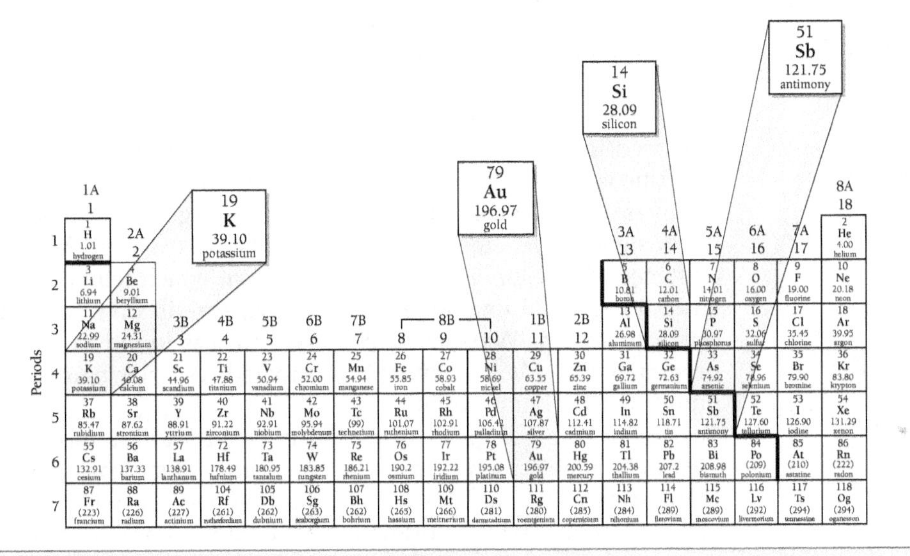

▶ SKILLBUILDER 4.1 | Atomic Number, Atomic Symbol, and Element Name

Find the name and atomic number for each element.

(a) Na
(b) Ni
(c) P
(d) Ta

▶ FOR MORE PRACTICE Problems 41, 42, 45, 46, 47, 48, 49, 50.

4.6 Looking for Patterns: The Periodic Law and the Periodic Table

▶ Use the periodic table to classify elements by group.

The organization of the periodic table is credited primarily to the work of Russian chemist Dmitri Mendeleev (1834–1907); however, German chemist Julius Lothar Meyer (1830–1895) had independently suggested a similar organization.

▲ Dmitri Mendeleev, a Russian chemistry professor who arranged early versions of the periodic table.

In Mendeleev's time, chemists had discovered about 65 different elements along with their relative masses, chemical activity, and some of their physical properties. However, there was no systematic way of organizing them.

The properties (colors) of these elements form a repeating pattern.

1	2	3	4	5	6	7	8	9	10	11	12	13	14	15	16	17	18	19	20
H	He	Li	Be	B	C	N	O	F	Ne	Na	Mg	Al	Si	P	S	Cl	Ar	K	Ca

▲ FIGURE 4.10 **Recurring properties** These elements are listed in order of increasing atomic number (Mendeleev used relative mass, which is similar). The color of each element represents its properties. Notice that the properties (colors) of these elements form a repeating pattern.

In 1869, Mendeleev noticed that certain groups of elements had similar properties. He found that if he listed the elements in order of increasing relative mass, those similar properties recurred in a regular pattern (▲ FIGURE 4.10). Mendeleev summarized these observations in the **periodic law**:

> When the elements are arranged in order of increasing relative mass, certain sets of properties recur periodically.

***Periodic* means "recurring regularly."**

Mendeleev organized all the known elements in a table in which relative mass increased from left to right and elements with similar properties were aligned in the same vertical columns (◀ FIGURE 4.11). Because many elements had not yet been discovered, Mendeleev's table contained some gaps, which allowed him to predict the existence of yet-undiscovered elements. For example, Mendeleev predicted the existence of an element he called *eka-silicon,* which fell below silicon on the table and between gallium and arsenic. In 1886, eka-silicon was discovered by German chemist Clemens Winkler (1838–1904) and was found to have almost exactly the properties that Mendeleev had anticipated. Winkler named the element germanium, after his home country.

Mendeleev's original listing has evolved into the modern **periodic table**. In the modern table, elements are listed in order of increasing atomic number rather than increasing relative mass. The modern periodic table also contains more elements than Mendeleev's original table because many more have been discovered since his time.

Mendeleev's periodic law was based on observation. Like all scientific laws, the periodic law summarized many observations but did not give the underlying reason for the observation—only theories do that. For now, we accept the periodic law as it is, but in Chapter 9 we will examine a powerful theory that explains the law and gives the underlying reasons for it.

Elements with similar properties align in vertical columns.

1 H							2 He
3 Li	4 Be	5 B	6 C	7 N	8 O	9 F	10 Ne
11 Na	12 Mg	13 Al	14 Si	15 P	16 S	17 Cl	18 Ar
19 K	20 Ca						

▲ FIGURE 4.11 **Making a periodic table** If we place the elements from Figure 4.10 in a table, we can arrange them in rows so that similar properties align in the same vertical columns. This is similar to Mendeleev's first periodic table.

We can broadly classify the elements in the periodic table as metals, nonmetals, and metalloids (▶ FIGURE 4.12). **Metals** occupy the left side of the periodic table and have similar properties: They are good conductors of heat and electricity; they can be pounded into flat sheets (malleability); they can be drawn into wires (ductility); they are often shiny; and they tend to lose electrons when they undergo chemical changes. Good examples of metals are iron, magnesium, chromium, and sodium.

Nonmetals occupy the upper right side of the periodic table. The dividing line between metals and nonmetals is the zigzag diagonal line running from boron to astatine in Figure 4.12. Nonmetals have more varied properties—some are solids at room temperature, others are gases—but as a whole, they tend to be poor conductors of heat and electricity, and they all tend to gain electrons when they undergo chemical changes. Good examples of nonmetals are oxygen, nitrogen, chlorine, and iodine.

Most of the elements that lie along the zigzag diagonal line dividing metals and nonmetals are **metalloids,** or semimetals, and display mixed properties.

Metals
Nonmetals
Metalloids

	1A 1	2A 2	3B 3	4B 4	5B 5	6B 6	7B 7	8B 8	8B 9	8B 10	1B 11	2B 12	3A 13	4A 14	5A 15	6A 16	7A 17	8A 18
1	1 H																	2 He
2	3 Li	4 Be											5 B	6 C	7 N	8 O	9 F	10 Ne
3	11 Na	12 Mg											13 Al	14 Si	15 P	16 S	17 Cl	18 Ar
4	19 K	20 Ca	21 Sc	22 Ti	23 V	24 Cr	25 Mn	26 Fe	27 Co	28 Ni	29 Cu	30 Zn	31 Ga	32 Ge	33 As	34 Se	35 Br	36 Kr
5	37 Rb	38 Sr	39 Y	40 Zr	41 Nb	42 Mo	43 Tc	44 Ru	45 Rh	46 Pd	47 Ag	48 Cd	49 In	50 Sn	51 Sb	52 Te	53 I	54 Xe
6	55 Cs	56 Ba	57 La	72 Hf	73 Ta	74 W	75 Re	76 Os	77 Ir	78 Pt	79 Au	80 Hg	81 Tl	82 Pb	83 Bi	84 Po	85 At	86 Rn
7	87 Fr	88 Ra	89 Ac	104 Rf	105 Db	106 Sg	107 Bh	108 Hs	109 Mt	110 Ds	111 Rg	112 Cn	113 Nh	114 Fl	115 Mc	116 Lv	117 Ts	118 Og

Lanthanides	58 Ce	59 Pr	60 Nd	61 Pm	62 Sm	63 Eu	64 Gd	65 Tb	66 Dy	67 Ho	68 Er	69 Tm	70 Yb	71 Lu
Actinides	90 Th	91 Pa	92 U	93 Np	94 Pu	95 Am	96 Cm	97 Bk	98 Cf	99 Es	100 Fm	101 Md	102 No	103 Lr

▲ FIGURE 4.12 **Metals, nonmetals, and metalloids** The elements in the periodic table can be broadly classified as metals, nonmetals, or metalloids.

We also call metalloids **semiconductors** because of their intermediate electrical conductivity, which can be changed and controlled. This property makes semiconductors useful in the manufacture of the electronic devices that are central to computers, cell phones, and many other technological gadgets. Silicon, arsenic, and germanium are good examples of metalloids.

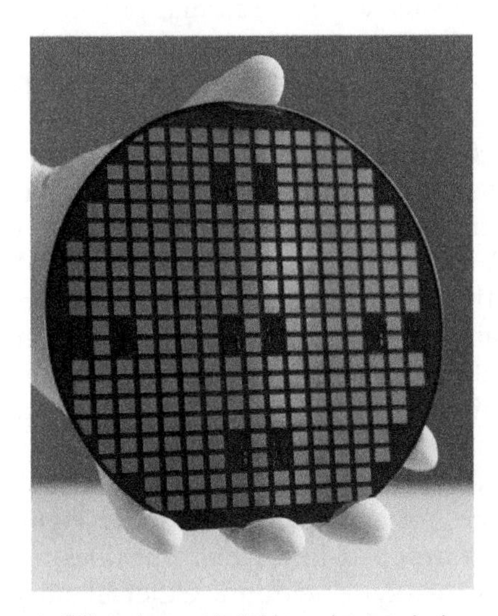

▲ Silicon is a metalloid used extensively in the computer and electronics industries.

EXAMPLE **4.2** **Classifying Elements as Metals, Nonmetals, or Metalloids**

Classify each element as a metal, nonmetal, or metalloid.

(a) Ba **(b)** I **(c)** O **(d)** Te

SOLUTION

(a) Barium is on the left side of the periodic table; it is a metal.
(b) Iodine is on the right side of the periodic table; it is a nonmetal.
(c) Oxygen is on the right side of the periodic table; it is a nonmetal.
(d) Tellurium is in the middle-right section of the periodic table, along the line that divides the metals from the nonmetals; it is a metalloid.

▶ **SKILLBUILDER 4.2 | Classifying Elements as Metals, Nonmetals, or Metalloids**

Classify each element as a metal, nonmetal, or metalloid.

(a) S **(b)** Cl **(c)** Ti **(d)** Sb

▶ **FOR MORE PRACTICE** Problems 51, 52, 53, 54.

We can also broadly divide the periodic table into **main-group elements**, whose properties tend to be more predictable based on their position in the periodic table, and **transition elements** or **transition metals**, whose properties are less

▶ FIGURE 4.13 **Main-group and transition elements** We broadly divide the periodic table into main-group elements, whose properties we can generally predict based on their position, and transition elements, whose properties tend to be less predictable based on their position.

Main-group elements | Transition elements | Main-group elements

Group number

Periods	1A	2A	3B	4B	5B	6B	7B	8B	8B	8B	1B	2B	3A	4A	5A	6A	7A	8A
1	1 H																	2 He
2	3 Li	4 Be											5 B	6 C	7 N	8 O	9 F	10 Ne
3	11 Na	12 Mg											13 Al	14 Si	15 P	16 S	17 Cl	18 Ar
4	19 K	20 Ca	21 Sc	22 Ti	23 V	24 Cr	25 Mn	26 Fe	27 Co	28 Ni	29 Cu	30 Zn	31 Ga	32 Ge	33 As	34 Se	35 Br	36 Kr
5	37 Rb	38 Sr	39 Y	40 Zr	41 Nb	42 Mo	43 Tc	44 Ru	45 Rh	46 Pd	47 Ag	48 Cd	49 In	50 Sn	51 Sb	52 Te	53 I	54 Xe
6	55 Cs	56 Ba	57 La	72 Hf	73 Ta	74 W	75 Re	76 Os	77 Ir	78 Pt	79 Au	80 Hg	81 Tl	82 Pb	83 Bi	84 Po	85 At	86 Rn
7	87 Fr	88 Ra	89 Ac	104 Rf	105 Db	106 Sg	107 Bh	108 Hs	109 Mt	110 Ds	111 Rg	112 Cn	113 Nh	114 Fl	115 Mc	116 Lv	117 Ts	118 Og

easily predictable based simply on their position in the periodic table (▲ FIGURE 4.13). Main-group elements are in columns labeled with a number and the letter A. Transition elements are in columns labeled with a number and the letter B. A competing numbering system does not use letters but only the numbers 1–18. We show both numbering systems in the periodic table in the inside front cover of this book.

CONCEPTUAL ✔ CHECKPOINT 4.2

Which element is a main-group metal?

(a) O (b) Ag (c) P (d) Pb

The noble gases are inert (unreactive) compared to other elements. However, some noble gases, especially the heavier ones, form a limited number of compounds with other elements under special conditions.

Each column within the periodic table is a **family** or **group** of elements. The elements within a family of main-group elements usually have similar properties, and some have a group name. For example, the Group 8A elements, the **noble gases**, are chemically inert gases. The most familiar noble gas is probably helium, used to fill balloons. Helium, like the other noble gases, is chemically stable—it won't combine with other elements to form compounds—and is therefore safe to put into balloons. Other noble gases include neon, often used in neon signs; argon, which makes up a small percentage of our atmosphere; krypton; and xenon. The Group 1A elements, the **alkali metals**, are all very reactive metals. A marble-sized piece of sodium can explode when dropped into water. Other alkali metals include lithium, potassium, and rubidium. The Group 2A elements, the **alkaline earth metals**, are also fairly reactive, although not quite as reactive as the alkali metals. Calcium, for example, reacts fairly vigorously when dropped into water but does not explode as readily as sodium. Other alkaline earth metals are magnesium, a common low-density structural metal; strontium; and barium. The Group 7A elements, the **halogens**, are very reactive nonmetals. Chlorine, a greenish-yellow gas with a pungent odor, is probably the most familiar halogen. Because of its reactivity, people often use chlorine as a sterilizing and disinfecting agent (it reacts with and kills bacteria and other microscopic organisms). Other halogens include bromine, a red-brown liquid that readily evaporates into a gas; iodine, a purple solid; and fluorine, a pale yellow gas.

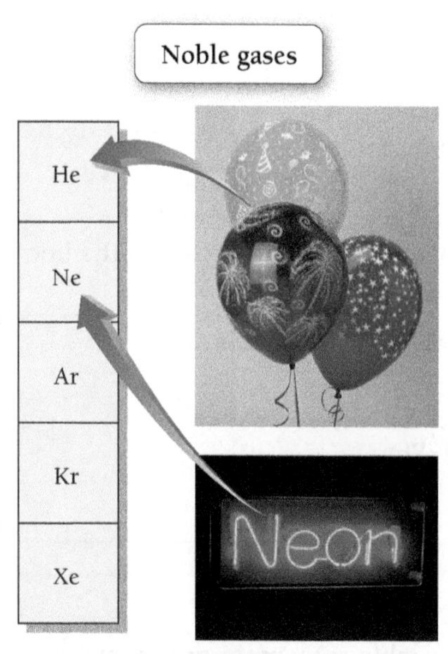

▲ The noble gases include helium (used in balloons), neon (used in neon signs), argon, krypton, and xenon.

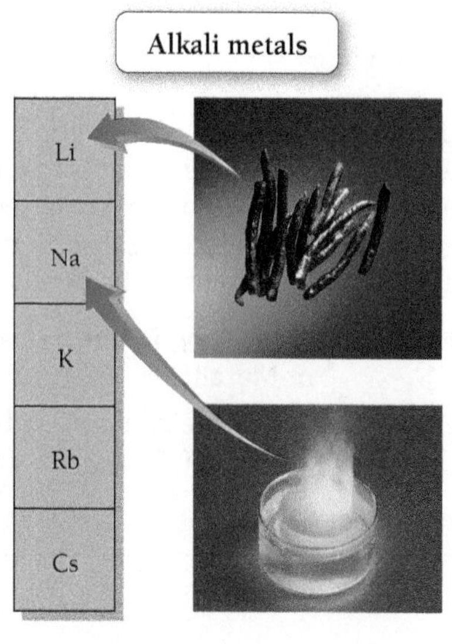

▲ The alkali metals include lithium, sodium (shown in the second photo reacting with water), potassium, rubidium, and cesium.

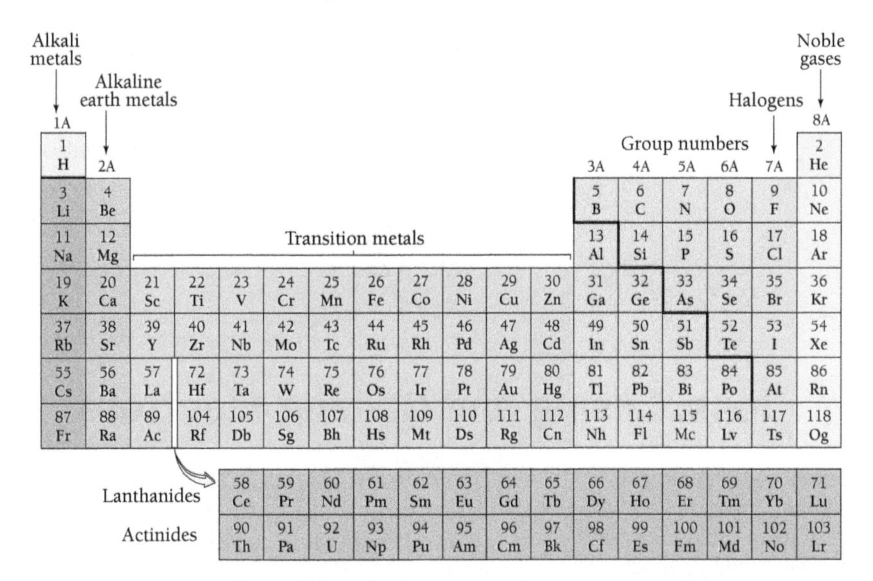

▶ The periodic table with Groups 1A, 2A, 7A, and 8A highlighted.

EXAMPLE 4.3 Groups and Families of Elements

To which group or family of elements does each element belong?

(a) Mg **(b)** N **(c)** K **(d)** Br

SOLUTION

(a) Mg is in Group 2A; it is an alkaline earth metal.
(b) N is in Group 5A.
(c) K is in Group 1A; it is an alkali metal.
(d) Br is in Group 7A; it is a halogen.

▶ **SKILLBUILDER 4.3 | Groups and Families of Elements**

To which group or family of elements does each element belong?

(a) Li **(b)** B **(c)** I **(d)** Ar

▶ **FOR MORE PRACTICE** Problems 57, 58, 59, 60, 61, 62, 63, 64.

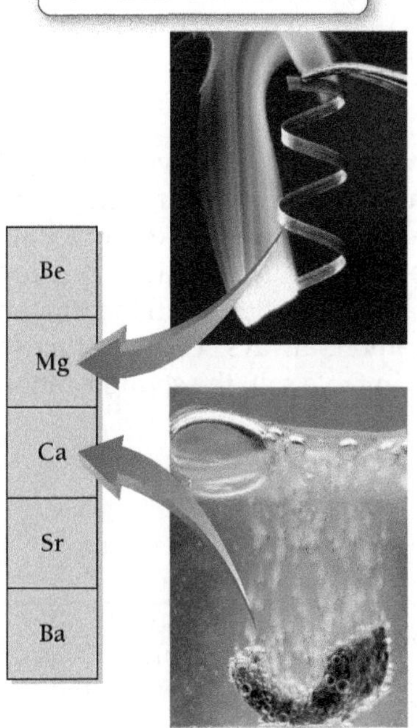

◀ The alkaline earth metals include beryllium, magnesium (shown burning in the first photo), calcium (shown reacting with water in the second photo), strontium, and barium.

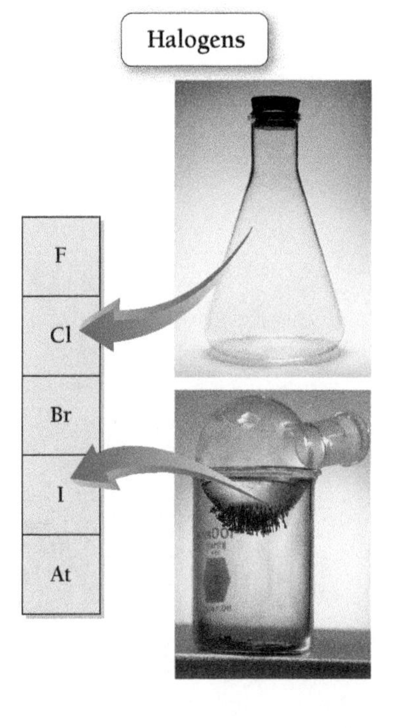

▶ The halogens include fluorine, chlorine, bromine, iodine, and astatine.

CONCEPTUAL CHECKPOINT 4.3

Which statement is NEVER true?

(a) An element can be both a transition element and a metal.

(b) An element can be both a transition element and a metalloid.

(c) An element can be both a metalloid and a halogen.

(d) An element can be both a main-group element and a halogen.

4.7 Ions: Losing and Gaining Electrons

- Determine ion charge from the number of protons and electrons.
- Determine the number of protons and electrons in an ion.

The charge of an ion is indicated in the upper right corner of the symbol.

In chemical reactions, atoms often lose or gain electrons to form charged particles called **ions.** For example, a neutral lithium (Li) atom contains 3 protons and 3 electrons; however, in reactions, a lithium atom loses 1 electron (e^-) to form a Li^+ ion.

$$Li \longrightarrow Li^+ + e^-$$

The Li^+ *ion* contains 3 protons but only 2 electrons, resulting in a net charge of 1+. We usually write ion charges with the magnitude of the charge first followed by the sign of the charge. For example, we write a positive two charge as 2+ and a negative two charge as 2−. The charge of an ion depends on how many electrons were gained or lost and is given by the formula:

$$\begin{aligned} \text{ion charge} &= \text{number of protons} - \text{number of electrons} \\ &= \#p^+ - \#e^- \end{aligned}$$

where p^+ stands for *proton* and e^- stands for *electron.*

For the Li^+ ion with 3 protons and 2 electrons, the charge is:

$$\text{ion charge} = 3 - 2 = 1+$$

A neutral fluorine (F) atom contains 9 protons and 9 electrons; however, in chemical reactions a fluorine atom gains 1 electron to form F^- ions:

$$F + e^- \longrightarrow F^-$$

The F^- *ion* contains 9 protons and 10 electrons, resulting in a 1− charge.

$$\begin{aligned} \text{ion charge} &= 9 - 10 \\ &= 1- \end{aligned}$$

Positively charged ions, such as Li^+, are **cations**, and negatively charged ions, such as F^-, are **anions**. Ions, on the one hand, behave very differently than the atoms from which they are formed. Neutral sodium atoms, for example, are extremely reactive, interacting violently with most things they contact. Sodium cations (Na^+), on the other hand, are relatively inert—we eat them all the time in sodium chloride (table salt). In nature, cations and anions always occur together so that, again, matter is charge-neutral. For example, in table salt, the sodium cation occurs together with the chloride anion (Cl^-).

CONCEPTUAL CHECKPOINT 4.4

What is the symbol for the ion that forms when oxygen gains two electrons?

$$O + 2e^- \longrightarrow \_\_\_\_$$

EXAMPLE **4.4** **Determining Ion Charge from Numbers of Protons and Electrons**

Determine the charge of each ion.

(a) a magnesium ion with 10 electrons
(b) a sulfur ion with 18 electrons
(c) an iron ion with 23 electrons

SOLUTION

To determine the charge of each ion, use the ion charge equation.

$$\text{ion charge} = \#p^+ - \#e^-$$

You are given the number of electrons in the problem. You can obtain the number of protons from the element's atomic number in the periodic table.

(a) Magnesium's atomic number is 12.

$$\text{ion charge} = 12 - 10 = 2+ \ (Mg^{2+})$$

(b) Sulfur's atomic number is 16.

$$\text{ion charge} = 16 - 18 = 2- \ (S^{2-})$$

(c) Iron's atomic number is 26.

$$\text{ion charge} = 26 - 23 = 3+ \ (Fe^{3+})$$

▶ **SKILLBUILDER 4.4 | Determining Ion Charge from Numbers of Protons and Electrons**

Determine the charge of each ion.

(a) a nickel ion with 26 electrons
(b) a bromine ion with 36 electrons
(c) a phosphorus ion with 18 electrons

▶ **FOR MORE PRACTICE** Example 4.10; Problems 75, 76.

EXAMPLE **4.5** **Determining the Number of Protons and Electrons in an Ion**

Determine the number of protons and electrons in the Ca^{2+} ion.

The periodic table indicates that the atomic number for calcium is 20, so calcium has 20 protons. You can find the number of electrons using the ion charge equation.

SOLUTION

$\text{ion charge} = \#p^+ - \#e^-$

$2+ = 20 - \#e^-$

$\#e^- = 20 - 2 = 18$

The number of electrons is 18. The Ca^{2+} ion has 20 protons and 18 electrons.

▶ **SKILLBUILDER 4.5 | Determining the Number of Protons and Electrons in an Ion**

Determine the number of protons and electrons in the S^{2-} ion.

▶ **FOR MORE PRACTICE** Example 4.11; Problems 77, 78.

Ions and the Periodic Table

For many main-group elements, we can use the periodic table to predict how many electrons tend to be lost or gained when an atom of that particular element ionizes. The number associated with the letter A above each *main-group* column in the periodic table—1 through 8—gives the number of *valence electrons* for the elements in

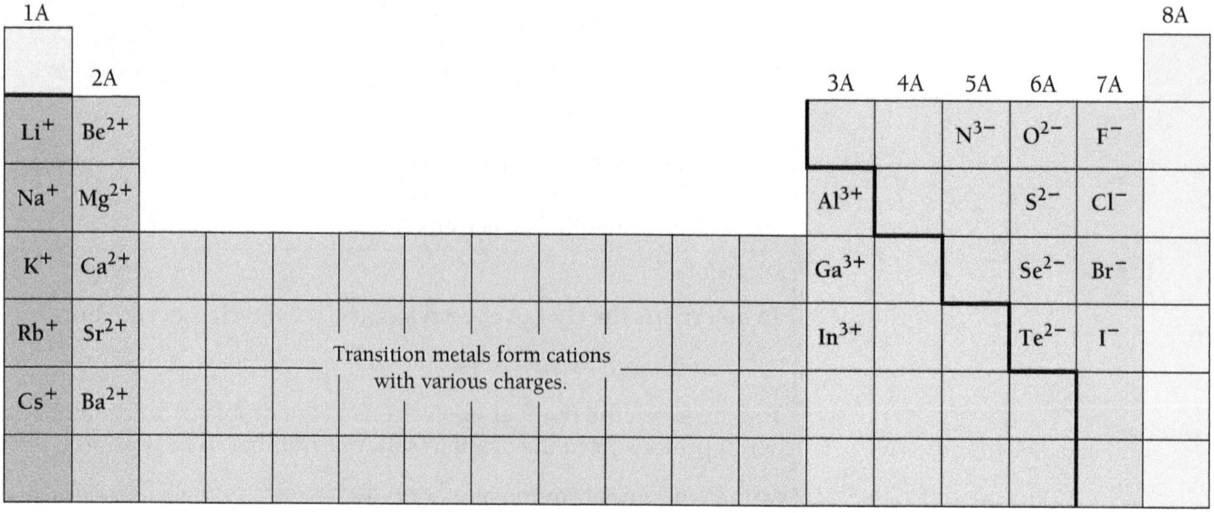

1A	2A		3A	4A	5A	6A	7A	8A
Li^+	Be^{2+}				N^{3-}	O^{2-}	F^-	
Na^+	Mg^{2+}		Al^{3+}			S^{2-}	Cl^-	
K^+	Ca^{2+}		Ga^{3+}			Se^{2-}	Br^-	
Rb^+	Sr^{2+}	Transition metals form cations with various charges.	In^{3+}			Te^{2-}	I^-	
Cs^+	Ba^{2+}							

▲ FIGURE 4.14 **Elements that form predictable ions**

that column. We will discuss the concept of valence electrons more fully in Chapter 9; for now, you can think of valence electrons as the outermost electrons in an atom. Because oxygen is in column 6A, we can deduce that it has 6 valence electrons; because magnesium is in column 2A, it has 2 valence electrons, and so on. An important exception to this rule is helium—it is in column 8A but has only 2 valence electrons. Valence electrons are particularly important because, as we shall see in Chapter 10, these electrons are the ones that are most important in chemical bonding.

We can predict the charge acquired by a particular element when it ionizes from its position in the periodic table relative to the noble gases.

Main-group elements tend to form ions that have the same number of valence electrons as the nearest noble gas.

For example, the closest noble gas to oxygen is neon. When oxygen ionizes, it *acquires* two additional electrons for a total of 8 valence electrons—the same number as neon. When determining the closest noble gas, we can move either forward or backward on the periodic table. For example, the closest noble gas to magnesium is also neon, even though neon (atomic number 10) falls before magnesium (atomic number 12) in the periodic table. Magnesium *loses* its 2 valence electrons to attain the same number of valence electrons as neon.

In accordance with this principle, the alkali metals (Group 1A) tend to lose 1 electron and form 1+ ions, while the alkaline earth metals (Group 2A) tend to lose 2 electrons and form 2+ ions. The halogens (Group 7A) tend to gain 1 electron and form 1– ions. The groups in the periodic table that form predictable ions are shown in ▲ FIGURE 4.14. Become familiar with these groups and the ions they form. In Chapter 9, we will examine a theory that more fully explains why these groups form ions as they do.

EXAMPLE 4.6 Charge of Ions from Position in Periodic Table

Based on their position in the periodic table, what ions do barium and iodine tend to form?

SOLUTION

Because barium is in Group 2A, it tends to form a cation with a 2+ charge (Ba^{2+}). Because iodine is in Group 7A, it tends to form an anion with a 1− charge (I^-).

▶ **SKILLBUILDER 4.6 | Charge of Ions from Position in Periodic Table**

Based on their position in the periodic table, what ions do potassium and selenium tend to form?

▶ **FOR MORE PRACTICE** Problems 81, 82.

CONCEPTUAL CHECKPOINT 4.5

Which pair of ions has the same total number of electrons?

(a) Na^+ and Mg^{2+} (b) F^- and Cl^-

(c) O^- and O^{2-} (d) Ga^{3+} and Fe^{3+}

4.8 Isotopes: When the Number of Neutrons Varies

- Determine atomic numbers, mass numbers, and isotope symbols for an isotope.
- Determine number of protons and neutrons from isotope symbols.

All atoms of a given element have the same number of protons; however, they do not necessarily have the same number of neutrons. Because neutrons and protons have nearly the same mass (approximately 1 amu), and the number of neutrons in the atoms of a given element can vary, all atoms of a given element *do not* have the same mass (contrary to what John Dalton originally proposed in his atomic theory). For example, all neon atoms in nature contain 10 protons, but they may have 10, 11, or 12 neutrons (▼ FIGURE 4.15). All three types of neon atoms exist, and each has a slightly different mass. Atoms with the same number of protons but different numbers of neutrons are **isotopes**. Some elements, such as beryllium (Be) and aluminum (Al), have only one naturally occurring isotope, while other elements, such as neon (Ne) and chlorine (Cl), have two or more.

Recent studies have shown that for some elements, the relative amounts of each different isotope vary depending on the history of the sample. However, these variations are usually small and beyond the scope of this book.

For a given element, the relative amounts of each different isotope in a naturally occurring sample of that element are always the same. For example, in any natural sample of neon atoms, 90.48% of the atoms are the isotope with 10 neutrons, 0.27% are the isotope with 11 neutrons, and 9.25% are the isotope with 12 neutrons as summarized in Table 4.2. This means that in a sample of 10,000 neon atoms, 9048 have 10 neutrons, 27 have 11 neutrons, and 925 have 12 neutrons. These percentages are the **percent natural abundance** of the isotopes. The preceding numbers are for neon only; each element has its own unique percent natural abundance of isotopes.

***Percent* means "per hundred." 90.48% means that 90.48 atoms out of 100 are the isotope with 10 neutrons.**

TABLE 4.2 Neon Isotopes

Symbol	Number of Protons	Number of Neutrons	A (Mass Number)	Percent Natural Abundance
Ne-20 or $^{20}_{10}Ne$	10	10	20	90.48%
Ne-21 or $^{21}_{10}Ne$	10	11	21	0.27%
Ne-22 or $^{22}_{10}Ne$	10	12	22	9.25%

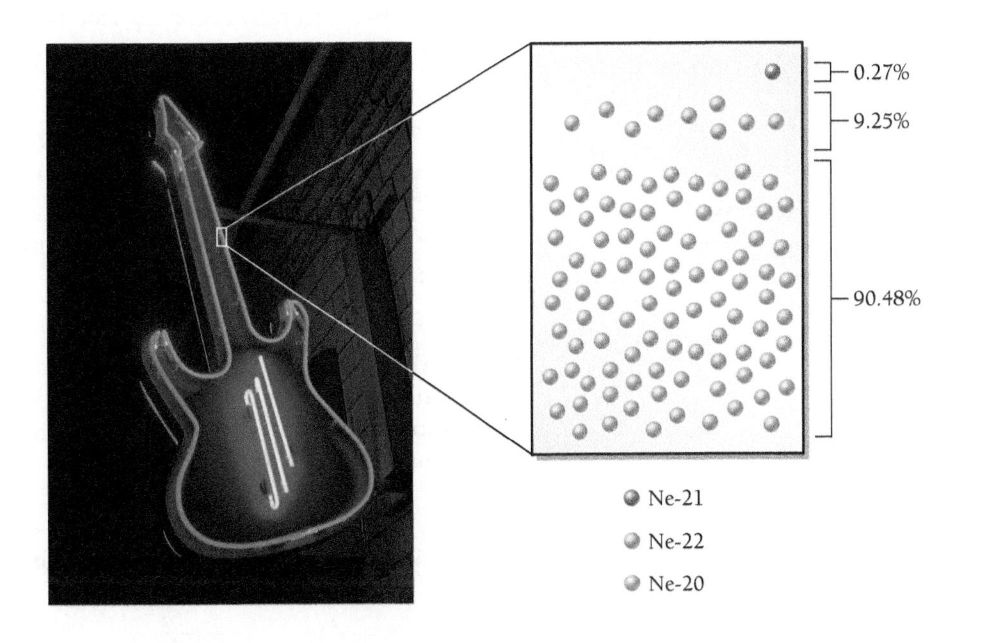

▶ FIGURE 4.15 **Isotopes of neon** Naturally occurring neon contains three different isotopes: Ne-20 (with 10 neutrons), Ne-21 (with 11 neutrons), and Ne-22 (with 12 neutrons).

The sum of the number of neutrons and protons in an atom is its **mass number** and is given the symbol **A**.

$$A = \text{number of protons} + \text{number of neutrons}$$

For neon, which has 10 protons, the mass numbers of the three different naturally occurring isotopes are 20, 21, and 22, corresponding to 10, 11, and 12 neutrons, respectively.

We often symbolize isotopes in the following way:

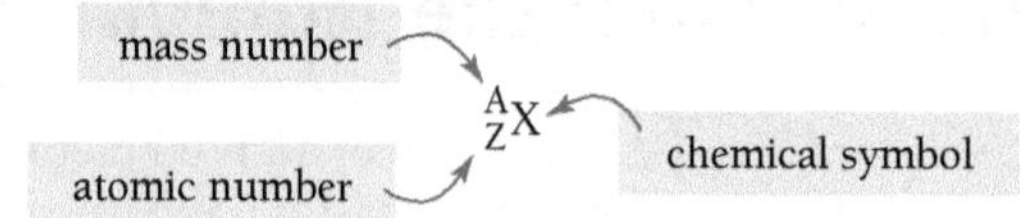

where X is the chemical symbol, A is the mass number, and Z is the atomic number.

For example, the symbols for the neon isotopes are:

$$^{20}_{10}Ne \quad ^{21}_{10}Ne \quad ^{22}_{10}Ne$$

Notice that the chemical symbol, Ne, and the atomic number, 10, are redundant. If the atomic number is 10, the symbol must be Ne, and vice versa. The mass numbers, however, are different, reflecting the different number of neutrons in each isotope.

A second common notation for isotopes is the chemical symbol (or chemical name) followed by a hyphen and the mass number of the isotope:

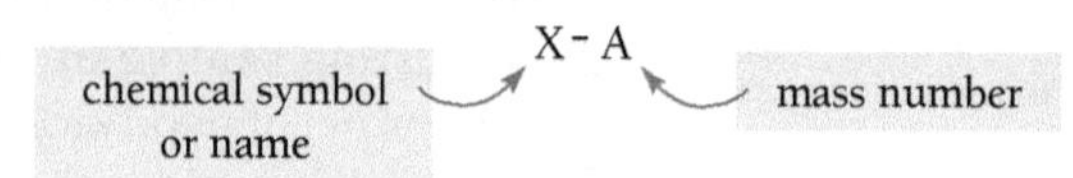

In this notation, the neon isotopes are:

Ne-20	neon-20
Ne-21	neon-21
Ne-22	neon-22

In general, mass number increases with increasing atomic number.

Notice that all isotopes of a given element have the same number of protons (otherwise they would be a different element). Notice also that the mass number is the *sum* of the number of protons and the number of neutrons. The number of neutrons in an isotope is the difference between the mass number and the atomic number.

Carbon has two naturally occurring isotopes: $^{12}_{6}C$ and $^{13}_{6}C$. Using circles to represent protons and squares to represent neutrons, draw the nucleus of each isotope.

EXAMPLE 4.7 Atomic Numbers, Mass Numbers, and Isotope Symbols

What are the atomic number (Z), mass number (A), and symbols of the carbon isotope that has 7 neutrons?

SOLUTION

You can determine that the atomic number (Z) of carbon is 6 (from the periodic table). This means that carbon atoms have 6 protons. The mass number (A) for the isotope with 7 neutrons is the sum of the number of protons and the number of neutrons.

$$A = 6 + 7 = 13$$

So, Z = 6, A = 13, and the symbols for the isotope are C-13 and $^{13}_{6}C$.

▶ **SKILLBUILDER 4.7 | Atomic Numbers, Mass Numbers, and Isotope Symbols**
What are the atomic number, mass number, and symbols for the chlorine isotope with 18 neutrons?

▶ **FOR MORE PRACTICE** Example 4.12; Problems 87, 89, 91, 92.

Interactive Worked Example Video 4.8

EXAMPLE **4.8** **Numbers of Protons and Neutrons from Isotope Symbols**

How many protons and neutrons are in the chromium isotope $^{52}_{24}Cr$?

	SOLUTION
The number of protons is equal to Z (lower left number).	$\#p^+ = Z = 24$
The number of neutrons is equal to A (upper left number) − Z (lower left number).	$\#n = A - Z$ $= 52 - 24$ $= 28$

▶ **SKILLBUILDER 4.8 | Numbers of Protons and Neutrons from Isotope Symbols**
How many protons and neutrons are in the potassium isotope $^{39}_{19}K$?

▶ **FOR MORE PRACTICE** Example 4.13; Problems 93, 94.

CONCEPTUAL ✔ CHECKPOINT 4.7

If an atom with a mass number of 27 has 14 neutrons, it is an isotope of which element?

(a) silicon
(b) aluminum
(c) cobalt
(d) niobium

CONCEPTUAL ✔ CHECKPOINT 4.8

Throughout this book, we represent atoms as spheres. For example, we represent a carbon atom by a black sphere as shown here. In light of the nuclear theory of the atom, when represented this way, would C-12 and C-13 look different? Why or why not?

4.9 Atomic Mass: The Average Mass of an Element's Atoms

▶ Calculate atomic mass from percent natural abundances and isotopic masses.

An important part of Dalton's atomic theory was that all atoms of a given element have the same mass. But as we just learned, the atoms of a given element may have different masses (because of isotopes). So Dalton was not completely correct. We can, however, calculate an average mass—called the **atomic mass**—for each element. You can find the atomic mass of each element in the periodic table directly beneath the element's symbol; it represents the average mass of the atoms that compose that element. For example, the periodic table lists the atomic mass of chlorine as 35.45 amu. Naturally occurring chlorine consists of 75.77% chlorine-35 (mass 34.97 amu) and 24.23% chlorine-37 (mass 36.97 amu). Its atomic mass is:

Some books use the term *average atomic mass* or *atomic weight* instead of simply *atomic mass*.

$$\text{atomic mass} = (0.7577 \times 34.97 \text{ amu}) + (0.2423 \times 36.97 \text{ amu})$$
$$= 35.45 \text{ amu}$$

CHEMISTRY IN THE ENVIRONMENT

Radioactive Isotopes at Hanford, Washington

The nuclei of the isotopes of a given element are not all equally stable. For example, naturally occurring lead is composed primarily of Pb-206, Pb-207, and Pb-208. Other isotopes of lead also exist, but their nuclei are unstable. Scientists can make some of these other isotopes, such as Pb-185, in the laboratory. However, within seconds Pb-185 atoms emit a few energetic subatomic particles from their nuclei and change into different isotopes of different elements (which are themselves unstable). These emitted subatomic particles are called **nuclear radiation**, and we call the isotopes that emit them **radioactive**. Nuclear radiation, always associated with unstable nuclei, can be harmful to humans and other living organisms because the emitted energetic particles interact with and damage biological molecules. Some isotopes, such as Pb-185, emit significant amounts of radiation only for a very short time. Others remain radioactive for a long time—in some cases millions or even billions of years.

The nuclear power and nuclear weapons industries produce by-products containing unstable isotopes of several different elements. Many of these isotopes emit nuclear radiation for a long time, and their disposal is an environmental problem. For example, in Hanford, Washington, which for 50 years produced fuel for nuclear weapons, 177 underground storage tanks contain 55 million gallons of highly radioactive nuclear waste. Certain radioactive isotopes within that waste will produce nuclear radiation for the foreseeable future. Unfortunately, some of the underground storage tanks in Hanford are aging, and leaks have allowed some of the waste to seep into the environment. While the danger from short-term external exposure to this waste is minimal, ingestion of the waste through contamination of drinking water or food supplies would pose significant health risks. Consequently, Hanford is now the site of the largest environmental cleanup project in U.S. history, involving 11,000 workers. The U.S. government expects the project to last for decades, and current costs are about $3 billion per year.

Radioactive isotopes are not always harmful, however, and many have beneficial uses. For example, physicians give technetium-99 (Tc-99) to patients to diagnose disease. The radiation emitted by Tc-99 helps doctors image internal organs or detect infection.

B4.2 CAN YOU ANSWER THIS? *Give the number of neutrons in each of the following isotopes: Pb-206, Pb-207, Pb-208, Pb-185, Tc-99.*

◀ Storage tanks at Hanford, Washington, contain 55 million gallons of high-level nuclear waste. Each tank pictured here holds 1 million gallons.

Notice that the atomic mass of chlorine is closer to 35 than 37 because naturally occurring chlorine contains more chlorine-35 atoms than chlorine-37 atoms. Notice also that when we use percentages in these calculations, we must always convert them to their decimal value. To convert a percentage to its decimal value, we divide by 100. For example:

$$75.77\% = 75.77/100 = 0.7577$$
$$24.23\% = 24.23/100 = 0.2423$$

In general, we calculate atomic mass according to the following equation:

$$\begin{aligned}\text{atomic mass} = &(\text{fraction of isotope 1} \times \text{mass of isotope 1}) + \\ &(\text{fraction of isotope 2} \times \text{mass of isotope 2}) + \\ &(\text{fraction of isotope 3} \times \text{mass of isotope 3}) + \ldots\end{aligned}$$

where the fractions of each isotope are the percent natural abundances converted to their decimal values. Atomic mass is useful because it allows us to assign a characteristic mass to each element, and, as we will see in Chapter 6, it allows us to quantify the number of atoms in a sample of that element.

EXAMPLE 4.9 Calculating Atomic Mass

Gallium has two naturally occurring isotopes: Ga-69 with mass 68.9256 amu and a natural abundance of 60.11%, and Ga-71 with mass 70.9247 amu and a natural abundance of 39.89%. Calculate the atomic mass of gallium.

	SOLUTION
Remember to convert the percent natural abundances into decimal form by dividing by 100.	$\text{fraction Ga-69} = \frac{60.11}{100} = 0.6011$ $\text{fraction Ga-71} = \frac{39.89}{100} = 0.3989$
Use the fractional abundances and the atomic masses of the isotopes to calculate the atomic mass according to the atomic mass definition.	$\text{atomic mass} = (0.6011 \times 68.9256\ \text{amu}) + (0.3989 \times 70.9247\ \text{amu})$ $= 41.4312\ \text{amu} + 28.2919\ \text{amu}$ $= 69.7231 = 69.72\ \text{amu}$

▶ **SKILLBUILDER 4.9 | Calculating Atomic Mass**

Magnesium has three naturally occurring isotopes with masses of 23.99, 24.99, and 25.98 amu and natural abundances of 78.99%, 10.00%, and 11.01%. Calculate the atomic mass of magnesium.

▶ **FOR MORE PRACTICE** Example 4.14; Problems 97, 98.

CONCEPTUAL CHECKPOINT 4.9

A fictitious element is composed of isotopes A and B with masses of 61.9887 and 64.9846 amu, respectively. The atomic mass of the element is 64.52. What can you conclude about the natural abundances of the two isotopes?

(a) The natural abundance of isotope A must be greater than the natural abundance of isotope B.

(b) The natural abundance of isotope B must be greater than the natural abundance of isotope A.

(c) The natural abundances of both isotopes must be about equal.

(d) Nothing can be concluded about the natural abundances of the two isotopes from the given information.

Chapter 4 in Review

MasteringChemistry™ provides end-of-chapter exercises, feedback-enriched tutorial problems, animations, and interactive activities to encourage problem solving practice and deeper understanding of key concepts and topics.

Self-Assessment Quiz

PEARSON eText 2.0

Q1. Which statement is not part of Dalton's atomic theory?
- **(a)** Each element is composed of indestructible particles called atoms.
- **(b)** All atoms of a given element have the same mass and other properties.
- **(c)** Atoms are themselves composed of protons, neutrons, and electrons.
- **(d)** Atoms combine in simple whole-number ratios to form compounds.

Q2. Which statement best summarizes the nuclear model of the atom that emerged from Rutherford's gold foil experiment?
- **(a)** The atom is composed of a dense core that contains most of its mass and all of its positive charge, while low-mass negatively charged particles compose most of its volume.
- **(b)** The atom is composed of a sphere of positive charge with many negatively charged particles within the sphere.
- **(c)** Most of the mass of the atom is evenly distributed throughout its volume.
- **(d)** All of the particles that compose an atom have exactly the same mass.

Q3. An ion composed of which of these particles would have a mass of approximately 16 amu and a charge of 2−?
- **(a)** 8 protons and 8 electrons
- **(b)** 8 protons, 8 neutrons, and 10 electrons
- **(c)** 8 protons, 8 neutrons, and 8 electrons
- **(d)** 8 protons, 8 neutrons, and 6 electrons

Q4. Which element is a main-group metal with an even atomic number?
(a) K **(b)** Ca **(c)** Cr **(d)** Se

Q5. Which element is a halogen?
(a) Ne **(b)** O **(c)** Ca **(d)** I

Q6. Which pair of elements has the most similar properties?
(a) Sr and Ba **(b)** S and Ar
(c) H and He **(d)** K and Se

Q7. Which element is a row 4 noble gas?
(a) Ne **(b)** Br **(c)** Zr **(d)** Kr

Q8. How many electrons does the predictable (most common) ion of fluorine contain?
(a) 1 **(b)** 4 **(c)** 9 **(d)** 10

Q9. How many neutrons does the Fe-56 isotope contain?
(a) 26 **(b)** 30 **(c)** 56 **(d)** 112

Q10. Determine the number of protons, neutrons, and electrons in $^{32}_{16}S^{2-}$.
- **(a)** 16 protons; 32 neutrons; and 18 electrons
- **(b)** 16 protons; 16 neutrons; and 18 electrons
- **(c)** 32 protons; 16 neutrons; and 2 electrons
- **(d)** 16 protons; 48 neutrons; and 16 electrons

Q11. What is the charge of the Cr ion that contains 21 electrons?
(a) 2− **(b)** 3− **(c)** 2+ **(d)** 3+

Q12. An element has four naturally occurring isotopes; the table lists the mass and natural abundance of each isotope. Find the atomic mass of the element.

Isotope	Mass (amu)	Natural Abundance
A	203.9730	1.4%
B	205.9744	24.1%
C	206.9758	22.1%
D	207.9766	52.4%

- **(a)** 207.2 amu
- **(b)** 2.072×10^4 amu
- **(c)** 206.2 amu
- **(d)** 206.5 amu

Answers: 1:c, 2:a, 3:b, 4:b, 5:d, 6:a, 7:d, 8:d, 9:b, 10:b, 11:d, 12:a

Chemical Principles

The Atomic Theory

Democritus and Leucippus, ancient Greek philosophers, were the first to assert that matter is ultimately composed of small, indestructible particles. It was not until 2000 years later, however, that John Dalton introduced a formal atomic theory stating that matter is composed of atoms; atoms of a given element have unique properties that distinguish them from atoms of other elements; and atoms combine in simple, whole-number ratios to form compounds.

Relevance

The concept of atoms is important because it explains the physical world. Humans and everything we see are made of atoms. To understand the physical world, we must begin by understanding atoms. Atoms are the key concept—they determine the properties of matter.

Discovery of the Atom's Nucleus

Rutherford's gold foil experiment probed atomic structure, and his results led to the nuclear model of the atom, which, with minor modifications to accommodate neutrons, is still valid today. According to this model, the atom is composed of protons and neutrons—which compose most of the atom's mass and are grouped together in a dense nucleus—and electrons—which compose most of the atom's volume. Protons and neutrons have similar masses (1 amu), while electrons have a much smaller mass (0.00055 amu).

We can understand why this is relevant by asking, what if it were otherwise? What if matter were *not* mostly empty space? While we cannot know for certain, it seems probable that such matter would not form the diversity of substances required for life—and then, of course, we would not be around to ask the question.

Charge

Protons and electrons both have electrical charge; the charge of the proton is 1+, and the charge of the electron is 1−. The neutron has no charge. When protons and electrons combine in atoms, their charges cancel.

Electrical charge is relevant to much of our modern world. Many of the machines and computers we depend on are powered by electricity, which is the movement of electrical charge.

The Periodic Table

The periodic table tabulates all known elements in order of increasing atomic number. The periodic table is arranged so that similar elements are grouped in columns. Columns of elements in the periodic table have similar properties and are called groups or families. Elements on the left side of the periodic table are metals and tend to lose electrons in their chemical changes. Elements on the upper right side of the periodic table are nonmetals and tend to gain electrons in their chemical changes. Elements between the two are metalloids.

The periodic table helps us organize the elements in ways that allow us to predict their properties. Helium, for example, is not toxic in small amounts because it is an inert gas—it does not react with anything. The gases in the column below it on the periodic table are also inert gases and form a family or group of elements called the noble gases. By tabulating the elements and grouping similar ones together, we begin to understand their properties.

Atomic Number

The characteristic that defines an element is the number of protons in the nuclei of its atoms; this number is the atomic number (Z).

Elements are the fundamental building blocks from which all compounds are made.

Ions

When an atom gains or loses electrons, it becomes an ion. Positively charged ions are cations, and negatively charged ions are anions. Cations and anions occur together so that matter is ordinarily charge-neutral.

Ions occur in many common compounds, such as sodium chloride.

Isotopes

While all atoms of a given element have the same number of protons, they do not necessarily have the same number of neutrons. Atoms of the same element with different numbers of neutrons are isotopes. Isotopes are characterized by their mass number (A), the sum of the number of protons and the number of neutrons in the nucleus.

Each naturally occurring sample of an element has the same percent natural abundance of each isotope. We can use these percentages, together with the mass of each isotope, to calculate the atomic mass of the element, a weighted average of the masses of the individual isotopes.

Isotopes are relevant because they influence atomic masses. To understand these masses, we must understand the presence and abundance of isotopes. In nuclear processes—processes in which the nuclei of atoms actually change—the presence of different isotopes becomes even more important.

Some isotopes are not stable—they lose subatomic particles and transform into other elements. The emission of subatomic particles by unstable nuclei is called radioactive decay. In many situations, such as in diagnosing and treating certain diseases, nuclear radiation is extremely useful. In other situations, such as in the disposal of radioactive waste, it can pose environmental problems.

Chemical Skills

Examples

LO: Determine ion charge from the number of protons and electrons (Section 4.7).

- Refer to the periodic table or the alphabetical list of elements to find the atomic number of the element; this number is equal to the number of protons.
- Use the ion charge equation to calculate charge.

$$\text{Ion charge} = \#p^+ - \#e^-$$

EXAMPLE 4.10 Determining Ion Charge from Numbers of Protons and Electrons

Determine the charge of a selenium ion with 36 electrons.

SOLUTION

Selenium is atomic number 34; therefore, it has 34 protons.

$$\text{ion charge} = 34 - 36 = 2-$$

LO: Determine the number of protons and electrons in an ion (Section 4.7).

- Refer to the periodic table or the alphabetical list of elements to find the atomic number of the element; this number is equal to the number of protons.
- Use the ion charge equation and substitute in the known values.

$$\text{Ion charge} = \#p^+ - \#e^-$$

- Solve the equation for the number of electrons.

EXAMPLE 4.11 Determining the Number of Protons and Electrons in an Ion

Find the number of protons and electrons in the O^{2-} ion.

SOLUTION

The atomic number of O is 8; therefore, it has 8 protons.

$$\text{ion charge} = \#p^+ - \#e^-$$

$$2- = 8 - \#e^-$$

$$\#e^- = 8 + 2 = 10$$

The ion has 8 protons and 10 electrons.

LO: Determine atomic numbers, mass numbers, and isotope symbols for an isotope (Section 4.8).

- Refer to the periodic table or the alphabetical list of elements to find the atomic number of the element.
- The mass number (A) is equal to the atomic number plus the number of neutrons.
- Write the symbol for the isotope by writing the symbol for the element with the mass number in the upper left corner and the atomic number in the lower left corner.
- The other symbol for the isotope is simply the chemical symbol followed by a hyphen and the mass number.

EXAMPLE 4.12 Determining Atomic Numbers, Mass Numbers, and Isotope Symbols for an Isotope

What are the atomic number (Z), mass number (A), and symbols for the iron isotope with 30 neutrons?

SOLUTION

The atomic number of iron is 26.

$$A = 26 + 30 = 56$$

The mass number is 56.

$${}^{56}_{26}\text{Fe}$$

Fe-56

LO: Determine number of protons and neutrons from isotope symbols (Section 4.8).

- The number of protons is equal to Z (lower left number).
- The number of neutrons is equal to

A (upper left number) − Z (lower left number)

EXAMPLE 4.13 Determining Number of Protons and Neutrons from Isotope Symbols

How many protons and neutrons are in ${}^{62}_{28}\text{Ni}$?

SOLUTION

28 protons

$$\#n = 62 - 28 = 34 \text{ neutrons}$$

LO: Calculate atomic mass from percent natural abundances and isotopic masses (Section 4.9).

- Convert the natural abundances from percent to decimal values by dividing by 100.
- Find the atomic mass by multiplying the fractions of each isotope by its respective mass and summing them.
- Round to the correct number of significant figures.
- Check your work.

EXAMPLE **4.14** **Calculating Atomic Mass from Percent Natural Abundances and Isotopic Masses**

Copper has two naturally occurring isotopes: Cu-63 with mass 62.9395 amu and a natural abundance of 69.17%, and Cu-65 with mass 64.9278 amu and a natural abundance of 30.83%. Calculate the atomic mass of copper.

SOLUTION

$$\text{fraction Cu-63} = \frac{69.17}{100} = 0.6917$$

$$\text{fraction Cu-65} = \frac{30.83}{100} = 0.3083$$

$$\begin{aligned}\text{atomic mass} &= (0.6917 \times 62.9395 \text{ amu}) + (0.3083 \times 64.9278)\\ &= 43.5\underline{3}53 \text{ amu} + 20.0\underline{1}72 \text{ amu}\\ &= 63.5\underline{5}25 \text{ amu}\\ &= 63.55 \text{ amu}\end{aligned}$$

Key Terms

alkali metals [4.6]
alkaline earth metals [4.6]
anion [4.7]
atom [4.1]
atomic mass [4.9]
atomic mass unit (amu) [4.4]
atomic number (Z) [4.5]
cation [4.7]
charge [4.4]
chemical symbol [4.5]
electron [4.3]
family (of elements) [4.6]
group (of elements) [4.6]
halogens [4.6]
ion [4.7]
isotope [4.8]
main-group elements [4.6]
mass number (A) [4.8]
metalloids [4.6]
metals [4.6]
neutron [4.3]
noble gases [4.6]
nonmetals [4.6]
nuclear radiation [4.9]
nuclear theory of the atom [4.3]
nucleus (of an atom) [4.3]
percent natural abundance [4.8]
periodic law [4.6]
periodic table [4.6]
proton [4.3]
radioactive [4.9]
semiconductor [4.6]
transition elements [4.6]
transition metals [4.6]

Exercises

Questions

1. What did Democritus contribute to our modern understanding of matter?
2. What are three main ideas in Dalton's atomic theory?
3. Describe Rutherford's gold foil experiment and the results of that experiment. How did these results refute the plum-pudding model of the atom?
4. What are the main ideas in the nuclear theory of the atom?
5. List the three subatomic particles and their properties.
6. What is electrical charge?
7. Is matter usually charge-neutral? How would matter be different if it were not charge-neutral?
8. What does the atomic number of an element specify?
9. What is a chemical symbol?
10. List some examples of how elements were named.
11. What was Dmitri Mendeleev's main contribution to our modern understanding of chemistry?
12. What is the main idea in the periodic law?
13. How is the periodic table organized?
14. What are the properties of metals? Where are metals found on the periodic table?
15. What are the properties of nonmetals? Where are nonmetals found on the periodic table?
16. Where on the periodic table are metalloids found?
17. What is a family or group of elements?
18. Locate each group of elements on the periodic table and list its group number.
    (a) alkali metals
    (b) alkaline earth metals
    (c) halogens
    (d) noble gases
19. What is an ion?
20. What is an anion? What is a cation?

21. Locate each group on the periodic table and list the charge of the ions it tends to form.
    (a) Group 1A
    (b) Group 2A
    (c) Group 3A
    (d) Group 6A
    (e) Group 7A

22. What are isotopes?

23. What is the percent natural abundance of isotopes?

24. What is the mass number of an isotope?

25. What notations are commonly used to specify isotopes? What do each of the numbers in these symbols mean?

26. What is the atomic mass of an element?

Problems

ATOMIC AND NUCLEAR THEORY

27. Which statements are *inconsistent* with Dalton's atomic theory as it was originally stated? Explain your answers.
    (a) All carbon atoms are identical.
    (b) Helium atoms can be split into two hydrogen atoms.
    (c) An oxygen atom combines with 1.5 hydrogen atoms to form water molecules.
    (d) Two oxygen atoms combine with a carbon atom to form carbon dioxide molecules.

28. Which statements are *consistent* with Dalton's atomic theory as it was originally stated? Explain your answers.
    (a) Calcium and titanium atoms have the same mass.
    (b) Neon and argon atoms are the same.
    (c) All cobalt atoms are identical.
    (d) Sodium and chlorine atoms combine in a 1:1 ratio to form sodium chloride.

29. Which statements are *inconsistent* with Rutherford's nuclear theory as it was originally stated? Explain your answers.
    (a) Helium atoms have two protons in the nucleus and two electrons outside the nucleus.
    (b) Most of the volume of hydrogen atoms is due to the nucleus.
    (c) Aluminum atoms have 13 protons in the nucleus and 22 electrons outside the nucleus.
    (d) The majority of the mass of nitrogen atoms is due to their 7 electrons.

30. Which statements are *consistent* with Rutherford's nuclear theory as it was originally stated? Explain your answers.
    (a) Atomic nuclei are small compared to the size of atoms.
    (b) The volume of an atom is mostly empty space.
    (c) Neutral potassium atoms contain more protons than electrons.
    (d) Neutral potassium atoms contain more neutrons than protons.

31. If atoms are mostly empty space and atoms compose all ordinary matter, why does solid matter seem to have no space within it?

32. Rutherford's experiment indicated that matter was not as uniform as it appears. What part of his experimental results implied this idea? Explain.

PROTONS, NEUTRONS, AND ELECTRONS

33. Which statement about electrons is true?
    (a) Electrons attract one another.
    (b) Electrons are repelled by protons.
    (c) Some electrons have a charge of 1− and some have no charge.
    (d) Electrons are much lighter than neutrons.

34. Which statement about electrons is false?
    (a) Most atoms have more electrons than protons.
    (b) Electrons have a charge of 1−.
    (c) If an atom has an equal number of protons and electrons, it will be charge-neutral.
    (d) Electrons experience an attraction to protons.

35. Which statement about protons is true?
    (a) Protons have twice the mass of neutrons.
    (b) Protons have the same magnitude of charge as electrons but are opposite in sign.
    (c) Most atoms have more protons than electrons.
    (d) Protons have a charge of 1−.

36. Which statement about protons is false?
    (a) Protons have about the same mass as neutrons.
    (b) Protons have about the same mass as electrons.
    (c) All atoms have protons.
    (d) Protons have the same magnitude of charge as neutrons but are opposite in sign.

37. How many electrons would it take to equal the mass of a proton?

38. A helium nucleus has two protons and two neutrons. How many electrons would it take to equal the mass of a helium nucleus?

39. What mass of electrons is required to neutralize the charge of 1.0 g of protons?

40. What mass of protons is required to neutralize the charge of 1.0 g of electrons?

ELEMENTS, SYMBOLS, AND NAMES

41. Find the atomic number (Z) for each element.
(a) Fr
(b) Kr
(c) Pa
(d) Ge
(e) Al

42. Find the atomic number (Z) for each element.
(a) Si
(b) W
(c) Ni
(d) Rn
(e) Sr

43. How many protons are in the nucleus of an atom of each element?
(a) Ar
(b) Sn
(c) Xe
(d) O
(e) Tl

44. How many protons are in the nucleus of an atom of each element?
(a) Ti
(b) Li
(c) U
(d) Br
(e) F

45. List the symbol and atomic number of each element.
(a) carbon
(b) nitrogen
(c) sodium
(d) potassium
(e) copper

46. List the symbol and atomic number of each element.
(a) boron
(b) neon
(c) silver
(d) mercury
(e) curium

47. List the name and the atomic number of each element.
(a) Mn
(b) Ag
(c) Au
(d) Pb
(e) S

48. List the name and the atomic number of each element.
(a) Y
(b) N
(c) Ne
(d) K
(e) Mo

49. Fill in the blanks to complete the table.

Element Name	Element Symbol	Atomic Number
___	Au	79
Tin	___	___
___	As	___
Copper	___	29
___	Fe	___
___	___	80

50. Fill in the blanks to complete the table.

Element Name	Element Symbol	Atomic Number
___	Al	13
Iodine	___	___
___	Sb	___
Sodium	___	___
___	Rn	86
___	___	82

THE PERIODIC TABLE

51. Classify each element as a metal, nonmetal, or metalloid.
(a) Sr
(b) Mg
(c) F
(d) N
(e) As

52. Classify each element as a metal, nonmetal, or metalloid.
(a) Na
(b) Ge
(c) Si
(d) Br
(e) Ag

53. Which elements would you expect to lose electrons in chemical changes?
(a) potassium
(b) sulfur
(c) fluorine
(d) barium
(e) copper

54. Which elements would you expect to gain electrons in chemical changes?
(a) nitrogen
(b) iodine
(c) tungsten
(d) strontium
(e) gold

55. Which elements are main-group elements?
(a) Te
(b) K
(c) V
(d) Re
(e) Ag

56. Which elements are *not* main-group elements?
(a) Al
(b) Br
(c) Mo
(d) Cs
(e) Pb

57. Which elements are alkaline earth metals?
(a) sodium
(b) aluminum
(c) calcium
(d) barium
(e) lithium

58. Which elements are alkaline earth metals?
(a) rubidium
(b) tungsten
(c) magnesium
(d) cesium
(e) beryllium

59. Which elements are alkali metals?
(a) barium
(b) sodium
(c) gold
(d) tin
(e) rubidium

60. Which elements are alkali metals?
(a) scandium
(b) iron
(c) potassium
(d) lithium
(e) cobalt

61. Classify each element as a halogen, a noble gas, or neither.
(a) Cl
(b) Kr
(c) F
(d) Ga
(e) He

62. Classify each element as a halogen, a noble gas, or neither.
(a) Ne
(b) Br
(c) S
(d) Xe
(e) I

63. To what group number does each element belong?
(a) oxygen
(b) aluminum
(c) silicon
(d) tin
(e) phosphorus

64. To what group number does each element belong?
(a) germanium
(b) nitrogen
(c) sulfur
(d) carbon
(e) boron

65. Which element do you expect to be most like sulfur? Why?
(a) nitrogen
(b) oxygen
(c) fluorine
(d) lithium
(e) potassium

66. Which element do you expect to be most like magnesium? Why?
(a) potassium
(b) silver
(c) bromine
(d) calcium
(e) lead

67. Which pair of elements do you expect to be most similar? Why?
(a) Si and P
(b) Cl and F
(c) Na and Mg
(d) Mo and Sn
(e) N and Ni

68. Which pair of elements do you expect to be most similar? Why?
(a) Ti and Ga
(b) N and O
(c) Li and Na
(d) Ar and Br
(e) Ge and Ga

69. Which element is a main-group nonmetal?
(a) K
(b) Fe
(c) Sn
(d) S

70. Which element is a row 5 transition element?
(a) Sr
(b) Pd
(c) P
(d) V

71. Fill in the blanks to complete the table.

Chemical Symbol	Group Number	Group Name	Metal or Nonmetal
K	___	___	metal
Br	___	halogens	___
Sr	___	___	___
He	8A	___	___
Ar	___	___	___

72. Fill in the blanks to complete the table.

Chemical Symbol	Group Number	Group Name	Metal or Nonmetal
Cl	7A	___	___
Ca	___	___	metal
Xe	___	___	nonmetal
Na	___	alkali metal	___
F	___	___	___

IONS

73. Complete each ionization equation.
(a) $Na \longrightarrow Na^+ +$ ___
(b) $O + 2e^- \longrightarrow$ ___
(c) $Ca \longrightarrow Ca^{2+} +$ ___
(d) $Cl + e^- \longrightarrow$ ___

74. Complete each ionization equation.
(a) $Mg \longrightarrow$ ___ $+ 2e^-$
(b) $Ba \longrightarrow Ba^{2+} +$ ___
(c) $I + e^- \longrightarrow$ ___
(d) $Al \longrightarrow$ ___ $+ 3e^-$

75. Determine the charge of each ion.
(a) oxygen ion with 10 electrons
(b) aluminum ion with 10 electrons
(c) titanium ion with 18 electrons
(d) iodine ion with 54 electrons

76. Determine the charge of each ion.
(a) tungsten ion with 68 electrons
(b) tellurium ion with 54 electrons
(c) nitrogen ion with 10 electrons
(d) barium ion with 54 electrons

77. Determine the number of protons and electrons in each ion.
(a) Na^+
(b) Ba^{2+}
(c) O^{2-}
(d) Co^{3+}

78. Determine the number of protons and electrons in each ion.
(a) Al^{3+}
(b) S^{2-}
(c) I^-
(d) Ag^+

79. Determine whether each statement is true or false. If false, correct it.
(a) The Ti^{2+} ion contains 22 protons and 24 electrons.
(b) The I^- ion contains 53 protons and 54 electrons.
(c) The Mg^{2+} ion contains 14 protons and 12 electrons.
(d) The O^{2-} ion contains 8 protons and 10 electrons.

80. Determine whether each statement is true or false. If false, correct it.
(a) The Fe^{2+} ion contains 29 protons and 26 electrons.
(b) The Cs^+ ion contains 55 protons and 56 electrons.
(c) The Se^{2-} ion contains 32 protons and 34 electrons.
(d) The Li^+ ion contains 3 protons and 2 electrons.

81. Predict the ion formed by each element.
(a) Rb
(b) K
(c) Al
(d) O

82. Predict the ion formed by each element.
(a) F
(b) N
(c) Mg
(d) Na

83. Predict how many electrons each element will most likely gain or lose.
(a) Ga
(b) Li
(c) Br
(d) S

84. Predict how many electrons each element will most likely gain or lose.
(a) I
(b) Ba
(c) Cs
(d) Se

85. Fill in the blanks to complete the table.

Symbol	Ion Commonly Formed	Number of Electrons in Ion	Number of Protons in Ion
Te	___	54	___
In	___	___	49
Sr	Sr^{2+}	___	___
___	Mg^{2+}	___	12
Cl	___	___	___

86. Fill in the blanks to complete the table.

Symbol	Ion Commonly Formed	Number of Electrons in Ion	Number of Protons in Ion
F	___	___	9
___	Be^{2+}	2	___
Br	___	36	___
Al	___	___	13
O	___	___	___

ISOTOPES

87. Determine the atomic number and mass number for each isotope.
(a) the hydrogen isotope with 2 neutrons
(b) the chromium isotope with 28 neutrons
(c) the calcium isotope with 22 neutrons
(d) the tantalum isotope with 109 neutrons

88. How many neutrons are in an atom with each atomic number and mass number?
(a) $Z = 28, A = 59$
(b) $Z = 92, A = 235$
(c) $Z = 21, A = 46$
(d) $Z = 18, A = 42$

89. Write isotopic symbols in the form $^{A}_{Z}X$ for each isotope.
(a) the oxygen isotope with 8 neutrons
(b) the fluorine isotope with 10 neutrons
(c) the sodium isotope with 12 neutrons
(d) the aluminum isotope with 14 neutrons

90. Write isotopic symbols in the form X-A (for example, C-13) for each isotope.
(a) the iodine isotope with 74 neutrons
(b) the phosphorus isotope with 16 neutrons
(c) the uranium isotope with 234 neutrons
(d) the argon isotope with 22 neutrons

91. Write the symbol for each isotope in the form $^{A}_{Z}X$.
(a) cobalt-60
(b) neon-22
(c) iodine-131
(d) plutonium-244

92. Write the symbol for each isotope in the form $^{A}_{Z}X$.
(a) U-235
(b) V-52
(c) P-32
(d) Xe-144

93. Determine the number of protons and neutrons in each isotope.
(a) $^{23}_{11}Na$
(b) $^{266}_{88}Ra$
(c) $^{208}_{32}Pb$
(d) $^{14}_{7}N$

94. Determine the number of protons and neutrons in each isotope.
(a) $^{33}_{15}P$
(b) $^{40}_{19}K$
(c) $^{222}_{86}Rn$
(d) $^{99}_{43}Tc$

95. Carbon-14, present within living organisms and substances derived from living organisms, is often used to establish the age of fossils and artifacts. Determine the number of protons and neutrons in a carbon-14 isotope and write its symbol in the form $^{A}_{Z}X$.

96. Plutonium-239 is used in nuclear bombs. Determine the number of protons and neutrons in plutonium-239 and write its symbol in the form $^{A}_{Z}X$.

ATOMIC MASS

97. Rubidium has two naturally occurring isotopes: Rb-85 with mass 84.9118 amu and a natural abundance of 72.17%, and Rb-87 with mass 86.9092 amu and a natural abundance of 27.83%. Calculate the atomic mass of rubidium.

98. Silicon has three naturally occurring isotopes: Si-28 with mass 27.9769 amu and a natural abundance of 92.21%, Si-29 with mass 28.9765 amu and a natural abundance of 4.69%, and Si-30 with mass 29.9737 amu and a natural abundance of 3.10%. Calculate the atomic mass of silicon.

99. Bromine has two naturally occurring isotopes (Br-79 and Br-81) and an atomic mass of 79.904 amu.
(a) If the natural abundance of Br-79 is 50.69%, what is the natural abundance of Br-81?
(b) If the mass of Br-81 is 80.9163 amu, what is the mass of Br-79?

100. Silver has two naturally occurring isotopes (Ag-107 and Ag-109).
(a) Use the periodic table to find the atomic mass of silver.
(b) If the natural abundance of Ag-107 is 51.84%, what is the natural abundance of Ag-109?
(c) If the mass of Ag-107 is 106.905 amu, what is the mass of Ag-109?

101. An element has two naturally occurring isotopes. Isotope 1 has a mass of 120.9038 amu and a relative abundance of 57.4%, and isotope 2 has a mass of 122.9042 amu and a relative abundance of 42.6%. Find the atomic mass of this element and, referring to the periodic table, identify it.

102. Copper has two naturally occurring isotopes. Cu-63 has a mass of 62.939 amu and relative abundance of 69.17%. Use the atomic weight of copper to determine the mass of the other copper isotope.

Cumulative Problems

103. Electrical charge is sometimes reported in coulombs (C). On this scale, 1 electron has a charge of -1.6×10^{-19} C. Suppose your body acquires −125 mC (millicoulombs) of charge on a dry day. How many excess electrons has it acquired? (*Hint*: Use the charge of an electron in coulombs as a conversion factor between charge and number of electrons.)

104. How many excess protons are in a positively charged object with a charge of +398 mC (millicoulombs)? The charge of 1 proton is $+1.6 \times 10^{-19}$ C. (*Hint*: Use the charge of the proton in coulombs as a conversion factor between charge and number of protons.)

105. The hydrogen atom contains 1 proton and 1 electron. The radius of the proton is approximately 1.0 fm (femtometer), and the radius of the hydrogen atom is approximately 53 pm (picometers). Calculate the volume of the nucleus and the volume of the atom for hydrogen. What percentage of the hydrogen atom's volume does the nucleus occupy? (*Hint*: Convert both given radii to m, and then calculate their volumes using the formula for the volume of a sphere, which is $V = \frac{4}{3}\pi r^3$.)

106. Carbon-12 contains 6 protons and 6 neutrons. The radius of the nucleus is approximately 2.7 fm, and the radius of the atom is approximately 70 pm. Calculate the volume of the nucleus and the volume of the atom. What percentage of the carbon atom's volume does the nucleus occupy? (*Hint*: Convert both given radii to m, and then calculate their volumes using the formula for the volume of a sphere, which is $V = \frac{4}{3}\pi r^3$.)

107. Prepare a table like Table 4.2 for the four different isotopes of Sr that have the natural abundances and masses listed here.

Sr-84	0.56%	83.9134 amu
Sr-86	9.86%	85.9093 amu
Sr-87	7.00%	86.9089 amu
Sr-88	82.58%	87.9056 amu

Use your table and the listed atomic masses to calculate the atomic mass of strontium.

108. Determine the number of protons and neutrons in each isotope of chromium and use the listed natural abundances and masses to calculate its atomic mass.

Cr-50	4.345%	49.9460 amu
Cr-52	83.79%	51.9405 amu
Cr-53	9.50%	52.9407 amu
Cr-54	2.365%	53.9389 amu

109. Fill in the blanks to complete the table.

Symbol	Z	A	Number of Protons	Number of Electrons	Number of Neutrons	Charge
Zn^{2+}	___	___	___	___	34	2+
___	25	55	___	22	___	___
___	___	___	15	15	16	___
O^{2-}	___	16	___	___	___	2−
___	___	___	16	18	18	___

110. Fill in the blanks to complete the table.

Symbol	Z	A	Number of Protons	Number of Electrons	Number of Neutrons	Charge
Mg^{2+}	___	25	___	___	13	2+
___	22	48	___	18	___	___
___	16	___	___	___	16	2−
Ga^{3+}	___	71	___	___	___	___
___	___	___	82	80	125	___

111. Europium has two naturally occurring isotopes: Eu-151 with a mass of 150.9198 amu and a natural abundance of 47.8%, and Eu-153. Use the atomic mass of europium to find the mass and natural abundance of Eu-153.

112. Rhenium has two naturally occurring isotopes: Re-185 with a natural abundance of 37.40% and Re-187 with a natural abundance of 62.60%. The sum of the masses of the two isotopes is 371.9087 amu. Find the masses of the individual isotopes.

113. Chapter 1 describes the difference between observations, laws, and theories. Cite two examples of theories from Chapter 4 and explain why they are theories.

114. Chapter 1 describes the difference between observations, laws, and theories. Cite one example of a law from Chapter 4 and explain why it is a law.

115. The atomic mass of fluorine is 19.00 amu, and all fluorine atoms in a naturally occurring sample of fluorine have this mass. The atomic mass of chlorine is 35.45 amu, but no chlorine atoms in a naturally occurring sample of chlorine have this mass. Provide an explanation for the difference.

116. The atomic mass of germanium is 72.61 amu. Is it likely that any individual germanium atoms have a mass of 72.61 amu?

117. Copper has only two naturally occurring isotopes, Cu-63 and Cu-65. The mass of Cu-63 is 62.9396 amu, and the mass of Cu-65 is 64.9278 amu. Use the atomic mass of copper to determine the relative abundance of each isotope in a naturally occurring sample. (*Hint*: The relative abundances of the two isotopes sum to 100%.)

118. Gallium has only two naturally occurring isotopes, Ga-69 and Ga-71. The mass of Ga-69 is 68.9256 amu, and the mass of Ga-71 is 70.9247 amu. Use the atomic mass of gallium to determine the relative abundance of each isotope in a naturally occurring sample.

Highlight Problems

119. The figure shown here is a representation of 50 atoms of a fictitious element with the symbol Nt and atomic number 120. Nt has three isotopes represented by the following colors: Nt-304 (red), Nt-305 (blue), and Nt-306 (green).

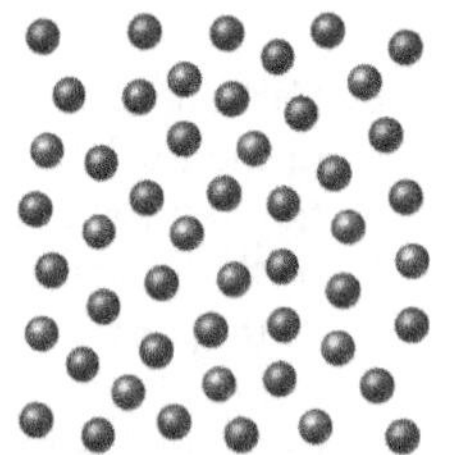

(a) Assuming that the figure is statistically representative of naturally occurring Nt, what is the percent natural abundance of each Nt isotope?

(b) Use the listed masses of each isotope to calculate the atomic mass of Nt. Then draw a box for the element similar to the boxes for each element shown in the periodic table in the inside front cover of this book. Make sure your box includes the atomic number, symbol, and atomic mass. (Assume that the percentages from part a are correct to four significant figures.)

Nt-304	303.956 amu
Nt-305	304.962 amu
Nt-306	305.978 amu

120. Neutron stars are believed to be composed of solid nuclear matter, primarily neutrons.

(a) If the radius of a neutron is 1.0×10^{-13} cm, calculate its density in g/cm^3.

(volume of a sphere $= \frac{4}{3}\pi r^3$)

(b) Assuming that a neutron star has the same density as a neutron, calculate the mass in kilograms of a small piece of a neutron star the size of a spherical pebble with a radius of 0.10 mm.

Questions for Group Work

Discuss these questions with the group and record your consensus answer.

121. Complete the following table.

Particle	Mass (amu)	Charge	In the nucleus? (yes/no)	# in ^{32}S atom	# in ^{79}Br$^-$ ion
Proton					
Neutron					
Electron					

122. Make a sketch of an oxygen atom. Include the correct number of protons, electrons, and neutrons for the most abundant isotope. Use the following symbols: proton = •, neutron = o, electron = •.

123. The table at right includes data similar to that used by Mendeleev when he made the periodic table. Write on a small card the symbol, atomic mass, and a stable compound formed by each element. Arrange your cards in order of increasing atomic mass. Do you observe any repeating patterns? Describe any patterns you observe. (*Hint*: There is one missing element somewhere in the pattern.)

Element	Atomic Mass	Stable Compound	Element	Atomic Mass	Stable Compound
Be	9	$BeCl_2$	O	16	H_2O
S	32	H_2S	Ga	69.7	GaH_3
F	19	F_2	As	75	AsF_3
Ca	40	$CaCl_2$	C	12	CH_4
Li	7	LiCl	K	39	KCl
Si	28	SiH_4	Mg	24.3	$MgCl_2$
Cl	35.4	Cl_2	Se	79	H_2Se
B	10.8	BH_3	Al	27	AlH_3
Ge	72.6	GeH_4	Br	80	Br_2
N	14	NF_3	Na	23	NaCl

124. Arrange the cards from Question 123 so that mass increases from left to right and elements with similar properties are above and below each other. Copy the periodic table you have invented onto a piece of paper. There is one element missing. Predict its mass and a stable compound it might form.

Data Interpretation and Analysis

125. The graph at the right shows the atomic radius for the 19 elements in the periodic table.
   (a) Describe the trend in atomic radius in going from H (atomic number 1) to K (atomic number 19).
   (b) Find the three elements represented with blue dots on a periodic table. What do their placements in the table have in common?
   (c) Find the three elements represented with red dots on a periodic table. What do their placements in the table have in common?
   (d) Based on the graph, what is the radius of C?

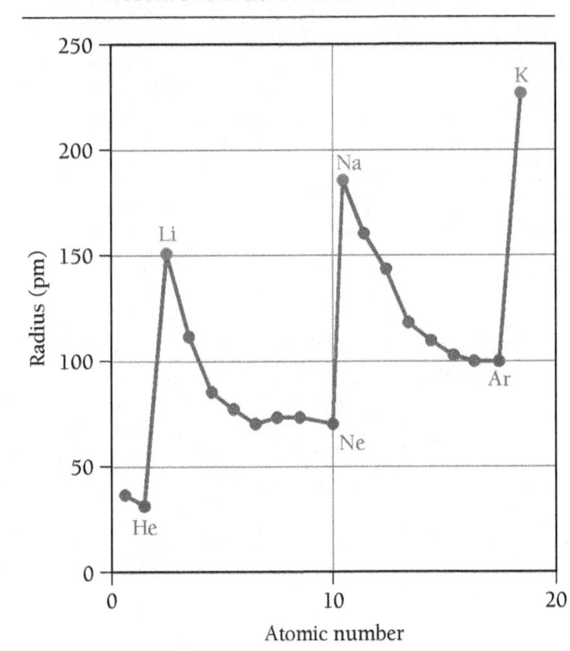

Answers to Skillbuilder Exercises

Skillbuilder 4.1 (a) sodium, 11
(b) nickel, 28
(c) phosphorus, 15
(d) tantalum, 73

Skillbuilder 4.2 (a) nonmetal
(b) nonmetal
(c) metal
(d) metalloid

Skillbuilder 4.3 (a) alkali metal, Group 1A
(b) Group 3A
(c) halogen, Group 7A
(d) noble gas, Group 8A

Skillbuilder 4.4 (a) 2+
(b) 1−
(c) 3−

Skillbuilder 4.5 16 protons, 18 electrons

Skillbuilder 4.6 K^+ and Se^{2-}

Skillbuilder 4.7 Z = 17, A = 35, Cl-35, and $^{35}_{17}Cl$

Skillbuilder 4.8 19 protons, 20 neutrons

Skillbuilder 4.9 24.31 amu

Answers to Conceptual Checkpoints

4.1 (c) The mass in amu is approximately equal to the number of protons plus the number of neutrons. In order to be charge-neutral, the number of protons must equal the number of electrons.

4.2 (d) Lead is a metal (see Figure 4.12) and a main-group element (see Figure 4.13).

4.3 (b) All of the metalloids are main-group elements (see Figures 4.12 and 4.13).

4.4 O^{2-}

4.5 (a) Both of these ions have 10 electrons.

4.6

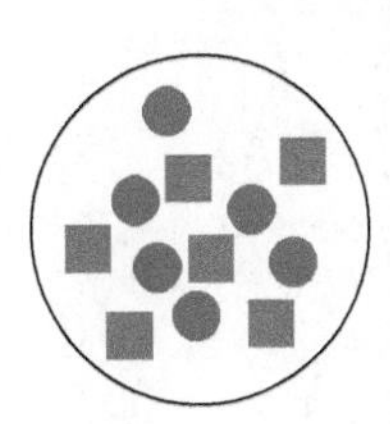
C-12 nucleus

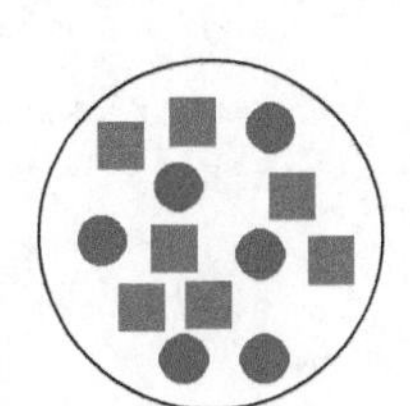
C-13 nucleus

4.7 (b) This atom must have (27 − 14) = 13 protons; the element with an atomic number of 13 is Al.

4.8 The isotopes C-12 and C-13 would not look different in this representation of atoms because the only difference between the two isotopes is that C-13 has an extra neutron in the nucleus. The illustration represents the whole atom and does not attempt to illustrate its nucleus. Because the nucleus of an atom is miniscule compared to the size of the atom itself, the extra neutron would not affect the size of the atom.

4.9 (b) The natural abundance of isotope B must be greater than the natural abundance of isotope A because the atomic mass is closer to the mass of isotope B than to the mass of isotope A.

Sucrose
AND
H_2O
SUCROSE
$C_{12}H_{22}O_{11}$
O_2
CAFFEINE
23RD
DUE
MASTERING
CHEMISTRY
CALL
Sucrose
Sucrose

5 Molecules and Compounds

Almost all aspects of life are engineered at the molecular level, and without understanding molecules, we can only have a very sketchy understanding of life itself.

—Francis Harry Compton Crick (1916–2004)

CHAPTER OUTLINE

5.1 Sugar and Salt

Sodium, a shiny metal (◀ FIGURE 5.1) that dulls almost instantly upon exposure to air, is extremely reactive and poisonous. If you were to consume any appreciable amount of elemental sodium, you would need immediate medical help. Chlorine, a pale yellow gas (▼ FIGURE 5.2), is equally reactive and poisonous. Yet the compound formed from these two elements, sodium chloride, is the relatively harmless flavor enhancer that we call table salt (▼ FIGURE 5.3). When elements combine to form compounds, their properties completely change.

▲ FIGURE 5.1 **Elemental sodium** Sodium is an extremely reactive metal that dulls almost instantly upon exposure to air.

▶ FIGURE 5.2 **Elemental chlorine** Chlorine is a yellow gas with a pungent odor. It is highly reactive and poisonous.

▲ FIGURE 5.3 **Sodium chloride** The compound formed by sodium and chlorine is table salt.

◀ Ordinary table sugar is a compound called sucrose. A sucrose molecule, such as the one shown here, contains carbon, hydrogen, and oxygen atoms. The properties of sucrose are, however, very different from those of carbon (also shown in the form of graphite), hydrogen, and oxygen. The properties of a compound are, in general, different from the properties of the elements that compose it.

Consider also ordinary sugar. Sugar is a compound composed of carbon, hydrogen, and oxygen. Each of these elements has its own unique properties. Carbon is most familiar to us as the graphite found in pencils or as the diamonds in jewelry. Hydrogen is an extremely flammable gas used as a fuel for rocket engines, and oxygen is one of the gases that compose air. When these three elements combine to form sugar, however, a sweet, white, crystalline solid results.

In Chapter 4, you learned how protons, neutrons, and electrons combine to form different elements, each with its own properties and its own chemistry, each different from the other. In this chapter, you will learn how these elements combine with each other to form different compounds, each with its own properties and its own chemistry, each different from all the others and different from the elements that compose it. This is the great wonder of nature: how from such simplicity—protons, neutrons, and electrons—we get such great complexity. It is exactly this complexity that makes life possible. Life could not exist with just 91 different elements if they did not combine to form compounds. It takes compounds in all of their diversity to make living organisms.

5.2 Compounds Display Constant Composition

▶ Restate and apply the law of constant composition.

Although some of the substances you encounter in everyday life are elements, most are not—they are compounds. Free atoms are rare in nature. As you learned in Chapter 3, a compound is different from a mixture of elements. In a compound, the elements combine in fixed, definite proportions, whereas in a mixture, they can have any proportions whatsoever. Consider the difference between a mixture of hydrogen and oxygen gas (▼ FIGURE 5.4) and the compound water (▼ FIGURE 5.5). A mixture of hydrogen and oxygen gas can contain any proportions of hydrogen and oxygen. Water, on the other hand, is composed of water molecules that consist of two hydrogen atoms bonded to one oxygen atom. Consequently, water has a definite proportion of hydrogen to oxygen.

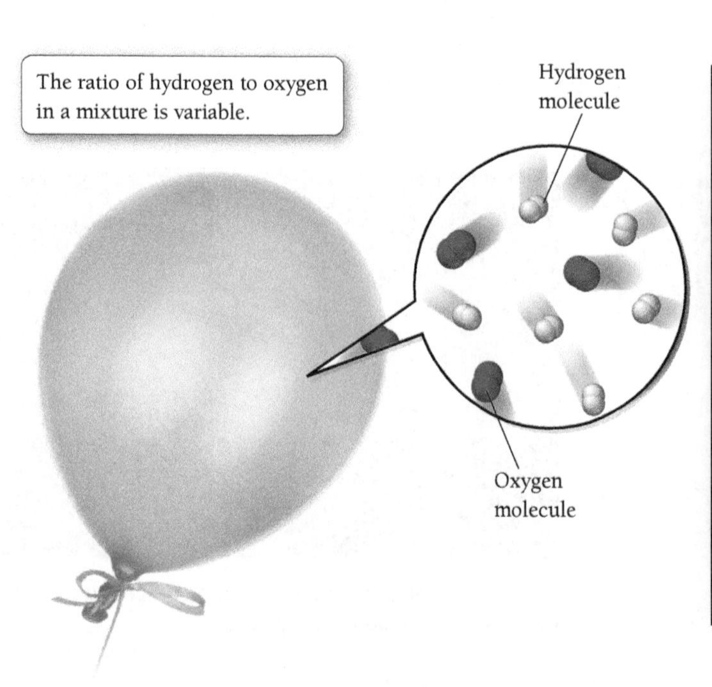
The ratio of hydrogen to oxygen in a mixture is variable.
Hydrogen molecule
Oxygen molecule

▲ FIGURE 5.4 **A mixture** This balloon is filled with a mixture of hydrogen and oxygen gas. The relative amounts of hydrogen and oxygen are variable. We could easily add either more hydrogen or more oxygen to the balloon.

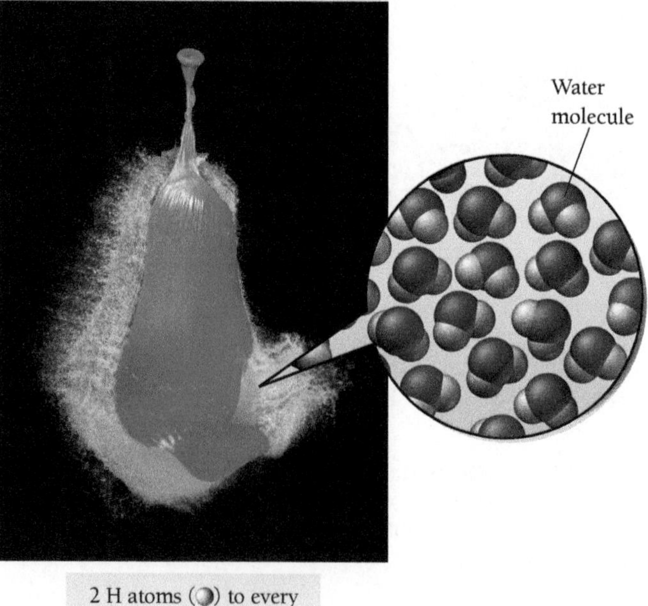
The ratio of hydrogen to oxygen in water is fixed.
Water molecule
2 H atoms ( ) to every
1 O atom ( )

▲ FIGURE 5.5 **A chemical compound** This balloon is filled with water, composed of molecules that have a fixed ratio of hydrogen to oxygen.

The first chemist to formally state the idea that elements combine in fixed proportions to form compounds was Joseph Proust (1754–1826) in the **law of constant composition**, which states:

> All samples of a given compound have the same proportions of their constituent elements.

For example, if we decompose an 18.0 g sample of water, we get 16.0 g of oxygen and 2.0 g of hydrogen, or an oxygen-to-hydrogen mass ratio of

Even though atoms combine in whole-number ratios, their mass ratios are not necessarily whole numbers.

$$\text{mass ratio} = \frac{16.0\ \text{g O}}{2.0\ \text{g H}} = 8.0 \quad \text{or} \quad 8.0:1$$

This is true of any sample of pure water, no matter what its origin. The law of constant composition applies not only to water but to every compound. If we decompose a 17.0 g sample of ammonia, a compound composed of nitrogen and hydrogen, we get 14.0 g of nitrogen and 3.0 g of hydrogen, or a nitrogen-to-hydrogen mass ratio of:

$$\text{mass ratio} = \frac{14.0\ \text{g N}}{3.0\ \text{g H}} = 4.7 \quad \text{or} \quad 4.7:1$$

Again, this ratio is the same for every sample of ammonia—the composition of each compound is constant.

EXAMPLE 5.1 Constant Composition of Compounds

Two samples of carbon dioxide, obtained from different sources, are decomposed into their constituent elements. One sample produces 4.8 g of oxygen and 1.8 g of carbon, and the other sample produces 17.1 g of oxygen and 6.4 g of carbon. Show that these results are consistent with the law of constant composition.

Calculate the mass ratio of one element to the other by dividing the larger mass by the smaller one.

SOLUTION

For the first sample:

$$\frac{\text{mass oxygen}}{\text{mass carbon}} = \frac{4.8\ \text{g}}{1.8\ \text{g}} = 2.7$$

For the second sample:

$$\frac{\text{mass oxygen}}{\text{mass carbon}} = \frac{17.1\ \text{g}}{6.4\ \text{g}} = 2.7$$

Because the ratios are the same for the two samples, these results are consistent with the law of constant composition.

▶ **SKILLBUILDER 5.1 | Constant Composition of Compounds**

Two samples of carbon monoxide, obtained from different sources, are decomposed into their constituent elements. One sample produces 4.3 g of oxygen and 3.2 g of carbon, and the other sample produces 7.5 g of oxygen and 5.6 g of carbon. Are these results consistent with the law of constant composition?

▶ **FOR MORE PRACTICE** Example 5.16; Problems 25, 26.

CONCEPTUAL CHECKPOINT 5.1

A compound composed of two elements A and B has a ratio of $\frac{\text{mass A}}{\text{mass B}} = 3.0$.

Decomposition of the compound produces 9.0 g of element A. What mass of element B is produced?

(a) 27.0 g B **(b)** 9.0 g B **(c)** 3.0 g B **(d)** 1.0 g B

5.3 Chemical Formulas: How to Represent Compounds

- ▶ Write chemical formulas.
- ▶ Determine the total number of each type of atom in a chemical formula.

We represent a compound with a **chemical formula**, which indicates the elements present in the compound and the relative number of atoms of each element. For example, H_2O is the chemical formula for water; it indicates that water consists of hydrogen and oxygen atoms in a 2:1 ratio. (Note that the ratio in a chemical

Compounds have constant composition with respect to mass (as you learned in the previous section) because they are composed of atoms in fixed ratios.

formula is a ratio of atoms, not a ratio of masses.) The formula contains the symbol for each element, accompanied by a subscript indicating the number of atoms of that element. By convention, a subscript of 1 is omitted.

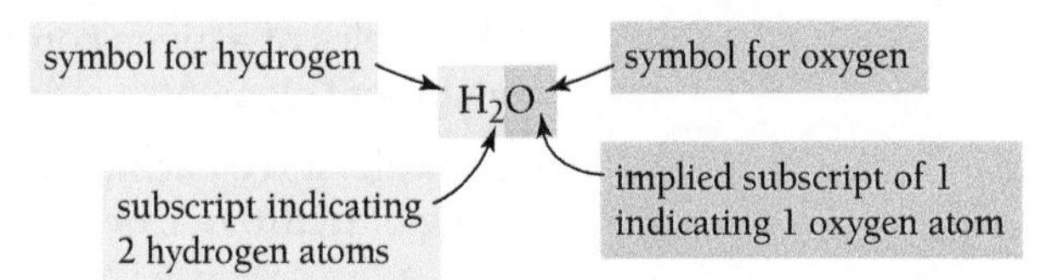

Other common chemical formulas include NaCl for table salt, indicating sodium and chlorine atoms in a 1:1 ratio; CO_2 for carbon dioxide, indicating carbon and oxygen atoms in a 1:2 ratio; and $C_{12}H_{22}O_{11}$ for table sugar (sucrose), indicating carbon, hydrogen, and oxygen atoms in a 12:22:11 ratio. The subscripts in a chemical formula are part of the compound's definition—if the subscripts change, the formula no longer specifies the same compound. For example, CO is the chemical formula for carbon monoxide, an air pollutant with adverse health effects on humans. When inhaled, carbon monoxide interferes with the blood's ability to carry oxygen, which can be fatal. CO is the primary substance responsible for the deaths of people who inhale too much automobile exhaust. If we change the subscript of the O in CO from 1 to 2, however, we get the formula for a totally different compound. CO_2 is the chemical formula for carbon dioxide, the relatively harmless product of combustion and human respiration. We breathe small amounts of CO_2 all the time with no harmful effects. So, remember that:

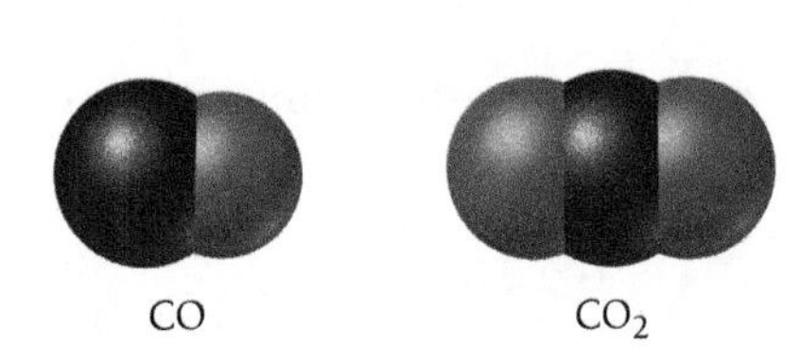

The subscripts in a chemical formula represent the relative numbers of each type of atom in a chemical compound; they never change for a given compound.

Chemical formulas normally list the most metallic elements first. Therefore, the formula for table salt is NaCl, not ClNa. In compounds that do not include a metal, we list the more metal-like element first. Recall from Chapter 4 that metals occupy the left side of the periodic table and nonmetals the upper right side. Among nonmetals, those to the left in the periodic table are more metal-like than those to the right and are normally listed first. Therefore, we write CO_2 and NO, not O_2C and ON. Within a single column in the periodic table, elements toward the bottom are more metal-like than elements toward the top. So, we write SO_2, not O_2S. Table 5.1 lists the specific order for listing nonmetal elements in a chemical formula.

There are a few historical exceptions in which the most metallic element is not listed first, such as the hydroxide ion, which we write as OH^-.

TABLE 5.1 Order of Listing Nonmetal Elements in a Chemical Formula

C	P	N	H	S	I	Br	Cl	O	F

Elements on the left are generally listed before elements on the right.

EXAMPLE 5.2 Writing Chemical Formulas

Write a chemical formula for each compound.

(a) the compound containing two aluminum atoms to every three oxygen atoms
(b) the compound containing three oxygen atoms to every sulfur atom
(c) the compound containing four chlorine atoms to every carbon atom

	SOLUTION
Aluminum is the metal, so list it first.	**(a)** Al_2O_3
Sulfur is below oxygen on the periodic table and occurs before oxygen in Table 5.1, so list it first.	**(b)** SO_3
Carbon is to the left of chlorine on the periodic table and occurs before chlorine in Table 5.1, so list it first.	**(c)** CCl_4

▶ **SKILLBUILDER 5.2 | Writing Chemical Formulas**

Write a chemical formula for each compound.

(a) the compound containing two silver atoms to every sulfur atom
(b) the compound containing two nitrogen atoms to every oxygen atom
(c) the compound containing two oxygen atoms to every titanium atom

▶ **FOR MORE PRACTICE** Example 5.17; Problems 31, 32, 33, 34.

Polyatomic Ions in Chemical Formulas

Some chemical formulas contain groups of atoms that act as a unit. When more than one group of the same kind is present, we set their formula off in parentheses with a subscript to indicate the number of units of that group. Many of these groups of atoms have a charge associated with them and are called **polyatomic ions**. For example, NO_3^- is a polyatomic ion with a 1− charge. We describe polyatomic ions in more detail in Section 5.5.

To determine the total number of each type of atom in a compound containing a group within parentheses, we multiply the subscript outside the parentheses by the subscript for each atom inside the parentheses. For example, $Mg(NO_3)_2$ indicates a compound containing one magnesium atom (present as the Mg^{2+} ion) and two NO_3^- groups.

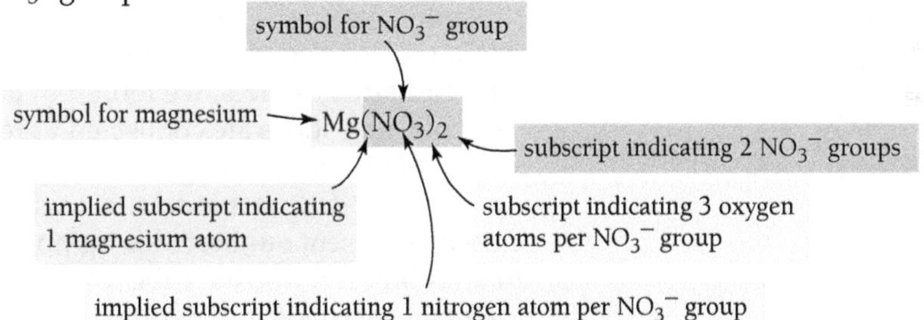

Therefore, the formula $Mg(NO_3)_2$ has the following numbers of each type of atom.

Mg: 1 Mg
N: $1 \times 2 = 2$ N (implied 1 inside parentheses times 2 outside parentheses)
O: $3 \times 2 = 6$ O (3 inside parentheses times 2 outside parentheses)

EXAMPLE **5.3** **Total Number of Each Type of Atom in a Chemical Formula**

Determine the number of each type of atom in $Mg_3(PO_4)_2$.

SOLUTION

Mg: There are three Mg atoms (present as Mg^{2+} ions), as indicated by the subscript 3.

P: There are two P atoms. We determine this by multiplying the subscript outside the parentheses (2) by the subscript for P inside the parentheses, which is 1 (implied).

O: There are eight O atoms. We determine this by multiplying the subscript outside the parentheses (2) by the subscript for O inside the parentheses (4).

▶ **SKILLBUILDER 5.3 | Total Number of Each Type of Atom in a Chemical Formula**

Determine the number of each type of atom in K_2SO_4.

▶ **SKILLBUILDER PLUS**

Determine the number of each type of atom in $Al_2(SO_4)_3$.

▶ **FOR MORE PRACTICE** Example 5.18; Problems 35, 36, 37, 38.

CONCEPTUAL CHECKPOINT 5.2

Which formula represents the greatest total number of atoms?

(a) $Al(C_2H_3O_2)_3$ **(b)** $Al_2(Cr_2O_7)_3$ **(c)** $Pb(HSO_4)_4$

(d) $Pb_3(PO_4)_4$ **(e)** $(NH_4)_3PO_4$

Types of Chemical Formulas

We categorize chemical formulas into three types: empirical, molecular, and structural. An **empirical formula** is the simplest whole-number ratio of atoms of each element in a compound. A **molecular formula** is the *actual* number of atoms of each element in a molecule of the compound. For example, the molecular formula for hydrogen peroxide is H_2O_2, and its empirical formula is HO. The molecular formula is always a whole-number multiple of the empirical formula. For many compounds, the molecular and empirical formulas are the same. For example, the empirical and molecular formula for water is H_2O because water molecules contain two hydrogen atoms and one oxygen atom; no simpler whole-number ratio can express the number of hydrogen atoms relative to oxygen atoms.

Hydrogen
Carbon
Nitrogen
Oxygen
Fluorine
Phosphorus
Sulfur
Chlorine

A **structural formula** uses lines to represent chemical bonds and shows how the atoms in a molecule are connected to each other. The structural formula for hydrogen peroxide is H—O—O—H. We can also use **molecular models**—three-dimensional representations of molecules—to represent compounds. In this book, we use two types of molecular models: ball-and-stick and space-filling. In **ball-and-stick models**, we represent atoms as balls and chemical bonds as sticks. The balls and sticks are connected to represent the molecule's shape. The balls are color coded, and we assign each element a color as shown in the margin.

In **space-filling models**, atoms fill the space between each other to more closely represent our best idea for how a molecule might appear if we could scale it to a visible size. Consider the following ways to represent a molecule of methane, the main component of natural gas:

CH_4

Molecular formula

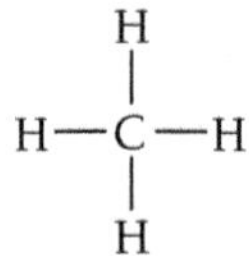

Structural formula

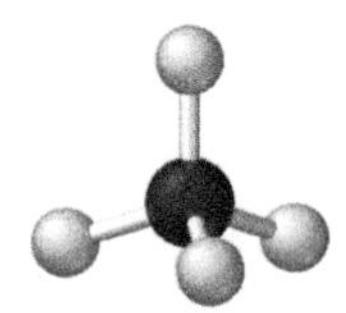

Ball-and-stick model

Space-filling model

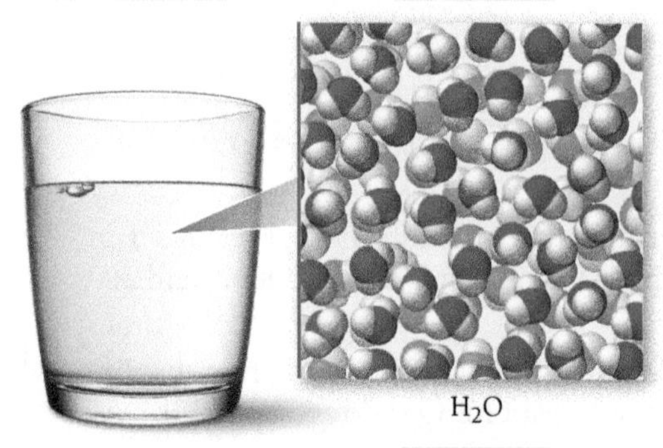

The molecular formula of methane indicates that methane has one carbon atom and four hydrogen atoms. The structural formula shows how the atoms are connected: each hydrogen atom is bonded to the central carbon atom. The ball-and-stick model and the space-filling model illustrate the *geometry* of the molecule: how the atoms are arranged in three dimensions.

Throughout this book, you have seen and will continue to see images that show the connection between the *macroscopic world* (what we see), the *atomic and molecular world* (the particles that compose matter), and the *symbolic way* that chemists represent the atomic and molecular world. For example, at left is a representation of water using this kind of image.

The main goal of these images is to help you visualize the main theme of this book: *the connection between the world around us and the world of atoms and molecules.*

CONCEPTUAL CHECKPOINT 5.3

Write a formula for the compound represented by this space-filling model.

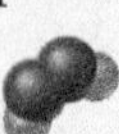

5.4 A Molecular View of Elements and Compounds

- Classify elements as atomic or molecular.
- Classify compounds as ionic or molecular.

Recall from Chapter 3 that we can categorize pure substances as either elements or compounds. We can further subcategorize elements and compounds according to the basic units that compose them (▼ FIGURE 5.6). Pure substances may be either elements or compounds. Elements may be either atomic or molecular. Compounds may be either molecular or ionic.

Atomic Elements

Atomic elements have single atoms as their basic units. Most elements fall into this category. For example, helium is composed of helium atoms, copper is composed of copper atoms, and mercury of mercury atoms (▼ FIGURE 5.7).

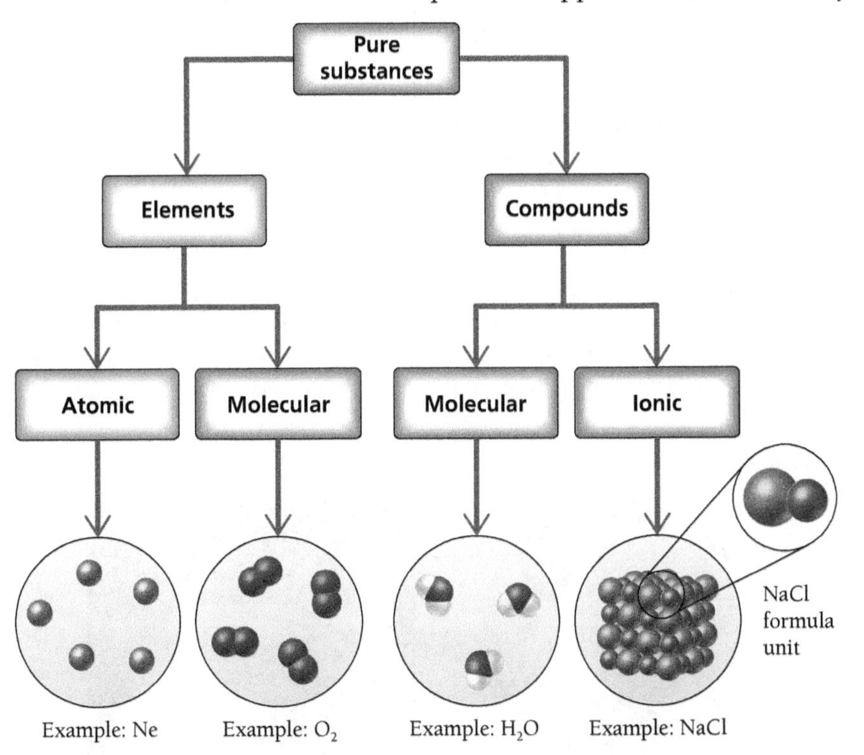

▲ FIGURE 5.6 **A molecular view of elements and compounds**

Molecular Elements

Molecular elements do not normally exist in nature with single atoms as their basic units. Instead, these elements exist as *diatomic molecules*—two atoms of that element bonded together—as their basic units. For example, hydrogen is composed of H_2 molecules, oxygen is composed of O_2 molecules, and chlorine of Cl_2 molecules (▼ FIGURE 5.8). Table 5.2 and ▶ FIGURE 5.9, on the next page, list elements that exist as diatomic molecules.

A few molecular elements, such as S_8 and P_4, are composed of molecules containing several atoms.

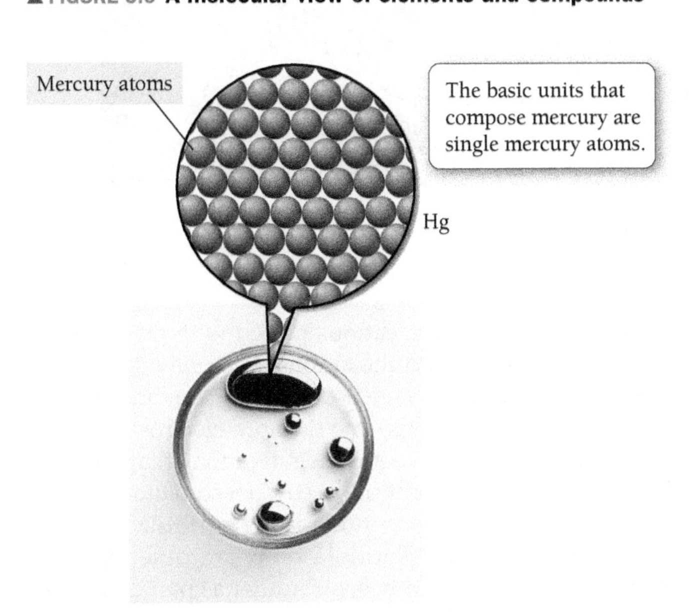

▲ FIGURE 5.7 **An atomic element**

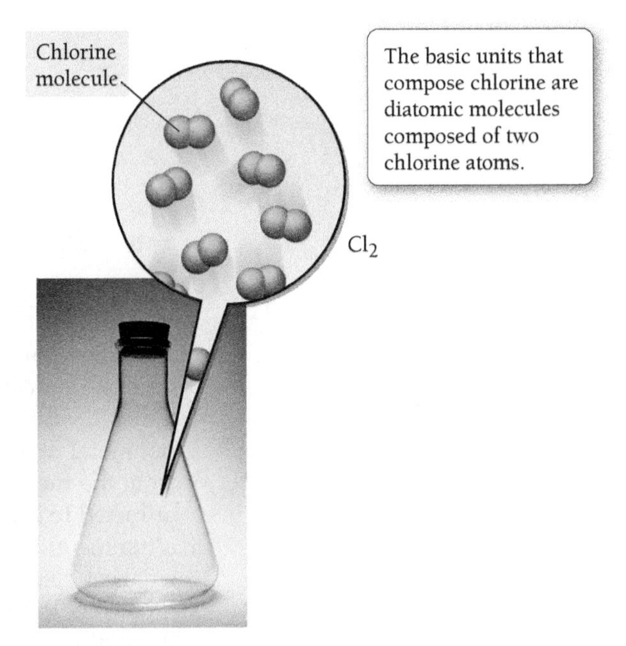

▲ FIGURE 5.8 **A molecular element**

TABLE 5.2 Elements That Occur as Diatomic Molecules

Name of Element	Formula of Basic Unit
hydrogen	H_2
nitrogen	N_2
oxygen	O_2
fluorine	F_2
chlorine	Cl_2
bromine	Br_2
iodine	I_2

Main group | Transition metals | Main group

☐ Elements that exist as diatomic molecules

Periods	1A 1	2A 2	3B 3	4B 4	5B 5	6B 6	7B 7	8B 8	8B 9	8B 10	1B 11	2B 12	3A 13	4A 14	5A 15	6A 16	7A 17	8A 18
1	1 H																	2 He
2	3 Li	4 Be											5 B	6 C	7 N	8 O	9 F	10 Ne
3	11 Na	12 Mg											13 Al	14 Si	15 P	16 S	17 Cl	18 Ar
4	19 K	20 Ca	21 Sc	22 Ti	23 V	24 Cr	25 Mn	26 Fe	27 Co	28 Ni	29 Cu	30 Zn	31 Ga	32 Ge	33 As	34 Se	35 Br	36 Kr
5	37 Rb	38 Sr	39 Y	40 Zr	41 Nb	42 Mo	43 Tc	44 Ru	45 Rh	46 Pd	47 Ag	48 Cd	49 In	50 Sn	51 Sb	52 Te	53 I	54 Xe
6	55 Cs	56 Ba	57 La	72 Hf	73 Ta	74 W	75 Re	76 Os	77 Ir	78 Pt	79 Au	80 Hg	81 Tl	82 Pb	83 Bi	84 Po	85 At	86 Rn
7	87 Fr	88 Ra	89 Ac	104 Rf	105 Db	106 Sg	107 Bh	108 Hs	109 Mt	110 Ds	111 Rg	112 Cn	113 Nh	114 Fl	115 Mc	116 Lv	117 Ts	118 Og

Lanthanides	58 Ce	59 Pr	60 Nd	61 Pm	62 Sm	63 Eu	64 Gd	65 Tb	66 Dy	67 Ho	68 Er	69 Tm	70 Yb	71 Lu
Actinides	90 Th	91 Pa	92 U	93 Np	94 Pu	95 Am	96 Cm	97 Bk	98 Cf	99 Es	100 Fm	101 Md	102 No	103 Lr

▲ FIGURE 5.9 **Elements that form diatomic molecules** Elements that normally exist as diatomic molecules are highlighted in yellow on this periodic table. Note that they are all nonmetals and include four of the halogens.

Molecular Compounds

Molecular compounds are composed of two or more nonmetals. The basic units of molecular compounds are molecules composed of the constituent atoms. For example, water is composed of H_2O molecules, dry ice is composed of CO_2 molecules (▼ FIGURE 5.10), and acetone (finger nail–polish remover) of C_3H_6O molecules.

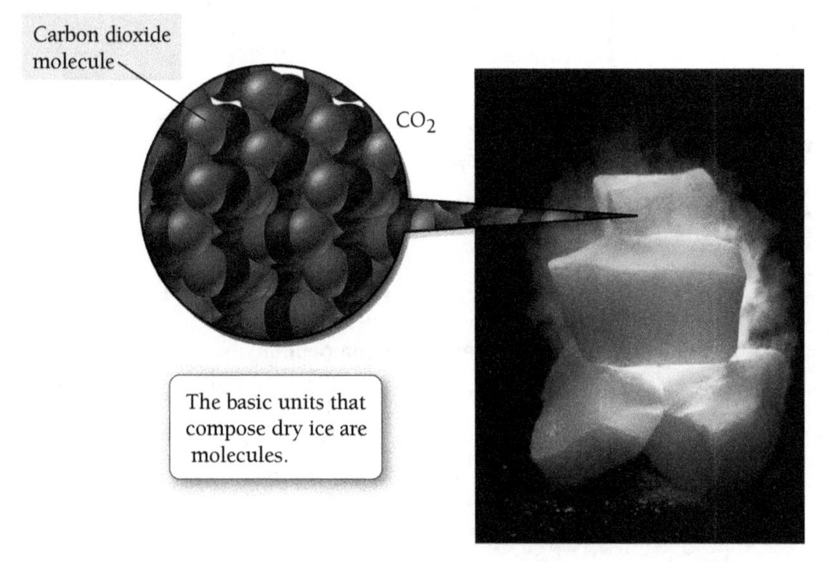

▲ FIGURE 5.10 **A molecular compound**

Ionic Compounds

Ionic compounds are composed of one or more cations paired with one or more anions. In most cases, the cations are metals and the anions are nonmetals. When a metal, which has a tendency to lose electrons (see Section 4.6), combines with a nonmetal, which has a tendency to gain electrons, one or more electrons transfer from the metal to the nonmetal, creating positive and negative ions that are then attracted to each other. We can assume that a compound composed of a metal and a nonmetal is ionic. The basic unit of ionic compounds is the **formula unit**, the smallest electrically neutral collection of ions. Formula units are different from molecules in that they do not exist as discrete entities, but rather as part of a larger three-dimensional array. For example, salt (NaCl) is composed of Na^+ and Cl^- ions

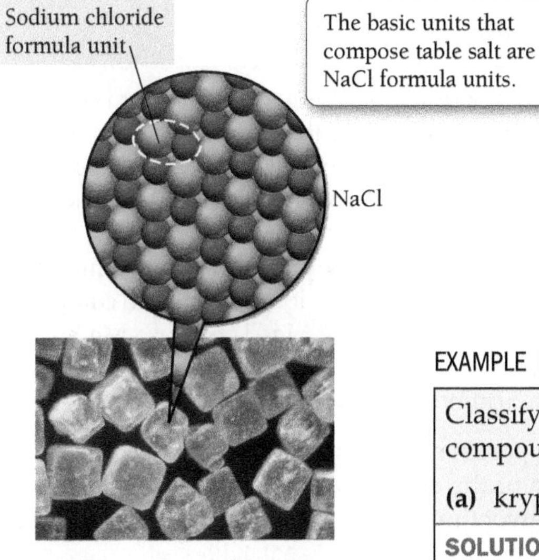
Sodium chloride formula unit
The basic units that compose table salt are NaCl formula units.
NaCl

▲ FIGURE 5.11 **An ionic compound** Unlike molecular compounds, ionic compounds do not contain individual molecules but rather sodium and chloride ions in an alternating three-dimensional array.

in a 1:1 ratio. In table salt, Na^+ and Cl^- ions exist in an alternating three-dimensional array (◀ FIGURE 5.11). However, any one Na^+ ion does not pair with one specific Cl^- ion. Sometimes chemists refer to formula units as molecules, but this is not strictly correct since ionic compounds do not contain distinct molecules.

EXAMPLE 5.4 Classifying Substances as Atomic Elements, Molecular Elements, Molecular Compounds, or Ionic Compounds

Classify each substance as an atomic element, molecular element, molecular compound, or ionic compound.

(a) krypton **(b)** $CoCl_2$ **(c)** nitrogen **(d)** SO_2 **(e)** KNO_3

SOLUTION

(a) Krypton is an element that is not listed as diatomic in Table 5.2; therefore, it is an atomic element.

(b) $CoCl_2$ is a compound composed of a metal (left side of periodic table) and nonmetal (right side of the periodic table); therefore, it is an ionic compound.

(c) Nitrogen is an element that is listed as diatomic in Table 5.2; therefore, it is a molecular element.

(d) SO_2 is a compound composed of two nonmetals; therefore, it is a molecular compound.

(e) KNO_3 is a compound composed of a metal and two nonmetals; therefore, it is an ionic compound.

▶ **SKILLBUILDER 5.4 | Classifying Substances as Atomic Elements, Molecular Elements, Molecular Compounds, or Ionic Compounds**

Classify each substance as an atomic element, molecular element, molecular compound, or ionic compound.

(a) chlorine **(b)** NO **(c)** Au **(d)** Na_2O **(e)** $CrCl_3$

▶ **FOR MORE PRACTICE** Example 5.19, Example 5.20; Problems 43, 44, 45, 46.

PEARSON eText 2.0

CONCEPTUAL CHECKPOINT 5.4

Which image represents a molecular compound?

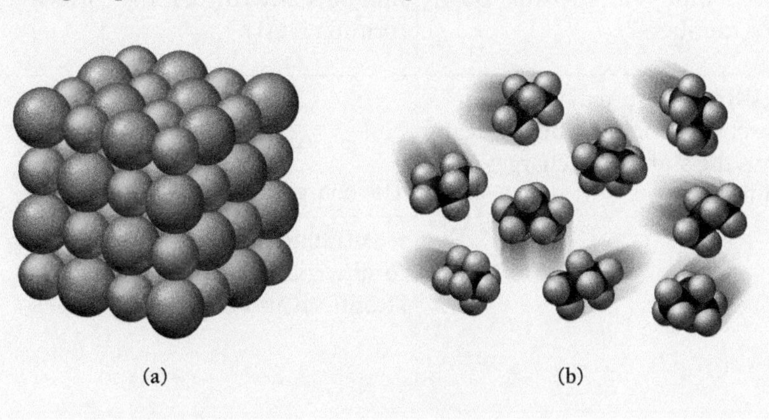
(a)
(b)

5.5 Writing Formulas for Ionic Compounds

▶ Write formulas for ionic compounds.

Revisit Section 4.7 and Figure 4.14 to review the elements that form ions with a predictable charge.

Because ionic compounds must be charge-neutral and many elements form only one type of ion with a predictable charge, we can determine the formulas for many ionic compounds based on their constituent elements. For example, on the one hand, the formula for the ionic compound composed of sodium and chlorine must be NaCl and not anything else because in ionic compounds Na always forms 1+ cations and Cl always forms 1− anions. In order for the compound to be charge-neutral, it must contain one Na^+ cation to every Cl^- anion. The formula for the ionic compound composed of magnesium and chlorine, on the other hand, must be $MgCl_2$ because Mg always forms 2+ cations and Cl always forms 1− anions. In order for the compound to be charge-neutral, it must contain one Mg^{2+} cation to every two Cl^- anions. In general:

- Ionic compounds always contain positive and negative ions.
- In the chemical formula, the sum of the charges of the positive ions (cations) must always exactly cancel the sum of the charges of the negative ions (anions).

Writing Formulas for Ionic Compounds Containing Only Monoatomic Ions

To write the formula for an ionic compound containing monoatomic ions (no polyatomic ions), follow the procedure in the left column of the following examples. The center and right columns provide two examples of how to apply the procedure.

PEARSON eText 2.0 Interactive Worked Example Video 5.5

Writing Formulas for Ionic Compounds	EXAMPLE 5.5 Write a formula for the ionic compound that forms from aluminum and oxygen.	EXAMPLE 5.6 Write a formula for the ionic compound that forms from magnesium and oxygen.
1. Write the symbol for the metal and its charge followed by the symbol of the nonmetal and its charge. For many elements, you can determine these charges from their group number in the periodic table (refer to Figure 4.14).	**SOLUTION** Al^{3+} O^{2-}	**SOLUTION** Mg^{2+} O^{2-}
2. Use the magnitude of the charge on each ion (without the sign) as the subscript for the other ion.	Al^{3+} O^{2-} Al_2O_3	Mg^{2+} O^{2-} Mg_2O_2
3. If possible, reduce the subscripts to give a ratio with the smallest whole numbers.	In this case, you cannot reduce the numbers any further; the correct formula is Al_2O_3.	To reduce the subscripts, divide both subscripts by 2. $Mg_2O_2 \div 2 = MgO$
4. Check to make sure that the sum of the charges of the cations exactly cancels the sum of the charges of the anions.	Cations: $2(3+) = 6+$ Anions: $3(2-) = 6-$ The charges cancel.	Cations: 2+ Anions: 2− The charges cancel.
	▶ **SKILLBUILDER 5.5** | Write a formula for the compound that forms from strontium and chlorine.	▶ **SKILLBUILDER 5.6** | Write a formula for the compound that forms from aluminum and nitrogen.
		▶ **FOR MORE PRACTICE** Problems 53, 54, 57.

Writing Formulas for Ionic Compounds Containing Polyatomic Ions

As noted previously, some ionic compounds contain polyatomic ions (ions that are themselves composed of a group of atoms with an overall charge). Table 5.3 lists the most common polyatomic ions. You need to be able to recognize polyatomic ions in a chemical formula, so it is good idea to become familiar with Table 5.3. To write a formula for ionic compounds containing polyatomic ions, use the formula and charge of the polyatomic ion as demonstrated in Example 5.7.

TABLE 5.3 Some Common Polyatomic Ions

Name	Formula	Name	Formula
acetate	$C_2H_3O_2^-$	hypochlorite	ClO^-
carbonate	CO_3^{2-}	chlorite	ClO_2^-
hydrogen carbonate (or bicarbonate)	HCO_3^-	chlorate	ClO_3^-
hydroxide	OH^-	perchlorate	ClO_4^-
nitrate	NO_3^-	permanganate	MnO_4^-
nitrite	NO_2^-	sulfate	SO_4^{2-}
chromate	CrO_4^{2-}	sulfite	SO_3^{2-}
dichromate	$Cr_2O_7^{2-}$	hydrogen sulfite (or bisulfite)	HSO_3^-
phosphate	PO_4^{3-}	hydrogen sulfate (or bisulfate)	HSO_4^-
hydrogen phosphate	HPO_4^{2-}	peroxide	O_2^{2-}
ammonium	NH_4^+	cyanide	CN^-

EXAMPLE 5.7 Writing Formulas for Ionic Compounds Containing Polyatomic Ions

Write a formula for the compound that forms from calcium and nitrate ions.

	SOLUTION
1. Write the symbol for the metal ion followed by the symbol for the polyatomic ion and their charges. You can deduce these charges from the group numbers in the periodic table. For polyatomic ions, look up the charges and names in Table 5.3.	Ca^{2+} NO_3^-
2. Use the magnitude of the charge on each ion as the subscript for the other ion.	$Ca_1(NO_3)_2$
3. Check to see if the subscripts can be reduced to simpler whole numbers. You should drop subscripts of 1 because they are implied.	In this case, you cannot further reduce the subscripts, but you can drop the subscript of 1. $Ca(NO_3)_2$
4. Confirm that the sum of the charges of the cations exactly cancels the sum of the charges of the anions.	Cations: 2+ Anions: $2(1-) = 2-$

▶ **SKILLBUILDER 5.7** | Write the formula for the compound that forms between aluminum and phosphate ions.

▶ **SKILLBUILDER PLUS** | Write the formula for the compound that forms between sodium and sulfite ions.

▶ **FOR MORE PRACTICE** Example 5.21; Problems 55, 56, 58.

CONCEPTUAL CHECKPOINT 5.5

Some metals can form ions of different charges in different compounds. Deduce the charge of the Cr ion in the compound $Cr(NO_3)_3$.

(a) 1+ (b) 2+ (c) 3+ (d) 1−

5.6 Nomenclature: Naming Compounds

- Distinguish between common and systematic names for compounds.

Because there are so many different compounds, chemists have developed systematic ways to name them. These naming rules can help you examine a compound's formula and determine its name, or vice versa. Many compounds also have a common name. For example, H_2O has the common name *water* and the systematic name *dihydrogen monoxide*. A common name is like a nickname for a compound, used by those who are familiar with it. Since water is such a familiar compound, everyone uses its common name and not its systematic name. In the sections that follow, you'll learn how to systematically name simple ionic and molecular compounds. Keep in mind, however, that some compounds also have common names that are often used instead of the systematic name. Common names can be learned only through familiarity.

5.7 Naming Ionic Compounds

- Name binary ionic compounds containing a metal that forms only one type of ion.
- Name binary ionic compounds containing a metal that forms more than one type of ion.
- Name ionic compounds containing a polyatomic ion.

The first step in naming an ionic compound is identifying it as one. Remember, any time we have a metal and one or more nonmetals together in a chemical formula, we can assume the compound is ionic. Ionic compounds are categorized into two types (▼ FIGURE 5.12) depending on the metal in the compound. The first type (sometimes called Type I) contains a metal with an invariant charge—one that does not vary from one compound to another. Sodium, for instance, has a 1+ charge in all of its compounds. ▼ FIGURE 5.13 lists examples of metals whose charge is invariant from one compound to another. The charge of most of these metals can be inferred from their group number in the periodic table (see Figure 4.14).

The second type of ionic compound (sometimes called Type II) contains a metal with a charge that can differ in different compounds. In other words, the

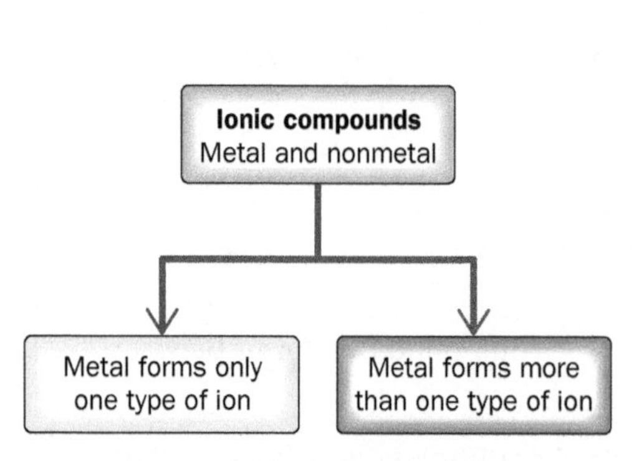

▲ FIGURE 5.12 **Classification of ionic compounds** We classify ionic compounds into two types, depending on the metal in the compound.

Metals Whose Charge Is Invariant from One Compound to Another

Periods	1A 1	2A 2	3B 3	4B 4	5B 5	6B 6	7B 7	8B 8	8B 9	8B 10	1B 11	2B 12	3A 13	4A 14	5A 15	6A 16	7A 17	8A 18
1	1 H																	2 He
2	3 Li 1+	4 Be											5 B	6 C	7 N	8 O	9 F	10 Ne
3	11 Na 1+	12 Mg 2+											13 Al 3+	14 Si	15 P	16 S	17 Cl	18 Ar
4	19 K 1+	20 Ca 2+	21 Sc 3+	22 Ti	23 V	24 Cr	25 Mn	26 Fe	27 Co	28 Ni	29 Cu	30 Zn 2+	31 Ga	32 Ge	33 As	34 Se	35 Br	36 Kr
5	37 Rb 1+	38 Sr 2+	39 Y	40 Zr	41 Nb	42 Mo	43 Tc	44 Ru	45 Rh	46 Pd	47 Ag 1+	48 Cd	49 In	50 Sn	51 Sb	52 Te	53 I	54 Xe
6	55 Cs 1+	56 Ba 2+	57 La	72 Hf	73 Ta	74 W	75 Re	76 Os	77 Ir	78 Pt	79 Au	80 Hg	81 Tl	82 Pb	83 Bi	84 Po	85 At	86 Rn
7	87 Fr	88 Ra	89 Ac	104 Rf	105 Db	106 Sg	107 Bh	108 Hs	109 Mt	110 Ds	111 Rg	112 Cn	113 Nh	114 Fl	115 Mc	116 Lv	117 Ts	118 Og

▲ FIGURE 5.13 **Metals with invariant charges** The metals highlighted in this periodic table always form an ion with the same charge in all of their compounds.

TABLE 5.4 Some Metals That Form More Than One Type of Ion and Their Common Charges (*This list is not exhaustive but meant to show examples.*)

Metal	Symbol Ion	Name	Older Name*
chromium	Cr^{2+}	chromium(II)	chromous
	Cr^{3+}	chromium(III)	chromic
iron	Fe^{2+}	iron(II)	ferrous
	Fe^{3+}	iron(III)	ferric
cobalt	Co^{2+}	cobalt(II)	cobaltous
	Co^{3+}	cobalt(III)	cobaltic
copper	Cu^{+}	copper(I)	cuprous
	Cu^{2+}	copper(II)	cupric
tin	Sn^{2+}	tin(II)	stannous
	Sn^{4+}	tin(IV)	stannic
mercury	Hg_2^{2+}	mercury(I)	mercurous
	Hg^{2+}	mercury(II)	mercuric
lead	Pb^{2+}	lead(II)	plumbous
	Pb^{4+}	lead(IV)	plumbic

* An older naming system substitutes the names found in this column for the name of the metal and its charge. Under this system, chromium(II) oxide is named chromous oxide. We do *not* use this older system in this text.

metal in this second type of ionic compound can form more than one kind of cation (depending on the compound). Iron, for instance, has a 2+ charge in some of its compounds and a 3+ charge in others. The best way to remember these is by elimination. For our purposes, you can assume that any metal *not* highlighted in Figure 5.13 is of this second type. Table 5.4 lists some examples of metals of this type. Metals that form cations whose charges can vary in different compounds are usually (but not always) found in the **transition metals** section of the periodic table (◀ FIGURE 5.14). The exceptions are Zn and Ag, which are transition metals, but form cations with the same charge in all of their compounds (as you can see from Figure 5.13). Two other exceptions are Pb and Sn, which are *not* transition metals but still form cations whose charges can vary in different compounds.

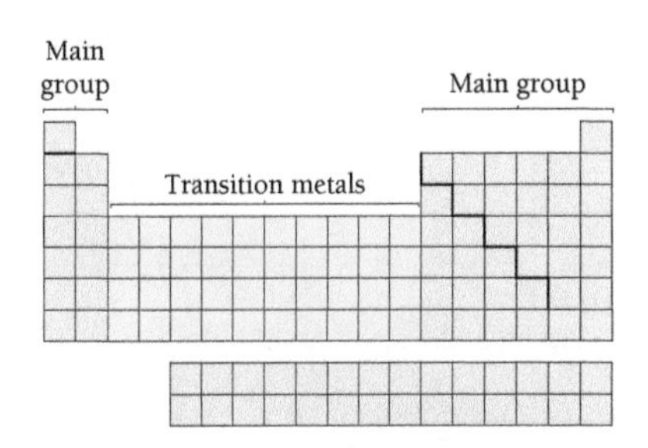

▲ FIGURE 5.14 **The transition metals** The metals that form more than one type of ion are usually (but not always) transition metals.

Naming Binary Ionic Compounds Containing a Metal That Forms Only One Type of Cation

Binary compounds contain only two different elements. The names for binary ionic compounds containing a metal that forms only one type of ion have the form:

name of cation (metal)	base name of anion (nonmetal) + *-ide*

Because the charge of the metal is always the same for these types of compounds, we do not need to specify it in the compound's name. For example, the name for NaCl consists of the name of the cation, *sodium*, followed by the base name of the anion, *chlor*, with the ending *-ide*. The full name is *sodium chloride*.

The name of the cation in ionic compounds is the same as the name of the metal.

NaCl sodium chloride

The name for $CaBr_2$ consists of the name of the cation, *calcium*, followed by the base name of the anion, *brom*, with the ending *-ide*. The full name is *calcium bromide*.

$CaBr_2$ calcium bromide

Table 5.5 (on the next page) contains the base names for various nonmetals and their most common charges in ionic compounds.

TABLE 5.5 Some Common Anions

Nonmetal	Symbol for Ion	Base Name	Anion Name
fluorine	F^-	fluor-	fluoride
chlorine	Cl^-	chlor-	chloride
bromine	Br^-	brom-	bromide
iodine	I^-	iod-	iodide
oxygen	O^{2-}	ox-	oxide
sulfur	S^{2-}	sulf-	sulfide
nitrogen	N^{3-}	nitr-	nitride

EXAMPLE **5.8** **Naming Ionic Compounds Containing a Metal That Forms Only One Type of Cation**

Name the compound MgF_2.

SOLUTION

The cation is magnesium. The anion is fluorine, which becomes *fluoride*. Its correct name is *magnesium fluoride*.

▶ **SKILLBUILDER 5.8 | Naming Ionic Compounds Containing a Metal That Forms Only One Type of Ion**

Name the compound KBr.

▶ **SKILLBUILDER PLUS** Name the compound Zn_3N_2.

▶ **FOR MORE PRACTICE** Example 5.22; Problems 59, 60.

Naming Binary Ionic Compounds Containing a Metal That Forms More Than One Type of Cation

Because the charge of the metal cation in these types of compounds is not always the same, we must specify the charge in the metal's name. We specify the charge with a roman numeral (in parentheses) following the name of the metal. For example, we distinguish between Cu^+ and Cu^{2+} by writing a (I) to indicate the 1+ ion or a (II) to indicate the 2+ ion:

Cu^+ copper(I)

Cu^{2+} copper(II)

The full names for these types of compounds have the form:

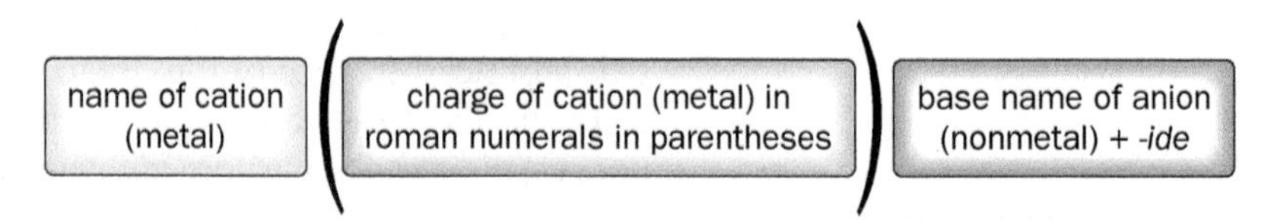

We can determine the charge of the metal from the chemical formula of the compound—remember that the sum of all the charges must be zero. For example, the charge of iron in $FeCl_3$ must be 3+ in order for the compound to be charge-neutral with the three Cl^- anions. The name for $FeCl_3$ is therefore the name of the cation, *iron*, followed by the charge of the cation in parentheses *(III)*, followed by the base name of the anion, *chlor*, with the ending *-ide*. The full name is *iron(III) chloride*.

$FeCl_3$ iron(III) chloride

Likewise, the name for CrO consists of the name of the cation, *chromium*, followed by the charge of the cation in parentheses *(II)*, followed by the base name of the anion, *ox-*, with the ending *-ide*. The full name is *chromium(II) oxide*.

CrO chromium(II) oxide

The charge of chromium must be 2+ in order for the compound to be charge-neutral with one O^{2-} anion.

EXAMPLE **5.9** **Naming Ionic Compounds Containing a Metal That Forms More Than One Type of Cation**

Name the compound $PbCl_4$.

SOLUTION

The name for $PbCl_4$ consists of the name of the cation, *lead*, followed by the charge of the cation in parentheses *(IV)*, followed by the base name of the anion, *chlor-*, with the ending *-ide*. The full name is *lead(IV) chloride*. We know the charge on Pb is 4+ because the charge on Cl is 1−. Since there are 4 Cl^- anions, the Pb cation must be Pb^{4+}.

$PbCl_4$ lead(IV) chloride

▶ **SKILLBUILDER 5.9 | Naming Ionic Compounds Containing a Metal That Forms More Than One Type of Cation**

Name the compound PbO.

▶ **FOR MORE PRACTICE** Example 5.23; Problems 61, 62.

CONCEPTUAL ✔ CHECKPOINT 5.6

Explain why CaO is NOT named calcium(II) oxide.

Naming Ionic Compounds Containing a Polyatomic Ion

We name ionic compounds containing polyatomic ions using the same procedure we applied to other ionic compounds, except that we use the name of the polyatomic ion whenever it occurs (see Table 5.3). For example, we name KNO_3 using its cation, K^+, *potassium*, and its polyatomic anion, NO_3^-, *nitrate*. The full name is *potassium nitrate*.

KNO_3 potassium nitrate

We name $Fe(OH)_2$ according to its cation, *iron*, its charge *(II)*, and its polyatomic ion, *hydroxide*. Its full name is *iron(II) hydroxide*.

$Fe(OH)_2$ iron(II) hydroxide

If the compound contains both a polyatomic cation and a polyatomic anion, we use the names of both polyatomic ions. For example, NH_4NO_3 is *ammonium nitrate*.

NH_4NO_3 ammonium nitrate

Most polyatomic ions are **oxyanions**, anions containing oxygen. Notice in Table 5.3 that when a series of oxyanions contain different numbers of oxygen atoms, we name them systematically according to the number of oxygen atoms in the ion. If there are two ions in the series, we give the one with more oxygen atoms the ending *-ate* and we give the one with fewer oxygen atoms the ending *-ite*. For example, NO_3^- is *nitrate* and NO_2^- is *nitrite*.

NO_3^- nitrate
NO_2^- nitrite

If there are more than two ions in the series, then we use the prefixes *hypo-*, meaning "less than," and *per-*, meaning "more than." So we call ClO^- *hypochlorite*, meaning "less oxygen than chlorite," and we call ClO_4^- *perchlorate*, meaning "more oxygen than chlorate."

ClO^- hypochlorite
ClO_2^- chlorite
ClO_3^- chlorate
ClO_4^- perchlorate

EXAMPLE **5.10** **Naming Ionic Compounds Containing a Polyatomic Ion**

Name the compound K_2CrO_4.

SOLUTION

The name for K_2CrO_4 consists of the name of the cation, *potassium*, followed by the name of the polyatomic ion, *chromate*.

K_2CrO_4 potassium chromate

▶ **SKILLBUILDER 5.10 | Naming Ionic Compounds Containing a Polyatomic Ion**
Name the compound $Mn(NO_3)_2$.

▶ **FOR MORE PRACTICE** Example 5.24; Problems 65, 66.

CONCEPTUAL ✔ CHECKPOINT 5.7

We just saw that the anion ClO_3^- is named chlorate. What is the name of the anion IO_3^-?

EVERYDAY CHEMISTRY

Polyatomic Ions

A glance at the labels of household products reveals the importance of polyatomic ions in everyday compounds. For example, the active ingredient in household bleach is sodium hypochlorite, which acts to decompose color-causing molecules in clothes (bleaching action) and to kill bacteria (disinfection). A box of baking soda contains sodium bicarbonate (sodium hydrogen carbonate), which acts as an antacid when consumed in small quantities and as a source of carbon dioxide gas in baking. The pockets of carbon dioxide gas make baked goods fluffy rather than flat.

Calcium carbonate is the active ingredient in many antacids such as Tums™ and Alka-Mints™. It neutralizes stomach acids, relieving the symptoms of indigestion and heartburn. Too much calcium carbonate, however, can cause constipation, so Tums should not be overused. Sodium nitrite is a common food additive used to preserve packaged meats such as ham, hot dogs, and bologna. Sodium nitrite inhibits the growth of bacteria, especially those that cause botulism, an often fatal type of food poisoning.

▲ Compounds containing polyatomic ions are present in many consumer products.

▲ The active ingredient in bleach is sodium hypochlorite.

B5.1 CAN YOU ANSWER THIS? *Write a formula for each of these compounds that contain polyatomic ions: sodium hypochlorite, sodium bicarbonate, calcium carbonate, sodium nitrite.*

5.8 Naming Molecular Compounds

▶ Name molecular compounds.

Key Concept Video
Naming Molecular Compounds

The first step in naming a molecular compound is identifying it as one. Remember, nearly all molecular compounds form from two or more nonmetals. In this section, we discuss how to name binary (two-element) molecular compounds. Their names have the form:

When writing the name of a molecular compound, as when writing the formula, the first element is the more metal-like one (see Table 5.1). The prefixes given to each element indicate the number of atoms present.

mono- 1	*hexa-* 6
di- 2	*hepta-* 7
tri- 3	*octa-* 8
tetra- 4	*nona-* 9
penta- 5	*deca-* 10

If there is only one atom of the *first element* in the formula, the prefix *mono-* is normally omitted. For example, the name for CO_2 begins with *carbon*, without a prefix because *mono-* is omitted for the first element, followed by the prefix *di-*, to indicate two oxygen atoms, followed by the base name of the second element, *ox*, with the ending *-ide*.

carbon di- ox -ide

The full name is *carbon dioxide*.

CO_2 carbon dioxide

When the prefix ends with a vowel and the base name starts with a vowel, we sometimes drop the first vowel, especially in the case of mono oxide, which becomes monoxide.

The name for the compound N_2O, also called laughing gas, begins with the first element, *nitrogen*, with the prefix *di-*, to indicate that there are two of them, followed by the base name of the second element, *ox*, prefixed by *mono-*, to indicate one, and the suffix *-ide*. Because *mono-* ends with a vowel and *oxide* begins with one, we drop an *o* and combine the two as *monoxide*. The entire name is *dinitrogen monoxide*.

N_2O dinitrogen monoxide

EXAMPLE 5.11 Naming Molecular Compounds

Name each compound.

(a) CCl_4 **(b)** BCl_3 **(c)** SF_6

SOLUTION

(a) The name of the compound is the name of the first element, *carbon*, followed by the base name of the second element, *chlor*, prefixed by *tetra-* to indicate four, and the suffix *-ide*.

CCl_4 carbon tetrachloride

(b) The name of the compound is the name of the first element, *boron*, followed by the base name of the second element, *chlor*, prefixed by *tri-* to indicate three, and the suffix *-ide*.

BCl_3 boron trichloride

(c) The name of the compound is the name of the first element, *sulfur*, followed by the base name of the second element, *fluor*, prefixed by *hexa-* to indicate six, and the suffix *-ide*. The entire name is *sulfur hexafluoride*.

SF_6 sulfur hexafluoride

▶ SKILLBUILDER 5.11 | Naming Molecular Compounds

Name the compound N_2O_4.

▶ FOR MORE PRACTICE Example 5.25; Problems 71, 72.

CONCEPTUAL CHECKPOINT 5.8

The compound NCl_3 is named nitrogen trichloride, while $AlCl_3$ is simply aluminum chloride. Why the difference?

5.9 Naming Acids

- Name binary acids.
- Name oxyacids containing an oxyanion ending in *-ate*.
- Name oxyacids containing an oxyanion ending in *-ite*.

Acids are molecular compounds that produce H^+ ions when dissolved in water. They are composed of hydrogen, which we usually write first in their formula, and one or more nonmetals, which we usually write second. Acids are characterized by their sour taste and their ability to dissolve some metals. For example, HCl(*aq*) is an acid—the (*aq*) indicates that the compound is "aqueous" or "dissolved in water." HCl(*aq*) has a characteristically sour taste. Since HCl(*aq*) is present in stomach fluids, its sour taste becomes painfully obvious during vomiting. HCl(*aq*) also dissolves some metals. If we drop a strip of zinc into a beaker of HCl(*aq*), it will slowly disappear as the acid converts the zinc metal into dissolved Zn^{2+} cations.

HCl(g) refers to HCl molecules in the gas state.

Acids are present in many foods, such as lemons and limes, and they are used in some household products such as toilet bowl cleaners and Lime-A-Way. In this section, we only learn how to name them, but in Chapter 14 we will learn more about the properties of acids. We categorize acids into two groups: **binary acids**, those containing only hydrogen and a nonmetal, and oxyacids, those containing hydrogen, a nonmetal, and oxygen (▼ FIGURE 5.15).

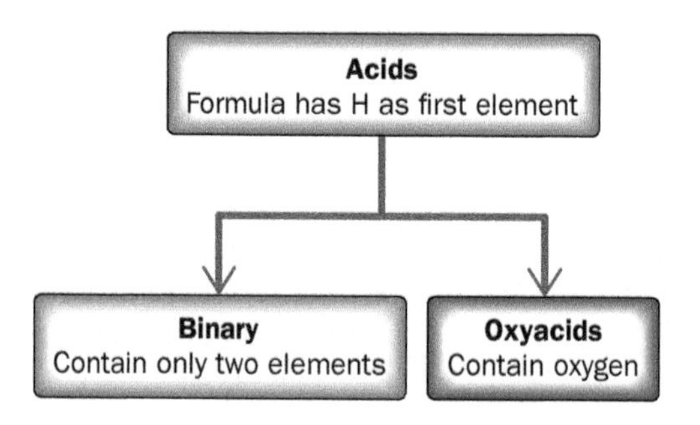

▶ FIGURE 5.15 **Classification of acids** We classify acids into two types, depending on the number of elements in the acid. If the acid contains only two elements, it is a binary acid. If it contains oxygen, it is an oxyacid.

Naming Binary Acids

Binary acids are composed of hydrogen and a nonmetal. The names for binary acids have the following form:

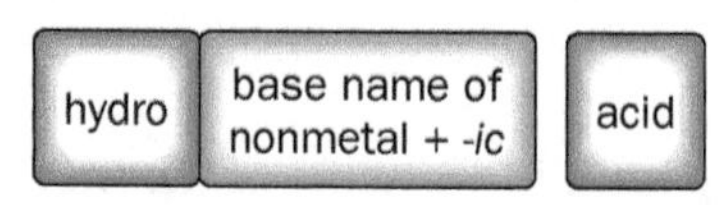

For example, HCl(*aq*) is hydro*chlor*ic acid and HBr(*aq*) is hydro*brom*ic acid.

HCl(*aq*) hydrochloric acid HBr(*aq*) hydrobromic acid

EXAMPLE **5.12** **Naming Binary Acids**

Give the name of H_2S(*aq*).

	SOLUTION
The base name of S is *sulfur*, so the name is *hydrosulfuric acid*.	H_2S(*aq*) hydrosulfuric acid

▶ **SKILLBUILDER 5.12 | Naming Binary Acids**

Name HF(*aq*).

▶ **FOR MORE PRACTICE** Example 5.26; Problems 77b, 78d.

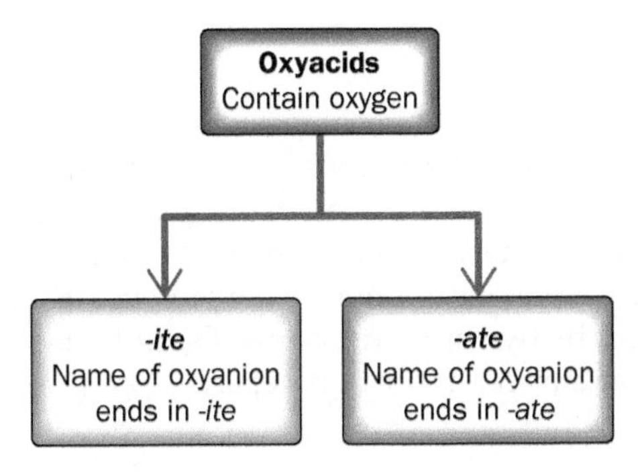

▲ FIGURE 5.16 **Classification of oxyacids** Oxyacids are classified into two types, depending on the endings of the oxyanions that they contain.

Naming Oxyacids

Oxyacids are acids that contain oxyanions, which are listed in the table of polyatomic ions (Table 5.3). For example, $HNO_3(aq)$ contains the nitrate (NO_3^-) ion, $H_2SO_3(aq)$ contains the sulfite (SO_3^{2-}) ion, and $H_2SO_4(aq)$ contains the sulfate (SO_4^{2-}) ion. All of these acids are a combination of one or more H^+ ions with an oxyanion. The number of H^+ ions depends on the charge of the oxyanion, so that the formula is always charge-neutral. The names of oxyacids depend on the ending of the oxyanion (◀ FIGURE 5.16).

The names of acids containing oxyanions ending with *-ite* take this form:

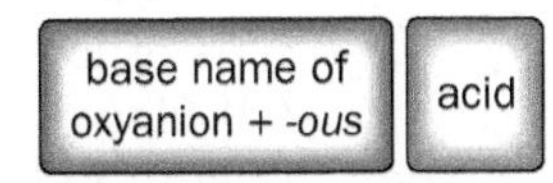

The names of acids containing oxyanions ending with *-ate* take this form:

You can use the saying, "*ic* I *ate* an acid" to remember the association of *-ic* with *-ate*.

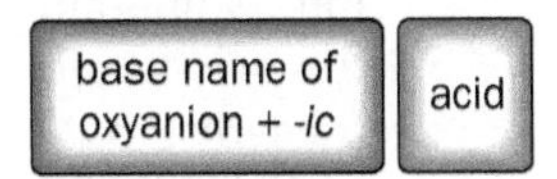

So H_2SO_3 is *sulfurous acid* (oxyanion is sulfite), and HNO_3 is *nitric acid* (oxyanion is nitrate).

$H_2SO_3(aq)$ sulfurous acid $\qquad$ $HNO_3(aq)$ nitric acid

Table 5.6 lists some common oxyacids and their oxyanions.

EXAMPLE 5.13 Naming Oxyacids

Name $HC_2H_3O_2(aq)$.

The oxyanion is acetate, which ends in *-ate*; therefore, the name of the acid is *acetic acid*.

SOLUTION

$HC_2H_3O_2(aq)$ acetic acid

▶ **SKILLBUILDER 5.13 | Naming Oxyacids**

Name $HNO_2(aq)$.

▶ **FOR MORE PRACTICE** Examples 5.27, 5.28; Problems 77acd, 78abc.

TABLE 5.6 Names of Some Common Oxyacids and Their Oxyanions

Acid Formula	Acid Name	Oxyanion Name	Oxyanion Formula
HNO_2	nitrous acid	nitrite	NO_2^-
HNO_3	nitric acid	nitrate	NO_3^-
H_2SO_3	sulfurous acid	sulfite	SO_3^{2-}
H_2SO_4	sulfuric acid	sulfate	SO_4^{2-}
$HClO_2$	chlorous acid	chlorite	ClO_2^-
$HClO_3$	chloric acid	chlorate	ClO_3^-
$HC_2H_3O_2$	acetic acid	acetate	$C_2H_3O_2^-$
H_2CO_3	carbonic acid	carbonate	CO_3^{2-}

5.10 Nomenclature Summary

▶ Recognize and name chemical compounds.

Acids are technically a subclass of molecular compounds; that is, they are molecular compounds that form H^+ ions when dissolved in water.

Naming compounds requires several steps. The flowchart in ▼ FIGURE 5.17 summarizes the different categories of compounds that we have covered in the chapter and how to identify and name them. The first step is to decide whether the compound is ionic, molecular, or an acid. We can recognize ionic compounds by the presence of a metal and a nonmetal, molecular compounds by two or more nonmetals, and acids by the presence of hydrogen (written first) and one or more nonmetals.

Ionic Compounds

For an ionic compound, we must next decide whether the metal forms only one type of ion or more than one type of ion. Group 1A (alkali) metals, Group 2A (alkaline earth) metals, and aluminum always form only one type of ion (Figure 5.13). Most of the transition metals (except Zn, Sc, and Ag) form more than one type of ion. Once we have identified the type of ionic compound, we name it according to the scheme in the chart. If the ionic compound contains a polyatomic ion—something we must learn to recognize by familiarity—we insert the name of the polyatomic ion in place of the metal (positive polyatomic ion) or the nonmetal (negative polyatomic ion).

Molecular Compounds

We have learned how to name only one type of molecular compound, the binary (two-element) compound. If we identify a compound as molecular, we name it according to the scheme in Figure 5.17.

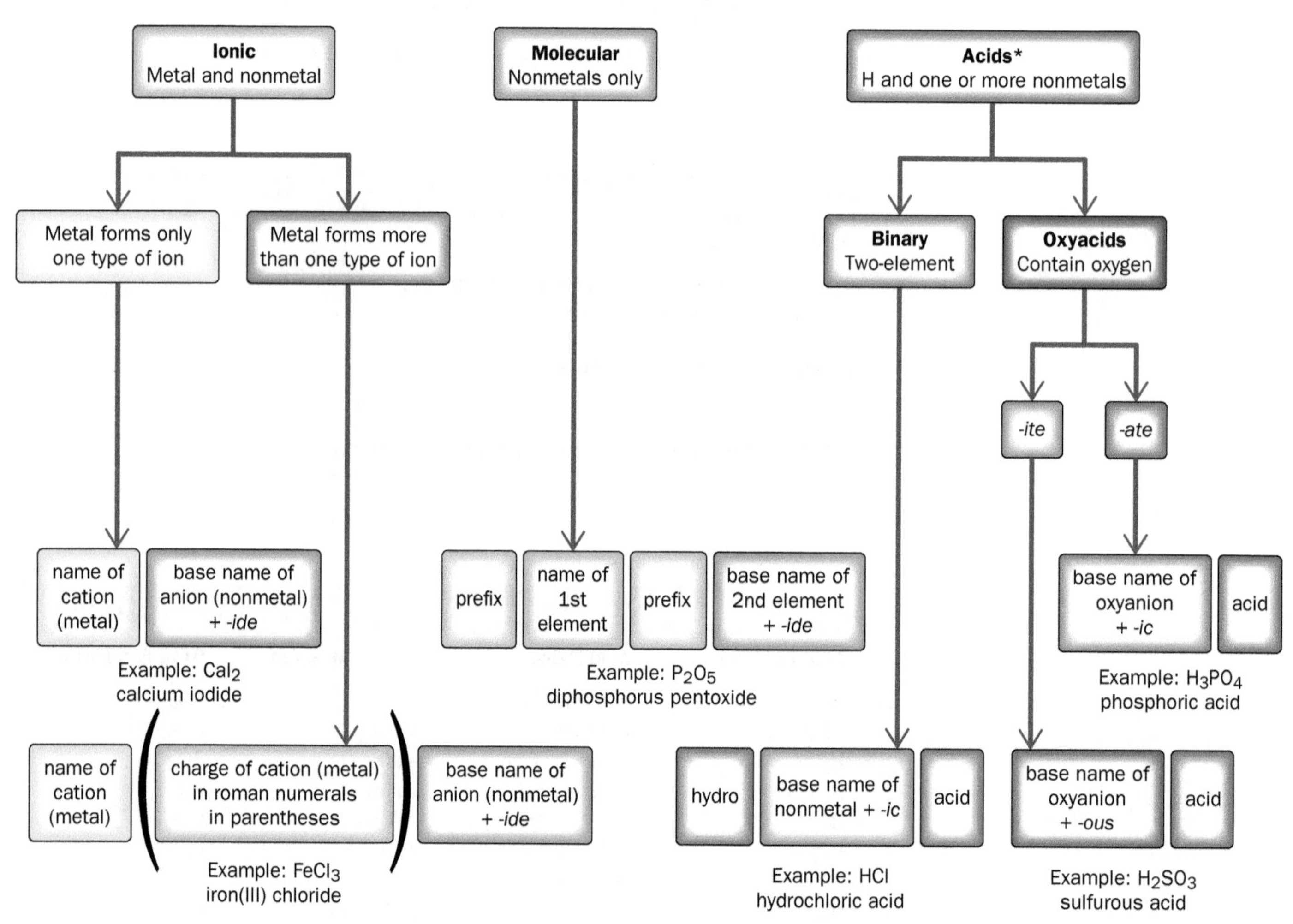

▲ FIGURE 5.17 **Nomenclature flowchart**

Acids

To name an acid, we must first decide whether it is a binary (two-element) acid or an oxyacid (an acid containing oxygen). We name binary acids according to the scheme in Figure 5.17. We must further subdivide oxyacids based on the name of their corresponding oxyanion. If the oxyanion ends in *-ite*, we use one scheme; if it ends with *-ate*, we use the other.

EXAMPLE **5.14** **Nomenclature Using Figure 5.17**

Name each compound: CO, CaF_2, HF(*aq*), $Fe(NO_3)_3$, $HClO_4$(*aq*), H_2SO_3(*aq*).

SOLUTION

The table illustrates how to use Figure 5.17 to arrive at a name for each compound.

Formula	Flowchart Path	Name
CO	molecular	carbon monoxide
CaF_2	ionic ⟶ one type of ion ⟶	calcium fluoride
HF(*aq*)	acid ⟶ binary ⟶	hydrofluoric acid
$Fe(NO_3)_3$	ionic ⟶ more than one type of ion ⟶	iron(III) nitrate
$HClO_4$(*aq*)	acid ⟶ oxyacid ⟶ *-ate* ⟶	perchloric acid
H_2SO_3(*aq*)	acid ⟶ oxyacid ⟶ *-ite* ⟶	sulfurous acid

▶ **FOR MORE PRACTICE** Problems 93, 94.

5.11 Formula Mass: The Mass of a Molecule or Formula Unit

▶ Calculate formula mass.

The terms *molecular mass* and *molecular weight*, which are also commonly used, have the same meaning as formula mass.

In Chapter 4, we discussed atoms and elements and defined the average mass of the atoms that compose an element as the *atomic mass* for that element. Similarly, in this chapter, which introduces molecules and compounds, we designate the average mass of the molecules (or formula units) that compose a compound as the **formula mass**.

For any compound, the *formula mass* is the sum of the atomic masses of all the atoms in its chemical formula:

$$\text{formula mass} = \left(\boxed{\begin{array}{c}\#\text{ atoms of 1st}\\ \text{element in}\\ \text{chemical formula}\end{array}} \times \boxed{\begin{array}{c}\text{atomic mass}\\ \text{of}\\ \text{1st element}\end{array}}\right) + \left(\boxed{\begin{array}{c}\#\text{ atoms of 2nd}\\ \text{element in}\\ \text{chemical formula}\end{array}} \times \boxed{\begin{array}{c}\text{atomic mass}\\ \text{of}\\ \text{2nd element}\end{array}}\right) + \ldots$$

Like atomic mass for atoms, formula mass characterizes the average mass of a molecule or formula unit. For example, the formula mass of water, H_2O, is:

$$\begin{aligned}\text{formula mass} &= 2(1.01\text{ amu}) + 16.00\text{ amu}\\ &= 18.02\text{ amu}\end{aligned}$$

and that of sodium chloride, NaCl, is:

$$\begin{aligned}\text{formula mass} &= 22.99\text{ amu} + 35.45\text{ amu}\\ &= 58.44\text{ amu}\end{aligned}$$

In addition to giving a characteristic mass to the molecules or formula units of a compound, formula mass—as we will discuss in Chapter 6—allows us to quantify the number of molecules or formula units in a sample of a given mass.

EXAMPLE **5.15** **Calculating Formula Mass**

Calculate the formula mass of carbon tetrachloride, CCl_4.

SOLUTION

To find the formula mass, sum the atomic masses of each atom in the chemical formula.

$$\begin{aligned} \text{formula mass} &= 1 \times (\text{atomic mass C}) + 4 \times (\text{atomic mass Cl}) \\ &= 12.01 \text{ amu} + 4(35.45 \text{ amu}) \\ &= 12.01 \text{ amu} + 141.8\underline{0} \text{ amu} \\ &= 153.8 \text{ amu} \end{aligned}$$

▶ **SKILLBUILDER 5.15 | Calculating Formula Masses**

Calculate the formula mass of dinitrogen monoxide, N_2O, also called laughing gas.

▶ **FOR MORE PRACTICE** Example 5.29; Problems 83, 84.

CONCEPTUAL CHECKPOINT 5.9

Which substance has the greatest formula mass?

(a) O_2 (b) O_3 (c) H_2O (d) H_2O_2

Chapter 5 in Review

MasteringChemistry™ provides end-of-chapter exercises, feedback-enriched tutorial problems, animations, and interactive activities to encourage problem solving practice and deeper understanding of key concepts and topics.

Self-Assessment Quiz

Q1. Carbon tetrachloride has a chlorine-to-carbon mass ratio of 11.8:1. If a sample of carbon tetrachloride contains 35 g of chlorine, what mass of carbon does it contain?
(a) 0.34 g C
(b) 1.0 g C
(c) 3.0 g C
(d) 11.8 g C

Q2. Write a chemical formula for a compound that contains two chlorine atoms to every one oxygen atom.
(a) Cl_2O
(b) ClO_2
(c) 2ClO
(d) $Cl(O_2)_2$

Q3. How many oxygen atoms are in the chemical formula $Fe_2(SO_4)_3$?
(a) 2
(b) 3
(c) 4
(d) 12

Q4. Which element is a molecular element?
(a) copper
(b) iodine
(c) krypton
(d) potassium

Q5. Which compound is ionic?
(a) BrF_5
(b) HNO_3
(c) $MgSO_4$
(d) NI_3

Q6. Write a formula for the compound that forms between Sr and Br.
(a) SrBr
(b) Sr_2Br
(c) $SrBr_2$
(d) Sr_2Br_2

Q7. Write a formula for the compound that forms between sodium and chlorite ions.
(a) $NaClO_2$
(b) Na_2ClO_2
(c) $Na(ClO)_2$
(d) $NaClO_3$

Q8. Name the compound Li_3N.
(a) trilithium mononitride
(b) trilithium nitride
(c) lithium(I) nitride
(d) lithium nitride

Q9. Name the compound $CrCl_3$.
(a) monochromium trichloride
(b) chromium trichloride
(c) chromium chloride
(d) chromium(III) chloride

Q10. Name the compound $BaSO_4$.
(a) barium sulfate
(b) barium(II) sulfate
(c) barium monosulfur tetraoxygen
(d) barium tetrasulfate

Q11. Name the compound PF_5.
(a) monophosphorus pentafluoride
(b) phosphorus pentafluoride
(c) phosphorus fluoride
(d) phosphorus(III) fluoride

Q12. What is the formula for manganese(III) oxide?
(a) MnO
(b) Mn_3O
(c) Mn_2O_3
(d) MnO_3

Q13. Name the acid $H_3PO_4(aq)$.
(a) hydrogen phosphate
(b) phosphoric acid
(c) phosphorus acid
(d) hydrophosphic acid

Q14. What is the formula for hydrobromic acid?
(a) HBr
(b) $HBrO$
(c) $HBrO_2$
(d) $HBrO_3$

Q15. Determine the formula mass of CCl_2F_2.
(a) 66.46 amu
(b) 108.9 amu
(c) 132.92 amu
(d) 120.91 amu

Answers: 1:c, 2:a, 3:d, 4:b, 5:c, 6:c, 7:a, 8:d, 9:d, 10:a, 11:b, 12:c, 13:b, 14:a, 15:d

Chemical Principles

Relevance

Compounds

Matter is ultimately composed of atoms, and those atoms are often combined in compounds. The most important characteristic of a compound is its constant composition. The elements that make up a particular compound are in fixed, definite proportions in all samples of the compound.

Most of the matter you encounter is in the form of compounds. Water, salt, and carbon dioxide are all examples of common simple compounds. More complex compounds include caffeine, aspirin, acetone, and testosterone.

Chemical Formulas

Chemical formulas represent compounds. Formulas indicate the elements present in the compound and the relative number of atoms of each. These formulas represent the basic units that make up a compound. Pure substances can be categorized according to the basic units that compose them. Elements can be composed of atoms or molecules. Compounds can be molecular, in which case their basic units are molecules, or ionic, in which case their basic units are formula units (composed of cations and anions). We can write the formulas for many ionic compounds simply by knowing the elements in the compound.

To understand compounds, you must understand their composition, which is represented by a chemical formula. The connection between the molecular world and the macroscopic world hinges on the particles that compose matter. Since most matter is in the form of compounds, the properties of most matter depend on the molecules or ions that compose it. Molecular matter does what its molecules do; ionic matter does what its ions do. The world you see and experience is governed by what these particles are doing.

Chemical Nomenclature

We can write the names of simple ionic compounds, molecular compounds, and acids, by examining their chemical formulas. The nomenclature flowchart (Figure 5.17) shows the basic procedure for determining these names.

Because there are so many compounds, there must be a systematic way to name them. By learning these few simple rules, you will be able to name thousands of different compounds. The next time you look at the label on a consumer product, try to identify as many of the compounds as you can by examining their names.

Formula Mass

The formula mass of a compound is the sum of the atomic masses of all the atoms in the chemical formula for the compound. Like atomic mass for elements, formula mass characterizes the average mass of a molecule or formula unit.

Besides being the characteristic mass of a molecule or formula unit, formula mass is important in many calculations involving the composition of compounds and quantities in chemical reactions.

Chemical Skills

Examples

LO: Restate and apply the law of constant composition (Section 5.2).

The law of constant composition states that all samples of a given compound have the same ratio of their constituent elements.

To determine whether experimental data are consistent with the law of constant composition, calculate the ratios of the masses of each element in all samples. When calculating these ratios, it is most convenient to put the larger number in the numerator (top) and the smaller one in the denominator (bottom); that way, the ratio is greater than 1. If the ratios are the same, then the data are consistent with the law of constant composition.

EXAMPLE 5.16 Constant Composition of Compounds

Two samples said to be carbon disulfide (CS_2) are decomposed into their constituent elements. One sample produces 8.08 g S and 1.51 g C, while the other produces 31.3 g S and 3.85 g C. Are these results consistent with the law of constant composition?

SOLUTION

Sample 1

$$\frac{\text{mass S}}{\text{mass C}} = \frac{8.08\text{ g}}{1.51\text{ g}} = 5.35$$

Sample 2

$$\frac{\text{mass S}}{\text{mass C}} = \frac{31.3\text{ g}}{3.85\text{ g}} = 8.13$$

These results are not consistent with the law of constant composition, so the information that the two samples are the same substance must therefore be in error.

LO: Write chemical formulas (Section 5.3).

Chemical formulas indicate the elements present in a compound and the relative number of atoms of each. When writing formulas, put the more metallic element first.

EXAMPLE 5.17 Writing Chemical Formulas

Write a chemical formula for the compound containing one nitrogen atom for every two oxygen atoms.

SOLUTION

NO_2

LO: Determine the total number of each type of atom in a chemical formula (Section 5.3).

The numbers of atoms not enclosed in parentheses are given directly by their subscript.

Find the numbers of atoms within parentheses by multiplying their subscript within the parentheses by their subscript outside the parentheses.

EXAMPLE 5.18 Determining the Total Number of Each Type of Atom in a Chemical Formula

Determine the number of each type of atom in $Pb(ClO_3)_2$.

SOLUTION

one Pb atom

two Cl atoms

six O atoms

LO: Classify elements as atomic or molecular (Section 5.4).

Most elements exist as atomic elements; their basic units in nature are individual atoms. However, several elements (H_2, N_2, O_2, F_2, Cl_2, Br_2, and I_2) exist as molecular elements; their basic units in nature are diatomic molecules.

EXAMPLE 5.19 Classifying Elements as Atomic or Molecular

Classify each element as atomic or molecular: sodium, iodine, and nitrogen.

SOLUTION

sodium: atomic

iodine: molecular (I_2)

nitrogen: molecular (N_2)

LO: Classify compounds as ionic or molecular (Section 5.4).

Compounds containing a metal and a nonmetal are ionic. If the metal is a transition metal, it is likely to form more than one type of ion (see exceptions in Figure 5.13). If the metal is not a transition metal, it is likely to form only one type of ion (see exceptions in Table 5.4).

Compounds composed of only nonmetals are molecular.

EXAMPLE 5.20 Classifying Compounds as Ionic or Molecular

Classify each compound as ionic or molecular. If they are ionic, determine whether the metal forms only one type of ion or more than one type of ion.

$FeCl_3$, K_2SO_4, CCl_4

SOLUTION

$FeCl_3$: ionic, metal forms more than one type of ion

K_2SO_4: ionic, metal forms only one type of ion

CCl_4: molecular

LO: Write formulas for ionic compounds (Section 5.5).

1. Write the symbol for the metal ion followed by the symbol for the nonmetal ion (or polyatomic ion) and their charges. These charges can be deduced from the group numbers in the periodic table. (In the case of polyatomic ions, the charges come from Table 5.3.)
2. Use the magnitude of the charge on each ion as the subscript for the other ion.
3. Check to see if you can reduce the subscripts to simpler whole numbers. Drop subscripts of 1; they are implied.
4. Confirm that the sum of the charges of the cations exactly cancels the sum of the charges of the anions.

EXAMPLE 5.21 Writing Formulas for Ionic Compounds

Write a formula for the compound that forms from lithium and sulfate ions.

SOLUTION

Li^+ SO_4^{2-}

$Li_2(SO_4)$

In this case, the subscripts cannot be further reduced.

Li_2SO_4

Cations	Anions
2(1+) = 2+	2−

LO: Name binary ionic compounds containing a metal that forms only one type of ion (Section 5.7).

The name of the metal is unchanged. The name of the nonmetal is its base name with the ending *-ide*.

EXAMPLE 5.22 Naming Binary Ionic Compounds Containing a Metal That Forms Only One Type of Ion

Name the compound Al_2O_3.

SOLUTION

aluminum oxide

LO: Name binary ionic compounds containing a metal that forms more than one type of ion (Section 5.7).

Because the names of these compounds include the charge of the metal ion, first determine that charge by calculating the total charge of the nonmetal ions.

The total charge of the metal ions must equal the total charge of the nonmetal ions, but have the opposite sign.

The name of the compound is the name of the metal ion, followed by the charge of the metal ion, followed by the base name of the nonmetal + *-ide*.

EXAMPLE 5.23 Naming Binary Ionic Compounds Containing a Metal That Forms More Than One Type of Ion

Name the compound Fe_2S_3.

SOLUTION

3 sulfide ions × (2−) = 6−

2 iron ions × (*ion charge*) = 6+

ion charge = 3+

charge of each iron ion = 3+

iron(III) sulfide

LO: Name compounds containing a polyatomic ion (Section 5.7).

Name ionic compounds containing a polyatomic ion in the normal way, except substitute the name of the polyatomic ion (from Table 5.3) in place of the nonmetal.

Because the metal in this example forms more than one type of ion, you need to determine the charge on the metal ion. The charge of the metal ion must be equal in magnitude to the sum of the charges of the polyatomic ions but opposite in sign.

The name of the compound is the name of the metal ion, followed by the charge of the metal ion, followed by the name of the polyatomic ion.

EXAMPLE 5.24 Naming Compounds Containing a Polyatomic Ion

Name the compound $Co(ClO_4)_2$.

SOLUTION

2 perchlorate ions $\times$ (1−) = 2−

charge of cobalt ion = 2+

cobalt(II) perchlorate

LO: Name molecular compounds (Section 5.8).

The name consists of a prefix indicating the number of atoms of the first element, followed by the name of the first element, and a prefix for the number of atoms of the second element, followed by the base name of the second element plus the suffix *-ide*. The prefix *-mono* is normally dropped on the first element.

EXAMPLE 5.25 Naming Molecular Compounds

Name the compound NO_2.

SOLUTION

nitrogen dioxide

LO: Name binary acids (Section 5.9).

The name begins with *hydro-*, followed by the base name of the nonmetal, plus the suffix *-ic*, and the word *acid*.

EXAMPLE 5.26 Naming Binary Acids

Name the acid $HI(aq)$.

SOLUTION

hydroiodic acid

LO: Name oxyacids containing an oxyanion ending in *–ate* (Section 5.9).

The name is the base name of the oxyanion + *-ic*, followed by the word *acid* (sulfate violates the rule somewhat; in strict terms, the base name would be *sulf*).

EXAMPLE 5.27 Naming Oxyacids Containing an Oxyanion Ending in *-ate*

Name the acid $H_2SO_4(aq)$.

SOLUTION

The oxyanion is sulfate. The name of the acid is *sulfuric acid*.

LO: Name oxyacids containing an oxyanion ending in *–ite* (Section 5.9).

The name is the base name of the oxyanion + *-ous*, followed by the word *acid*.

EXAMPLE 5.28 Naming Oxyacids Containing an Oxyanion Ending in *-ite*

Name the acid $HClO_2(aq)$.

SOLUTION

The oxyanion is chlorite. The name of the acid is *chlorous acid*.

LO: Calculate formula mass (Section 5.11).

The formula mass is the sum of the atomic masses of all the atoms in the chemical formula. In determining the number of each type of atom, multiply subscripts inside parentheses by subscripts outside parentheses.

EXAMPLE 5.29 Calculating Formula Mass

Calculate the formula mass of $Mg(NO_3)_2$.

SOLUTION

$$\text{formula mass} = 24.31 + 2(14.01) + 6(16.00) = 148.33 \text{ amu}$$

Key Terms

acid [5.9]
atomic element [5.4]
ball-and-stick model [5.3]
binary acid [5.9]
binary compound [5.7]
chemical formula [5.3]
empirical formula [5.3]
formula mass [5.11]
formula unit [5.4]
ionic compound [5.4]
law of constant composition [5.2]
molecular compound [5.4]
molecular element [5.4]
molecular formula [5.3]
molecular model [5.3]
oxyacid [5.9]
oxyanion [5.7]
polyatomic ion [5.3]
space-filling model [5.3]
structural formula [5.3]
transition metals [5.7]

Exercises

Questions

1. Do the properties of an element change when it combines with another element to form a compound? Explain.
2. How might the world be different if elements did not combine to form compounds?
3. What is the law of constant composition? Who discovered it?
4. What is a chemical formula? List some examples.
5. In a chemical formula, which element is listed first?
6. In a chemical formula, how do you calculate the number of atoms of an element within parentheses? Provide an example.
7. Explain the difference between a molecular formula and an empirical formula.
8. What is a structural formula? What is the difference between a structural formula and a molecular model?
9. What is the difference between a molecular element and an atomic element? List the elements that occur as diatomic molecules.
10. What is the difference between an ionic compound and a molecular compound?
11. What is the difference between a common name for a compound and a systematic name?
12. List the metals that form only one type of ion (that is, metals whose charge is invariant from one compound to another). What are the group numbers of these metals?
13. Identify the block in the periodic table of metals that tend to form more than one type of ion.
14. What is the basic form for the names of ionic compounds containing a metal that forms only one type of ion?
15. What is the basic form for the names of ionic compounds containing a metal that forms more than one type of ion?
16. Why are roman numerals needed in the names of ionic compounds containing a metal that forms more than one type of ion?
17. How are compounds containing a polyatomic ion named?
18. Which polyatomic ions have a 2− charge? Which polyatomic ions have a 3− charge?
19. What is the basic form for the names of molecular compounds?
20. How many atoms does each prefix specify? *mono-*, *di-*, *tri-*, *tetra-*, *penta-*, *hexa-*.
21. What is the basic form for the names of binary acids?
22. What is the basic form for the name of oxyacids whose oxyanions end with *-ate*?
23. What is the basic form for the name of oxyacids whose oxyanions end with *-ite*?
24. What is the formula mass of a compound?

Problems

CONSTANT COMPOSITION OF COMPOUNDS

25. Two samples of sodium chloride are decomposed into their constituent elements. One sample produces 4.65 g of sodium and 7.16 g of chlorine, and the other sample produces 7.45 g of sodium and 11.5 g of chlorine. Are these results consistent with the law of constant composition? Explain your answer.

26. Two samples of carbon tetrachloride are decomposed into their constituent elements. One sample produces 32.4 g of carbon and 373 g of chlorine, and the other sample produces 12.3 g of carbon and 112 g of chlorine. Are these results consistent with the law of constant composition? Explain your answer.

27. Upon decomposition, one sample of magnesium fluoride produced 1.65 kg of magnesium and 2.57 kg of fluorine. A second sample produced 1.32 kg of magnesium. How much fluorine (in grams) did the second sample produce? Remember that, according to the law of constant composition, the ratio of the masses of the two elements must be the same in both samples.

28. The mass ratio of sodium to fluorine in sodium fluoride is 1.21:1. A sample of sodium fluoride produces 34.5 g of sodium upon decomposition. How much fluorine (in grams) forms? *Hint:* the ratio $\frac{\text{mass sodium}}{\text{mass fluorine}} = 1.21$.

29. Use the law of constant composition to complete the table summarizing the amounts of nitrogen and oxygen produced upon the decomposition of several samples of dinitrogen monoxide. Remember that, according to the law of constant composition, the ratio of the masses of the two elements $\left(\frac{\text{mass nitrogen}}{\text{mass oxygen}}\right)$ must be the same in all samples.

	Mass N_2O	Mass N	Mass O
Sample A	2.85 g	1.82 g	1.03 g
Sample B	4.55 g	____	____
Sample C	____	____	1.35 g
Sample D	____	1.11 g	____

30. Use the law of constant composition to complete the table summarizing the amounts of iron and chlorine produced upon the decomposition of several samples of iron(III) chloride. Remember that, according to the law of constant composition, the ratio of the masses of the two elements $\left(\frac{\text{mass chlorine}}{\text{mass iron}}\right)$ must be the same in all samples.

	Mass $FeCl_3$	Mass Fe	Mass Cl
Sample A	3.785 g	1.302 g	2.483 g
Sample B	2.175 g	____	____
Sample C	____	2.012 g	____
Sample D	____	____	2.329 g

CHEMICAL FORMULAS

31. Write a chemical formula for the compound containing one nitrogen atom for every three iodine atoms.

32. Write a chemical formula for the compound containing one carbon atom for every four bromine atoms.

33. Write chemical formulas for compounds containing:
(a) three iron atoms for every four oxygen atoms
(b) one phosphorus atom for every three chlorine atoms
(c) one phosphorus atom for every five chlorine atoms
(d) two silver atoms for every oxygen atom

34. Write chemical formulas for compounds containing:
(a) one calcium atom for every two iodine atoms
(b) two nitrogen atoms for every four oxygen atoms
(c) one silicon atom for every two oxygen atoms
(d) one zinc atom for every two chlorine atoms

35. How many oxygen atoms are in each chemical formula?
(a) H_3PO_4 (b) Na_2HPO_4
(c) $Ca(HCO_3)_2$ (d) $Ba(C_2H_3O_2)_2$

36. How many hydrogen atoms are in each of the formulas in Question 35?

37. Determine the number of each type of atom in each formula.
(a) $MgCl_2$ (b) $NaNO_3$
(c) $Ca(NO_2)_2$ (d) $Sr(OH)_2$

38. Determine the number of each type of atom in each formula.
(a) NH_4Cl (b) $Mg_3(PO_4)_2$
(c) NaCN (d) $Ba(HCO_3)_2$

39. Complete the table.

Formula	Number of $C_2H_3O_2^-$ Units	Number of Carbon Atoms	Number of Hydrogen Atoms	Number of Oxygen Atoms	Number of Metal Atoms
$Mg(C_2H_3O_2)_2$	____	____	____	____	____
$NaC_2H_3O_2$	____	____	____	____	____
$Cr_2(C_2H_3O_2)_4$	____	____	____	____	____

40. Complete the table.

Formula	Number of SO_4^{2-} Units	Number of Sulfur Atoms	Number of Oxygen Atoms	Number of Metal Atoms
$CaSO_4$	____	____	____	____
$Al_2(SO_4)_3$	____	____	____	____
K_2SO_4	____	____	____	____

41. Give the empirical formula that corresponds to each molecular formula.
(a) C_2H_6 (b) N_2O_4 (c) $C_4H_6O_2$ (d) NH_3

42. Give the empirical formula that corresponds to each molecular formula.
(a) C_2H_2 (b) CO_2 (c) $C_6H_{12}O_6$ (d) B_2H_6

MOLECULAR VIEW OF ELEMENTS AND COMPOUNDS

43. Classify each element as atomic or molecular.
(a) chlorine (b) argon
(c) cobalt (d) hydrogen

44. Which elements have molecules as their basic units?
(a) helium (b) oxygen
(c) iron (d) bromine

45. Classify each compound as ionic or molecular.
(a) CS_2 (b) CuO (c) KI (d) PCl_3

46. Classify each compound as ionic or molecular.
(a) PtO_2 (b) CF_2Cl_2 (c) CO (d) SO_3

47. Match the substances on the left with the basic units that compose them on the right. Remember that atomic elements are composed of atoms, molecular elements are composed of diatomic molecules, molecular compounds are composed of molecules, and ionic compounds are composed of formula units.

helium	molecules
CCl_4	formula units
K_2SO_4	diatomic molecules
bromine	single atoms

48. Match the substances on the left with the basic units that compose them on the right. Remember that atomic elements are composed of atoms, molecular elements are composed of diatomic molecules, molecular compounds are composed of molecules, and ionic compounds are composed of formula units.

NI_3	molecules
copper metal	single atoms
$SrCl_2$	diatomic molecules
nitrogen	formula units

49. What are the basic units—single atoms, molecules, or formula units—that compose each substance?
(a) $BaBr_2$ (b) Ne (c) I_2 (d) CO

50. What are the basic units—single atoms, molecules, or formula units—that compose each substance?
(a) Rb_2O (b) N_2 (c) $Fe(NO_3)_2$ (d) N_2F_4

51. Classify each compound as ionic or molecular. If it is ionic, determine whether the metal forms only one type of ion or more than one type of ion.
(a) KCl **(b)** CBr_4 **(c)** NO_2 **(d)** $Sn(SO_4)_2$

52. Classify each compound as ionic or molecular. If it is ionic, determine whether the metal forms only one type of ion or more than one type of ion.
(a) $CoCl_2$ **(b)** CF_4 **(c)** $BaSO_4$ **(d)** NO

WRITING FORMULAS FOR IONIC COMPOUNDS

53. Write a formula for the ionic compound that forms from each pair of elements.
(a) sodium and sulfur
(b) strontium and oxygen
(c) aluminum and sulfur
(d) magnesium and chlorine

54. Write a formula for the ionic compound that forms from each pair of elements.
(a) aluminum and oxygen
(b) beryllium and iodine
(c) calcium and sulfur
(d) calcium and iodine

55. Write a formula for the compound that forms from potassium and
(a) acetate **(b)** chromate
(c) phosphate **(d)** cyanide

56. Write a formula for the compound that forms from calcium and
(a) hydroxide **(b)** carbonate
(c) phosphate **(d)** hydrogen phosphate

57. Write formulas for the compounds formed from the element on the left and each of the elements on the right.
(a) Li N, O, F
(b) Ba N, O, F
(c) Al N, O, F

58. Write formulas for the compounds formed from the element on the left and each polyatomic ion on the right.
(a) Rb NO_3^-, SO_4^{2-}, PO_4^{3-}
(b) Sr NO_3^-, SO_4^{2-}, PO_4^{3-}
(c) In NO_3^-, SO_4^{2-}, PO_4^{3-}
(Assume In charge is 3+.)

NAMING IONIC COMPOUNDS

59. Name each ionic compound. In each of these compounds, the metal forms only one type of ion.
(a) $CsCl$ **(b)** $SrBr_2$ **(c)** K_2O **(d)** LiF

60. Name each ionic compound. In each of these compounds, the metal forms only one type of ion.
(a) LiI **(b)** MgS **(c)** BaF_2 **(d)** NaF

61. Name each ionic compound. In each of these compounds, the metal forms more than one type of ion.
(a) $CrCl_2$ **(b)** $CrCl_3$ **(c)** SnO_2 **(d)** PbI_2

62. Name each ionic compound. In each of these compounds, the metal forms more than one type of ion.
(a) $HgBr_2$ **(b)** Fe_2O_3 **(c)** CuI_2 **(d)** $SnCl_4$

63. Determine whether the metal in each ionic compound forms only one type of ion or more than one type of ion and name the compound accordingly.
(a) Cr_2O_3 **(b)** NaI **(c)** $CaBr_2$ **(d)** SnO

64. Determine whether the metal in each ionic compound forms only one type of ion or more than one type of ion and name the compound accordingly.
(a) FeI_3 **(b)** $PbCl_4$ **(c)** SrI_2 **(d)** BaO

65. Name each ionic compound containing a polyatomic ion.
(a) $Ba(NO_3)_2$ **(b)** $Pb(C_2H_3O_2)_2$
(c) NH_4I **(d)** $KClO_3$
(e) $CoSO_4$ **(f)** $NaClO_4$

66. Name each ionic compound containing a polyatomic ion.
(a) $Ba(OH)_2$ **(b)** $Fe(OH)_3$
(c) $Cu(NO_2)_2$ **(d)** $PbSO_4$
(e) $KClO$ **(f)** $Mg(C_2H_3O_2)_2$

67. Name each polyatomic ion.
(a) BrO^- **(b)** BrO_2^- **(c)** BrO_3^- **(d)** BrO_4^-

68. Name each polyatomic ion.
(a) IO^- **(b)** IO_2^- **(c)** IO_3^- **(d)** IO_4^-

69. Write a formula for each ionic compound.
(a) copper(II) bromide
(b) silver nitrate
(c) potassium hydroxide
(d) sodium sulfate
(e) potassium hydrogen sulfate
(f) sodium hydrogen carbonate

70. Write a formula for each ionic compound.
(a) copper(I) chlorate
(b) potassium permanganate
(c) lead(II) chromate
(d) calcium fluoride
(e) iron(II) phosphate
(f) lithium hydrogen sulfite

NAMING MOLECULAR COMPOUNDS

71. Name each molecular compound.
(a) SO_2 (b) NI_3 (c) BrF_5
(d) NO (e) N_4Se_4

72. Name each molecular compound.
(a) XeF_4 (b) PI_3 (c) SO_3
(d) $SiCl_4$ (e) I_2O_5

73. Write a formula for each molecular compound.
(a) carbon monoxide
(b) disulfur tetrafluoride
(c) dichlorine monoxide
(d) phosphorus pentafluoride
(e) boron tribromide
(f) diphosphorus pentasulfide

74. Write a formula for each molecular compound.
(a) chlorine monoxide
(b) xenon tetroxide
(c) xenon hexafluoride
(d) carbon tetrabromide
(e) diboron tetrachloride
(f) tetraphosphorus triselenide

75. Determine whether the name shown for each molecular compound is correct. If not, provide the compound's correct name.
(a) PBr_5 phosphorus(V) pentabromide
(b) P_2O_3 phosphorus trioxide
(c) SF_4 monosulfur hexafluoride
(d) NF_3 nitrogen trifluoride

76. Determine whether the name shown for each molecular compound is correct. If not, provide the compound's correct name.
(a) NCl_3 nitrogen chloride
(b) CI_4 carbon(IV) iodide
(c) CO carbon oxide
(d) SCl_4 sulfur tetrachloride

NAMING ACIDS

77. Determine whether each acid is a binary acid or an oxyacid and name each acid. If the acid is an oxyacid, provide the name of the oxyanion.
(a) $HNO_2(aq)$ (b) $HI(aq)$
(c) $H_2SO_4(aq)$ (d) $HNO_3(aq)$

78. Determine whether each acid is a binary acid or an oxyacid and name each acid. If the acid is an oxyacid, provide the name of the oxyanion.
(a) $H_2CO_3(aq)$ (b) $HC_2H_3O_2(aq)$
(c) $H_3PO_4(aq)$ (d) $HCl(aq)$

79. Name each acid.
(a) $HClO$ (b) $HClO_2$ (c) $HClO_3$ (d) $HClO_4$

80. Name each acid. (*Hint:* The names of the oxyanions are analogous to the names of the oxyanions of chlorine.)
(a) $HBrO_3$ (b) HIO_3

81. Write a formula for each acid.
(a) phosphoric acid (b) hydrobromic acid
(c) sulfurous acid

82. Write a formula for each acid.
(a) hydrofluoric acid (b) hydrocyanic acid
(c) chlorous acid

FORMULA MASS

83. Calculate the formula mass for each compound.
(a) HNO_3 (b) $CaBr_2$ (c) CCl_4 (d) $Sr(NO_3)_2$

84. Calculate the formula mass for each compound.
(a) CS_2 (b) $C_6H_{12}O_6$ (c) $Fe(NO_3)_3$ (d) C_7H_{16}

85. Arrange the compounds in order of decreasing formula mass.
Ag_2O, PtO_2, $Al(NO_3)_3$, PBr_3

86. Arrange the compounds in order of decreasing formula mass.
WO_2, Rb_2SO_4, $Pb(C_2H_3O_2)_2$, RbI

Cumulative Problems

87. Write a molecular formula for each molecular model. (White = hydrogen; red = oxygen; black = carbon; blue = nitrogen; yellow = sulfur)

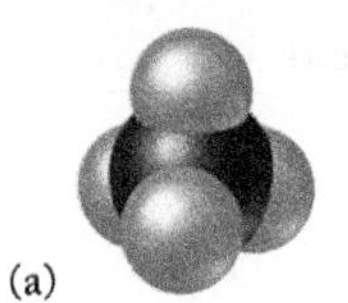
(a)

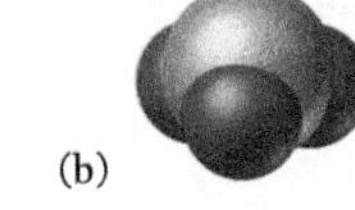
(b)

(c)

88. Write a molecular formula for each molecular model. (White = hydrogen; red = oxygen; black = carbon; blue = nitrogen; yellow = sulfur)

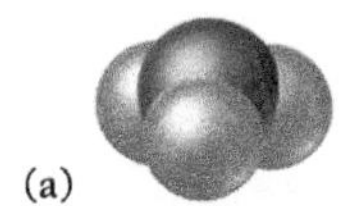
(a)

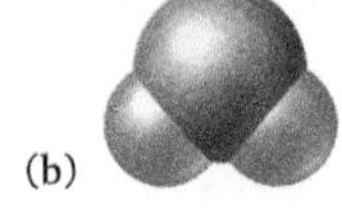
(b)

(c)

89. How many chlorine atoms are in each set?
(a) three carbon tetrachloride molecules
(b) two calcium chloride formula units
(c) four phosphorus trichloride molecules
(d) seven sodium chloride formula units

90. How many oxygen atoms are in each set?
(a) four dinitrogen monoxide molecules
(b) two calcium carbonate formula units
(c) three sulfur dioxide molecules
(d) five perchlorate ions

91. Specify the number of hydrogen atoms (white) represented in each set of molecular models:

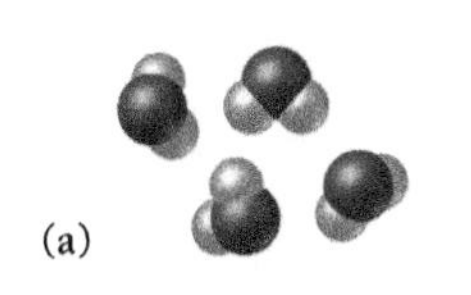
(a)
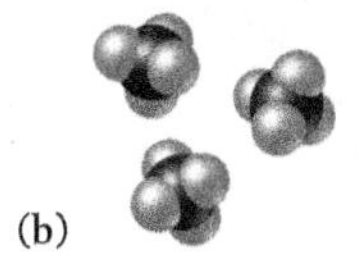
(b)
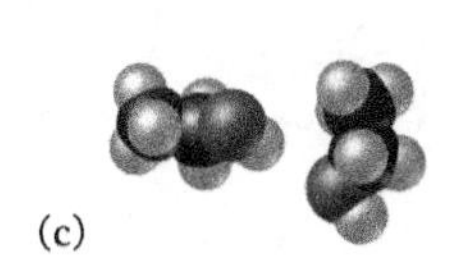
(c)

92. Specify the number of oxygen atoms (red) represented in each set of molecular models:

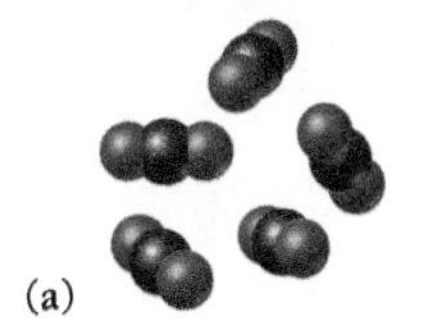
(a)
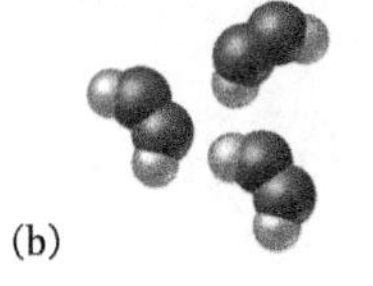
(b)
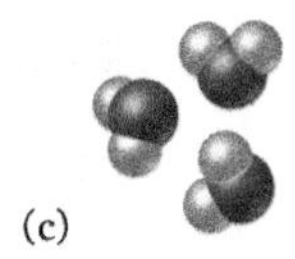
(c)

93. Complete the table:

Formula	Type of Compound (Ionic, Molecular, Acid)	Name
N_2H_4	molecular	________
________	________	potassium chloride
$H_2CrO_4(aq)$	________	________
________	________	cobalt(III) cyanide

94. Complete the table:

Formula	Type of Compound (Ionic, Molecular, Acid)	Name
$K_2Cr_2O_7$	ionic	________
$HBr(aq)$	________	hydrobromic acid
________	________	dinitrogen pentoxide
PbO_2	________	________

95. Is each name correct for the given formula? If not, provide the correct name.
(a) $Ca(NO_2)_2$ calcium nitrate
(b) K_2O dipotassium monoxide
(c) PCl_3 phosphorus chloride
(d) $PbCO_3$ lead(II) carbonate
(e) KIO_2 potassium hypoiodite

96. Is each name correct for the given formula? If not, provide the correct name.
(a) $HNO_3(aq)$ hydrogen nitrate
(b) NaClO sodium hypochlorite
(c) CaI_2 calcium diiodide
(d) $SnCrO_4$ tin chromate
(e) $NaBrO_3$ sodium bromite

97. For each compound, list the correct formula and calculate the formula mass.
(a) tin(IV) sulfate
(b) nitrous acid
(c) sodium bicarbonate
(d) phosphorus pentafluoride

98. For each compound, list the correct formula and calculate the formula mass.
(a) barium bromide
(b) dinitrogen trioxide
(c) copper(I) sulfate
(d) hydrobromic acid

99. Name each compound and calculate its formula mass.
(a) PtO_2
(b) N_2O_5
(c) $Al(ClO_3)_3$
(d) PBr_5

100. Name each compound and calculate its formula mass.
(a) $Al_2(SO_4)_3$
(b) P_2O_3
(c) $HClO(aq)$
(d) $Cr(C_2H_3O_2)_3$

101. A compound contains only carbon and hydrogen and has a formula mass of 28.06 amu. What is its molecular formula?

102. A compound contains only nitrogen and oxygen and has a formula mass of 44.02 amu. What is its molecular formula?

103. Carbon has two naturally occurring isotopes: carbon-12 (mass = 12.00 amu) and carbon-13 (mass = 13.00 amu). Chlorine also has two naturally occurring isotopes: chlorine-35 (mass = 34.97 amu) and chlorine-37 (mass = 36.97 amu). How many CCl_4 molecules of different masses can exist? Determine the mass (in amu) of each of them.

104. Nitrogen has two naturally occurring isotopes: nitrogen-14 (mass = 14.00 amu) and nitrogen-15 (mass = 15.00 amu). Bromine also has two naturally occurring isotopes: bromine-79 (mass = 78.92 amu) and bromine-81 (mass = 80.92 amu). How many types of NBr_3 molecules of different masses can exist? Determine the mass (in amu) of each of them.

Highlight Problems

105. Examine each substance and the corresponding molecular view and classify it as an atomic element, a molecular element, a molecular compound, or an ionic compound.

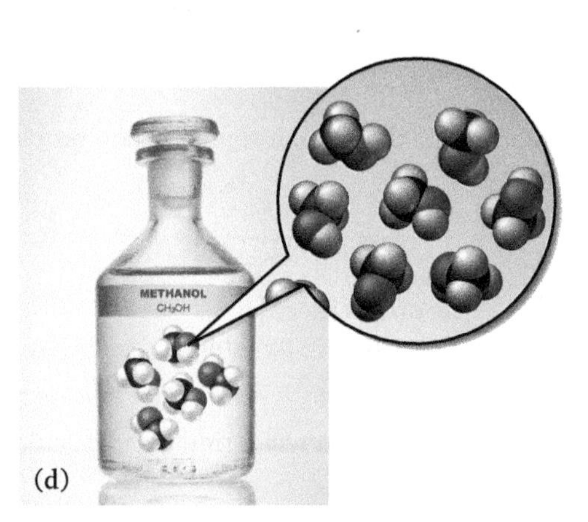

106. Molecules can be as small as two atoms or as large as thousands of atoms. In 1962, Max F. Perutz and John C. Kendrew were awarded the Nobel Prize for their discovery of the structure of hemoglobin, a very large molecule that transports oxygen from the lungs to cells through the bloodstream. The chemical formula of hemoglobin is $C_{2952}H_{4664}O_{832}N_{812}S_8Fe_4$. Calculate the formula mass of hemoglobin.

▲ Max Perutz and John C. Kendrew won a Nobel Prize in 1962 for determining the structure of hemoglobin by X-ray diffraction.

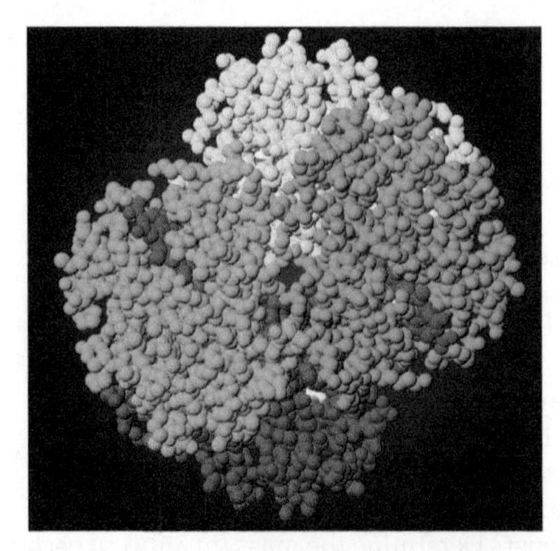

▲ Computer-generated model of hemoglobin.

107. Examine each consumer product label. Write chemical formulas for as many of the compounds as possible based on what you have learned in this chapter.

(a)

Active Ingredient:
Sodium Hypochlorite 6.0%
Other Ingredients: 94.0%
Total: 100.0%
(Yields 5.7% available chlorine)

KEEP OUT OF REACH OF CHILDREN

DANGER: CORROSIVE.

FIRST AID: IF IN EYES: Hold eye open and rinse slowly and gently with water for 15–20 minutes. Remove contact lenses, if present, after the first 5 minutes, then continue rinsing eye. IF ON SKIN OR CLOTHING: Take off contaminated clothing. Rinse skin immediately with plenty of water for 15–20 minutes. IN EITHER CASE, CALL A POISON CONTROL CENTER OR DOCTOR IMMEDIATELY FOR TREATMENT ADVICE. See back panel for additional precautionary labeling.

Kills Methicillin Resistant Staphylococcus aureus (MRSA) *Staphylococcus aureus, Streptococcus pyogenes, Salmonella enterica, Escherichia coli O157:H7 and Influenza A2

3 QT (96 FL OZ) 2.83 L

(b)

Drug Facts

Active ingredients (in each 5 mL teaspoon)	Purposes
Aluminum hydroxide (equivalent to dried gel, USP) 400 mg	Antacid
Magnesium hydroxide 400 mg	Antacid
Simethicone 40 mg	Antigas

Use relieves: ■ heartburn ■ acid indigestion ■ sour stomach ■ upset stomach due to these symptoms ■ pressure and bloating commonly referred to as gas

Warnings
Ask a doctor before use if you have
■ kidney disease ■ a magnesium-restricted diet

Ask a doctor or pharmacist if you are taking a prescription drug. Antacids may interact with certain prescription drugs.

Stop use and ask a doctor if symptoms last more than 2 weeks.

Keep out of reach of children.

Directions ■ shake well ■ adults/children 12 years and older: take 2-4 teaspoonfuls between meals, at bedtime, or as directed by a doctor ■ do not take more than 12 teaspoonfuls in a 24-hour period, or use the maximum dosage for more than 2 weeks ■ children under 12 years: ask a doctor

Other information ■ each teaspoon contains: magnesium 171 mg ■ do not use if breakaway band on plastic cap is broken or missing ■ does not meet USP requirements for preservative effectiveness ■ do not freeze

Inactive ingredients butylparaben, carboxymethylcellulose sodium, flavors, hypromellose, microcrystalline cellulose, propylparaben, purified water, sodium saccharin, sorbitol

Questions or comments?
1-800-469-5268 (English) or
1-888-466-8746 (Spanish)

Johnson-Johnson - MERCK
Consumer Pharmaceuticals Co.
FORT WASHINGTON, PA 19054 USA

(c)

NOTICE! PROTECTIVE INNER SEAL BENEATH CAP. IF MISSING OR DAMAGED, DO NOT USE CONTENTS.

Drug Facts

Active ingredients (in each tablet)	Purpose
Calcium carbonate 1000 mg	Antacid
Simethicone 60 mg	Antigas

Uses for the relief of
• acid indigestion
• heartburn
• sour stomach
• upset stomach associated with these symptoms
• bloating and pressure commonly referred to as gas

Warning
Do not take more than 8 tablets in a 24-hour period or use the maximum dosage for more than 2 weeks except under the advice and supervision of a physician

Ask a doctor before use if you have
• kidney stones • a calcium-restricted diet

Ask a doctor or pharmacist before use if your are presently taking a prescription drug. Antacids may interact with certain prescription drugs.

When using this product
• at maximum dose, constipation may occur

Stop use and ask a doctor if
• symptoms last more than two weeks

Keep out of reach of children.

Directions

(d)

Nutrition Facts
Serving Size 1/8 tsp (0.6g)
Servings Per Container about 472
Amount Per Serving
Calories 0
% Daily Value*
Total Fat 0g 0%
Sodium 65mg 3%
Total Carb. 0g 0%
Protein 0g
Calcium 2%
Not a significant source of calories from fat, saturated fat, trans fat, cholesterol, dietary fiber, sugars, vitamin A, vitamin C and iron.
*Percent Daily Values are based on a 2,000 calorie diet.
Ingredients: Cornstarch, Sodium Bicarbonate, Sodium Aluminum Sulfate, Monocalcium Phosphate.
CLABBER GIRL CORPORATION
TERRE HAUTE, IN 47808
davisbakingpowder.com
MADE IN USA

Questions for Group Work

Discuss these questions with the group and record your consensus answer.

108. Write the correct formula for each species: carbon monoxide, carbon dioxide, the carbonate ion. List as many similarities and differences between these three species as you can. Try to get at least one contribution from each group member.

109. What questions do you need to ask about a substance in order to determine whether it is (1) an atomic element, (2) a molecular element, (3) a molecular compound, or (4) an ionic compound? Write a detailed set of instructions describing how to determine the classification of a substance based on the answers to your questions.

110. Name each compound: $Ca(NO_3)_2$, N_2O, NaS, $CrCl_3$. For each compound, include a detailed step-by-step description of the process you used to determine the name. (*Tip:* Each group member could name one compound and present it to the whole group.)

111. Calculate the formula mass for each compound in Group Work Question 110.

Data Interpretation and Analysis

112. Climate scientists have become increasingly concerned that rising levels of carbon dioxide in the atmosphere (produced by the burning of fossil fuels) will affect the global climate in harmful ways such as increased temperatures, rising sea levels, and coastal flooding. The graph at right shows the concentration of carbon dioxide in the atmosphere from the mid-1800s to the present time. Study the graph and answer the questions that follow.

(a) What were the carbon dioxide concentrations in 1950 and 2000? How much did the carbon dioxide concentration increase during these 50 years?

(b) What was the average yearly increase between 1950 and 2000?

(c) Beginning from the carbon dioxide concentration in 2010 (390 ppm), and assuming the average yearly increase you calculated in part b, what will the carbon dioxide concentration be in 2050?

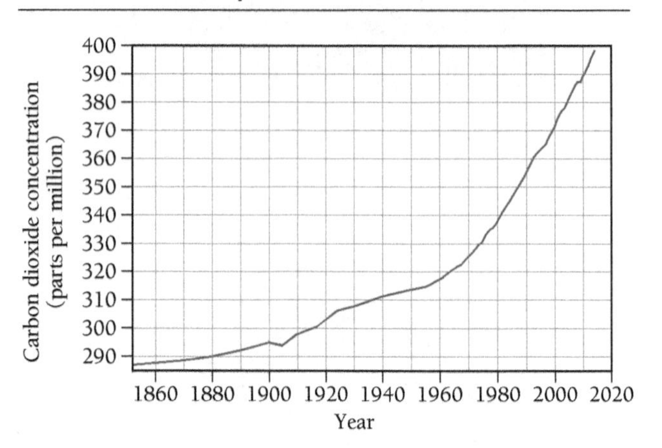

Answers to Skillbuilder Exercises

Skillbuilder 5.1 .... Yes, because in both cases

$$\frac{\text{Mass O}}{\text{Mass C}} = 1.3$$

Skillbuilder 5.2 .... **(a)** Ag_2S
(b) N_2O
(c) TiO_2

Skillbuilder 5.3 .... two K atoms, one S atom, four O atoms

Skillbuilder Plus, p. 137 .... two Al atoms, three S atoms, twelve O atoms

Skillbuilder 5.4 .... **(a)** molecular element
(b) molecular compound
(c) atomic element
(d) ionic compound
(e) ionic compound

Skillbuilder 5.5 .... $SrCl_2$

Skillbuilder 5.6 .... AlN

Skillbuilder 5.7 .... $AlPO_4$

Skillbuilder Plus, p. 143 .... Na_2SO_3

Skillbuilder 5.8 .... potassium bromide

Skillbuilder Plus, p. 146 .... zinc nitride

Skillbuilder 5.9 .... lead(II) oxide

Skillbuilder 5.10 .... manganese(II) nitrate

Skillbuilder 5.11 .... dinitrogen tetroxide

Skillbuilder 5.12 .... hydrofluoric acid

Skillbuilder 5.13 .... nitrous acid

Skillbuilder 5.15 .... 44.02 amu

Answers to Conceptual Checkpoints

5.1 (c) The ratio of A/B is 3.0, and A is 9.0 g, so B must be 3.0 g.

5.2 (b) This formula represents 2 Al atoms + 3(2 Cr atoms +7 O atoms) = 29 atoms.

5.3 H_2O_2

5.4 (b) The figure represents a molecular compound because the compound exists as individual molecules. Figure (a) represents an ionic compound with formula units in a lattice structure.

5.5 (c) Because the nitrate ion has a charge of 1−, the three nitrate ions together have a charge of 3−. Because the compound must be charge-neutral, Cr must have a charge of 3+.

5.6 Because calcium forms only one type of ion (Ca^{2+}); therefore, the charge of the ion is not included in the name (it is always the same, 2+).

5.7 Iodate

5.8 This question addresses one of the most common errors in nomenclature: the failure to correctly categorize the compound. NCl_3 is a molecular compound (two or more nonmetals) and therefore requires prefixes to indicate the number of each type of atom. $AlCl_3$ is an ionic compound (metal and nonmetal) and therefore requires no such prefixes.

5.9 (b)

Moonlight Diner
Specials
Breaksfast Special
Two eggs, Three strips
of bacon, Sausage link,
Toast ,NaCl
$5.95
Omelettes
Served with side of NaCl
Denver Omelette
BP Delight
Atherosclerosis
Sodium Chloride
Hypertension
On the Healthy Side
120/80 mmHg

Lunch Special
1500 milligrams a day

142/93 mmHg
Symptoms:
Chest pain
Confusion
Ear noise or buzzing
Irregular heartbeat
Nosebleed

6 Chemical Composition

In science, you don't ask why, you ask how much.
—Erwin Chargaff (1905–2002)

CHAPTER OUTLINE

6.1 How Much Sodium?

Sodium is an important dietary mineral that we eat in food, primarily as sodium chloride (table salt). Sodium helps regulate body fluids, and eating too much of it can lead to high blood pressure. High blood pressure, in turn, increases the risk of stroke and heart attack. Consequently, people with high blood pressure should limit their sodium intake. The FDA recommends that a person consume less than 2.4 g (2400 mg) of sodium per day. However, sodium is usually consumed as sodium chloride, so the mass of sodium that we eat is not the same as the mass of sodium chloride that we eat. How many grams of sodium chloride can we consume and still stay below the FDA recommendation for sodium?

To answer this question, we need to know the *chemical composition* of sodium chloride. From Chapter 5, we are familiar with its formula, $NaCl$, which indicates that there is one sodium ion to every chloride ion. However, because the masses of sodium and chlorine are different, the relationship between the mass of sodium and the mass of sodium chloride is not clear from the chemical formula alone. In this chapter, we learn how to use the information in a chemical formula, together with atomic and formula masses, to calculate the amount of a constituent element in a given amount of a compound (or vice versa).

▲ The mining of iron requires knowing how much iron is in a given amount of iron ore.

Chemical composition is important not just for assessing dietary sodium intake but for addressing many other questions as well. A company that mines iron, for example, wants to know how much iron it can extract from a given amount of iron ore; an organization interested in developing hydrogen as a potential fuel would want to know how much hydrogen it can extract from a given amount of water. Many environmental issues also require knowledge of chemical composition. An estimate of the threat of

◀ Ordinary table salt is a compound called sodium chloride. The sodium within sodium chloride is linked to high blood pressure. In this chapter, we learn how to determine how much sodium is in a given amount of sodium chloride.

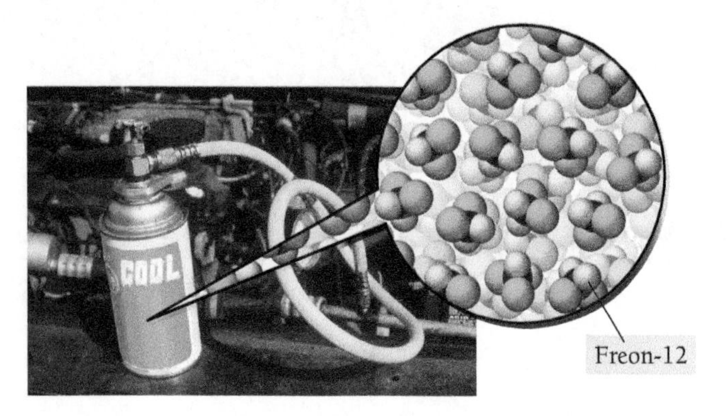

ozone depletion requires knowing how much chlorine is in a given amount of a particular chlorofluorocarbon such as freon-12. To determine these kinds of quantities, we must understand the relationships inherent in a chemical formula and the relationship between numbers of atoms or molecules and their masses. In this chapter, we examine these relationships.

◀ Estimating the threat of ozone depletion requires knowing the amount of chlorine in a given amount of a chlorofluorocarbon.

6.2 Counting Nails by the Pound

▶ Recognize that we use the mass of atoms to count them because they are too small and numerous to count individually.

Some hardware stores sell nails by the pound, which is easier than selling them by the nail because customers often need hundreds of nails and counting them takes too long. However, a customer may still want to know the number of nails contained in a given weight of nails. This problem is similar to asking how many atoms are in a given mass of an element. With atoms, we *must* use their mass as a way to count them because atoms are too small and too numerous to count individually. Even if you could see atoms and counted them 24 hours a day for as long as you lived, you would barely begin to count the number of atoms in something as small as a grain of sand. However, just as the hardware store customer wants to know the number of nails in a given weight, we want to know the number of atoms in a given mass. How do we do that?

Suppose the hardware store customer buys 2.60 lb of medium-sized nails and a dozen nails weigh 0.150 lb. How many nails did the customer buy? This calculation requires two conversions: one between pounds and dozens and another between dozens and number of nails. The conversion factor for the first part is the weight per dozen nails.

$$0.150 \text{ lb nails} = 1 \text{ doz nails}$$

3.4 lb nails

The conversion factor for the second part is the number of nails in one dozen.

$$1 \text{ doz nails} = 12 \text{ nails}$$

The solution map for the problem is:

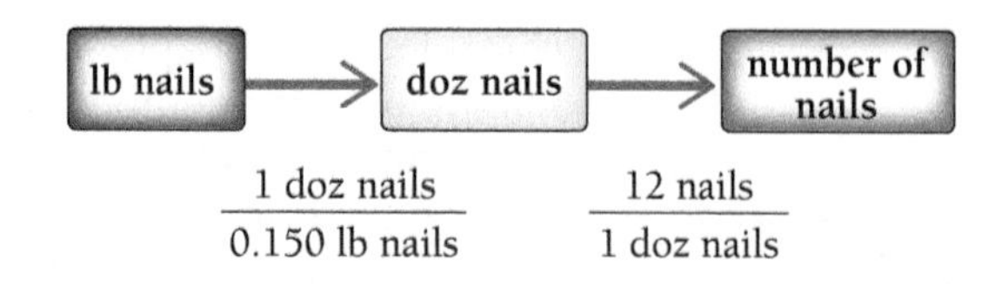

$$\frac{1 \text{ doz nails}}{0.150 \text{ lb nails}} \qquad \frac{12 \text{ nails}}{1 \text{ doz nails}}$$

8.25 grams carbon

▲ Asking how many nails are in a given weight of nails is similar to asking how many atoms are in a given mass of an element. In both cases, we count the objects by weighing them.

Beginning with 2.60 lb and using the solution map as a guide, we convert from lb to number of nails.

$$2.60 \text{ lb nails} \times \frac{1 \text{ doz nails}}{0.150 \text{ lb nails}} \times \frac{12 \text{ nails}}{1 \text{ doz nails}} = 208 \text{ nails}$$

The customer who bought 2.60 lb of nails has 208 nails. She counted the nails by weighing them. If the customer purchased a different size of nail, the first conversion factor—relating pounds to dozens—would change, but the second conversion factor would not. One dozen corresponds to 12 nails, regardless of their size.

CONCEPTUAL CHECKPOINT 6.1

A certain type of nail weighs 0.50 lb per dozen. How many nails are contained in 3.5 lb of these nails?

(a) 84 **(b)** 21 **(c)** 0.58 **(d)** 12

6.3 Counting Atoms by the Gram

- Convert between moles and number of atoms.
- Convert between grams and moles.
- Convert between grams and number of atoms.

Determining the number of atoms in a sample with a certain mass is similar to determining the number of nails in a sample with a certain weight. With nails, we used a dozen as a convenient number in our conversions, but a dozen is too small to use with atoms. We need a larger number because atoms are so small. The chemist's "dozen" is called the **mole (mol)** and has a value of 6.022×10^{23}.

$$1 \text{ mol} = 6.022 \times 10^{23}$$

This is **Avogadro's number**, named after Amadeo Avogadro (1776–1856).

The first thing to understand about the mole is that it can specify Avogadro's number of anything. *One mole of anything is* 6.022×10^{23} *units of that thing.* For example, 1 mol of marbles corresponds to 6.022×10^{23} marbles, and 1 mol of sand grains corresponds to 6.022×10^{23} sand grains. One mole of atoms, ions, or molecules generally makes up objects of reasonable size. For example, 22 copper pennies contain approximately 1 mol of copper (Cu) atoms, and a couple of large helium balloons contain approximately 1 mol of helium (He) atoms.

1 mol of copper atoms

▲ Twenty-two *copper* pennies contain approximately one mole of copper atoms. Pennies were mostly copper until 1982, at which point the U.S. Mint started making them out of zinc with a copper coating (because copper became too valuable).

The second thing to understand about the mole is how it gets its specific value. *The numerical value of the mole is defined as being equal to the number of atoms in exactly 12 g of pure carbon-12.*

This definition of the mole establishes a relationship between mass (grams of carbon) and number of atoms (Avogadro's number). This relationship, as we will see shortly, allows us to count atoms by weighing them.

Converting between Moles and Number of Atoms

Converting between moles and number of atoms is similar to converting between dozens and number of nails. To convert between moles of atoms and number of atoms, we use these conversion factors:

$$\frac{1 \text{ mol}}{6.022 \times 10^{23} \text{ atoms}} \quad \text{or} \quad \frac{6.022 \times 10^{23} \text{ atoms}}{1 \text{ mol}}$$

For example, suppose we want to convert 3.5 mol of helium to a number of helium atoms. We set up the problem in the standard way.

1 mol of helium atoms

▲ Two large helium balloons contain approximately one mole of helium atoms.

GIVEN: 3.5 mol He

FIND: He atoms

RELATIONSHIPS USED $1 \text{ mol He} = 6.022 \times 10^{23} \text{ He atoms}$

SOLUTION MAP We draw a solution map showing the conversion from moles of He to He atoms.

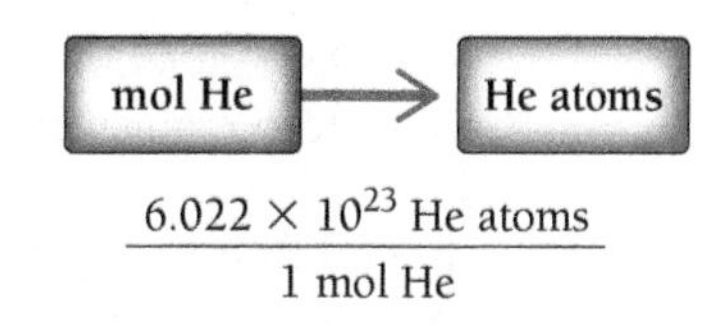

SOLUTION

Beginning with 3.5 mol He, we use the conversion factor to get to He atoms.

$$3.5 \text{ mol He} \times \frac{6.022 \times 10^{23} \text{ He atoms}}{1 \text{ mol He}} = 2.1 \times 10^{24} \text{ He atoms}$$

EXAMPLE 6.1 Converting between Moles and Number of Atoms

A silver ring contains 1.1×10^{22} silver atoms. How many moles of silver are in the ring?

SORT You are given the number of silver atoms and asked to find the number of moles.	**GIVEN:** 1.1×10^{22} Ag atoms **FIND:** mol Ag
STRATEGIZE Draw a solution map, beginning with silver atoms and ending at moles. The conversion factor is Avogadro's number.	**SOLUTION MAP** Ag atoms → mol Ag $\frac{1 \text{ mol Ag}}{6.022 \times 10^{23} \text{ Ag atoms}}$ **RELATIONSHIPS USED** 1 mol Ag = 6.022×10^{23} Ag atoms (Avogadro's number)
SOLVE Follow the solution map to solve the problem. Beginning with 1.1×10^{22} Ag atoms, use the conversion factor to determine the moles of Ag.	**SOLUTION** $1.1 \times 10^{22} \text{ Ag atoms} \times \frac{1 \text{ mol Ag}}{6.022 \times 10^{23} \text{ Ag atoms}} = 1.8 \times 10^{-2} \text{ mol Ag}$
CHECK Are the units correct? Does the answer make physical sense?	The units, mol Ag, are the desired units. The magnitude of the answer is orders of magnitude smaller than the given quantity because it takes many atoms to make a mole, so you expect the answer to be orders of magnitude smaller than the given quantity.

▶ **SKILLBUILDER 6.1 | Converting between Moles and Number of Atoms**

How many gold atoms are in a pure gold ring containing 8.83×10^{-2} mol Au?

▶ **FOR MORE PRACTICE** Example 6.14; Problems 17, 18, 19, 20.

Converting between Grams and Moles of an Element

We just explained how to convert between moles and number of atoms, which is like converting between dozens and number of nails. We need one more conversion factor to convert from the mass of a sample to the number of atoms in the sample. For nails, we used the weight of one dozen nails; for atoms, we use the mass of 1 mol of atoms.

> The mass of 1 mol of atoms of an element is its **molar mass**. The value of an element's molar mass in grams per mole is numerically equal to the element's atomic mass in atomic mass units.

Recall that Avogadro's number, the number of atoms in a mole, is defined as the number of atoms in exactly 12 g of carbon-12. The atomic mass unit is defined as one-twelfth of the mass of a carbon-12 atom, so it follows that the molar mass of any element—the mass of 1 mol of atoms in grams of that element—is equal to the atomic mass of that element expressed in atomic mass units. For example, copper has an atomic mass of 63.55 amu; therefore, 1 mol of copper atoms has a mass of 63.55 g, and the molar mass of copper is 63.55 g/mol. Just as the weight of 1 doz nails changes for different types of nails, so the mass of 1 mol of atoms changes for different elements: 1 mol of sulfur atoms (sulfur atoms are lighter than copper atoms) has a mass of 32.06 g; 1 mol of carbon atoms (lighter than sulfur) has a mass of 12.01 g; and 1 mol of lithium atoms (lighter yet) has a mass of 6.94 g.

$$32.06 \text{ g sulfur} = 1 \text{ mol sulfur} = 6.022 \times 10^{23} \text{ S atoms}$$

$$12.01 \text{ g carbon} = 1 \text{ mol carbon} = 6.022 \times 10^{23} \text{ C atoms}$$

$$6.94 \text{ g lithium} = 1 \text{ mol lithium} = 6.022 \times 10^{23} \text{ Li atoms}$$

The lighter the atom, the less mass in 1 mol of that atom (▼ FIGURE 6.1).

The molar mass of any element is a conversion factor between grams of that element and moles of that element. For carbon:

$$12.01 \text{ g C} = 1 \text{ mol C} \quad \text{or} \quad \frac{12.01 \text{ g C}}{1 \text{ mol C}} \quad \text{or} \quad \frac{1 \text{ mol C}}{12.01 \text{ g C}}$$

1 dozen large nails

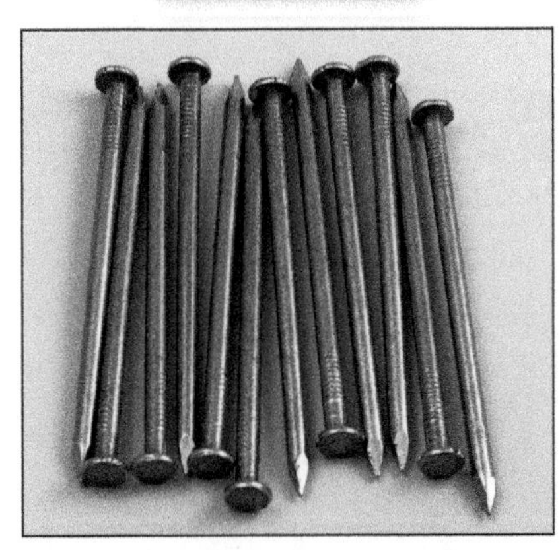

1 dozen small nails

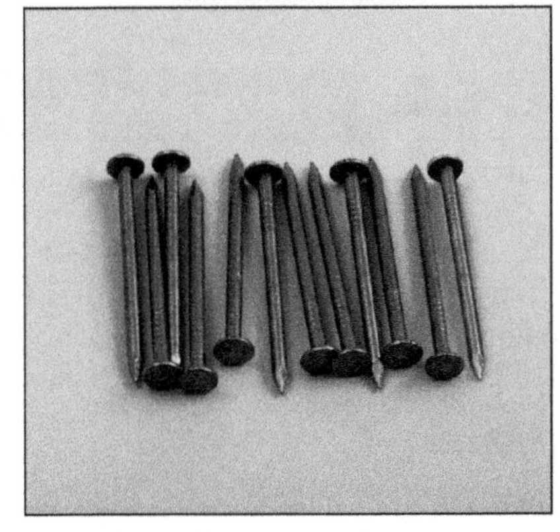

(a)

1 mole S (32.06 g)

1 mole C (12.01 g)

(b)

▶ FIGURE 6.1 **The mass of 1 mol**
(a) Each of these pictures shows the same number of nails: 12. As you can see, 12 large nails have more weight and occupy more space than 12 small nails. The same is true for atoms. **(b)** Each of these samples has the same number of atoms: 6.022×10^{23}. Because sulfur atoms are more massive and larger than carbon atoms, 1 mol of S atoms is heavier and occupies more space than 1 mol of C atoms.

A 0.58-g diamond is about a three-carat diamond.

Suppose we want to calculate the number of moles of carbon in a 0.58-g diamond (pure carbon).

We first sort the information in the problem.

GIVEN: 0.58 g C

FIND: mol C

SOLUTION MAP We then strategize by drawing a solution map showing the conversion from grams of C to moles of C. The conversion factor is the molar mass of carbon.

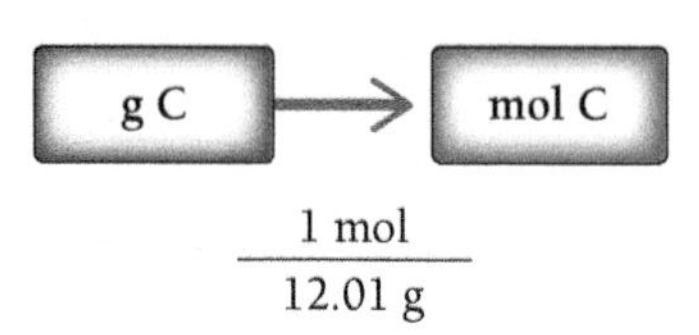

RELATIONSHIPS USED

12.01 g C = 1 mol C (molar mass of carbon, from periodic table)

SOLUTION

Finally, we solve the problem by following the solution map.

$$0.58\ \text{g C} \times \frac{1\ \text{mol C}}{12.01\ \text{g C}} = 4.8 \times 10^{-2}\ \text{mol C}$$

EXAMPLE **6.2** **The Mole Concept—Converting between Grams and Moles**

Calculate the number of moles of sulfur in 57.8 g of sulfur.	
SORT Begin by sorting the information in the problem. You are given the mass of sulfur and asked to find the number of moles.	**GIVEN:** 57.8 g S **FIND:** mol S
STRATEGIZE Draw a solution map showing the conversion from g S to mol S. The conversion factor is the molar mass of sulfur.	**SOLUTION MAP** g S → mol S $\frac{1\ \text{mol S}}{32.06\ \text{g S}}$ **RELATIONSHIPS USED** 32.06 g S = 1 mol S (molar mass of sulfur, from periodic table)
SOLVE Follow the solution map to solve the problem. Begin with 57.8 g S and use the conversion factor to determine mol S.	**SOLUTION** $57.8\ \text{g S} \times \frac{1\ \text{mol C}}{32.06\ \text{g S}} = 1.80\ \text{mol S}$
CHECK Check your answer. Are the units correct? Does the answer make physical sense?	The units (mol S) are correct. The magnitude of the answer makes sense because 1 mol of S has a mass of 32.06 g; therefore, 57.8 g of S should be close to 2 mols.

▶ **SKILLBUILDER 6.2 | The Mole Concept—Converting between Grams and Moles**

Calculate the number of grams of sulfur in 2.78 mol of sulfur.

▶ **FOR MORE PRACTICE** Example 6.15; Problems 25, 26, 27, 28, 29, 30.

Converting between Grams of an Element and Number of Atoms

Suppose we want to know the number of carbon *atoms* in the 0.58-g diamond. We first convert from grams to moles and then from moles to number of atoms. The solution map is:

g C → mol C → number of C atoms

$$\frac{1 \text{ mol C}}{12.01 \text{ g C}} \qquad \frac{6.022 \times 10^{23} \text{ C atoms}}{1 \text{ mol C}}$$

Notice the similarity between this solution map and the one we used for nails:

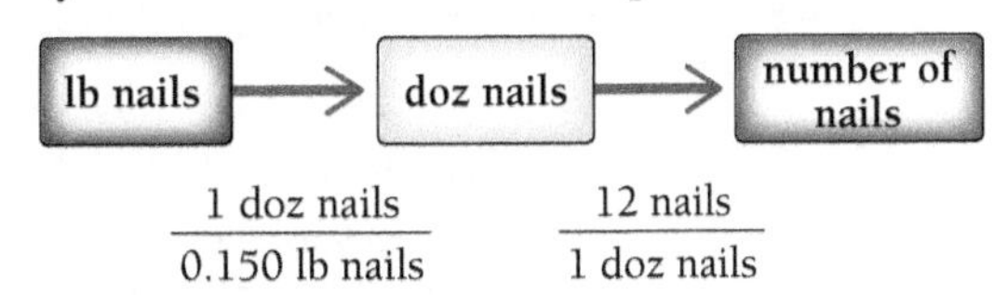

Beginning with 0.58 g carbon and using the solution map as a guide, we convert to the number of carbon atoms.

Interactive Worked Example Video 6.3

$$0.58 \text{ g C} \times \frac{1 \text{ mol C}}{12.01 \text{ g C}} \times \frac{6.022 \times 10^{23} \text{ C atoms}}{1 \text{ mol C}} = 2.9 \times 10^{22} \text{ C atoms}$$

EXAMPLE **6.3** **The Mole Concept—Converting between Grams and Number of Atoms**

How many aluminum atoms are in an aluminum can with a mass of 16.2 g?

SORT You are given the mass of aluminum and asked to find the number of aluminum atoms.	**GIVEN:** 16.2 g Al **FIND:** Al atoms
STRATEGIZE The solution map has two steps. In the first step, convert from g Al to mol Al. In the second step, convert from mol Al to the number of Al atoms. The required conversion factors are the molar mass of aluminum and the number of atoms in a mole.	**SOLUTION MAP** 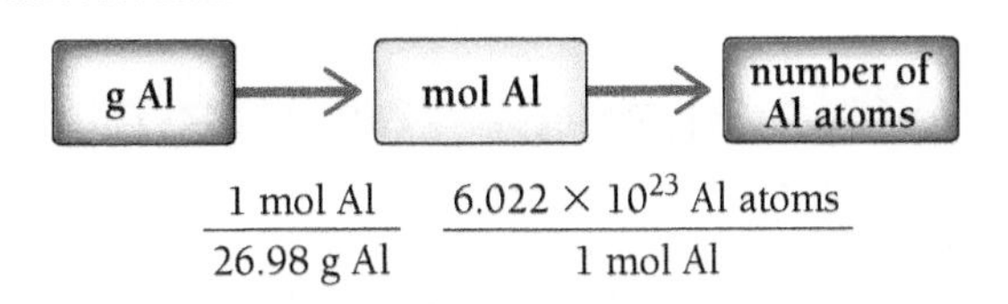 **RELATIONSHIPS USED** 26.98 g Al = 1 mol Al (molar mass of aluminum, from periodic table) $6.022 \times 10^{23} = 1$ mol (Avogadro's number)
SOLVE Follow the solution map to solve the problem, beginning with 16.2 g Al and multiplying by the appropriate conversion factors to arrive at Al atoms.	**SOLUTION** $16.2 \text{ g Al} \times \frac{1 \text{ mol Al}}{26.98 \text{ g Al}} \times \frac{6.022 \times 10^{23} \text{ Al atoms}}{1 \text{ mol Al}} = 3.62 \times 10^{23} \text{ Al atoms}$
CHECK Are the units correct? Does the answer make physical sense?	The units, Al atoms, are correct. The answer makes sense because the number of atoms in any macroscopic-sized sample of matter is very large.

▶ **SKILLBUILDER 6.3** | **The Mole Concept—Converting between Grams and Number of Atoms**
Calculate the mass of 1.23×10^{24} helium atoms.

▶ **FOR MORE PRACTICE** Example 6.16; Problems 35, 36, 37, 38, 39, 40, 41, 42.

Before we move on, notice that numbers with large exponents, such as 6.022×10^{23}, are almost unimaginably large. Twenty-two copper pennies contain 6.022×10^{23} or 1 mol of copper atoms; 6.022×10^{23} pennies would cover Earth's entire surface to a depth of 300 m. Even objects that are small by everyday standards occupy a huge space when we have a mole of them. For example, one crystal of granulated sugar has a mass of less than 1 mg and a diameter of less than 0.1 mm, yet 1 mol of sugar crystals would cover the state of Texas to a depth of several feet. For every increase of 1 in the exponent of a number, the number increases by 10. So a number with an exponent of 23 is incredibly large. A mole has to be a large number because atoms are so small.

CONCEPTUAL CHECKPOINT 6.2

Which statement is *always* true for samples of atomic elements, regardless of the type of element present in the samples?

(a) If two samples of different elements contain the same number of atoms, they contain the same number of moles.

(b) If two samples of different elements have the same mass, they contain the same number of moles.

(c) If two samples of different elements have the same mass, they contain the same number of atoms.

CONCEPTUAL CHECKPOINT 6.3

Without doing any calculations, determine which sample contains the most atoms.

(a) one gram of cobalt

(b) one gram of carbon

(c) one gram of lead

6.4 Counting Molecules by the Gram

- Convert between grams and moles of a compound.
- Convert between mass of a compound and number of molecules.

The calculations we just performed for atoms can also be applied to molecules for covalent compounds or formula units for ionic compounds. We first convert between the mass of a compound and moles of the compound, and then we calculate the number of molecules (or formula units) from moles.

Converting between Grams and Moles of a Compound

Remember, ionic compounds do not contain individual molecules. The smallest electrically neutral collection of ions is a formula unit.

For elements, the molar mass is the mass of 1 mol of atoms of that element. For compounds, the molar mass is the mass of 1 mol of molecules or formula units of that compound. The molar mass of a compound in grams per mole is numerically equal to the formula mass of the compound in atomic mass units. For example, the formula mass of CO_2 is:

Remember, the formula mass for a compound is the sum of the atomic masses of all the atoms in a chemical formula.

$$\begin{aligned}\text{formula mass} &= 1(\text{atomic mass of C}) + 2(\text{atomic mass of O})\\ &= 1(12.01\text{ amu}) + 2(16.00\text{ amu})\\ &= 44.01\text{ amu}\end{aligned}$$

The molar mass of CO_2 is therefore:

$$\text{molar mass} = 44.01\text{ g/mol}$$

Just as the molar mass of an element serves as a conversion factor between grams and moles of that element, the molar mass of a compound serves as a conversion factor between grams and moles of that compound. For example, suppose we want to find the number of moles in a 22.5-g sample of dry ice (solid CO_2). We begin by sorting the information.

GIVEN: 22.5 g CO_2

FIND: mol CO_2

SOLUTION MAP

We then strategize by drawing a solution map that shows how the molar mass converts grams of the compound to moles of the compound.

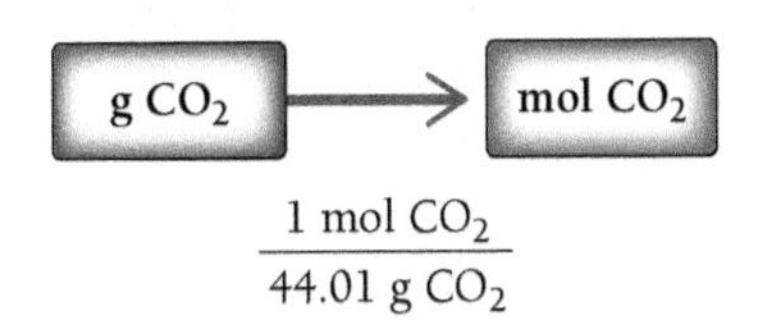

RELATIONSHIPS USED

44.01 g CO_2 = 1 mol CO_2 (molar mass of CO_2)

SOLUTION

Finally, we solve the problem.

$$22.5 \text{ g} \times \frac{1 \text{ mol } CO_2}{44.01 \text{ g}} = 0.511 \text{ mol } CO_2$$

EXAMPLE 6.4 The Mole Concept—Converting between Grams and Moles for Compounds

Calculate the mass (in grams) of 1.75 mol of water.

SORT You are given moles of water and asked to find the mass.	**GIVEN:** 1.75 mol H_2O **FIND:** g H_2O
STRATEGIZE Draw a solution map showing the conversion from mol H_2O to g H_2O. The conversion factor is the molar mass of water, which you can determine by summing the atomic masses of all the atoms in the chemical formula.	**SOLUTION MAP** mol H_2O → g H_2O $\frac{18.02 \text{ g } H_2O}{1 \text{ mol } H_2O}$ **RELATIONSHIPS USED** H_2O molar mass = 2(atomic mass H) + 1(atomic mass O) = 2(1.01) + 1(16.00) = 18.02 g/mol
SOLVE Follow the solution map to solve the problem. Begin with 1.75 mol of water and use the molar mass to convert to grams of water.	**SOLUTION** $1.75 \text{ mol } H_2O \times \frac{18.02 \text{ g } H_2O}{\text{mol } H_2O} = 31.5 \text{ g } H_2O$
CHECK Check your answer. Are the units correct? Does the answer make physical sense?	The units (g H_2O) are the desired units. The magnitude of the answer makes sense because 1 mol of water has a mass of 18.02 g; therefore, 1.75 mol should have a mass that is slightly less than 36 g.

▶ SKILLBUILDER 6.4 | The Mole Concept—Converting between Grams and Moles

Calculate the number of moles of NO_2 in 1.18 g of NO_2.

▶ FOR MORE PRACTICE Problems 47, 48, 49, 50.

Converting between Grams of a Compound and Number of Molecules

Suppose that we want to find the *number of* CO_2 *molecules* in a sample of dry ice (solid CO_2) with a mass of 22.5 g. The solution map for the problem is:

g CO_2 → mol CO_2 → CO_2 molecules

$$\frac{1 \text{ mol } CO_2}{44.01 \text{ g } CO_2} \qquad \frac{6.022 \times 10^{23} \ CO_2 \text{ molecules}}{1 \text{ mol } CO_2}$$

Notice that the first part of the solution map is identical to calculating the number of moles of CO_2 in 22.5 g of dry ice. The second part of the solution map shows the conversion from moles to number of molecules. Following the solution map, we calculate:

$$22.5 \text{ g } CO_2 \times \frac{1 \text{ mol } CO_2}{44.01 \text{ g } CO_2} \times \frac{6.022 \times 10^{23} \ CO_2 \text{ molecules}}{\text{mol } CO_2}$$

$$= 3.08 \times 10^{23} \ CO_2 \text{ molecules}$$

EXAMPLE 6.5 The Mole Concept—Converting between Mass of a Compound and Number of Molecules

What is the mass of 4.78×10^{24} NO_2 molecules?

SORT You are given the number of NO_2 molecules and asked to find the mass.	**GIVEN:** 4.78×10^{24} NO_2 molecules **FIND:** g NO_2
STRATEGIZE The solution map has two steps. In the first step, convert from molecules of NO_2 to moles of NO_2. In the second step, convert from moles of NO_2 to mass of NO_2. The required conversion factors are the molar mass of NO_2 and the number of molecules in a mole.	**SOLUTION MAP** NO_2 molecules → mol NO_2 → g NO_2 $\frac{1 \text{ mol } NO_2}{6.022 \times 10^{23} \ NO_2 \text{ molecules}}$ $\frac{46.01 \text{ g } NO_2}{1 \text{ mol } NO_2}$ **RELATIONSHIPS USED** 6.022×10^{23} molecules = 1 mol (Avogadro's number) NO_2 molar mass = 1(atomic mass N) + 2(atomic mass O) = 14.01 + 2(16.00) = 46.01 g/mol
SOLVE Using the solution map as a guide, begin with molecules of NO_2 and multiply by the appropriate conversion factors to arrive at g NO_2.	**SOLUTION** $4.78 \times 10^{24} \ NO_2 \text{ molecules} \times \frac{1 \text{ mol } NO_2}{6.022 \times 10^{23} \ NO_2 \text{ molecules}} \times \frac{46.01 \text{ g } NO_2}{1 \text{ mol } NO_2} = 365 \text{ g } NO_2$
CHECK Check your answer. Are the units correct? Does the answer make physical sense?	The units, g NO_2, are correct. Because the number of NO_2 molecules is more than one mole, the answer should be more than one molar mass (more than 46.01 g), which it is; therefore, the magnitude of the answer is reasonable.

▶ **SKILLBUILDER 6.5 | The Mole Concept—Converting between Mass and Number of Molecules**

How many H_2O molecules are in a sample of water with a mass of 3.64 g?

▶ **FOR MORE PRACTICE** Problems 51, 52, 53, 54.

CONCEPTUAL CHECKPOINT 6.4

Compound A has a molar mass of 100 g/mol, and Compound B has a molar mass of 200 g/mol. If you have samples of equal mass of both compounds, which sample contains the greater number of molecules?

6.5 Chemical Formulas as Conversion Factors

- Convert between moles of a compound and moles of a constituent element.
- Convert between grams of a compound and grams of a constituent element.

We are almost ready to address the sodium problem posed in Section 6.1. To determine how much of a particular element (such as sodium) is in a given amount of a particular compound (such as sodium chloride), we must understand the numerical relationships inherent in a chemical formula. We can understand these relationships with a straightforward analogy: Asking how much sodium is in a given amount of sodium chloride is similar to asking how many leaves are on a given number of clovers. For example, suppose we want to know the number of leaves on 14 clovers. We need a conversion factor between leaves and clovers. For clovers, the conversion factor comes from our everyday knowledge about them—we know that each clover has three leaves. We can express that relationship as a ratio between clovers and leaves.

▲ We know that each clover has three leaves. We can express that as a ratio: 3 leaves : 1 clover.

3 leaves : 1 clover

Like other conversion factors, this ratio gives the relationship between leaves and clovers. With this ratio, we can write a conversion factor to determine the number of leaves in 14 clovers. The solution map is:

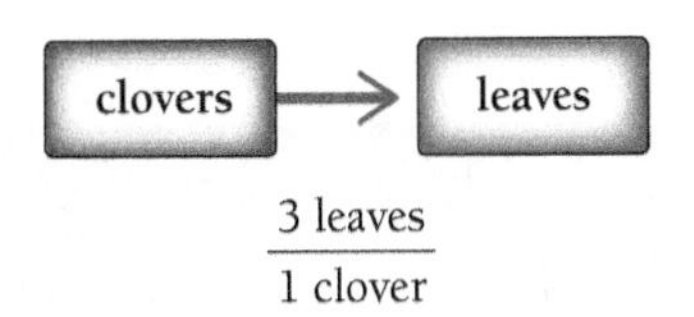

$$\frac{3 \text{ leaves}}{1 \text{ clover}}$$

We solve the problem by beginning with clovers and converting to leaves.

$$14 \text{ clovers} \times \frac{3 \text{ leaves}}{1 \text{ clover}} = 42 \text{ leaves}$$

Similarly, a chemical formula gives us ratios between elements and molecules for a particular compound. For example, the formula for carbon dioxide (CO_2) indicates that there are two O atoms per CO_2 molecule. We write this as:

$$2 \text{ O atoms} : 1 \text{ CO}_2 \text{ molecule}$$

Just as 3 leaves : 1 clover can also be written as 3 dozen leaves: 1 dozen clovers, for molecules we can write:

$$2 \text{ doz O atoms} : 1 \text{ doz CO}_2 \text{ molecules}$$

However, for atoms and molecules, we normally work in moles.

$$2 \text{ mol O} : 1 \text{ mol CO}_2$$

Chemical formulas are discussed in Chapter 5.

With conversion factors such as these—which come directly from the chemical formula—we can determine the amounts of the constituent elements present in a given amount of a compound.

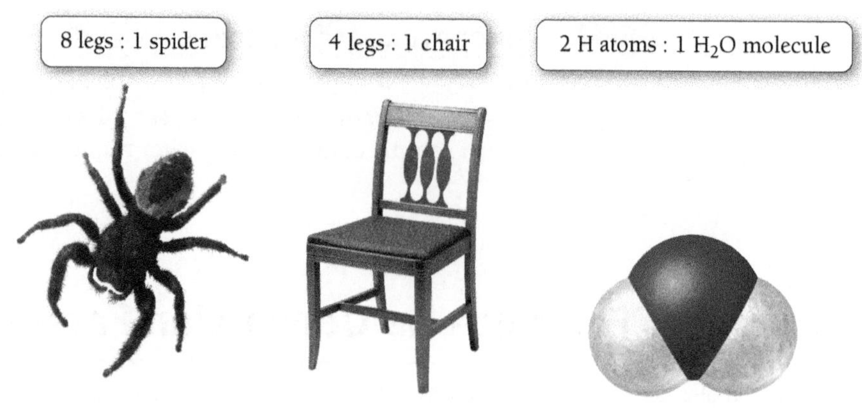

▲ Each of these shows a ratio.

Converting between Moles of a Compound and Moles of a Constituent Element

Suppose we want to know the number of moles of O in 18 mol of CO_2. Our solution map is:

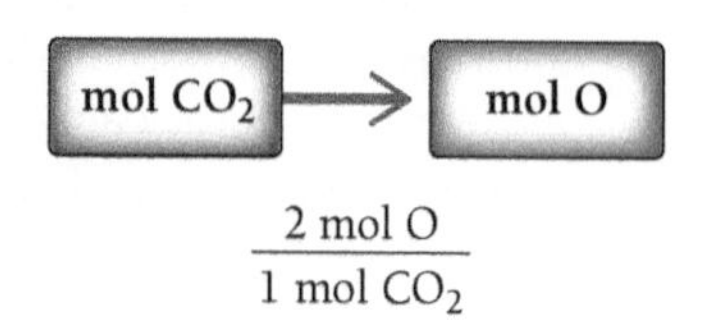

We can then calculate the moles of O.

$$18 \text{ mol } CO_2 \times \frac{2 \text{ mol O}}{1 \text{ mol } CO_2} = 36 \text{ mol O}$$

EXAMPLE 6.6 Chemical Formulas as Conversion Factors—Converting between Moles of a Compound and Moles of a Constituent Element

Determine the number of moles of O in 1.7 mol of $CaCO_3$.

SORT You are given the number of moles of $CaCO_3$ and asked to find the number of moles of O.	**GIVEN:** 1.7 mol $CaCO_3$ **FIND:** mol O
STRATEGIZE The solution map begins with moles of calcium carbonate and ends with moles of oxygen. Determine the conversion factor from the chemical formula, which indicates three O atoms for every $CaCO_3$ unit.	**SOLUTION MAP** mol $CaCO_3$ → mol O $\frac{3 \text{ mol O}}{1 \text{ mol } CaCO_3}$ **RELATIONSHIPS USED** 3 mol O : 1 mol $CaCO_3$ (from chemical formula)
SOLVE Follow the solution map to solve the problem. The subscripts in a chemical formula are exact, so they never limit significant figures.	**SOLUTION** $1.7 \text{ mol } CaCO_3 \times \frac{3 \text{ mol O}}{1 \text{ mol } CaCO_3} = 5.1 \text{ mol O}$
CHECK Check your answer. Are the units correct? Does the answer make physical sense?	The units (mol O) are correct. The magnitude is reasonable as the number of moles of oxygen should be larger than the number of moles of $CaCO_3$ (because each $CaCO_3$ unit contains 3 O atoms).

▶ **SKILLBUILDER 6.6 | Chemical Formulas as Conversion Factors—Converting between Moles of a Compound and Moles of a Constituent Element**

Determine the number of moles of O in 1.4 mol of H_2SO_4.

▶ **FOR MORE PRACTICE** Example 6.17; Problems 63, 64.

CONCEPTUAL CHECKPOINT 6.5

How many moles of H are present in 12 moles of CH_4?

Converting between Grams of a Compound and Grams of a Constituent Element

Now, we have the tools we need to solve our sodium problem from the beginning of the chapter. Suppose we want to know the mass of sodium in 15 g of NaCl. The chemical formula gives us the relationship between moles of Na and moles of NaCl:

$$1 \text{ mol Na} : 1 \text{ mol NaCl}$$

To use this relationship, we need *mol* NaCl, but we have *g* NaCl. We can use the *molar mass* of NaCl to convert from g NaCl to mol NaCl. Then we use the conversion factor from the chemical formula to convert to mol Na. Finally, we use the molar mass of Na to convert to g Na. The solution map is:

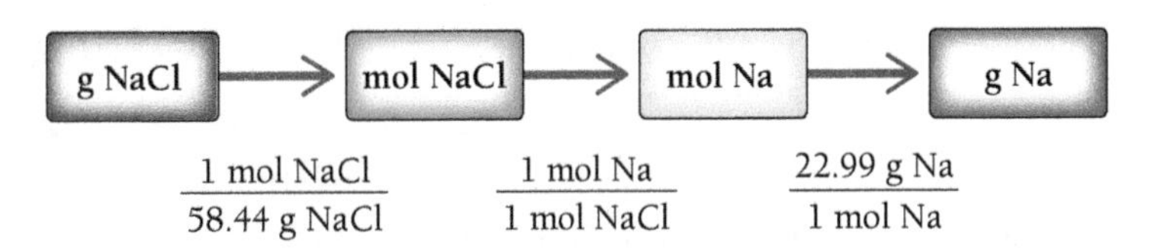

$$\frac{1 \text{ mol NaCl}}{58.44 \text{ g NaCl}} \qquad \frac{1 \text{ mol Na}}{1 \text{ mol NaCl}} \qquad \frac{22.99 \text{ g Na}}{1 \text{ mol Na}}$$

Notice that we must convert from g NaCl to mol NaCl *before* we can use the chemical formula as a conversion factor.

The chemical formula gives us a relationship between moles of substances, not between grams.

We follow the solution map to solve the problem.

$$15 \text{ g NaCl} \times \frac{1 \text{ mol NaCl}}{58.44 \text{ g NaCl}} \times \frac{1 \text{ mol Na}}{1 \text{ mol NaCl}} \times \frac{22.99 \text{ g Na}}{1 \text{ mol Na}} = 5.9 \text{ g Na}$$

The general form for solving problems where you are asked to find the mass of an element present in a given mass of a compound is:

mass compound ⟶ **moles** compound ⟶ **moles** element ⟶ **mass** element

Use the atomic or molar mass to convert between mass and moles, and use the relationships inherent in the chemical formula to convert between moles and moles (▼ FIGURE 6.2).

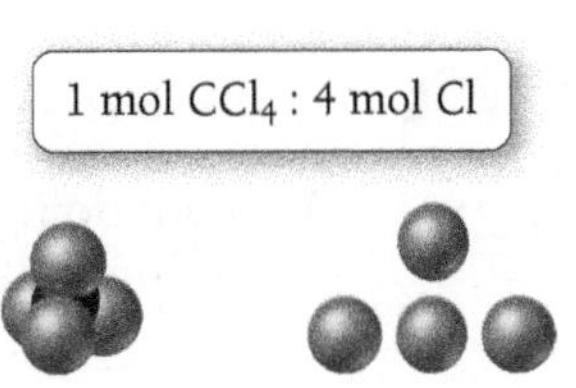

▶ FIGURE 6.2 **Mole relationships from a chemical formula** The relationships inherent in a chemical formula allow us to convert between moles of the compound and moles of a constituent element (or vice versa).

Interactive Worked Example Video 6.7

EXAMPLE 6.7 Chemical Formulas as Conversion Factors—Converting between Grams of a Compound and Grams of a Constituent Element

Carvone ($C_{10}H_{14}O$) is the main component of spearmint oil. It has a pleasant aroma and mint flavor. Carvone is added to chewing gum, liqueurs, soaps, and perfumes. Calculate the mass of carbon in 55.4 g of carvone.

SORT

You are given the mass of carvone and asked to find the mass of one of its constituent elements.

GIVEN: 55.4 g $C_{10}H_{14}O$

FIND: g C

STRATEGIZE

Base the solution map on:
grams ⟶ mole ⟶ mole ⟶ grams

SOLUTION MAP

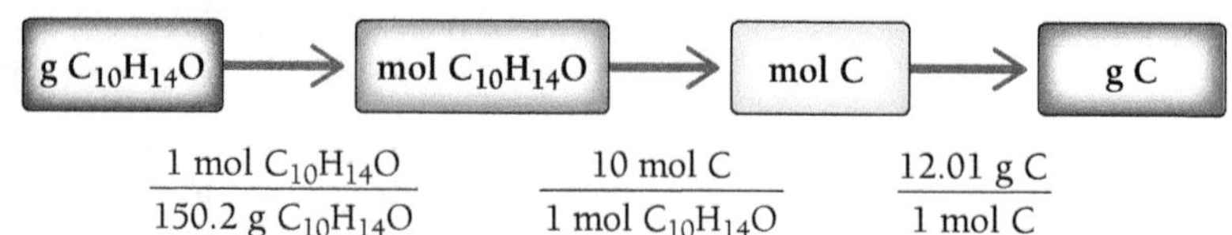

$$\frac{1 \text{ mol } C_{10}H_{14}O}{150.2 \text{ g } C_{10}H_{14}O} \qquad \frac{10 \text{ mol C}}{1 \text{ mol } C_{10}H_{14}O} \qquad \frac{12.01 \text{ g C}}{1 \text{ mol C}}$$

You need three conversion factors. The first is the molar mass of carvone.

RELATIONSHIPS USED

$$\begin{aligned}\text{molar mass carvone} &= 10(12.01) + 14(1.01) + 1(16.00)\\ &= 120.1 + 14.14 + 16.00\\ &= 150.2 \text{ g/mol}\end{aligned}$$

The second conversion factor is the relationship between moles of carbon and moles of carvone from the molecular formula.

10 mol C : 1 mol $C_{10}H_{14}O$ (from chemical formula)

The third conversion factor is the molar mass of carbon.

1 mol C = 12.01 g C (molar mass C, from periodic table)

SOLVE

Follow the solution map to solve the problem, beginning with g $C_{10}H_{14}O$ and multiplying by the appropriate conversion factors to arrive at g C.

SOLUTION

$$55.4 \text{ g } C_{10}H_{14}O \times \frac{1 \text{ mol } C_{10}H_{14}O}{150.2 \text{ g } C_{10}H_{14}O} \times \frac{10 \text{ mol C}}{1 \text{ mol } C_{10}H_{14}O} \times \frac{12.01 \text{ g C}}{1 \text{ mol C}} = 44.3 \text{ g C}$$

CHECK

Check your answer. Are the units correct? Does the answer make physical sense?

The units, g C, are correct. The magnitude of the answer is reasonable because the mass of carbon with the compound must be less than the mass of the compound itself. If you had arrived at a mass of carbon that was greater than the mass of the compound, you would know that you had made a mistake; the mass of a constituent element can never be greater than the mass of the compound itself.

▶ **SKILLBUILDER 6.7 | Chemical Formulas as Conversion Factors—Converting between Grams of a Compound and Grams of a Constituent Element**

Determine the mass of oxygen in a 5.8-g sample of sodium bicarbonate ($NaHCO_3$).

▶ **SKILLBUILDER PLUS** Determine the mass of oxygen in a 7.20-g sample of $Al_2(SO_4)_3$.

▶ **FOR MORE PRACTICE** Example 6.18; Problems 67, 68, 69, 70.

CONCEPTUAL CHECKPOINT 6.6

Without doing any detailed calculations, determine which sample contains the most fluorine atoms.

(a) 25 g of HF **(b)** 1.5 mol of CH_3F **(c)** 1.0 mol of F_2

6.6 Mass Percent Composition of Compounds

▶ Use mass percent composition as a conversion factor.

Another way to express how much of an element is in a given compound is to use the element's mass percent composition for that compound. The **mass percent composition** or simply **mass percent** of an element is the element's percentage of the total mass of the compound. For example, the mass percent composition of sodium in sodium chloride is 39%. This information tells us that a 100-g sample of sodium chloride contains 39 g of sodium. We can determine the mass percent composition for a compound from experimental data using the formula:

$$\text{mass percent of element } X = \frac{\text{mass of } X \text{ in a sample of the compound}}{\text{mass of the sample of the compound}} \times 100\%$$

Suppose a 0.358-g sample of chromium reacts with oxygen to form 0.523 g of the metal oxide. Then the mass percent of chromium is:

$$\begin{aligned}\text{mass percent Cr} &= \frac{\text{mass Cr}}{\text{mass metal oxide}} \times 100\% \\ &= \frac{0.358\text{ g}}{0.523\text{ g}} \times 100\% = 68.5\%\end{aligned}$$

We can use mass percent composition as a conversion factor between grams of a constituent element and grams of the compound. For example, we just saw that the mass percent composition of sodium in sodium chloride is 39%. This can be written as:

$$39 \text{ g sodium} : 100 \text{ g sodium chloride}$$

or in fractional form:

$$\frac{39 \text{ g Na}}{100 \text{ g NaCl}} \quad \text{or} \quad \frac{100 \text{ g NaCl}}{39 \text{ g Na}}$$

These fractions are conversion factors between g Na and g NaCl, as shown in Example 6.8.

EXAMPLE 6.8 Using Mass Percent Composition as a Conversion Factor

The FDA recommends that adults consume less than 2.4 g of sodium per day. How many grams of sodium chloride can you consume and still be within the FDA guidelines? Sodium chloride is 39% sodium by mass.

SORT

You are given the mass of sodium and the mass percent of sodium in sodium chloride. When mass percent is given, write it as a fraction. *Percent means per hundred*, so 39% sodium indicates that there are 39 g Na per 100 g NaCl. You are asked to find the mass of sodium chloride that contains the given mass of sodium.

GIVEN: 2.4 g Na

$$\frac{39 \text{ g Na}}{100 \text{ g NaCl}}$$

FIND: g NaCl

STRATEGIZE

Draw a solution map that starts with the mass of sodium and uses the mass percent as a conversion factor to get to the mass of sodium chloride.

SOLUTION MAP

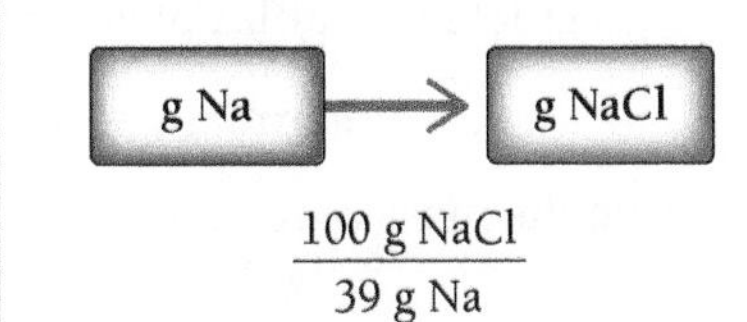

$$\frac{100 \text{ g NaCl}}{39 \text{ g Na}}$$

RELATIONSHIPS USED

39 g Na : 100 g NaCl (given in the problem)

continued on page 184 ▶

continued from page 183

SOLVE	**SOLUTION**
Follow the solution map to solve the problem, beginning with grams Na and ending with grams of NaCl. The amount of salt you can consume and still be within the FDA guideline is 6.2 g NaCl.	$2.4 \text{ g Na} \times \frac{100 \text{ g NaCl}}{39 \text{ g Na}} = 6.2 \text{ g NaCl}$
CHECK Check your answer. Are the units correct? Does the answer make physical sense?	The units, g NaCl, are correct. The answer makes physical sense because the mass of NaCl should be *larger* than the mass of Na. The mass of a compound containing a given mass of a particular element is always larger than the mass of the element itself.  ▲ Twelve and a half packets of salt contain 6.2 g NaCl.

▶ **SKILLBUILDER 6.8 | Using Mass Percent Composition as a Conversion Factor**

If a woman consumes 22 g of sodium chloride, how much sodium does she consume? Sodium chloride is 39% sodium by mass.

▶ **FOR MORE PRACTICE** Example 6.19; Problems 75, 76, 77, 78.

6.7 Mass Percent Composition from a Chemical Formula

▶ Determine mass percent composition from a chemical formula.

In the previous section, we demonstrated how to calculate mass percent composition from experimental data and how to use mass percent composition as a conversion factor. We can also calculate the mass percent of any element in a compound from the chemical formula for the compound. Based on the chemical formula, the mass percent of element *X* in a compound is:

$$\text{mass percent of element } X = \frac{\text{mass of element } X \text{ in 1 mol of compound}}{\text{mass of 1 mol of compound}} \times 100\%$$

Suppose, for example, that we want to calculate the mass percent composition of Cl in the chlorofluorocarbon CCl_2F_2. The mass percent of Cl is given by:

$$CCl_2F_2$$

$$\text{mass percent Cl} = \frac{2 \times \text{molar mass Cl}}{\text{molar mass } CCl_2F_2} \times 100\%$$

We must multiply the molar mass of Cl by 2 because the chemical formula has a subscript of 2 for Cl, meaning that 1 mol of CCl_2F_2 contains 2 mol of Cl atoms. We calculate the molar mass of CCl_2F_2 as follows:

$$\text{molar mass} = 1(12.01) + 2(35.45) + 2(19.00) = 120.91 \text{ g/mol}$$

So the mass percent of Cl in CCl_2F_2 is:

$$\text{mass percent Cl} = \frac{2 \times \text{molar mass Cl}}{\text{molar mass } CCl_2F_2} \times 100\% = \frac{2 \times 35.45 \text{ g}}{120.91 \text{ g}} \times 100\%$$

$$= 58.64\%$$

Interactive Worked Example Video 6.9

EXAMPLE 6.9 Mass Percent Composition

Calculate the mass percent of Cl in freon-114 ($C_2Cl_4F_2$).

SORT You are given the molecular formula of freon-114 and asked to find the mass percent of Cl.	**GIVEN:** $C_2Cl_4F_2$ **FIND:** Mass % Cl
STRATEGIZE You can use the information in the chemical formula to substitute into the mass percent equation and obtain the mass percent Cl.	**SOLUTION MAP** 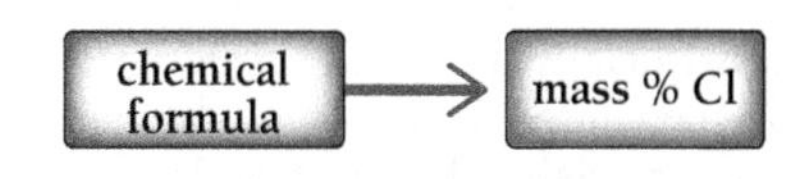 $\text{mass \% Cl} = \frac{4 \times \text{molar mass Cl}}{\text{molar mass } C_2Cl_4F_2} \times 100\%$ **RELATIONSHIPS USED** mass percent of element X = $\frac{\text{mass of element } X \text{ in 1 mol of compound}}{\text{mass of 1 mol of compound}} \times 100\%$ (mass percent equation, introduced in this section)
SOLVE Calculate the molar mass of freon-114 and substitute the values into the equation to find mass percent Cl.	**SOLUTION** $4 \times \text{molar mass Cl} = 4(35.45 \text{ g}) = 141.8 \text{ g}$ $\text{molar mass } C_2Cl_4F_2 = 2(12.01) + 4(35.45) + 2(19.00)$ $= 24.02 + 141.8 + 38.00$ $= \frac{203.8 \text{ g}}{\text{mol}}$ $\text{mass \% Cl} = \frac{4 \times \text{molar mass Cl}}{\text{molar mass } C_2Cl_4F_2} \times 100\%$ $= \frac{141.8 \text{ g}}{203.8 \text{ g}} \times 100\%$ $= 69.58\%$
CHECK Check your answer. Are the units correct? Does the answer make physical sense?	The units (%) are correct. The answer makes physical sense. Mass percent composition should never exceed 100%. If your answer is greater than 100%, you have made an error.

▶ **SKILLBUILDER 6.9 | Mass Percent Composition**

Acetic acid ($HC_2H_3O_2$) is the active ingredient in vinegar. Calculate the mass percent composition of O in acetic acid.

▶ **FOR MORE PRACTICE** Example 6.20; Problems 79, 80, 81, 82, 85, 86.

CONCEPTUAL ✔ CHECKPOINT 6.7

Which compound has the highest mass percent of O? (You should not have to perform any detailed calculations to answer this question.)

(a) CrO

(b) CrO_2

(c) Cr_2O_3

CHEMISTRY AND HEALTH

Fluoridation of Drinking Water

In the early 1900s, scientists discovered that people whose drinking water naturally contained fluoride (F^-) ions had fewer cavities than people whose water did not. At appropriate levels, fluoride strengthens tooth enamel, which prevents tooth decay. In an effort to improve public health, fluoride has been artificially added to drinking water supplies since 1945. In the United States today, about 62% of the population drinks artificially fluoridated drinking water. The American Dental Association and public health agencies estimate that water fluoridation reduces tooth decay by 40 to 65%.

The fluoridation of public drinking water, however, is often controversial. Some opponents argue that fluoride is available from other sources—such as toothpaste, mouthwash, drops, and pills—and therefore should not be added to drinking water. Anyone who wants fluoride can get it from these optional sources, they argue, and the government should not impose fluoride on the population. Other opponents argue that the risks associated with fluoridation are too great. Indeed, too much fluoride can cause teeth to become brown and spotted, a condition known as dental fluorosis. Extremely high levels can lead to skeletal fluorosis, a condition in which the bones become brittle and arthritic.

The scientific consensus is that, like many minerals, fluoride shows some health benefits at certain levels—about 1–4 mg/day for adults—but can have detrimental effects at higher levels. Consequently, most major cities fluoridate their drinking water at a level of about 0.7 mg/L. Most adults drink between 1 and 2 L of water per day, so they receive beneficial amounts of fluoride from the water. Bottled water does not normally contain fluoride. Fluoridated bottled water can sometimes be found in the infant section of supermarkets.

B6.1 CAN YOU ANSWER THIS? *Fluoride is often added to water as sodium fluoride (NaF). What is the mass percent composition of F^- in NaF? How many grams of NaF must be added to 1500 L of water to fluoridate it at a level of 0.7 mg F^-/L?*

6.8 Calculating Empirical Formulas for Compounds

- ▶ Determine an empirical formula from experimental data.
- ▶ Calculate an empirical formula from reaction data.

In Section 6.7, we learned how to calculate mass percent composition from a chemical formula. But can we go the other way? Can we calculate a chemical formula from mass percent composition? This is important because laboratory analyses of compounds do not often give chemical formulas directly; rather, they give the relative masses of each element present in a compound. For example, if we decompose water into hydrogen and oxygen in the laboratory, we could measure the masses of hydrogen and oxygen produced. Can we determine the chemical formula for water from this kind of data?

▶ We just learned how to go from the chemical formula of a compound to its mass percent composition. Can we also go the other way?

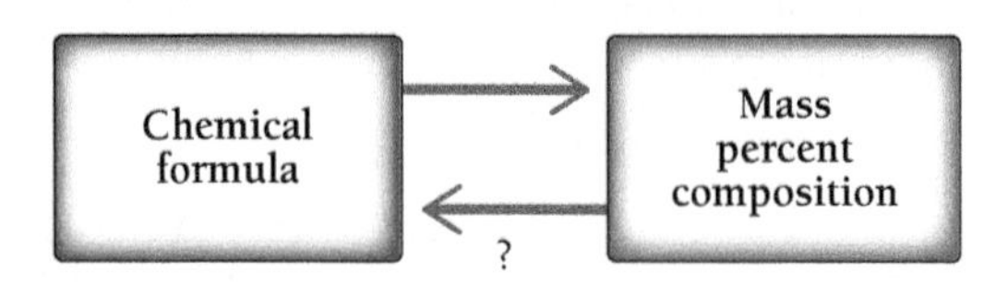

The answer is a qualified yes. We can determine a chemical formula, but it is the **empirical formula**, not the molecular formula. As we saw in Section 5.3, an empirical formula gives the smallest whole-number ratio of each type of atom in a compound, not the specific number of each type of atom in a molecule. Recall that the **molecular formula** is always a whole-number multiple of the empirical formula: Molecular formula = empirical $\times$ n, where $n = 1, 2, 3 \ldots$.

A chemical formula represents a ratio of atoms or moles of atoms, not a ratio of masses.

For example, the molecular formula for hydrogen peroxide is H_2O_2, and its empirical formula is HO.

$$HO \times 2 \longrightarrow H_2O_2$$

Calculating an Empirical Formula from Experimental Data

Suppose we decompose a sample of water in the laboratory and find that it produces 3.0 g of hydrogen and 24 g of oxygen. How do we determine an empirical formula from these data?

▲ Water can be decomposed by an electric current into hydrogen and oxygen. How can we find the empirical formula for water from the masses of its component elements?

We know that an empirical formula represents a ratio of atoms or a ratio of moles of atoms, but it *does not* represent a ratio of masses. So the first thing we must do is convert our data from grams to moles. How many moles of each element formed during the decomposition? To convert to moles, we divide each mass by the molar mass of that element.

$$\text{mol H} = 3.0\ \cancel{\text{g H}} \times \frac{1\ \text{mol H}}{1.01\ \cancel{\text{g H}}} = 3.0\ \text{mol H}$$

$$\text{mol O} = 24\ \cancel{\text{g O}} \times \frac{1\ \text{mol O}}{16.00\ \cancel{\text{g O}}} = 1.5\ \text{mol O}$$

From these data, we know there are 3 mol of H for every 1.5 mol of O. We can now write a pseudoformula for water:

$$H_3O_{1.5}$$

To get whole-number subscripts in our formula, we divide all the subscripts by the smallest one, in this case 1.5.

$$H_{\frac{3}{1.5}}O_{\frac{1.5}{1.5}} = H_2O$$

Our empirical formula for water, which in this case also happens to be the molecular formula, is H_2O. The following procedure can be used to obtain the empirical formula of any compound from experimental data. The left column outlines the procedure, and the center and right columns contain two examples of how to apply the procedure.

PEARSON eText 2.0 Interactive Worked Example Video 6.11

Obtaining an Empirical Formula from Experimental Data	EXAMPLE 6.10	EXAMPLE 6.11
	You decompose a compound containing nitrogen and oxygen in the laboratory and produce 24.5 g of nitrogen and 70.0 g of oxygen. Calculate the empirical formula of the compound.	A laboratory analysis of aspirin determines the following mass percent composition: C 60.00% H 4.48% O 35.53% Find the empirical formula.
1. Write down (or calculate) the masses of each element present in a sample of the compound. If you are given mass percent composition, assume a 100-g sample and calculate the masses of each element from the given percentages.	**GIVEN:** 24.5 g N 70.0 g O **FIND:** empirical formula	**GIVEN:** In a 100-g sample: 60.00 g C 4.48 g H 35.53 g O **FIND:** empirical formula
2. Convert each of the masses in Step 1 to moles by using the appropriate molar mass for each element as a conversion factor.	**SOLUTION** $24.5 \text{ g N} \times \frac{1 \text{ mol N}}{14.01 \text{ g N}} = 1.75 \text{ mol N}$ $70.0 \text{ g O} \times \frac{1 \text{ mol O}}{16.00 \text{ g O}} = 4.38 \text{ mol O}$	**SOLUTION** $60.00 \text{ g C} \times \frac{1 \text{ mol C}}{12.01 \text{ g C}} = 4.996 \text{ mol C}$ $4.48 \text{ g H} \times \frac{1 \text{ mol H}}{1.01 \text{ g H}} = 4.44 \text{ mol H}$ $35.53 \text{ g O} \times \frac{1 \text{ mol O}}{16.00 \text{ g O}} = 2.221 \text{ mol O}$
3. Write down a pseudoformula for the compound, using the moles of each element (from Step 2) as subscripts.	$N_{1.75}O_{4.38}$	$C_{4.996}H_{4.44}O_{2.221}$
4. Divide all the subscripts in the formula by the smallest subscript.	$N_{\frac{1.75}{1.75}}O_{\frac{4.38}{1.75}} \longrightarrow N_1O_{2.5}$	$C_{\frac{4.996}{2.221}}H_{\frac{4.44}{2.221}}O_{\frac{2.221}{2.221}} \longrightarrow C_{2.25}H_2O_1$
5. If the subscripts are not whole numbers, multiply all the subscripts by a small whole number (see the following table) to arrive at whole-number subscripts.	$N_1O_{2.5} \times 2 \longrightarrow N_2O_5$ The correct empirical formula is N_2O_5.	$C_{2.25}H_2O_1 \times 4 \longrightarrow C_9H_8O_4$ The correct empirical formula is $C_9H_8O_4$.
	▶ **SKILLBUILDER 6.10** | A sample of a compound is decomposed in the laboratory and produces 165 g of carbon, 27.8 g of hydrogen, and 220.2 g O. Calculate the empirical formula of the compound.	▶ **SKILLBUILDER 6.11** | Ibuprofen, an aspirin substitute, has the mass percent composition: C 75.69%; H 8.80%; O 15.51%. Calculate the empirical formula of ibuprofen.
	▶ **FOR MORE PRACTICE** Problems 87, 88, 89, 90.	▶ **FOR MORE PRACTICE** Example 6.21; Problems 91, 92, 93, 94.

Fractional Subscript	Multiply by This Number to Get Whole-Number Subscripts
_.10	10
_.20	5
_.25	4
_.33	3
_.50	2
_.66	3
_.75	4

EXAMPLE **6.12** **Calculating an Empirical Formula from Reaction Data**

A 3.24-g sample of titanium reacts with oxygen to form 5.40 g of the metal oxide. What is the empirical formula of the metal oxide?

You are given the mass of titanium and the mass of the metal oxide that forms. You are asked to find the empirical formula. You need to recognize this problem as one requiring a special procedure and apply that procedure, which is outlined below.	**GIVEN:** 3.24 g Ti 5.40 g metal oxide **FIND:** empirical formula
1. Write down (or calculate) the masses of each element present in a sample of the compound. In this case, you are given the mass of the initial Ti sample and the mass of its oxide after the sample reacts with oxygen. The mass of oxygen is the difference between the mass of the oxide and the mass of titanium.	**SOLUTION** 3.24 g Ti mass O = mass oxide − mass titanium = 5.40 g − 3.24 g = 2.16 g O
2. Convert each of the masses in Step 1 to moles by using the appropriate molar mass for each element as a conversion factor.	$3.24\ \text{g Ti} \times \frac{1\ \text{mol Ti}}{47.88\ \text{g Ti}} = 0.0677\ \text{mol Ti}$ $2.16\ \text{g O} \times \frac{1\ \text{mol O}}{16.00\ \text{g O}} = 0.135\ \text{mol O}$
3. Write down a pseudoformula for the compound, using the moles of each element obtained in Step 2 as subscripts.	$Ti_{0.0677}O_{0.135}$
4. Divide all the subscripts in the formula by the smallest subscript.	$Ti_{\frac{0.0677}{0.0677}}O_{\frac{0.135}{0.0677}} \longrightarrow TiO_2$
5. If the subscripts are not whole numbers, multiply all the subscripts by a small whole number to arrive at whole-number subscripts.	As the subscripts are already whole numbers, this last step is unnecessary. The correct empirical formula is TiO_2.

▶ **SKILLBUILDER 6.12 | Calculating an Empirical Formula from Reaction Data**

A 1.56-g sample of copper reacts with oxygen to form 1.95 g of the metal oxide. What is the formula of the metal oxide?

▶ **FOR MORE PRACTICE** Problems 95, 96, 97, 98.

6.9 Calculating Molecular Formulas for Compounds

▶ Calculate a molecular formula from an empirical formula and molar mass.

We can determine the *molecular* formula of a compound from the empirical formula if we also know the molar mass of the compound. Recall from Section 6.8 that the molecular formula is always a whole-number multiple of the empirical formula.

$$\text{molecular formula} = \text{empirical formula} \times n, \text{ where } n = 1, 2, 3 \ldots$$

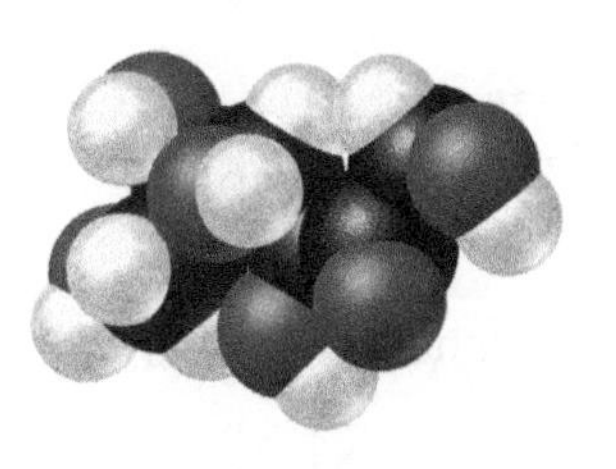

▲ Fructose, a sugar found in fruit.

Suppose we want to find the molecular formula for fructose (a sugar found in fruit) from its empirical formula, CH_2O, and its molar mass, 180.2 g/mol. We know that the molecular formula is a whole-number multiple of CH_2O.

$$\text{molecular formula} = CH_2O \times n$$

We also know that the molar mass is a whole-number multiple of the **empirical formula molar mass**, the sum of the masses of all the atoms in the empirical formula.

$$\text{molar mass} = \text{empirical formula molar mass} \times n$$

For a particular compound, the value of n in both cases is the same. Therefore, we can find n by calculating the ratio of the molar mass to the empirical formula molar mass.

$$n = \frac{\text{molar mass}}{\text{empirical formula molar mass}}$$

For fructose, the empirical formula molar mass is:

$$\text{empirical formula molar mass} = 1(12.01) + 2(1.01) + 16.00 = 30.03\ \text{g/mol}$$

Therefore, n is:

$$n = \frac{180.2\ \text{g/mol}}{30.03\ \text{g/mol}} = 6$$

We can then use this value of n to find the molecular formula.

$$\text{molecular formula} = CH_2O \times 6 = C_6H_{12}O_6$$

EXAMPLE 6.13 Calculating Molecular Formula from Empirical Formula and Molar Mass

Naphthalene is a compound containing carbon and hydrogen that is used in mothballs. Its empirical formula is C_5H_4 and its molar mass is 128.16 g/mol. What is its molecular formula?

SORT

You are given the empirical formula and the molar mass of a compound and asked to find its molecular formula.

GIVEN: empirical formula = C_5H_4
molar mass = 128.16 g/mol

FIND: molecular formula

STRATEGIZE

In the first step, use the molar mass (which is given) and the empirical formula molar mass (which you can calculate based on the empirical formula) to determine n (the integer by which you must multiply the empirical formula to determine the molecular formula).

In the second step, multiply the subscripts in the empirical formula by n to arrive at the molecular formula.

SOLUTION MAP

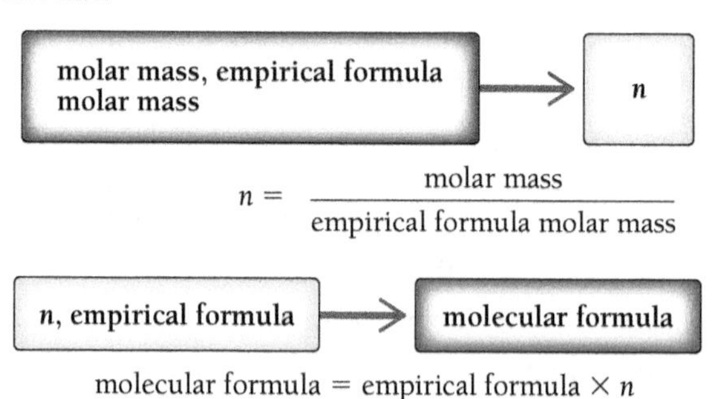

SOLVE

First find the empirical formula molar mass. Next follow the solution map. Find n by dividing the molar mass by the empirical formula molar mass (which you just calculated). Multiply the empirical formula by n to determine the molecular formula.

SOLUTION

$$\text{empirical formula molar mass} = 5(12.01) + 4(1.01) = 64.09\ \text{g/mol}$$

$$n = \frac{\text{molar mass}}{\text{empirical formula mass}} = \frac{128.16\ \text{g/mol}}{64.09\ \text{g/mol}} = 2$$

$$\text{molecular formula} = C_5H_4 \times 2 = C_{10}H_8$$

CHECK

Check your answer. Does the answer make physical sense?

The answer makes physical sense because it is a whole-number multiple of the empirical formula. Any answer containing fractional subscripts would be an error.

▶ **SKILLBUILDER 6.13 | Calculating Molecular Formula from Empirical Formula and Molar Mass**

Butane is a compound containing carbon and hydrogen used as a fuel in butane lighters. Its empirical formula is C_2H_5, and its molar mass is 58.12 g/mol. Find its molecular formula.

▶ **SKILLBUILDER PLUS** A compound with the following mass percent composition has a molar mass of 60.10 g/mol. Find its molecular formula.

C 39.97% H 13.41% N 46.62%

▶ **FOR MORE PRACTICE** Example 6.22; Problems 99, 100, 101, 102.

Chapter 6 in Review

MasteringChemistry™ provides end-of-chapter exercises, feedback-enriched tutorial problems, animations, and interactive activities to encourage problem solving practice and deeper understanding of key concepts and topics.

Self-Assessment Quiz

Q1. How many atoms are there in 5.8 mol helium?
(a) 23.2 atoms
(b) 9.6×10^{-24} atoms
(c) 5.8×10^{23} atoms
(d) 3.5×10^{24} atoms

Q2. A sample of pure silver has a mass of 155 g. How many moles of silver are in the sample?
(a) 1.44 mol
(b) 1.67×10^{4} mol
(c) 0.696 mol
(d) 155 mol

Q3. How many carbon atoms are there in a 12.5-kg sample of carbon?
(a) 6.27×10^{20} atoms
(b) 9.04×10^{28} atoms
(c) 6.27×10^{26} atoms
(d) 1.73×10^{-21} atoms

Q4. Which sample contains the greatest number of atoms?
(a) 15 g Ne
(b) 15 g Ar
(c) 15 g Kr
(d) None of the above (all contain the same number of atoms).

Q5. What is the average mass (in grams) of a single carbon dioxide molecule?
(a) 3.8×10^{-26} g
(b) 7.31×10^{-23} g
(c) 2.65×10^{25} g
(d) 44.01 g

Q6. How many moles of O are in 1.6 mol of $Ca(NO_3)_2$?
(a) 1.6 mol O
(b) 3.2 mol O
(c) 4.8 mol O
(d) 9.6 mol O

Q7. How many grams of Cl are in 25.8 g CF_2Cl_2?
(a) 7.56 g
(b) 3.78 g
(c) 15.1 g
(d) 0.427 g

Q8. Which sample contains the greatest number of F atoms?
(a) 2.0 mol HF
(b) 1.5 mol F_2
(c) 1.0 mol CF_4
(d) 0.5 mol CH_2F_2

Q9. The compound A_2X is 35.8% A by mass. What mass of the compound contains 55.1 g A?
(a) 308 g
(b) 154 g
(c) 19.7 g
(d) 35.8 g

Q10. Which compound has the highest mass percent C?
(a) CO
(b) CO_2
(c) H_2CO_3
(d) H_2CO

Q11. What is the mass percent N in $C_2H_8N_2$?
(a) 23.3% N
(b) 16.6% N
(c) 215% N
(d) 46.6% N

Q12. A compound is 52.14% C, 13.13% H, and 34.73% O by mass. What is the empirical formula of the compound?
(a) C_4HO_3
(b) C_2H_6O
(c) $C_2H_8O_3$
(d) C_3HO_6

Q13. A compound has the empirical formula CH_2O and a formula mass of 120.10 amu. What is the molecular formula of the compound?
(a) CH_2O
(b) $C_2H_4O_2$
(c) $C_3H_6O_3$
(d) $C_4H_8O_4$

Q14. A compound is decomposed in the laboratory and produces 1.40 g N and 0.20 g H. What is the empirical formula of the compound?
(a) NH (b) N_2H
(c) NH_2 (d) N_7H

Answers: 1:d, 2:a, 3:c, 4:a, 5:b, 6:d, 7:c, 8:c, 9:b, 10:a, 11:d, 12:b, 13:d, 14:c

Chemical Principles

The Mole Concept

The mole is a specific number (6.022×10^{23}) that allows us to count atoms or molecules by weighing them. One mole of any element has a mass equivalent to its atomic mass in grams, and a mole of any compound has a mass equivalent to its formula mass in grams. The mass of 1 mol of an element or compound is its molar mass.

Relevance: The mole concept allows us to determine the number of atoms or molecules in a sample from its mass. Just as a hardware store customer wants to know the number of nails in a certain weight of nails, we want to know the number of atoms in a certain mass of atoms. Because atoms are too small to count, we use their mass.

Chemical Formulas and Chemical Composition

Chemical formulas indicate the relative number of each kind of element in a compound. These numbers are based on atoms or moles. By using molar masses, we can use the information in a chemical formula to determine the relative masses of each kind of element in a compound. We can then relate the mass of a sample of a compound to the masses of the elements contained in the compound.

Relevance: The chemical composition of compounds is important because it lets us determine how much of a particular element is contained within a particular compound. For example, to assess the threat to the Earth's ozone layer from chlorofluorocarbons (CFCs), we need to know how much chlorine is in a particular CFC.

Empirical and Molecular Formulas from Laboratory Data

The relative masses of the elements within a compound allow us to determine the empirical formula of the compound. If we know the molar mass of the compound, we can also determine its molecular formula.

Relevance: The first thing we want to know about an unknown compound is its chemical formula because the formula reveals the compound's composition. Chemists often arrive at formulas by analyzing compounds in the laboratory—either by decomposing them or by synthesizing them—to determine the relative masses of the elements they contain.

Chemical Skills

LO: Convert between moles and number of atoms (Section 6.3).

Examples

EXAMPLE **6.14** **Converting between Moles and Number of Atoms**

Calculate the number of atoms in 4.8 mol of copper.

SORT

You are given moles of copper and asked to find the number of copper atoms.

GIVEN: 4.8 mol Cu

FIND: Cu atoms

STRATEGIZE

To convert between moles and number of atoms, use Avogadro's number, 6.022×10^{23} atoms = 1 mol, as a conversion factor.

SOLUTION MAP

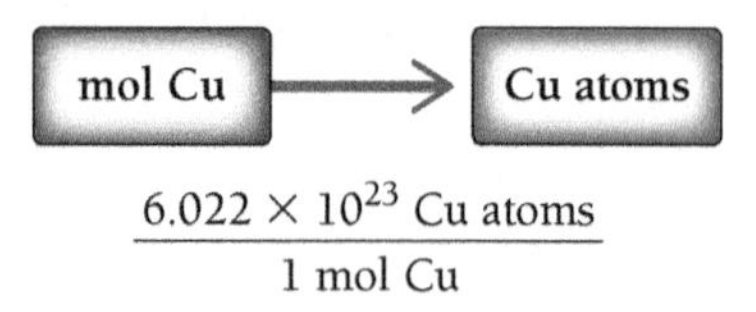

$$\frac{6.022 \times 10^{23} \text{ Cu atoms}}{1 \text{ mol Cu}}$$

RELATIONSHIPS USED

1 mol Cu = 6.022×10^{23} Cu atoms (Avogadro's number, from inside back cover)

SOLVE

Follow the solution map to solve the problem.

SOLUTION

$$4.8 \text{ mol Cu} \times \frac{6.022 \times 10^{23} \text{ Cu atoms}}{1 \text{ mol Cu}} = 2.9 \times 10^{24} \text{ Cu atoms}$$

CHECK

Check your answer. Are the units correct? Does the answer make physical sense?

The units, Cu atoms, are correct. The answer makes physical sense because the number is very large, as you would expect for nearly 5 mol of atoms.

LO: Convert between grams and moles (Section 6.3).

EXAMPLE **6.15** **Converting between Grams and Moles**

Calculate the mass of aluminum (in grams) of 6.73 mol of aluminum.

SORT
You are given the number of moles of aluminum and asked to find the mass of aluminum in grams.

GIVEN: 6.73 mol Al

FIND: g Al

STRATEGIZE
Use the molar mass of aluminum to convert between moles and grams.

SOLUTION MAP

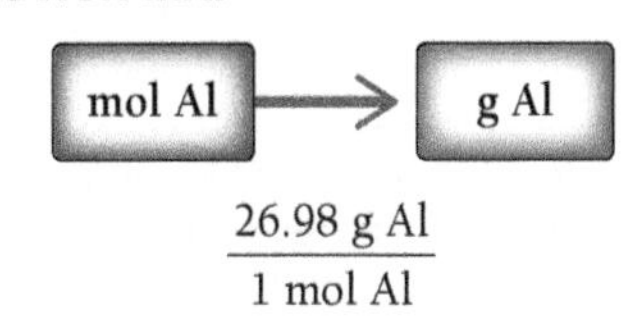

$$\frac{26.98 \text{ g Al}}{1 \text{ mol Al}}$$

RELATIONSHIPS USED
26.98 g Al = 1 mol Al (molar mass of Al from periodic table)

SOLVE
Follow the solution map to solve the problem.

SOLUTION

$$6.73 \text{ mol Al} \times \frac{26.98 \text{ g Al}}{1 \text{ mol Al}} = 182 \text{ g Al}$$

CHECK
Check your answer. Are the units correct? Does the answer make physical sense?

The units, g Al, are correct. The answer makes physical sense because each mole has a mass of about 27 g; therefore, nearly 7 mol should have a mass of nearly 190 g.

LO: Convert between grams and number of atoms or molecules (Section 6.3).

EXAMPLE **6.16** **Converting between Grams and Number of Atoms or Molecules**

Determine the number of atoms in a 48.3-g sample of zinc.

SORT
You are given the mass of a zinc sample and asked to find the number of Zn atoms that it contains.

GIVEN: 48.3 g Zn

FIND: Zn atoms

STRATEGIZE
Use the molar mass of the element to convert from grams to moles, and then use Avogadro's number to convert moles to number of atoms.

SOLUTION MAP

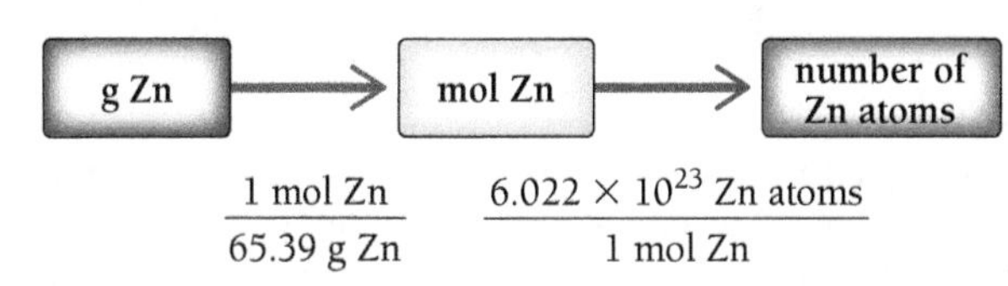

$$\frac{1 \text{ mol Zn}}{65.39 \text{ g Zn}} \qquad \frac{6.022 \times 10^{23} \text{ Zn atoms}}{1 \text{ mol Zn}}$$

RELATIONSHIPS USED
65.39 g Zn = 1 mol Zn (molar mass of Zn from periodic table)

1 mol = 6.022×10^{23} atoms (Avogadro's number, from inside back cover)

SOLVE
Follow the solution map to solve the problem.

SOLUTION

$$48.3 \text{ g Zn} \times \frac{1 \text{ mol Zn}}{65.39 \text{ g Zn}} \times \frac{6.022 \times 10^{23} \text{ Zn atoms}}{1 \text{ mol Zn}}$$

$$= 4.45 \times 10^{23} \text{ Zn atoms}$$

CHECK
Check your answer. Are the units correct? Does the answer make physical sense?

The units, Zn atoms, are correct. The answer makes physical sense because the number of atoms in any macroscopic-sized sample should be very large.

LO: Convert between moles of a compound and moles of a constituent element (Section 6.5).

EXAMPLE **6.17** **Converting between Moles of a Compound and Moles of a Constituent Element**

Determine the number of moles of oxygen in 7.20 mol of H_2SO_4.

SORT
You are given the number of moles of sulfuric acid and asked to find the number of moles of oxygen.

GIVEN: 7.20 mol H_2SO_4

FIND: mol O

STRATEGIZE
To convert between moles of a compound and moles of a constituent element, use the chemical formula of the compound to determine a ratio between the moles of the element and the moles of the compound.

SOLUTION MAP

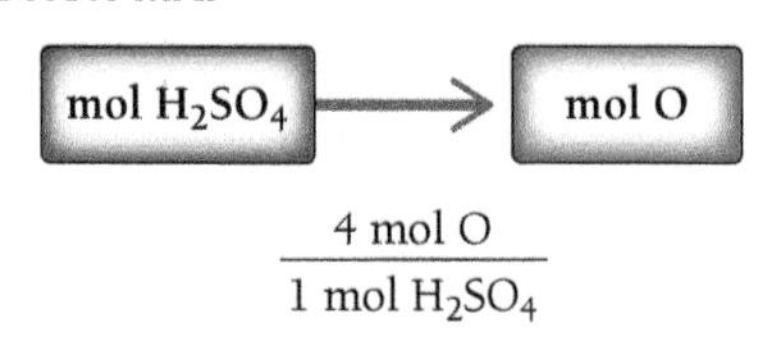

$$\frac{4 \text{ mol O}}{1 \text{ mol } H_2SO_4}$$

RELATIONSHIPS USED

4 mol O : 1 mol H_2SO_4

SOLVE
Follow the solution map to solve the problem.

SOLUTION

$$7.20 \text{ mol } H_2SO_4 \times \frac{4 \text{ mol O}}{1 \text{ mol } H_2SO_4} = 28.8 \text{ mol O}$$

CHECK
Check your answer. Are the units correct? Does the answer make physical sense?

The units, mol O, are correct. The answer makes physical sense because the number of moles of an element in a compound is equal to or greater than the number of moles of the compound itself.

LO: Convert between grams of a compound and grams of a constituent element (Section 6.5).

EXAMPLE **6.18** **Converting between Grams of a Compound and Grams of a Constituent Element**

Find the grams of iron in 79.2 g of Fe_2O_3.

SORT
You are given the mass of iron(III) oxide and asked to find the mass of iron contained within it.

GIVEN: 79.2 g Fe_2O_3

FIND: g Fe

STRATEGIZE
Use the molar mass of the compound to convert from grams of the compound to moles of the compound. Then use the chemical formula to obtain a conversion factor to convert from moles of the compound to moles of the constituent element. Finally, use the molar mass of the constituent element to convert from moles of the element to grams of the element.

SOLUTION MAP

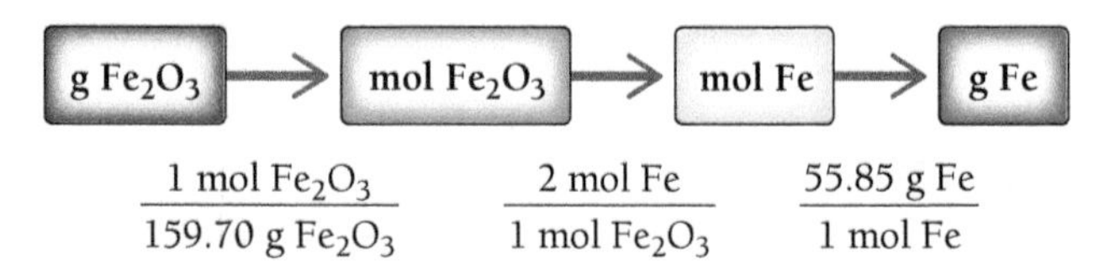

$$\frac{1 \text{ mol } Fe_2O_3}{159.70 \text{ g } Fe_2O_3} \qquad \frac{2 \text{ mol Fe}}{1 \text{ mol } Fe_2O_3} \qquad \frac{55.85 \text{ g Fe}}{1 \text{ mol Fe}}$$

RELATIONSHIPS USED

molar mass Fe_2O_3
$= 2(55.85) + 3(16.00)$
$= 159.70 \text{ g/mol}$

2 mol Fe : 1 mol Fe_2O_3 (from given chemical formula)

SOLVE
Follow the solution map to solve the problem.

SOLUTION

$$79.2 \text{ g } Fe_2O_3 \times \frac{1 \text{ mol } Fe_2O_3}{159.70 \text{ g } Fe_2O_3} \times \frac{2 \text{ mol Fe}}{1 \text{ mol } Fe_2O_3} \times \frac{55.85 \text{ g Fe}}{1 \text{ mol Fe}} = 55.4 \text{ g Fe}$$

CHECK
Check your answer. Are the units correct? Does the answer make physical sense?

The units, g Fe, are correct. The answer makes physical sense because the mass of a constituent element within a compound should be less than the mass of the compound itself.

LO: **Use mass percent composition as a conversion factor (Section 6.6).**

EXAMPLE **6.19** **Using Mass Percent Composition as a Conversion Factor**

Determine the mass of titanium in 57.2 g of titanium(IV) oxide. The mass percent of titanium in titanium(IV) oxide is 59.9%.

SORT

You are given the mass of titanium(IV) oxide and the mass percent titanium in the oxide. You are asked to find the mass of titanium in the sample.

GIVEN: 57.2 g TiO_2

$$\frac{59.9 \text{ g Ti}}{100 \text{ g TiO}_2}$$

FIND: g Ti

STRATEGIZE

Use the percent composition as a conversion factor between grams of titanium(IV) oxide and grams of titanium.

SOLUTION MAP

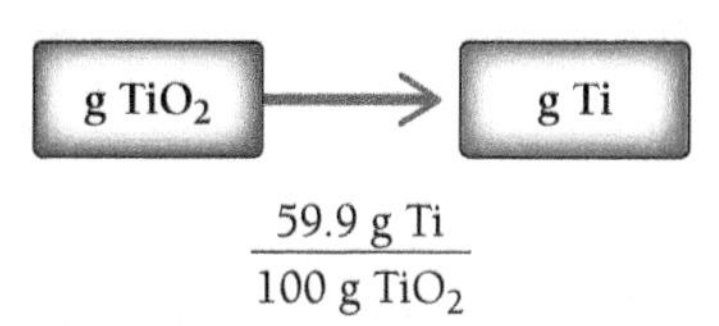

$$\frac{59.9 \text{ g Ti}}{100 \text{ g TiO}_2}$$

RELATIONSHIPS USED

59.9 g Ti : 100 g TiO_2

SOLVE

Follow the solution map to solve the problem.

SOLUTION

$$57.2 \text{ g TiO}_2 \times \frac{59.9 \text{ g Ti}}{100 \text{ g TiO}_2} = 34.3 \text{ g Ti}$$

CHECK

Check your answer. Are the units correct? Does the answer make physical sense?

The units, g Ti, are correct. The answer makes physical sense because the mass of an element within a compound should be less than the mass of the compound itself.

LO: **Determine mass percent composition from a chemical formula (Section 6.7).**

EXAMPLE **6.20** **Determining Mass Percent Composition from a Chemical Formula**

Calculate the mass percent composition of potassium in potassium oxide (K_2O).

SORT

You are given the formula of potassium oxide and asked to determine the mass percent of potassium within it.

GIVEN: K_2O

FIND: mass % K

STRATEGIZE

The solution map shows how you can substitute the information derived from the chemical formula into the mass percent equation to yield the mass percent of the element.

SOLUTION MAP

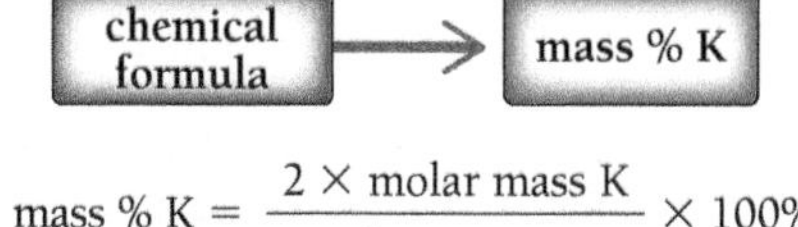

$$\text{mass \% K} = \frac{2 \times \text{molar mass K}}{\text{molar mass K}_2\text{O}} \times 100\%$$

RELATIONSHIPS USED

mass percent of element X

$$= \frac{\text{mass of element } X \text{ in 1 mol of compound}}{\text{mass of 1 mol of compound}} \times 100\%$$

(mass percent equation, from Section 6.6)

SOLVE

Calculate the molar mass of potassium oxide and follow the solution map to solve the problem.

SOLUTION

$$\text{molar mass K}_2\text{O} = 2(39.10) + 16.00$$
$$= 94.20 \text{ g/mol}$$

$$\text{mass \% K} = \frac{2(39.10 \text{ g K})}{94.20 \text{ g K}_2\text{O}} \times 100\% = 83.01\% \text{ K}$$

CHECK

Check your answer. Are the units correct? Does the answer make physical sense?

The units, % K, are correct. The answer makes physical sense because it should be below 100%.

LO: Determine an empirical formula from experimental data (Section 6.8).

You need to recognize this problem as one requiring a special procedure. Follow these steps to solve the problem.

1. Write down (or calculate) the masses of each element present in a sample of the compound. If you are given mass percent composition, assume a 100-g sample and calculate the masses of each element from the given percentages.
2. Convert each of the masses in Step 1 to moles by using the appropriate molar mass for each element as a conversion factor.
3. Write down a pseudoformula for the compound using the moles of each element (from Step 2) as subscripts.
4. Divide all the subscripts in the formula by the smallest subscript.
5. If the subscripts are not whole numbers, multiply all the subscripts by a small whole number to arrive at whole-number subscripts.

EXAMPLE 6.21 Determining an Empirical Formula from Experimental Data

A laboratory analysis of vanillin, the favoring agent in vanilla, determined the mass percent composition: C, 63.15%; H, 5.30%; O, 31.55%. Determine the empirical formula of vanillin.

GIVEN: 63.15% C, 5.30% H, and 31.55% O

FIND: empirical formula

SOLUTION

In a 100-g sample:

63.15 g C

5.30 g H

31.55 g O

$$63.15\ \cancel{g\ C} \times \frac{1\ mol\ C}{12.01\ \cancel{g\ C}} = 5.258\ mol\ C$$

$$5.30\ \cancel{g\ H} \times \frac{1\ mol\ H}{1.01\ \cancel{g\ H}} = 5.25\ mol\ H$$

$$31.55\ \cancel{g\ O} \times \frac{1\ mol\ O}{16.01\ \cancel{g\ O}} = 1.972\ mol\ O$$

$C_{5.258}H_{5.25}O_{1.972}$

$$C_{\frac{5.258}{1.972}}H_{\frac{5.25}{1.972}}O_{\frac{1.972}{1.972}} \longrightarrow C_{2.67}H_{2.66}O_1$$

$$C_{2.67}H_{2.66}O_1 \times 3 \longrightarrow C_8H_8O_3$$

The correct empirical formula is $C_8H_8O_3$.

LO: Calculate a molecular formula from an empirical formula and molar mass (Section 6.9).

EXAMPLE 6.22 Calculating a Molecular Formula from an Empirical Formula and Molar Mass

Acetylene, a gas used in welding torches, has the empirical formula CH and a molar mass of 26.04 g/mol. Find its molecular formula.

SORT

You are given the empirical formula and molar mass of acetylene and asked to find the molecular formula.

GIVEN: empirical formula = CH

molar mass = 26.04 g/mol

FIND: molecular formula

STRATEGIZE

In the first step, use the molar mass (which is given) and the empirical formula molar mass (which you can calculate based on the empirical formula) to determine n (the integer by which you must multiply the empirical formula to arrive at the molecular formula).

In the second step, multiply the coefficients in the empirical formula by n to arrive at the molecular formula.

SOLUTION MAP

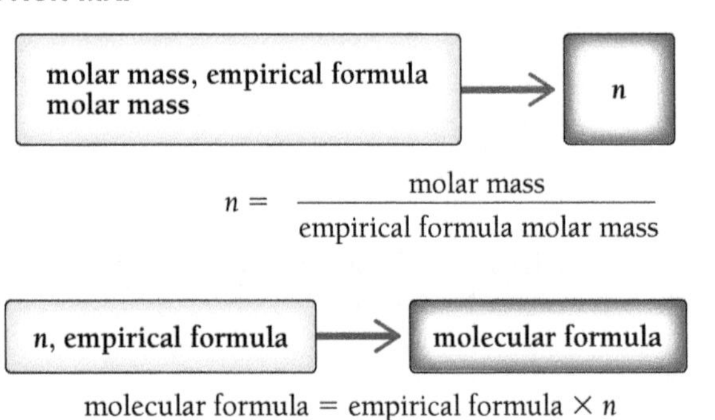

$$n = \frac{\text{molar mass}}{\text{empirical formula molar mass}}$$

molecular formula = empirical formula × n

SOLVE Follow the solution map to solve the problem. Calculate the empirical formula molar mass, which is the sum of the masses of all the atoms in the empirical formula.	**SOLUTION** empirical formula molar mass $= 12.01 + 1.01$ $= 13.02\ \text{g/mol}$
Next, find n, the ratio of the molar mass to empirical mass.	$n = \dfrac{\text{molar mass}}{\text{empirical formula molar mass}}$ $= \dfrac{26.04\ \text{g/mol}}{13.02\ \text{g/mol}} = 2$
Finally, multiply the empirical formula by n to get the molecular formula.	molecular formula $= CH \times 2 \longrightarrow C_2H_2$
CHECK Check your answer. Does the answer make physical sense?	The answer makes physical sense because the formula subscripts are all whole numbers. Any answer with non-whole numbers is incorrrect.

Key Terms

Avogadro's number [6.3]
empirical formula [6.8]
empirical formula molar mass [6.9]
mass percent (composition) [6.6]
molar mass [6.3]
mole (mol) [6.3]
molecular formula [6.8]

Exercises

Questions

1. Why is chemical composition important?
2. How can you efficiently determine the number of atoms in a sample of an element? Why is counting them not an option?
3. How many atoms are in 1 mol of atoms?
4. How many molecules are in 1 mol of molecules?
5. What is the mass of 1 mol of atoms for an element?
6. What is the mass of 1 mol of molecules for a compound?
7. What is the mass of 1 mol of atoms of each element?
   (a) P **(b)** Pt **(c)** C **(d)** Cr
8. What is the mass of 1 mol of molecules of each compound?
   (a) CO_2 **(b)** CH_2Cl_2 **(c)** $C_{12}H_{22}O_{11}$ **(d)** SO_2
9. The subscripts in a chemical formula give relationships between moles of the constituent elements and moles of the compound. Explain why these subscripts *do not* give relationships between grams of the constituent elements and grams of the compound.
10. Write the conversion factors between moles of each constituent element and moles of the compound for $C_{12}H_{22}O_{11}$.
11. You can use mass percent composition as a conversion factor between grams of a constituent element and grams of the compound. Write the conversion factor (including units) inherent in each mass percent composition.
    (a) Water is 11.19% hydrogen by mass.
    (b) Fructose, also known as fruit sugar, is 53.29% oxygen by mass.
    (c) Octane, a component of gasoline, is 84.12% carbon by mass.
    (d) Ethanol, the alcohol in alcoholic beverages, is 52.14% carbon by mass.
12. What is the mathematical formula for calculating mass percent composition from a chemical formula?
13. How are the empirical formula and the molecular formula of a compound related?
14. Why is it important to be able to calculate an empirical formula from experimental data?
15. What is the empirical formula mass of a compound?
16. How are the molar mass and empirical formula mass for a compound related?

Problems

THE MOLE CONCEPT

17. How many mercury atoms are in 5.8 mol of mercury?
18. How many moles of gold atoms do 3.45×10^{24} gold atoms constitute?
19. How many atoms are in each elemental sample?
    (a) 3.4 mol Cu **(b)** 9.7×10^{-3} mol C
    (c) 22.9 mol Hg **(d)** 0.215 mol Na
20. How many moles of atoms are in each elemental sample?
    (a) 4.6×10^{24} Pb atoms **(b)** 2.87×10^{22} He atoms
    (c) 7.91×10^{23} K atoms **(d)** 4.41×10^{21} Ca atoms

21. Complete the table.

Element	Moles	Number of Atoms
Ne	0.552	____
Ar	____	3.25×10^{24}
Xe	1.78	____
He	____	1.08×10^{20}

22. Complete the table.

Element	Moles	Number of Atoms
Cr	____	9.61×10^{23}
Fe	1.52×10^{-5}	____
Ti	0.0365	____
Hg	____	1.09×10^{23}

23. Consider these definitions.
1 doz = 12   1 gross = 144
1 ream = 500   1 mol = 6.022×10^{23}
Suppose you have 872 sheets of paper. How many ______ of paper sheets do you have?
(a) dozens (b) gross (c) reams (d) moles

24. A pure copper penny contains approximately 3.0×10^{22} copper atoms. Use the definitions in the previous problem to determine how many ______ of copper atoms are in a penny.
(a) dozens (b) gross (c) reams (d) moles

25. How many moles of tin atoms are in a pure tin cup with a mass of 38.1 g?

26. A lead fishing weight contains 0.12 mol of lead atoms. What is its mass?

27. A pure gold coin contains 0.145 mol of gold. What is its mass?

28. A helium balloon contains 0.46 g of helium. How many moles of helium does it contain?

29. How many moles of atoms are in each elemental sample?
(a) 1.34 g Zn (b) 24.9 g Ar
(c) 72.5 g Ta (d) 0.0223 g Li

30. What is the mass in grams of each elemental sample?
(a) 6.64 mol W (b) 0.581 mol Ba
(c) 68.1 mol Xe (d) 1.57 mol S

31. Complete the table.

Element	Moles	Mass
Ne	____	22.5 g
Ar	0.117	____
Xe	____	1.00 kg
He	1.44×10^{-4}	____

32. Complete the table.

Element	Moles	Mass
Cr	0.00442	____
Fe	____	73.5 mg
Ti	1.009×10^{-3}	____
Hg	____	1.78 kg

33. A pure silver ring contains 0.0134 mmol (millimol) Ag. How many silver atoms does it contain?

34. A pure gold ring contains 0.0102 mmol (millimol) Au. How many gold atoms does it contain?

35. How many aluminum atoms are in 3.78 g of aluminum?

36. What is the mass of 4.91×10^{21} platinum atoms?

37. How many atoms are in each elemental sample?
(a) 16.9 g Sr
(b) 26.1 g Fe
(c) 8.55 g Bi
(d) 38.2 g P

38. Calculate the mass in grams of each elemental sample.
(a) 1.32×10^{20} uranium atoms
(b) 2.55×10^{22} zinc atoms
(c) 4.11×10^{23} lead atoms
(d) 6.59×10^{24} silicon atoms

39. How many carbon atoms are in a diamond (pure carbon) with a mass of 38 mg?

40. How many helium atoms are in a helium blimp containing 495 kg of helium?

41. How many titanium atoms are in a pure titanium bicycle frame with a mass of 1.28 kg?

42. How many copper atoms are in a pure copper statue with a mass of 133 kg?

43. Complete the table.

Element	Mass	Moles	Number of Atoms
Na	38.5 mg	___	___
C	___	1.12	___
V	___	___	214
Hg	1.44 kg	___	___

44. Complete the table.

Element	Mass	Moles	Number of Atoms
Pt	___	0.0449	___
Fe	___	___	1.14×10^{25}
Ti	23.8 mg	___	___
Hg	___	2.05	___

45. Which sample contains the greatest number of atoms?
(a) 27.2 g Cr **(b)** 55.1 g Ti **(c)** 205 g Pb

46. Which sample contains the greatest number of atoms?
(a) 10.0 g He **(b)** 25.0 g Ne **(c)** 115 g Xe

47. Determine the number of moles of molecules (or formula units) in each sample.
(a) 38.2 g sodium chloride
(b) 36.5 g nitrogen monoxide
(c) 4.25 kg carbon dioxide
(d) 2.71 mg carbon tetrachloride

48. Determine the mass of each sample.
(a) 1.32 mol carbon tetrafluoride
(b) 0.555 mol magnesium fluoride
(c) 1.29 mmol carbon disulfide
(d) 1.89 kmol sulfur trioxide

49. Complete the table.

Compound	Mass	Moles	Number of Molecules
H_2O	112 kg	___	___
N_2O	6.33 g	___	___
SO_2	___	2.44	___
CH_2Cl_2	___	0.0643	___

50. Complete the table.

Compound	Mass	Moles	Number of Molecules
CO_2	___	0.0153	___
CO	___	0.0150	___
BrI	23.8 mg	___	___
CF_2Cl_2	1.02 kg	___	___

51. A mothball, composed of naphthalene ($C_{10}H_8$), has a mass of 1.32 g. How many naphthalene molecules does it contain?

52. Calculate the mass in grams of a single water molecule.

53. How many molecules are in each sample?
(a) 3.5 g H_2O
(b) 56.1 g N_2
(c) 89 g CCl_4
(d) 19 g $C_6H_{12}O_6$

54. Calculate the mass in grams of each sample.
(a) 5.94×10^{20} H_2O_2 molecules
(b) 2.8×10^{22} SO_2 molecules
(c) 4.5×10^{25} O_3 molecules
(d) 9.85×10^{19} CH_4 molecules

55. A sugar crystal contains approximately 1.8×10^{17} sucrose ($C_{12}H_{22}O_{11}$) molecules. What is its mass in milligrams?

56. A salt crystal has a mass of 0.12 mg. How many NaCl formula units does it contain?

57. How much money, in dollars, does 1 mol of pennies represent? If this amount of money were evenly distributed to the entire world's population (about 7.1 billion people), how much would each person get? Would each person be a millionaire? Billionaire? Trillionaire?

58. A typical dust particle has a diameter of about 10.0 mm. If 1.0 mol of dust particles were laid end to end along the equator, how many times would they encircle the planet? The circumference of the Earth at the equator is 40,076 km.

CHEMICAL FORMULAS AS CONVERSION FACTORS

59. Determine the number of moles of Cl in 2.7 mol $CaCl_2$.

60. How many moles of O are in 12.4 mol $Fe(NO_3)_3$?

61. Which sample contains the greatest number of moles of O?
(a) 2.3 mol H_2O **(b)** 1.2 mol H_2O_2
(c) 0.9 mol $NaNO_3$ **(d)** 0.5 mol $Ca(NO_3)_2$

62. Which sample contains the greatest number of moles of Cl?
(a) 3.8 mol HCl **(b)** 1.7 mol CH_2Cl_2
(c) 4.2 mol $NaClO_3$ **(d)** 2.2 mol $Mg(ClO_4)_2$

63. Determine the number of moles of C in each sample.
(a) 2.5 mol CH_4
(b) 0.115 mol C_2H_6
(c) 5.67 mol C_4H_{10}
(d) 25.1 mol C_8H_{18}

64. Determine the number of moles of H in each sample.
(a) 4.67 mol H_2O
(b) 8.39 mol NH_3
(c) 0.117 mol N_2H_4
(d) 35.8 mol $C_{10}H_{22}$

65. For each set of molecular models, write a relationship between moles of hydrogen and moles of molecules. Then determine the total number of hydrogen atoms present. (H—white; O—red; C—black; N—blue)

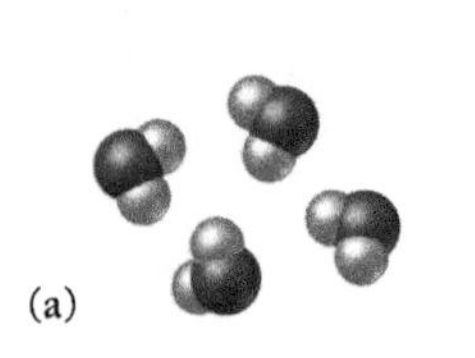
(a)

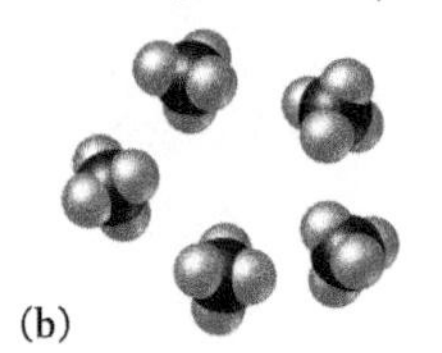
(b)

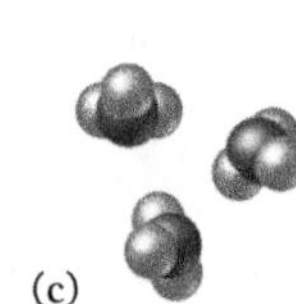
(c)

66. For each set of molecular models, write a relationship between moles of oxygen and moles of molecules. Then determine the total number of oxygen atoms present. (H— white; O—red; C—black; S—yellow)

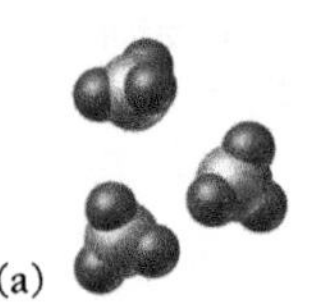
(a)

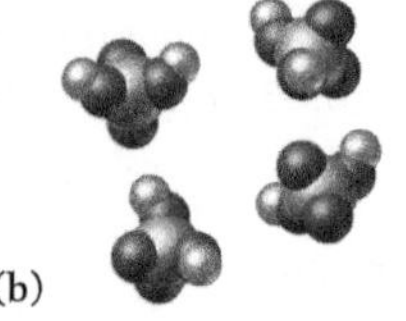
(b)

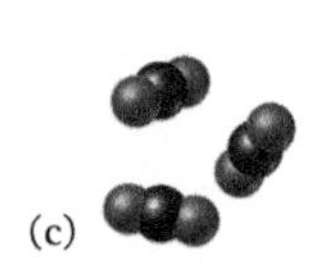
(c)

67. How many grams of Cl are in 38.0 g of each sample of chlorofluorocarbons (CFCs)?
(a) CF_2Cl_2
(b) $CFCl_3$
(c) $C_2F_3Cl_3$
(d) CF_3Cl

68. Calculate the number of grams of sodium in 1.00 g of each sodium-containing food additive.
(a) NaCl (table salt)
(b) Na_3PO_4 (sodium phosphate)
(c) $NaC_7H_5O_2$ (sodium benzoate)
(d) $Na_2C_6H_6O_7$ (sodium hydrogen citrate)

69. Iron is found in Earth's crust as several different iron compounds. Calculate the mass (in kg) of each compound that contains 1.0×10^3 kg of iron.
(a) Fe_2O_3 (hematite)
(b) Fe_3O_4 (magnetite)
(c) $FeCO_3$ (siderite)

70. Lead is found in Earth's crust as several lead compounds. Calculate the mass (in kg) of each compound that contains 1.0×10^3 kg of lead.
(a) PbS (galena)
(b) $PbCO_3$ (cerussite)
(c) $PbSO_4$ (anglesite)

MASS PERCENT COMPOSITION

71. A 2.45-g sample of strontium completely reacts with oxygen to form 2.89 g of strontium oxide. Use this data to calculate the mass percent composition of strontium in strontium oxide.

72. A 4.78-g sample of aluminum completely reacts with oxygen to form 6.67 g of aluminum oxide. Use this data to calculate the mass percent composition of aluminum in aluminum oxide.

73. A 1.912-g sample of calcium chloride is decomposed into its constituent elements and found to contain 0.690 g Ca and 1.222 g Cl. Calculate the mass percent composition of Ca and Cl in calcium chloride.

74. A 0.45-g sample of aspirin is decomposed into its constituent elements and found to contain 0.27 g C, 0.020 g H, and 0.16 g O. Calculate the mass percent composition of C, H, and O in aspirin.

75. Copper(II) fluoride contains 37.42% F by mass. Use this percentage to calculate the mass of fluorine in grams contained in 28.5 g of copper(II) fluoride.

76. Silver chloride, used in silver plating, contains 75.27% Ag. Calculate the mass of silver chloride in grams required to make 4.8 g of silver plating.

77. In small amounts, the fluoride ion (often consumed as NaF) prevents tooth decay. According to the American Dental Association, an adult female should consume 3.0 mg of fluorine per day. Calculate the amount of sodium fluoride (45.24% F) that a woman should consume to get the recommended amount of fluorine.

78. The iodide ion, usually consumed as potassium iodide, is a dietary mineral essential to good nutrition. In countries where potassium iodide is added to salt, iodine deficiency or goiter has been almost completely eliminated. The recommended daily allowance (RDA) for iodine is 150 μg/day. How much potassium iodide (76.45% I) should you consume to meet the RDA?

MASS PERCENT COMPOSITION FROM CHEMICAL FORMULA

79. Calculate the mass percent composition of nitrogen in each compound.
(a) N_2O (b) NO (c) NO_2 (d) N_2O_5

80. Calculate the mass percent composition of carbon in each compound.
(a) C_2H_2 (b) C_3H_6 (c) C_2H_6 (d) C_2H_6O

81. Calculate the mass percent composition of each element in each compound.
(a) $C_2H_4O_2$ (b) CH_2O_2 (c) C_3H_9N (d) $C_4H_{12}N_2$

82. Calculate the mass percent composition of each element in each compound.
(a) $FeCl_3$ (b) TiO_2 (c) H_3PO_4 (d) HNO_3

83. Calculate the mass percent composition of O in each compound.
(a) calcium nitrate (b) iron(II) sulfate
(c) carbon dioxide

84. Calculate the mass percent composition of Cl in each compound.
(a) carbon tetrachloride (b) calcium hypochlorite
(c) perchloric acid

85. Various iron ores have different amounts of iron per kilogram of ore. Calculate the mass percent composition of iron for each iron ore: Fe_2O_3 (hematite), Fe_3O_4 (magnetite), $FeCO_3$ (siderite). Which ore has the highest iron content?

86. Plants need nitrogen to grow, so many fertilizers consist of nitrogen-containing compounds. Calculate the mass percent composition of nitrogen in each fertilizer: NH_3, $CO(NH_2)_2$, NH_4NO_3, $(NH_4)_2SO_4$. Which fertilizer has the highest nitrogen content?

CALCULATING EMPIRICAL FORMULAS

87. A compound containing nitrogen and oxygen is decomposed in the laboratory and produces 1.78 g of nitrogen and 4.05 g of oxygen. Calculate the empirical formula of the compound.

88. A compound containing selenium and fluorine is decomposed in the laboratory and produces 2.231 g of selenium and 3.221 g of fluorine. Calculate the empirical formula of the compound.

89. Samples of several compounds are decomposed, and the masses of their constituent elements are measured. Calculate the empirical formula for each compound.
(a) 1.245 g Ni, 5.381 g I
(b) 1.443 g Se, 5.841 g Br
(c) 2.128 g Be, 7.557 g S, 15.107 g O

90. Samples of several compounds are decomposed, and the masses of their constituent elements are measured. Calculate the empirical formula for each compound.
(a) 2.677 g Ba, 3.115 g Br
(b) 1.651 g Ag, 0.1224 g O
(c) 0.672 g Co, 0.569 g As, 0.486 g O

91. The rotten smell of a decaying animal carcass is partially due to a nitrogen-containing compound called putrescine. Elemental analysis of putrescine indicates that it consists of 54.50% C, 13.73% H, and 31.77% N. Calculate the empirical formula of putrescine. (*Hint*: Begin by assuming a 100-g sample and determining the mass of each element in that 100-g sample.)

92. Citric acid, the compound responsible for the sour taste of lemons, has the elemental composition: C, 37.51%; H, 4.20%; O, 58.29%. Calculate the empirical formula of citric acid. (*Hint*: Begin by assuming a 100-g sample and determining the mass of each element in that 100-g sample.)

93. These compounds are found in many natural favors and scents. Calculate the empirical formula for each compound.
(a) ethyl butyrate (pineapple oil): C, 62.04%; H, 10.41%; O, 27.55%
(b) methyl butyrate (apple flavor): C, 58.80%; H, 9.87%; O, 31.33%
(c) benzyl acetate (oil of jasmine): C, 71.98%; H, 6.71%; O, 21.31%

94. Calculate the empirical formula for each over-the-counter pain reliever
(a) acetaminophen (Tylenol): C, 63.56%; H, 6.00%; N, 9.27%; O, 21.17%
(b) naproxen (Aleve): C, 73.03%; H, 6.13%; O, 20.84%

95. A 1.45-g sample of phosphorus burns in air and forms 2.57 g of a phosphorus oxide. Calculate the empirical formula of the oxide. (*Hint*: Determine the mass of oxygen in the 2.57 g of phosphorus oxide by determining the difference in mass before and after the phosphorus burns in air.)

96. A 2.241-g sample of nickel reacts with oxygen to form 2.852 g of the metal oxide. Calculate the empirical formula of the oxide.

97. A 0.77-mg sample of nitrogen reacts with chlorine to form 6.61 mg of the chloride. What is the empirical formula of the nitrogen chloride?

98. A 45.2-mg sample of phosphorus reacts with selenium to form 131.6 mg of the selenide. What is the empirical formula of the phosphorus selenide?

CALCULATING MOLECULAR FORMULAS

99. A compound containing carbon and hydrogen has a molar mass of 56.11 g/mol and an empirical formula of CH_2. Determine its molecular formula.

100. A compound containing phosphorus and oxygen has a molar mass of 219.9 g/mol and an empirical formula of P_2O_3. Determine its molecular formula.

101. The molar masses and empirical formulas of several compounds containing carbon and chlorine are listed here. Find the molecular formula of each compound.
(a) 284.77 g/mol, CCl
(b) 131.39 g/mol, C_2HCl_3
(c) 181.44 g/mol, C_2HCl

102. The molar masses and empirical formulas of several compounds containing carbon and nitrogen are listed here. Find the molecular formula of each compound.
(a) 163.26 g/mol, $C_{11}H_{17}N$
(b) 186.24 g/mol, C_6H_7N
(c) 312.29 g/mol, C_3H_2N

Cumulative Problems

103. A pure copper cube has an edge length of 1.42 cm. How many copper atoms does it contain? (volume of a cube = (edge length)^3; density of copper = 8.96 g/cm^3) *Hint*: Start by calculating the volume of the copper cube.

104. A pure silver sphere has a radius of 0.886 cm. How many silver atoms does it contain? (volume of a sphere = $\frac{4}{3}\pi r^3$; density of silver = 10.5 g/cm^3) *Hint*: Start by calculating the volume of the silver sphere.

105. A drop of water has a volume of approximately 0.05 mL. How many water molecules does it contain? (density of water = 1.0 g/cm^3)

106. Fingernail–polish remover is primarily acetone (C_3H_6O). How many acetone molecules are in a bottle of acetone with a volume of 325 mL? (density of acetone = 0.788 g/cm^3)

107. Complete the table.

Substance	Mass	Moles	Number of Particles (atoms or molecules)
Ar	____	4.5×10^{-4}	____
NO_2	____	____	1.09×10^{20}
K	22.4 mg	____	____
C_8H_{18}	3.76 kg	____	____

108. Complete the table.

Substance	Mass	Moles	Number of Particles (atoms or molecules)
$C_6H_{12}O_6$	15.8 g	____	____
Pb	____	____	9.04×10^{21}
CF_4	22.5 mg	____	____
C	____	0.0388	____

109. Determine the chemical formula of each compound and refer to the formula to calculate the mass percent composition of each constituent element.
(a) copper(II) iodide
(b) sodium nitrate
(c) lead(II) sulfate
(d) calcium fluoride

110. Determine the chemical formula of each compound and refer to the formula to calculate the mass percent composition of each constituent element.
(a) nitrogen triiodide
(b) xenon tetrafluoride
(c) phosphorus trichloride
(d) carbon monoxide

111. The rock in a particular iron ore deposit contains 78% Fe_2O_3 by mass. How many kilograms of the rock must a mining company process to obtain 1.0×10^3 kg of iron?

112. The rock in a lead ore deposit contains 84% PbS by mass. How many kilograms of the rock must a mining company process to obtain 1.0 kg of Pb?

113. A leak in the air conditioning system of an office building releases 12 kg of CHF_2Cl per month. If the leak continues, how many kilograms of Cl are emitted into the atmosphere each year?

114. A leak in the air conditioning system of an older car releases 55 g of CF_2Cl_2 per month. How much Cl is emitted into the atmosphere each year by this car?

115. Hydrogen is a possible future fuel. However, elemental hydrogen is rare, so it must be obtained from a hydrogen-containing compound such as water. If hydrogen were obtained from water, how much hydrogen, in grams, could be obtained from 1.0 L of water? (density of water = 1.0 g/cm^3)

116. Hydrogen, a possible future fuel mentioned in Problem 115, can also be obtained from ethanol. Ethanol can be made from the fermentation of crops such as corn. How much hydrogen, in grams, could be obtained from 1.0 kg of ethanol (C_2H_5OH)?

117. Complete the table of compounds that contain only carbon and hydrogen.

Formula	Molar Mass	% C (by mass)	% H (by mass)
C_2H_4	___	___	___
___	58.12	82.66%	___
C_4H_8	___	___	___
___	44.09	___	18.29%

118. Complete the table of compounds that contain only chromium and oxygen.

Formula	Name	Molar Mass	% Cr (by mass)	% O (by mass)
___	Chromium(III) oxide	___	___	___
___	___	84.00	61.90%	___
___	___	100.00	___	48.00%

119. Butanedione, a component of butter and body odor, has a cheesy smell. Elemental analysis of butanedione gave the mass percent composition: C, 55.80%; H, 7.03%; O, 37.17%. The molar mass of butanedione is 86.09 g/mol. Determine the molecular formula of butanedione.

120. Caffeine, a stimulant found in coffee and soda, has the mass percent composition: C, 49.48%; H, 5.19%; N, 28.85%; O, 16.48%. The molar mass of caffeine is 194.19 g/mol. Find the molecular formula of caffeine.

121. Nicotine, a stimulant found in tobacco, has the mass percent composition: C, 74.03%; H, 8.70%; N, 17.27%. The molar mass of nicotine is 162.26 g/mol. Find the molecular formula of nicotine.

122. Estradiol is a female sexual hormone that causes maturation and maintenance of the female reproductive system. Elemental analysis of estradiol gave the mass percent composition: C, 79.37%; H, 8.88%; O, 11.75%. The molar mass of estradiol is 272.37 g/mol. Find the molecular formula of estradiol.

123. A sample contains both KBr and KI in unknown quantities. If the sample has a total mass of 5.00 g and contains 1.51 g K, what are the percentages of KBr and KI in the sample by mass?

124. A sample contains both CO_2 and Ne in unknown quantities. If the sample contains a combined total of 1.75 mol and has a total mass of 65.3 g, what are the percentages of CO_2 and Ne in the sample by mole?

125. Ethanethiol (C_2H_6S) is a compound with a disagreeable odor that is used to impart an odor to natural gas. When ethanethiol is burned, the sulfur reacts with oxygen to form SO_2. What mass of SO_2 forms upon the complete combustion of 28.7 g of ethanethiol?

126. Methanethiol (CH_4S) has a disagreeable odor and is often a component of bad breath. When methanethiol is burned, the sulfur reacts with oxygen to form SO_2. What mass of SO_2 forms upon the complete combustion of 1.89 g of methanethiol?

127. An iron ore contains 38% Fe_2O_3 by mass. What is the maximum mass of iron that can be recovered from 10.0 kg of this ore?

128. Seawater contains approximately 3.5% NaCl by mass and has a density of 1.02 g/mL. What volume of seawater contains 1.0 g of sodium?

Highlight Problems

129. You can use the concepts in this chapter to obtain an estimate of the number of atoms in the universe. These steps will guide you through this calculation.
   (a) Begin by calculating the number of atoms in the sun. Assume that the sun is pure hydrogen with a density of 1.4 g/cm^3. The radius of the sun is 7×10^8 m, and the volume of a sphere is $V = \frac{4}{3}\pi r^3$.
   (b) The sun is an average-sized star, and stars are believed to compose most of the mass of the visible universe (planets are so small they can be ignored), so we can estimate the number of atoms in a galaxy by assuming that every star in the galaxy has the same number of atoms as our sun. The Milky Way galaxy is believed to contain 1×10^{11} stars. Use your answer from part a to calculate the number of atoms in the Milky Way galaxy.
   (c) Astronomers estimate that the universe contains approximately 1×10^{11} galaxies. If each of these galaxies contains the same number of atoms as the Milky Way galaxy, what is the total number of atoms in the universe?

▲ Our sun is one of the 100 billion stars in the Milky Way galaxy. The universe is estimated to contain about 100 billion galaxies.

130. Because of increasing evidence of damage to the ozone layer, chlorofluorocarbon (CFC) production was banned in 1996. However, about 100 million auto air conditioners still use CFC-12 (CF_2Cl_2). These air conditioners are recharged from stockpiled supplies of CFC-12. If each of the 100 million automobiles contains 1.1 kg of CFC-12 and leaks 25% of its CFC-12 into the atmosphere per year, how much Cl in kilograms do auto air conditioners add to the atmosphere each year? (Assume two significant figures in your calculations.)

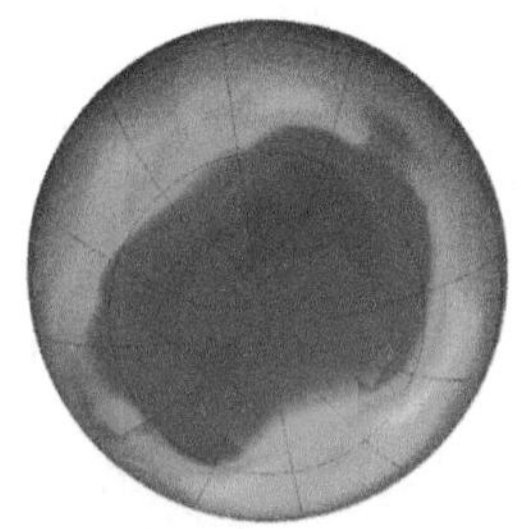

▲ The ozone hole over Antarctica on September 24, 2009. The dark blue and purple areas over the South Pole represent depressed ozone concentrations.

131. In 1996, the media reported that possible evidence of life on Mars was found on a meteorite called Allan Hills 84001 (AH 84001). The meteorite was discovered in Antarctica in 1984 and is believed to have originated on Mars. Elemental analysis of substances within its crevices revealed carbon-containing compounds that normally derive only from living organisms. Suppose that one of those compounds had a molar mass of 202.23 g/mol and the mass percent composition: C, 95.02%; H, 4.98%. What is the molecular formula for the carbon-containing compound?

▲ The Allan Hills 84001 meteorite. Elemental analysis of the substances within the crevices of this meteorite revealed carbon-containing compounds that normally originate from living organisms.

Questions for Group Work

Discuss these questions with the group and record your consensus answer.

132. Using grammatically correct English sentences, describe the relationship between the number 6.022×10^{23} and the gram.

133. Imagine that all our balances displayed mass in units of slugs instead of grams. Would we still want to use the number 6.022×10^{23} for Avogadro's number? Why or why not? If not, what number would we want to use instead? (A slug is equal to 14,954 g.)

134. Amylose is a "polysaccharide" that plants use to store energy. It is made of repeating subunits of $C_6H_{10}O_5$. If a particular amylose molecule has 2537 of these subunits, what is its molecular formula? What is its molar mass? What is the empirical formula for amylose?

Data Interpretation and Analysis

135. Public water systems often add fluoride to drinking water because, in the proper amounts, fluoride improves dental health and prevents cavities. Too much fluoride, however, can cause fluorosis, which stains teeth. In 2015, the U.S. Public Health Service (PHS) revised its 1962 recommendations for the amount of fluoride in public water systems. The 1962 recommendations depended on the average temperature for the region in question as shown here.

1962 Fluoride Recommendations

Annual Average of Maximum Daily Air Temperatures (°C)	Recommended Fluoride Concentration (mg/L)			Maximum Allowable Fluoride Concentration (mg/L)
	Lower	Optimum	Upper	
10.0–12.0	0.9	1.2	1.7	2.4
12.1–14.6	0.8	1.1	1.5	2.2
14.7–17.7	0.8	1.0	1.3	2.0
17.8–21.4	0.7	0.9	1.2	1.8
21.5–26.3	0.7	0.8	1.0	1.6
26.4–31.5	0.6	0.7	0.8	1.4

The new recommendation is simply for municipalities to fluoridate public water systems at a level of 0.7 mg/L. Notice that this level is at the lower end of previous recommendations. The recommended level was lowered because U.S. citizens are now getting fluoride from other sources, including toothpaste and mouthwash. The recommended level balances the need for fluoride to improve dental health with the risk of developing fluorosis from too much fluoride. Examine the data in the table and answer the following questions:

(a) Determine the percent change in optimum recommended fluoride concentration for a water system with annual average maximum daily temperatures of 17.8–21.4 °C. *Hint*: the percent change is given by

$$\% \text{ change} = \frac{\text{final value} - \text{initial value}}{\text{initial value}} \times 100\%.$$

(b) Sodium fluoride (NaF) and sodium fluorosilicate (Na_2SiF_6) are commercially available in 100.0-lb bags. Calculate the mass in kg of fluoride in a 100.0-lb bag for each of these compounds. If the compounds cost about the same per 1000.0 lb, which compound would be the better choice from an economic point of view?

(c) The National Institutes of Health (NIH) recommends a fluoride intake of 3.1 mg/day for adult females and 3.8 mg/day for adult males. If drinking water contains 0.7 mg/L, how much water should a person consume daily to meet the NIH recommendation for women? For men?

Answers to Skillbuilder Exercises

Skillbuilder 6.1 .... 5.32×10^{22} Au atoms
Skillbuilder 6.2 .... 89.1 g S
Skillbuilder 6.3 .... 8.17 g He
Skillbuilder 6.4 .... 2.56×10^{-2} mol NO_2
Skillbuilder 6.5 .... 1.22×10^{23} H_2O molecules
Skillbuilder 6.6 .... 5.6 mol O
Skillbuilder 6.7 .... 3.3 g O
Skillbuilder Plus, p. 182 ... 4.04 g O
Skillbuilder 6.8 .... 8.6 g Na
Skillbuilder 6.9 .... 53.28% O
Skillbuilder 6.10 .... CH_2O
Skillbuilder 6.11 .... $C_{13}H_{18}O_2$
Skillbuilder 6.12 .... CuO
Skillbuilder 6.13 .... C_4H_{10}
Skillbuilder Plus, p. 190 ... $C_2H_8N_2$

Answers to Conceptual Checkpoints

6.1 (a) $3.5 \text{ lb} \times \frac{1 \text{ doz}}{0.50 \text{ lb}} \times \frac{12 \text{ nails}}{\text{doz}} = 84 \text{ nails}$

6.2 (a) The mole is a counting unit; it represents a definite number (Avogadro's number, 6.022×10^{23}). Therefore, a given number of atoms always represents a precise number of moles, regardless of what atom is involved. Atoms of different elements have different masses. So if samples of different elements have the same mass, they *cannot* contain the same number of atoms or moles.

6.3 (b) Because carbon has lower molar mass than cobalt or lead, a 1-g sample of carbon contains more atoms than 1 g of cobalt or lead.

6.4 Sample A would have the greater number of molecules. Sample A has a lower molar mass than sample B, so a given mass of sample A has more moles and therefore more molecules than the same mass of sample B.

6.5 48 mol of H

6.6 (c) 1.0 mol of F_2 contains 2.0 mol of F atoms. Each of the other two options contains less than 2 mol of F atoms.

6.7 (b) This compound has the highest ratio of oxygen atoms to chromium atoms and therefore has the greatest mass percent of oxygen.

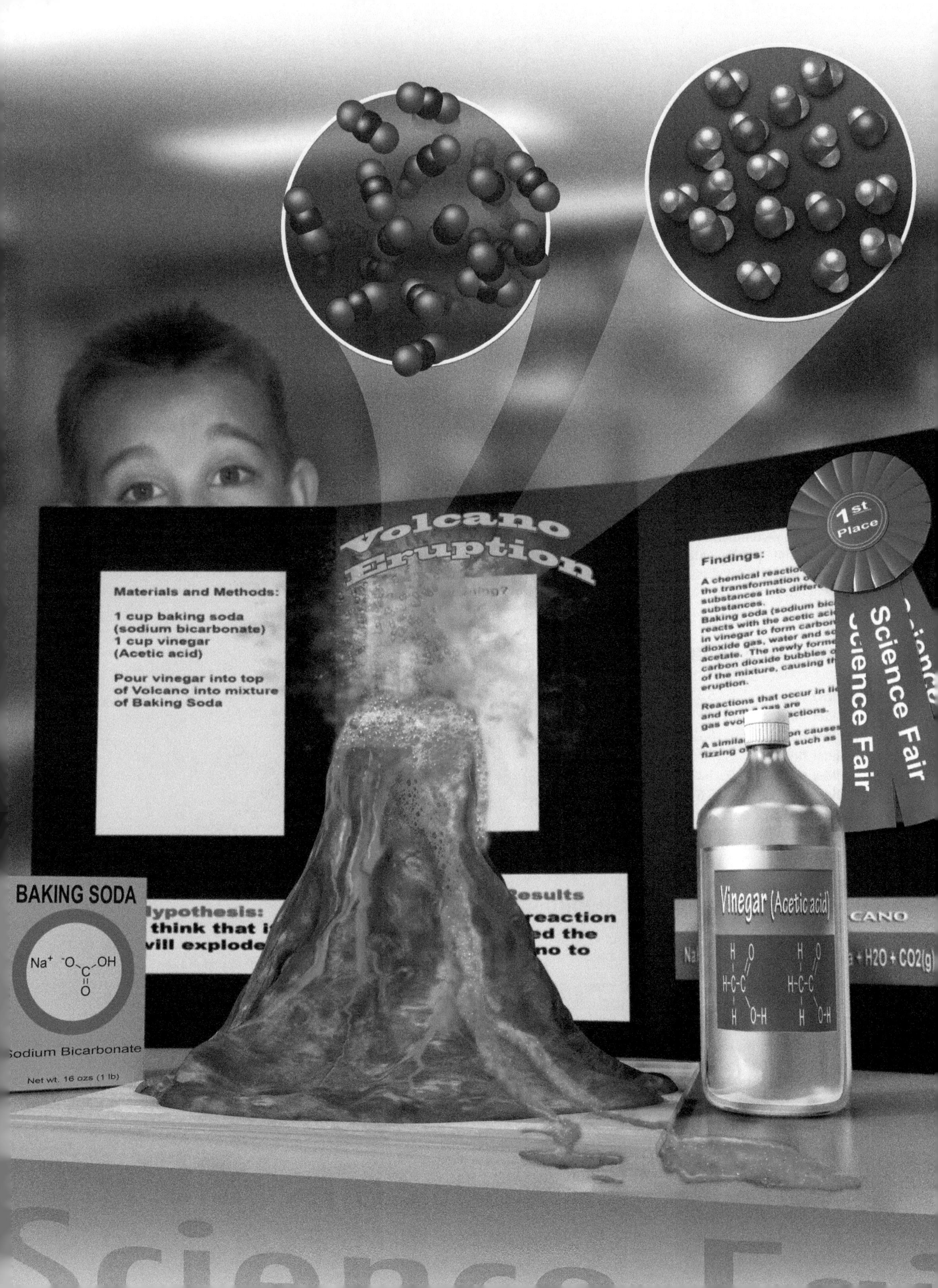

Volcano Eruption
Materials and Methods:
1 cup baking soda
(sodium bicarbonate)
1 cup vinegar
(Acetic acid)
Pour vinegar into top
of Volcano into mixture
of Baking Soda
Findings:
1st Place
Science Fair
Science Fair
BAKING SODA
Net wt. 16 ozs (1 lb)
Vinegar (Acetic acid)

7 Chemical Reactions

Chemistry . . . is one of the broadest branches of science if for no other reason than, when we think about it, everything is chemistry.

—Luciano Caglioti (1933–)

CHAPTER OUTLINE

7.1 Grade School Volcanoes, Automobiles, and Laundry Detergents

Did you make a clay volcano in grade school that erupted when filled with vinegar and baking soda? Have you pushed the gas pedal of a car and felt the acceleration as the car moved forward? Have you wondered why laundry detergents work better than hand soap to clean your clothes? Each of these processes involves a *chemical reaction*—the transformation of one or more substances into different substances.

In the classic grade school volcano, baking soda (which is sodium bicarbonate) reacts with acetic acid in vinegar to form carbon dioxide gas, water, and sodium acetate. The newly formed carbon dioxide bubbles out of the mixture, causing the eruption. Reactions that occur in liquids and form gases are *gas evolution reactions*. A similar reaction causes the fizzing of antacids such as Alka-Seltzer™.

When you drive a car, hydrocarbons such as octane (in gasoline) react with oxygen from the air to form carbon dioxide gas and water (▼ FIGURE 7.1).

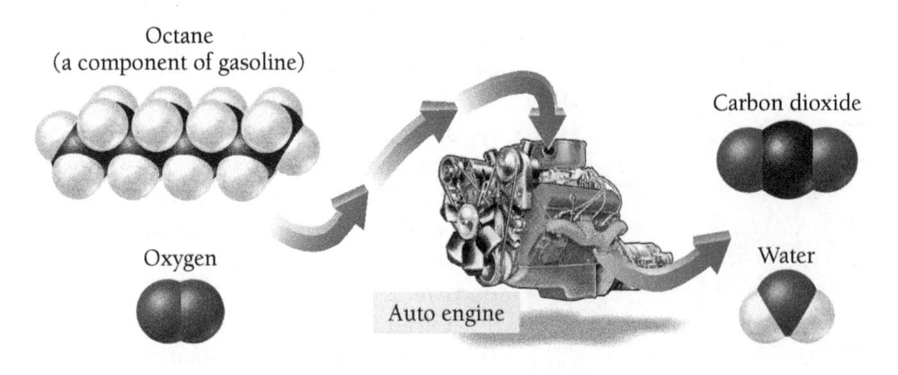

▶ FIGURE 7.1 **A combustion reaction** In an automobile engine, hydrocarbons such as octane (C_8H_{18}) from gasoline combine with oxygen from the air and react to form carbon dioxide and water.

◀ Schoolchildren sometimes make clay volcanoes that erupt by combining vinegar and baking soda, which react to produce the bubbling and splattering.

▲ FIGURE 7.2 **Soap and water** Soap forms suds with pure water (left) but reacts with the ions in hard water (right) to form a gray residue that adheres to clothes.

This reaction produces heat, which expands the gases in the car's cylinders, accelerating it forward. Reactions such as this one—in which a substance reacts with oxygen, emitting heat and forming one or more oxygen-containing compounds—are **combustion reactions**. Combustion reactions are a subcategory of *oxidation–reduction reactions*, in which electrons are transferred from one substance to another. The formation of rust and the dulling of automobile paint are other examples of oxidation–reduction reactions.

Laundry detergent works better than hand soap to wash clothes because it contains substances that soften hard water. Hard water contains dissolved calcium (Ca^{2+}) and magnesium (Mg^{2+}) ions. These ions react with soap to form a gray, slimy substance called *curd* or *soap scum* (◀ FIGURE 7.2). If you have ever tried to do your laundry with hand soap, you may have noticed gray soap scum residue on your clothes.

Laundry detergents prevent curd formation by removing Ca^{2+} and Mg^{2+} from water. Why? Because they contain substances that react with Ca^{2+} and Mg^{2+}. For example, many laundry detergents contain the carbonate ($CO_3{}^{2-}$) ion. Carbonate ions react with calcium and magnesium ions in the hard water to form solid calcium carbonate ($CaCO_3$) and solid magnesium carbonate ($MgCO_3$). These solids simply settle to the bottom of the laundry mixture, resulting in the removal of the ions from the water. In other words, laundry detergents contain substances that react with the ions in hard water to immobilize them. Reactions such as these—that form solid substances in water—are *precipitation reactions*. Precipitation reactions are also used to remove dissolved toxic metals in industrial wastes.

Chemical reactions take place all around us and even inside us. They are involved in many of the products we use daily and in many of our experiences. Chemical reactions can be relatively simple, like the combination of hydrogen and oxygen to form water, or they can be complex, like the synthesis of a protein molecule from thousands of simpler molecules. In some cases, such as the neutralization reaction that occurs in a swimming pool when acid is added to adjust the water's acidity level, chemical reactions are not noticeable to the naked eye. In other cases, such as the combustion reaction that produces a pillar of smoke and fire under a rocket during liftoff, chemical reactions are very obvious. In all cases, however, chemical reactions produce changes in the arrangements of the molecules and atoms that compose matter. Often, these molecular changes cause macroscopic changes that we can experience directly.

7.2 Evidence of a Chemical Reaction

▶ Identify evidence of a chemical reaction.

If we could see the atoms and molecules that compose matter, we could easily identify a chemical reaction. Do the atoms combine with other atoms to form compounds? Do new molecules form? Do the original molecules decompose? Do the atoms in one molecule change places with atoms in another? If the answer to one or more of these questions is yes, a chemical reaction has occurred.

Although we can't see atoms, many chemical reactions do produce easily detectable changes as they occur. For example, when the color-causing molecules in a brightly colored shirt decompose with repeated exposure to sunlight, the color of the shirt fades. Similarly, when the molecules embedded in the plastic of a child's temperature-sensitive spoon transform upon warming, the color of the spoon changes. These *color changes* are evidence that a chemical reaction has occurred.

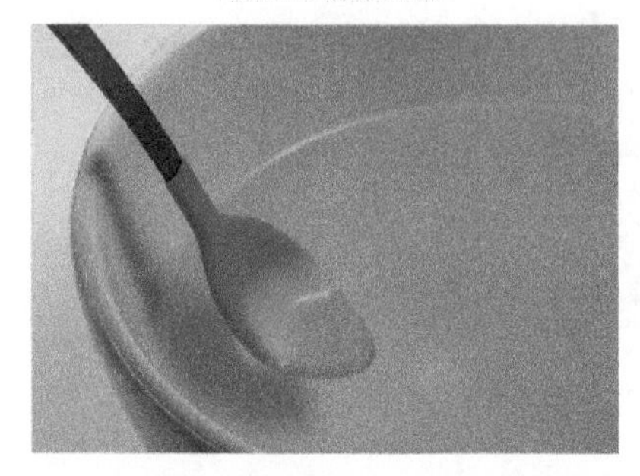

▲ A child's temperature-sensitive spoon changes color upon warming due to a reaction induced by the higher temperature.

Other changes that identify chemical reactions include the *formation of a solid* (▶ FIGURE 7.3) or *the formation of a gas* (▶ FIGURE 7.4). Dropping Alka-Seltzer tablets into water or combining baking soda and vinegar (as in our opening example of the grade school volcano) are both good examples of chemical reactions that produce a gas—the gas is visible as bubbles in the liquid.

Solid formation

▲ FIGURE 7.3 **A precipitation reaction** The formation of a solid in a previously clear solution is evidence of a chemical reaction.

Gas formation

▲ FIGURE 7.4 **A gas evolution reaction** The formation of a gas is evidence of a chemical reaction.

Recall from Section 3.9 that a reaction that emits heat is an *exothermic* reaction and one that absorbs heat is an *endothermic* reaction.

Heat absorption and *emission*, as well as *light emission*, are also evidence of reactions. For example, a natural gas flame produces heat and light. A chemical cold pack becomes cold when the plastic barrier separating two substances is broken. Both of these changes suggest that a chemical reaction is occurring.

Heat absorption

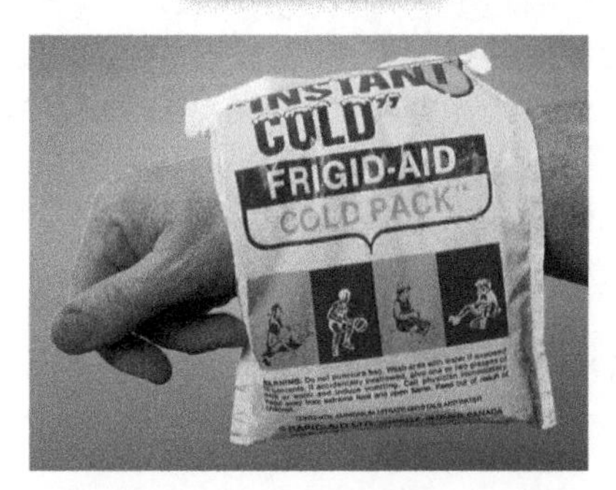

▲ A change in temperature due to absorption or emission of heat is evidence of a chemical reaction. This chemical cold pack becomes cold when the barrier separating two substances is broken and the substances combine.

▲ FIGURE 7.5 **Boiling: A physical change** When water boils, bubbles are formed and a gas is evolved. However, no chemical change has occurred because the gas, like the liquid water, is also composed of water molecules.

While these changes provide evidence of a chemical reaction, they are not *definitive* evidence. Only chemical analysis showing that the initial substances have changed into other substances conclusively proves that a chemical reaction has occurred. We can be fooled. For example, when water boils, bubbles form, but no chemical reaction has occurred. Boiling water forms gaseous steam, but both water and steam are composed of water molecules—no chemical change has occurred (◀ FIGURE 7.5). On the other hand, chemical reactions may occur without any obvious signs, yet chemical analysis may show that a reaction has indeed occurred. The changes occurring at the atomic and molecular level determine whether a chemical reaction has taken place.

In summary, each of the following provides evidence of a chemical reaction:

- a *color change*

- the *emission of light*

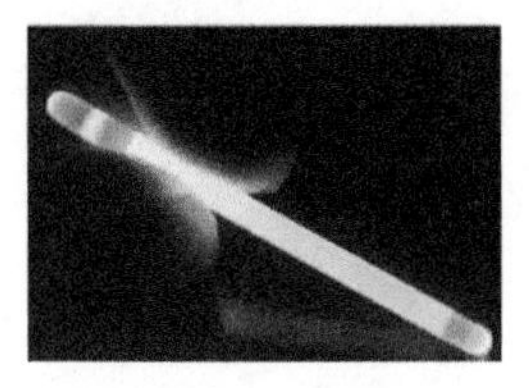

- the *formation of a solid* in a previously clear (unclouded) solution

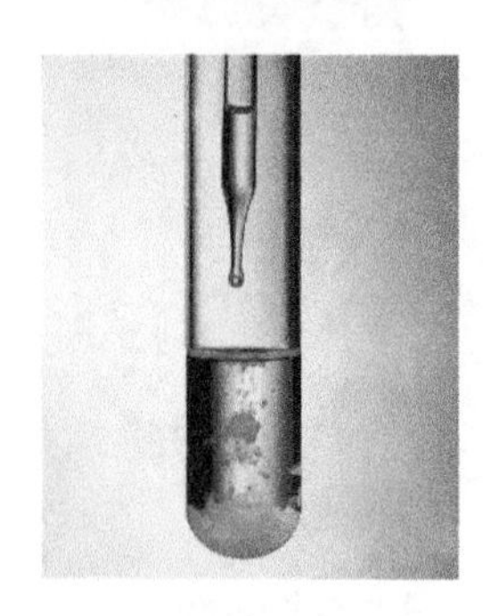

- the *emission* or *absorption of heat*

- the *formation of a gas* when we add a substance to a solution

EXAMPLE 7.1 Evidence of a Chemical Reaction

Which changes involve a chemical reaction? Explain your answers.

(a) ice melting upon warming
(b) an electric current passing through water, resulting in the formation of hydrogen and oxygen gas that appears as bubbles rising in the water
(c) iron rusting
(d) bubbles forming when a soda can is opened

SOLUTION

(a) not a chemical reaction; melting ice forms water, but both the ice and water are composed of water molecules.
(b) chemical reaction; water decomposes into hydrogen and oxygen, as evidenced by the bubbling.
(c) chemical reaction; iron changes into iron oxide, changing color in the process.
(d) not a chemical reaction; even though there is bubbling, it is just carbon dioxide coming out of the liquid.

▶ SKILLBUILDER 7.1 | Evidence of a Chemical Reaction

Which changes involve a chemical reaction? Explain your answers.

(a) butane burning in a butane lighter
(b) butane evaporating out of a butane lighter
(c) wood burning
(d) dry ice subliming

▶ FOR MORE PRACTICE Example 7.16; Problems 25, 26, 27, 28, 29, 30.

CONCEPTUAL CHECKPOINT 7.1

These images portray molecular views of various substances before and after a change. Determine whether a chemical reaction has occurred in each case.

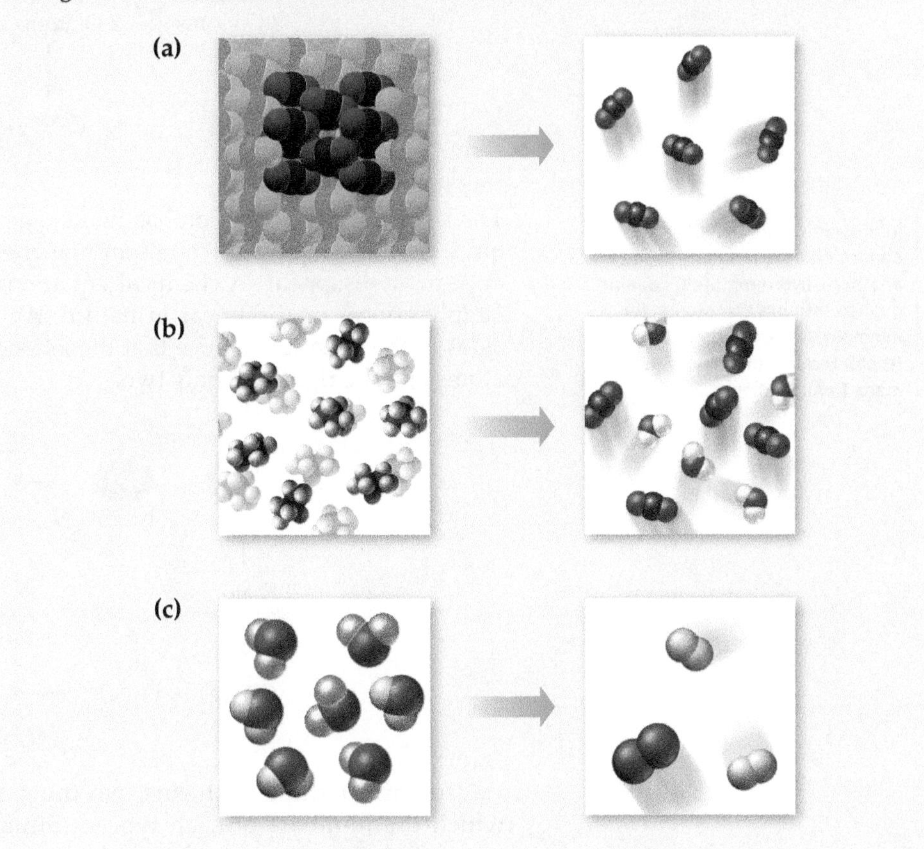

7.3 The Chemical Equation

▶ Identify balanced chemical equations.

Key Concept Video
Writing and Balancing Chemical Equations

TABLE 7.1 Abbreviations Indicating the States of Reactants and Products in Chemical Equations

Abbreviation	State
(*g*)	gas
(*l*)	liquid
(*s*)	solid
(*aq*)	aqueous (water solution)*

*The (*aq*) designation stands for *aqueous*, which indicates that a substance is dissolved in water. When a substance dissolves in water, the mixture is called a *solution*. We discuss solutions in greater detail in Section 7.5.

Recall from Section 3.6 that we represent chemical reactions with *chemical equations*. For example, the reaction occurring in a natural-gas flame, such as the flame on a kitchen stove, is methane (CH_4) reacting with oxygen (O_2) to form carbon dioxide (CO_2) and water (H_2O). We represent this reaction with the equation:

$$\underbrace{CH_4 + O_2}_{\text{reactants}} \longrightarrow \underbrace{CO_2 + H_2O}_{\text{products}}$$

The substances on the left side of the equation are the *reactants*, and the substances on the right side are the *products*. We often specify the state of each reactant or product in parentheses next to the formula. If we add states to our equation, it becomes:

$$CH_4(g) + O_2(g) \longrightarrow CO_2(g) + H_2O(g)$$

The (g) indicates that these substances are gases in the reaction. Table 7.1 summarizes the common states of reactants and products and the symbols used in chemical reactions.

Let's look more closely at the equation for the burning of natural gas. How many oxygen atoms are on each side of the equation?

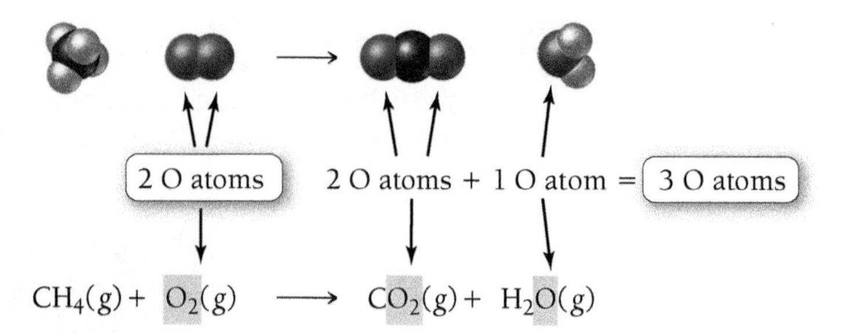

In chemical equations, atoms cannot change from one type to another—hydrogen atoms cannot change into oxygen atoms, for example. Nor can atoms disappear (recall the law of conservation of mass from Section 3.7).

The left side of the equation has two oxygen atoms, and the right side has three. Since chemical equations represent real chemical reactions, atoms cannot simply appear or disappear in chemical equations because, as we know, atoms don't simply appear or disappear in nature. We must account for the atoms on both sides of the equation. Notice that the left side of the equation has four hydrogen atoms and the right side only two.

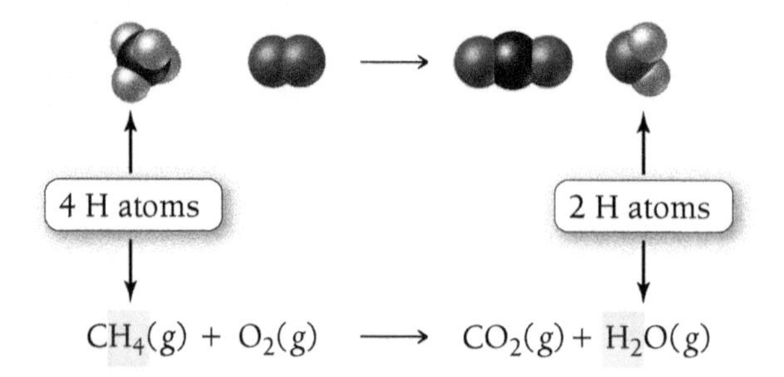

To correct these problems, we must create a **balanced equation**, one in which the numbers of each type of atom on both sides of the equation are equal. To balance an equation, we insert coefficients—not subscripts—in front of the chemical formulas as needed to make the number of each type of atom in the reactants equal to the number of each type of atom in the products. New atoms do not form during a reaction, nor do atoms vanish—matter must be conserved.

When we balance chemical equations by inserting coefficients in front of the formulas of the reactants and products, it changes the number of molecules in the equation, but it does not change the *kinds* of molecules. To balance the preceding equation, for example, we put the coefficient 2 before O_2 in the reactants, and the coefficient 2 before H_2O in the products:

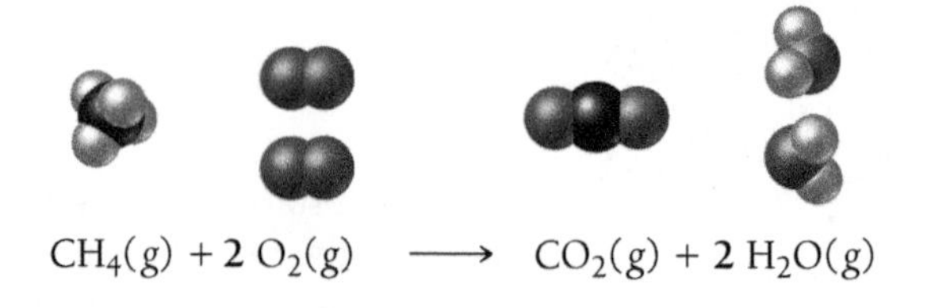

The equation is now balanced because the numbers of each type of atom on both sides of the equation are equal. We can verify this by summing the number of each type of atom.

We determine the number of a particular type of atom within a chemical formula in an equation by multiplying the subscript for the atom by the coefficient for the chemical formula. If there is no coefficient or subscript, a 1 is implied. The balanced equation for the combustion of natural gas is:

$$CH_4(g) + 2\,O_2(g) \longrightarrow CO_2(g) + 2\,H_2O(g)$$

Reactants	Products
1 C atom ($1 \times CH_4$)	1 C atom ($1 \times CO_2$)
4 H atoms ($1 \times CH_4$)	4 H atoms ($2 \times H_2O$)
4 O atoms ($2 \times O_2$)	4 O atoms ($1 \times CO_2 + 2 \times H_2O$)

The numbers of each type of atom on both sides of the equation are equal—the equation is balanced.

▶ A balanced chemical equation represents a chemical reaction. In this equation, methane molecules combine with oxygen to form carbon dioxide and water. The photo shows a stove burning methane and the balanced equation for the reaction above it.

CONCEPTUAL CHECKPOINT 7.2

In photosynthesis, plants make the sugar glucose, $C_6H_{12}O_6$, from carbon dioxide and water. The equation for the reaction is:

$$6\,CO_2 + 6\,H_2O \longrightarrow C_6H_{12}O_6 + x\,O_2$$

In order for this equation to be balanced, the coefficient x must be

(a) 3 **(b)** 6 **(c)** 9 **(d)** 12

7.4 How to Write Balanced Chemical Equations

▶ Write balanced chemical equations.

The following procedure details the steps for writing balanced chemical equations. As in other procedures in the book, we show the steps in the left column and examples of applying each step in the center and right columns. Remember, change only the *coefficients* to balance a chemical equation; *never change the subscripts because changing the subscripts changes the kinds of molecules, not the number of molecules.*

PEARSON eText 2.0 Interactive Worked Example Video 7.2

Writing Balanced Chemical Equations	EXAMPLE 7.2	EXAMPLE 7.3
	Write a balanced equation for the reaction between solid silicon dioxide and solid carbon to produce solid silicon monocarbide and carbon monoxide gas.	Write a balanced equation for the combustion reaction between liquid octane (C_8H_{18}), a component of gasoline, and gaseous oxygen to form gaseous carbon dioxide and gaseous water.
1. Write the unbalanced equation by writing chemical formulas for each of the reactants and products. Review Chapter 5 for nomenclature rules. (If the unbalanced equation is provided in the problem, skip this step and go to Step 2.)	**SOLUTION** $SiO_2(s) + C(s) \longrightarrow SiC(s) + CO(g)$	**SOLUTION** $C_8H_{18}(l) + O_2(g) \longrightarrow CO_2(g) + H_2O(g)$
2. If an element occurs in only one compound on both sides of the equation, balance it first. If there is more than one such element, balance metals before nonmetals.	**BEGIN WITH SI** $SiO_2(s) + C(s) \longrightarrow SiC(s) + CO(g)$ **1 Si atom ⟶ 1 Si atom** Si is already balanced. **BALANCE O NEXT** $SiO_2(s) + C(s) \longrightarrow SiC(s) + CO(g)$ **2 O atoms ⟶ 1 O atom** To balance O, put a 2 before $CO(g)$. $SiO_2(s) + C(s) \longrightarrow SiC(s) + 2\,CO(g)$ **2 O atoms ⟶ 2 O atoms**	**BEGIN WITH C** $C_8H_{18}(l) + O_2(g) \longrightarrow CO_2(g) + H_2O(g)$ **8 C atoms ⟶ 1 C atom** To balance C, put an 8 before $CO_2(g)$. $C_8H_{18}(l) + O_2(g) \longrightarrow 8\,CO_2(g) + H_2O(g)$ **8 C atoms ⟶ 8 C atoms** **BALANCE H NEXT** $C_8H_{18}(l) + O_2(g) \longrightarrow 8\,CO_2(g) + H_2O(g)$ **18 H atoms ⟶ 2 H atoms** To balance H, put a 9 before $H_2O(g)$. $C_8H_{18}(l) + O_2(g) \longrightarrow 8\,CO_2(g) + 9\,H_2O(g)$ **18 H atoms ⟶ 18 H atoms**
3. If an element occurs as a free element (not as part of a compound) on either side of the chemical equation, balance it last. Always balance free elements by adjusting the coefficient *on the free element.*	**BALANCE C** $SiO_2(s) + C(s) \longrightarrow SiC(s) + 2\,CO(g)$ **1 C atom ⟶** 1 C + 2 C = **3 C atoms** To balance C, put a 3 before $C(s)$. $SiO_2(s) + 3\,C(s) \longrightarrow SiC(s) + 2\,CO(g)$ **3 C atoms ⟶** 1 C + 2 C = **3 C atoms**	**BALANCE O** $C_8H_{18}(l) + O_2(g) \longrightarrow 8\,CO_2(g) + 9\,H_2O(g)$ **2 O atoms ⟶** 16 O + 9 O = **25 O atoms** To balance O, put a $\frac{25}{2}$ before $O_2(g)$. $C_8H_{18}(l) + \frac{25}{2}\,O_2(g) \longrightarrow 8\,CO_2(g) + 9\,H_2O(g)$ **25 O atoms ⟶** 16 O + 9 O = **25 O atoms**
4. If the balanced equation contains coefficient fractions, change these into whole numbers by multiplying the entire equation by the appropriate factor.	This step is not necessary in this example. Proceed to Step 5.	$[C_8H_{18}(l) + \frac{25}{2}\,O_2(g) \longrightarrow 8\,CO_2(g) + 9\,H_2O(g)] \times 2$ $2\,C_8H_{18}(l) + 25\,O_2(g) \longrightarrow 16\,CO_2(g) + 18\,H_2O(g)$

5. Check to make certain the equation is balanced by summing the total number of each type of atom on both sides of the equation.

$SiO_2(s) + 3\,C(s) \longrightarrow SiC(s) + 2\,CO(g)$

Reactants		Products
1 Si atom	⟶	1 Si atom
2 O atoms	⟶	2 O atoms
3 C atoms	⟶	3 C atoms

The equation is balanced.

▶ **SKILLBUILDER 7.2** | Write a balanced equation for the reaction between solid chromium(III) oxide and solid carbon to produce solid chromium and carbon dioxide gas.

$2\,C_8H_{18}(l) + 25\,O_2(g) \longrightarrow 16\,CO_2(g) + 18\,H_2O(g)$

Reactants		Products
16 C atoms	⟶	16 C atoms
36 H atoms	⟶	36 H atoms
50 O atoms	⟶	50 O atoms

The equation is balanced.

▶ **SKILLBUILDER 7.3** | Write a balanced equation for the combustion reaction of gaseous C_4H_{10} and gaseous oxygen to form gaseous carbon dioxide and gaseous water.

▶ **FOR MORE PRACTICE** Example 7.17; Problems 35, 36, 37, 38.

EXAMPLE 7.4 Balancing Chemical Equations

Write a balanced equation for the reaction of solid aluminum with aqueous sulfuric acid to form aqueous aluminum sulfate and hydrogen gas.

Use your knowledge of chemical nomenclature from Chapter 5 to write a skeletal equation containing formulas for each of the reactants and products. The formulas for each compound MUST BE CORRECT before you begin to balance the equation.

SOLUTION

$Al(s) + H_2SO_4(aq) \longrightarrow Al_2(SO_4)_3(aq) + H_2(g)$

Since both aluminum and hydrogen occur as free elements, balance those last. Sulfur and oxygen occur in only one compound on each side of the equation, so balance these first. Sulfur and oxygen are also part of a polyatomic ion that stays intact on both sides of the equation. *Balance polyatomic ions such as these as a unit.* There are 3 SO_4^{2-} ions on the right side of the equation, so put a 3 in front of H_2SO_4.

$Al(s) + 3\,H_2SO_4(aq) \longrightarrow Al_2(SO_4)_3(aq) + H_2(g)$

Balance Al next. Since there are 2 Al atoms on the right side of the equation, place a 2 in front of Al on the left side of the equation.

$2\,Al(s) + 3\,H_2SO_4(aq) \longrightarrow Al_2(SO_4)_3(aq) + H_2(g)$

Balance H next. Since there are 6 H atoms on the left side, place a 3 in front of $H_2(g)$ on the right side.

$2\,Al(s) + 3\,H_2SO_4(aq) \longrightarrow Al_2(SO_4)_3(aq) + 3\,H_2(g)$

Finally, sum the number of atoms on each side to make sure that the equation is balanced.

$2\,Al(s) + 3\,H_2SO_4(aq) \longrightarrow Al_2(SO_4)_3(aq) + 3\,H_2(g)$

Reactants		Products
2 Al atoms	⟶	2 Al atoms
6 H atoms	⟶	6 H atoms
3 S atoms	⟶	3 S atoms
12 O atoms	⟶	12 O atoms

▶ **SKILLBUILDER 7.4** | **Balancing Chemical Equations**

Write a balanced equation for the reaction of aqueous lead(II) acetate with aqueous potassium iodide to form solid lead(II) iodide and aqueous potassium acetate.

▶ **FOR MORE PRACTICE** Problems 39, 40, 41, 42, 43, 44.

EXAMPLE **7.5** **Balancing Chemical Equations**

Balance the chemical equation.

$Fe(s) + HCl(aq) \longrightarrow FeCl_3(aq) + H_2(g)$

	SOLUTION
Since Cl occurs in only one compound on each side of the equation, balance it first. One Cl atom is on the left side of the equation, and 3 Cl atoms are on the right side. To balance Cl, place a 3 in front of HCl.	$Fe(s) + 3\,HCl(aq) \longrightarrow FeCl_3(aq) + H_2(g)$
Since H and Fe occur as free elements, balance them last. There is 1 Fe atom on the left side of the equation and 1 Fe atom on the right, so Fe is balanced. There are 3 H atoms on the left and 2 H atoms on the right. Balance H by placing a $\frac{3}{2}$ in front of H_2. (That way you don't alter other elements that are already balanced.)	$Fe(s) + 3\,HCl(aq) \longrightarrow FeCl_3(aq) + \frac{3}{2}H_2(g)$
The equation now contains a coefficient fraction; clear it by multiplying the entire equation (both sides) by 2.	$[Fe(s) + 3\,HCl(aq) \longrightarrow FeCl_3(aq) + \frac{3}{2}H_2(g)] \times 2$ $2\,Fe(s) + 6\,HCl(aq) \longrightarrow 2\,FeCl_3(aq) + 3\,H_2(g)$
Finally, sum the number of atoms on each side to check that the equation is balanced.	$2\,Fe(s) + 6\,HCl(aq) \longrightarrow 2\,FeCl_3(aq) + 3\,H_2(g)$

Reactants		Products
2 Fe atoms	⟶	2 Fe atoms
6 Cl atoms	⟶	6 Cl atoms
6 H atoms	⟶	6 H atoms

▶ **SKILLBUILDER 7.5 | Balancing Chemical Equations**

Balance the chemical equation.

$HCl(g) + O_2(g) \longrightarrow H_2O(l) + Cl_2(g)$

▶ **FOR MORE PRACTICE** Problems 45, 46, 47, 48, 49, 50.

CONCEPTUAL CHECKPOINT 7.3

Which quantity must always be the same on both sides of a balanced chemical equation?

(a) the number of each type of atom

(b) the number of each type of molecule

(c) the sum of all of the coefficients

7.5 Aqueous Solutions and Solubility: Compounds Dissolved in Water

▶ Determine whether a compound is soluble.

In the previous section, we balanced chemical equations that represent chemical reactions. We now turn to investigating several types of reactions.

Aqueous Solutions

Since many of these reactions occur in water, we must first understand *aqueous solutions*. Reactions occurring in aqueous solutions are among the most common and important. An **aqueous solution** is a homogeneous mixture of a substance with water. For example, a sodium chloride (NaCl) solution (also called a saline solution) is composed of sodium chloride dissolved in water. Sodium chloride

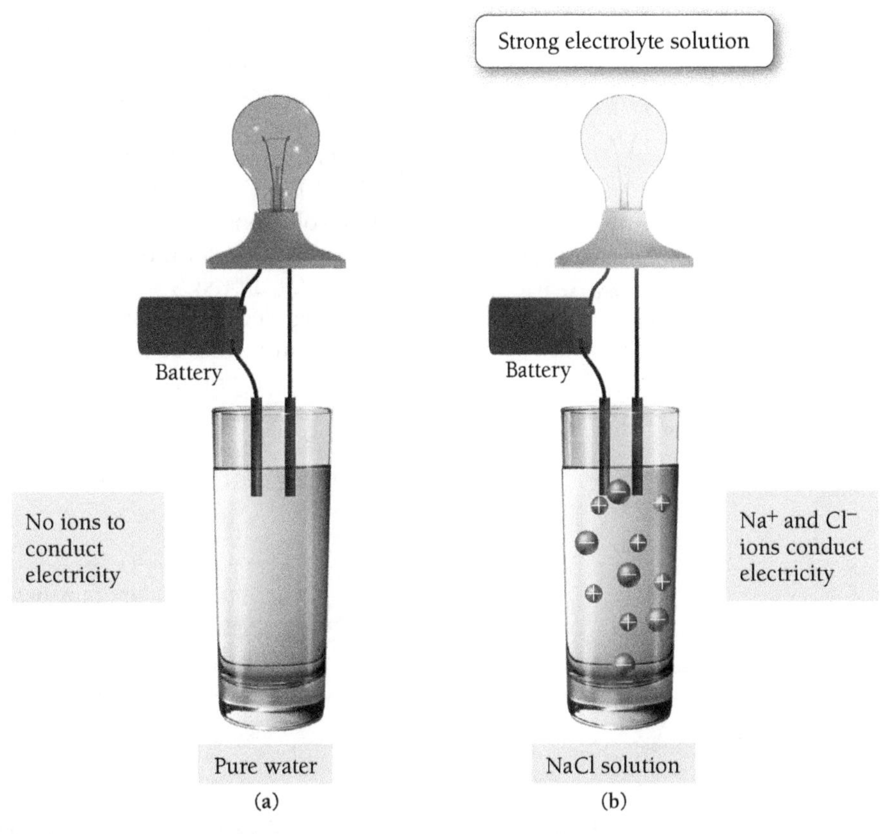

▲ FIGURE 7.6 **Ions as conductors** (a) Pure water does not conduct electricity. (b) Ions in a sodium chloride solution conduct electricity, causing the bulb to light. Solutions such as NaCl are called strong electrolyte solutions.

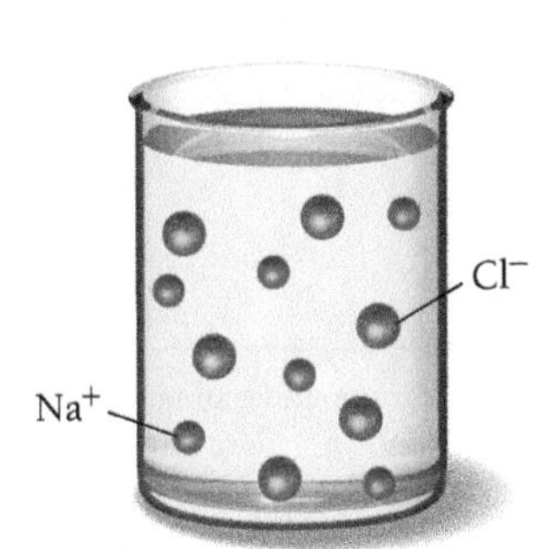

A sodium chloride solution contains independent Na^+ and Cl^- ions.

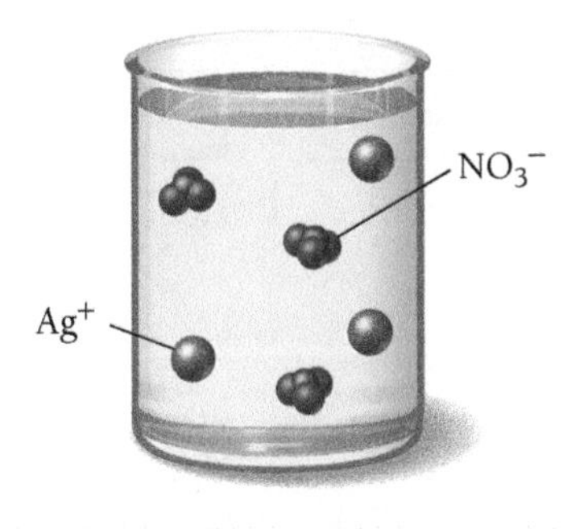

A silver nitrate solution contains independent Ag^+ and NO_3^- ions.

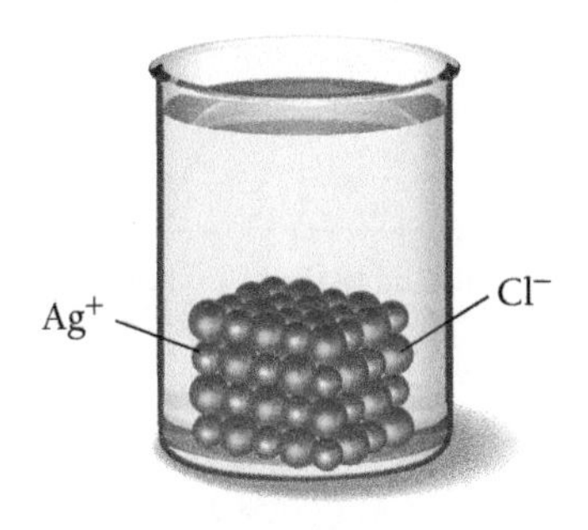

When silver chloride is added to water, it remains as solid AgCl—it does not dissolve into independent ions.

solutions are common both in the oceans and in living cells. You can form a sodium chloride solution yourself by adding table salt to water. As you stir the salt into the water, it seems to disappear. However, you know the salt is still there because if you taste the water, it has a salty flavor. How does sodium chloride dissolve in water?

When ionic compounds such as NaCl dissolve in water, they usually dissociate into their component ions. A sodium chloride solution, represented as NaCl(*aq*), does not contain any NaCl units; only dissolved Na^+ ions and Cl^- ions are present.

We know that NaCl is present as independent sodium and chloride ions in solution because sodium chloride solutions conduct electricity, which requires the presence of freely moving charged particles. Substances (such as NaCl) that completely dissociate into ions in solution are *strong electrolytes*, and the resultant solutions are **strong electrolyte solutions** (▲ FIGURE 7.6). Similarly, a silver nitrate solution, represented as $AgNO_3(aq)$, does not contain any $AgNO_3$ units, but only dissolved Ag^+ ions and NO_3^- ions. It, too, is a strong electrolyte solution. When compounds containing polyatomic ions such as NO_3^- dissolve, the polyatomic ions dissolve as intact units.

Not all ionic compounds, however, dissolve in water. AgCl, for example, does not. If we add AgCl to water, it remains as solid AgCl and appears as a white solid at the bottom of the beaker.

Solubility

A compound is **soluble** in a particular liquid if it dissolves in that liquid; a compound is **insoluble** if it does not dissolve in the liquid. NaCl, for example, is soluble in water. If we mix solid sodium chloride into water, it dissolves and forms a strong electrolyte solution. AgCl, on the other hand, is insoluble in water. If we mix solid silver chloride into water, it remains as a solid within the liquid water.

There is no easy way to predict whether a particular compound will be soluble or insoluble in water. For ionic compounds, however, empirical rules have been deduced from observations of many compounds. These **solubility rules** are summarized in Table 7.2 and ▼ FIGURE 7.7. For example, the solubility rules indicate that compounds containing the lithium ion are *soluble*. That means that compounds such as LiBr, $LiNO_3$, Li_2SO_4, LiOH, and Li_2CO_3 all dissolve in water to form strong electrolyte solutions. If a compound contains Li^+, it is soluble. Similarly, the solubility rules state that compounds containing the NO_3^- ion are soluble. Compounds such as $AgNO_3$, $Pb(NO_3)_2$, $NaNO_3$, $Ca(NO_3)_2$, and $Sr(NO_3)_2$ all dissolve in water to form strong electrolyte solutions.

The solubility rules apply only to the solubility of the compounds in water.

The solubility rules also state that, with some exceptions, compounds containing the CO_3^{2-} ion are *insoluble*. Compounds such as $CuCO_3$, $CaCO_3$, $SrCO_3$, and $FeCO_3$ do not dissolve in water. Note that the solubility rules have many exceptions. For example, compounds containing CO_3^{2-} are *soluble when paired with* Li^+, Na^+, K^+, or NH_4^+. Thus Li_2CO_3, Na_2CO_3, K_2CO_3, and $(NH_4)_2CO_3$ are all soluble.

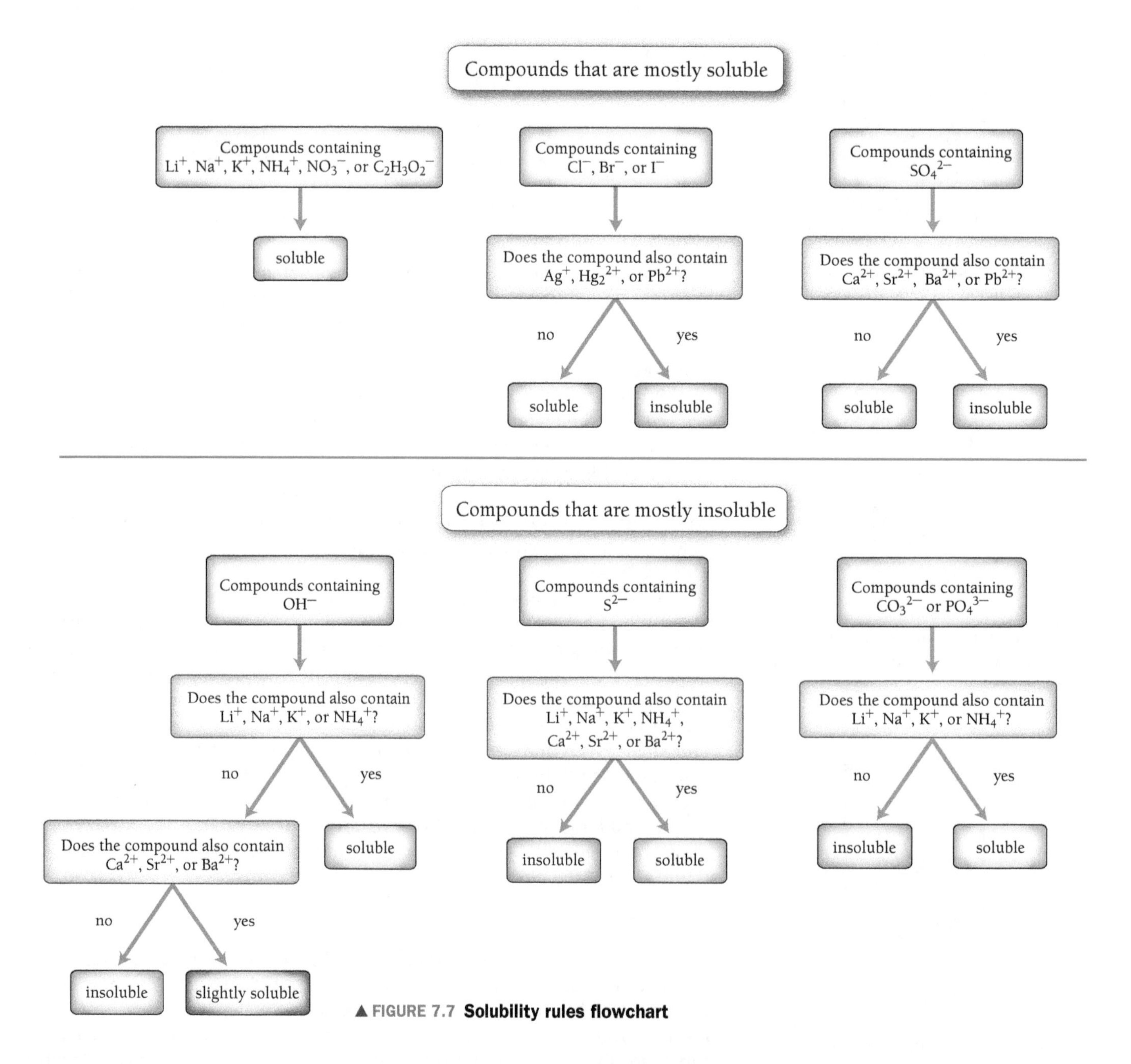

▲ FIGURE 7.7 **Solubility rules flowchart**

TABLE 7.2 Solubility Rules

Compounds Containing the Following Ions Are Mostly Soluble	Exceptions
Li^+, Na^+, K^+, NH_4^+	None
NO_3^-, $C_2H_3O_2^-$	None
Cl^-, Br^-, I^-	When any of these ions pair with Ag^+, Hg_2^{2+}, or Pb^{2+}, the compound is insoluble.
SO_4^{2-}	When SO_4^{2-} pairs with Ca^{2+}, Sr^{2+}, Ba^{2+}, or Pb^{2+}, the compound is insoluble.
Compounds Containing the Following Ions Are Mostly Insoluble	**Exceptions**
OH^-, S^{2-}	When either of these ions pairs with Li^+, Na^+, K^+, or NH_4^+, the compound is soluble. When S^{2-} pairs with Ca^{2+}, Sr^{2+}, or Ba^{2+}, the compound is soluble. When OH^- pairs with Ca^{2+}, Sr^{2+}, or Ba^{2+}, the compound is slightly soluble.*
CO_3^{2-}, PO_4^{3-}	When either of these ions pairs with Li^+, Na^+, K^+, or NH_4^+, the compound is soluble.

*For many purposes these can be considered insoluble.

EXAMPLE 7.6 Determining Whether a Compound is Soluble

Is each compound soluble or insoluble?

(a) AgBr (b) $CaCl_2$ (c) $Pb(NO_3)_2$ (d) $PbSO_4$

SOLUTION

(a) Insoluble; compounds containing Br^- are normally soluble, but Ag^+ is an exception.
(b) Soluble; compounds containing Cl^- are normally soluble, and Ca^{2+} is not an exception.
(c) Soluble; compounds containing NO_3^- are always soluble.
(d) Insoluble; compounds containing SO_4^{2-} are normally soluble, but Pb^{2+} is an exception.

▶ **SKILLBUILDER 7.6 | Determining Whether a Compound Is Soluble**

Is each compound soluble or insoluble?

(a) CuS (b) $FeSO_4$ (c) $PbCO_3$ (d) NH_4Cl

▶ **FOR MORE PRACTICE** Example 7.18; Problems 57, 58, 59, 60, 61, 62.

PEARSON eText 2.0

CONCEPTUAL CHECKPOINT 7.4

Which image best depicts a mixture of $BaCl_2$ and water?

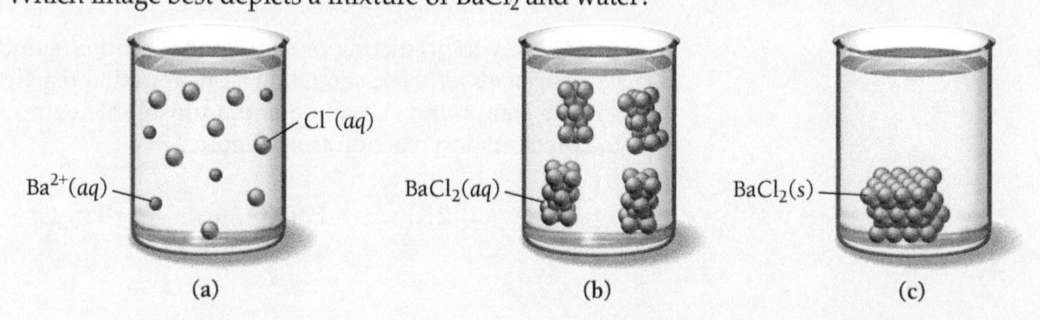

7.6 Precipitation Reactions: Reactions in Aqueous Solution That Form a Solid

▶ Predict and write equations for precipitation reactions.

Recall from Section 7.1 that sodium carbonate in laundry detergent reacts with dissolved Mg^{2+} and Ca^{2+} ions to form solids that precipitate (come out of) solution. This reaction is an example of a **precipitation reaction**—a reaction that forms a solid, called a **precipitate**, when two aqueous solutions are mixed.

Precipitation reactions are common in chemistry. Potassium iodide and lead nitrate, for example, both form colorless, strong electrolyte solutions when dissolved in water (see the solubility rules in Section 7.5). When the two solutions are combined, however, a brilliant yellow precipitate forms (▼ FIGURE 7.8). We describe this precipitation reaction with the chemical equation:

$$2\,KI(aq) + Pb(NO_3)_2(aq) \longrightarrow PbI_2(s) + 2\,KNO_3(aq)$$

Precipitation reactions do not always occur when two aqueous solutions mix. For example, when we combine solutions of KI(*aq*) and NaCl(*aq*), nothing happens (▼ FIGURE 7.9).

$$KI(aq) + NaCl(aq) \longrightarrow \text{NO REACTION}$$

$$2\,KI(aq) + Pb(NO_3)_2(aq) \longrightarrow PbI_2(s) + 2\,KNO_3(aq)$$

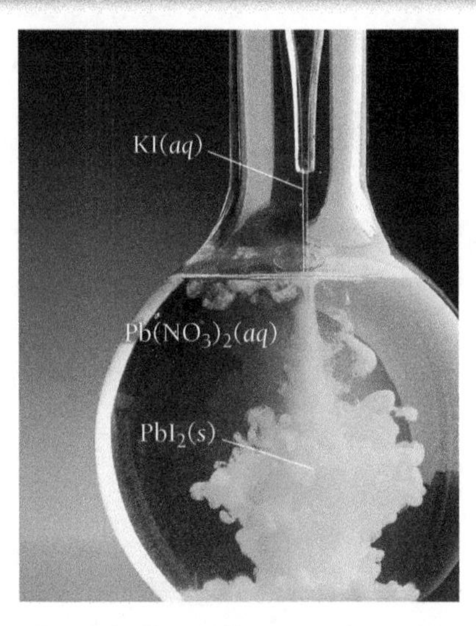

▲ FIGURE 7.8 **Precipitation** When we mix a potassium iodide solution with a lead(II) nitrate solution, a brilliant yellow precipitate of $PbI_2(s)$ forms.

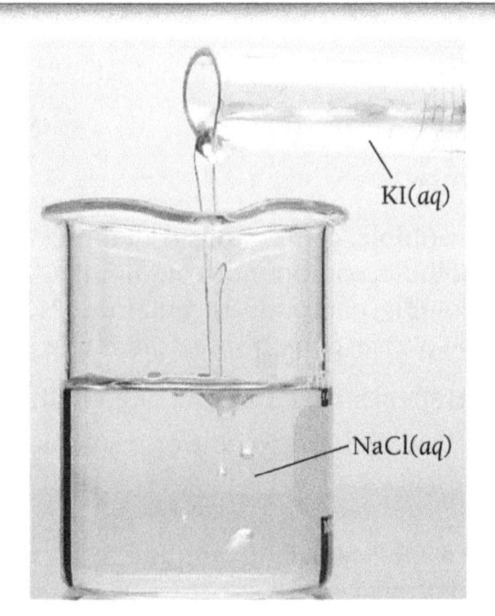

▲ FIGURE 7.9 **No reaction** When we mix a potassium iodide solution with a sodium chloride solution, no reaction occurs.

The key to predicting precipitation reactions is understanding that *only insoluble compounds form precipitates*. In a precipitation reaction, two solutions containing soluble compounds combine and an insoluble compound precipitates. Consider the precipitation reaction from Figure 7.8:

$$\underset{\text{soluble}}{2\,KI(aq)} + \underset{\text{soluble}}{Pb(NO_3)_2(aq)} \longrightarrow \underset{\text{insoluble}}{PbI_2(s)} + \underset{\text{soluble}}{2\,KNO_3(aq)}$$

KI and $Pb(NO_3)_2$ are both soluble, but the precipitate, PbI_2, is *insoluble*. Before mixing, KI(*aq*) and $Pb(NO_3)_2(aq)$ are each dissociated in their respective solutions.

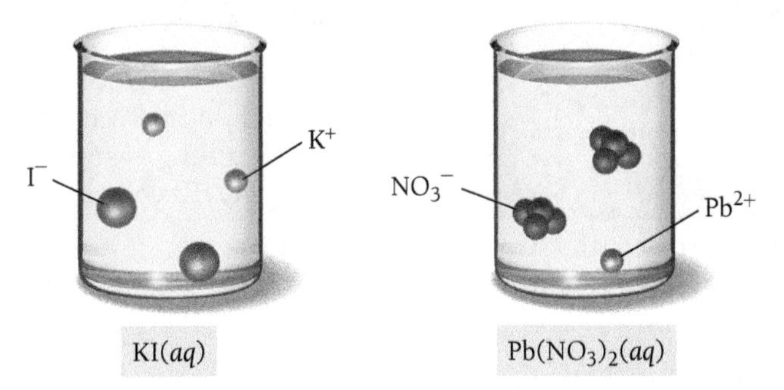

The instant that the solutions are mixed, all four ions are present.

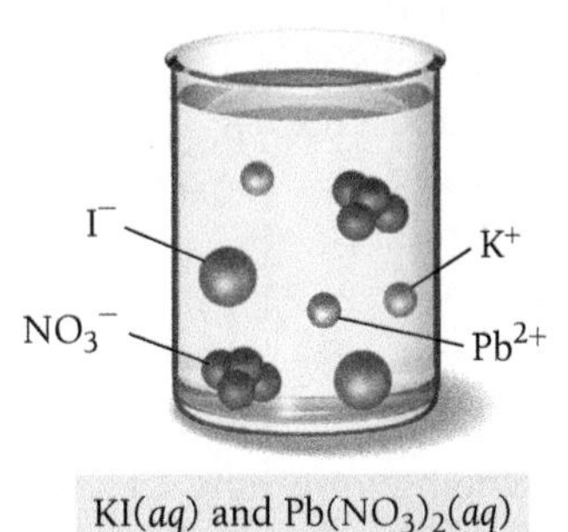

However, new compounds—potentially insoluble ones—are now possible. Specifically, the cation from one compound can pair with the anion from the other compound to form new (and potentially insoluble) products:

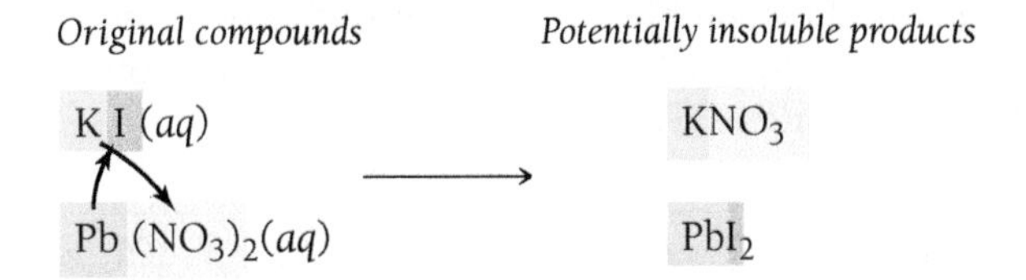

If, on the one hand, the *potentially insoluble* products are both *soluble*, no reaction occurs. If, on the other hand, one or both of the potentially insoluble products are *indeed insoluble*, a precipitation reaction occurs. In this case, KNO_3 is soluble, but PbI_2 is insoluble. Consequently, PbI_2 precipitates.

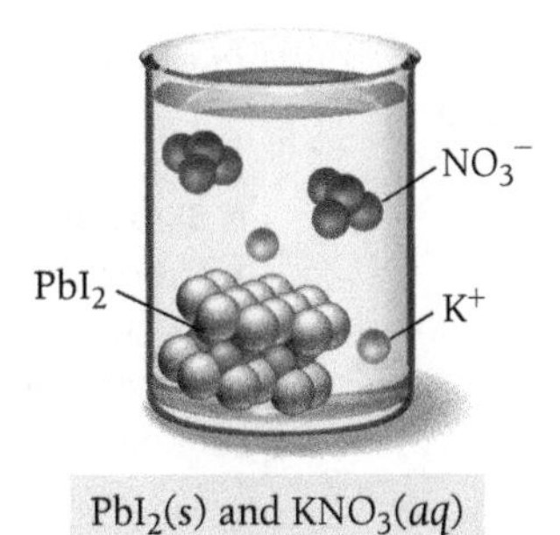

To predict whether a precipitation reaction occurs when two solutions are mixed and to write an equation for the reaction, we follow the steps in the procedure that accompanies Examples 7.7 and 7.8. As usual, the steps are shown in the left column, and two examples of applying the procedure are shown in the center and right columns.

PEARSON eText 2.0 Interactive Worked Example Video 7.7

Writing Equations for Precipitation Reactions	EXAMPLE **7.7** Write an equation for the precipitation reaction that occurs (if any) when solutions of sodium carbonate and copper(II) chloride are mixed.	EXAMPLE **7.8** Write an equation for the precipitation reaction that occurs (if any) when solutions of lithium nitrate and sodium sulfate are mixed.
1. Write the formulas of the two compounds being mixed as reactants in a chemical equation.	**SOLUTION** $Na_2CO_3(aq) + CuCl_2(aq) \longrightarrow$	**SOLUTION** $LiNO_3(aq) + Na_2SO_4(aq) \longrightarrow$
2. Below the equation, write the formulas of the potentially insoluble products that could form from the reactants. Obtain these by combining the cation from one reactant with the anion from the other. Make sure to write correct (charge-neutral) formulas for these ionic compounds as described in Section 5.5.	$Na_2CO_3(aq) + CuCl_2(aq) \longrightarrow$ **Potentially insoluble products** NaCl $CuCO_3$	$LiNO_3(aq) + Na_2SO_4(aq) \longrightarrow$ **Potentially insoluble products** $NaNO_3$ Li_2SO_4
3. Use the solubility rules from Section 7.5 to determine whether any of the potential new products are insoluble.	NaCl is *soluble* (compounds containing Cl^- are usually soluble, and Na^+ is not an exception). $CuCO_3$ is *insoluble* (compounds containing CO_3^{2-} are usually insoluble, and Cu^{2+} is not an exception).	$NaNO_3$ is *soluble* (compounds containing NO_3^- are soluble, and Na^+ is not an exception). Li_2SO_4 is *soluble* (compounds containing SO_4^{2-} are soluble, and Li^+ is not an exception).
4. If all of the potentially insoluble products are soluble, there will be no precipitate. Write NO REACTION next to the arrow.	Because this example has an insoluble product, you proceed to the next step.	$LiNO_3(aq) + Na_2SO_4(aq) \longrightarrow$ NO REACTION
5. If one or both of the potentially insoluble products are insoluble, write their formula(s) as the product(s) of the reaction, using (*s*) to indicate solid. Follow any soluble products with (*aq*) to indicate aqueous.	$Na_2CO_3(aq) + CuCl_2(aq) \longrightarrow CuCO_3(s) + NaCl(aq)$	
6. Balance the equation. Remember to adjust only coefficients, not subscripts.	$Na_2CO_3(aq) + CuCl_2(aq) \longrightarrow CuCO_3(s) + 2\,NaCl(aq)$	
	▶ **SKILLBUILDER 7.7** | Write an equation for the precipitation reaction that occurs (if any) when solutions of potassium hydroxide and nickel(II) bromide are mixed.	▶ **SKILLBUILDER 7.8** | Write an equation for the precipitation reaction that occurs (if any) when solutions of ammonium chloride and iron(III) nitrate are mixed.
		▶ **FOR MORE PRACTICE** Example 7.19; Problems 63, 64, 65, 66.

EXAMPLE 7.9 Predicting and Writing Equations for Precipitation Reactions

Write an equation for the precipitation reaction (if any) that occurs when solutions of lead(II) acetate and sodium sulfate are mixed. If no reaction occurs, write *NO REACTION*.

	SOLUTION
1. Write the formulas of the two compounds being mixed as reactants in a chemical equation.	$Pb(C_2H_3O_2)_2(aq) + Na_2SO_4(aq) \longrightarrow$
2. Below the equation, write the formulas of the potentially insoluble products that could form from the reactants. Determine these by combining the cation from one reactant with the anion from the other reactant. Make sure to adjust the subscripts so that all formulas are charge-neutral.	$Pb(C_2H_3O_2)_2(aq) + Na_2SO_4(aq) \longrightarrow$ **Potentially insoluble products** $NaC_2H_3O_2$ $PbSO_4$
3. Use the solubility rules from Section 7.5 to determine whether any of the potentially insoluble products are insoluble.	$NaC_2H_3O_2$ is *soluble* (compounds containing Na^+ are always soluble). $PbSO_4$ is *insoluble* (compounds containing SO_4^{2-} are normally soluble, but Pb^{2+} is an exception).
4. If all of the potentially insoluble products are soluble, there will be no precipitate. Write *NO REACTION* next to the arrow.	This reaction has an insoluble product so you proceed to the next step.
5. If one or both of the potentially insoluble products are insoluble, write their formula(s) as the product(s) of the reaction, using (*s*) to indicate solid. Follow any soluble products with (*aq*) to indicate aqueous.	$Pb(C_2H_3O_2)_2(aq) + Na_2SO_4(aq) \longrightarrow PbSO_4(s) + NaC_2H_3O_2(aq)$
6. Balance the equation.	$Pb(C_2H_3O_2)_2(aq) + Na_2SO_4(aq) \longrightarrow PbSO_4(s) + 2\,NaC_2H_3O_2(aq)$

▶ **SKILLBUILDER 7.9 | Predicting and Writing Equations for Precipitation Reactions**

Write an equation for the precipitation reaction (if any) that occurs when solutions of potassium sulfate and strontium nitrate are mixed. If no reaction occurs, write *NO REACTION*.

▶ **FOR MORE PRACTICE** Problems 67, 68.

CONCEPTUAL ✓ CHECKPOINT 7.5

Which reaction results in the formation of a precipitate?

(a) $NaNO_3(aq) + CaS(aq)$

(b) $MgSO_4(aq) + CaS(aq)$

(c) $NaNO_3(aq) + MgSO_4(aq)$

7.7 Writing Chemical Equations for Reactions in Solution: Molecular, Complete Ionic, and Net Ionic Equations

▶ Write molecular, complete ionic, and net ionic equations.

Consider the following equation for a precipitation reaction:

$$AgNO_3(aq) + NaCl(aq) \longrightarrow AgCl(s) + NaNO_3(aq)$$

This equation is a **molecular equation**, an equation showing the complete neutral formulas for every compound in the reaction. We can also write equations for reactions occurring in aqueous solution to show that aqueous ionic compounds normally dissociate in solution. For example, we can write the previous equation as:

$$Ag^+(aq) + NO_3^-(aq) + Na^+(aq) + Cl^-(aq) \longrightarrow AgCl(s) + Na^+(aq) + NO_3^-(aq)$$

When writing complete ionic equations, separate only aqueous ionic compounds into their constituent ions. Do NOT separate solid, liquid, or gaseous compounds.

Equations such as this one, which show the reactants and products as they are actually present in solution, are **complete ionic equations.**

Notice that in the complete ionic equation, some of the ions in solution appear unchanged on both sides of the equation. These ions are called **spectator ions** because they do not participate in the reaction.

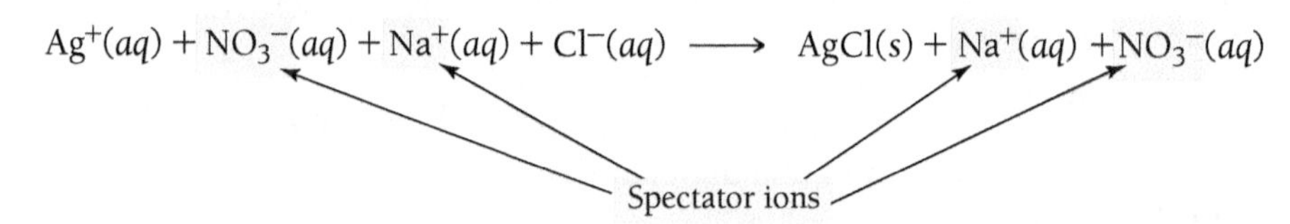

To simplify the equation and to more clearly show what is happening, spectator ions can be omitted.

$$Ag^+(aq) + Cl^-(aq) \longrightarrow AgCl(s)$$

Equations such as this one, which show only the *species* that actually participate in the reaction, are **net ionic equations.**

As another example, consider the reaction between $HCl(aq)$ and $NaOH(aq)$.

$$HCl(aq) + NaOH(aq) \longrightarrow H_2O(l) + NaCl(aq)$$

HCl, NaOH, and NaCl exist in solution as independent ions. The *complete ionic equation* for this reaction is:

$$H^+(aq) + Cl^-(aq) + Na^+(aq) + OH^-(aq) \longrightarrow H_2O(l) + Na^+(aq) + Cl^-(aq)$$

To write the *net ionic equation*, we remove the spectator ions, those that are unchanged on both sides of the equation.

$$H^+(aq) + \cancel{Cl^-}(aq) + \cancel{Na^+}(aq) + OH^-(aq) \longrightarrow H_2O(l) + \cancel{Na^+}(aq) + \cancel{Cl^-}(aq)$$

Spectator ions

The net ionic equation is $H^+(aq) + OH^-(aq) \longrightarrow H_2O(l)$.

To summarize:

- A molecular equation is a chemical equation showing the complete, neutral formulas for every compound in a reaction.
- A complete ionic equation is a chemical equation showing all of the species as they are actually present in solution.
- A net ionic equation is an equation showing only the species that actually participate in the reaction.

CONCEPTUAL CHECKPOINT 7.6

Which chemical equation is a net ionic equation?

(a) $K_2SO_4(aq) + BaCl_2(aq) \longrightarrow BaSO_4(s) + 2\,KCl(aq)$

(b) $2\,K^+(aq) + SO_4^{2-}(aq) + Ba^{2+}(aq) + 2\,Cl^-(aq) \longrightarrow BaSO_4(s) + 2\,K^+(aq) + 2\,Cl^-(aq)$

(c) $Ba^{2+}(aq) + SO_4^{2-}(aq) \longrightarrow BaSO_4(s)$

EXAMPLE **7.10** **Writing Complete Ionic and Net Ionic Equations**

Consider this precipitation reaction occurring in aqueous solution.

$$Pb(NO_3)_2(aq) + 2\,LiCl(aq) \longrightarrow PbCl_2(s) + 2\,LiNO_3(aq)$$

Write a complete ionic equation and a net ionic equation for the reaction.

Write the complete ionic equation by separating aqueous ionic compounds into their constituent ions. The $PbCl_2(s)$ remains as one unit because it does not dissociate in solution (it is insoluble).

SOLUTION

Complete ionic equation

$$Pb^{2+}(aq) + 2\,NO_3^-(aq) + 2\,Li^+(aq) + 2\,Cl^-(aq) \longrightarrow PbCl_2(s) + 2\,Li^+(aq) + 2\,NO_3^-(aq)$$

Write the net ionic equation by eliminating the spectator ions, those that do not change during the reaction.

Net ionic equation

$$Pb^{2+}(aq) + 2\,Cl^-(aq) \longrightarrow PbCl_2(s)$$

▶ **SKILLBUILDER 7.10 | Writing Complete Ionic and Net Ionic Equations**

Write a complete ionic equation and a net ionic equation for this reaction occurring in aqueous solution.

$$2\,HBr(aq) + Ca(OH)_2(aq) \longrightarrow 2\,H_2O(l) + CaBr_2(aq)$$

▶ **FOR MORE PRACTICE** Example 7.20; Problems 69, 70, 71, 72.

7.8 Acid–Base and Gas Evolution Reactions

- Identify and write equations for acid–base reactions.
- Identify and write equations for gas evolution reactions.

Two other kinds of reactions that occur in solution are **acid–base reactions**—reactions that form water upon mixing of an acid and a base—and **gas evolution reactions**—reactions that evolve a gas. Like precipitation reactions, these reactions occur when the cation of one reactant combines with the anion of another. As we will see in Section 7.9, many gas evolution reactions also happen to be acid–base reactions.

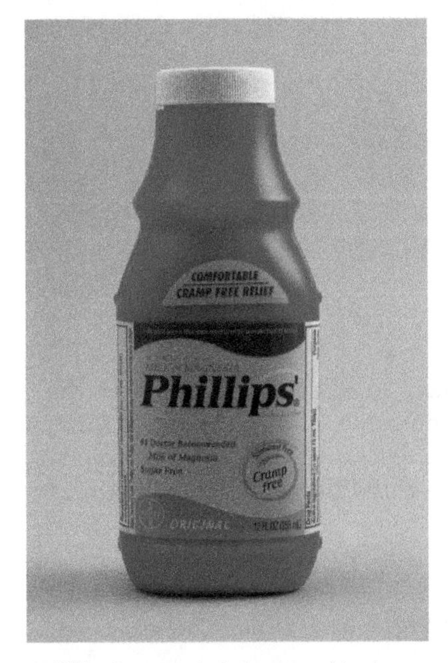

▲ Milk of magnesia is basic and tastes bitter.

Even though coffee is acidic overall, it contains some naturally occurring bases (such as caffeine) that give it a bitter taste.

Acid–Base (Neutralization) Reactions

As we saw in Chapter 5, an acid is a compound characterized by its sour taste, its ability to dissolve some metals, and its tendency to form H^+ ions in solution. A base is a compound characterized by its bitter taste, its slippery feel, and its tendency to form OH^- ions in solution. Table 7.3 lists some common acids and bases. Acids and bases are also found in many everyday substances. Foods such as lemons, limes, and vinegar contain acids. Soap, coffee, and milk of magnesia all contain bases.

When an acid and a base are mixed, the $H^+(aq)$ from the acid combines with the $OH^-(aq)$ from the base to form $H_2O(l)$. Consider the reaction between hydrochloric acid and sodium hydroxide mentioned earlier.

$$\underset{\text{Acid}}{HCl(aq)} + \underset{\text{Base}}{NaOH(aq)} \longrightarrow \underset{\text{Water}}{H_2O(l)} + \underset{\text{Salt}}{NaCl(aq)}$$

TABLE 7.3 Some Common Acids and Bases

Acid	Formula	Base	Formula
hydrochloric acid	HCl	sodium hydroxide	$NaOH$
hydrobromic acid	HBr	lithium hydroxide	$LiOH$
nitric acid	HNO_3	potassium hydroxide	KOH
sulfuric acid	H_2SO_4	calcium hydroxide	$Ca(OH)_2$
perchloric acid	$HClO_4$	barium hydroxide	$Ba(OH)_2$
acetic acid	$HC_2H_3O_2$		

▶ Common foods and everyday substances such as oranges, lemons, vinegar, and vitamin C contain acids.

Acid–base reactions (also called **neutralization reactions**) generally form water and an ionic compound—called a **salt**—that usually remains dissolved in the solution. The net ionic equation for many acid–base reactions is:

$$H^+(aq) + OH^-(aq) \longrightarrow H_2O(l)$$

Another acid–base reaction is the reaction that occurs between sulfuric acid and potassium hydroxide.

$$\underset{\text{Acid}}{H_2SO_4(aq)} + \underset{\text{Base}}{2\,KOH} \longrightarrow \underset{\text{Water}}{2\,H_2O(l)} + \underset{\text{Salt}}{K_2SO_4(aq)}$$

Notice the pattern of acid and base reacting to form water and a salt.

$$\text{Acid} + \text{Base} \longrightarrow \text{Water} + \text{Salt} \quad \text{(acid–base reactions)}$$

When writing equations for acid–base reactions, write the formula of the salt using the procedure for writing formulas of ionic compounds presented in Section 5.5.

EXAMPLE **7.11** **Writing Equations for Acid–Base Reactions**

Write a molecular and a net ionic equation for the reaction between aqueous HNO_3 and aqueous $Ca(OH)_2$.

	SOLUTION
You must recognize these substances as an acid and a base. Write the skeletal reaction following the general pattern of acid plus base produces water plus salt.	$\underset{\text{Acid}}{HNO_3(aq)} + \underset{\text{Base}}{Ca(OH)_2(aq)} \longrightarrow \underset{\text{Water}}{H_2O(l)} + \underset{\text{Salt}}{Ca(NO_3)_2(aq)}$
Next, balance the equation.	$2\,HNO_3(aq) + Ca(OH)_2(aq) \longrightarrow 2\,H_2O(l) + Ca(NO_3)_2(aq)$
Write the net ionic equation by eliminating the ions that remain the same on both sides of the equation.	$2\,H^+(aq) + 2\,OH^-(aq) \longrightarrow 2\,H_2O(l)$ or simply $H^+(aq) + OH^-(aq) \longrightarrow H_2O(l)$

▶ **SKILLBUILDER 7.11 | Writing Equations for Acid–Base Reactions**

Write a molecular and a net ionic equation for the reaction that occurs between aqueous H_2SO_4 and aqueous KOH.

▶ **FOR MORE PRACTICE** Example 7.21; Problems 77, 78, 79, 80.

Gas Evolution Reactions

Some aqueous reactions form a gas as a product. These reactions, as we learned in Section 7.1, are gas evolution reactions. Some gas evolution reactions form a gaseous product directly when the cation of one reactant reacts with the anion of the other. For example, when sulfuric acid reacts with lithium sulfide, dihydrogen monosulfide gas forms:

$$H_2SO_4(aq) + Li_2S(aq) \longrightarrow \underset{\text{Gas}}{H_2S(g)} + Li_2SO_4(aq)$$

Many gas evolution reactions such as this one are also acid–base reactions. In Chapter 14 we learn how ions such as HCO_3^- act as bases in aqueous solution.

Other gas evolution reactions form an intermediate product that then decomposes into a gas. For example, when aqueous hydrochloric acid is mixed with aqueous sodium bicarbonate, the following reaction occurs:

$$HCl(aq) + NaHCO_3(aq) \longrightarrow H_2CO_3(aq) + NaCl(aq) \longrightarrow H_2O(l) + \underset{\text{Gas}}{CO_2(g)} + NaCl(aq)$$

Gas evolution reaction

$HC_2H_3O_2(aq) + NaHCO_3(aq) \longrightarrow$
$H_2O(l) + CO_2(g) + NaC_2H_3O_2(aq)$

▲ In this gas evolution reaction, vinegar (a dilute solution of acetic acid) and baking soda (sodium bicarbonate) produce carbon dioxide.

The intermediate product, H_2CO_3, is not stable and decomposes to form H_2O and gaseous CO_2. This reaction is almost identical to the reaction in the grade school volcano of Section 7.1, which involves the mixing of acetic acid and sodium bicarbonate:

$$HC_2H_3O_2(aq) + NaHCO_3(aq) \longrightarrow H_2CO_3(aq) + NaC_2H_3O_2(aq) \longrightarrow H_2O(l) + CO_2(g) + NaC_2H_3O_2(aq)$$

The bubbling of the classroom volcano is caused by the newly formed carbon dioxide gas.

Other important gas evolution reactions form either H_2SO_3 or NH_4OH as intermediate products:

$$HCl(aq) + NaHSO_3(aq) \longrightarrow H_2SO_3(aq) + NaCl(aq) \longrightarrow H_2O(l) + SO_2(g) + NaCl(aq)$$

$$NH_4Cl(aq) + NaOH(aq) \longrightarrow NH_4OH(aq) + NaCl(aq) \longrightarrow H_2O(l) + NH_3(g) + NaCl(aq)$$

Table 7.4 lists the main types of compounds that form gases in aqueous reactions, as well as the gases that they form.

TABLE 7.4 Types of Compounds That Undergo Gas Evolution Reactions

Reactant Type	Intermediate Product	Gas Evolved	Example
sulfides	none	H_2S	$2\ HCl(aq) + K_2S(aq) \longrightarrow H_2S(g) + 2\ KCl(aq)$
carbonates and bicarbonates	H_2CO_3	CO_2	$2\ HCl(aq) + K_2CO_3(aq) \longrightarrow H_2O(l) + CO_2(g) + 2\ KCl(aq)$
sulfites and bisulfites	H_2SO_3	SO_2	$2\ HCl(aq) + K_2SO_3(aq) \longrightarrow H_2O(l) + SO_2(g) + 2\ KCl(aq)$
ammonium	NH_4OH	NH_3	$NH_4Cl(aq) + KOH(aq) \longrightarrow H_2O(l) + NH_3(g) + KCl(aq)$

EXAMPLE 7.12 Writing Equations for Gas Evolution Reactions

Write a molecular equation for the gas evolution reaction that occurs when you mix aqueous nitric acid and aqueous sodium carbonate.

	SOLUTION
Begin by writing a skeletal equation that includes the reactants and products that form when the cation of each reactant combines with the anion of the other.	$HNO_3(aq) + Na_2CO_3(aq) \longrightarrow H_2CO_3(aq) + NaNO_3(aq)$
You must recognize that $H_2CO_3(aq)$ decomposes into $H_2O(l)$ and $CO_2(g)$ and write the corresponding equation.	$HNO_3(aq) + Na_2CO_3(aq) \longrightarrow H_2O(l) + CO_2(g) + NaNO_3(aq)$
Finally, balance the equation.	$2\ HNO_3(aq) + Na_2CO_3(aq) \longrightarrow H_2O(l) + CO_2(g) + 2\ NaNO_3(aq)$

▶ **SKILLBUILDER 7.12 | Writing Equations for Gas Evolution Reactions**

Write a molecular equation for the gas evolution reaction that occurs when you mix aqueous hydrobromic acid and aqueous potassium sulfite.

▶ **SKILLBUILDER PLUS** | Write a net ionic equation for the previous reaction.

▶ **FOR MORE PRACTICE** Example 7.22; Problems 81, 82.

CHEMISTRY AND HEALTH

Neutralizing Excess Stomach Acid

Heartburn is a painful burning sensation in the esophagus (the tube that joins the throat and stomach). The pain is casued by stomach acid, which helps to break down food during digestion. Sometimes—especially after a large meal—some of that stomach acid can work its way up into the esophagus, causing the pain. You can relieve mild heartburn simply by swallowing repeatedly. Saliva contains the bicarbonate ion (HCO_3^-), which acts as a base to neutralize the acid. You can treat more severe heartburn with antacids, over-the-counter medicines that work by reacting with and neutralizing stomach acid. Antacids employ different bases as neutralizing agents. Tums™, for example, contains $CaCO_3$; milk of magnesia contains $Mg(OH)_2$; and Mylanta™ contains $Al(OH)_3$. They all, however, have the same effect of neutralizing stomach acid and relieving heartburn.

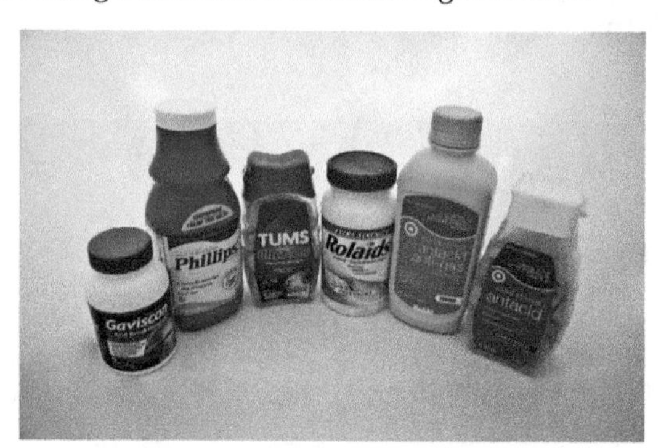

▲ Antacids contain bases such as $Mg(OH)_2$, $Al(OH)_3$, and $NaHCO_3$.

B7.1 CAN YOU ANSWER THIS? *Assume that stomach acid is HCl and write equations showing how each of these antacids neutralizes stomach acid.*

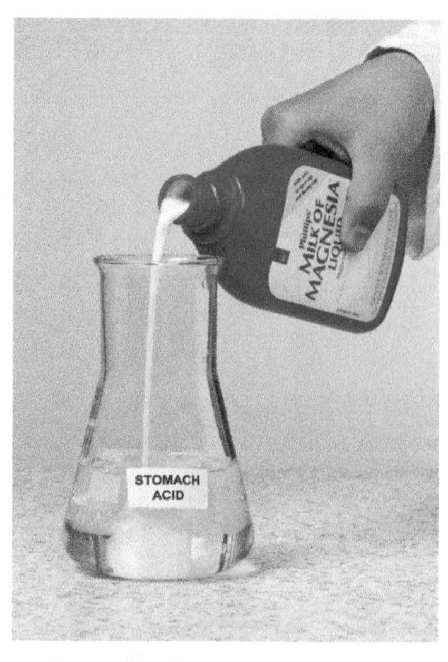

▲ The base in an antacid neutralizes excess stomach acid, relieving heartburn and acid stomach.

7.9 Oxidation–Reduction Reactions

- Identify redox reactions.
- Identify and write equations for combustion reactions.

We will cover oxidation–reduction reactions in more detail in Chapter 16.

Reactions involving the transfer of electrons are **oxidation–reduction reactions** or **redox reactions.** Redox reactions are responsible for the rusting of iron, the bleaching of hair, and the production of electricity in batteries. Many redox reactions involve the reaction of a substance with oxygen.

$$2\,H_2(g) + O_2(g) \longrightarrow 2\,H_2O(g)$$
(reaction that powers the space shuttle)

$$4\,Fe(s) + 3\,O_2(g) \longrightarrow 2\,Fe_2O_3(s)$$
(rusting of iron)

$$CH_4(g) + 2\,O_2(g) \longrightarrow CO_2(g) + 2\,H_2O(g)$$
(combustion of natural gas)

However, redox reactions don't always have to involve oxygen. Consider, for example, the reaction between sodium and chlorine to form table salt (NaCl).

$$2\,Na(s) + Cl_2(g) \longrightarrow 2\,NaCl(s)$$

The reaction between sodium and oxygen also forms other oxides besides Na_2O.

This reaction is similar to the reaction between sodium and oxygen, which can form sodium oxide.

$$4\,Na(s) + O_2(g) \longrightarrow 2\,Na_2O(s)$$

Helpful mnemonics:
OIL RIG—Oxidation Is Loss; Reduction Is Gain.
LEO GER—Lose Electrons Oxidation; Gain Electrons Reduction.

What do these two reactions have in common? In both cases, sodium (a metal with a tendency to lose electrons) reacts with a nonmetal (that has a tendency to gain electrons). In both cases, sodium atoms lose electrons to nonmetal atoms. A fundamental definition of oxidation is *the loss of electrons*, and a fundamental definition of reduction is *the gain of electrons*.

Notice that oxidation and reduction must occur together. If one substance loses electrons (oxidation), then another substance must gain electrons (reduction). For now, you simply need to be able to identify redox reactions. In Chapter 16 we will examine them more thoroughly.

Redox reactions are those in which:

A reaction can be classified as a redox reaction if it meets any one of these requirements.

- A substance reacts with elemental oxygen.
- A metal reacts with a nonmetal.
- More generally, one substance transfers electrons to another substance.

EXAMPLE **7.13** **Identifying Redox Reactions**

Which of these are redox reactions?

(a) $2\,Mg(s) + O_2(g) \longrightarrow 2\,MgO(s)$
(b) $2\,HBr(aq) + Ca(OH)_2(aq) \longrightarrow 2\,H_2O(l) + CaBr_2(aq)$
(c) $Ca(s) + Cl_2(g) \longrightarrow CaCl_2(s)$
(d) $Zn(s) + Fe^{2+}(aq) \longrightarrow Zn^{2+}(aq) + Fe(s)$

SOLUTION

(a) Redox reaction; Mg reacts with elemental oxygen.
(b) Not a redox reaction; it is an acid–base reaction.
(c) Redox reaction; a metal reacts with a nonmetal.
(d) Redox reaction; Zn transfers two electrons to Fe^{2+}.

▶ **SKILLBUILDER 7.13 | Identifying Redox Reactions**

Which of these are redox reactions?

(a) $2\,Li(s) + Cl_2(g) \longrightarrow 2\,LiCl(s)$
(b) $2\,Al(s) + 3\,Sn^{2+}(aq) \longrightarrow 2\,Al^{3+}(aq) + 3\,Sn(s)$
(c) $Pb(NO_3)_2(aq) + 2\,LiCl(aq) \longrightarrow PbCl_2(s) + 2\,LiNO_3(aq)$
(d) $C(s) + O_2(g) \longrightarrow CO_2(g)$

▶ **FOR MORE PRACTICE** Example 7.23; Problems 83, 84.

Combustion reactions are a type of redox reaction. They are important because most of our society's energy is derived from combustion reactions. Combustion reactions are characterized by the reaction of a substance with O_2 to form one or more oxygen-containing compounds, often including water. Combustion reactions are exothermic (they emit heat). For example, as we saw in Section 7.3, natural gas (CH_4) reacts with oxygen to form carbon dioxide and water.

The water formed in combustion reactions may be gaseous (*g*) or liquid (*l*) depending on the reaction conditions.

$$CH_4(g) + 2\,O_2(g) \longrightarrow CO_2(g) + 2\,H_2O(g)$$

As mentioned in Section 7.1, combustion reactions power automobiles. For example, octane, a component of gasoline, reacts with oxygen to form carbon dioxide and water.

$$2\,C_8H_{18}(l) + 25\,O_2(g) \longrightarrow 16\,CO_2(g) + 18\,H_2O(g)$$

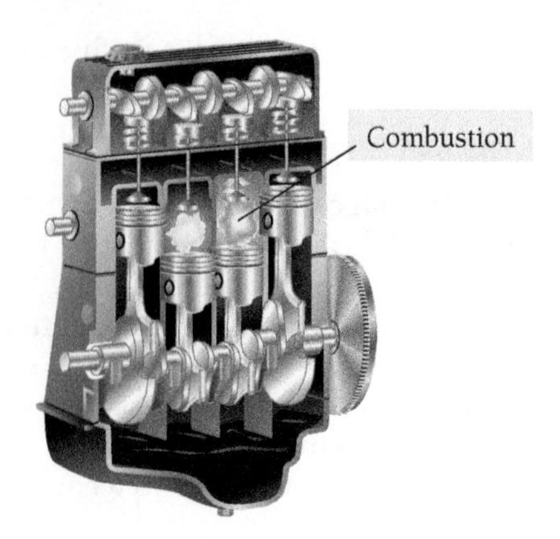

▲ Combustion of octane occurs in the cylinders of an automobile engine.

Ethanol, the alcohol in alcoholic beverages, also reacts with oxygen in a combustion reaction to form carbon dioxide and water.

$$C_2H_5OH(l) + 3\,O_2(g) \longrightarrow 2\,CO_2(g) + 3\,H_2O(g)$$

Compounds containing carbon and hydrogen—or carbon, hydrogen, and oxygen—always form carbon dioxide and water upon combustion. Other combustion reactions include the reaction of carbon with oxygen to form carbon dioxide:

$$C(s) + O_2(g) \longrightarrow CO_2(g)$$

and the reaction of hydrogen with oxygen to form water:

$$2\,H_2(g) + O_2(g) \longrightarrow 2\,H_2O(g)$$

EXAMPLE **7.14** **Writing Combustion Reactions**

Write a balanced equation for the combustion of liquid methyl alcohol (CH_3OH).

	SOLUTION
Write a skeletal equation showing the reaction of CH_3OH with O_2 to form CO_2 and H_2O.	$CH_3OH(l) + O_2(g) \longrightarrow CO_2(g) + H_2O(g)$
Balance the skeletal equation using the rules in Section 7.4.	$2CH_3OH(l) + 3O_2(g) \longrightarrow 2CO_2(g) + 4H_2O(g)$

▶ **SKILLBUILDER 7.14 | Writing Combustion Reactions**
Write a balanced equation for the combustion of liquid pentane (C_5H_{12}), a component of gasoline.

▶ **SKILLBUILDER PLUS** | Write a balanced equation for the combustion of liquid propanol (C_3H_7OH).

▶ **FOR MORE PRACTICE** Example 7.24; Problems 85, 86.

7.10 Classifying Chemical Reactions

▶ Classify chemical reactions.

Throughout this chapter, we have examined different types of chemical reactions. We have seen examples of precipitation reactions, acid–base reactions, gas evolution reactions, oxidation–reduction reactions, and combustion reactions. We can organize these different types of reactions with the following flowchart:

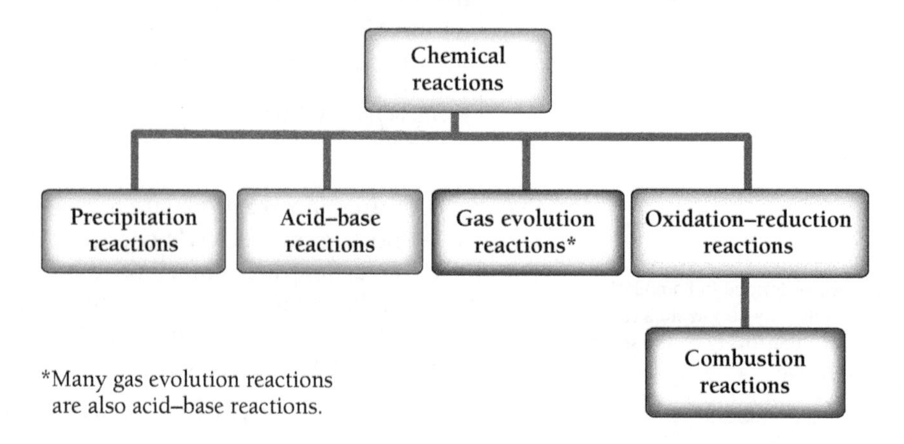

This classification scheme focuses on the type of chemistry or phenomenon that is occurring during the reaction (such as the formation of a precipitate or the transfer of electrons). However, an alternative way to classify chemical reactions is by what atoms or groups of atoms do during the reaction.

Classifying Chemical Reactions by What Atoms Do

In an alternative way of classifying reactions, we focus on the pattern of the reaction by classifying it into one of the following four categories. In this classification scheme, the letters (A, B, C, D) represent atoms or groups of atoms.

Type of Reaction	Generic Equation
synthesis or combination	$A + B \longrightarrow AB$
decomposition	$AB \longrightarrow A + B$
single-displacement	$A + BC \longrightarrow AC + B$
double-displacement	$AB + CD \longrightarrow AD + CB$

Synthesis or Combination Reactions

In a **synthesis** or **combination reaction**, simple substances combine to form more complex substances. The simpler substances may be elements, such as sodium and chlorine combining to form sodium chloride.

$$2\,Na(s) + Cl_2(g) \longrightarrow 2\,NaCl(s)$$

The simpler substances may also be compounds, such as calcium oxide and carbon dioxide combining to form calcium carbonate.

$$CaO(s) + CO_2(g) \longrightarrow CaCO_3(s)$$

In either case, a synthesis reaction follows the general equation:

$$A + B \longrightarrow AB$$

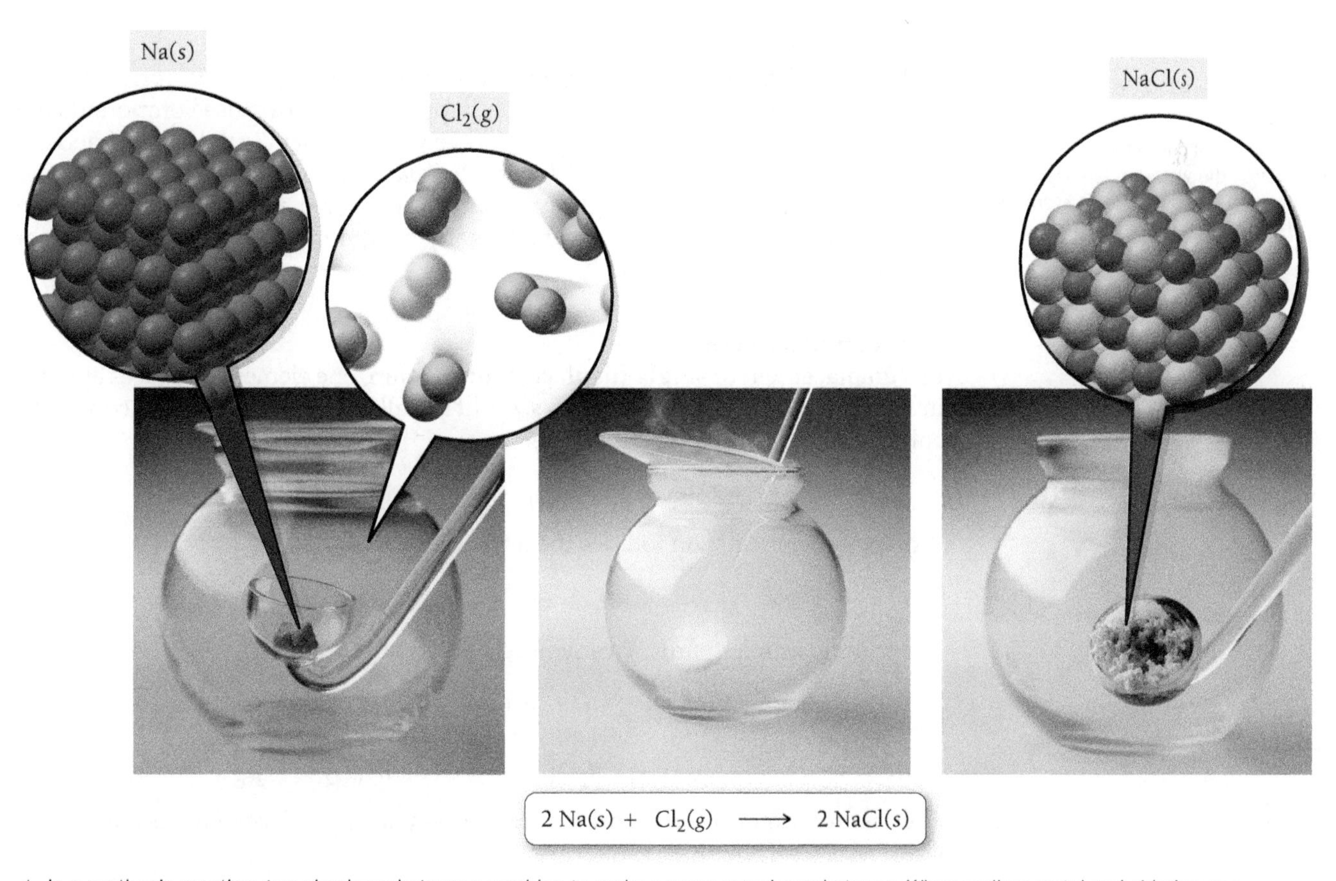

▲ In a synthesis reaction, two simpler substances combine to make a more complex substance. When sodium metal and chlorine gas combine, a chemical reaction occurs that forms sodium chloride.

Note that the first two of these reactions are also redox reactions.

Other examples of synthesis reactions include:

$$2H_2(g) + O_2(g) \longrightarrow 2H_2O(l)$$

$$2Mg(s) + O_2(g) \longrightarrow 2MgO(s)$$

$$SO_3(g) + H_2O(l) \longrightarrow H_2SO_4(aq)$$

$$2H_2O(l) \longrightarrow 2H_2(g) + O_2(g)$$

▲ When electrical current is passed through water, the water undergoes a decomposition reaction to form hydrogen gas and oxygen gas.

Decomposition Reactions

In a **decomposition reaction**, a complex substance decomposes to form simpler substances. The simpler substances may be elements, such as the hydrogen and oxygen gases that form upon the decomposition of water when electrical current passes through it.

$$2H_2O(l) \xrightarrow[\text{electrical current}]{} 2H_2(g) + O_2(g)$$

The simpler substances may also be compounds, such as the calcium oxide and carbon dioxide that form upon heating calcium carbonate.

$$CaCO_3(s) \xrightarrow[\text{heat}]{} CaO(s) + CO_2(g)$$

In either case, a decomposition reaction follows the general equation:

$$AB \longrightarrow A + B$$

Other examples of decomposition reactions include:

$$2HgO(s) \xrightarrow[\text{heat}]{} 2Hg(l) + O_2(g)$$

$$2KClO_3(s) \xrightarrow[\text{heat}]{} 2KCl(s) + 3O_2(g)$$

$$CH_3I(g) \xrightarrow[\text{light}]{} CH_3(g) + I(g)$$

Notice that these decomposition reactions require energy in the form of heat, electrical current, or light to make them happen. This is because compounds are normally stable and energy is required to decompose them. A number of decomposition reactions require *ultraviolet* or *UV light*, which is light in the ultraviolet region of the spectrum. UV light carries more energy than visible light and can therefore initiate the decomposition of many compounds. (We will discuss light in more detail in Chapter 9.)

Displacement Reactions

In a **displacement** or **single-displacement reaction**, one element displaces another in a compound. For example, when we add metallic zinc to a solution of copper(II) chloride, the zinc replaces the copper.

$$Zn(s) + CuCl_2(aq) \longrightarrow ZnCl_2(aq) + Cu(s)$$

A displacement reaction follows the general equation:

$$A + BC \longrightarrow AC + B$$

Other examples of displacement reactions include:

$$Mg(s) + 2HCl(aq) \longrightarrow MgCl_2(aq) + H_2(g)$$

$$2Na(s) + 2H_2O(l) \longrightarrow 2NaOH(aq) + H_2(g)$$

The last reaction can be identified more easily as a displacement reaction if we write water as HOH(l).

$$2Na(s) + 2HOH(l) \longrightarrow 2NaOH(aq) + H_2(g)$$

▶ In a single-displacement reaction, one element displaces another in a compound. When zinc metal is immersed in a copper(II) chloride solution, the zinc atoms displace the copper ions in solution and the copper ions coat onto the zinc metal.

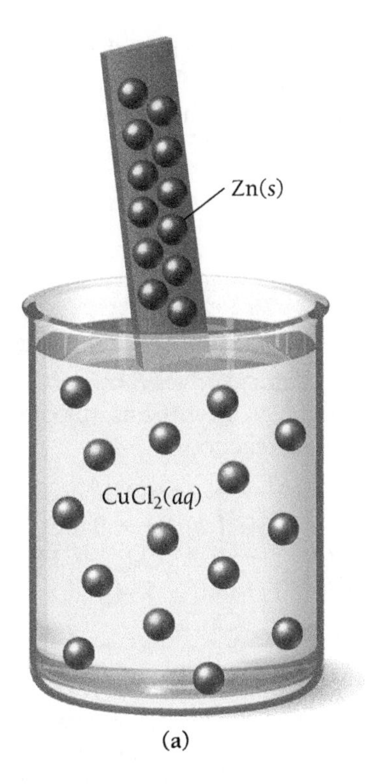

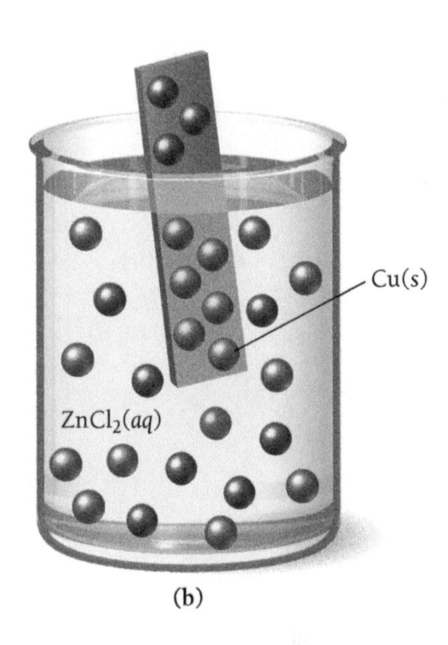

Double-Displacement Reactions

In a **double-displacement reaction**, two elements or groups of elements in two different compounds exchange places to form two new compounds. For example, in aqueous solution, the silver in silver nitrate changes places with the sodium in sodium chloride and solid silver chloride and aqueous sodium nitrate form.

This double-displacement reaction is also a precipitation reaction.

$$AgNO_3(aq) + NaCl(aq) \longrightarrow AgCl(s) + NaNO_3(aq)$$

A double-displacement reaction follows the general form:

$$AB + CD \longrightarrow AD + CB$$

Other examples of double-displacement reactions include:

These double-displacement reactions are also acid–base reactions.

$$HCl(aq) + NaOH(aq) \longrightarrow H_2O(l) + NaCl(aq)$$

$$2\,HCl(aq) + Na_2CO_3(aq) \longrightarrow H_2CO_3(aq) + 2\,NaCl(aq)$$

As we learned in Section 7.8, $H_2CO_3(aq)$ is not stable and decomposes to form $H_2O(l) + CO_2(g)$, so the overall equation is:

This double-displacement reaction is also a gas evolution reaction and an acid–base reaction.

$$2\,HCl(aq) + Na_2CO_3(aq) \longrightarrow H_2O(l) + CO_2(g) + 2\,NaCl(aq)$$

Classification Flowchart

A flowchart for this classification scheme of chemical reactions is as follows:

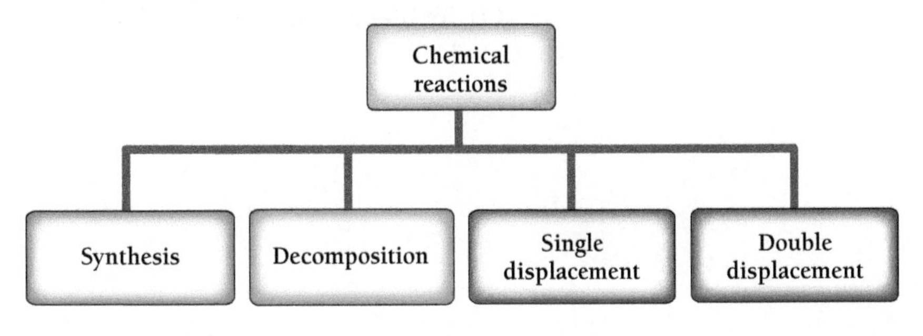

No single classification scheme is perfect because all chemical reactions are unique in some sense. However, both classification schemes—one that focuses on the type of chemistry occurring and the other that focuses on what atoms or groups of atoms are doing—are helpful because they help us see differences and similarities among chemical reactions.

EXAMPLE 7.15 Classifying Chemical Reactions According to What Atoms Do

Classify each reaction as a synthesis, decomposition, single-displacement, or double-displacement reaction.

(a) $Na_2O(s) + H_2O(l) \longrightarrow 2\,NaOH(aq)$
(b) $Ba(NO_3)_2(aq) + K_2SO_4(aq) \longrightarrow BaSO_4(s) + 2\,KNO_3(aq)$
(c) $2\,Al(s) + Fe_2O_3(s) \longrightarrow Al_2O_3(s) + 2\,Fe(l)$
(d) $2\,H_2O_2(aq) \longrightarrow 2\,H_2O(l) + O_2(g)$
(e) $Ca(s) + Cl_2(g) \longrightarrow CaCl_2(s)$

SOLUTION

(a) Synthesis; a more complex substance forms from two simpler ones.
(b) Double-displacment; Ba and K switch places to form two new compounds.
(c) Single-displacement; Al displaces Fe in Fe_2O_3.
(d) Decomposition; a complex substance decomposes into simpler ones.
(e) Synthesis; a more complex substance forms from two simpler ones.

▶ SKILLBUILDER 7.15 | Classifying Chemical Reactions According to What Atoms Do

Classify each reaction as a synthesis, decomposition, single-displacement, or double-displacement reaction.

(a) $2\,Al(s) + 2\,H_3PO_4(aq) \longrightarrow 2\,AlPO_4(aq) + 3\,H_2(g)$
(b) $CuSO_4(aq) + 2\,KOH(aq) \longrightarrow Cu(OH)_2(s) + K_2SO_4(aq)$
(c) $2\,K(s) + Br_2(l) \longrightarrow 2\,KBr(s)$
(d) $CuCl_2(aq) \xrightarrow[\text{electrical current}]{} Cu(s) + Cl_2(g)$

▶ FOR MORE PRACTICE Example 7.25; Problems 89, 90, 91, 92.

PEARSON eText 2.0

CONCEPTUAL ✔ CHECKPOINT 7.7

Both precipitation reactions and acid–base reactions can also be classified as:

(a) synthesis reactions
(b) decomposition reactions
(c) single-displacement reactions
(d) double-displacement reactions

Chapter 7 in Review

MasteringChemistry™ provides end-of-chapter exercises, feedback-enriched tutorial problems, animations, and interactive activities to encourage problem solving practice and deeper understanding of key concepts and topics.

Self-Assessment Quiz

PEARSON eText 2.0

Q1. Which process is a chemical reaction?
(a) Gasoline evaporating from a gasoline tank
(b) Iron rusting when left outdoors
(c) Dew condensing on grass during the night
(d) Water boiling on a stove top

Q2. How many oxygen atoms are on the reactant side of this chemical equation?

$$K_2CO_3(aq) + Pb(NO_3)_2(aq) \longrightarrow 2\,KNO_3(aq) + PbCO_3(s)$$

(a) 3
(b) 6
(c) 9
(d) 12

Q3. What is the coefficient for hydrogen in the balanced equation for the reaction of solid iron(III) oxide with gaseous hydrogen to form solid iron and liquid water?
(a) 2
(b) 3
(c) 4
(d) 6

Q4. Determine the correct set of coefficients to balance the chemical equation.

$$\_C_6H_6(l) + \_O_2(g) \longrightarrow \_CO_2(g) + \_H_2O(g)$$

(a) 2, 15, 12, 6
(b) 1, 15, 6, 3
(c) 2, 15, 12, 12
(d) 1, 7, 6, 3

Q5. Which compound is soluble in water?
(a) $Fe(OH)_2$
(b) CuS
(c) $AgCl$
(d) $CuCl_2$

Q6. Name the precipitate that forms (if any) when aqueous solutions of barium nitrate and potassium sulfate are mixed.
(a) $BaK(s)$
(b) $NO_3SO_4(s)$
(c) $BaSO_4(s)$
(d) $KNO_3(s)$

Q7. Which set of reactants forms a solid precipitate when mixed?
(a) $NaNO_3(aq)$ and $KCl(aq)$
(b) $KOH(aq)$ and $Na_2CO_3(aq)$
(c) $CuCl_2(aq)$ and $NaC_2H_3O_2(aq)$
(d) $Na_2CO_3(aq)$ and $CaCl_2(aq)$

Q8. What is the net ionic equation for the reaction between $Pb(C_2H_3O_2)_2(aq)$ and $KBr(aq)$?
(a) $Pb(C_2H_3O_2)_2(aq) + 2\,KBr(aq) \longrightarrow PbBr_2(s) + 2\,KC_2H_3O_2(aq)$
(b) $Pb^{2+}(aq) + 2\,Br^-(aq) \longrightarrow PbBr_2(s)$
(c) $K^+(aq) + C_2H_3O_2^-(aq) \longrightarrow KC_2H_3O_2(s)$
(d) $Pb(C_2H_3O_2)_2(aq) + 2\,KBr(aq) \longrightarrow 2\,KC_2H_3O_2(s) + PbBr_2(aq)$

Q9. Complete the equation:
$HBr(aq) + NaOH\,(aq) \longrightarrow$ ______________
(a) $NaH(s) + BrOH(aq)$
(b) $NaBr(s) + NaOH(aq)$
(c) $H_2O(l) + NaBr(aq)$
(d) No reaction occurs.

Q10. Complete the equation:
$HNO_3(aq) + KHCO_3(aq) \longrightarrow$ ______________
(a) $H_2O(l) + CO_2(g) + KNO_3(aq)$
(b) $KNO_3(s) + H_2CO_3(aq)$
(c) $HK(aq) + NO_3HCO_3$
(d) No reaction occurs.

Q11. What are the products of the balanced equation for the combustion of C_4H_9OH?
(a) $C_4H_9(s) + NaOH(aq)$
(b) $4\,CO_2(g) + 5\,H_2O(l)$
(c) $4\,O_2(g) + 5\,H_2O(l)$
(d) $2\,C_2H_4(g) + H_2O(l)$

Q12. Precipitation reactions are best classified as which type of reaction?
(a) Single displacement
(b) Double displacement
(c) Decomposition
(d) None of the above

Answers: 1:b, 2:c, 3:b, 4:a, 5:d, 6:c, 7:d, 8:b, 9:c, 10:a, 11:b, 12:b

Chemical Principles	Relevance
Chemical Reactions In a chemical reaction, one or more substances—either elements or compounds—change into a different substance.	Chemical reactions are central to many processes, including transportation, energy generation, manufacturing of household products, vision, and life itself.
Evidence of a Chemical Reaction The only absolute evidence for a chemical reaction is chemical analysis showing that one or more substances have changed into another substance. However, at least one of the following observations is often evidence of a chemical reaction: a color change; the formation of a solid or precipitate; the formation of a gas; the emission of light; and the emission or absorption of heat.	We can often perceive the changes that accompany chemical reactions. In fact, we often employ chemical reactions for the changes they produce. For example, we use the heat emitted by the combustion of fossil fuels to warm our homes, drive our cars, and generate electricity.
Chemical Equations Chemical equations represent chemical reactions. They include formulas for the reactants (the substances present before the reaction) and for the products (the new substances formed by the reaction). Chemical equations must be balanced to reflect the conservation of matter in nature; atoms do not spontaneously appear or disappear.	Chemical equations allow us to represent and understand chemical reactions. For example, the equations for the combustion reactions of fossil fuels let us see that carbon dioxide, a gas that contributes to global warming, is one of the products of these reactions.
Aqueous Solutions and Solubility Aqueous solutions are mixtures of a substance dissolved in water. If a substance dissolves in water, it is soluble; otherwise, it is insoluble.	Aqueous solutions are common. Oceans, lakes, and most of the fluids in our bodies are aqueous solutions.
Some Specific Types of Reactions **Precipitation reaction:** A solid or precipitate forms upon mixing two aqueous solutions. **Acid–base reaction:** Water form(s) upon mixing an acid and base. **Gas evolution reaction:** A gas forms upon mixing two aqueous solutions. **Redox reaction:** Electrons transfer from one substance to another. **Combustion reaction:** A substance reacts with oxygen, emitting heat, and forming an oxygen-containing compound and, in many cases, water.	Many of the specific types of reactions discussed in this chapter occur in aqueous solutions and are therefore important to living organisms. Acid–base reactions, for example, constantly occur in the blood of living organisms to maintain constant blood acidity levels. In humans, a small change in blood acidity levels would result in death, so our bodies carry out chemical reactions to prevent this. Combustion reactions are important because they are the main energy source for our society.
Classifying Chemical Reactions We can classify many chemical reactions into one of the following four categories according to what atoms or groups of atoms do: • synthesis: $(A + B \longrightarrow AB)$ • decomposition: $(AB \longrightarrow A + B)$ • single-displacement: $(A + BC \longrightarrow AC + B)$ • double-displacement: $(AB + CD \longrightarrow AD + CB)$	We classify chemical reactions to better understand them and to recognize similarities and differences among reactions.

Chemical Skills

Examples

LO: Identify evidence of a chemical reaction (Section 7.2).

To identify a chemical reaction, determine whether one or more of the initial substances changed into a different substance. If so, a chemical reaction occurred. One or more of the following often accompanies a chemical reaction: a color change; the formation of a solid or precipitate; the formation of a gas; the emission of light; and the emission or absorption of heat.

EXAMPLE 7.16 Identifying a Chemical Reaction

Which of these are chemical reactions?

(a) Copper turns green on exposure to air.
(b) When sodium bicarbonate is combined with hydrochloric acid, bubbling is observed.
(c) Liquid water freezes to form solid ice.
(d) A pure copper penny forms bubbles of a dark brown gas when dropped into nitric acid. The nitric acid solution turns blue.

SOLUTION

(a) Chemical reaction, as evidenced by the color change.
(b) Chemical reaction, as evidenced by the evolution of a gas.
(c) Not a chemical reaction; solid ice is still water.
(d) Chemical reaction, as evidenced by the evolution of a gas and by a color change.

LO: Write balanced chemical equations (Sections 7.3, 7.4).

To write balanced chemical equations, follow these steps.

1. Write a skeletal equation by writing chemical formulas for each of the reactants and products. (If a skeletal equation is provided, proceed to Step 2.)
2. If an element occurs in only one compound on both sides of the equation, balance that element first. If there is more than one such element, and the equation contains both metals and nonmetals, balance metals before nonmetals.
3. If an element occurs as a free element on either side of the chemical equation, balance that element last.
4. If the balanced equation contains coefficient fractions, clear these by multiplying the entire equation by the appropriate factor.
5. Check to make certain the equation is balanced by summing the total number of each type of atom on both sides of the equation.

Reminders

- Change only the *coefficients* to balance a chemical equation, *never the subscripts*. Changing the subscripts would change the compounds themselves.
- If the equation contains polyatomic ions that stay intact on both sides of the equation, balance the polyatomic ions as a group.

EXAMPLE 7.17 Writing Balanced Chemical Equations

Write a balanced chemical equation for the reaction of solid vanadium(V) oxide with hydrogen gas to form solid vanadium(III) oxide and liquid water.

$$V_2O_5(s) + H_2(g) \longrightarrow V_2O_3(s) + H_2O(l)$$

SOLUTION

A skeletal equation is given. Proceed to Step 2.

Vanadium occurs in only one compound on both sides of the equation. However, it is balanced, so you can proceed and balance oxygen by placing a 2 in front of H_2O on the right side.

$$V_2O_5(s) + H_2(g) \longrightarrow V_2O_3(s) + 2\,H_2O(l)$$

Hydrogen occurs as a free element; balance it last by placing a 2 in front of H_2 on the left side.

$$V_2O_5(s) + 2\,H_2(g) \longrightarrow V_2O_3(s) + 2\,H_2O(l)$$

The equation does not contain coefficient fractions. Proceed to Step 5.

Check the equation.

$$V_2O_5(s) + 2\,H_2(g) \longrightarrow V_2O_3(s) + 2\,H_2O(l)$$

Reactants		Products
2 V atoms	⟶	2 V atoms
5 O atoms	⟶	5 O atoms
4 H atoms	⟶	4 H atoms

LO: Determine whether a compound is soluble (Section 7.5).

To determine whether or not a compound is soluble, refer to the solubility rules in Table 7.2. It is simplest to begin by looking for those ions that always form soluble compounds (Li^+, Na^+, K^+, NH_4^+, NO_3^-, and $C_2H_3O_2^-$). If a compound contains one of those, it is soluble. If it does not, determine if the anion is mostly soluble (Cl^-, Br^-, I^-, or SO_4^{2-}) or mostly insoluble (OH^-, S^{2-}, CO_3^{2-}, or PO_4^{3-}). Look at the cation as well to determine whether it is one of the exceptions.

EXAMPLE 7.18 Determining Whether a Compound is Soluble

Is each compound soluble or insoluble?

(a) $CuCO_3$
(b) $BaSO_4$
(c) $Fe(NO_3)_3$

SOLUTION

(a) Insoluble; compounds containing CO_3^{2-} are insoluble, and Cu^{2+} is not an exception.
(b) Insoluble; compounds containing SO_4^{2-} are usually soluble, but Ba^{2+} is an exception.
(c) Soluble; all compounds containing NO_3^- are soluble.

LO: Predict and write equations for precipitation reactions (Section 7.6).

To predict whether a precipitation reaction occurs when two solutions are mixed and to write an equation for the reaction, follow these steps.

1. Write the formulas of the two compounds being mixed as reactants in a chemical equation.
2. Below the equation, write the formulas of the potentially insoluble products that could form from the reactants. Determine these by combining the cation from one reactant with the anion from the other. Make sure to adjust the subscripts so that all formulas are charge-neutral.
3. Use the solubility rules to determine whether any of the potentially insoluble products are indeed insoluble.
4. If all of the potentially insoluble products are soluble, there will be no precipitate. Write *NO REACTION* next to the arrow.
5. If one or both of the potentially insoluble products are insoluble, write their formula(s) as the product(s) of the reaction using (*s*) to indicate *solid*. Write any soluble products with (*aq*) to indicate *aqueous*.
6. Balance the equation.

EXAMPLE 7.19 Predicting Precipitation Reactions

Write an equation for the precipitation reaction that occurs, if any, when solutions of sodium phosphate and cobalt(II) chloride are mixed.

SOLUTION

$$Na_3PO_4(aq) + CoCl_2(aq) \longrightarrow$$

Potentially Insoluble Products:

$$NaCl \qquad Co_3(PO_4)_2$$

NaCl is soluble.

$Co_3(PO_4)_2$ is insoluble.

Reaction contains an insoluble product; proceed to Step 5.

$$Na_3PO_4(aq) + CoCl_2(aq) \longrightarrow Co_3(PO_4)_2(s) + NaCl(aq)$$

$$2\,Na_3PO_4(aq) + 3\,CoCl_2(aq) \longrightarrow Co_3(PO_4)_2(s) + 6\,NaCl(aq)$$

LO: Write molecular, complete ionic, and net ionic equations (Section 7.7).

To write a molecular equation, include the complete, neutral formulas for every compound in the reaction.

To write a complete ionic equation from a molecular equation, separate all aqueous ionic compounds into independent ions. Do not separate solid, liquid, or gaseous compounds.

To write a net ionic equation from a complete ionic equation, eliminate all species that do not change (spectator ions) in the course of the reaction.

EXAMPLE 7.20 Writing Complete Ionic and Net Ionic Equations

Write a complete ionic and a net ionic equation for the reaction.

$$2\,NH_4Cl(aq) + Hg_2(NO_3)_2(aq) \longrightarrow Hg_2Cl_2(s) + 2\,NH_4NO_3(aq)$$

SOLUTION

Complete ionic equation:

$$2\,NH_4^+(aq) + 2\,Cl^-(aq) + Hg_2^{2+}(aq) + 2\,NO_3^-(aq) \longrightarrow Hg_2Cl_2(s) + 2\,NH_4^+(aq) + 2\,NO_3^-(aq)$$

Net ionic equation:

$$2\,Cl^-(aq) + Hg_2^{2+}(aq) \longrightarrow Hg_2Cl_2(s)$$

LO: Identify and write equations for acid–base reactions (Section 7.8).

When you see an acid and a base (see Table 7.3) as reactants in an equation, write a reaction in which the acid and the base react to form water and a salt.

EXAMPLE **7.21** **Writing Equations for Acid–Base Reactions**

Write an equation for the reaction that occurs when aqueous hydroiodic acid is mixed with aqueous barium hydroxide.

SOLUTION

$$\underset{\text{Acid}}{2\,HI(aq)} + \underset{\text{Base}}{Ba(OH)_2(aq)} \longrightarrow \underset{\text{Water}}{2\,H_2O(l)} + \underset{\text{Salt}}{BaI_2(aq)}$$

LO: Identify and write equations for gas evolution reactions (Section 7.8).

Refer to Table 7.4 to identify gas evolution reactions.

EXAMPLE **7.22** **Writing Equations for Gas Evolution Reactions**

Write an equation for the reaction that occurs when aqueous hydrobromic acid is mixed with aqueous potassium bisulfite.

SOLUTION

$$HBr(aq) + KHSO_3(aq) \longrightarrow H_2SO_3(aq) + KBr(aq) \longrightarrow H_2O(l) + SO_2(g) + KBr(aq)$$

LO: Identify redox reactions (Section 7.9).

Redox reactions are those in which any of the following occurs:

- A substance reacts with elemental oxygen.
- A metal reacts with a nonmetal.
- One substance transfers electrons to another substance.

EXAMPLE **7.23** **Identifying Redox Reactions**

Which of these reactions is a redox reaction?

(a) $4\,Fe(s) + 3\,O_2(g) \longrightarrow 2\,Fe_2O_3(s)$
(b) $CaO(s) + CO_2(g) \longrightarrow CaCO_3(s)$
(c) $AgNO_3(aq) + NaCl(aq) \longrightarrow AgCl(s) + NaNO_3(aq)$

SOLUTION

Only **(a)** is a redox reaction.

LO: Identify and write equations for combustion reactions (Section 7.9).

In a combustion reaction, a substance reacts with O_2 to form one or more oxygen-containing compounds and, in many cases, water.

EXAMPLE **7.24** **Writing Equations for Combustion Reactions**

Write a balanced equation for the combustion of gaseous ethane (C_2H_6), a minority component of natural gas.

SOLUTION

The skeletal equation is:

$$C_2H_6(g) + O_2(g) \longrightarrow CO_2(g) + H_2O(g)$$

The balanced equation is:

$$2\,C_2H_6(g) + 7\,O_2(g) \longrightarrow 4\,CO_2(g) + 6\,H_2O(g)$$

LO: Classify chemical reactions (Section 7.10).

You can classify chemical reactions by inspection. The four major categories are:

Synthesis or combination

$$A + B \longrightarrow AB$$

Decomposition

$$AB \longrightarrow A + B$$

Single-displacement

$$A + BC \longrightarrow AC + B$$

Double-displacement

$$AB + CD \longrightarrow AD + CB$$

EXAMPLE **7.25** **Classifying Chemical Reactions**

Classify each chemical reaction as a synthesis, decomposition, single-displacement, or double-displacement reaction.

(a) $2\,K(s) + Br_2(g) \longrightarrow 2\,KBr(s)$
(b) $Fe(s) + 2\,AgNO_3(aq) \longrightarrow Fe(NO_3)_2(aq) + 2\,Ag(s)$
(c) $CaSO_3(s) \longrightarrow CaO(s) + SO_2(g)$
(d) $CaCl_2(aq) + Li_2SO_4(aq) \longrightarrow CaSO_4(s) + 2\,LiCl(aq)$

SOLUTION

(a) Synthesis; KBr, a more complex substance, is formed from simpler substances.
(b) Single-displacement; Fe displaces Ag in $AgNO_3$.
(c) Decomposition; $CaSO_3$ decomposes into simpler substances.
(d) Double-displacement; Ca and Li switch places to form new compounds.

Key Terms

acid–base reaction [7.8]
aqueous solution [7.5]
balanced equation [7.3]
combination reaction [7.10]
combustion reaction [7.9]
complete ionic equation [7.7]
decomposition reaction [7.10]
displacement reaction [7.10]
double-displacement reaction [7.10]
gas evolution reaction [7.8]
insoluble [7.5]
molecular equation [7.7]
net ionic equation [7.7]
neutralization reaction [7.8]
oxidation–reduction (redox) reaction [7.9]
precipitate [7.6]
precipitation reaction [7.6]
salt [7.8]
single-displacement reaction [7.10]
solubility rules [7.5]
soluble [7.5]
spectator ion [7.7]
strong electrolyte solution [7.5]
synthesis reaction [7.10]

Exercises

Questions

1. What is a chemical reaction? List some examples.
2. If you could observe atoms and molecules with the naked eye, what would you look for as conclusive evidence of a chemical reaction?
3. What are the main indications that a chemical reaction has occurred?
4. What is a chemical equation? Provide an example and identify the reactants and products.
5. What does each abbreviation, often used in chemical equations, represent?
   (a) (g) **(b)** (l) **(c)** (s) **(d)** (aq)
6. To balance a chemical equation, adjust the ______ as necessary to make the numbers of each type of atom on both sides of the equation equal. Never adjust the ______ to balance a chemical equation.
7. Is the chemical equation balanced? Why or why not?

$$2\,Ag_2O(s) + C(s) \longrightarrow CO_2(g) + 4\,Ag(s)$$

8. What is an aqueous solution? List two examples.
9. What does it mean if a compound is referred to as soluble? insoluble?
10. Explain what happens to an ionic substance when it dissolves in water.
11. Do polyatomic ions dissociate when they dissolve in water, or do they remain intact?
12. What is a strong electrolyte solution?
13. What are the solubility rules, and how are they useful?
14. What is a precipitation reaction? Provide an example and identify the precipitate.
15. Is the precipitate in a precipitation reaction always a compound that is soluble or insoluble? Explain.
16. Describe the differences between a molecular equation, a complete ionic equation, and net ionic equation. Give an example of each to illustrate the differences.
17. What is an acid–base reaction? List an example and identify the acid and the base.
18. What are the distinguishing properties of acids and bases?
19. What is a gas evolution reaction? Give an example.
20. What is a redox reaction? Give an example.
21. What is a combustion reaction? Give an example.
22. What are two different ways to classify chemical reactions presented in Section 7.10? Explain the differences between the two methods.
23. Explain the difference between a synthesis reaction and a decomposition reaction and provide an example of each.
24. Explain the difference between a single-displacement reaction and a double-displacement reaction and provide an example of each.

Problems

EVIDENCE OF CHEMICAL REACTIONS

25. Which observation is consistent with a chemical reaction occurring? Why?
    (a) Solid copper deposits on a piece of aluminum foil when the foil is placed in a blue copper nitrate solution. The blue color of the solution fades.
    (b) Liquid ethyl alcohol turns into a solid when placed in a low-temperature freezer.
    (c) A white precipitate forms when solutions of barium nitrate and sodium sulfate are mixed.
    (d) A mixture of sugar and water bubbles when yeasts are added. After several days, the sugar is gone and ethyl alcohol is found in the water.
26. Which observation is consistent with a chemical reaction occurring? Why?
    (a) Propane forms a flame and emits heat as it burns.
    (b) Acetone feels cold as it evaporates from the skin.
    (c) Bubbling occurs when potassium carbonate and hydrochloric acid solutions are mixed.
    (d) Heat is felt when a warm object is placed in your hand.

27. Vinegar forms bubbles when it is poured onto the calcium deposits on a faucet, and some of the calcium dissolves. Has a chemical reaction occurred? Explain your answer.

28. When a chemical drain opener is added to a clogged sink, bubbles form and the water in the sink gets warmer. Has a chemical reaction occurred? Explain your answer.

29. When a commercial hair bleaching mixture is applied to brown hair, the hair turns blond. Has a chemical reaction occurred? Explain your answer.

30. When water is boiled in a pot, it bubbles. Has a chemical reaction occurred? Explain your answer.

WRITING AND BALANCING CHEMICAL EQUATIONS

31. For each chemical equation (which may or may not be balanced), list the number of each type of atom on each side of the equation, and determine if the equation is balanced.
   (a) $Pb(NO_3)_2(aq) + 2\,NaCl(aq) \longrightarrow PbCl_2(s) + 2\,NaNO_3(aq)$
   (b) $C_3H_8(g) + O_2(g) \longrightarrow 3\,CO_2(g) + 4\,H_2O(g)$

32. For each chemical equation (which may or may not be balanced), list the number of each type of atom on each side of the equation, and determine if the equation is balanced.
   (a) $MgS(aq) + 2\,CuCl_2(aq) \longrightarrow 2\,CuS(s) + MgCl_2(aq)$
   (b) $2\,C_6H_{14}(l) + 19\,O_2(g) \longrightarrow 12\,CO_2(g) + 14\,H_2O(g)$

33. Consider the unbalanced chemical equation.

$$H_2O(l) \xrightarrow[\text{electrical current}]{} H_2(g) + O_2(g)$$

A chemistry student tries to balance the equation by placing the subscript 2 after the oxygen atom in H_2O. Explain why this is not correct. What is the correct balanced equation?

34. Consider the unbalanced chemical equation.

$$Al(s) + Cl_2(g) \longrightarrow AlCl_3(s)$$

A student tries to balance the equation by changing the subscript 2 on Cl to a 3. Explain why this is not correct. What is the correct balanced equation?

35. Write a balanced chemical equation for each chemical reaction.
   (a) Solid lead(II) sulfide reacts with aqueous hydrochloric acid to form solid lead(II) chloride and dihydrogen sulfide gas.
   (b) Gaseous carbon monoxide reacts with hydrogen gas to form gaseous methane (CH_4) and liquid water.
   (c) Solid iron(III) oxide reacts with hydrogen gas to form solid iron and liquid water.
   (d) Gaseous ammonia (NH_3) reacts with gaseous oxygen to form gaseous nitrogen monoxide and gaseous water.

36. Write a balanced chemical equation for each chemical reaction.
   (a) Solid copper reacts with solid sulfur to form solid copper(I) sulfide.
   (b) Sulfur dioxide gas reacts with oxygen gas to form sulfur trioxide gas.
   (c) Aqueous hydrochloric acid reacts with solid manganese(IV) oxide to form aqueous manganese(II) chloride, liquid water, and chlorine gas.
   (d) Liquid benzene (C_6H_6) reacts with gaseous oxygen to form carbon dioxide and liquid water.

37. Write a balanced chemical equation for each chemical reaction.
   (a) Solid magnesium reacts with aqueous copper(I) nitrate to form aqueous magnesium nitrate and solid copper.
   (b) Gaseous dinitrogen pentoxide decomposes to form nitrogen dioxide and oxygen gas.
   (c) Solid calcium reacts with aqueous nitric acid to form aqueous calcium nitrate and hydrogen gas.
   (d) Liquid methanol (CH_3OH) reacts with oxygen gas to form gaseous carbon dioxide and gaseous water.

38. Write a balanced chemical equation for each chemical reaction.
   (a) Gaseous acetylene (C_2H_2) reacts with oxygen gas to form gaseous carbon dioxide and gaseous water.
   (b) Chlorine gas reacts with aqueous potassium iodide to form solid iodine and aqueous potassium chloride.
   (c) Solid lithium oxide reacts with liquid water to form aqueous lithium hydroxide.
   (d) Gaseous carbon monoxide reacts with oxygen gas to form carbon dioxide gas.

39. When solid sodium is added to liquid water, it reacts with the water to produce hydrogen gas and aqueous sodium hydroxide. Write a balanced chemical equation for this reaction.

40. When iron rusts, solid iron reacts with gaseous oxygen to form solid iron(III) oxide. Write a balanced chemical equation for this reaction.

41. Sulfuric acid in acid rain forms when gaseous sulfur dioxide pollutant reacts with gaseous oxygen and liquid water to form aqueous sulfuric acid. Write a balanced chemical equation for this reaction.

42. Nitric acid in acid rain forms when gaseous nitrogen dioxide pollutant reacts with gaseous oxygen and liquid water to form aqueous nitric acid. Write a balanced chemical equation for this reaction.

43. Write a balanced chemical equation for the reaction of solid vanadium(V) oxide with hydrogen gas to form solid vanadium(III) oxide and liquid water.

44. Write a balanced chemical equation for the reaction of gaseous nitrogen dioxide with hydrogen gas to form gaseous ammonia and liquid water.

45. Write a balanced chemical equation for the fermentation of sugar ($C_{12}H_{22}O_{11}$) by yeasts in which the aqueous sugar reacts with water to form aqueous ethyl alcohol (C_2H_5OH) and carbon dioxide gas.

46. Write a balanced chemical equation for the photosynthesis reaction in which gaseous carbon dioxide and liquid water react in the presence of chlorophyll to produce aqueous glucose ($C_6H_{12}O_6$) and oxygen gas.

47. Balance each chemical equation.
(a) $Na_2S(aq) + Cu(NO_3)_2(aq) \longrightarrow NaNO_3(aq) + CuS(s)$
(b) $HCl(aq) + O_2(g) \longrightarrow H_2O(l) + Cl_2(g)$
(c) $H_2(g) + O_2(g) \longrightarrow H_2O(l)$
(d) $FeS(s) + HCl(aq) \longrightarrow FeCl_2(aq) + H_2S(g)$

48. Balance each chemical equation.
(a) $N_2H_4(l) \longrightarrow NH_3(g) + N_2(g)$
(b) $H_2(g) + N_2(g) \longrightarrow NH_3(g)$
(c) $Cu_2O(s) + C(s) \longrightarrow Cu(s) + CO(g)$
(d) $H_2(g) + Cl_2(g) \longrightarrow HCl(g)$

49. Balance each chemical equation.
(a) $BaO_2(s) + H_2SO_4(aq) \longrightarrow BaSO_4(s) + H_2O_2(aq)$
(b) $Co(NO_3)_3(aq) + (NH_4)_2S(aq) \longrightarrow Co_2S_3(s) + NH_4NO_3(aq)$
(c) $Li_2O(s) + H_2O(l) \longrightarrow LiOH(aq)$
(d) $Hg_2(C_2H_3O_2)_2(aq) + KCl(aq) \longrightarrow Hg_2Cl_2(s) + KC_2H_3O_2(aq)$

50. Balance each chemical equation.
(a) $MnO_2(s) + HCl(aq) \longrightarrow Cl_2(g) + MnCl_2(aq) + H_2O(l)$
(b) $Co_2(g) + CaSiO_3(s) + H_2O(l) \longrightarrow SiO_2(s) + Ca(HCO_3)_2(aq)$
(c) $Fe(s) + S(l) \longrightarrow Fe_2S_3(s)$
(d) $NO_2(g) + H_2O(l) \longrightarrow HNO_3(aq) + NO(g)$

51. Is each chemical equation correctly balanced? If not, correct it.
(a) $Rb(s) + H_2O(l) \longrightarrow RbOH(aq) + H_2(g)$
(b) $2N_2H_4(g) + N_2O_4(g) \longrightarrow 3N_2(g) + 4H_2O(g)$
(c) $NiS(s) + O_2(g) \longrightarrow NiO(s) + SO_2(g)$
(d) $PbO(s) + 2NH_3(g) \longrightarrow Pb(s) + N_2(g) + H_2O(l)$

52. Is each chemical equation correctly balanced? If not, correct it.
(a) $SiO_2(s) + 4HF(aq) \longrightarrow SiF_4(g) + 2H_2O(l)$
(b) $2Cr(s) + 3O_2(g) \longrightarrow Cr_2O_3(s)$
(c) $Al_2S_3(s) + H_2O(l) \longrightarrow 2Al(OH)_3(s) + 3H_2S(g)$
(d) $Fe_2O_3(s) + CO(g) \longrightarrow 2Fe(s) + CO_2(g)$

53. Human cells obtain energy from a reaction called cellular respiration. Balance the skeletal equation for cellular respiration.

$$C_6H_{12}O_6(aq) + O_2(g) \longrightarrow CO_2(g) + H_2O(l)$$

54. Propane camping stoves produce heat by the combustion of gaseous propane (C_3H_8). Balance the skeletal equation for the combustion of propane.

$$C_3H_8(g) + O_2(g) \longrightarrow CO_2(g) + H_2O(g)$$

55. Catalytic converters work to remove nitrogen oxides and carbon monoxide from exhaust. Balance the skeletal equation for one of the reactions that occurs in a catalytic converter.

$$NO(g) + CO(g) \longrightarrow N_2(g) + CO_2(g)$$

56. Billions of pounds of urea are produced annually for use as fertilizer. Balance the skeletal equation for the synthesis of urea.

$$NH_3(g) + CO_2(g) \longrightarrow CO(NH_2)_2(s) + H_2O(l)$$

SOLUBILITY

57. Is each compound soluble or insoluble? For the soluble compounds, identify the ions present in solution.
(a) $NaC_2H_3O_2$
(b) $Sn(NO_3)_2$
(c) AgI
(d) $Na_3(PO_4)$

58. Is each compound soluble or insoluble? For the soluble compounds, identify the ions present in solution.
(a) $(NH_4)_2S$
(b) $CuCO_3$
(c) ZnS
(d) $Pb(C_2H_3O_2)_2$

59. Pair each cation on the left with an anion on the right that will form an *insoluble* compound with it and write a formula for the insoluble compound. Use each anion only once.

Ag^+	SO_4^{2-}
Ba^{2+}	Cl^-
Cu^{2+}	CO_3^{2-}
Fe^{3+}	S^{2-}

60. Pair each cation on the left with an anion on the right that will form a *soluble* compound with it and write a formula for the soluble compound. Use each anion only once.

Na^+	NO_3^-
Sr^{2+}	SO_4^{2-}
Co^{2+}	S^{2-}
Pb^{2+}	CO_3^{2-}

61. Move any misplaced compounds to the correct column.

Soluble	Insoluble
K_2S	K_2SO_4
$PbSO_4$	Hg_2I_2
BaS	$Cu_3(PO_4)_2$
$PbCl_2$	MgS
Hg_2Cl_2	$CaSO_4$
NH_4Cl	SrS
Na_2CO_3	Li_2S

62. Move any misplaced compounds to the correct column.

Soluble	Insoluble
$LiOH$	$CaCl_2$
Na_2CO_3	$Cu(OH)_2$
$AgCl$	$Ca(C_2H_3O_2)_2$
K_3PO_4	$SrSO_4$
CuI_2	Hg_2Br_2
$Pb(NO_3)_2$	$PbBr_2$
$CoCO_3$	PbI_2

PRECIPITATION REACTIONS

63. Complete and balance each equation. If no reaction occurs, write *NO REACTION*.
(a) $KI(aq) + BaS(aq) \longrightarrow$
(b) $K_2SO_4(aq) + BaBr_2(aq) \longrightarrow$
(c) $NaCl(aq) + Hg_2(C_2H_3O_2)_2(aq) \longrightarrow$
(d) $NaC_2H_3O_2(aq) + Pb(NO_3)_2(aq) \longrightarrow$

64. Complete and balance each equation. If no reaction occurs, write *NO REACTION*.
(a) $NaOH(aq) + FeBr_3(aq) \longrightarrow$
(b) $BaCl_2(aq) + AgNO_3(aq) \longrightarrow$
(c) $Na_2CO_3(aq) + CoCl_2(aq) \longrightarrow$
(d) $K_2S(aq) + BaCl_2(aq) \longrightarrow$

65. Write a molecular equation for the precipitation reaction that occurs (if any) when each pair of solutions is mixed. If no reaction occurs, write *NO REACTION*.
(a) sodium carbonate and lead(II) nitrate
(b) potassium sulfate and lead(II) acetate
(c) copper(II) nitrate and barium sulfide
(d) calcium nitrate and sodium iodide

66. Write a molecular equation for the precipitation reaction that occurs (if any) when each pair of solutions is mixed. If no reaction occurs, write *NO REACTION*.
(a) potassium chloride and lead(II) acetate
(b) lithium sulfate and strontium chloride
(c) potassium bromide and calcium sulfide
(d) chromium(III) nitrate and potassium phosphate

67. Correct any incorrect equations. If no reaction occurs, write *NO REACTION*.
(a) $Ba(NO_3)_2(aq) + (NH_4)_2SO_4(aq) \longrightarrow BaSO_4(s) + 2\,NH_4NO_3(aq)$
(b) $BaS(aq) + 2\,KCl(aq) \longrightarrow BaCl_2(s) + K_2S(aq)$
(c) $2\,KI(aq) + Pb(NO_3)_2(aq) \longrightarrow PbI_2(s) + 2\,KNO_3(aq)$
(d) $Pb(NO_3)_2(aq) + 2\,LiCl(aq) \longrightarrow 2\,LiNO_3(s) + PbCl_2(aq)$

68. Correct any incorrect equations. If no reaction occurs, write *NO REACTION*.
(a) $AgNO_3(aq) + NaCl(aq) \longrightarrow NaCl(s) + AgNO_3(aq)$
(b) $K_2SO_4(aq) + Co(NO_3)_2(aq) \longrightarrow CoSO_4(s) + 2\,KNO_3(aq)$
(c) $Cu(NO_3)_2(aq) + (NH_4)_2S(aq) \longrightarrow CuS(s) + 2\,NH_4NO_3(aq)$
(d) $Hg_2(NO_3)_2(aq) + 2\,LiCl(aq) \longrightarrow Hg_2Cl_2(s) + 2\,LiNO_3(aq)$

IONIC AND NET IONIC EQUATIONS

69. Identify the spectator ions in the complete ionic equation.

$$2\,K^+(aq) + S^{2-}(aq) + Pb^{2+}(aq) + 2\,NO_3^-(aq) \longrightarrow PbS(s) + 2\,K^+(aq) + 2\,NO_3^-(aq)$$

70. Identify the spectator ions in the complete ionic equation.

$$Ba^{2+}(aq) + 2\,I^-(aq) + 2\,Na^+(aq) + SO_4^{2-}(aq) \longrightarrow BaSO_4(s) + 2\,I^-(aq) + 2\,Na^+(aq)$$

71. Write balanced complete ionic and net ionic equations for each reaction.
(a) $AgNO_3(aq) + KCl(aq) \longrightarrow AgCl(s) + KNO_3(aq)$
(b) $CaS(aq) + CuCl_2(aq) \longrightarrow CuS(s) + CaCl_2(aq)$
(c) $NaOH(aq) + HNO_3(aq) \longrightarrow H_2O(l) + NaNO_3(aq)$
(d) $2\,K_3PO_4(aq) + 3\,NiCl_2(aq) \longrightarrow Ni_3(PO_4)_2(s) + 6\,KCl(aq)$

72. Write balanced complete ionic and net ionic equations for each reaction.
(a) $HI(aq) + KOH(aq) \longrightarrow H_2O(l) + KI(aq)$
(b) $Na_2SO_4(aq) + CaI_2(aq) \longrightarrow CaSO_4(s) + 2\,NaI(aq)$
(c) $2\,HC_2H_3O_2(aq) + Na_2CO_3(aq) \longrightarrow H_2O(l) + CO_2(g) + 2\,NaC_2H_3O_2(aq)$
(d) $NH_4Cl(aq) + NaOH(aq) \longrightarrow H_2O(l) + NH_3(g) + NaCl(aq)$

73. Mercury(I) ions (Hg_2^{2+}) can be removed from solution by precipitation with Cl^-. Suppose a solution contains aqueous $Hg_2(NO_3)_2$. Write complete ionic and net ionic equations to show the reaction of aqueous $Hg_2(NO_3)_2$ with aqueous sodium chloride to form solid Hg_2Cl_2 and aqueous sodium nitrate.

74. Lead ions can be removed from solution by precipitation with sulfate ions. Suppose a solution contains lead(II) nitrate. Write a complete ionic and net ionic equation to show the reaction of aqueous lead(II) nitrate with aqueous potassium sulfate to form solid lead(II) sulfate and aqueous potassium nitrate.

75. Write complete ionic and net ionic equations for each of the reactions in Problem 67.

76. Write complete ionic and net ionic equations for each of the reactions in Problem 68.

ACID–BASE AND GAS EVOLUTION REACTIONS

77. When a hydrochloric acid solution is combined with a potassium hydroxide solution, an acid–base reaction occurs. Write a balanced molecular equation and a net ionic equation for this reaction.

78. A beaker of nitric acid is neutralized with calcium hydroxide. Write a balanced molecular equation and a net ionic equation for this reaction.

79. Complete and balance each acid–base reaction.
   (a) $HCl(aq) + Ba(OH)_2(aq) \longrightarrow$
   (b) $H_2SO_4(aq) + KOH(aq) \longrightarrow$
   (c) $HClO_4(aq) + NaOH(aq) \longrightarrow$

80. Complete and balance each acid–base reaction.
   (a) $HC_2H_3O_2(aq) + Ca(OH)_2(aq) \longrightarrow$
   (b) $HBr(aq) + LiOH(aq) \longrightarrow$
   (c) $H_2SO_4(aq) + Ba(OH)_2(aq) \longrightarrow$

81. Complete and balance each gas evolution reaction.
   (a) $HBr(aq) + NaHCO_3(aq) \longrightarrow$
   (b) $NH_4I(aq) + KOH(aq) \longrightarrow$
   (c) $HNO_3(aq) + K_2SO_3(aq) \longrightarrow$
   (d) $HI(aq) + Li_2S(aq) \longrightarrow$

82. Complete and balance each gas evolution reaction.
   (a) $HClO_4(aq) + K_2CO_3(aq) \longrightarrow$
   (b) $HC_2H_3O_2(aq) + LiHSO_3(aq) \longrightarrow$
   (c) $(NH_4)_2SO_4(aq) + Ca(OH)_2(aq) \longrightarrow$
   (d) $HCl(aq) + ZnS(s) \longrightarrow$

OXIDATION–REDUCTION AND COMBUSTION

83. Which reactions are redox reactions?
   (a) $Ba(NO_3)_2(aq) + K_2SO_4(aq) \longrightarrow BaSO_4(s) + 2\,KNO_3(aq)$
   (b) $Ca(s) + Cl_2(g) \longrightarrow CaCl_2(s)$
   (c) $HCl(aq) + NaOH(aq) \longrightarrow H_2O(l) + NaCl(aq)$
   (d) $Zn(s) + Fe^{2+}(aq) \longrightarrow Zn^{2+}(aq) + Fe(s)$

84. Which reactions are redox reactions?
   (a) $Al(s) + 3\,Ag^{+}(aq) \longrightarrow Al^{3+}(aq) + 3\,Ag(s)$
   (b) $4\,K(s) + O_2(g) \longrightarrow 2\,K_2O(s)$
   (c) $SO_3(g) + H_2O(l) \longrightarrow H_2SO_4(aq)$
   (d) $Mg(s) + Br_2(l) \longrightarrow MgBr_2(s)$

85. Complete and balance each combustion reaction.
   (a) $C_2H_6(g) + O_2(g) \longrightarrow$
   (b) $Ca(s) + O_2(g) \longrightarrow$
   (c) $C_3H_8O(l) + O_2(g) \longrightarrow$
   (d) $C_4H_{10}S(l) + O_2(g) \longrightarrow$

86. Complete and balance each combustion reaction.
   (a) $S(s) + O_2(g) \longrightarrow$
   (b) $C_7H_{16}(l) + O_2(g) \longrightarrow$
   (c) $C_4H_{10}O(l) + O_2(g) \longrightarrow$
   (d) $CS_2(l) + O_2(g) \longrightarrow$

87. Write a balanced chemical equation for the synthesis reaction of $Br_2(g)$ with each metal.
   (a) $Ag(s)$ (b) $K(s)$ (c) $Al(s)$ (d) $Ca(s)$

88. Write a balanced chemical equation for the synthesis reaction of $Cl_2(g)$ with each metal.
   (a) $Zn(s)$ (b) $Ga(s)$ (c) $Rb(s)$ (d) $Mg(s)$

CLASSIFYING CHEMICAL REACTIONS BY WHAT ATOMS DO

89. Classify each chemical reaction as a synthesis, decomposition, single-displacement, or double-displacement reaction.
   (a) $K_2S(aq) + Co(NO_3)_2(aq) \longrightarrow 2\,KNO_3(aq) + CoS(s)$
   (b) $3\,H_2(g) + N_2(g) \longrightarrow 2\,NH_3(g)$
   (c) $Zn(s) + CoCl_2(aq) \longrightarrow ZnCl_2(aq) + Co(s)$
   (d) $CH_3Br(g) \xrightarrow[\text{UV light}]{} CH_3(g) + Br(g)$

90. Classify each chemical reaction as a synthesis, decomposition, single-displacement, or double-displacement reaction.
   (a) $CaSO_4(g) \xrightarrow[\text{heat}]{} CaO(s) + SO_3(g)$
   (b) $2\,Na(s) + O_2(g) \longrightarrow Na_2O_2(s)$
   (c) $Pb(s) + 2\,AgNO_3(aq) \longrightarrow Pb(NO_3)_2(aq) + 2\,Ag(s)$
   (d) $HI(aq) + NaOH(aq) \longrightarrow H_2O(l) + NaI(aq)$

91. NO is a pollutant emitted by motor vehicles. It is formed by the reaction:
   (a) $N_2(g) + O_2(g) \longrightarrow 2\,NO(g)$
   Once in the atmosphere, NO (through a series of reactions) adds one oxygen atom to form NO_2. NO_2 then interacts with UV light according to the reaction:
   (b) $NO_2(g) \xrightarrow[\text{UV light}]{} NO(g) + O(g)$
   These freshly formed oxygen atoms then react with O_2 in the air to form ozone (O_3), a main component of smog:
   (c) $O(g) + O_2(g) \longrightarrow O_3(g)$
   Classify each of the preceding reactions (a, b, c) as a synthesis, decomposition, single-displacement, or double-displacement reaction.

92. A main source of sulfur oxide pollutants are smelters where sulfide ores are converted into metals. The first step in this process is the reaction of the sulfide ore with oxygen in reactions such as:
   (a) $2\,PbS(s) + 3\,O_2(g) \xrightarrow[\text{UV light}]{} 2\,PbO(s) + 2\,SO_2(g)$
   Sulfur dioxide can then react with oxygen in air to form sulfur trioxide:
   (b) $2\,SO_2(g) + O_2(g) \longrightarrow 2\,SO_3(g)$
   Sulfur trioxide can then react with water from rain to form sulfuric acid that falls as acid rain:
   (c) $SO_3(g) + H_2O(l) \longrightarrow H_2SO_4(aq)$
   Classify each of the preceding reactions (a, b, c) as a synthesis, decomposition, single-displacement, or double-displacement reaction.

Cumulative Problems

93. Predict the products of each reaction and write balanced complete ionic and net ionic equations for each. If no reaction occurs, write *NO REACTION.*
(a) $NaI(aq) + Hg_2(NO_3)_2(aq) \longrightarrow$
(b) $HClO_4(aq) + Ba(OH)_2(aq) \longrightarrow$
(c) $Li_2CO_3(aq) + NaCl(aq) \longrightarrow$
(d) $HCl(aq) + Li_2CO_3(aq) \longrightarrow$

94. Predict the products of each reaction and write balanced complete ionic and net ionic equations for each. If no reaction occurs, write *NO REACTION.*
(a) $LiCl(aq) + AgNO_3(aq) \longrightarrow$
(b) $H_2SO_4(aq) + Li_2SO_3(aq) \longrightarrow$
(c) $HC_2H_3O_2(aq) + Ca(OH)_2(aq) \longrightarrow$
(d) $HCl(aq) + KBr(aq) \longrightarrow$

95. Predict the products of each reaction and write balanced complete ionic and net ionic equations for each. If no reaction occurs, write *NO REACTION.*
(a) $BaS(aq) + NH_4Cl(aq) \longrightarrow$
(b) $NaC_2H_3O_2(aq) + KCl(aq) \longrightarrow$
(c) $KHSO_3(aq) + HNO_3(aq) \longrightarrow$
(d) $MnCl_3(aq) + K_3PO_4(aq) \longrightarrow$

96. Predict the products of each reaction and write balanced complete ionic and net ionic equations for each. If no reaction occurs, write *NO REACTION.*
(a) $H_2SO_4(aq) + HNO_3(aq) \longrightarrow$
(b) $NaOH(aq) + LiOH(aq) \longrightarrow$
(c) $Cr(NO_3)_3(aq) + LiOH(aq) \longrightarrow$
(d) $HCl(aq) + Hg_2(NO_3)_2(aq) \longrightarrow$

97. Predict the type of reaction (if any) that occurs between each pair of substances. Write balanced molecular equations for each. If no reaction occurs, write *NO REACTION.*
(a) aqueous potassium hydroxide and aqueous acetic acid
(b) aqueous hydrobromic acid and aqueous potassium carbonate
(c) gaseous hydrogen and gaseous oxygen
(d) aqueous ammonium chloride and aqueous lead(II) nitrate

98. Predict the type of reaction (if any) that occurs between each pair of substances. Write balanced molecular equations for each. If no reaction occurs, write *NO REACTION.*
(a) aqueous hydrochloric acid and aqueous copper(II) nitrate
(b) liquid pentanol ($C_5H_{12}O$) and gaseous oxygen
(c) aqueous ammonium chloride and aqueous calcium hydroxide
(d) aqueous strontium sulfide and aqueous copper(II) sulfate

99. Classify each reaction in as many ways as possible.
(a) $2\,Al(s) + 3\,Cu(NO_3)_2(aq) \longrightarrow 2\,Al(NO_3)_3(aq) + 3\,Cu(s)$
(b) $HBr(aq) + KHSO_3(aq) \longrightarrow H_2O(l) + SO_2(g) + NaBr(aq)$
(c) $2\,HI(aq) + Na_2S(aq) \longrightarrow H_2S(g) + 2\,NaI(aq)$
(d) $K_2CO_3(aq) + FeBr_2(aq) \longrightarrow FeCO_3(s) + 2\,KBr(aq)$

100. Classify each reaction in as many ways as possible.
(a) $NaCl(aq) + AgNO_3(aq) \longrightarrow AgCl(s) + NaNO_3(aq)$
(b) $2\,Rb(s) + Br_2(g) \longrightarrow 2\,RbBr(s)$
(c) $Zn(s) + NiBr_2(aq) \longrightarrow Ni(s) + ZnBr_2(aq)$
(d) $Ca(s) + 2H_2O(l) \longrightarrow Ca(OH)_2(aq) + H_2(g)$

101. Hard water often contains dissolved Ca^{2+} and Mg^{2+} ions. One way to soften water is to add phosphates. The phosphate ion forms insoluble precipitates with calcium and magnesium ions, removing them from solution. Suppose that a solution contains aqueous calcium chloride and aqueous magnesium nitrate. Write molecular, complete ionic, and net ionic equations showing how the addition of sodium phosphate precipitates the calcium and magnesium ions.

102. Lakes that have been acidified by acid rain (HNO_3 and H_2SO_4) can be neutralized by a process called *liming*, in which limestone ($CaCO_3$) is added to the acidified water. Write ionic and net ionic equations to show how limestone reacts with HNO_3 and H_2SO_4 to neutralize them. How would you be able to tell if the neutralization process was working?

103. What solution can you add to each cation mixture to precipitate one cation while keeping the other cation in solution? Write a net ionic equation for the precipitation reaction that occurs.
(a) $Fe^{2+}(aq)$ and $Pb^{2+}(aq)$
(b) $K^+(aq)$ and $Ca^{2+}(aq)$
(c) $Ag^+(aq)$ and $Ba^{2+}(aq)$
(d) $Cu^{2+}(aq)$ and $Hg_2^{2+}(aq)$

104. What solution can you add to each cation mixture to precipitate one cation while keeping the other cation in solution? Write a net ionic equation for the precipitation reaction that occurs.
(a) $Sr^{2+}(aq)$ and $Hg_2^{2+}(aq)$
(b) $NH_4^+(aq)$ and $Ca^{2+}(aq)$
(c) $Ba^{2+}(aq)$ and $Mg^{2+}(aq)$
(d) $Ag^+(aq)$ and $Zn^{2+}(aq)$

105. A solution contains an unknown amount of dissolved calcium. Addition of 0.112 mol of K_3PO_4 causes complete precipitation of all of the calcium. How many moles of calcium were dissolved in the solution? What mass of calcium was dissolved in the solution?

106. A solution contains an unknown amount of dissolved magnesium. Addition of 0.0877 mol of Na_2CO_3 causes complete precipitation of all of the magnesium. What mass of magnesium was dissolved in the solution?

107. A solution contains 0.133 g of dissolved lead. How many moles of sodium chloride must be added to the solution to completely precipitate all of the dissolved lead? What mass of sodium chloride must be added?

108. A solution contains 1.77 g of dissolved silver. How many moles of potassium chloride must be added to the solution to completely precipitate all of the silver? What mass of potassium chloride must be added?

Highlight Problems

109. Shown here are molecular views of two different possible mechanisms by which an automobile airbag might function. One of these mechanisms involves a chemical reaction and the other does not. By looking at the molecular views, can you tell which mechanism operates via a chemical reaction?

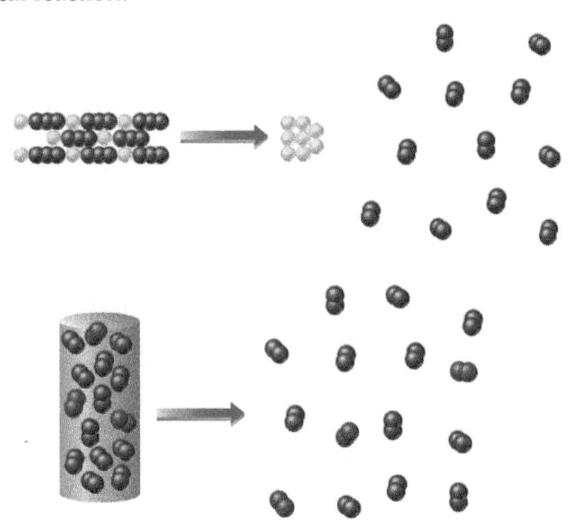

▲ When an airbag is detonated, the bag inflates. These figures show two possible ways in which the inflation may happen.

110. Precipitation reactions often produce brilliant colors. Look at the photographs of each precipitation reaction and write molecular, complete ionic, and net ionic equations for each one.

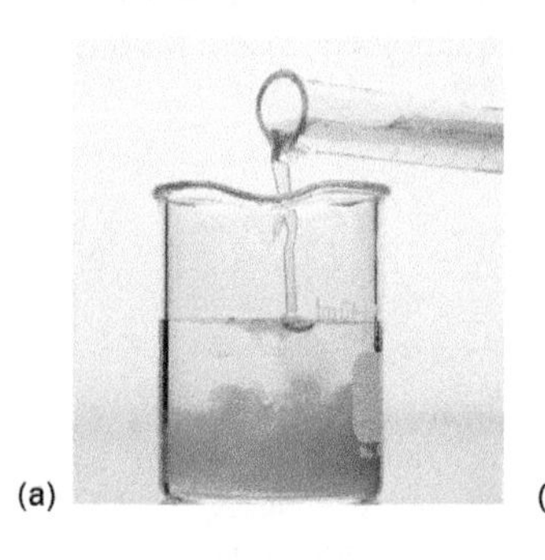

(a)

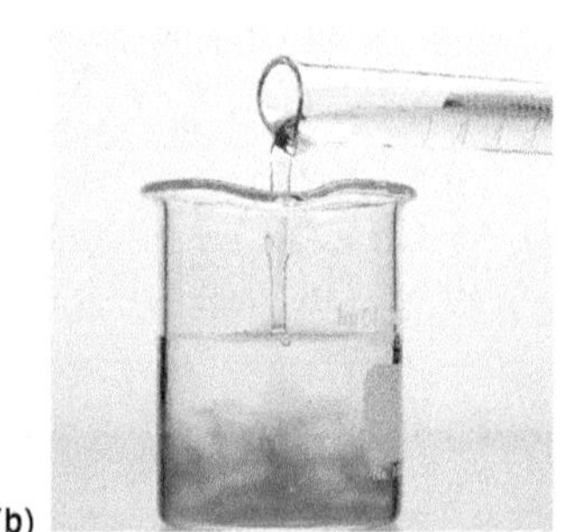

(b)

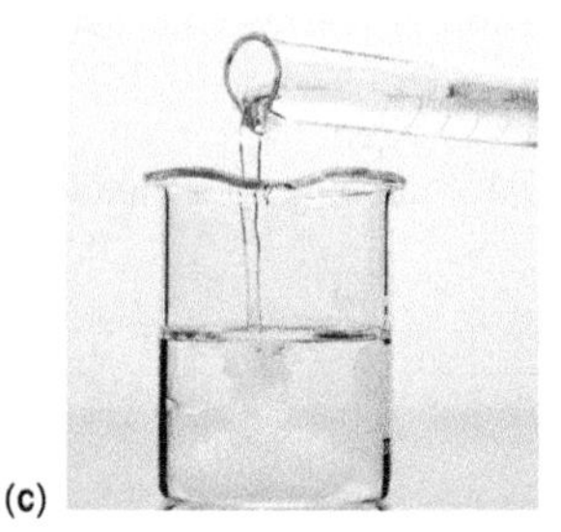

(c)

▲ **(a)** The precipitation reaction that occurs when aqueous iron(III) nitrate is added to aqueous sodium hydroxide. **(b)** The precipitation reaction that occurs when aqueous cobalt(II) chloride is added to aqueous potassium hydroxide. **(c)** The precipitation reaction that occurs when aqueous $AgNO_3$ is added to aqueous sodium iodide.

Questions for Group Work

Discuss these questions with the group and record your consensus answer.

111. Use group members to represent atoms or ions and act out a specific balanced chemical reaction.

112. Memorize the solubility rules. Without referring back to the rules, have each group member list two ionic compounds that are expected to be soluble and two that are expected to be insoluble. Include at least one exception. Check the work of the other members of your group.

113. Define and give an example of each of the following classes of reactions: precipitation, acid–base, gas evolution, redox (noncombustion), and combustion. Have each group member define one type and provide an example, and then present his or her reaction to the group.

Data Interpretation and Analysis

114. Water samples often contain dissolved ions such as Ca^{2+} and Fe^{2+}. The presence of these ions can often be detected by adding a precipitation agent—a substance that causes one of the dissolved ions to precipitate. For example, if sodium chloride is added to a water sample containing dissolved Ag^+, a white precipitate forms. If no precipitate forms, then the water sample does not contain dissolved Ag^+. Use the solubility rules from Section 7.5 and the data provided to determine the ions present in the water samples.

(a) Water Sample A may contain Ag^+, Ca^{2+}, and Cu^{2+}. Use the tabulated data to determine the ions present in Sample A.

Substance Added to Water Sample A	Observation
NaCl	No precipitate forms
Na_2SO_4	Precipitate forms
Na_2CO_3 (after filtering off precipitate from previous step)	Precipitate forms

(b) Water Sample B may contain Hg_2^{2+}, Ba^{2+}, and Fe^{2+}. Use the tabulated data to determine the ions present in Sample B.

Substance Added to Water Sample B	Observation
KCl	Precipitate forms
K_2SO_4 (after filtering off precipitate from previous step)	No precipitate forms
K_2CO_3	Precipitate forms

Answers to Skillbuilder Exercises

Skillbuilder 7.1

(a) Chemical reaction; heat and light are emitted.

(b) Not a chemical reaction; gaseous and liquid butane are both butane.

(c) Chemical reaction; heat and light are emitted.

(d) Not a chemical reaction; solid dry ice is made of carbon dioxide, which sublimes (evaporates) as carbon dioxide gas.

Skillbuilder 7.2

$2\,Cr_2O_3(s) + 3\,C(s) \longrightarrow 4\,Cr(s) + 3\,CO_2(g)$

Skillbuilder 7.3

$2\,C_4H_{10}(g) + 13\,O_2(g) \longrightarrow 8\,CO_2(g) + 10\,H_2O(g)$

Skillbuilder 7.4

$Pb(C_2H_3O_2)_2(aq) + 2\,KI(aq) \longrightarrow PbI_2(s) + 2\,KC_2H_3O_2(aq)$

Skillbuilder 7.5

$4\,HCl(g) + O_2(g) \longrightarrow 2\,H_2O(l) + 2\,Cl_2(g)$

Skillbuilder 7.6

(a) insoluble

(b) soluble

(c) insoluble

(d) soluble

Skillbuilder 7.7

$2\,KOH(aq) + NiBr_2(aq) \longrightarrow Ni(OH)_2(s) + 2\,KBr(aq)$

Skillbuilder 7.8

$NH_4Cl(aq) + Fe(NO_3)_3(aq) \longrightarrow$ NO REACTION

Skillbuilder 7.9

$K_2SO_4(aq) + Sr(NO_3)_2(aq) \longrightarrow SrSO_4(s) + 2\,KNO_3(aq)$

Skillbuilder 7.10

Complete ionic equation:

$2\,H^+(aq) + 2\,Br^-(aq) + Ca^{2+}(aq) + 2\,OH^-(aq) \longrightarrow 2\,H_2O(l) + Ca^{2+}(aq) + 2\,Br^-(aq)$

Net ionic equation:

$2\,H^+(aq) + 2\,OH^-(aq) \longrightarrow 2\,H_2O(l)$, or simply

$H^+(aq) + OH^-(aq) \longrightarrow H_2O(l)$

Skillbuilder 7.11

Molecular equation:

$H_2SO_4(aq) + 2\,KOH(aq) \longrightarrow 2\,H_2O(l) + K_2SO_4(aq)$

Net ionic equation:

$H^+(aq) + OH^-(aq) \longrightarrow H_2O(l)$

Skillbuilder 7.12

$2\,HBr(aq) + K_2SO_3(aq) \longrightarrow H_2O(l) + SO_2(g) + 2\,KBr(aq)$

Skillbuilder Plus, p. 227

$2\,H^+(aq) + SO_3^{2-}(aq) \longrightarrow H_2O(l) + SO_2(g)$

Skillbuilder 7.13

(a), **(b)**, and **(d)** are all redox reactions; **(c)** is a precipitation reaction.

Skillbuilder 7.14

$C_5H_{12}(l) + 8\,O_2(g) \longrightarrow 5\,CO_2(g) + 6\,H_2O(g)$

Skillbuilder Plus, p. 230

$2\,C_3H_7OH(l) + 9\,O_2(g) \longrightarrow 6\,CO_2(g) + 8\,H_2O(g)$

Skillbuilder 7.15

(a) single-displacement **(b)** double-displacement **(c)** synthesis **(d)** decomposition

Answers to Conceptual Checkpoints

7.1 (a) No reaction occurred. The molecules are the same before and after the change.

(b) A reaction occurred; the molecules have changed.

(c) A reaction occurred; the molecules have changed.

7.2 (b) There are 18 oxygen atoms on the left side of the equation, so the same number is needed on the right: $6 + 6(2) = 18$.

7.3 (a) The number of each type of atom must be the same on both sides of a balanced chemical equation. Since molecules change during a chemical reaction, their number is not the same on both sides (b), nor is the sum of all of the coefficients the same (c).

7.4 (a) Since chlorides are usually soluble and Ba^{2+} is not an exception, $BaCl_2$ is soluble and will dissolve in water. When it dissolves, it dissociates into its component ions, as shown in (a).

7.5 (b) Both of the possible products, MgS and $CaSO_4$, are insoluble. The possible products of the other reactions—Na_2S, $Ca(NO_3)_2$, Na_2SO_4, and $Mg(NO_3)_2$—are all soluble.

7.6 (c) The net ionic equation shows only the species that actually participate in the reaction.

7.7 (d) In a precipitation reaction, cations and anions "exchange partners" to produce at least one product. In an acid–base reaction, H^+ and OH^- combine to form water, and their partners pair off to a salt.

Octane
2,2,4 ISO

8 Quantities in Chemical Reactions

Man masters nature not by force but by understanding. That is why science has succeeded where magic failed: because it has looked for no spell to cast.

—Jacob Bronowski (1908–1974)

CHAPTER OUTLINE

8.1 Climate Change: Too Much Carbon Dioxide

Outgoing heat is trapped by atmospheric greenhouse gases.

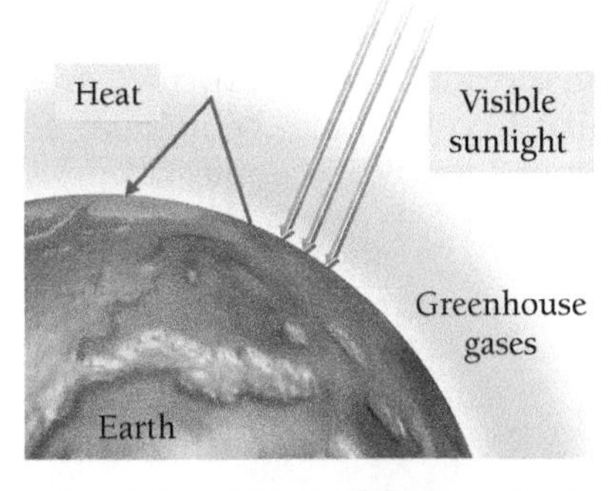

▲ **FIGURE 8.1 The greenhouse effect** Greenhouse gases act like glass in a greenhouse, allowing visible-light energy to enter the atmosphere but preventing heat energy from escaping.

Average global temperatures depend on the balance between incoming sunlight, which warms Earth, and outgoing heat lost to space, which cools it. Certain gases in Earth's atmosphere, called **greenhouse gases**, affect that balance by acting like glass in a greenhouse. They allow sunlight into the atmosphere to warm Earth but prevent heat from escaping (◀ FIGURE 8.1). Without greenhouse gases, more heat would escape, and Earth's average temperature would be about 60 °F colder. Caribbean tourists would freeze at an icy 21 °F (−6 °C), instead of baking at a tropical 81 °F (27 °C). On the other hand, if the concentration of greenhouse gases in the atmosphere were to increase, Earth's average temperature would rise.

In recent decades, scientists have become concerned because the atmospheric concentration of carbon dioxide (CO_2)—Earth's most significant greenhouse gas in terms of its contribution to climate—is rising. This rise in CO_2 concentration enhances the atmosphere's ability to hold heat and therefore leads to **climate change**, seen most clearly as an increase in Earth's average temperature. Since 1880, atmospheric CO_2 levels have risen by 38%, and Earth's average temperature has increased by 0.8 °C (about 1.4 °F) (▶ FIGURE 8.2, on the next page).

The primary cause of rising atmospheric CO_2 concentration is the burning of fossil fuels. Fossil fuels—natural gas, petroleum, and coal—provide approximately 90% of our society's energy. Combustion of fossil fuels, however, produces CO_2. As an example, consider the combustion of octane (C_8H_{18}), a component of gasoline:

$$2\,C_8H_{18}(l) + 25\,O_2(g) \longrightarrow 16\,CO_2(g) + 18\,H_2O(g)$$

The balanced chemical equation shows that 16 mol of CO_2 are produced for every 2 mol of octane burned. Because we know the world's annual fossil fuel consumption, we can estimate the world's annual CO_2 production. A simple calculation

◀ The combustion of fossil fuels such as octane (shown here) produces water and carbon dioxide as products. Carbon dioxide is a greenhouse gas that most climate scientists believe is responsible for climate change.

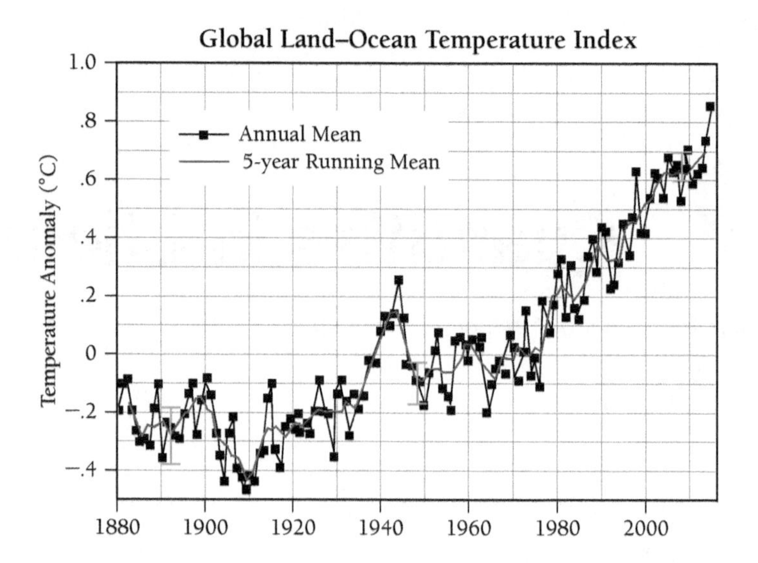

▶ **FIGURE 8.2 Climate change** Yearly temperature differences from the 120-year average temperature. Earth's average temperature has increased by about 0.8 °C since 1880.

shows that the world's annual CO_2 production—from fossil fuel combustion—matches the measured annual atmospheric CO_2 increase. This implies that fossil fuel combustion is indeed responsible for increased atmospheric CO_2 levels.

The numerical relationship between chemical quantities in a balanced chemical equation is called reaction **stoichiometry**. Stoichiometry allows us to predict the amounts of products that form in a chemical reaction based on the amounts of reactants. Stoichiometry also allows us to predict how much of the reactants is necessary to form a given amount of product or how much of one reactant is required to completely react with another reactant. These calculations are central to chemistry, allowing chemists to plan and carry out chemical reactions to obtain products in desired quantities.

8.2 Making Pancakes: Relationships between Ingredients

▶ Recognize the numerical relationship between chemical quantities in a balanced chemical equation.

Key Concept Video Reaction Stoichiometry

The concepts of stoichiometry are similar to the concepts we use in following a cooking recipe. Calculating the amount of carbon dioxide produced by the combustion of a given amount of a fossil fuel is similar to calculating the number of pancakes that we can make from a given number of eggs. For example, suppose we use the following pancake recipe:

For the sake of simplicity, this recipe omits liquid ingredients.

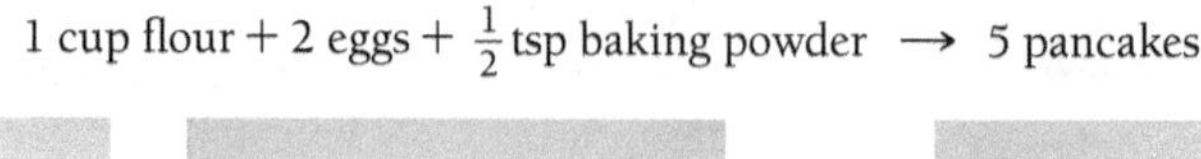

$$1 \text{ cup flour} + 2 \text{ eggs} + \tfrac{1}{2} \text{ tsp baking powder} \longrightarrow 5 \text{ pancakes}$$

▲ A recipe gives numerical relationships between the ingredients and the number of pancakes.

The recipe shows the numerical relationships between the pancake ingredients. It says that if we have 2 eggs—and enough of everything else—we can make 5 pancakes. We can write this relationship as a ratio:

2 eggs : 5 pancakes

What if we have 8 eggs? Assuming that we have enough of everything else, how many pancakes can we make? Using the preceding ratio as a conversion factor, we can determine that 8 eggs are sufficient to make 20 pancakes.

$$8 \text{ eggs} \times \frac{5 \text{ pancakes}}{2 \text{ eggs}} = 20 \text{ pancakes}$$

The pancake recipe contains numerical conversion factors between the pancake ingredients and the number of pancakes. Other conversion factors from this recipe include:

1 cup flour : 5 pancakes

$\frac{1}{2}$ tsp baking powder : 5 pancakes

The recipe also gives us relationships among the ingredients themselves. For example, how much baking powder is required to go with 3 cups of flour? From the recipe:

1 cup flour : $\frac{1}{2}$ tsp baking powder

With this ratio, we can form the conversion factor to calculate the appropriate amount of baking powder.

$$3 \text{ cups flour} \times \frac{\frac{1}{2} \text{ tsp baking powder}}{1 \text{ cup flour}} = \frac{3}{2} \text{ tsp baking powder}$$

8.3 Making Molecules: Mole-to-Mole Conversions

▶ Carry out mole-to-mole conversions between reactants and products in a balanced chemical equation.

A balanced chemical equation is like a "recipe" for how reactants combine to form products. For example, the following equation shows how hydrogen and nitrogen combine to form ammonia (NH_3):

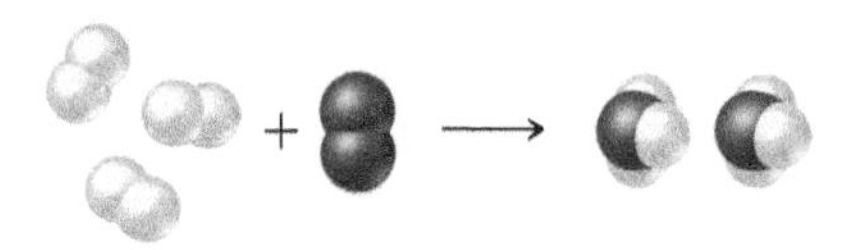

$$3H_2(g) + N_2(g) \longrightarrow 2NH_3(g)$$

The balanced equation shows that 3 H_2 molecules react with 1 N_2 molecule to form 2 NH_3 molecules. We can express these relationships as the following ratios:

3 H_2 molecules : 1 N_2 molecule : 2 NH_3 molecules

Since we do not ordinarily deal with individual molecules, we can express the same ratios in moles.

3 mol H_2 : 1 mol N_2 : 2 mol NH_3

If we have 3 mol of N_2, and more than enough H_2, how much NH_3 can we make? We first sort the information in the problem.

GIVEN: 3 mol N_2

FIND: mol NH_3

SOLUTION MAP

We then strategize by drawing a solution map that begins with mol N_2 and ends with mol NH_3. The conversion factor comes from the balanced chemical equation.

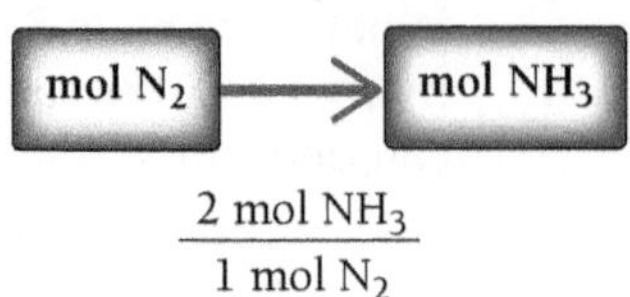

$$\frac{2 \text{ mol } NH_3}{1 \text{ mol } N_2}$$

RELATIONSHIPS USED

1 mol N_2:2 mol NH_3 (from balanced equation)

SOLUTION

We can then do the conversion.

$$3 \cancel{\text{mol } N_2} \times \frac{2 \text{ mol } NH_3}{1 \cancel{\text{mol } N_2}} = 6 \text{ mol } NH_3$$

We have enough N_2 to make 6 mol of NH_3.

EXAMPLE 8.1 Mole-to-Mole Conversions

Sodium chloride, NaCl, forms in this reaction between sodium and chlorine.

$$2 \text{ Na}(s) + Cl_2(g) \longrightarrow 2 \text{ NaCl}(s)$$

How many moles of NaCl result from the complete reaction of 3.4 mol of Cl_2? Assume that there is more than enough Na.

SORT You are given the number of moles of a reactant (Cl_2) and asked to find the number of moles of product (NaCl) that will form if the reactant completely reacts.	**GIVEN:** 3.4 mol Cl_2 **FIND:** mol NaCl
STRATEGIZE Draw the solution map beginning with moles of chlorine and using the stoichiometric conversion factor to calculate moles of sodium chloride. The conversion factor comes from the balanced chemical equation.	**SOLUTION MAP** 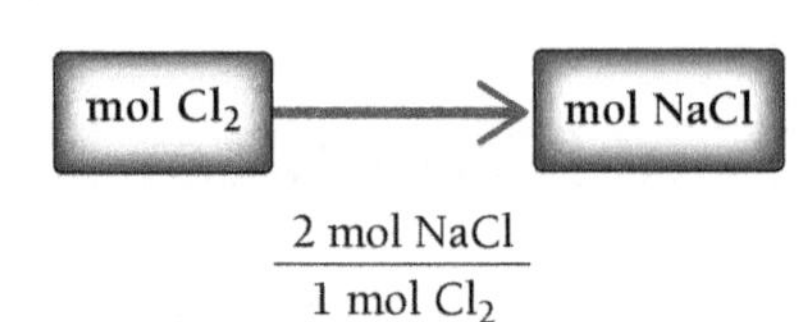 $\frac{2 \text{ mol NaCl}}{1 \text{ mol } Cl_2}$ **RELATIONSHIPS USED** 1 mol Cl_2:2 mol NaCl (from balanced chemical equation)
SOLVE Follow the solution map to solve the problem.	**SOLUTION** $3.4 \cancel{\text{mol } Cl_2} \frac{2 \text{ mol NaCl}}{1 \cancel{\text{mol } Cl_2}} = 6.8 \text{ mol NaCl}$ There is enough Cl_2 to produce 6.8 mol of NaCl.
CHECK Check your answer. Are the units correct? Does the answer make physical sense?	The answer has the correct units, moles. The answer is reasonable because each mole of Cl_2 makes two moles of NaCl.

▶ **SKILLBUILDER 8.1 | Mole-to-Mole Conversions**

Water forms when hydrogen gas reacts explosively with oxygen gas according to the balanced equation:

$$O_2(g) + 2H_2(g) \longrightarrow 2H_2O(g)$$

How many mol of H_2O result from the complete reaction of 24.6 mol of O_2? Assume that there is more than enough H_2.

▶ **FOR MORE PRACTICE** Example 8.8; Problems 15, 16, 17, 18.

PEARSON eText 2.0

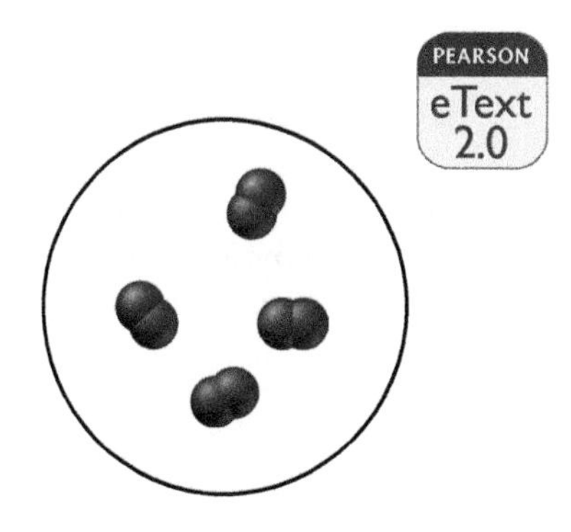

CONCEPTUAL CHECKPOINT 8.1

Methane (CH_4) undergoes combustion according to this reaction.

$$CH_4(g) + 2\,O_2(g) \longrightarrow CO_2(g) + 2\,H_2O(g)$$

If the figure shown in the left margin represents the amount of oxygen available to react, which of the following figures best represents the amount of CH_4 required to completely react with all of the oxygen?

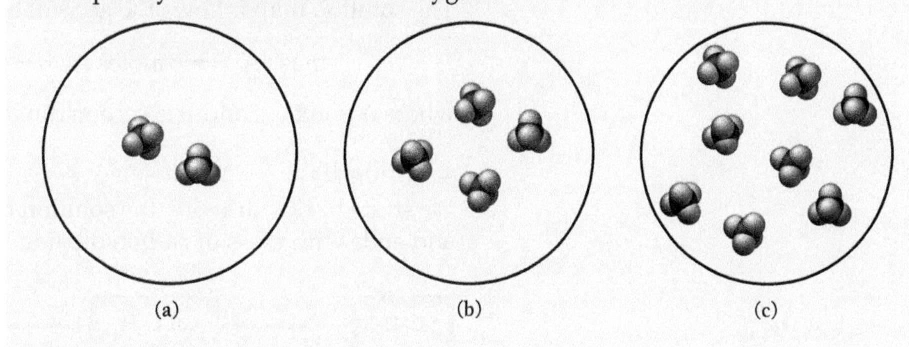

8.4 Making Molecules: Mass-to-Mass Conversions

▶ Carry out mass-to-mass conversions between reactants and products in a balanced chemical equation and molar masses.

In Chapter 6, we learned how a chemical *formula* contains conversion factors for converting between moles of a compound and moles of its constituent elements. In this chapter, we have seen how a chemical *equation* contains conversion factors between moles of reactants and moles of products. However, we are often interested in relationships between *mass* of reactants and *mass* of products. For example, we might want to know the mass of carbon dioxide emitted by an automobile per kilogram of gasoline used. Or we might want to know the mass of each reactant required to obtain a certain mass of a product in a synthesis reaction.

These calculations are similar to calculations covered in Section 6.5, where we converted between mass of a compound and mass of a constituent element. The general outline for these types of calculations is:

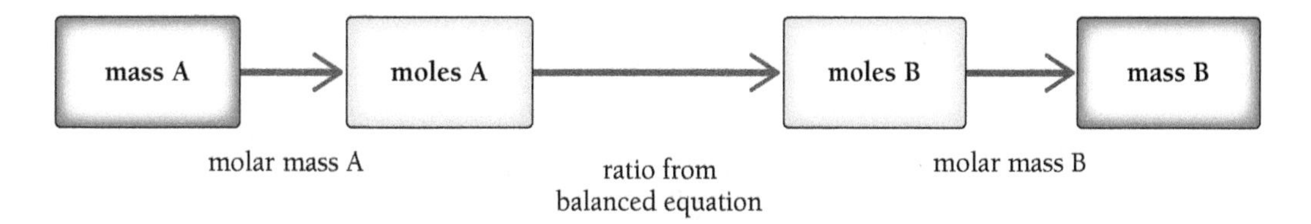

where A and B are two different substances involved in the reaction. We use the molar mass of A to convert from mass of A to moles of A. We use the ratio from the balanced equation to convert from moles of A to moles of B, and we use the molar mass of B to convert moles of B to mass of B.

For example, suppose we want to calculate the mass of CO_2 emitted upon the combustion of 5.0×10^2 g of pure octane. The balanced chemical equation for octane combustion is:

$$2\,C_8H_{18}(l) + 25\,O_2(g) \longrightarrow 16\,CO_2(g) + 18\,H_2O(g)$$

We begin by sorting the information in the problem.

GIVEN: 5.0×10^2 g C_8H_{18}

FIND: g CO_2

Notice that we are given g C_8H_{18} and asked to find g CO_2. The balanced chemical equation, however, gives us a relationship between moles of C_8H_{18} and moles of CO_2. Consequently, before using that relationship, we must convert from grams to moles.

The solution map follows the general outline:

$$\text{mass A} \longrightarrow \text{moles A} \longrightarrow \text{moles B} \longrightarrow \text{mass B}$$

where A is octane and B is carbon dioxide.

SOLUTION MAP

We strategize by drawing the solution map, which begins with mass of octane and ends with mass of carbon dioxide.

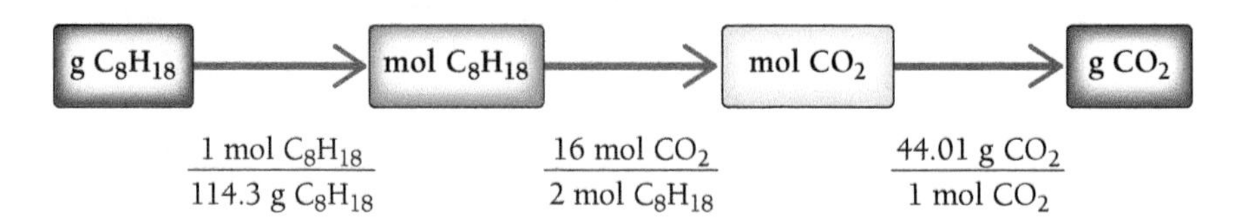

$$\frac{1 \text{ mol } C_8H_{18}}{114.3 \text{ g } C_8H_{18}} \qquad \frac{16 \text{ mol } CO_2}{2 \text{ mol } C_8H_{18}} \qquad \frac{44.01 \text{ g } CO_2}{1 \text{ mol } CO_2}$$

RELATIONSHIPS USED

2 mol C_8H_{18} : 16 mol CO_2 (from chemical equation)

molar mass C_8H_{18} = 114.3 g/mol

molar mass CO_2 = 44.01 g/mol

SOLUTION

We then follow the solution map to solve the problem, beginning with g C_8H_{18} and canceling units to arrive at g CO_2.

$$5.0 \times 10^2 \text{ g } \cancel{C_8H_{18}} \times \frac{1 \cancel{\text{mol } C_8H_{18}}}{114.3 \text{ g } \cancel{C_8H_{18}}} \times \frac{16 \cancel{\text{mol } CO_2}}{2 \cancel{\text{mol } C_8H_{18}}} \times \frac{44.01 \text{ g } CO_2}{1 \cancel{\text{mol } CO_2}} = 1.5 \times 10^3 \text{ g } CO_2$$

Upon combustion, 5.0×10^2 g of octane produces 1.5×10^3 g of carbon dioxide.

CONCEPTUAL CHECKPOINT 8.2

Consider the reaction $A + 2B \rightarrow 3C$. If the molar mass of C is twice the molar mass of A, what mass of C is produced by the complete reaction of 10.0 g A?

(a) 10.0 g

(b) 30.0 g

(c) 60.0 g

Interactive Worked Example Video 8.2

EXAMPLE 8.2 Mass-to-Mass Conversions

In photosynthesis, plants convert carbon dioxide and water into glucose ($C_6H_{12}O_6$) according to the reaction:

$$6\ CO_2(g) + 6\ H_2O(l) \xrightarrow{\text{sunlight}} 6\ O_2(g) + C_6H_{12}O_6(aq)$$

How many grams of glucose can be synthesized from 58.5 g of CO_2? Assume that there is more than enough water present to react with all of the CO_2.

SORT

You are given the mass of carbon dioxide and asked to find the mass of glucose that can form if the carbon dioxide completely reacts.

GIVEN: 58.5 g CO_2

FIND: g $C_6H_{12}O_6$

STRATEGIZE

The solution map uses the general outline:

$$\text{mass A} \longrightarrow \text{moles A} \longrightarrow \text{moles B} \longrightarrow \text{mass B}$$

where A is carbon dioxide and B is glucose.

SOLUTION MAP

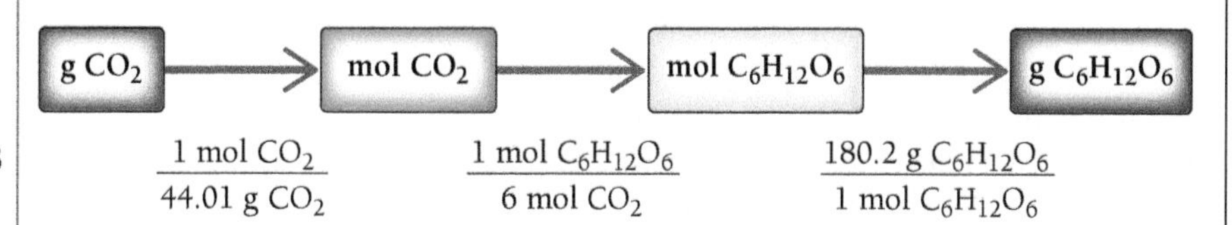

The main conversion factor is the stoichiometric relationship between moles of carbon dioxide and moles of glucose. This conversion factor comes from the balanced equation. The other conversion factors are the molar masses of carbon dioxide and glucose.

RELATIONSHIPS USED

6 mol CO_2 : 1 mol $C_6H_{12}O_6$ (from balanced chemical equation)

molar mass CO_2 = 44.01 g/mol

molar mass $C_6H_{12}O_6$ = 180.2 g/mol

SOLVE

Follow the solution map to solve the problem. Begin with grams of carbon dioxide and multiply by the appropriate factors to arrive at grams of glucose.

SOLUTION

$$58.5\ \cancel{\text{g}\ CO_2} \times \frac{1\ \cancel{\text{mol}\ CO_2}}{44.01\ \cancel{\text{g}\ CO_2}} \times \frac{1\ \cancel{\text{mol}\ C_6H_{12}O_6}}{6\ \cancel{\text{mol}\ CO_2}} \times \frac{180.2\ \text{g}\ C_6H_{12}O_6}{1\ \cancel{\text{mol}\ C_6H_{12}O_6}}$$

$$= 39.9\ \text{g}\ C_6H_{12}O_6$$

CHECK

Are the units correct? Does the answer make physical sense?

The units, g $C_6H_{12}O_6$, are correct. The magnitude of the answer seems reasonable because it is of the same order of magnitude as the given mass of carbon dioxide. An answer that is orders of magnitude different would immediately be suspect.

▶ **SKILLBUILDER 8.2 | Mass-to-Mass Conversions**

Magnesium hydroxide, the active ingredient in milk of magnesia, neutralizes stomach acid, primarily HCl, according to the reaction:

$$Mg(OH)_2(aq) + 2\ HCl(aq) \longrightarrow 2\ H_2O(l) + MgCl_2(aq)$$

How much HCl in grams can be neutralized by 5.50 g of $Mg(OH)_2$?

▶ **FOR MORE PRACTICE** Example 8.9; Problems 31, 32, 33, 34.

EXAMPLE 8.3 Mass-to-Mass Conversions

One of the components of acid rain (rain that becomes acidified due to air pollution) is nitric acid, which forms when NO_2, a pollutant, reacts with oxygen and rainwater according to the following simplified reaction:

$$4\,NO_2(g) + O_2(g) + 2\,H_2O(l) \longrightarrow 4\,HNO_3(aq)$$

Assuming that there is more than enough O_2 and H_2O, how much HNO_3 in kilograms forms from 1.5×10^3 kg of NO_2 pollutant?

SORT
You are given the mass of nitrogen dioxide (a reactant) and asked to find the mass of nitric acid that can form if the nitrogen dioxide completely reacts.

GIVEN: 1.5×10^3 kg NO_2

FIND: kg HNO_3

STRATEGIZE
The solution map follows the general format of:
mass ⟶ moles ⟶
moles ⟶ mass
However, because the original quantity of NO_2 is given in kilograms, you must first convert to grams. The final quantity is requested in kilograms, so you must convert back to kilograms at the end. The main conversion factor is the stoichiometric relationship between moles of nitrogen dioxide and moles of nitric acid. This conversion factor comes from the balanced equation. The other conversion factors are the molar masses of nitrogen dioxide and nitric acid and the relationship between kilograms and grams.

SOLUTION MAP

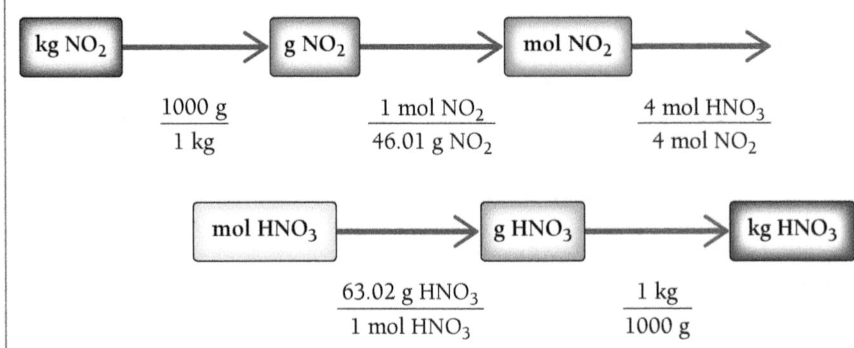

RELATIONSHIPS USED

4 mol NO_2 : 4 mol HNO_3 (from balanced chemical equation)

molar mass NO_2 = 46.01 g/mol

molar mass HNO_3 = 63.02 g/mol

1 kg = 1000 g

SOLVE
Follow the solution map to solve the problem. Begin with kilograms of nitrogen dioxide and multiply by the appropriate conversion factors to arrive at kilograms of nitric acid.

SOLUTION

$$1.5 \times 10^3\ \text{kg } NO_2 \times \frac{1000\ \text{g}}{1\ \text{kg}} \times \frac{1\ \text{mol } NO_2}{46.01\ \text{g } NO_2} \times \frac{4\ \text{mol } HNO_3}{4\ \text{mol } NO_2} \times \frac{63.02\ \text{g } HNO_3}{1\ \text{mol } HNO_3} \times \frac{1\ \text{kg}}{1000\ \text{g}} = 2.1 \times 10^3\ \text{kg } HNO_3$$

CHECK
Are the units correct? Does the answer make physical sense?

The units, kg HNO_3, are correct. The magnitude of the answer seems reasonable because it is of the same order of magnitude as the given mass of nitrogen dioxide. An answer that is orders of magnitude different would immediately be suspect.

▶ **SKILLBUILDER 8.3 | Mass-to-Mass Conversions**

Another component of acid rain is sulfuric acid, which forms when SO_2, also a pollutant, reacts with oxygen and rainwater according to the following reaction:

$$2\,SO_2(g) + O_2(g) + 2\,H_2O(l) \longrightarrow 2\,H_2SO_4(aq)$$

Assuming that there is more than enough O_2 and H_2O, how much H_2SO_4 in kilograms forms from 2.6×10^3 kg of SO_2?

▶ **FOR MORE PRACTICE** Problems 35, 36, 37, 38.

8.5 More Pancakes: Limiting Reactant, Theoretical Yield, and Percent Yield

▶ Calculate limiting reactant, theoretical yield, and percent yield in a balanced chemical equation.

PEARSON eText 2.0 **Key Concept Video** Limiting Reactant, Theoretical Yield, and Percent Yield

Let's return to our pancake analogy to understand two more concepts important in reaction stoichiometry: limiting reactant and percent yield. Recall our pancake recipe:

$$1 \text{ cup flour} + 2 \text{ eggs} + \tfrac{1}{2} \text{ tsp baking powder} \longrightarrow 5 \text{ pancakes}$$

Suppose we have 3 cups flour, 10 eggs, and 4 tsp baking powder. How many pancakes can we make? We have enough flour to make:

$$3 \text{ cups flour} \times \frac{5 \text{ pancakes}}{1 \text{ cup flour}} = 15 \text{ pancakes}$$

We have enough eggs to make:

$$10 \text{ eggs} \times \frac{5 \text{ pancakes}}{2 \text{ eggs}} = 25 \text{ pancakes}$$

We have enough baking powder to make:

$$4 \text{ tsp baking powder} \times \frac{5 \text{ pancakes}}{\frac{1}{2} \text{ tsp baking powder}} = 40 \text{ pancakes}$$

We have enough flour for 15 pancakes, enough eggs for 25 pancakes, and enough baking powder for 40 pancakes. Consequently, unless we get more ingredients, *we can make only 15 pancakes*. The amount of flour we have *limits* the number of pancakes we can make. If this were a chemical reaction, the flour would be the *limiting reactant*, the reactant that limits the amount of product in a chemical reaction. Notice that the **limiting reactant** is simply the reactant that makes *the least amount of product*. If this were a chemical reaction, 15 pancakes would be the **theoretical yield**, the amount of product that can be made in a chemical reaction based on the amount of limiting reactant.

The term *limiting reagent* is sometimes used in place of limiting reactant.

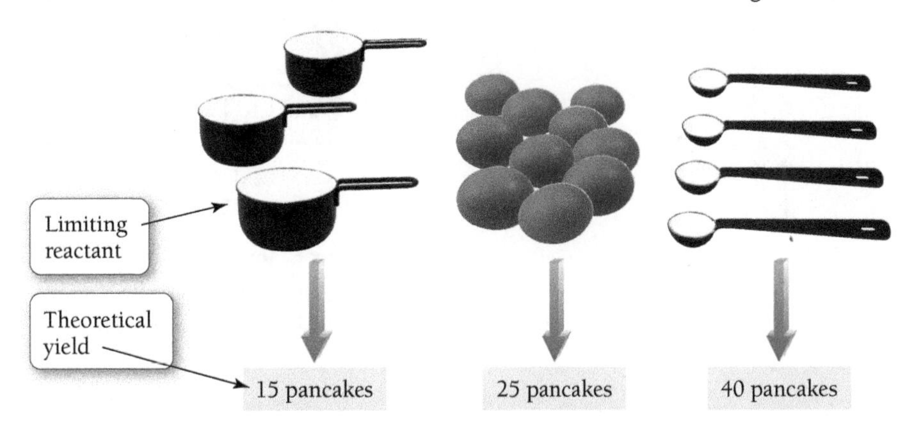

▶ If this were a chemical reaction, the flour would be the limiting reactant and 15 pancakes would be the theoretical yield.

Let us carry this analogy one step further. Suppose we go on to make our pancakes. We accidentally burn three of them and one falls on the floor. So even though we had enough flour for 15 pancakes, we finished with only 11 pancakes. If this were a chemical reaction, the 11 pancakes would be our **actual yield**, the amount of product actually produced by a chemical reaction. Finally, our **percent yield**, the percentage of the theoretical yield that was actually attained, would be:

$$\text{percent yield} = \frac{11 \text{ pancakes}}{15 \text{ pancakes}} \times 100\% = 73\%$$

Since four of the pancakes were ruined, we ended up with only 73% of our theoretical yield. In a chemical reaction, the actual yield is almost always less than 100% because at least some of the product does not form or is lost in the process of recovering it (in analogy to some of the pancakes being burned).

The actual yield of a chemical reaction, which must be determined experimentally, often depends in various ways on the reaction conditions. We will explore some of the factors involved in Chapter 15.

To summarize:

- **Limiting reactant (or limiting reagent)**—the reactant that is completely consumed in a chemical reaction.
- **Theoretical yield**—the amount of product that can be made in a chemical reaction based on the amount of limiting reactant.
- **Actual yield**—the amount of product actually produced by a chemical reaction.
- **Percent yield** $= \dfrac{\textbf{Actual yield}}{\textbf{Theoretical yield}} \times \textbf{100\%}$

Consider this reaction:

$$Ti(s) + 2Cl_2(g) \longrightarrow TiCl_4(s)$$

If we begin with 1.8 mol of titanium and 3.2 mol of chlorine, what is the limiting reactant and theoretical yield of $TiCl_4$ in mol? We begin by sorting the information in the problem according to our standard problem-solving procedure.

GIVEN: 1.8 mol Ti
3.2 mol Cl_2

FIND: limiting reactant
theoretical yield

SOLUTION MAP

As in our pancake analogy, we determine the limiting reactant by calculating how much product can be made from each reactant. The reactant that makes the *least amount of product* is the limiting reactant.

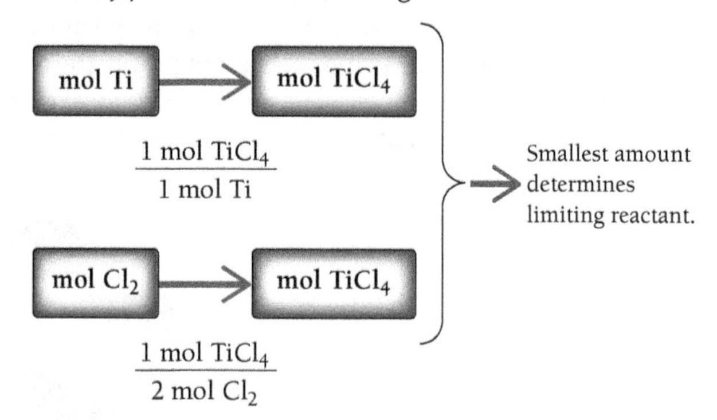

RELATIONSHIPS USED

The conversion factors come from the balanced chemical equation and give the relationships between moles of each of the reactants and moles of product.

$$1 \text{ mol Ti} : 1 \text{ mol } TiCl_4$$

$$2 \text{ mol } Cl_2 : 1 \text{ mol } TiCl_4$$

SOLUTION

$$1.8 \text{ mol Ti} \times \frac{1 \text{ mol } TiCl_4}{1 \text{ mol Ti}} = 1.8 \text{ mol } TiCl_4$$

$$3.2 \text{ mol } Cl_2 \times \frac{1 \text{ mol } TiCl_4}{2 \text{ mol } Cl_2} = 1.6 \text{ mol } TiCl_4$$

Limiting reactant (3.2 mol Cl_2); Least amount of product (1.6 mol $TiCl_4$)

In many industrial applications, the more costly reactant or the reactant that is most difficult to remove from the product mixture is chosen to be the limiting reactant.

Since the 3.2 mol of Cl_2 make the least amount of $TiCl_4$, Cl_2 is the limiting reactant. Notice that we began with more moles of Cl_2 than Ti, but because the reaction requires 2 Cl_2 for each Ti, Cl_2 is still the limiting reactant. The theoretical yield is 1.6 mol of $TiCl_4$.

EXAMPLE 8.4 Limiting Reactant and Theoretical Yield from Initial Moles of Reactants

Consider this reaction:

$$2\ \text{Al}(s) + 3\ \text{Cl}_2(g) \longrightarrow 2\ \text{AlCl}_3(s)$$

If you begin with 0.552 mol of aluminum and 0.887 mol of chlorine, what is the limiting reactant and theoretical yield of $AlCl_3$ in moles?

SORT
You are given the number of moles of aluminum and chlorine and asked to find the limiting reactant and theoretical yield of aluminum chloride.

GIVEN: 0.552 mol Al
0.887 mol Cl_2

FIND: limiting reactant
theoretical yield of $AlCl_3$

STRATEGIZE
Draw a solution map that shows how to get from moles of each reactant to moles of $AlCl_3$. The reactant that makes the *least amount of $AlCl_3$* is the limiting reactant. The conversion factors are the stoichiometric relationships (from the balanced equation).

SOLUTION MAP

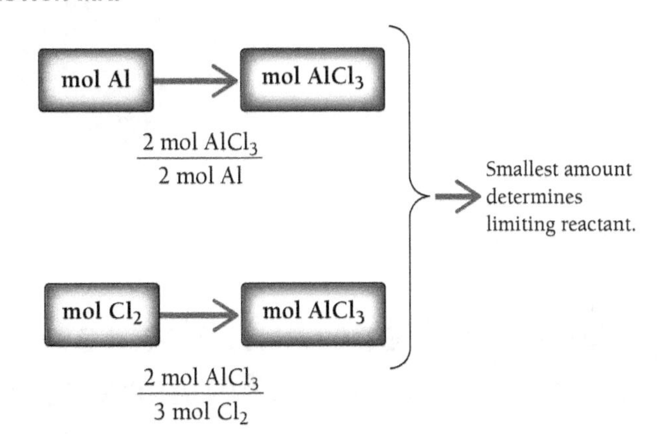

RELATIONSHIPS USED

2 mol Al : 2 mol $AlCl_3$ (from balanced equation)

3 mol Cl_2 : 2 mol $AlCl_3$ (from balanced equation)

SOLVE
Follow the solution map to solve the problem.

SOLUTION

$$0.552\ \text{mol Al} \times \frac{2\ \text{mol AlCl}_3}{2\ \text{mol Al}} = 0.552\ \text{mol AlCl}_3$$

Limiting reactant — Least amount of product

$$0.887\ \text{mol Cl}_2 \times \frac{2\ \text{mol AlCl}_3}{3\ \text{mol Cl}_2} = 0.591\ \text{mol AlCl}_3$$

Because the 0.552 mol of Al makes the least amount of $AlCl_3$, Al is the limiting reactant. The theoretical yield is 0.552 mol of $AlCl_3$.

CHECK
Are the units correct? Does the answer make physical sense?

The units, mol $AlCl_3$, are correct. The magnitude of the answer seems reasonable because it is of the same order of magnitude as the given number of moles of Al and Cl_2. An answer that is orders of magnitude different would immediately be suspect.

▶ SKILLBUILDER 8.4 | Limiting Reactant and Theoretical Yield from Initial Moles of Reactants

Consider the reaction:

$$2\ \text{Na}(s) + \text{F}_2(g) \longrightarrow 2\ \text{NaF}(s)$$

If you begin with 4.8 mol of sodium and 2.6 mol of fluorine, what is the limiting reactant and theoretical yield of NaF in mol?

▶ FOR MORE PRACTICE Problems 43, 44, 45, 46, 47, 48, 49, 50.

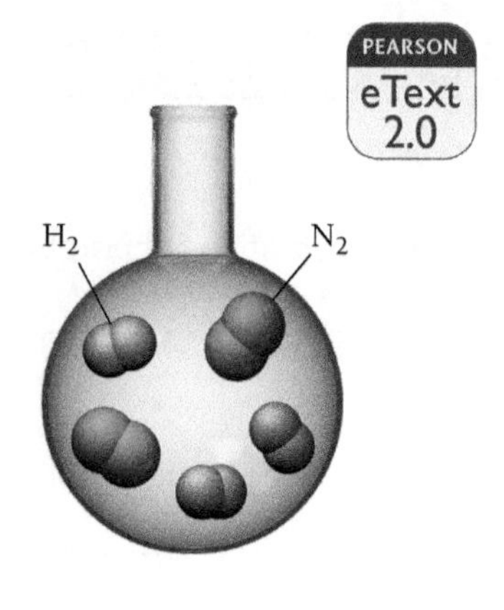

PEARSON eText 2.0

CONCEPTUAL CHECKPOINT 8.3

Consider the reaction:

$$N_2(g) + 3\,H_2(g) \longrightarrow 2NH_3(g)$$

If the flask in the left margin represents the mixture before the reaction, which flask represents the products after the limiting reactant has completely reacted?

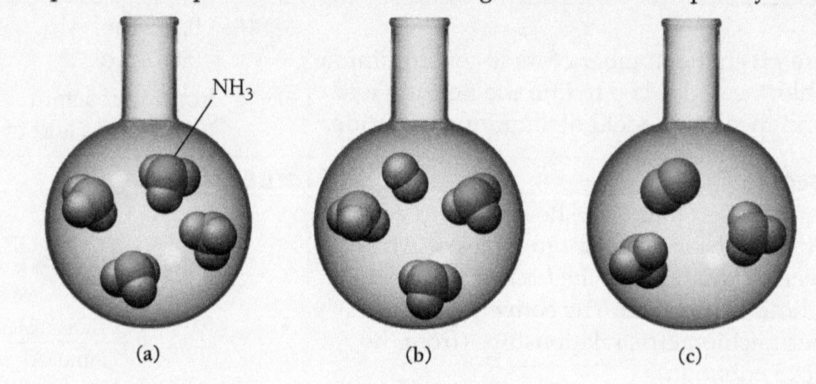

8.6 Limiting Reactant, Theoretical Yield, and Percent Yield from Initial Masses of Reactants

▶ Calculate limiting reactant, theoretical yield, and percent yield in a balanced chemical equation.

When working in the laboratory, we normally measure the initial amounts of reactants in grams. To find limiting reactants and theoretical yields from initial masses, we must add two steps to our calculations. Consider, for example, the synthesis reaction:

$$2\,Na(s) + Cl_2(g) \longrightarrow 2\,NaCl(s)$$

If we have 53.2 g of Na and 65.8 g of Cl_2, what is the limiting reactant and theoretical yield? We begin by sorting the information in the problem.

GIVEN: 53.2 g Na
65.8 g Cl_2

FIND: limiting reactant
theoretical yield

SOLUTION MAP

Again, we find the limiting reactant by calculating how much product can be made from each reactant. Since we are given the initial amounts in grams, we must first convert to moles. After we convert to moles of product, we convert back to grams of product. The reactant that makes the *least amount of product* is the limiting reactant.

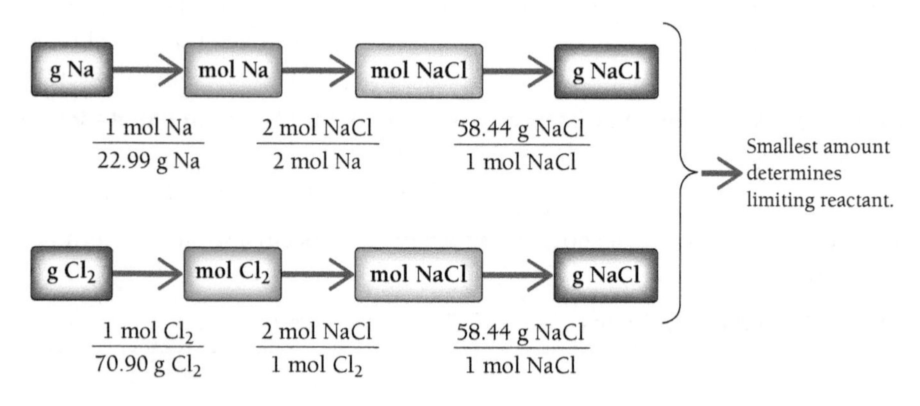

RELATIONSHIPS USED

From the balanced chemical equation, we know:

$$2\,mol\,Na : 2\,mol\,NaCl$$

$$1\,mol\,Cl_2 : 2\,mol\,NaCl$$

We also use these molar masses:

$$\text{molar mass Na} = \frac{22.99\text{ g Na}}{1\text{ mol Na}}$$

$$\text{molar mass } Cl_2 = \frac{70.90\text{ g } Cl_2}{1\text{ mol } Cl_2}$$

$$\text{molar mass NaCl} = \frac{58.44\text{ g NaCl}}{1\text{ mol NaCl}}$$

SOLUTION

Beginning with the actual amounts of each reactant, we follow the solution map to calculate how much product can be made from each.

We can also find the limiting reactant by calculating the number of moles of NaCl (rather than grams) that can be made from each reactant. However, since theoretical yields are normally calculated in grams, we take the calculation all the way to grams to determine limiting reactant.

$$53.2\text{ g Na} \times \frac{1\text{ mol Na}}{22.99\text{ g Na}} \times \frac{2\text{ mol NaCl}}{2\text{ mol Na}} \times \frac{58.44\text{ g NaCl}}{1\text{ mol NaCl}} = 135\text{ g NaCl}$$

$$65.8\text{ g } Cl_2 \times \frac{1\text{ mol } Cl_2}{70.90\text{ g } Cl_2} \times \frac{2\text{ mol NaCl}}{1\text{ mol } Cl_2} \times \frac{58.44\text{ g NaCl}}{1\text{ mol NaCl}} = 108\text{ g NaCl}$$

Limiting reactant

Least amount of product

Since Cl_2 makes the least amount of product, it is the limiting reactant. Notice that the limiting reactant is not necessarily the reactant with the least mass. In this case, we had fewer grams of Na than Cl_2, yet Cl_2 was the limiting reactant because it made less NaCl. The theoretical yield is therefore 108 g of NaCl, the amount of product possible based on the limiting reactant.

Now suppose that when the synthesis is carried out, the actual yield of NaCl is 86.4 g. What is the percent yield? The percent yield is:

The actual yield is always less than the theoretical yield because at least a small amount of product is usually lost or does not form during a reaction.

$$\text{percent yield} = \frac{\text{actual yield}}{\text{theoretical yield}} \times 100\% = \frac{86.4\text{ g}}{108\text{ g}} \times 100\% = 80.0\%$$

PEARSON eText 2.0

CONCEPTUAL CHECKPOINT 8.4

Consider the reaction $A + 2\,B \longrightarrow 3\,C$. The molar mass of B is twice the molar mass of A. Equal masses of A and B are in a reaction vessel. Which reactant, A or B, is the limiting reactant?

Interactive Worked Example Video 8.5

EXAMPLE 8.5 Finding Limiting Reactant and Theoretical Yield

Ammonia, NH_3, can be synthesized by this reaction:

$$2\,NO(g) + 5\,H_2(g) \longrightarrow 2\,NH_3(g) + 2\,H_2O(g)$$

What maximum amount of ammonia in grams can be synthesized from 45.8 g of NO and 12.4 g of H_2?

SORT

You are given the masses of two reactants and asked to find the maximum mass of ammonia that forms. Although this problem does not specifically ask for the limiting reactant, you must know it to determine the theoretical yield, which is the maximum amount of ammonia that can be synthesized.

GIVEN: 45.8 g NO, 12.4 g H_2

FIND: maximum amount of NH_3 in g (this is the theoretical yield)

STRATEGIZE

Identify the limiting reactant by calculating how much product can be made from each reactant. The reactant that makes the *least amount of product* is the limiting reactant. The mass of ammonia formed by the limiting reactant is the maximum amount of ammonia that can be synthesized.

SOLUTION MAP

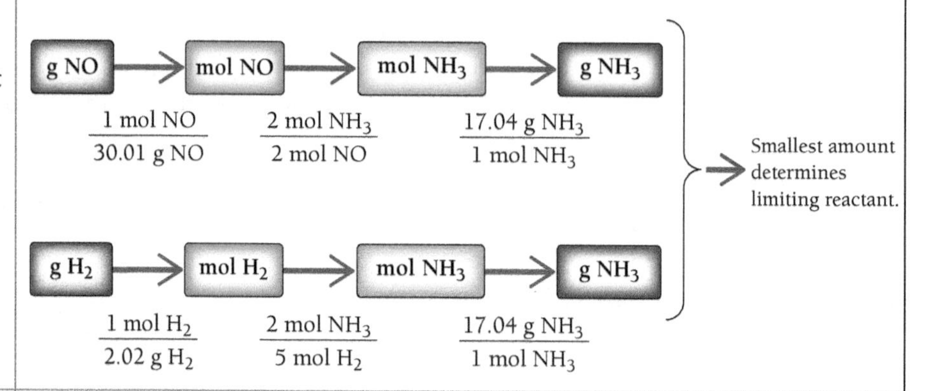

The main conversion factors come from the stoichiometric relationship between moles of each reactant and moles of ammonia. The other conversion factors are the molar masses of nitrogen monoxide, hydrogen gas, and ammonia.

RELATIONSHIPS USED

2 mol NO : 2 mol NH_3 5 mol H_2 : 2 mol NH_3

$$\text{molar mass NO} = \frac{30.01\,\text{g NO}}{1\,\text{mol NO}} \qquad \text{molar mass } H_2 = \frac{2.02\,\text{g } H_2}{1\,\text{mol } H_2}$$

$$\text{molar mass } NH_3 = \frac{17.04\,\text{g } NH_3}{1\,\text{mol } NH_3}$$

SOLVE

Follow the solution map, beginning with the actual amount of each reactant given, to calculate the amount of product that can be made from each reactant.

SOLUTION

$$45.8\,\text{g NO} \times \frac{1\,\text{mol NO}}{30.01\,\text{g NO}} \times \frac{2\,\text{mol } NH_3}{2\,\text{mol NO}} \times \frac{17.04\,\text{g } NH_3}{1\,\text{mol } NH_3} = 26.0\,\text{g } NH_3$$

Limiting reactant   Least amount of product

$$12.4\,\text{g } H_2 \times \frac{1\,\text{mol } H_2}{2.02\,\text{g } H_2} \times \frac{2\,\text{mol } NH_3}{5\,\text{mol } H_2} \times \frac{17.04\,\text{g } NH_3}{1\,\text{mol } NH_3} = 41.8\,\text{g } NH_3$$

There is enough NO to make 26.0 g of NH_3 and enough H_2 to make 41.8 g of NH_3. Therefore, NO is the limiting reactant, and the maximum amount of ammonia that can possibly be made is 26.0 g, which is the theoretical yield.

CHECK Are the units correct? Does the answer make physical sense?	The units of the answer, g NH_3, are correct. The magnitude of the answer seems reasonable because it is of the same order of magnitude as the given masses of NO and H_2. An answer that is orders of magnitude different would immediately be suspect.

▶ **SKILLBUILDER 8.5 | Finding Limiting Reactant and Theoretical Yield**

Ammonia can also be synthesized by this reaction:

$$3\,H_2(g) + N_2(g) \longrightarrow 2\,NH_3(g)$$

What maximum amount of ammonia in grams can be synthesized from 25.2 g of N_2 and 8.42 g of H_2?

▶ **SKILLBUILDER PLUS** What maximum amount of ammonia in kilograms can be synthesized from 5.22 kg of H_2 and 31.5 kg of N_2?

▶ **FOR MORE PRACTICE** Problems 55, 56, 57, 58.

EXAMPLE 8.6 Finding Limiting Reactant, Theoretical Yield, and Percent Yield

Consider this reaction:

$$Cu_2O(s) + C(s) \longrightarrow 2\,Cu(s) + CO(g)$$

When 11.5 g of C reacts with 114.5 g of Cu_2O, 87.4 g of Cu are obtained. Determine the limiting reactant, theoretical yield, and percent yield.

SORT You are given the mass of the reactants, carbon and copper(I) oxide, as well as the mass of copper formed by the reaction. You are asked to find the limiting reactant, theoretical yield, and percent yield.	**GIVEN:** 11.5 g C 114.5 g Cu_2O 87.4 g Cu produced **FIND:** limiting reactant theoretical yield percent yield
STRATEGIZE The solution map shows how to find the mass of Cu formed by the initial masses of Cu_2O and C. The reactant that makes the *least amount of product* is the limiting reactant and determines the theoretical yield.	**SOLUTION MAP** 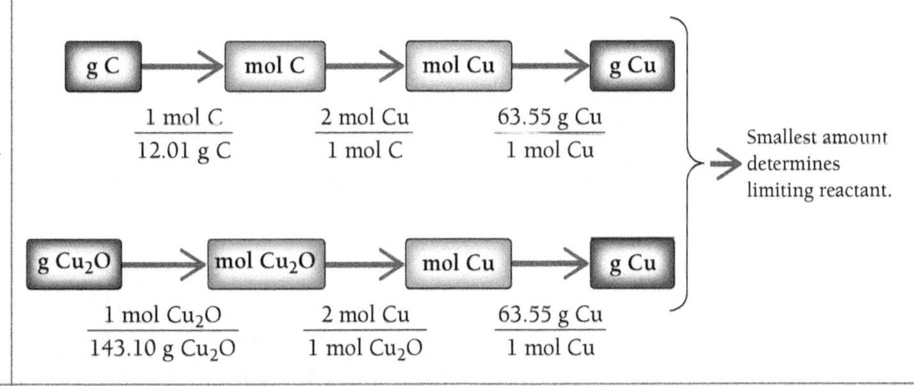
The main conversion factors are the stoichiometric relationships between moles of each reactant and moles of copper. The other conversion factors are the molar masses of copper(I) oxide, carbon, and copper.	**RELATIONSHIPS USED** 1 mol Cu_2O : 2 mol Cu 1 mol C : 2 mol Cu molar mass Cu_2O = 143.10 g/mol molar mass C = 12.01 g/mol molar mass Cu = 63.55 g/mol

continued on page 264 ▶

continued from page 263

SOLVE Follow the solution map, beginning with the actual given amount of each reactant, to calculate the amount of product that can be made from each reactant. Since Cu_2O makes the least amount of product, Cu_2O is the limiting reactant. The theoretical yield is then the amount of product made by the limiting reactant. The percent yield is the actual yield (87.4 g Cu) divided by the theoretical yield (101.7 g Cu) multiplied by 100%.	**SOLUTION** $11.5 \text{ g C} \times \frac{1 \text{ mol C}}{12.01 \text{ g C}} \times \frac{2 \text{ mol Cu}}{1 \text{ mol C}} \times \frac{63.55 \text{ g Cu}}{1 \text{ mol Cu}} = 122 \text{ g Cu}$ $114.5 \text{ g } Cu_2O \times \frac{1 \text{ mol } Cu_2O}{143.10 \text{ g } Cu_2O} \times \frac{2 \text{ mol Cu}}{1 \text{ mol } Cu_2O} \times \frac{63.55 \text{ g Cu}}{1 \text{ mol Cu}} = 101.7 \text{ g Cu}$ Limiting reactant — Least amount of product theoretical yield = 101.7 g Cu $\text{percent yield} = \frac{\text{actual yield}}{\text{theoretical yield}} \times 100\%$ $= \frac{87.4 \text{ g}}{101.7 \text{ g}} \times 100\% = 85.9\%$
CHECK Are the units correct? Does the answer make physical sense?	The theoretical yield has the right units (g Cu). The magnitude of the theoretical yield seems reasonable because it is of the same order of magnitude as the given masses of C and Cu_2O. The theoretical yield is reasonable because it is less than 100%. Any calculated theoretical yield above 100% is incorrect.

▶ **SKILLBUILDER 8.6 | Finding Limiting Reactant, Theoretical Yield, and Percent Yield**

This reaction is used to obtain iron from iron ore:

$$Fe_2O_3(s) + 3\,CO(g) \longrightarrow 2\,Fe(s) + 3\,CO_2(g)$$

The reaction of 185 g of Fe_2O_3 with 95.3 g of CO produces 87.4 g of Fe. Determine the limiting reactant, theoretical yield, and percent yield.

▶ **FOR MORE PRACTICE** Example 8.10; Problems 61, 62, 63, 64, 65, 66.

CONCEPTUAL CHECKPOINT 8.5

Ammonia can be synthesized by the reaction of nitrogen monoxide and hydrogen gas.

$$2\,NO(g) + 5\,H_2(g) \longrightarrow 2\,NH_3(g) + 2\,H_2O(g)$$

A reaction vessel initially contains 4.0 mol of NO and 15.0 mol of H_2. What is in the reaction vessel once the reaction has occurred to the fullest extent possible?

(a) 2 mol NO; 5 mol H_2; 2 mol NH_3; and 2 mol H_2O

(b) 0 mol NO; 0 mol H_2; 6 mol NH_3; and 6 mol H_2O

(c) 2 mol NO; 0 mol H_2; 4 mol NH_3; and 2 mol H_2O

(d) 0 mol NO; 5 mol H_2; 4 mol NH_3; and 4 mol H_2O

8.7 Enthalpy: A Measure of the Heat Evolved or Absorbed in a Reaction

▶ Calculate the amount of thermal energy emitted or absorbed by a chemical reaction.

Chapter 3 (see Section 3.9) describes how chemical reactions can be *exothermic* (in which case they *emit* thermal energy when they occur) or *endothermic* (in which case they *absorb* thermal energy when they occur). The *amount* of thermal energy emitted or absorbed by a chemical reaction, under conditions of constant pressure (which are common for most everyday reactions), can be quantified with a function called **enthalpy**. Specifically, we define a quantity called the **enthalpy of reaction** (ΔH_{rxn}) as the amount of thermal energy (or heat) that is emitted or absorbed when a reaction occurs at constant pressure.

Sign of ΔH_{rxn}

The *sign* of ΔH_{rxn} (positive or negative) depends on the *direction* in which thermal energy flows when the reaction occurs. If thermal energy flows out of the reaction and into the surroundings (as in an exothermic reaction), then ΔH_{rxn} is negative.

For example, we can specify the enthalpy of reaction for the combustion of CH_4, the main component in natural gas, as:

$$CH_4(g) + 2\,O_2(g) \longrightarrow CO_2(g) + 2\,H_2O(g) \qquad \Delta H_{rxn} = -802.3\text{ kJ}$$

EVERYDAY CHEMISTRY

Bunsen Burners

In the laboratory, we often use Bunsen burners as heat sources. These burners are normally fueled by methane. The balanced equation for methane (CH_4) combustion is:

$$CH_4(g) + 2\,O_2(g) \longrightarrow CO_2(g) + 2\,H_2O(g)$$

Most Bunsen burners have a mechanism to adjust the amount of air (and therefore of oxygen) that is mixed with the methane. If you light the burner with the air completely closed off, you get a yellow, smoky flame that is not very hot. As you increase the amount of air going into the burner, the flame becomes bluer, less smoky, and hotter. When you reach the optimum adjustment, the flame has a sharp, inner blue triangle, no smoke, and is hot enough to melt glass easily. Continuing to increase the air beyond this point causes the flame to become cooler again and may actually extinguish it.

B8.1 CAN YOU ANSWER THIS? *Can you use the concepts from this chapter to explain the changes in the Bunsen burner flame as the air intake is adjusted?*

(a) No air

(b) Small amount of air

(c) Optimum

(d) Too much air

▲ Bunsen burner flame at various stages of air-intake adjustment.

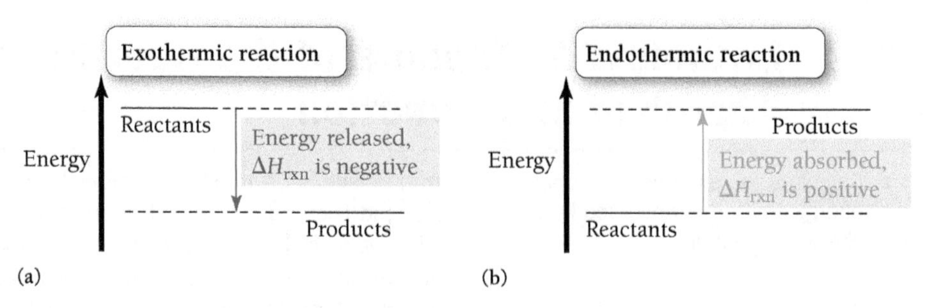

▲ FIGURE 8.3 **Exothermic and endothermic reactions** **(a)** In an exothermic reaction, energy is released into the surroundings. **(b)** In an endothermic reaction, energy is absorbed from the surroundings.

This reaction is exothermic and therefore has a negative enthalpy of reaction. The magnitude of ΔH_{rxn} tells us that 802.3 kJ of heat are emitted when 1 mol of CH_4 reacts with 2 mol of O_2.

If, by contrast, thermal energy flows into the reaction and out of the surroundings (as in an endothermic reaction), then ΔH_{rxn} is positive. For example, we specify the enthalpy of reaction for the reaction between nitrogen and oxygen gas to form nitrogen monoxide as:

$$N_2(g) + O_2(g) \longrightarrow 2\,NO(g) \qquad \Delta H_{rxn} = +182.6\,kJ$$

This reaction is endothermic and therefore has a positive enthalpy of reaction. When 1 mol of N_2 reacts with 1 mol of O_2, 182.6 kJ of heat are absorbed from the surroundings.

We can think of the energy of a chemical system in the same way that we think about the balance in a checking account. Energy flowing *out* of the chemical system is like a withdrawal and carries a negative sign as shown in ▲ FIGURE 8.3a. Energy flowing *into* the system is like a deposit and carries a positive sign as shown in ▲ FIGURE 8.3b.

Stoichiometry of ΔH_{rxn}

The amount of heat emitted or absorbed when a chemical reaction occurs depends on the *amounts* of reactants that actually react. As we have just seen, we usually specify ΔH_{rxn} in combination with the balanced chemical equation for the reaction. The magnitude of ΔH_{rxn} is for the stoichiometric amounts of reactants and products for the reaction *as written.*

For example, the balanced equation and ΔH_{rxn} for the combustion of propane (the fuel used in LP gas) is:

$$C_3H_8(g) + 5\,O_2(g) \longrightarrow 3\,CO_2(g) + 4\,H_2O(g) \qquad \Delta H_{rxn} = -2044\,kJ$$

This means that when 1 mol of C_3H_8 reacts with 5 mol of O_2 to form 3 mol of CO_2 and 4 mol of H_2O, 2044 kJ of heat are emitted. We can write these relationships in the same way that we express stoichiometric relationships: as ratios between two quantities. For the reactants in this reaction, we write:

$$1\text{ mol } C_3H_8 : -2044\text{ kJ} \quad \text{or} \quad 5\text{ mol } O_2 : -2044\text{ kJ}$$

The ratios mean that 2044 kJ of thermal energy are evolved when 1 mol of C_3H_8 and 5 mol of O_2 completely react. We can use these ratios to construct conversion factors between amounts of reactants or products and the quantity of heat emitted (for exothermic reactions) or absorbed (for endothermic reactions). To find out how much heat is emitted upon the combustion of a certain mass in grams of C_3H_8, we use the following solution map:

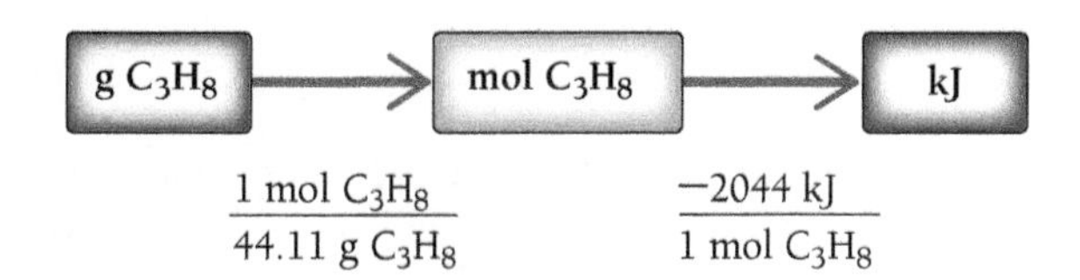

We use the molar mass to convert between grams and moles, and the stoichiometric relationship between moles of C_3H_8 and kilojoules to convert between moles and kilojoules, as shown in Example 8.7.

PEARSON eText 2.0 **Interactive Worked Example Video 8.7**

EXAMPLE 8.7 Stoichiometry Involving ΔH_{rxn}

An LP gas tank in a home barbecue contains 1.18×10^4 g of propane (C_3H_8). Calculate the heat (in kJ) associated with the complete combustion of all of the propane in the tank.

$$C_3H_8(g) + 5\,O_2(g) \longrightarrow 3\,CO_2(g) + 4\,H_2O(g) \qquad \Delta H_{rxn} = -2044 \text{ kJ}$$

SORT
You are given the mass of propane and asked to find the heat evolved (in kJ) in its combustion.

GIVEN: 1.18×10^4 g C_3H_8

FIND: kJ

STRATEGIZE
Start with the given mass of propane and use its molar mass to find the number of moles. Use the stoichiometric relationship between moles of propane and kilojoules of heat to find the heat evolved.

SOLUTION MAP

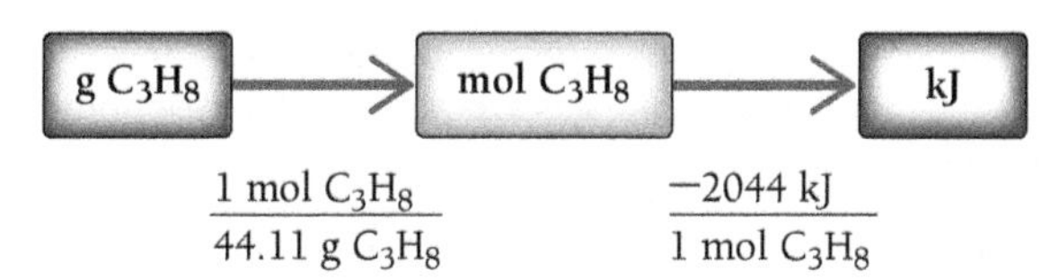

RELATIONSHIPS USED

1 mol C_3H_8: −2044 kJ (from balanced equation)

molar mass C_3H_8 = 44.11 g/mol

SOLVE
Follow the solution map to solve the problem. Begin with 11.8×10^4 g C_3H_8 and multiply by the appropriate conversion factors to arrive at kJ.

SOLUTION

$$1.18 \times 10^4 \text{ g } C_3H_8 \times \frac{1 \text{ mol } C_3H_8}{44.11 \text{ g } C_3H_8} \times \frac{-2044 \text{ kJ}}{1 \text{ mol } C_3H_8} = -5.47 \times 10^5 \text{ kJ}$$

CHECK
Check your answer. Are the units correct? Does the answer make physical sense?

The units, kJ, are correct. The answer is negative, as it should be when heat is evolved by a reaction.

▶ **SKILLBUILDER 8.7 | Stoichiometry Involving ΔH**
Ammonia reacts with oxygen according to the equation:

$$4\,NH_3(g) + 5\,O_2(g) \longrightarrow 4\,NO(g) + 6\,H_2O \qquad \Delta H_{rxn} = -906 \text{ kJ}$$

Calculate the heat (in kJ) associated with the complete reaction of 155 g of NH_3.

▶ **SKILLBUILDER PLUS** What mass of butane in grams is necessary to produce 1.5×10^3 kJ of heat? What mass of CO_2 is produced?

$$C_4H_{10}(g) + \tfrac{13}{2}\,O_2(g) \longrightarrow 4\,CO_2(g) + 5\,H_2O(g) \qquad \Delta H_{rxn} = -2658 \text{ kJ}$$

▶ **FOR MORE PRACTICE** Example 8.11; Problems 71, 72, 73, 74, 75, 76.

PEARSON eText 2.0

CONCEPTUAL CHECKPOINT 8.6

Consider the generic reaction:

$$2\,A + 3\,B \longrightarrow 2\,C \qquad \Delta H_{rxn} = -100\text{ kJ}$$

If a reaction mixture initially contains 5 mol of A and 6 mol of B, how much heat (in kJ) will have evolved once the reaction has occurred to the greatest extent possible?

(a) 100 kJ **(b)** 150 kJ **(c)** 200 kJ **(d)** 300 kJ

Chapter 8 in Review

MasteringChemistry™ provides end-of-chapter exercises, feedback-enriched tutorial problems, animations, and interactive activities to encourage problem solving practice and deeper understanding of key concepts and topics.

Self-Assessment Quiz

PEARSON eText 2.0

Q1. Sulfur and fluorine react to form sulfur hexafluoride according to the reaction shown here. How many mol of F_2 are required to react completely with 2.55 mol of S?

$$S(s) + 3\,F_2(g) \longrightarrow SF_6(g)$$

(a) 0.85 mol F_2 **(b)** 2.55 mol F_2
(c) 7.65 mol F_2 **(d)** 15.3 mol F_2

Q2. Hydrogen chloride gas and oxygen gas react to form gaseous water and chlorine gas according to the reaction shown here.

$$4\,HCl(g) + O_2(g) \longrightarrow 2\,H_2O(g) + 2\,Cl_2(g)$$

If the first image below represents the amount of HCl available for the reaction, which image represents the amount of oxygen required to react completely with the amount of available HCl?

$= O_2$

$= HCl$

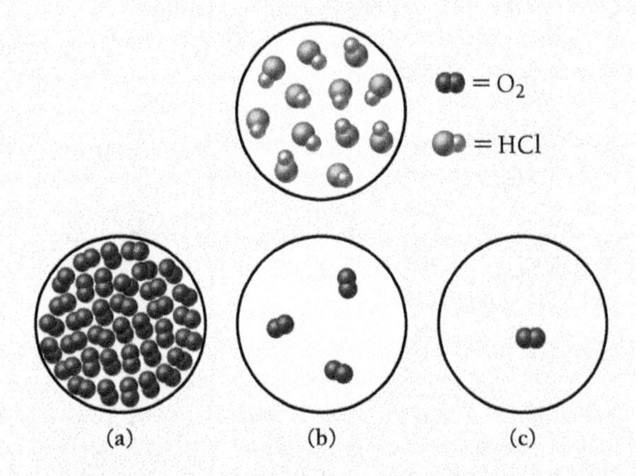

Q3. Sodium reacts with fluorine to form sodium fluoride. What mass of sodium fluoride forms from the complete reaction of 12.5 g of fluorine with enough sodium to completely react with it?

$$2\,Na(s) + F_2(g) \longrightarrow 2\,NaF(s)$$

(a) 0.658 g NaF **(b)** 13.8 g NaF
(c) 6.91 g NaF **(d)** 27.6 g NaF

Q4. Consider the hypothetical reaction shown here. If 11 mol of A combine with 16 mol of B and the reaction occurs to the greatest extent possible, how many mol of C form?

$$3\,A + 4\,B \longrightarrow 2\,C$$

(a) 7.3 mol C **(b)** 8.0 mol C
(c) 17 mol C **(d)** 32 mol C

Q5. Consider the generic reaction:

$$2\,A + 3\,B + C \longrightarrow 2\,D$$

A reaction mixture contains 6 mol A, 8 mol B, and 10 mol C. What is the limiting reactant?

(a) A **(b)** B **(c)** C **(d)** D

Q6. Methanol (CH_3OH) reacts with oxygen to form carbon dioxide and water according to the reaction shown here.

$$2\,CH_3OH(g) + 3\,O_2(g) \longrightarrow 2\,CO_2(g) + 4\,H_2O(g)$$

If the first image below represents a reaction mixture of methanol and oxygen, which image represents the reaction mixture after the reaction has occurred to the maximum extent possible?

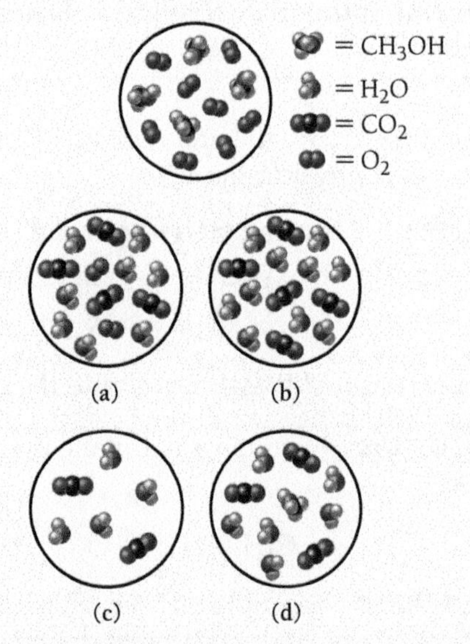

Q7. Sodium and chlorine react to form sodium chloride.

$$2\,Na(s) + Cl_2(g) \longrightarrow 2\,NaCl(s)$$

What is the theoretical yield of sodium chloride for the reaction of 55.0 g Na with 67.2 g Cl_2?

(a) 1.40×10^2 g NaCl **(b)** 111 g NaCl
(c) 55.4 g NaCl **(d)** 222 g NaCl

Q8. A reaction has a theoretical yield of 22.8 g. When the reaction is carried out, 15.1 g of the product is obtained. What is the percent yield?

(a) 151%
(b) 66.2%
(c) 344%
(d) 88.2%

Q9. Titanium can be obtained from its oxide by the reaction shown here. When 42.0 g of TiO_2 react with 11.5 g C, 18.7 g Ti are obtained. What is the percent yield for the reaction?

$$TiO_2(s) + 2\,C(s) \longrightarrow Ti(s) + CO(g)$$

(a) 74.2%
(b) 81.6%
(c) 122%
(d) 41%

Q10. Which statement best describes an exothermic reaction?

(a) An exothermic reaction gives off heat.
(b) An exothermic reaction absorbs heat.
(c) An exothermic reaction produces only small amounts of products.
(d) none of the above

Q11. Consider the generic reaction:

$$A + 2\,B \longrightarrow AB_2 \quad \Delta H_{rxn} = -155\,kJ$$

If a reaction mixture contains 5 mol A and 8 mol B, how much heat is emitted or absorbed once the reaction has occurred to the greatest extent possible?

(a) 775 kJ emitted
(b) 775 kJ absorbed
(c) 620 kJ emitted
(d) 620 kJ absorbed

Q12. Hydrogen gas reacts with oxygen to form water.

$$2\,H_2(g) + O_2(g) \longrightarrow 2\,H_2O(g) \quad \Delta H = -483.5\,kJ$$

Determine the minimum mass of hydrogen gas required to produce 226 kJ of heat.

(a) 8.63 g
(b) 1.88 g
(c) 0.942 g
(d) 0.935 g

Answers: 1:c, 2:b, 3:d, 4:a, 5:b, 6:a, 7:b, 8:b, 9:b, 10:a, 11:c, 12:b

Chemical Principles

Stoichiometry

A balanced chemical equation indicates quantitative relationships between the amounts of reactants and products. For example, the reaction $2H_2 + O_2 \longrightarrow 2H_2O$ tells us that 2 mol of H_2 reacts with 1 mol of O_2 to form 2 mol of H_2O. We can use these relationships to calculate quantities such as the amount of product possible with a certain amount of reactant, or the amount of one reactant required to completely react with a certain amount of another reactant. The quantitative relationship between reactants and products in a chemical reaction is reaction stoichiometry.

Relevance: Reaction stoichiometry is important because we often want to know the numerical relationship between the reactants and products in a chemical reaction. For example, we might want to know how much carbon dioxide, a greenhouse gas, is formed when a certain amount of a particular fossil fuel burns.

Limiting Reactant, Theoretical Yield, and Percent Yield

The limiting reactant in a chemical reaction is the reactant that limits the amount of product that can be made. The theoretical yield in a chemical reaction is the amount of product that can be made based on the amount of the limiting reactant. The actual yield in a chemical reaction is the amount of product actually produced. The percent yield in a chemical reaction is the actual yield divided by theoretical yield times 100%.

Relevance: Calculations of limiting reactant, theoretical yield, and percent yield are central to chemistry because they allow for quantitative understanding of chemical reactions. Just as we need to know relationships between ingredients to follow a recipe, so we must know relationships between reactants and products to carry out a chemical reaction. The percent yield in a chemical reaction is often used as a measure of the success of the reaction. Imagine following a recipe and making only 1% of the final product—your cooking would be a failure. Similarly, low percent yields in chemical reactions are usually considered poor, and high percent yields are considered good.

Chemical Principles

Enthalpy of Reaction

The amount of heat released or absorbed by a chemical reaction under conditions of constant pressure is the enthalpy of reaction (ΔH_{rxn}).

Relevance

The enthalpy of reaction describes the relationship between the amount of reactant that undergoes reaction and the amount of thermal energy produced. This is important, for example, in determining quantities such as the amount of fuel needed to produce a given amount of energy.

Chemical Skills

LO: Carry out mole-to-mole conversions between reactants and products in a balanced chemical equation (Section 8.3).

SORT

You are given the number of moles of sodium and asked to find the number of moles of sodium oxide formed by the reaction.

STRATEGIZE

Draw a solution map beginning with the number of moles of the given substance and then use the conversion factor from the balanced chemical equation to determine the number of moles of the substance you are trying to find.

SOLVE

Follow the solution map to get to the number of moles of the substance you are trying to find.

CHECK

Are the units correct? Does the answer make physical sense?

Examples

EXAMPLE 8.8 Mole-to-Mole Conversions

How many mol of sodium oxide can be synthesized from 4.8 mol of sodium? Assume that more than enough oxygen is present. The balanced equation is:

$$4\,Na(s) + O_2(g) \longrightarrow 2\,Na_2O(s)$$

GIVEN: 4.8 mol Na

FIND: mol Na_2O

SOLUTION MAP

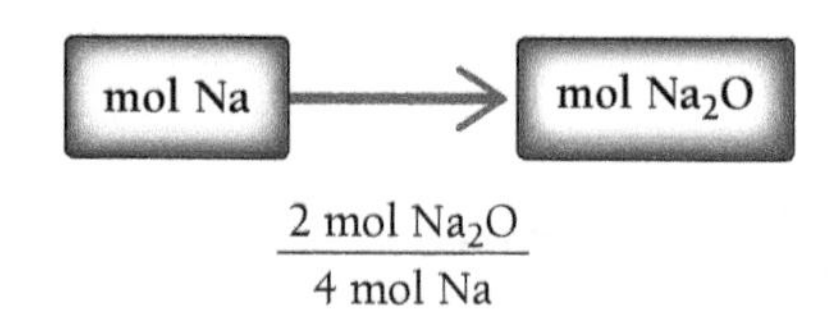

$$\frac{2\text{ mol } Na_2O}{4\text{ mol Na}}$$

RELATIONSHIPS USED

4 mol Na : 2 mol Na_2O

SOLUTION

$$4.8\text{ mol Na} \times \frac{2\text{ mol } Na_2O}{4\text{ mol Na}} = 2.4\text{ mol } Na_2O$$

The units of the answer, mol Na_2O, are correct. The magnitude of the answer seems reasonable because it is of the same order of magnitude as the given number of mol of Na.

LO: Carry out mass-to-mass conversions between reactants and products in a balanced chemical equation and molar masses (Section 8.4).

EXAMPLE **8.9** **Mass-to-Mass Conversions**

How many grams of sodium oxide can be synthesized from 17.4 g of sodium? Assume that more than enough oxygen is present. The balanced equation is:

$$4\,Na(s) + O_2(g) \longrightarrow 2\,Na_2O(s)$$

SORT
You are given the mass of sodium and asked to find the mass of sodium oxide that forms upon reaction.

GIVEN: 17.4 g Na

FIND: g Na_2O

STRATEGIZE
Draw the solution map by beginning with the mass of the given substance. Convert to moles using the molar mass and then convert to moles of the substance you are trying to find, using the conversion factor obtained from the balanced chemical equation.

Finally, convert to the mass of the substance you are trying to find, using its molar mass.

SOLUTION MAP

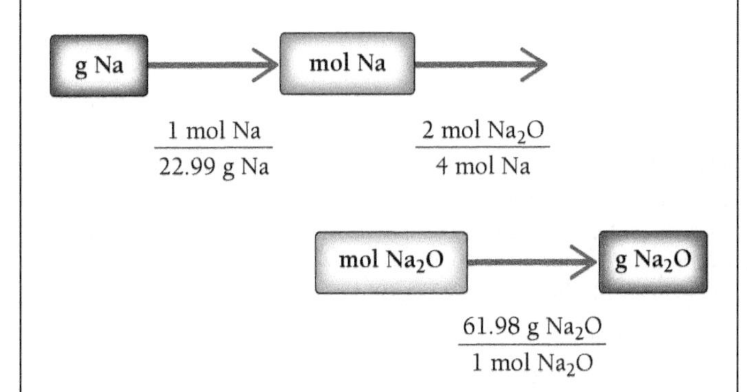

RELATIONSHIPS USED

4 mol Na : 2 mol Na_2O (from balanced equation)

molar mass Na = 22.99 g/mol

molar mass Na_2O = 61.98 g/mol

SOLVE
Follow the solution map and calculate the answer by beginning with the mass of the given substance and multiplying by the appropriate conversion factors to determine the mass of the substance you are trying to find.

SOLUTION

$$17.4\ \text{g Na} \times \frac{1\ \text{mol Na}}{22.99\ \text{g Na}} \times \frac{2\ \text{mol Na}_2\text{O}}{4\ \text{mol Na}} \times \frac{61.98\ \text{g Na}_2\text{O}}{1\ \text{mol Na}_2\text{O}} = 23.5\ \text{g Na}_2\text{O}$$

CHECK
Are the units correct? Does the answer make physical sense?

The units of the answer, g Na_2O, are correct. The magnitude of the answer seems reasonable because it is of the same order of magnitude as the given number of mol of Na.

LO: Calculate limiting reactant, theoretical yield, and percent yield in a balanced chemical equation (Sections 8.5, 8.6).

EXAMPLE **8.10** **Limiting Reactant, Theoretical Yield, and Percent Yield**

10.4 g of As reacts with 11.8 g of S to produce 14.2 g of As_2S_3. Find the limiting reactant, theoretical yield, and percent yield for this reaction. The balanced chemical equation is:

$$2\,As(s) + 3\,S(l) \longrightarrow As_2S_3(s)$$

SORT

You are given the masses of arsenic and sulfur as well as the mass of arsenic sulfide formed by the reaction. You are asked to find the limiting reactant, theoretical yield, and percent yield.

GIVEN: 10.4 g As
11.8 g S
14.2 g As_2S_3

FIND: limiting reactant
theoretical yield
percent yield

STRATEGIZE

The solution map for limiting-reactant problems shows how to convert from the mass of each of the reactants to the mass of the product for each reactant. These are mass-to-mass conversions with the basic outline of

$$\text{mass} \longrightarrow \text{moles} \longrightarrow \text{moles} \longrightarrow \text{mass}$$

The reactant that forms the least amount of product is the limiting reactant.

The conversion factors you need are the stoichiometric relationships between each of the reactants and the product. You also need the molar masses of each reactant and product.

SOLUTION MAP

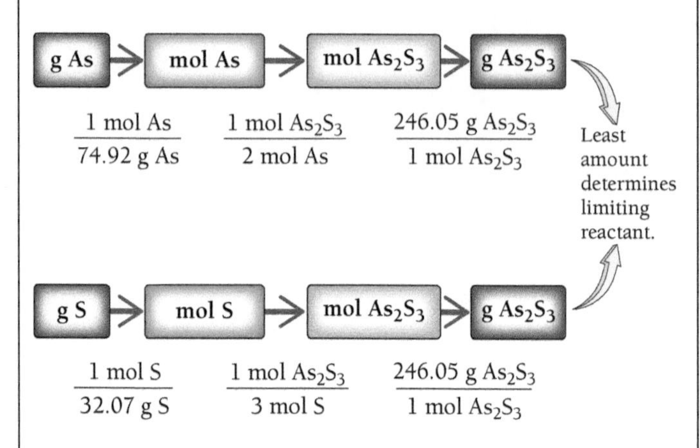

RELATIONSHIPS USED

2 mol As : 1 mol As_2S_3
3 mol S : 1 mol As_2S_3
molar mass As = 74.92 g/mol
molar mass S = 32.07 g/mol
molar mass As_2S_3 = 246.05 g/mol

SOLVE

To calculate the amount of product formed by each reactant, begin with the given amount of each reactant and multiply by the appropriate conversion factors, as shown in the solution map, to arrive at the mass of product for each reactant. The reactant that forms the least amount of product is the limiting reactant.

SOLUTION

$$10.4 \text{ g As} \times \frac{1 \text{ mol As}}{74.92 \text{ g As}} \times \frac{1 \text{ mol } As_2S_3}{2 \text{ mol As}} \times \frac{246.05 \text{ g } As_2S_3}{1 \text{ mol } As_2S_3} = 17.1 \text{ g } As_2S_3$$

Limiting reactant (As); Least amount of product (17.1 g As_2S_3)

$$11.8 \text{ g S} \times \frac{1 \text{ mol S}}{32.07 \text{ g S}} \times \frac{1 \text{ mol } As_2S_3}{3 \text{ mol S}} \times \frac{246.05 \text{ g } As_2S_3}{1 \text{ mol } As_2S_3} = 30.2 \text{ g } As_2S_3$$

The limiting reactant is As.

The theoretical yield is the amount of product formed by the limiting reactant.
The percent yield is the actual yield divided by the theoretical yield times 100%.

The theoretical yield is 17.1 g of As_2S_3.

$$\text{percent yield} = \frac{\text{actual yield}}{\text{theoretical yield}} \times 100\%$$

$$= \frac{14.2\,g}{17.1\,g} \times 100\% = 83.0\%$$

The percent yield is 83.0%.

CHECK
Check your answer. Are the units correct? Does the answer make physical sense?

The theoretical yield has the right units (g As_2S_3). The magnitude of the theoretical yield seems reasonable because it is of the same order of magnitude as the given masses of As and S. The theoretical yield is reasonable because it is less than 100%. Any calculated theoretical yield above 100% would be suspect.

LO: Calculate the amount of thermal energy emitted or absorbed by a chemical reaction (Section 8.7).

EXAMPLE **8.11** **Stoichiometry Involving ΔH_{rxn}**

Calculate the heat evolved (in kJ) upon complete combustion of 25.0 g of methane (CH_4).

$CH_4(g) + 2O_2(g) \longrightarrow CO_2(g) + 2H_2O(g)$ $\Delta H_{rxn} = -802\,kJ$

SORT
You are given the mass of methane and asked to find the quantity of heat in kJ emitted upon combustion.
Draw the solution map by beginning with the mass of the given substance. Convert to moles using molar mass and to kJ using ΔH_{rxn}.

GIVEN: 25 g CH_4

FIND: kJ

SOLUTION MAP

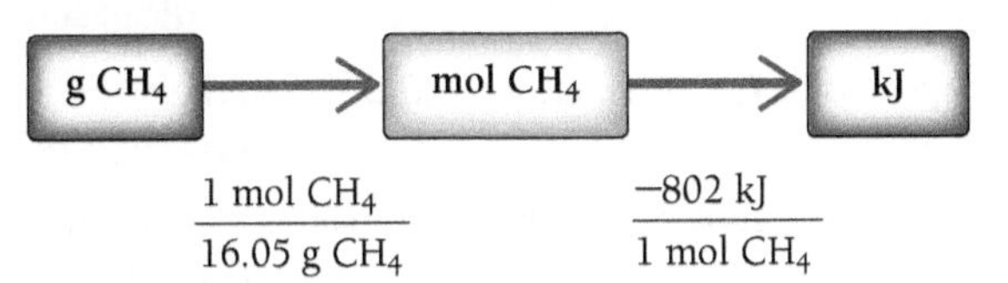

$$\frac{1\text{ mol }CH_4}{16.05\text{ g }CH_4} \qquad \frac{-802\text{ kJ}}{1\text{ mol }CH_4}$$

RELATIONSHIPS USED

1 mol CH_4 : −802 kJ (from balanced equation)

molar mass CH_4 = 16.05 g/mol

SOLVE
Follow the solution map to solve the problem. Begin with the mass of the given substance and multiply by the appropriate conversion factors to arrive at kJ. A negative answer indicates that heat is evolved into the surroundings. A positive answer indicates that heat is absorbed from the surroundings.

SOLUTION

$$25.0\,g\,CH_4 \times \frac{1\text{ mol }CH_4}{16.05\,g\,CH_4} \times \frac{-802\,kJ}{1\text{ mol }CH_4} = -1.25 \times 10^3\,kJ$$

CHECK
Are the units correct? Does the answer make physical sense?

The units, kJ, are correct. The answer is negative, as it should be since heat is evolved by the reaction.

Key Terms

actual yield [8.5]
climate change [8.1]
enthalpy [8.7]
enthalpy of reaction (ΔH_{rxn}) [8.7]
greenhouse gases [8.1]
limiting reactant [8.5]
percent yield [8.5]
stoichiometry [8.1]
theoretical yield [8.5]

Exercises

Questions

1. Why is reaction stoichiometry important? Cite some examples in your answer.
2. Nitrogen and hydrogen can react to form ammonia:

$$N_2(g) + 3H_2(g) \longrightarrow 2NH_3(g)$$

(a) Write ratios showing the relationships between moles of each of the reactants and products in the reaction.
(b) How many molecules of H_2 are required to completely react with two molecules of N_2?
(c) How many moles of H_2 are required to completely react with 2 mol of N_2?

3. Write the conversion factor that you would use to convert from moles of Cl_2 to moles of NaCl in the reaction:

$$2Na(s) + Cl_2(g) \longrightarrow 2NaCl(s)$$

4. What is wrong with this statement in reference to the reaction in the previous problem? "2 g of Na react with 1 g of Cl_2 to form 2 g of NaCl." Correct the statement to make it true.
5. What is the general form of the solution map for problems in which you are given the mass of a reactant in a chemical reaction and asked to find the mass of the product that can be made from the given amount of reactant?
6. Consider the recipe for making tomato and garlic pasta.

$$2 \text{ cups noodles} + 12 \text{ tomatoes} + 3 \text{ cloves garlic} \longrightarrow 4 \text{ servings pasta}$$

If you have 7 cups of noodles, 27 tomatoes, and 9 cloves of garlic, how many servings of pasta can you make? Which ingredient limits the amount of pasta that it is possible to make?

7. In a chemical reaction, what is the limiting reactant?
8. In a chemical reaction, what is the theoretical yield?
9. In a chemical reaction, what are the actual yield and the percent yield?
10. If you are given a chemical equation and specific amounts for each reactant in grams, how do you determine the maximum amount of product that can be made?
11. Consider the generic chemical reaction:

$$A + 2B \longrightarrow C + D$$

Suppose you have 12 g of A and 24 g of B. Which statement is true?
(a) A will definitely be the limiting reactant.
(b) B will definitely be the limiting reactant.
(c) A will be the limiting reactant if its molar mass is less than B.
(d) A will be the limiting reactant if its molar mass is greater than B.

12. Consider the generic chemical equation:

$$A + B \longrightarrow C$$

Suppose 25 g of A were allowed to react with 8 g of B. Analysis of the final mixture showed that A was completely used up and 4 g of B remained. What was the limiting reactant?

13. What is the enthalpy of reaction (ΔH_{rxn})? Why is this quantity important?
14. Explain the relationship between the sign of ΔH_{rxn} and whether a reaction is exothermic or endothermic.

Problems

MOLE-TO-MOLE CONVERSIONS

15. Consider the generic chemical reaction:

$$A + 2B \longrightarrow C$$

How many moles of C are formed upon complete reaction of:
(a) 2 mol of A
(b) 2 mol of B
(c) 3 mol of A
(d) 3 mol of B

16. Consider the generic chemical reaction:

$$2A + 3B \longrightarrow 3C$$

How many moles of B are required to completely react with:
(a) 6 mol of A
(b) 2 mol of A
(c) 7 mol of A
(d) 11 mol of A

17. For the reaction shown, calculate how many moles of NO_2 form when each amount of reactant completely reacts.

$$2N_2O_5(g) \longrightarrow 4NO_2(g) + O_2(g)$$

(a) 1.3 mol N_2O_5
(b) 5.8 mol N_2O_5
(c) 4.45×10^3 mol N_2O_5
(d) 1.006×10^{-3} mol N_2O_5

18. For the reaction shown, calculate how many moles of NH_3 form when each amount of reactant completely reacts.

$$3N_2H_4(l) \longrightarrow 4NH_3(g) + N_2(g)$$

(a) 5.3 mol N_2H_4
(b) 2.28 mol N_2H_4
(c) 5.8×10^{-2} mol N_2H_4
(d) 9.76×10^7 mol N_2H_4

19. Dihydrogen monosulfide reacts with sulfur dioxide according to the balanced equation:

$$2\,H_2S(g) + SO_2(g) \longrightarrow 3\,S(s) + 2\,H_2O(g)$$

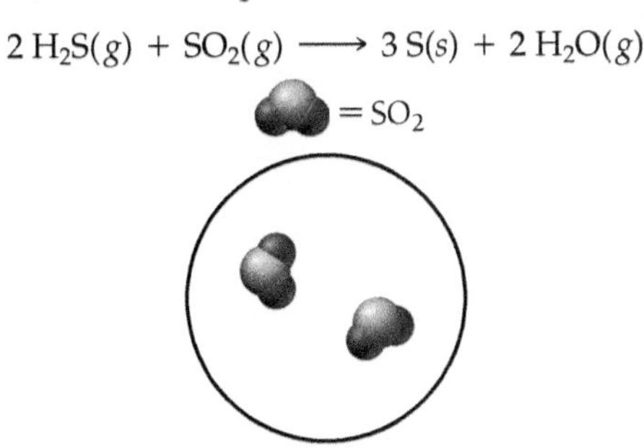

If the first figure represents the amount of SO_2 available to react, which figure best represents the amount of H_2S required to completely react with all of the SO_2?

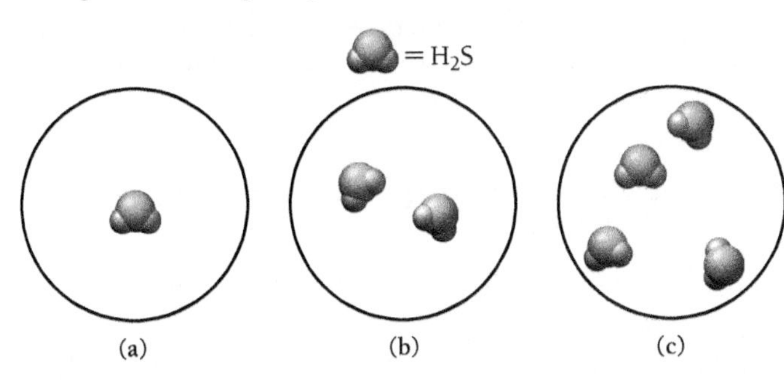

20. Chlorine gas reacts with fluorine gas according to the balanced equation:

$$Cl_2(g) + 3\,F_2(g) \longrightarrow 2\,ClF_3(g)$$

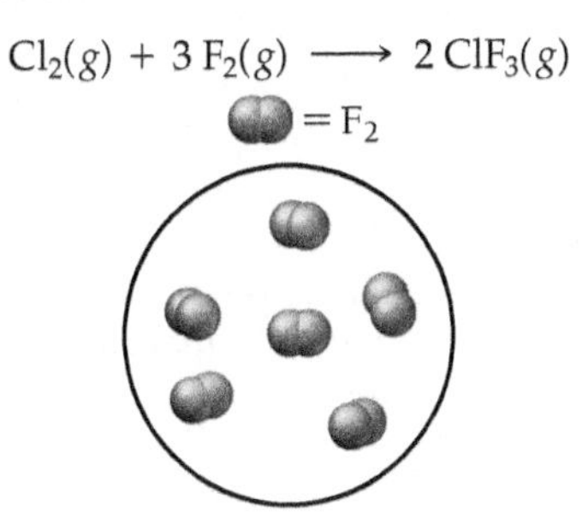

If the first figure represents the amount of fluorine available to react, and assuming that there is more than enough chlorine, which figure best represents the amount of chlorine trifluoride that would form upon complete reaction of all of the fluorine?

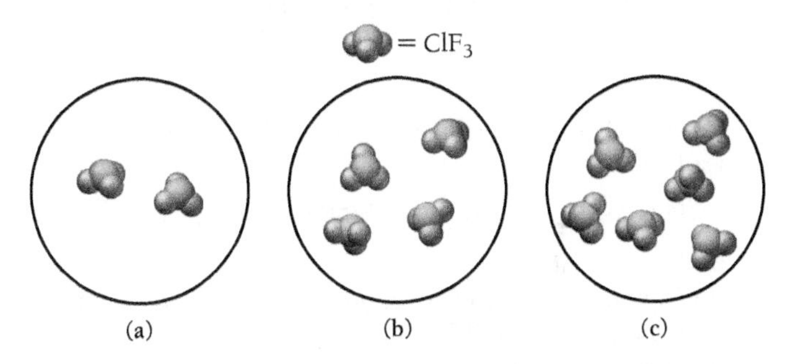

21. For each reaction, calculate how many moles of product form when 1.75 mol of the reactant in color completely reacts. Assume there is more than enough of the other reactant.
(a) $H_2(g) + Cl_2(g) \longrightarrow 2\,HCl(g)$
(b) $2\,H_2(g) + O_2(g) \longrightarrow 2\,H_2O(l)$
(c) $2\,Na(s) + O_2(g) \longrightarrow Na_2O_2(s)$
(d) $2\,S(s) + 3\,O_2(g) \longrightarrow 2\,SO_3(g)$

22. For each reaction, calculate how many moles of the product form when 0.112 mol of the reactant in color completely reacts. Assume there is more than enough of the other reactant.
(a) $2\,Ca(s) + O_2(g) \longrightarrow 2\,CaO(s)$
(b) $4\,Fe(s) + 3\,O_2(g) \longrightarrow 2\,Fe_2O_3(s)$
(c) $4\,K(s) + O_2(g) \longrightarrow 2\,K_2O(s)$
(d) $4\,Al(s) + 3\,O_2(g) \longrightarrow 2\,Al_2O_3(s)$

23. For the reaction shown, calculate how many moles of each product form when the given amount of each reactant completely reacts. Assume there is more than enough of the other reactant.

$$2\,PbS(s) + 3\,O_2(g) \longrightarrow 2\,PbO(s) + 2\,SO_2(g)$$

(a) 2.4 mol PbS
(b) 2.4 mol O_2
(c) 5.3 mol PbS
(d) 5.3 mol O_2

24. For the reaction shown, calculate how many moles of each product form when the given amount of each reactant completely reacts. Assume there is more than enough of the other reactant.

$$C_3H_8(g) + 5\,O_2(g) \longrightarrow 3\,CO_2(g) + 4\,H_2O(g)$$

(a) 4.6 mol C_3H_8
(b) 4.6 mol O_2
(c) 0.0558 mol C_3H_8
(d) 0.0558 mol O_2

25. Consider the balanced equation:

$$2\,N_2H_4(g) + N_2O_4(g) \longrightarrow 3\,N_2(g) + 4\,H_2O(g)$$

Complete the table with the appropriate number of moles of reactants and products. If the number of moles of a reactant is provided, fill in the required amount of the other reactant, as well as the moles of each product formed. If the number of moles of a product is provided, fill in the required amount of each reactant to make that amount of product, as well as the amount of the other product that is made.

mol N_2H_4	mol N_2O_4	mol N_2	mol H_2O
____	2	____	____
6	____	____	____
____	____	____	8
____	5.5	____	____
3	____	____	____
____	____	12.4	____

26. Consider the balanced equation:

$$SiO_2(s) + 3C(s) \longrightarrow SiC(s) + 2CO(g)$$

Complete the table with the appropriate number of moles of reactants and products. If the number of moles of a reactant is provided, fill in the required amount of the other reactant, as well as the moles of each product formed. If the number of moles of a product is provided, fill in the required amount of each reactant to make that amount of product, as well as the amount of the other product that is made.

mol SiO_2	mol C	mol SiC	mol CO
____	6	____	____
3	____	____	____
____	____	____	10
____	9.5	____	____
3.2	____	____	____

27. Consider the unbalanced equation for the combustion of butane:

$$C_4H_{10}(g) + O_2(g) \longrightarrow CO_2(g) + H_2O(g)$$

Balance the equation and determine how many moles of O_2 are required to react completely with 4.9 mol of C_4H_{10}.

28. Consider the unbalanced equation for the neutralization of acetic acid:

$$HC_2H_3O_2(aq) + Ca(OH)_2(aq) \longrightarrow H_2O(l) + Ca(C_2H_3O_2)_2(aq)$$

Balance the equation and determine how many moles of $Ca(OH)_2$ are required to completely neutralize 1.07 mol of $HC_2H_3O_2$.

29. Consider the unbalanced equation for the reaction of solid lead with silver nitrate:

$$Pb(s) + AgNO_3(aq) \longrightarrow Pb(NO_3)_2(aq) + Ag(s)$$

(a) Balance the equation.
(b) How many moles of silver nitrate are required to completely react with 9.3 mol of lead?
(c) How many moles of Ag are formed by the complete reaction of 28.4 mol of Pb?

30. Consider the unbalanced equation for the reaction of aluminum with sulfuric acid:

$$Al(s) + H_2SO_4(aq) \longrightarrow Al_2(SO_4)_3(aq) + H_2(g)$$

(a) Balance the equation.
(b) How many moles of H_2SO_4 are required to completely react with 8.3 mol of Al?
(c) How many moles of H_2 are formed by the complete reaction of 0.341 mol of Al?

MASS-TO-MASS CONVERSIONS

31. For the reaction shown, calculate how many grams of oxygen form when each quantity of reactant completely reacts.

$$2HgO(s) \longrightarrow 2Hg(l) + O_2(g)$$

(a) 2.13 g HgO
(b) 6.77 g HgO
(c) 1.55 kg HgO
(d) 3.87 mg HgO

32. For the reaction shown, calculate how many grams of oxygen form when each quantity of reactant completely reacts.

$$2KClO_3(s) \longrightarrow 2KCl(s) + 3O_2(g)$$

(a) 2.72 g $KClO_3$
(b) 0.361 g $KClO_3$
(c) 83.6 kg $KClO_3$
(d) 22.4 mg $KClO_3$

33. For each of the reactions, calculate how many grams of the product form when 2.4 g of the reactant in color completely reacts. Assume there is more than enough of the other reactant.

(a) $2Na(s) + Cl_2(g) \longrightarrow 2NaCl(s)$
(b) $CaO(s) + CO_2(g) \longrightarrow CaCO_3(s)$
(c) $2Mg(s) + O_2(g) \longrightarrow 2MgO(s)$
(d) $Na_2O(s) + H_2O(l) \longrightarrow 2NaOH(aq)$

34. For each of the reactions, calculate how many grams of the product form when 17.8 g of the reactant in color completely reacts. Assume there is more than enough of the other reactant.

(a) $Ca(s) + Cl_2(g) \longrightarrow CaCl_2(s)$
(b) $2K(s) + Br_2(l) \longrightarrow 2KBr(s)$
(c) $4Cr(s) + 3O_2(g) \longrightarrow 2Cr_2O_3(s)$
(d) $2Sr(s) + O_2(g) \longrightarrow 2SrO(s)$

35. For the reaction shown, calculate how many grams of each product form when the given amount of each reactant completely reacts to form products. Assume there is more than enough of the other reactant.

$$2Al(s) + Fe_2O_3(s) \longrightarrow Al_2O_3(s) + 2Fe(l)$$

(a) 4.7 g Al
(b) 4.7 g Fe_2O_3

36. For the reaction shown, calculate how many grams of each product form when the given amount of each reactant completely reacts to form products. Assume there is more than enough of the other reactant.

$$2HCl(aq) + Na_2CO_3(aq) \longrightarrow 2NaCl(aq) + H_2O(l) + CO_2(g)$$

(a) 10.8 g HCl
(b) 10.8 g Na_2CO_3

37. Consider the balanced equation for the combustion of methane, a component of natural gas:

$$CH_4(g) + 2\,O_2(g) \longrightarrow CO_2(g) + 2\,H_2O(g)$$

Complete the table with the appropriate masses of reactants and products. If the mass of a reactant is provided, fill in the mass of other reactants required to completely react with the given mass, as well as the mass of each product formed. If the mass of a product is provided, fill in the required masses of each reactant to make that amount of product, as well as the mass of the other product that forms.

Mass CH_4	Mass O_2	Mass CO_4	Mass H_2O
____	2.57 g	____	____
22.32 g	____	____	____
____	____	____	11.32 g
____	____	2.94 g	____
3.18 kg	____	____	____
____	____	2.35×10^3 kg	____

38. Consider the balanced equation for the combustion of butane, a fuel often used in lighters:

$$2\,C_4H_{10}(g) + 13\,O_2(g) \longrightarrow 8\,CO_2(g) + 10\,H_2O(g)$$

Complete the table showing the appropriate masses of reactants and products. If the mass of a reactant is provided, fill in the mass of other reactants required to completely react with the given mass, as well as the mass of each product formed. If the mass of a product is provided, fill in the required masses of each reactant to make that amount of product, as well as the mass of the other product that forms.

Mass C_4H_{10}	Mass O_2	Mass CO_2	Mass H_2O
____	1.11 g	____	____
5.22 g	____	____	____
____	____	10.12 g	____
____	____	____	9.04 g
232 mg	____	____	____
____	____	118 mg	____

39. For each acid–base reaction, calculate how many grams of acid are necessary to completely react with and neutralize 2.5 g of the base.

(a) $HCl(aq) + NaOH(aq) \longrightarrow H_2O(l) + NaCl(aq)$

(b) $2\,HNO_3(aq) + Ca(OH)_2(aq) \longrightarrow 2\,H_2O(l) + Ca(NO_3)_2(aq)$

(c) $H_2SO_4(aq) + 2\,KOH(aq) \longrightarrow 2\,H_2O(l) + K_2SO_4(aq)$

40. For each precipitation reaction, calculate how many grams of the first reactant are necessary to completely react with 17.3 g of the second reactant.

(a) $2\,KI(aq) + Pb(NO_3)_2(aq) \longrightarrow PbI_2(s) + 2\,KNO_3(aq)$

(b) $Na_2CO_3(aq) + CuCl_2(aq) \longrightarrow CuCO_3(s) + 2\,NaCl(aq)$

(c) $K_2SO_4(aq) + Sr(NO_3)_2(aq) \longrightarrow SrSO_4(s) + 2\,KNO_3(aq)$

41. Sulfuric acid can dissolve aluminum metal according to the reaction:

$$2\,Al(s) + 3\,H_2SO_4(aq) \longrightarrow Al_2(SO_4)_3(aq) + 3\,H_2(g)$$

Suppose you wanted to dissolve an aluminum block with a mass of 22.5 g. What minimum amount of H_2SO_4 in grams do you need? How many grams of H_2 gas will be produced by the complete reaction of the aluminum block?

42. Hydrochloric acid can dissolve solid iron according to the reaction:

$$Fe(s) + 2\,HCl(aq) \longrightarrow FeCl_2(aq) + H_2(g)$$

What minimum mass of HCl in grams dissolves a 2.8-g iron bar on a padlock? How much H_2 is produced by the complete reaction of the iron bar?

LIMITING REACTANT, THEORETICAL YIELD, AND PERCENT YIELD

43. Consider the generic chemical equation:

$$2\,A + 4\,B \longrightarrow 3\,C$$

What is the limiting reactant when each of the initial quantities of A and B is allowed to react?

(a) 2 mol A; 5 mol B
(b) 1.8 mol A; 4 mol B
(c) 3 mol A; 4 mol B
(d) 22 mol A; 40 mol B

44. Consider the generic chemical equation:

$$A + 3\,B \longrightarrow C$$

What is the limiting reactant when each of the initial quantities of A and B is allowed to react?

(a) 1 mol A; 4 mol B
(b) 2 mol A; 3 mol B
(c) 0.5 mol A; 1.6 mol B
(d) 24 mol A; 75 mol B

45. Determine the theoretical yield of C when each of the initial quantities of A and B is allowed to react in the generic reaction:

$$A + 2\,B \longrightarrow 3\,C$$

(a) 1 mol A; 1 mol B
(b) 2 mol A; 2 mol B
(c) 1 mol A; 3 mol B
(d) 32 mol A; 68 mol B

46. Determine the theoretical yield of C when each of the initial quantities of A and B is allowed to react in the generic reaction:

$$2\,A + 3\,B \longrightarrow 2\,C$$

(a) 2 mol A; 4 mol B
(b) 3 mol A; 3 mol B
(c) 5 mol A; 6 mol B
(d) 4 mol A; 5 mol B

47. For the reaction shown, find the limiting reactant for each of the initial quantities of reactants.

$$2\,K(s) + Cl_2(g) \longrightarrow 2\,KCl(s)$$

(a) 1 mol K; 1 mol Cl_2
(b) 1.8 mol K; 1 mol Cl_2
(c) 2.2 mol K; 1 mol Cl_2
(d) 14.6 mol K; 7.8 mol Cl_2

48. For the reaction shown, find the limiting reactant for each of the initial quantities of reactants.

$$4\,Cr(s) + 3\,O_2(g) \longrightarrow 2\,Cr_2O_3(s)$$

(a) 1 mol Cr; 1 mol O_2
(b) 4 mol Cr; 2.5 mol O_2
(c) 12 mol Cr; 10 mol O_2
(d) 14.8 mol Cr; 10.3 mol O_2

49. For the reaction shown, calculate the theoretical yield of product in moles for each of the initial quantities of reactants.

$$2\,Mn(s) + 3\,O_2(g) \longrightarrow 2\,MnO_3(s)$$

(a) 2 mol Mn; 2 mol O_2
(b) 4.8 mol Mn; 8.5 mol O_2
(c) 0.114 mol Mn; 0.161 mol O_2
(d) 27.5 mol Mn; 43.8 mol O_2

50. For the reaction shown, calculate the theoretical yield of the product in moles for each of the initial quantities of reactants.

$$Ti(s) + 2\,Cl_2(g) \longrightarrow TiCl_4(s)$$

(a) 2 mol Ti; 2 mol Cl_2
(b) 5 mol Ti; 9 mol Cl_2
(c) 0.483 mol Ti; 0.911 mol Cl_2
(d) 12.4 mol Ti; 15.8 mol Cl_2

51. Consider the generic reaction between reactants A and B:

$$3\,A + 4\,B \longrightarrow 2\,C$$

If a reaction vessel initially contains 9 mol A and 8 mol B, how many moles of A, B, and C will be in the reaction vessel after the reactants have reacted as much as possible? (Assume 100% actual yield.)

52. Consider the reaction between reactants S and O_2:

$$2\,S(s) + 3\,O_2(g) \longrightarrow 2\,SO_3(g)$$

If a reaction vessel initially contains 5 mol S and 9 mol O_2, how many moles of S, O_2, and SO_3 will be in the reaction vessel after the reactants have reacted as much as possible? (Assume 100% actual yield.)

53. Consider the reaction:

$$4\,HCl(g) + O_2(g) \longrightarrow 2\,H_2O(g) + 2\,Cl_2(g)$$

Each molecular diagram represents an initial mixture of the reactants. How many molecules of Cl_2 are formed by complete reaction in each case? (Assume 100% actual yield.)

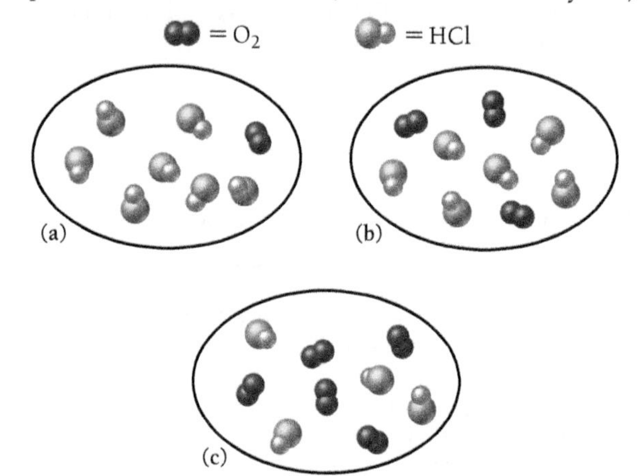

54. Consider the reaction:

$$2\,CH_3OH(g) + 3\,O_2(g) \longrightarrow 2\,CO_2(g) + 4\,H_2O(g)$$

Each molecular diagram represents an initial mixture of the reactants. How many CO_2 molecules are formed by complete reaction in each case? (Assume 100% actual yield.)

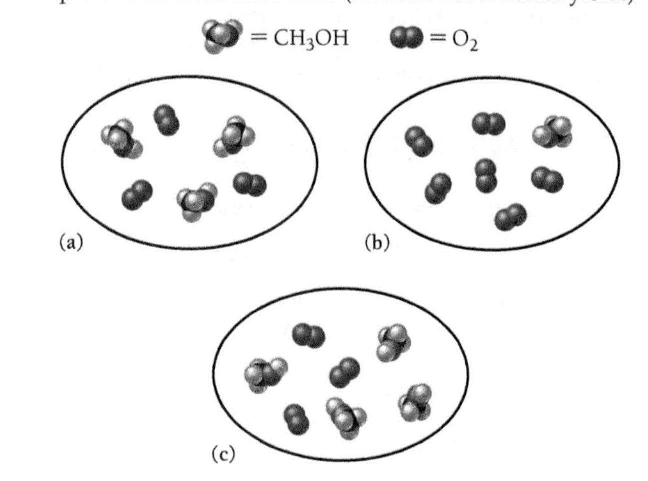

55. For the reaction shown, find the limiting reactant for each of the initial quantities of reactants.

$$2\,Li(s) + F_2(g) \longrightarrow 2\,LiF(s)$$

(a) 1.0 g Li; 1.0 g F_2
(b) 10.5 g Li; 37.2 g F_2
(c) 2.85×10^3 g Li; 6.79×10^3 g F_2

56. For the reaction shown, find the limiting reactant for each of the initial quantities of reactants.

$$4\,Al(s) + 3\,O_2(g) \longrightarrow 2\,Al_2O_3(s)$$

(a) 1.0 g Al; 1.0 g O_2
(b) 2.2 g Al; 1.8 g O_2
(c) 0.353 g Al; 0.482 g O_2

57. For the reaction shown, calculate the theoretical yield of the product in grams for each of the initial quantities of reactants.

$$2\ Al(s) + 3\ Cl_2(g) \longrightarrow 2\ AlCl_3(s)$$

(a) 1.0 g Al; 1.0 g Cl_2
(b) 5.5 g Al; 19.8 g Cl_2
(c) 0.439 g Al; 2.29 g Cl_2

58. For the reaction shown, calculate the theoretical yield of the product in grams for each of the initial quantities of reactants.

$$Ti(s) + 2\ F_2(g) \longrightarrow TiF_4(s)$$

(a) 1.0 g Ti; 1.0 g F_2
(b) 4.8 g Ti; 3.2 g F_2
(c) 0.388 g Ti; 0.341 g F_2

59. If the theoretical yield of a reaction is 24.8 g and the actual yield is 18.5 g, what is the percent yield?

60. If the theoretical yield of a reaction is 0.118 g and the actual yield is 0.104 g, what is the percent yield?

61. Consider the reaction between calcium oxide and carbon dioxide:

$$CaO(s) + CO_2(g) \longrightarrow CaCO_3(s)$$

A chemist allows 14.4 g of CaO and 13.8 g of CO_2 to react. When the reaction is finished, the chemist collects 19.4 g of $CaCO_3$. Determine the limiting reactant, theoretical yield, and percent yield for the reaction.

62. Consider the reaction between sulfur trioxide and water:

$$SO_3(g) + H_2O(l) \longrightarrow H_2SO_4(aq)$$

A chemist allows 61.5 g of SO_3 and 11.2 g of H_2O to react. When the reaction is finished, the chemist collects 54.9 g of H_2SO_4. Determine the limiting reactant, theoretical yield, and percent yield for the reaction.

63. Consider the reaction between NiS_2 and O_2:

$$2\ NiS_2(s) + 5\ O_2(g) \longrightarrow 2\ NiO(s) + 4\ SO_2(g)$$

When 11.2 g of NiS_2 react with 5.43 g of O_2, 4.86 g of NiO are obtained. Determine the limiting reactant, theoretical yield of NiO, and percent yield for the reaction.

64. Consider the reaction between HCl and O_2:

$$4\ HCl(g) + O_2(g) \longrightarrow 2\ H_2O(l) + 2\ Cl_2(g)$$

When 63.1 g of HCl react with 17.2 g of O_2, 49.3 g of Cl_2 are collected. Determine the limiting reactant, theoretical yield of Cl_2, and percent yield for the reaction.

65. Lead ions can be precipitated from solution with NaCl according to the reaction:

$$Pb^{2+}(aq) + 2\ NaCl(aq) \longrightarrow PbCl_2(s) + 2\ Na^+(aq)$$

When 135.8 g of NaCl are added to a solution containing 195.7 g of Pb^{2+}, a $PbCl_2$ precipitate forms. The precipitate is filtered and dried and found to have a mass of 252.4 g. Determine the limiting reactant, theoretical yield of $PbCl_2$, and percent yield for the reaction.

66. Magnesium oxide can be produced by heating magnesium metal in the presence of oxygen. The balanced equation for the reaction is:

$$2\ Mg(s) + O_2(g) \longrightarrow 2\ MgO(s)$$

When 10.1 g of Mg react with 10.5 g of O_2, 11.9 g of MgO are collected. Determine the limiting reactant, theoretical yield, and percent yield for the reaction.

67. Consider the reaction between TiO_2 and C:

$$TiO_2(s) + 2\ C(s) \longrightarrow Ti(s) + 2\ CO(g)$$

A reaction vessel initially contains 10.0 g of each of the reactants. Calculate the masses of TiO_2, C, Ti, and CO that will be in the reaction vessel after the reactants have reacted as much as possible. Assume 100% yield. *Hint*: The limiting reactant is completely consumed, but the reactant in excess is not. Use the amount of limiting reactant to determine the amount of products that form and the amount of the reactant in excess that remains after complete reaction.

68. Consider the reaction between N_2H_4 and N_2O_4:

$$2\ N_2H_4(g) + N_2O_4(g) \longrightarrow 3\ N_2(g) + 4\ H_2O(g)$$

A reaction vessel initially contains 27.5 g N_2H_4 and 74.9 g of N_2O_4. Calculate the masses of N_2H_4, N_2O_4, N_2, and H_2O that will be in the reaction vessel after the reactants have reacted as much as possible. Assume 100% yield. *Hint*: The limiting reactant is completely consumed, but the reactant in excess is not. Use the amount of limiting reactant to determine the amount of products that form and the amount of the reactant in excess that remains after complete reaction.

ENTHALPY AND STOICHIOMETRY OF ΔH_{rxn}

69. Classify each process as exothermic or endothermic and indicate the sign of ΔH_{rxn}.
(a) butane gas burning in a lighter
(b) the reaction that occurs in the chemical cold packs used to ice athletic injuries
(c) the burning of wax in a candle

70. Classify each process as exothermic or endothermic and indicate the sign of ΔH_{rxn}.
(a) ice melting
(b) a sparkler burning
(c) acetone evaporating from skin

71. Consider the generic reaction:

$$A + 2\,B \longrightarrow C \quad \Delta H_{rxn} = -55\ kJ$$

Determine the amount of heat emitted when each amount of reactant completely reacts (assume that there is more than enough of the other reactant).
(a) 1 mol A
(b) 2 mol A
(c) 1 mol B
(d) 2 mol B

72. Consider the generic reaction:

$$2\,A + 3\,B \longrightarrow C \quad \Delta H_{rxn} = -125\ kJ$$

Determine the amount of heat emitted when each amount of reactant completely reacts (assume that there is more than enough of the other reactant).
(a) 2 mol A
(b) 3 mol A
(c) 3 mol B
(d) 5 mol B

73. Consider the equation for the combustion of acetone (C_3H_6O), the main ingredient in nail polish remover:

$$C_3H_6O(l) + 4\,O_2(g) \longrightarrow 3\,CO_2(g) + 3\,H_2O(g) \quad \Delta H_{rxn} = -1790\ kJ$$

If a bottle of nail polish remover contains 155 g of acetone, how much heat is released by its complete combustion?

74. The equation for the combustion of CH_4 (the main component of natural gas) is shown below. How much heat is produced by the complete combustion of 237 g of CH_4?

$$CH_4(g) + 2\,O_2(g) \longrightarrow CO_2(g) + 2\,H_2O(g) \quad \Delta H_{rxn} = -802.3\ kJ$$

75. Octane (C_8H_{18}) is a component of gasoline that burns according to the equation:

$$C_8H_{18}(l) + \tfrac{25}{2}\,O_2(g) \longrightarrow 8\,CO_2(g) + 9\,H_2O(g) \quad \Delta H_{rxn} = -5074.1\ kJ$$

What mass of octane (in g) is required to produce 1.55×10^3 kJ of heat?

76. The evaporation of water is endothermic:

$$H_2O(l) \longrightarrow H_2O(g) \quad \Delta H_{rxn} = +44.01\ kJ$$

What minimum mass of water (in g) has to evaporate to absorb 175 kJ of heat?

Cumulative Problems

77. Consider the reaction:

$$2\,N_2(g) + 5\,O_2(g) + 2\,H_2O(g) \longrightarrow 4\,HNO_3(g)$$

If a reaction mixture contains 28 g of N_2, 150 g of O_2, and 36 g of H_2O, what is the limiting reactant? (Try to do this problem in your head without any written calculations.)

78. Consider the reaction:

$$2\,CO(g) + O_2(g) \longrightarrow 2\,CO_2(g)$$

If a reaction mixture contains 28 g of CO and 32 g of O_2, what is the limiting reactant? (Try to do this problem in your head without any written calculations.)

79. A solution contains an unknown mass of dissolved barium ions. When sodium sulfate is added to the solution, a white precipitate forms. The precipitate is filtered and dried and found to have a mass of 258 mg. What mass of barium was in the original solution? (Assume that all of the barium was precipitated out of solution by the reaction.)

80. A solution contains an unknown mass of dissolved silver ions. When potassium chloride is added to the solution, a white precipitate forms. The precipitate is filtered and dried and found to have a mass of 212 mg. What mass of silver was in the original solution? (Assume that all of the silver was precipitated out of solution by the reaction.)

81. Sodium bicarbonate is often used as an antacid to neutralize excess hydrochloric acid in an upset stomach. How much hydrochloric acid in grams can be neutralized by 3.5 g of sodium bicarbonate? (*Hint*: Begin by writing a balanced equation for the reaction between aqueous sodium bicarbonate and aqueous hydrochloric acid.)

82. Toilet bowl cleaners often contain hydrochloric acid to dissolve the calcium carbonate deposits that accumulate within a toilet bowl. How much calcium carbonate in grams can be dissolved by 5.8 g of HCl? (*Hint*: Begin by writing a balanced equation for the reaction between hydrochloric acid and calcium carbonate.)

83. The combustion of gasoline produces carbon dioxide and water. Assume gasoline to be pure octane (C_8H_{18}) and calculate how many kilograms of carbon dioxide are added to the atmosphere per 1.0 kg of octane burned. (*Hint*: Begin by writing a balanced equation for the combustion reaction.)

84. Many home barbecues are fueled with propane gas (C_3H_8). How much carbon dioxide in kilograms is produced upon the complete combustion of 18.9 L of propane (approximate contents of one 5-gal tank)? Assume that the density of the liquid propane in the tank is 0.621 g/mL. (*Hint*: Begin by writing a balanced equation for the combustion reaction.)

85. A hard-water solution contains 4.8 g of calcium chloride. How much sodium phosphate in grams should be added to the solution to completely precipitate all of the calcium?

86. Magnesium ions can be precipitated from seawater by the addition of sodium hydroxide. How much sodium hydroxide in grams must be added to a sample of seawater to completely precipitate the 88.4 mg of magnesium present?

87. Hydrogen gas can be prepared in the laboratory by a single-displacement reaction in which solid zinc reacts with hydrochloric acid. How much zinc in grams is required to make 14.5 g of hydrogen gas through this reaction?

88. Sodium peroxide (Na_2O_2) reacts with water to form sodium hydroxide and oxygen gas. Write a balanced equation for the reaction and determine how much oxygen in grams is formed by the complete reaction of 35.23 g of Na_2O_2.

89. Ammonium nitrate reacts explosively upon heating to form nitrogen gas, oxygen gas, and gaseous water. Write a balanced equation for this reaction and determine how much oxygen in grams is produced by the complete reaction of 1.00 kg of ammonium nitrate.

90. Pure oxygen gas can be prepared in the laboratory by the decomposition of solid potassium chlorate to form solid potassium chloride and oxygen gas. How much oxygen gas in grams can be prepared from 45.8 g of potassium chlorate?

91. Aspirin can be made in the laboratory by reacting acetic anhydride ($C_4H_6O_3$) with salicylic acid ($C_7H_6O_3$) to form aspirin ($C_9H_8O_4$) and acetic acid ($C_2H_4O_2$). The balanced equation is:

$$C_4H_6O_3 + C_7H_6O_3 \longrightarrow C_9H_8O_4 + C_2H_4O_2$$

In a laboratory synthesis, a student begins with 5.00 mL of acetic anhydride (density = 1.08 g/mL) and 2.08 g of salicylic acid. Once the reaction is complete, the student collects 2.01 g of aspirin. Determine the limiting reactant, theoretical yield of aspirin, and percent yield for the reaction.

92. The combustion of liquid ethanol (C_2H_5OH) produces carbon dioxide and water. After 3.8 mL of ethanol (density = 0.789 g/mL) is allowed to burn in the presence of 12.5 g of oxygen gas, 3.10 mL of water (density = 1.00 g/mL) is collected. Determine the limiting reactant, theoretical yield of H_2O, and percent yield for the reaction. (*Hint*: Write a balanced equation for the combustion of ethanol.)

93. Urea (CH_4N_2O), a common fertilizer, can be synthesized by the reaction of ammonia (NH_3) with carbon dioxide:

$$2\,NH_3(aq) + CO_2(aq) \longrightarrow CH_4N_2O(aq) + H_2O(l)$$

An industrial synthesis of urea produces 87.5 kg of urea upon reaction of 68.2 kg of ammonia with 105 kg of carbon dioxide. Determine the limiting reactant, theoretical yield of urea, and percent yield for the reaction.

94. Silicon, which occurs in nature as SiO_2, is the material from which most computer chips are made. If SiO_2 is heated until it melts into a liquid, it reacts with solid carbon to form liquid silicon and carbon monoxide gas. In an industrial preparation of silicon, 52.8 kg of SiO_2 reacts with 25.8 kg of carbon to produce 22.4 kg of silicon. Determine the limiting reactant, theoretical yield, and percent yield for the reaction.

95. The ingestion of lead from food, water, or other environmental sources can cause lead poisoning, a serious condition that affects the central nervous system, causing symptoms such as distractibility, lethargy, and loss of motor function. Lead poisoning is treated with chelating agents, substances that bind to lead and allow it to be eliminated in the urine. A modern chelating agent used for this purpose is succimer ($C_4H_6O_4S_2$). Suppose you are trying to determine the appropriate dose for succimer treatment of lead poisoning. Assume that a patient's blood lead levels are 0.550 mg/L, that total blood volume is 5.0 L, and that 1 mol of succimer binds 1 mol of lead. What minimum mass of succimer in milligrams is needed to bind all of the lead in this patient's bloodstream?

96. An emergency breathing apparatus placed in mines or caves works via the chemical reaction:

$$4\,KO_2(s) + 2\,CO_2(g) \longrightarrow 2\,K_2CO_3(s) + 3\,O_2(g)$$

If the oxygen supply becomes limited or if the air becomes poisoned, a worker can use the apparatus to breathe while exiting the mine. Notice that the reaction produces O_2, which can be breathed, and absorbs CO_2, a product of respiration. What minimum amount of KO_2 is required for the apparatus to produce enough oxygen to allow the user 15 minutes to exit the mine in an emergency? Assume that an adult consumes approximately 4.4 g of oxygen in 15 minutes of normal breathing.

97. The propane fuel (C_3H_8) used in gas barbecues burns according to the equation:

$$C_3H_8(g) + 5\,O_2(g) \longrightarrow 3\,CO_2(g) + 4\,H_2O(g) \qquad \Delta H_{rxn} = -2044 \text{ kJ}$$

If a pork roast must absorb 1.6×10^3 kJ to fully cook, and if only 10% of the heat produced by the barbecue is actually absorbed by the roast, what mass of CO_2 is emitted into the atmosphere during the grilling of the pork roast?

98. Charcoal is primarily carbon. Determine the mass of CO_2 produced by burning enough carbon to produce 5.00×10^2 kJ of heat.

$$C(s) + O_2(g) \longrightarrow CO_2(g) \qquad \Delta H_{rxn} = -393.5 \text{ kJ}$$

Highlight Problems

99. A loud classroom demonstration involves igniting a hydrogen-filled balloon. The hydrogen within the balloon reacts explosively with oxygen in the air to form water according to this reaction:

$$2\,H_2(g) + O_2(g) \longrightarrow 2\,H_2O(g)$$

If the balloon is filled with a mixture of hydrogen and oxygen, the explosion is even louder than if the balloon is filled with only hydrogen; the intensity of the explosion depends on the relative amounts of oxygen and hydrogen within the balloon. Consider the molecular views representing different amounts of hydrogen and oxygen in four different balloons. Based on the balanced chemical equation, which balloon will make the loudest explosion?

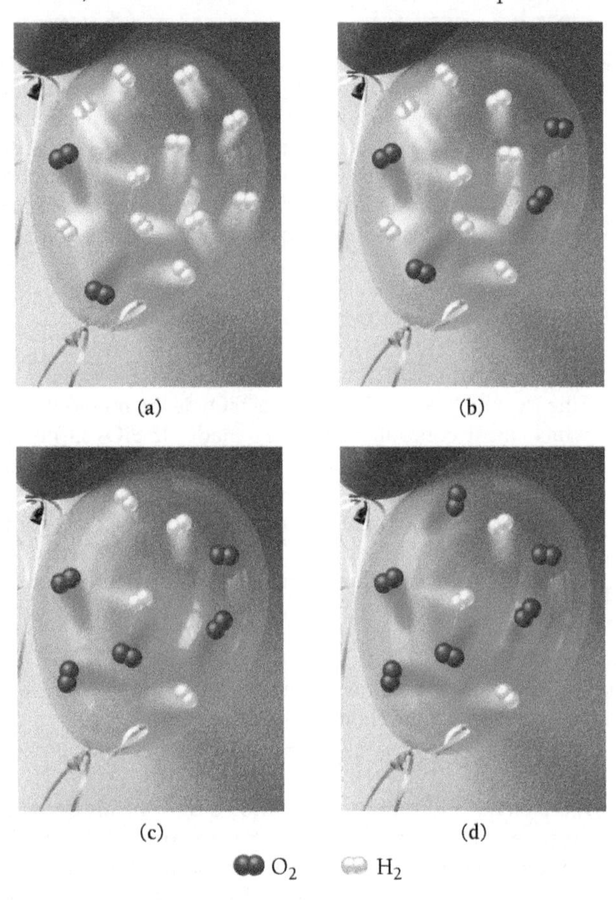

100. A hydrochloric acid solution will neutralize a sodium hydroxide solution. Consider the molecular views showing one beaker of HCl and four beakers of NaOH. Which NaOH beaker will just neutralize the HCl beaker? Begin by writing a balanced chemical equation for the neutralization reaction.

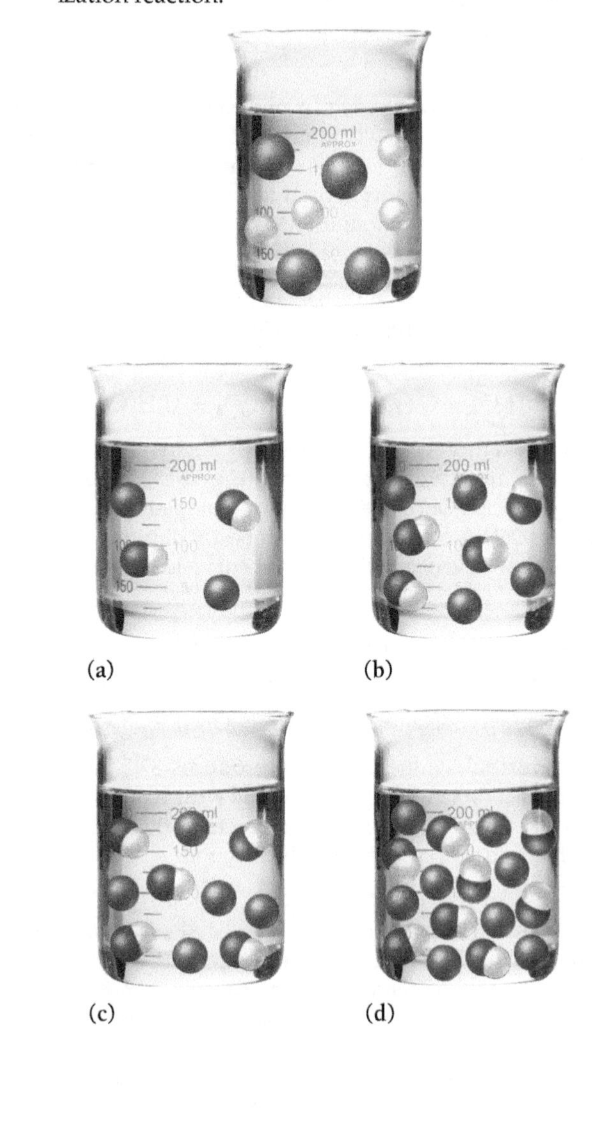

101. Scientists have grown progressively more worried about the potential for climate change caused by increasing atmospheric carbon dioxide levels. The world burns the fossil fuel equivalent of approximately 9.0×10^{12} kg of petroleum per year. Assume that all of this fossil fuel is in the form of octane (C_8H_{18}) and calculate how much CO_2 in kilograms is produced by world fossil fuel combustion per year. (*Hint*: Begin by writing a balanced equation for the combustion of octane.) If the atmosphere currently contains approximately 3.0×10^{15} kg of CO_2, how long will it take for the world's fossil fuel combustion to double the amount of atmospheric carbon dioxide?

102. Lakes that have been acidified by acid rain can be neutralized by the addition of limestone ($CaCO_3$). How much limestone in kilograms would be required to completely neutralize a 5.2×10^9-L lake containing 5.0×10^{-3} g of H_2SO_4 per liter?

Questions for Group Work

Discuss these questions with the group and record your consensus answer.

103. What volume of air is needed to burn an entire 55-L (approximately 15-gal) tank of gasoline? Assume that the gasoline is pure octane, C_8H_{18}. *Hint*: Air is 20% oxygen, 1 mol of a gas occupies about 25 L at room temperature, and the density of octane is 0.70 g/cm^3.

104. Have each member of your group choose a precipitation reaction from Chapter 7 and write a limiting reagent problem based on it. Provide the masses of reactants and the product. Trade problems within your group and solve them by determining the limiting reagent, the theoretical yield, and the percent yield.

105. Consider the combustion of propane:

$$C_3H_8(g) + O_2(g) \longrightarrow CO_2(g) + H_2O(g)$$

(a) Balance the reaction.
(b) Divide all coefficients by the coefficient on propane, so that you have the reaction for the combustion of 1 mol of propane.
(c) ΔH_{rxn} for the combustion of one mole of propane is −2219 kJ. What mass of propane would you need to burn to generate 5.0 MJ of heat?
(d) If propane costs about $0.67/L and has a density of 2.01 g/cm^3, how much would it cost to generate 5.0 MJ of heat by burning propane?

Data Interpretation and Analysis

106. A chemical reaction in which reactants A and B form the product C is studied in the laboratory. The researcher carries out the reaction with differing relative amounts of reactants and measures the amount of product produced. Examine the given tabulated data from the experiment and answer the questions.

(a) For which experiments is A the limiting reactant?
(b) For which experiments is B the limiting reactant?
(c) The molar mass of A is 50.0 g/mol, and the molar mass of B is 75.0 g/mol. What are the coefficients of A and B in the balanced chemical equation?
(d) The molar mass of C is 88.0 g/mol. What is the coefficient of C in the balanced chemical equation?
(e) Calculate an average percent yield for the reaction.

Experiment #	Mass A (g)	Mass B (g)	Mass C Obtained (g)
1	2.51	7.54	3.76
2	5.03	7.51	7.43
3	7.55	7.52	11.13
4	12.53	7.49	14.84
5	15.04	7.47	14.94
6	19.98	7.51	15.17
7	20.04	9.95	19.31
8	20.02	12.55	24.69

Answers to Skillbuilder Exercises

Skillbuilder 8.1.......... 49.2 mol H_2O
Skillbuilder 8.2.......... 6.88 g HCl
Skillbuilder 8.3.......... 4.0×10^3 kg H_2SO_4
Skillbuilder 8.4.......... Limiting reactant is Na; theoretical yield is 4.8 mol of NaF
Skillbuilder 8.5.......... 30.7 g NH_3
Skillbuilder Plus, p. 263... 29.4 kg NH_3
Skillbuilder 8.6.......... Limiting reactant is CO; theoretical yield = 127 g Fe; percent yield = 68.8%
Skillbuilder 8.7.......... -2.06×10^3 kJ
Skillbuilder Plus, p. 267... 33 g C_4H_{10} necessary; 99 g CO_2 produced

Answers to Conceptual Checkpoints

8.1 (a) Because the reaction requires two O_2 molecules to react with 1 CH_4 molecule, and there are four O_2 molecules available to react, two CH_4 molecules are required for complete reaction.

8.2 (c) One mole of A produces three moles of C. If the molar mass of C is three times the molar mass of A, then the mass of C produced is six times the mass of A that reacts.

8.3 (c) Hydrogen is the limiting reactant. The reaction mixture contains three H_2 molecules; therefore, two NH_3 molecules will form when the reactants have reacted as completely as possible. Nitrogen is in excess, and there is one leftover nitrogen molecule.

8.4 B is the limiting reactant because since B has a higher molar mass, a reaction mixture containing equal masses of both A and B has fewer moles of B. Since two moles of B are required to react with one mole of A and since fewer moles of B are present, B is the limiting reactant.

8.5 (d) NO is the limiting reactant. The reaction mixture initially contains 4 mol NO; therefore, 10 mol of H_2O will be consumed, leaving 5 mol H_2 unreacted. The products will be 4 mol NH_3 and 4 mol H_2O.

8.6 (c) B is the limiting reactant. If 4 mol B react, then 200 kJ of heat is produced.

TRO
Introductory Chemistry

9 Electrons in Atoms and the Periodic Table

Anyone who is not shocked by quantum mechanics has not understood it.

—Niels Bohr (1885–1962)

CHAPTER OUTLINE

9.1 Blimps, Balloons, and Models of the Atom

▲ The *Hindenburg* was filled with hydrogen, a reactive and flammable gas. **Question: What makes hydrogen reactive?**

You have probably seen one of the Goodyear blimps floating in the sky. A Goodyear blimp is often present at championship sporting events such as the Rose Bowl, the Indy 500, or the U.S. Open golf tournament. The blimp's inherent stability allows cameras to provide spectacular views of the world below for television and film.

The Goodyear blimp is similar to a large balloon. Unlike airplanes, which must be moving fast to stay in flight, a blimp or *airship* floats in air because it is filled with a gas that is less dense than air. The Goodyear blimp is filled with helium. Other airships in history, however, have used hydrogen for buoyancy. For example, the *Hindenburg*—the largest airship ever constructed—was filled with hydrogen. Hydrogen, a reactive and flammable gas, turned out to be a poor choice. On May 6, 1937, while landing in New Jersey after its first transatlantic crossing, the *Hindenburg* burst into flames. The fire destroyed the airship, killing 36 of its 97 passengers. Apparently, as the *Hindenburg* was landing, leaking hydrogen gas ignited, resulting in the explosion that destroyed the ship. A similar accident cannot happen to the Goodyear blimp because it uses helium—an inert gas—for buoyancy. A spark or even a flame would actually be *extinguished* by helium.

Why is helium inert? What is it about helium *atoms* that makes helium *gas* inert? By contrast, why is hydrogen so reactive? Recall from Chapter 5 that elemental hydrogen exists as a diatomic element. Hydrogen atoms are so reactive that they react with each other to form hydrogen molecules. What is it about hydrogen atoms that makes them so reactive? What is the difference between hydrogen and helium that accounts for their different reactivities?

◀ Modern blimps are filled with helium, an inert gas. The nucleus of the helium atom (inset) has two protons, so the neutral helium atom has two electrons—a highly stable configuration. In this chapter, we learn about models that explain the inertness of helium and the reactivity of other elements.

The periodic law stated here is a modification of Mendeleev's original formulation. Mendeleev listed elements in order of increasing *mass*; today we list them in order of increasing *atomic number*.

When we examine the properties of hydrogen and helium, we make observations about nature. Mendeleev's periodic law, first discussed in Chapter 4, summarizes the results of many similar observations on the properties of elements:

> When the elements are arranged in order of increasing atomic number, certain sets of properties recur periodically.

The reactivity exhibited by hydrogen (the first element in Group 1A) is also seen in other Group 1A elements, such as lithium and sodium. Likewise, the inertness of helium (a Group 8A element) is seen in other Group 8A elements such as neon and argon and the other noble gases. In this chapter, we consider models that help explain the observed behaviors of groups of elements such as the Group 1A metals and the noble gases. We examine two important models in particular—the **Bohr model** and the **quantum-mechanical model**—that propose explanations for the periodic law. These models explain how electrons exist in atoms and how those electrons affect the chemical and physical properties of elements.

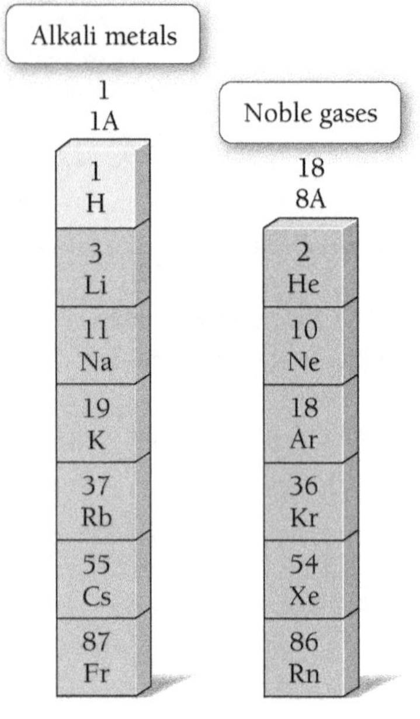

▲ The noble gases are chemically inert, and the alkali metals are chemically reactive. Why? (Hydrogen is a Group 1 element, but it is not considered an alkali metal.)

We have already learned much about the behavior of elements in this book. We know, for example, that sodium tends to form 1+ ions and that fluorine tends to form 1− ions. We know that some elements are metals and that others are nonmetals. And we know that the noble gases are, in general, chemically inert and that the alkali metals are chemically reactive. But we have not yet explored *why*. The models in this chapter explain why.

When the Bohr model and the quantum-mechanical model were developed in the early 1900s, they caused a revolution in the physical sciences, changing our fundamental view of matter at its most basic level. The scientists who devised these models—including Niels Bohr, Erwin Schrödinger, and Albert Einstein—were bewildered by their own discoveries. Bohr claimed, "Anyone who is not shocked by quantum mechanics has not understood it." Schrödinger lamented, "I don't like it, and I am sorry I ever had anything to do with it." Einstein disbelieved it, insisting that "God does not play dice with the universe." However, the quantum-mechanical model has such explanatory power that it is rarely questioned today. It forms the basis of the modern periodic table and our understanding of chemical bonding. Its applications include lasers, computers, and semiconductor devices, and it has led us to discover new ways to design drugs that cure disease. The quantum-mechanical model for the atom is, in many ways, the foundation of modern chemistry.

▲ Niels Bohr (left) and Erwin Schrödinger (right), along with Albert Einstein, played a role in the development of quantum mechanics, yet they were bewildered by their own findings.

9.2 Light: Electromagnetic Radiation

▶ Understand and explain the nature of electromagnetic radiation.

Before we explore models of the atom, we must understand a few things about light, because observations of the interaction of light with atoms helped to shape these models. Light is familiar to all of us—we see the world by it—but what is light? Unlike most of what we have encountered so far in this book, light is not matter—it has no mass. Light is a form of **electromagnetic radiation**, a

▲ When a water surface is disturbed, waves are created that radiate outward from the site.

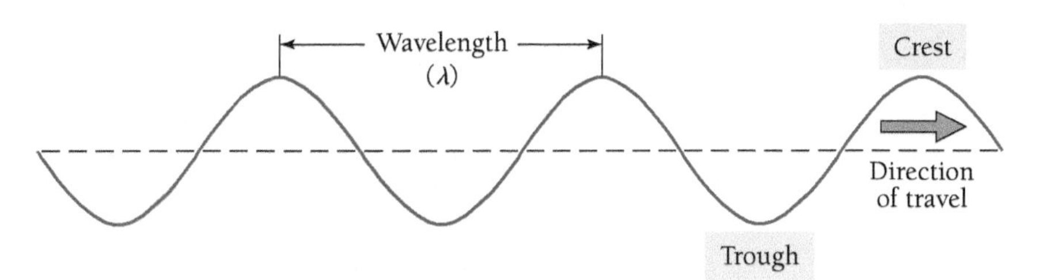

▲ FIGURE 9.1 **Wavelength** The wavelength of light (λ) is the distance between adjacent wave crests.

type of energy that travels through space at a constant speed of 3.0×10^8 m/s (186,000 mi/s). At this speed, a flash of light generated at the equator can travel around the world in one-seventh of a second. This extremely fast speed is part of the reason that we *see* a firework in the sky before we *hear* the sound of its explosion. The light from the exploding firework reaches our eyes almost instantaneously. The sound, traveling much more slowly, takes longer.

Before the advent of quantum mechanics, light was described exclusively as a wave of electromagnetic energy traveling through space. You are probably familiar with water waves (think of the waves created by a rock dropped into a still pond), or you may have created a wave on a rope by moving the end of the rope up and down in a quick motion. In either case, the wave carries energy as it moves through the water or along the rope.

The Greek letter *lambda* (λ) is pronounced "lam-duh."

Waves are generally characterized by **wavelength** (λ), the distance between adjacent wave crests (▲ FIGURE 9.1). For visible light, wavelength determines color. For example, orange light has a longer wavelength than blue light. White light, as produced by the sun or by a light bulb, contains a spectrum of wavelengths and therefore a spectrum of color. We see these colors—red, orange, yellow, green, blue, indigo, and violet—in a rainbow or when white light is passed through a prism (▼ FIGURE 9.2). Red light, with a wavelength of 750 nm (nanometers), has the longest wavelength of visible light. Violet light, with a wavelength of 400 nm, has the shortest ($1\text{ nm} = 10^{-9}\text{ m}$). The presence of color in white light is responsible for the colors we see in our everyday vision. For example, a red shirt is red because it reflects red light (▼ FIGURE 9.3). Our eyes see only the reflected light, making the shirt appear red.

Helpful mnemonic: ROY G BIV—Red, Orange, Yellow, Green, Blue, Indigo, Violet

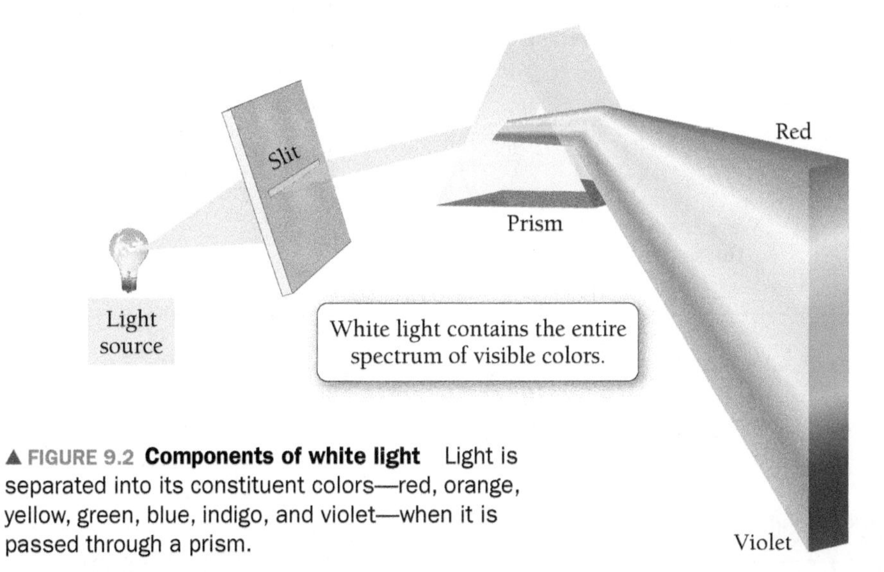

▲ FIGURE 9.2 **Components of white light** Light is separated into its constituent colors—red, orange, yellow, green, blue, indigo, and violet—when it is passed through a prism.

▲ FIGURE 9.3 **Color in objects** A red shirt appears red because it absorbs all colors except red, which it reflects.

Light waves are also often characterized by **frequency** (ν), the number of cycles or crests that pass through a stationary point in one second. Wavelength and frequency are inversely related—the shorter the wavelength, the higher the frequency. Blue light, for example, has a higher frequency than red light.

The Greek letter *nu* (ν) is pronounced "noo."

In the early twentieth century, scientists such as Albert Einstein discovered that the results of certain experiments could be explained only by describing light, not as waves, but as particles. In this description, the light leaving a flashlight, for example, is viewed as a stream of particles. A particle of light is called a **photon**, and we can think of a photon as a single packet of light energy. The amount of energy carried in the packet depends on the wavelength of the light—the shorter the wavelength, the greater the energy. Therefore, violet light (shorter wavelength) carries more energy per photon than red light (longer wavelength).

To summarize:

- Electromagnetic radiation is a form of energy that travels through space at a constant speed of 3.0×10^8 m/s and can exhibit wavelike or particle-like properties.
- The wavelength of electromagnetic radiation determines the amount of energy carried by one of its photons. The shorter the wavelength, the greater the energy of each photon.
- The frequency and energy of electromagnetic radiation are inversely related to its wavelength.

CONCEPTUAL CHECKPOINT 9.1

Which wavelength of light has the highest frequency?

(a) 350 nm **(b)** 500 nm **(c)** 750 nm

9.3 The Electromagnetic Spectrum

▶ Predict the relative wavelength, energy, and frequency of different types of light.

Electromagnetic radiation ranges in wavelength from 10^{-16} m (gamma rays) to 10^6 m (radio waves). Visible light composes only a tiny portion of that range. The entire range of electromagnetic radiation is called the **electromagnetic spectrum**. ▼ FIGURE 9.4 shows the electromagnetic spectrum, with short-wavelength, high-frequency radiation on the right and long-wavelength, low-frequency radiation on the left. Visible light is the small sliver in the middle.

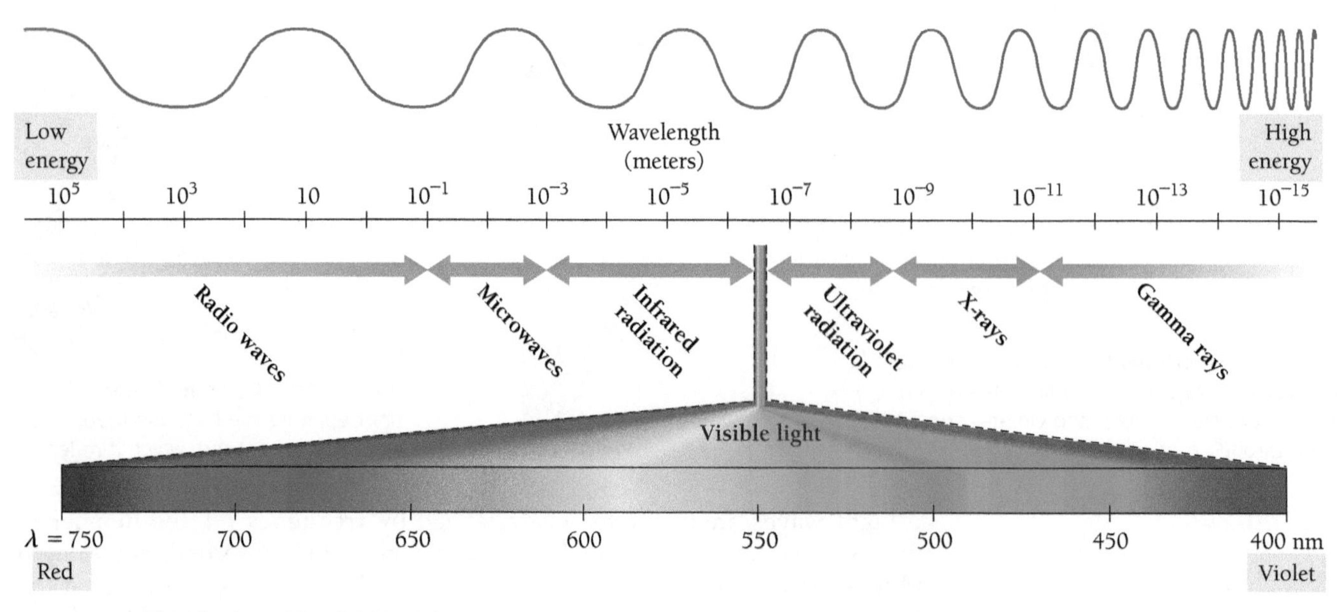

▲ FIGURE 9.4 **The electromagnetic spectrum**

Remember that the energy carried per photon is greater for short-wavelength light than for long-wavelength light. The shortest wavelength (and therefore most energetic) photons are those of **gamma rays**, shown on the far right of Figure 9.4. Gamma rays are produced by the sun, by stars, and by certain unstable atomic nuclei on Earth. Excessive human exposure to gamma rays is dangerous because the high energy of gamma-ray photons can damage biological molecules.

Next on the electromagnetic spectrum (to the left in Figure 9.4), with longer wavelengths (and lower energy) than gamma rays, are **X-rays**, familiar to us from their medical use. X-rays pass through many substances that block visible light and are therefore used to image internal bones and organs. Like gamma-ray photons, X-ray photons carry enough energy to damage biological molecules. While several yearly exposures to X-rays are relatively harmless, excessive exposure to X-rays increases cancer risk.

Sandwiched between X-rays and visible light in the electromagnetic spectrum is **ultraviolet** or **UV light**, most familiar to us as the component of sunlight that produces a sunburn or suntan. Though not as energetic as gamma-ray or X-ray photons, ultraviolet photons still carry enough energy to damage biological molecules. Excessive exposure to ultraviolet light increases the risk of skin cancer and cataracts and causes premature wrinkling of the skin. Next on the spectrum is **visible light**, ranging from violet (shorter wavelength, higher energy) to red (longer wavelength, lower energy). Photons of visible light do not damage biological molecules. They do, however, cause molecules in our eyes to rearrange, which sends a signal to our brains that results in vision.

Infrared light is next, with even longer wavelengths than visible light. The heat we feel when we place a hand near a hot object is infrared light. All warm objects, including human bodies, emit infrared light. While infrared light is invisible to our eyes, infrared sensors can detect it and are often used in night-vision technology to "see" in the dark. In the infrared region of the spectrum, warm objects—such as human bodies—glow, much as a light bulb glows in the visible region of the spectrum.

Beyond infrared light, at longer wavelengths still, are **microwaves**, used for radar and in microwave ovens. Although microwave light has longer wavelengths—and therefore lower energy per photon—than visible or infrared light, it is efficiently absorbed by water and therefore heats substances that contain water. For this reason substances that contain water, such as food, are warmed when placed in a microwave oven, but substances that do not contain water, such as plate and cups, are not.

Some dishes contain substances that absorb microwave radiation, but most do not.

The longest wavelengths of light are **radio waves**, which are used to transmit the signals used by AM and FM radio, cellular telephones, television, and other forms of communication.

Normal photograph

Infrared photograph

▲ Warm objects, such as human or animal bodies, give off infrared light that is easily detected with an infrared camera. In the infrared photograph, the warmest areas appear as red and the coolest as dark blue.

CHEMISTRY AND HEALTH

Radiation Treatment for Cancer

X-rays and gamma rays are sometimes called ionizing radiation because the high energy in their photons can ionize atoms and molecules. When ionizing radiation interacts with biological molecules, it can permanently change or even destroy them. Consequently we normally try to limit our exposure to ionizing radiation. However, doctors can use ionizing radiation to destroy molecules within unwanted cells such as cancer cells.

In radiation therapy (or radiotherapy), medical professionals aim X-ray or gamma-ray beams at cancerous tumors. The ionizing radiation damages the molecules within the tumor's cells that carry genetic information—information necessary for the cell to grow and divide—and the cells die or stop dividing. Ionizing radiation also damages molecules within healthy cells; however, cancerous cells divide more quickly than healthy cells, making cancerous cells more susceptible to genetic damage. Nonetheless, healthy cells often inadvertently sustain damage during treatments, resulting in side effects for patients such as fatigue, skin lesions, and hair loss. Doctors try to minimize the exposure of healthy cells by appropriate shielding and by targeting the tumor from multiple directions, minimizing the exposure of healthy cells while maximizing the exposure of cancerous cells (▶ FIGURE 9.5).

Another side effect of exposing healthy cells to radiation is that they too may become cancerous. In this way, a treatment for cancer may cause cancer. So why do we continue to use it? Radiation therapy, as most other disease therapies, has associated risks. However, we take risks all the time, many for lesser reasons. For example, every time we drive a car, we risk injury or even death. Why? Because we perceive the benefit—such as getting to the grocery store to buy food—to be worth the risk. The situation is similar in cancer therapy or any other therapy for that matter. The benefit of cancer therapy (possibly curing a cancer that will certainly kill) is worth the risk (a slight increase in the chance of developing a future cancer).

B9.1 CAN YOU ANSWER THIS? *Why is visible light not used to destroy cancerous tumors?*

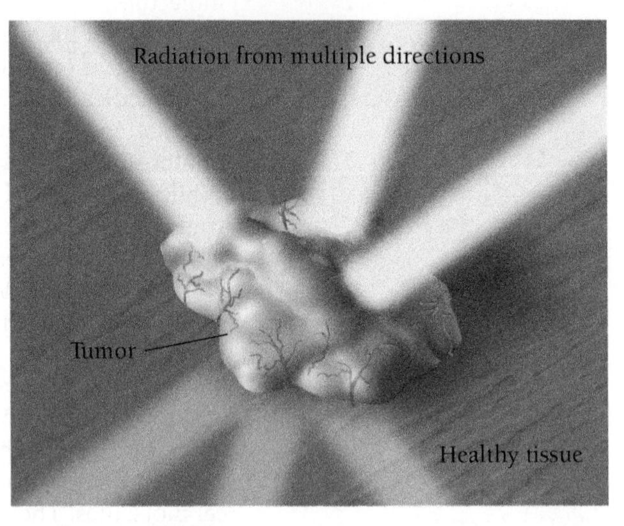

▲ FIGURE 9.5 **Radiation therapy** By targeting the tumor from various different directions, radiologists attempt to limit damage to healthy tissue.

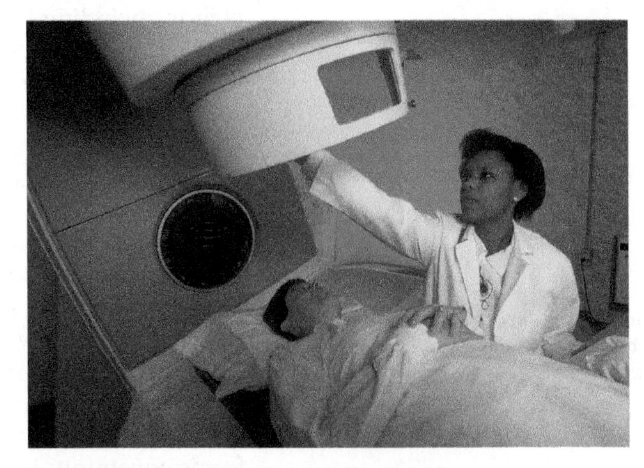

▲ Cancer patient undergoing radiation therapy.

EXAMPLE 9.1 Wavelength, Energy, and Frequency

Arrange these three types of electromagnetic radiation—visible light, X-rays, and microwaves—in order of increasing:

(a) wavelength (b) frequency (c) energy per photon

	SOLUTION
(a) wavelength Figure 9.4 indicates that X-rays have the shortest wavelength, followed by visible light and then microwaves.	X-rays, visible light, microwaves
(b) frequency Since frequency and wavelength are inversely proportional—the longer the wavelength, the shorter the frequency—the ordering with respect to frequency is exactly the reverse of the ordering with respect to wavelength.	microwaves, visible light, X-rays

(c) energy per photon Energy per photon decreases with increasing wavelength but increases with increasing frequency; therefore, the ordering with respect to energy per photon is the same as frequency.	microwaves, visible light, X-rays

▶ **SKILLBUILDER 9.1 | Wavelength, Energy, and Frequency**

Arrange these colors of visible light—green, red, and blue—in order of increasing:

(a) wavelength (b) frequency (c) energy per photon

▶ **FOR MORE PRACTICE** Example 9.9; Problems 31, 32, 33, 34, 35, 36, 37, 38.

CONCEPTUAL CHECKPOINT 9.2

Yellow light has a longer wavelength than violet light. Therefore:

(a) Yellow light has more energy per photon than violet light.

(b) Yellow light has less energy per photon than violet light.

(c) Both yellow light and violet light have the same energy per photon.

9.4 The Bohr Model: Atoms with Orbits

▶ Understand and explain the key characteristics of the Bohr model of the atom.

When an atom absorbs energy—in the form of heat, light, or electricity—it often reemits that energy as light. For example, a neon sign is composed of one or more glass tubes filled with gaseous neon atoms. When an electrical current passes through the tube, the neon atoms absorb some of the electrical energy and reemit it as the familiar red light of a neon sign (◀ FIGURE 9.6).

The absorption and emission of light by atoms is due to the interaction of the light with the electrons in the atom. If the atoms in the tube are different, the electrons have different energies, and the emitted light is a different color. In other words, atoms of each unique element emit light of a unique color (or unique wavelength). Mercury atoms, for example, emit light that appears blue, hydrogen atoms emit light that appears pink (▼ FIGURE 9.7), and helium atoms emit light that appears yellow-orange.

▲ FIGURE 9.6 **A neon sign** Neon atoms inside a glass tube absorb electrical energy and reemit the energy as light.

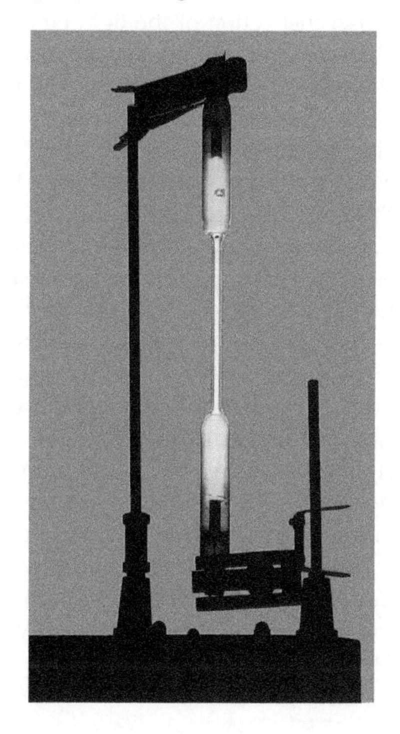

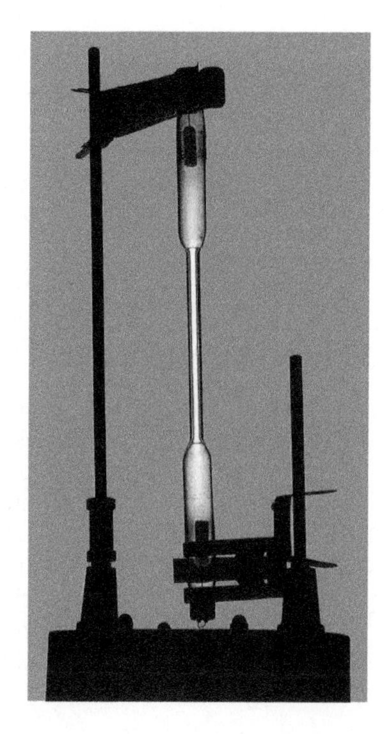

▶ FIGURE 9.7 **Light emission by different elements** Light emitted from a mercury lamp (left) appears blue, and light emitted from a hydrogen lamp (right) appears pink.

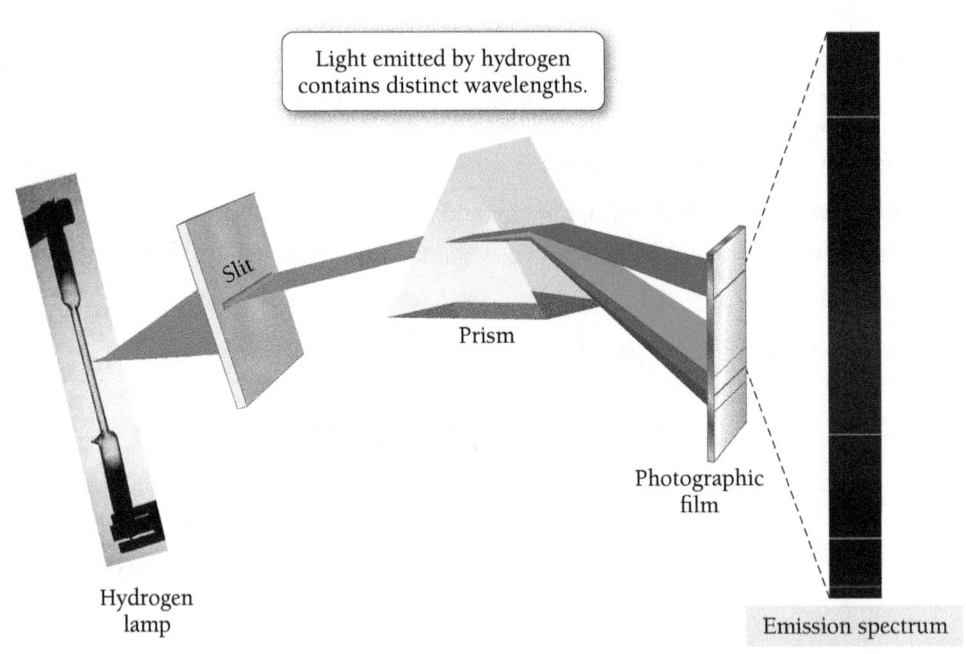

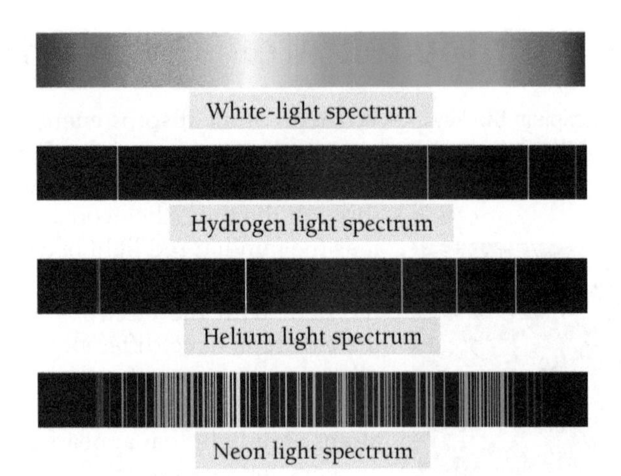

▶ FIGURE 9.8 **Emission spectra** A white-light spectrum is continuous, with some radiation emitted at every wavelength. The emission spectrum of an individual element, however, includes only certain specific wavelengths. (The different wavelengths appear as *lines* because the light from the source passes through a slit before entering the prism.) Each element produces its own unique and distinctive emission spectrum.

Closer inspection of the light emitted by hydrogen, helium, and neon atoms reveals that the light contains several distinct colors or wavelengths. Just as we can separate the white light from a light bulb into its constituent wavelengths by passing it through a prism, so we can separate the light emitted by glowing hydrogen, helium, or neon into its constituent wavelengths (▲ FIGURE 9.8) by passing it through a prism. The result is an **emission spectrum**. Notice the differences between a white-light spectrum and the emission spectra of hydrogen, helium, and neon. The white-light spectrum is *continuous*, meaning that the light intensity is uninterrupted or smooth across the entire visible range—there is some radiation at all wavelengths, with no gaps. The emission spectra of hydrogen, helium, and neon, however, are not continuous. They consist of bright spots or lines at specific wavelengths with complete darkness in between.

Since the emission of light in atoms is related to the motion of electrons within the atoms, a model for how electrons exist in atoms must account for these spectra. A major challenge in developing a model for electrons in atoms is the discrete or bright-line nature of the emission spectra. Why do atoms, when excited with energy, emit light only at particular wavelengths? Why do they *not* emit a continuous spectrum? Niels Bohr developed a simple model to explain these results. In his model, now called the Bohr model, electrons travel around the nucleus in circular orbits that are similar to planetary orbits around the sun. However, unlike planets revolving around the sun—which can theoretically orbit at any distance whatsoever from the sun—electrons in the Bohr model can orbit only at *specific, fixed* distances from the nucleus (◀ FIGURE 9.9).

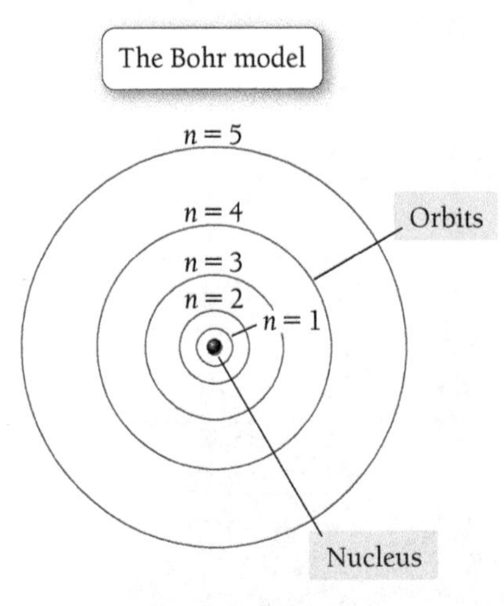

▲ FIGURE 9.9 **Bohr orbits**

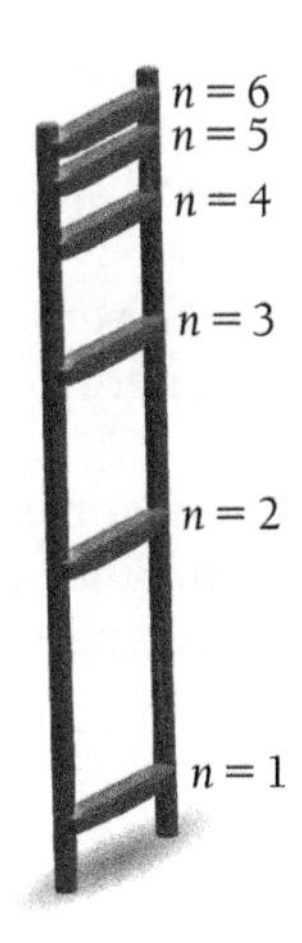

▲ FIGURE 9.10 **The Bohr energy ladder** Bohr orbits are like steps on a ladder. It is possible to stand on one step or another, but impossible to stand between steps.

The radii of Bohr orbits become increasingly further apart as n increases, but the energy levels get increasingly closer together.

The Bohr model is still important because it provides a logical foundation to the quantum-mechanical model and reveals the historical development of scientific understanding.

The *energy* of each Bohr orbit, specified by a **quantum number** $n = 1, 2, 3 \ldots,$ is also fixed, or **quantized**. The energy of each orbit increases with increasing value of n, but the energy levels become more closely spaced as n increases. Bohr orbits are like steps of a ladder (◀ FIGURE 9.10), each at a specific distance from the nucleus and each at a specific energy. Just as it is impossible to stand *between steps* on a ladder, so it is impossible for an electron to exist *between orbits* in the Bohr model. An electron in an $n = 3$ orbit, for example, is farther from the nucleus and has more energy than an electron in an $n = 2$ orbit. And an electron cannot exist at an intermediate distance or energy between the two orbits—the orbits are *quantized*. As long as an electron remains in a given orbit, it does not absorb or emit light, and its energy remains fixed and constant.

When an atom absorbs energy, an electron in one of these fixed orbits is *excited* or promoted to an orbit that is farther away from the nucleus (▼ FIGURE 9.11) and therefore higher in energy (this is analogous to moving up a step on the ladder). However, in this new configuration, the atom is less stable, and the electron quickly falls back or *relaxes* to a lower-energy orbit (this is analogous to moving down a step on the ladder). As it does so, it releases a photon of light containing the precise amount of energy—called a **quantum** of energy—that corresponds to the energy difference between the two orbits.

Since the amount of energy in a photon is directly related to its wavelength, the photon has a specific wavelength. *Consequently, the light emitted by excited atoms consists of specific lines at specific wavelengths, each corresponding to a specific transition between two orbits.* For example, the line at 486 nm in the hydrogen emission spectrum corresponds to an electron relaxing from the $n = 4$ orbit to the $n = 2$ orbit (▼ FIGURE 9.12). In the same way, the line at 657 nm (longer wavelength and therefore lower energy) corresponds to an electron relaxing from the $n = 3$ orbit to the $n = 2$ orbit. Notice that transitions between orbits that are closer together produce lower-energy (and therefore longer-wavelength) light than transitions between orbits that are farther apart.

The great success of the Bohr model of the atom was that it predicted the lines of the hydrogen emission spectrum. However, it failed to predict the emission spectra of other elements that contained more than one electron. For this, and other reasons, the Bohr model was replaced with a more sophisticated model called the quantum-mechanical or wave-mechanical model.

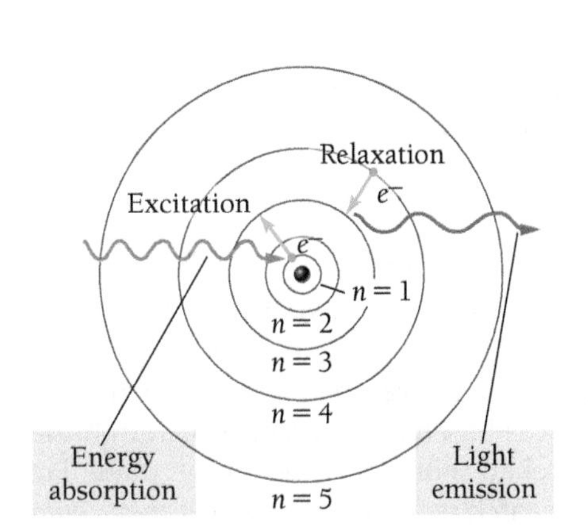

▲ FIGURE 9.11 **Excitation and emission** When a hydrogen atom absorbs energy, an electron is excited to a higher-energy orbit. The electron then relaxes back to a lower-energy orbit, emitting a photon of light.

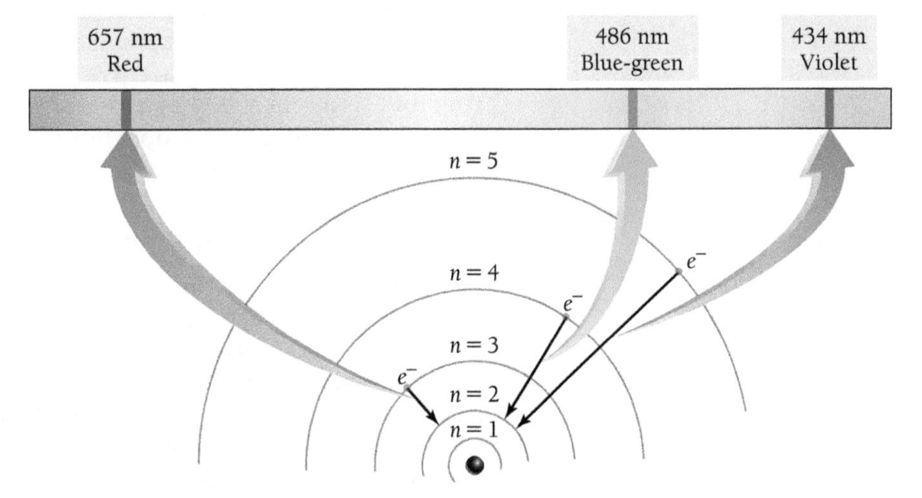

▲ FIGURE 9.12 **Hydrogen emission lines** The 657-nm line of the hydrogen emission spectrum corresponds to an electron relaxing from the $n = 3$ orbit to the $n = 2$ orbit. The 486-nm line corresponds to an electron relaxing from the $n = 4$ orbit to the $n = 2$ orbit, and the 434-nm line corresponds to an electron relaxing from $n = 5$ to $n = 2$.

To summarize:

- Electrons exist in quantized orbits at specific, fixed energies and specific, fixed distances from the nucleus.
- When energy is put into an atom, electrons are excited to higher-energy orbits.
- When an electron in an atom relaxes (or falls) from a higher-energy orbit to a lower-energy orbit, the atom emits light.
- The energy (and therefore the wavelength) of the emitted light corresponds to the energy difference between the two orbits in the transition. Since these energies are fixed and discrete, the energy (and therefore the wavelength) of the emitted light is fixed and discrete.

CONCEPTUAL CHECKPOINT 9.3

In one transition, an electron in a hydrogen atom falls from the $n = 3$ level to the $n = 2$ level. In a second transition, an electron in a hydrogen atom falls from the $n = 2$ level to the $n = 1$ level. Compared to the radiation emitted by the first of these transitions, the radiation emitted by the second has:

(a) a lower frequency
(b) a smaller energy per photon
(c) a shorter wavelength
(d) a longer wavelength

9.5 The Quantum-Mechanical Model: Atoms with Orbitals

▶ Understand and explain the key characteristics of the quantum-mechanical model of the atom.

The quantum-mechanical model of the atom replaced the Bohr model in the early twentieth century. In the quantum-mechanical model, *Bohr orbits* are replaced with *quantum-mechanical* **orbitals**. Orbitals are different from orbits in that they represent, not specific paths that electrons follow, but probability maps that show a statistical distribution of where the electron is likely to be found. The idea of an orbital is not easy to visualize. Quantum mechanics revolutionized physics and chemistry because in the quantum-mechanical model, electrons *do not* behave like particles flying through space. We cannot, in general, describe their exact paths. An orbital is a probability map that shows where the electron is *likely* to be found when the atom is probed; it does not represent the exact path that an electron takes as it travels through space.

Baseball Paths and Electron Probability Maps

To understand orbitals, let's contrast the behavior of a baseball with that of an electron. Imagine a baseball thrown from the pitcher's mound to a catcher at home plate (▶ FIGURE 9.13). The baseball's path can easily be traced as it travels from the pitcher to the catcher. The catcher can watch the baseball as it travels through the air, and he can predict exactly where the baseball will cross over home plate. He can even place his mitt in the correct place to catch it. *This sequence of events would be impossible for an electron.* Like photons, electrons exhibit a *wave–particle duality;* sometimes they act as particles, and other times as waves. This duality leads to behavior that makes it impossible to trace an electron's path. If an electron were "thrown" from the pitcher's mound to home plate, it would land in a different place every time, *even if it were thrown in exactly the same way*. Baseballs have predictable paths—electrons do not.

▶ FIGURE 9.13 **Baseballs follow predictable paths** A baseball follows a well-defined path as it travels from the pitcher to the catcher.

▲ FIGURE 9.14 **Electrons are unpredictable** To describe the behavior of a "pitched" electron, we would have to construct a probability map of where it would cross home plate.

In the quantum-mechanical world of the electron, the catcher cannot know exactly where the electron would cross the plate for any given throw. He has no way of putting his mitt in the right place to catch it. However, if the catcher were able to keep track of hundreds of electron throws, he could observe a reproducible, statistical pattern of where the electron crosses the plate. He could even draw maps in the strike zone showing the probability of an electron crossing a certain area (◀ FIGURE 9.14). These maps are called *probability maps*.

From Orbits to Orbitals

In the Bohr model, an *orbit* is a circular path—analogous to a baseball's path—that shows the electron's motion around an atomic nucleus. In the quantum-mechanical model, an *orbital* is a probability map, analogous to the probability map drawn by our catcher. It shows the relative likelihood of the electron being found at various locations when the atom is probed. Just as the Bohr model has different orbits with different radii, the quantum-mechanical model has different orbitals with different shapes.

9.6 Quantum-Mechanical Orbitals and Electron Configurations

▶ Write electron configurations and orbital diagrams for atoms.

In the Bohr model of the atom, a single quantum number (n) specifies each orbit. In the quantum-mechanical model, a number and a letter specify an orbital (or orbitals). In this section, we examine quantum-mechanical orbitals and electron configurations. An electron configuration is a compact way to specify the occupation of quantum-mechanical orbitals by electrons.

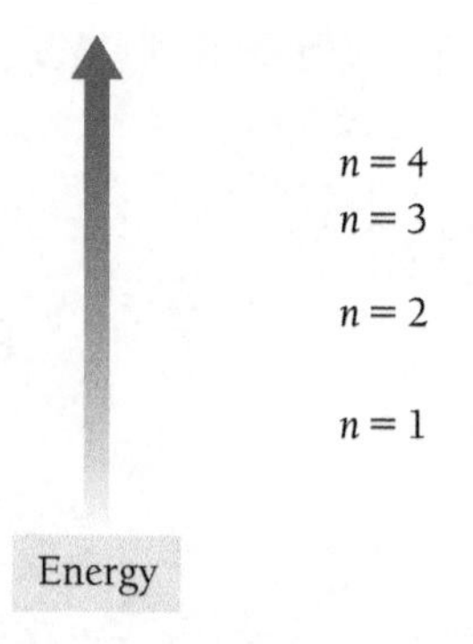

▲ FIGURE 9.15 **Principal quantum numbers** The principal quantum numbers ($n = 1, 2, 3 \ldots$) determine the energy of the hydrogen quantum-mechanical orbitals.

Quantum-Mechanical Orbitals

The lowest-energy orbital in the quantum-mechanical model—analogous to the $n = 1$ orbit in the Bohr model—is the *1s orbital*. We specify it by the number 1 and the letter *s*. The number is the **principal quantum number** (n) and specifies the **principal shell** of the orbital. The higher the principal quantum number, the higher the energy of the orbital. The possible principal quantum numbers are $n = 1, 2, 3 \ldots$, with energy increasing as n increases (◀ FIGURE 9.15). Because the 1*s* orbital has the lowest possible principal quantum number, it is in the lowest-energy shell and has the lowest possible energy.

The letter indicates the **subshell** of the orbital and specifies its shape. The possible letters are *s*, *p*, *d*, and *f*, and each letter corresponds to a different shape. For example, orbitals within the *s* subshell have a spherical shape. Unlike the $n = 1$ Bohr orbit, which shows the electron's circular path, the 1*s* quantum-mechanical orbital is a three-dimensional probability map. We sometimes represent orbitals with dots (▼ FIGURE 9.16), where the dot density is proportional to the probability of finding the electron.

We can understand the dot representation of an orbital better with another analogy. Imagine you could take a photograph of an electron in an atom every second for 10 or 15 minutes. One second the electron is very close to the nucleus; the next second it is farther away and so on. Each photo shows a dot representing the electron's position relative to the nucleus at that time. If we took hundreds of photos and superimposed all of them, we would have an image like Figure 9.16—a statistical representation of where the electron is found. Notice that the dot density for the 1*s* orbital is greatest near the nucleus and decreases farther away from the nucleus. This means that the electron is more likely to be found close to the nucleus than far away from it.

This analogy is purely hypothetical. It is impossible to photograph electrons in this way.

Orbitals can also be represented as geometric shapes that encompass most of the volume where the electron is likely to be found. For example, we represent the 1*s* orbital as a sphere (▼ FIGURE 9.17) that encompasses the volume within which the electron is found 90% of the time. If we superimpose the dot representation of the 1*s* orbital on the shape representation (▼ FIGURE 9.18), we can see that most of the dots are within the sphere, indicating that the electron is most likely to be found within the sphere when it is in the 1*s* orbital.

The single electron of an undisturbed hydrogen atom at room temperature is in the 1*s* orbital. This is the **ground state**, or lowest energy state, of the hydrogen atom. However, like the Bohr model, the quantum-mechanical model allows electrons to transition to higher-energy orbitals upon the absorption of energy. What are these higher-energy orbitals? What do they look like?

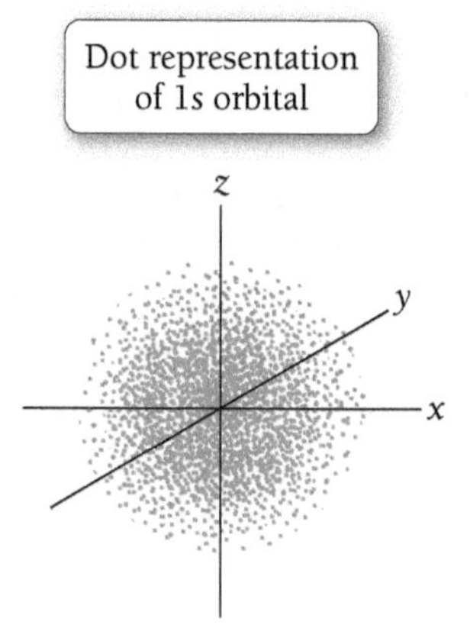

▲ FIGURE 9.16 **1s orbital** The dot density in this plot is proportional to the probability of finding the electron. The greater dot density near the middle indicates a higher probability of finding the electron near the nucleus.

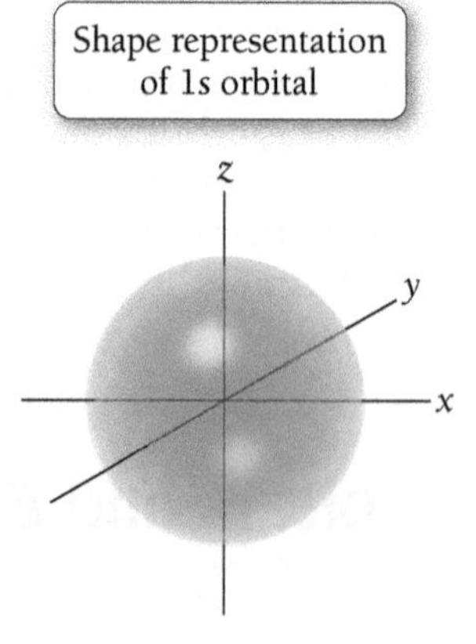

▲ FIGURE 9.17 **Shape representation of the 1s orbital** Because the distribution of electron density around the nucleus in Figure 9.16 is symmetrical—the same in all directions—we can represent the 1s orbital as a sphere.

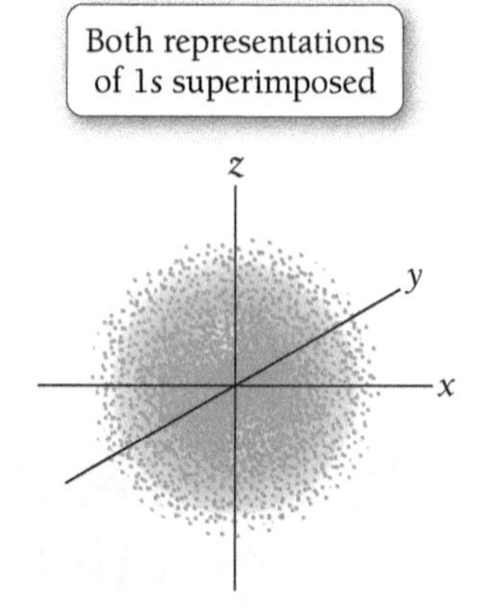

▲ FIGURE 9.18 **Orbital shape and dot representation for the 1s orbital** The shape representation of the 1s orbital superimposed on the dot density representation. We can see that when the electron is in the 1s orbital, it is most likely to be found within the sphere.

Shell	Number of subshells	Letters specifying subshells
$n = 4$	4	*s* *p* *d* *f*
$n = 3$	3	*s* *p* *d*
$n = 2$	2	*s* *p*
$n = 1$	1	*s*

▲ FIGURE 9.19 **Subshells** The number of subshells in a given principal shell is equal to the value of *n*.

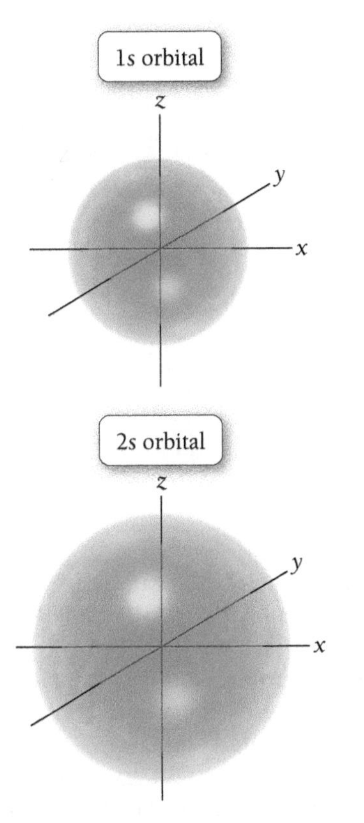

▲ FIGURE 9.20 **The 2s orbital** The 2s orbital is similar to the 1s orbital, but larger in size.

The next orbitals in the quantum-mechanical model are those with principal quantum number $n = 2$. Unlike the $n = 1$ principal shell, which contains only one subshell (specified by *s*), the $n = 2$ principal shell contains two subshells, specified by *s* and *p*.

The number of subshells in a given principal shell is equal to the value of *n*. Therefore, the $n = 1$ principal shell has one subshell, the $n = 2$ principal shell has two subshells, and so on (▲ FIGURE 9.19). The *s* subshell contains the 2*s* orbital, which is higher in energy than the 1*s* orbital and slightly larger (◀ FIGURE 9.20), but otherwise similar in shape. The *p* subshell contains three 2*p* orbitals. The three 2*p* orbitals all have the same dumbbell-like shape but each has a different orientation (▼ FIGURE 9.21).

The next principal shell, $n = 3$, contains three subshells specified by *s*, *p*, and *d*. The *s* and *p* subshells contain the 3*s* and 3*p* orbitals, similar in shape to the 2*s* and 2*p* orbitals, but slightly larger and higher in energy. The 3*d* subshell contains the five *d* orbitals shown in ▶ FIGURE 9.22, on the next page. The next principal shell, $n = 4$, contains four subshells specified by *s*, *p*, *d*, and *f*. The *s*, *p*, and *d* subshells for the $n = 4$ principal shell are similar to those in $n = 3$. The 4*f* subshell contains seven orbitals (called the 4*f* orbitals), whose shape we do not consider in this book.

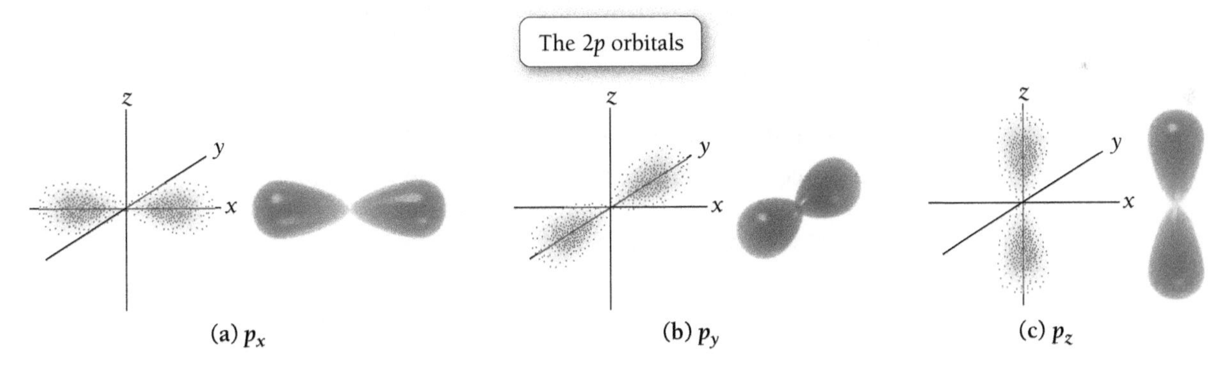

▲ FIGURE 9.21 **The 2*p* orbitals** This figure shows both the dot representation (left) and shape representation (right) for each *p* orbital.

PEARSON eText 2.0

CONCEPTUAL CHECKPOINT 9.4

Which subshells are in the $n = 3$ principal shell?

(a) *s* subshell (only)
(b) *s* and *p* subshells (only)
(c) *s*, *p*, and *d* subshells (only)

PEARSON eText 2.0

CONCEPTUAL CHECKPOINT 9.5

How many orbitals are in the *p* subshell?

(a) 1 **(b)** 3 **(c)** 5

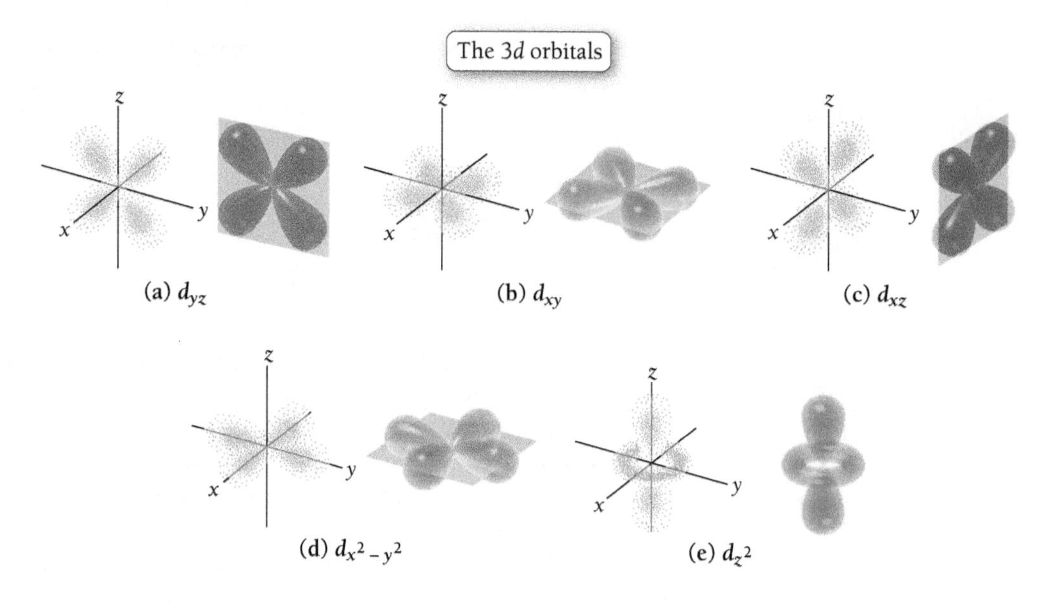

▶ **FIGURE 9.22 The 3*d* orbitals** This figure shows both the dot representation (left) and shape representation (right) for each *d* orbital.

As we have already discussed, hydrogen's single electron is usually in the 1*s* orbital because electrons generally occupy the lowest-energy orbital available. In hydrogen, the rest of the orbitals are normally empty. However, the absorption of energy by a hydrogen atom can cause the electron to jump (or make a transition) from the 1*s* orbital to a higher-energy orbital. When the electron is in a higher-energy orbital, we say that the hydrogen atom is in an **excited state**.

Because of their higher energy, excited states are unstable, and an electron in a higher-energy orbital will usually fall (or relax) back to a lower-energy orbital. In the process the electron emits energy, often in the form of light. As in the Bohr model, the energy difference between the two orbitals involved in the transition determines the wavelength of the emitted light (the greater the energy difference, the shorter the wavelength). The quantum-mechanical model predicts the bright-line spectrum of hydrogen as well as the Bohr model. However, unlike the Bohr model, it also predicts the bright-line spectra of other elements.

Electron Configurations: How Electrons Occupy Orbitals

An **electron configuration** illustrates the occupation of orbitals by electrons for a particular atom. For example, the electron configuration for a ground-state (or lowest energy) hydrogen atom is:

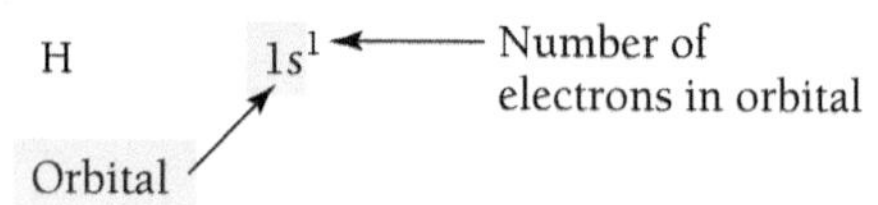

The electron configuration tells us that hydrogen's single electron is in the 1*s* orbital.

Another way to represent this information is with an **orbital diagram**, which gives similar information but shows the electrons as arrows in a box representing the orbital. The orbital diagram for a ground-state hydrogen atom is:

H [↿]
1*s*

The box represents the 1*s* orbital, and the arrow within the box represents the electron in the 1*s* orbital. In orbital diagrams, the direction of the arrow (pointing up or pointing down) represents **electron spin**, a fundamental property of electrons. All electrons have spin. The **Pauli exclusion principle** states that *orbitals may hold no more than two electrons with opposing spins*. We symbolize this as two arrows pointing in opposite directions:

⇃↾

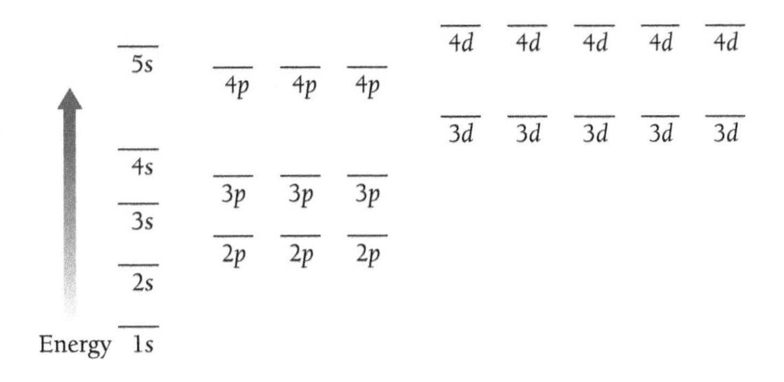

▶ FIGURE 9.23 **Energy ordering of orbitals for multi-electron atoms** Different subshells within the same principal shell have different energies.

A helium atom, for example, has two electrons. The electron configuration and orbital diagram for helium are:

	Electron configuration	Orbital diagram
He	$1s^2$	[⇅] 1s

Since we know that electrons occupy the lowest-energy orbitals available, and since we know that only two electrons (with opposing spins) are allowed in each orbital, we can continue to build ground-state electron configurations for the rest of the elements as long as we know the energy ordering of the orbitals. ▲ FIGURE 9.23 shows the energy ordering of a number of orbitals for multi-electron atoms.

In multi-electron atoms, the subshells within a principal shell do not have the same energy because of electron–electron interactions.

Notice that, for multi-electron atoms (in contrast to hydrogen which has only one electron), the subshells within a principal shell *do not* have the same energy. In elements other than hydrogen, the energy ordering is not determined by the principal quantum number alone. For example, in multi-electron atoms, the 4*s* subshell is lower in energy than the 3*d* subshell, even though its principal quantum number is higher. Using this relative energy ordering, we can write ground-state electron configurations and orbital diagrams for other elements.

For lithium, which has three electrons, the electron configuration and orbital diagram are:

Remember that the number of electrons in an atom is equal to its atomic number.

	Electron configuration	Orbital diagram
Li	$1s^2 2s^1$	[⇅] 1s [↑] 2s

For carbon, which has six electrons, the electron configuration and orbital diagram are:

	Electron configuration	Orbital diagram
C	$1s^2 2s^2 2p^2$	[⇅] 1s [⇅] 2s [↑ ↑] 2p

Notice that the 2*p* electrons occupy the *p* orbitals (of equal energy) singly rather than pairing in one orbital. This is the result of **Hund's rule**, which states that *when filling orbitals of equal energy, electrons fill them singly first, with parallel spins.*

Before we write electron configurations for other elements, let's summarize what we have learned so far:

- Electrons occupy orbitals so as to minimize the energy of the atom; therefore, lower-energy orbitals fill before higher-energy orbitals. Orbitals fill in the following order: 1*s* 2*s* 2*p* 3*s* 3*p* 4*s* 3*d* 4*p* 5*s* 4*d* 5*p* 6*s* (◀ FIGURE 9.24).
- Orbitals can hold no more than two electrons each. When two electrons occupy the same orbital, they must have opposing spins. This is known as the Pauli exclusion principle.
- When orbitals of identical energy are available, these are first occupied singly with parallel spins rather than in pairs. This is known as Hund's rule.

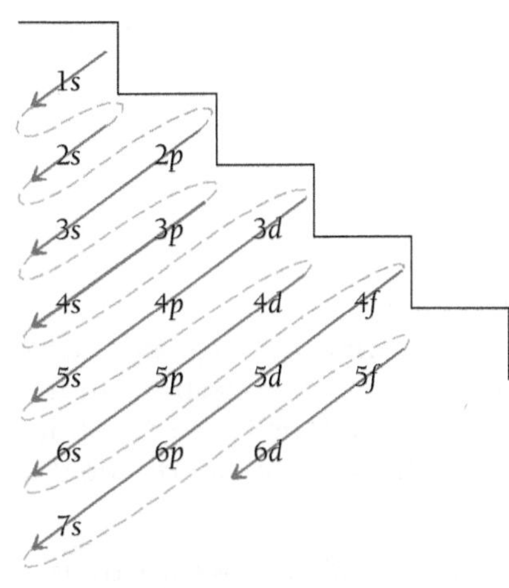

▲ FIGURE 9.24 **Orbital filling order** The arrows indicate the order in which orbitals fill.

Consider the electron configurations and orbital diagrams for elements with atomic numbers 3 through 10:

Symbol (# e^-)	Electron configuration	Orbital diagram 1s	2s	2p
Li (3)	$1s^22s^1$	⇅	↑	
Be (4)	$1s^22s^2$	⇅	⇅	
B (5)	$1s^22s^22p^1$	⇅	⇅	↑ _ _
C (6)	$1s^22s^22p^2$	⇅	⇅	↑ ↑ _
N (7)	$1s^22s^22p^3$	⇅	⇅	↑ ↑ ↑
O (8)	$1s^22s^22p^4$	⇅	⇅	⇅ ↑ ↑
F (9)	$1s^22s^22p^5$	⇅	⇅	⇅ ⇅ ↑
Ne (10)	$1s^22s^22p^6$	⇅	⇅	⇅ ⇅ ⇅

Notice how the *p* orbitals fill. As a result of Hund's rule, the *p* orbitals fill with single electrons before they fill with paired electrons. The electron configuration of neon represents the complete filling of the $n = 2$ principal shell. When writing electron configurations for elements beyond neon—or beyond any other noble gas—we often abbreviate the electron configuration of the previous noble gas by the symbol for the noble gas in brackets.

For example, the electron configuration of sodium is:

$$\text{Na} \quad 1s^22s^22p^63s^1$$

We can write this using the noble gas core notation as:

$$\text{Na} \quad [\text{Ne}]3s^1$$

where [Ne] represents $1s^22s^22p^6$, the electron configuration for neon.

To write an electron configuration for an element, we first find its atomic number from the periodic table—this number equals the number of electrons in the neutral atom. Then we use the order of filling from Figure 9.23 or 9.24 to distribute the electrons in the appropriate orbitals. Remember that each orbital can hold a maximum of two electrons. Consequently:

- the *s* subshell has only one orbital and therefore can hold only two electrons.
- the *p* subshell has three orbitals and therefore can hold six electrons.
- the *d* subshell has five orbitals and therefore can hold ten electrons.
- the *f* subshell has seven orbitals and therefore can hold 14 electrons.

EXAMPLE 9.2 Electron Configurations

Write electron configurations for each element.

(a) Mg (b) S (c) Ga

SOLUTION

(a) Magnesium has 12 electrons. Distribute two of these into the 1*s* orbital, two into the 2*s* orbital, six into the 2*p* orbitals, and two into the 3*s* orbital.
You can also write the electron configuration more compactly using the noble gas core notation. For magnesium, use [Ne] to represent $1s^22s^22p^6$.

Mg $1s^22s^22p^63s^2$
or
Mg $[Ne]3s^2$

(b) Sulfur has 16 electrons. Distribute two of these into the 1*s* orbital, two into the 2*s* orbital, six into the 2*p* orbitals, two into the 3*s* orbital, and four into the 3*p* orbitals.
You can write the electron configuration more compactly by using [Ne] to represent $1s^22s^22p^6$.

S $1s^22s^22p^63s^23p^4$
or
S $[Ne]3s^23p^4$

(c) Gallium has 31 electrons. Distribute two of these into the 1*s* orbital, two into the 2*s* orbital, six into the 2*p* orbitals, two into the 3*s* orbital, six into the 3*p* orbitals, two into the 4*s* orbital, ten into the 3*d* orbitals, and one into the 4*p* orbitals. Notice that the *d* subshell has five orbitals and can therefore hold 10 electrons.
You can write the electron configuration more compactly by using [Ar] to represent $1s^22s^22p^63s^23p^6$.

Ga $1s^22s^22p^63s^23p^64s^23d^{10}4p^1$
or
Ga $[Ar]4s^23d^{10}4p^1$

▶ **SKILLBUILDER 9.2 | Electron Configurations**

Write electron configurations for each element.

(a) Al (b) Br (c) Sr

▶ **SKILLBUILDER PLUS**

Write electron configurations for each ion. (*Hint:* To determine the number of electrons to include in the electron configuration of an ion, add or subtract electrons as needed to account for the charge of the ion.)

(a) Al^{3+} (b) Cl^- (c) O^{2-}

▶ **FOR MORE PRACTICE** Problems 49, 50, 53, 54, 55, 56.

EXAMPLE 9.3 Writing Orbital Diagrams

Write an orbital diagram for silicon.

SOLUTION

Since silicon is atomic number 14, it has 14 electrons. Draw a box for each orbital, putting the lowest-energy orbital (1*s*) on the far left and proceeding to orbitals of higher energy to the right.

1s	2s	2p	3s	3p
□	□	□□□	□	□□□

Distribute the 14 electrons into the orbitals, allowing a maximum of two electrons per orbital and remembering Hund's rule. The complete orbital diagram is:

	1s	2s	2p	3s	3p
Si	⇅	⇅	⇅ ⇅ ⇅	⇅	↑ ↑ □

▶ **SKILLBUILDER 9.3 | Writing Orbital Diagrams**

Write an orbital diagram for argon.

▶ **FOR MORE PRACTICE** Example 9.10; Problems 51, 52.

CONCEPTUAL CHECKPOINT 9.6

Which pair of elements has the same *total* number of electrons in its *p* orbitals?

(a) Na and K
(b) K and Kr
(c) P and N
(d) Ar and Ca

9.7 Electron Configurations and the Periodic Table

- Identify valence electrons and core electrons.
- Write electron configurations for elements based on their positions in the periodic table.

Valence electrons are the electrons in the outermost principal shell (the principal shell with the highest principal quantum number, *n*). These electrons are important because, as we will see in Chapter 10, they are held most loosely and are most easily lost or shared; therefore they are involved in chemical bonding. Electrons that are *not* in the outermost principal shell are **core electrons**. For example, silicon, with the electron configuration of $1s^22s^22p^63s^23p^2$, has four valence electrons (those in the $n = 3$ principal shell) and ten core electrons.

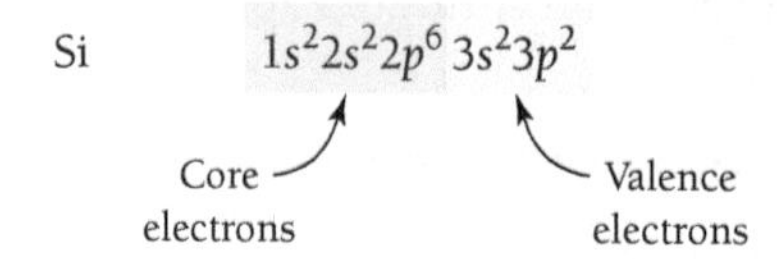

EXAMPLE **9.4** **Valence Electrons and Core Electrons**

Write an electron configuration for selenium and identify the valence electrons and the core electrons.

SOLUTION

Write the electron configuration for selenium by determining the total number of electrons from selenium's atomic number (34) and distributing them into the appropriate orbitals.

$$\text{Se}\quad 1s^22s^22p^63s^23p^64s^23d^{10}4p^4$$

The valence electrons are those in the outermost principal shell. For selenium, the outermost principal shell is the $n = 4$ shell, which contains six electrons (two in the 4*s* orbital and four in the three 4*p* orbitals). All other electrons, including those in the 3*d* orbitals, are core electrons.

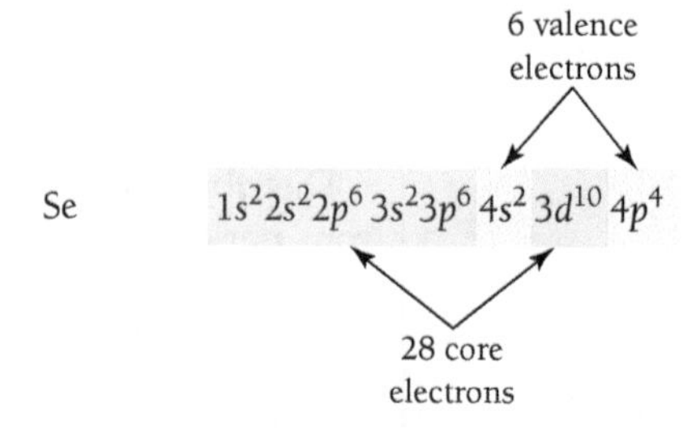

▶ **SKILLBUILDER 9.4 | Valence Electrons and Core Electrons**

Write an electron configuration for chlorine and identify the valence electrons and core electrons.

▶ **FOR MORE PRACTICE** Example 9.11; Problems 57, 58, 61, 62.

▶ FIGURE 9.25 **Outer electron configurations of the first 18 elements**

1A							8A
1 H $1s^1$	2A	3A	4A	5A	6A	7A	2 He $1s^2$
3 Li $2s^1$	4 Be $2s^2$	5 B $2s^22p^1$	6 C $2s^22p^2$	7 N $2s^22p^3$	8 O $2s^22p^4$	9 F $2s^22p^5$	10 Ne $2s^22p^6$
11 Na $3s^1$	12 Mg $3s^2$	13 Al $3s^23p^1$	14 Si $3s^23p^2$	15 P $3s^23p^3$	16 S $3s^23p^4$	17 Cl $3s^23p^5$	18 Ar $3s^23p^6$

▲ FIGURE 9.25 shows the first 18 elements in the periodic table with an outer electron configuration listed below each one. As we move across a row, the orbitals are simply filling in the correct order. As we move down a column, the highest principal quantum number increases, but the number of electrons in each subshell remains the same. Consequently, the elements within a column (or family) all have the same number of valence electrons and similar outer electron configurations.

A similar pattern exists for the entire periodic table (▼ FIGURE 9.26). Notice that, because of the filling order of orbitals, we can divide the periodic table into blocks representing the filling of particular subshells.

- The first two columns on the left side of the periodic table are the *s* block with outer electron configurations of ns^1 (first column) and ns^2 (second column).
- The six columns on the right side of the periodic table are the *p* block with outer electron configurations of: ns^2np^1, ns^2np^2, ns^2np^3, ns^2np^4, ns^2np^5 (halogens), and ns^2np^6 (noble gases).
- The transition metals are the *d* block.
- The lanthanides and actinides (also called the inner transition metals) are the *f* block.

Notice that, except for helium, the number of valence electrons for any main-group element is equal to the group number of its column. For example, we can tell that chlorine has seven valence electrons because it is in the column with group

s-block elements
p-block elements
d-block elements
f-block elements

Periods / Groups	1 1A	2 2A	3 3B	4 4B	5 5B	6 6B	7 7B	8 8B	9 8B	10 8B	11 1B	12 2B	13 3A	14 4A	15 5A	16 6A	17 7A	18 8A
1	1 H $1s^1$																	2 He $1s^2$
2	3 Li $2s^1$	4 Be $2s^2$											5 B $2s^22p^1$	6 C $2s^22p^2$	7 N $2s^22p^3$	8 O $2s^22p^4$	9 F $2s^22p^5$	10 Ne $2s^22p^6$
3	11 Na $3s^1$	12 Mg $3s^2$											13 Al $3s^23p^1$	14 Si $3s^23p^2$	15 P $3s^23p^3$	16 S $3s^23p^4$	17 Cl $3s^23p^5$	18 Ar $3s^23p^6$
4	19 K $4s^1$	20 Ca $4s^2$	21 Sc $4s^23d^1$	22 Ti $4s^23d^2$	23 V $4s^23d^3$	24 Cr $4s^13d^5$	25 Mn $4s^23d^5$	26 Fe $4s^23d^6$	27 Co $4s^23d^7$	28 Ni $4s^23d^8$	29 Cu $4s^13d^{10}$	30 Zn $4s^23d^{10}$	31 Ga $4s^24p^1$	32 Ge $4s^24p^2$	33 As $4s^24p^3$	34 Se $4s^24p^4$	35 Br $4s^24p^5$	36 Kr $4s^24p^6$
5	37 Rb $5s^1$	38 Sr $5s^2$	39 Y $5s^24d^1$	40 Zr $5s^24d^2$	41 Nb $5s^14d^4$	42 Mo $5s^14d^5$	43 Tc $5s^24d^5$	44 Ru $5s^14d^7$	45 Rh $5s^14d^8$	46 Pd $4d^{10}$	47 Ag $5s^14d^{10}$	48 Cd $5s^24d^{10}$	49 In $5s^25p^1$	50 Sn $5s^25p^2$	51 Sb $5s^25p^3$	52 Te $5s^25p^4$	53 I $5s^25p^5$	54 Xe $5s^25p^6$
6	55 Cs $6s^1$	56 Ba $6s^2$	57 La $6s^25d^1$	72 Hf $6s^25d^2$	73 Ta $6s^25d^3$	74 W $6s^25d^4$	75 Re $6s^25d^5$	76 Os $6s^25d^6$	77 Ir $6s^25d^7$	78 Pt $6s^15d^9$	79 Au $6s^15d^{10}$	80 Hg $6s^25d^{10}$	81 Tl $6s^26p^1$	82 Pb $6s^26p^2$	83 Bi $6s^26p^3$	84 Po $6s^26p^4$	85 At $6s^26p^5$	86 Rn $6s^26p^6$
7	87 Fr $7s^1$	88 Ra $7s^2$	89 Ac $7s^26d^1$	104 Rf $7s^26d^2$	105 Db $7s^26d^3$	106 Sg $7s^26d^4$	107 Bh	108 Hs	109 Mt	110 Ds	111 Rg	112 Cn	113 Nh	114 Fl	115 Mc	116 Lv	117 Ts	118 Og

Lanthanides	58 Ce $6s^25d^14f^1$	59 Pr $6s^24f^3$	60 Nd $6s^24f^4$	61 Pm $6s^24f^5$	62 Sm $6s^24f^6$	63 Eu $6s^24f^7$	64 Gd $6s^24f^75d$	65 Tb $6s^24f^9$	66 Dy $6s^24f^{10}$	67 Ho $6s^24f^{11}$	68 Er $6s^24f^{12}$	69 Tm $6s^24f^{13}$	70 Yb $6s^24f^{14}$	71 Lu $6s^24f^{14}5d^1$
Actinides	90 Th $7s^26d^2$	91 Pa $7s^25f^26d^1$	92 U $7s^25f^36d^1$	93 Np $7s^25f^46d^1$	94 Pu $7s^25f^6$	95 Am $7s^25f^7$	96 Cm $7s^25f^76d^1$	97 Bk $7s^25f^9$	98 Cf $7s^25f^{10}$	99 Es $7s^25f^{11}$	100 Fm $7s^25f^{12}$	101 Md $7s^25f^{13}$	102 No $7s^25f^{14}$	103 Lr $7s^25f^{14}6d^1$

▲ FIGURE 9.26 **Outer electron configurations of the elements**

Remember that main-group elements are those in the two far-left columns (1A, 2A) and the six far-right columns (3A–8A) of the periodic table (see Section 4.6).

number 7A. The row number in the periodic table is equal to the number of the highest principal shell (n value). For example, since chlorine is in row 3, its highest principal shell is the $n = 3$ shell.

The transition metals have electron configurations with trends that differ somewhat from main-group elements. As we move across a row in the *d* block, the *d* orbitals are filling (see Figure 9.26). However, the principal quantum number of the *d* orbital being filled across each row in the transition series is equal to the row number minus one (in the fourth row, the 3*d* orbitals fill; in the fifth row, the 4*d* orbitals fill; and so on). For the first transition series, the outer configuration is $4s^23d^x$ (x = number of *d* electrons) with two exceptions: Cr is $4s^13d^5$ and Cu is $4s^13d^{10}$. These exceptions occur because a half-filled *d* subshell and a completely filled *d* subshell are particularly stable. Otherwise, the number of outershell electrons in a transition series does not change as we move across a period. In other words, *the transition series represents the filling of core orbitals, and the number of outershell electrons is mostly constant.*

We can now see that the organization of the periodic table allows us to write the electron configuration for any element based simply on its position in the periodic table. For example, suppose we want to write an electron configuration for P. The inner electrons of P are those of the noble gas that precedes P in the periodic table, Ne. So we can represent the inner electrons with [Ne]. We obtain the outer electron configuration by tracing the elements between Ne and P and assigning electrons to the appropriate orbitals (▼ FIGURE 9.27). Remember that the highest n value is given by the row number (3 for phosphorus). So we begin with [Ne], then add in the two 3*s* electrons as we trace across the *s* block, followed by three 3*p* electrons as we trace across the *p* block to P, which is in the third column of the *p* block. The electron configuration is:

$$\text{P} \quad [\text{Ne}]3s^23p^3$$

Notice that P is in column 5A and therefore has five valence electrons and an outer electron configuration of ns^2np^3.

To summarize writing an electron configuration for an element based on its position in the periodic table:

- The inner electron configuration for any element is the electron configuration of the noble gas that immediately precedes that element in the periodic table. We represent the inner configuration with the symbol for the noble gas in brackets.
- We can determine the outer electrons from the element's position within a particular block (*s*, *p*, *d*, or *f*) in the periodic table. We trace the elements between the preceding noble gas and the element of interest, and assign electrons to the appropriate orbitals.
- The highest principal quantum number (highest n value) is equal to the row number of the element in the periodic table.
- For any element containing *d* electrons, the principal quantum number (n value) of the outermost *d* electrons is equal to the row number of the element minus 1.

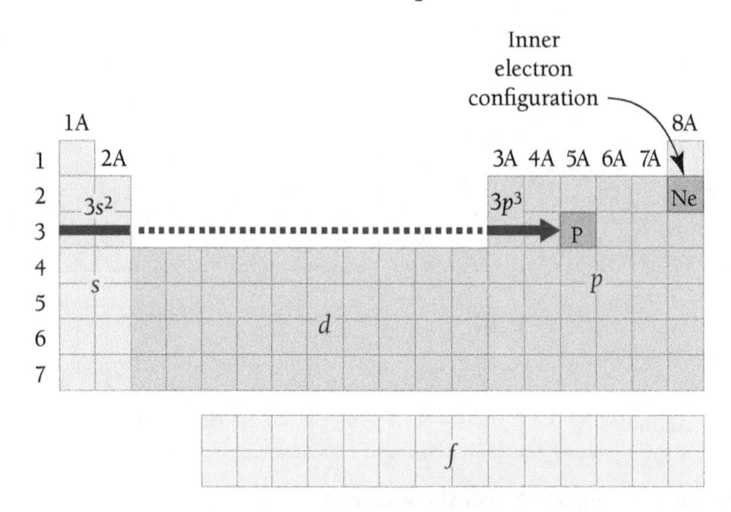

▶ FIGURE 9.27 **Electron configuration of phosphorus** Determining the electron configuration for P from its position in the periodic table.

Interactive Worked Example Video 9.5

EXAMPLE **9.5** **Writing Electron Configurations from the Periodic Table**

Write an electron configuration for arsenic based on its position in the periodic table.

SOLUTION

The noble gas that precedes arsenic in the periodic table is argon, so the inner electron configuration is [Ar]. Obtain the outer electron configuration by tracing the elements between Ar and As and assigning electrons to the appropriate orbitals.

Remember that the highest n value is given by the row number (4 for arsenic). So, begin with [Ar], then add in the two $4s$ electrons as you trace across the s block, followed by ten $3d$ electrons as you trace across the d block (the n value for d subshells is equal to the row number minus one), and finally the three $4p$ electrons as you trace across the p block to As, which is in the third column of the p block.

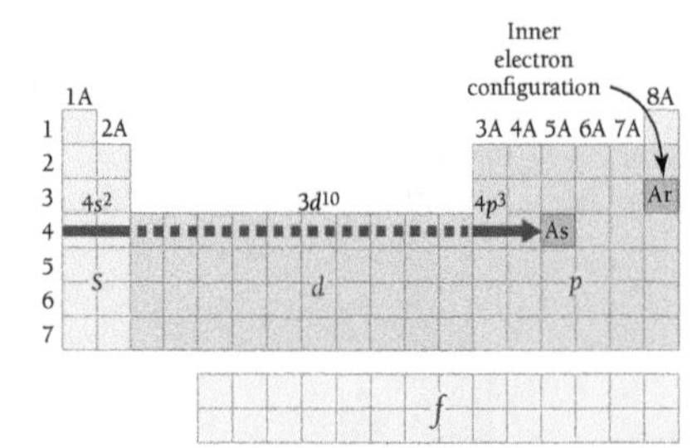

The electron configuration is:

$$\text{As} \quad [\text{Ar}]4s^2 3d^{10} 4p^3$$

▶ **SKILLBUILDER 9.5 | Writing Electron Configurations from the Periodic Table**

Use the periodic table to determine the electron configuration for tin.

▶ **FOR MORE PRACTICE** Example 9.12; Problems 65, 66, 67, 68.

CONCEPTUAL CHECKPOINT 9.7

Which element has the *fewest* valence electrons?

(a) B **(b)** Ca **(c)** O **(d)** K **(e)** Ga

9.8 The Explanatory Power of the Quantum-Mechanical Model

▶ Explain why the chemical properties of elements are largely determined by the number of valence electrons they contain.

Noble gases
18
8A

2 He $1s^2$
10 Ne $2s^2 2p^6$
18 Ar $3s^2 3p^6$
36 Kr $4s^2 4p^6$
54 Xe $5s^2 5p^6$
86 Rn $6s^2 6p^6$

▶ FIGURE 9.28 **Electron configurations of the noble gases** The noble gases (except for helium) all have eight valence electrons and completely full outer principal shells.

At the beginning of this chapter, we asserted that the quantum-mechanical model explains the chemical properties of the elements such as the inertness of helium, the reactivity of hydrogen, and the periodic law. We can now see why: *The chemical properties of elements are largely determined by the number of valence electrons they contain.* The properties of elements vary in a periodic fashion because the number of valence electrons is periodic.

Because elements within a column in the periodic table have the same number of valence electrons, they also have similar chemical properties. The noble gases, for example, all have eight valence electrons, except for helium, which has two (◀ FIGURE 9.28). Although we don't get into the quantitative (or numerical) aspects of the quantum-mechanical model in this book, calculations show that atoms with eight valence electrons (or two for helium) are particularly low in energy, and therefore stable. The noble gases are indeed chemically stable and thus relatively inert or nonreactive as accounted for by the quantum model.

Elements with electron configurations close to the noble gases are the most reactive because they can attain noble gas electron configurations by losing or gaining a small number of electrons. Alkali metals (Group 1) are among the most reactive metals since their outer electron configuration (ns^1) is one electron beyond a noble gas configuration (◀ FIGURE 9.29). If an alkali metal can react to lose its ns^1 electron, it attains a noble gas configuration. This explains why—as we learned in Chapter 4—the Group 1A metals tend to form 1+ cations. As an example, consider the electron configuration of sodium:

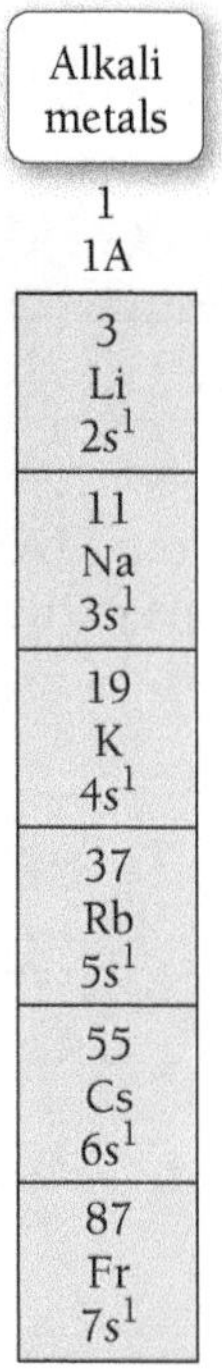

$$\text{Na} \quad 1s^2 2s^2 2p^6 3s^1$$

In reactions, sodium loses its 3s electron, forming a 1+ ion with the electron configuration of neon:

$$\text{Na}^+ \quad 1s^2 2s^2 2p^6$$

$$\text{Ne} \quad 1s^2 2s^2 2p^6$$

Similarly, alkaline earth metals, with an outer electron configuration of ns^2, also tend to be reactive metals. Each alkaline earth metal loses its two ns^2 electrons to form a 2+ cation (▼ FIGURE 9.30). For example, consider magnesium:

$$\text{Mg} \quad 1s^2 2s^2 2p^6 3s^2$$

In reactions, magnesium loses its two 3s electrons, forming a 2+ ion with the electron configuration of neon:

$$\text{Mg}^{2+} \quad 1s^2 2s^2 2p^6$$

On the other side of the periodic table, halogens are among the most reactive nonmetals because of their ns^2np^5 electron configurations (▼ FIGURE 9.31). Each halogen is only one electron away from a noble gas configuration and tends to react to gain that one electron, forming a 1− ion. For example, consider fluorine:

$$\text{F} \quad 1s^2 2s^2 2p^5$$

In reactions, fluorine gains one additional 2p electron, forming a 1− ion with the electron configuration of neon:

$$\text{F}^- \quad 1s^2 2s^2 2p^6$$

The elements that form predictable ions are shown in ▶ FIGURE 9.32 (first introduced in Chapter 4). Notice how the charge of these ions reflects their electron configurations—these elements form ions with noble gas electron configurations.

Atoms and/or ions that share the same electron configuration are termed isoelectronic.

Alkali metals 1 1A
3 Li $2s^1$
11 Na $3s^1$
19 K $4s^1$
37 Rb $5s^1$
55 Cs $6s^1$
87 Fr $7s^1$

▲ FIGURE 9.29 **Electron configurations of the alkali metals** The alkali metals all have ns^1 electron configurations and are therefore one electron beyond a noble gas configuration. In their reactions, they tend to lose that electron, forming 1+ ions and attaining a noble gas configuration.

Alkaline earth metals 2 2A
4 Be $2s^2$
12 Mg $3s^2$
20 Ca $4s^2$
38 Sr $5s^2$
56 Ba $6s^2$
88 Ra $7s^2$

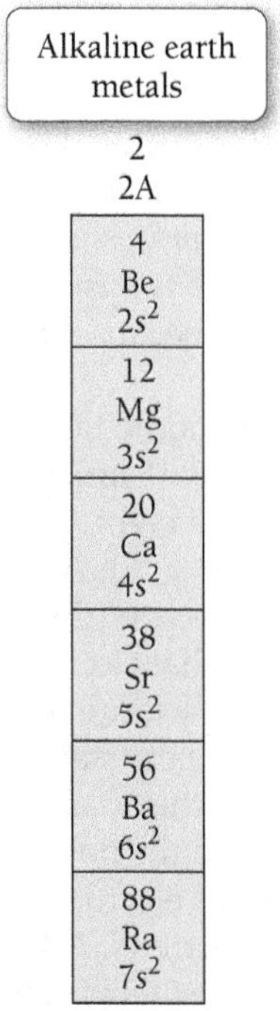

◀ FIGURE 9.30 **Electron configurations of the alkaline earth metals** The alkaline earth metals all have ns^2 electron configurations and are therefore two electrons beyond a noble gas configuration. In their reactions, they tend to lose two electrons, forming 2+ ions and attaining a noble gas configuration.

Halogens 17 7A
9 F $2s^2 2p^5$
17 Cl $3s^2 3p^5$
35 Br $4s^2 4p^5$
53 I $5s^2 5p^5$
85 At $6s^2 6p^5$

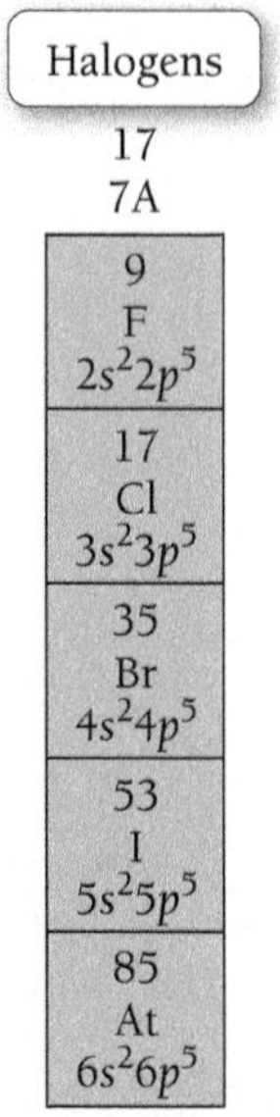

▶ FIGURE 9.31 **Electron configurations of the halogens** The halogens all have ns^2np^5 electron configurations and are therefore one electron short of a noble gas configuration. In their reactions, they tend to gain one electron, forming 1− ions and attaining a noble gas configuration.

1	2		3	4	5	6	7	8
Li^+					N^{3-}	O^{2-}	F^-	
Na^+	Mg^{2+}		Al^{3+}			S^{2-}	Cl^-	
K^+	Ca^{2+}		Ga^{3+}			Se^{2-}	Br^-	
Rb^+	Sr^{2+}	Most transition metals form cations with various charges	In^{3+}			Te^{2-}	I^-	
Cs^+	Ba^{2+}							

▲ FIGURE 9.32 **Elements that form predictable ions**

CONCEPTUAL CHECKPOINT 9.8

Shown here is the electron configuration of calcium:

$$\text{Ca} \quad 1s^22s^22p^63s^23p^64s^2$$

In its reactions, calcium tends to form the Ca^{2+} ion. Which electrons are lost upon ionization?

(a) all of the 4*s* electrons
(b) two of the 3*p* electrons
(c) all of the 3*s* electrons
(d) the 1*s* electrons

9.9 Periodic Trends: Atomic Size, Ionization Energy, and Metallic Character

▶ Identify and understand periodic trends in atomic size, ionization energy, and metallic character.

The quantum-mechanical model also explains other periodic trends such as atomic size, ionization energy, and metallic character. We examine these trends individually in this section of the chapter.

Atomic Size

The **atomic size** of an atom is determined by the distance between its outermost electrons and its nucleus. As we move across a period in the periodic table, we know that electrons occupy orbitals with the same principal quantum number, *n*. Since the principal quantum number largely determines the size of an orbital, electrons are therefore filling orbitals of approximately the same size, and we might expect atomic size to remain constant across a period. However, with each step across a period, the number of protons in the nucleus also increases. This increase in the number of protons results in a greater pull on the electrons from the nucleus, causing atomic size to actually decrease. Therefore:

> As we move to the right across a period, or row, in the periodic table, atomic size decreases, as shown in ▶ FIGURE 9.33, on the next page.

As we move down a column in the periodic table, the highest principal quantum number, *n*, increases. Because the size of an orbital increases with increasing principal quantum number, the electrons that occupy the outermost orbitals are farther from the nucleus as we move down a column. Therefore:

> As we move down a column, or family, in the periodic table, atomic size increases, as shown in Figure 9.33.

▶ **FIGURE 9.33 Periodic properties: atomic size** Atomic size decreases as we move to the right across a period and increases as we move down a column in the periodic table.

Relative atomic sizes of the main-group elements

Period	1A	2A	3A	4A	5A	6A	7A	8A
1	H							He
2	Li	Be	B	C	N	O	F	Ne
3	Na	Mg	Al	Si	P	S	Cl	Ar
4	K	Ca	Ga	Ge	As	Se	Br	Kr
5	Rb	Sr	In	Sn	Sb	Te	I	Xe
6	Cs	Ba	Tl	Pb	Bi	Po	At	Rn

Sizes of atoms tend to increase down a column.

Sizes of atoms tend to decrease across a period.

Interactive Worked Example Video 9.6

EXAMPLE 9.6 Atomic Size

Choose the larger atom in each pair.

(a) C or O **(b)** Li or K **(c)** C or Al **(d)** Se or I

SOLUTION

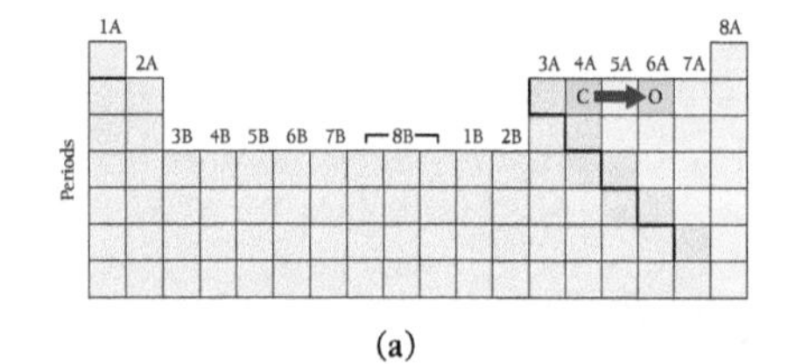

(a)

(a) C or O

Carbon atoms are larger than O atoms because, as you trace the path between C and O on the periodic table, you move to the right within the same period. Atomic size decreases as you go to the right.

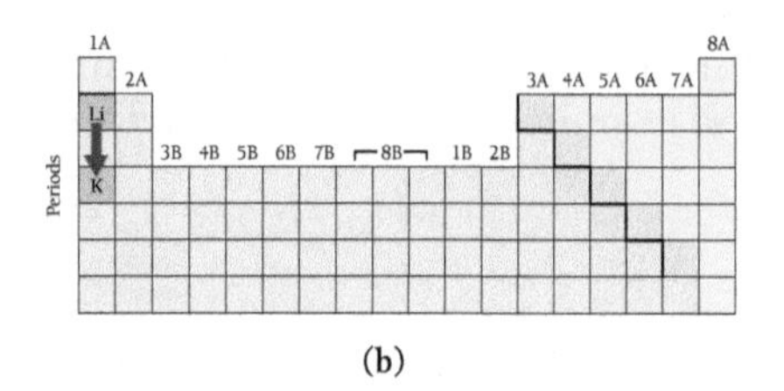

(b)

(b) Li or K

Potassium atoms are larger than Li atoms because, as you trace the path between Li and K on the periodic table, you move down a column. Atomic size increases as you go down a column.

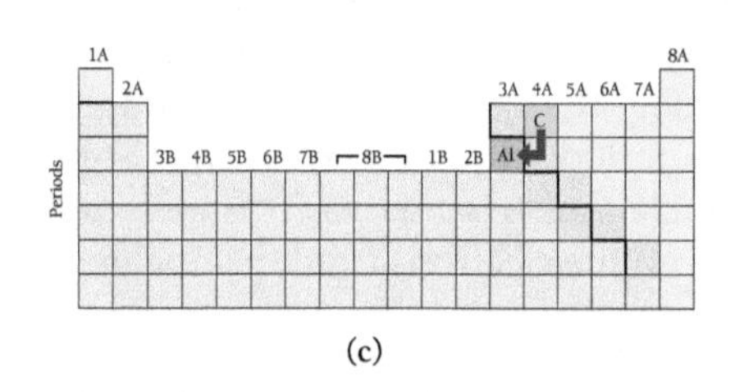

(c)

(c) C or Al

Aluminum atoms are larger than C atoms because, as you trace the path between C and Al on the periodic table, you move down a column (atomic size increases) and then to the left across a period (atomic size increases). These effects add together for an overall increase.

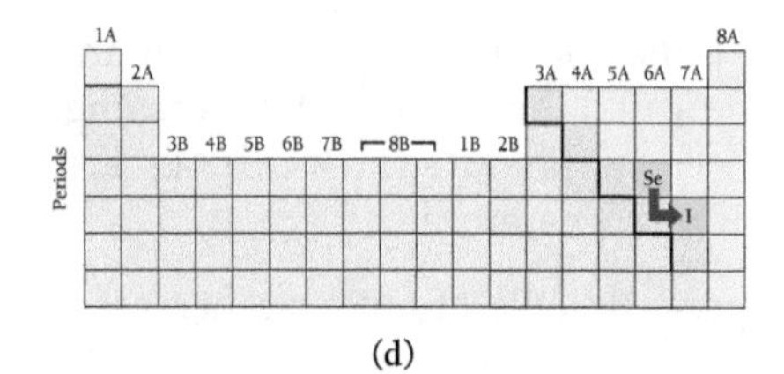

(d)

(d) Se or I

Based on periodic properties alone, you cannot tell which atom is larger because as you trace the path between Se and I, you go down a column (atomic size increases) and then to the right across a period (atomic size decreases). These effects tend to cancel one another.

▶ **SKILLBUILDER 9.6 | Atomic Size**

Choose the larger atom in each pair.

(a) Pb or Po **(b)** Rb or Na **(c)** Sn or Bi **(d)** F or Se

▶ **FOR MORE PRACTICE** Example 9.13a; Problems 81, 82, 83, 84.

CHEMISTRY AND HEALTH

Pumping Ions: Atomic Size and Nerve Impulses

No matter what you are doing at this moment, tiny pumps in each of the trillions of cells that make up your body are hard at work. These pumps, located in the cell membrane, move a number of different ions into and out of the cell. The most important of these ions are sodium (Na^+) and potassium (K^+), which happen to be pumped in opposite directions. Sodium ions are pumped *out of cells*, while potassium ions are pumped *into cells*. The result is a *chemical gradient* for each ion: The concentration of sodium is higher outside the cell than within, while exactly the opposite is true for potassium.

The ion pumps within the cell membrane are analogous to water pumps in a high-rise building that pump water against the force of gravity to a tank on the roof. Other structures within the membrane, called ion channels, are like the building's faucets. When they open momentarily, bursts of sodium and potassium ions, driven by their concentration gradients, flow back across the membrane—sodium flowing in and potassium flowing out. These ion pulses are the basis for the transmission of nerve signals in the brain, heart, and throughout the body. Consequently, every move you make or every thought you have is mediated by the flow of these ions.

How do the pumps and channels differentiate between sodium and potassium ions? How do the ion pumps selectively move sodium out of the cell and potassium into the cell? To answer this question, we must examine the sodium and potassium ions more closely. In what ways do they differ? Both are cations of Group I metals. All Group I metals tend to lose one electron to form cations with 1+ charge, so the magnitude of the charge cannot be the decisive factor. But potassium (atomic number 19) lies directly below sodium in the periodic table (atomic number 11) and based on periodic properties potassium is therefore larger than sodium. The potassium ion has a radius of 133 pm, while the sodium ion has a radius of 95 pm. (Recall from Chapter 2 that $1\,\text{pm} = 10^{-12}\,\text{m}$.) The pumps and channels within cell membranes are so sensitive that they distinguish between the sizes of these two ions and selectively allow only one or the other to pass. The result is the transmission of nerve signals that allows you to read this page.

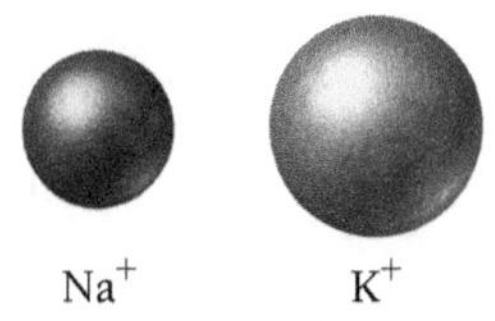

B9.2 CAN YOU ANSWER THIS? *Other ions, including calcium and magnesium, are also important to nerve signal transmission. Arrange these four ions in order of increasing size:* K^+, Na^+, Mg^{2+}, *and* Ca^{2+}.

Ionization Energy

The **ionization energy** of an atom is the energy required to remove an electron from the atom in the gaseous state. The ionization of sodium, for example, is represented with the equation:

$$\text{Na} + \text{ionization energy} \longrightarrow \text{Na}^+ + 1\text{e}^-$$

Based on what we know about electron configurations, what can we predict about ionization energy trends? Would it take more or less energy to remove an electron from Na than from Cl? We know that Na has an outer electron configuration of $3s^1$ and Cl has an outer electron configuration of $3s^23p^5$. Since removing an electron from Na gives it a noble gas configuration—and removing an electron from Cl does not—we would expect sodium to have a lower ionization energy, and that is the case. It is easier to remove an electron from sodium than it is from chlorine. We can generalize this idea in this statement:

> As we move across a period, or row, to the right in the periodic table, ionization energy increases (▶ FIGURE 9.34, on the next page).

What happens to ionization energy as we move down a column? As we have learned, the principal quantum number, n, increases as we move down a column. Within a given subshell, orbitals with higher principal quantum numbers are larger than orbitals with smaller principal quantum numbers. Consequently, electrons in the outermost principal shell are farther away from the positively charged nucleus—and therefore are held less tightly—as we move down a column. This

Ionization energy trends

Period	1 1A	2 2A	3 3B	4 4B	5 5B	6 6B	7 7B	8 8B	9 8B	10 8B	11 1B	12 2B	13 3A	14 4A	15 5A	16 6A	17 7A	18 8A
1	1 H																	2 He
2	3 Li	4 Be											5 B	6 C	7 N	8 O	9 F	10 Ne
3	11 Na	12 Mg											13 Al	14 Si	15 P	16 S	17 Cl	18 Ar
4	19 K	20 Ca	21 Sc	22 Ti	23 V	24 Cr	25 Mn	26 Fe	27 Co	28 Ni	29 Cu	30 Zn	31 Ga	32 Ge	33 As	34 Se	35 Br	36 Kr
5	37 Rb	38 Sr	39 Y	40 Zr	41 Nb	42 Mo	43 Tc	44 Ru	45 Rh	46 Pd	47 Ag	48 Cd	49 In	50 Sn	51 Sb	52 Te	53 I	54 Xe
6	55 Cs	56 Ba	57 La	72 Hf	73 Ta	74 W	75 Re	76 Os	77 Ir	78 Pt	79 Au	80 Hg	81 Tl	82 Pb	83 Bi	84 Po	85 At	86 Rn
7	87 Fr	88 Ra	89 Ac	104 Rf	105 Db	106 Sg	107 Bh	108 Hs	109 Mt	110 Ds	111 Rg	112 Cn	113 Nh	114 Fl	115 Mc	116 Lv	117 Ts	118 Og

Lanthanides	58 Ce	59 Pr	60 Nd	61 Pm	62 Sm	63 Eu	64 Gd	65 Tb	66 Dy	67 Ho	68 Er	69 Tm	70 Yb	71 Lu
Actinides	90 Th	91 Pa	92 U	93 Np	94 Pu	95 Am	96 Cm	97 Bk	98 Cf	99 Es	100 Fm	101 Md	102 No	103 Lr

Ionization energy decreases

Ionization energy increases

▶ **FIGURE 9.34 Periodic properties: ionization energy** Ionization energy increases as we move to the right across a period and decreases as we move down a column in the periodic table.

results in a lower ionization energy (if the electron is held less tightly, it is easier to pull away) as we move down a column. Therefore:

> As we move down a column (or family) in the periodic table, ionization energy decreases (see Figure 9.34).

Notice that the trends in ionization energy are consistent with the trends in atomic size. Smaller atoms are more difficult to ionize because their electrons are held more tightly. Therefore, as we move across a period, atomic size decreases and ionization energy increases. Similarly, as we move down a column, atomic size increases and ionization energy decreases since electrons are farther from the nucleus and are therefore less tightly held.

EXAMPLE 9.7 Ionization Energy

Choose the element with the higher ionization energy from each pair.

(a) Mg or P **(b)** As or Sb **(c)** N or Si **(d)** O or Cl

SOLUTION

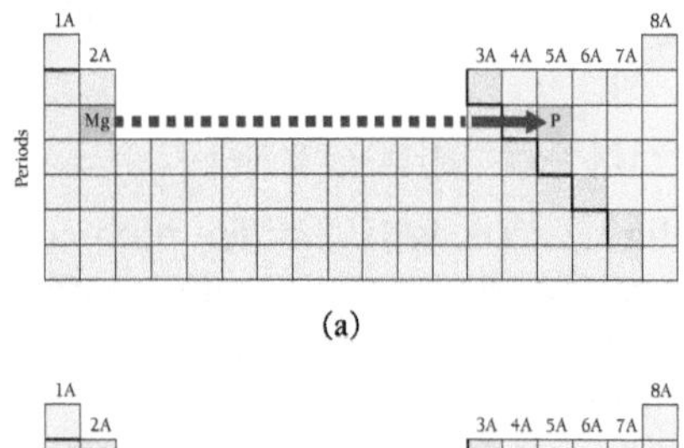

(a)

(a) Mg or P
P has a higher ionization than Mg because, as you trace the path between Mg and P on the periodic table, you move to the right within the same period. Ionization energy increases as you go to the right.

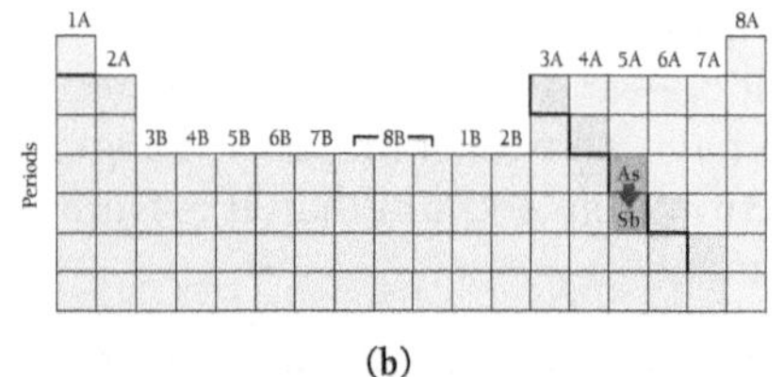

(b)

(b) As or Sb
As has a higher ionization energy than Sb because, as you trace the path between As and Sb on the periodic table, you move down a column. Ionization energy decreases as you go down a column.

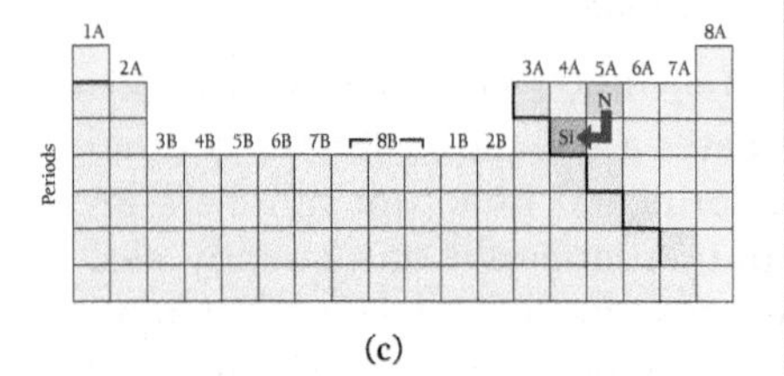

(c)

(c) N or Si
N has a higher ionization energy than Si because, as you trace the path between N and Si on the periodic table, you move down a column (ionization energy decreases) and then to the left across a period (ionization energy decreases). These effects sum together for an overall decrease.

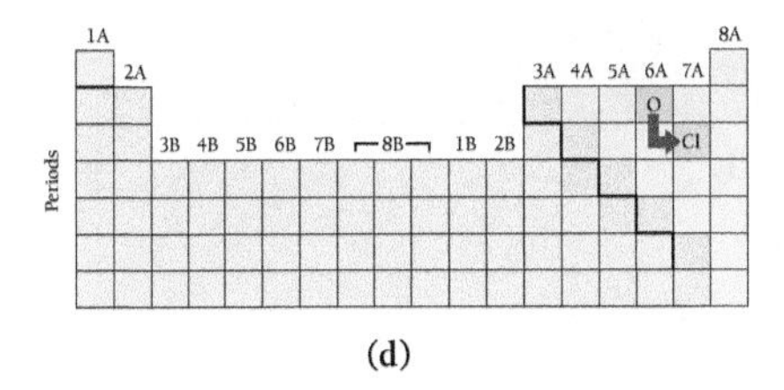

(d) O or Cl

Based on periodic properties alone, you cannot tell which has a higher ionization energy because, as you trace the path between O and Cl you move down a column (ionization energy decreases) and then to the right across a period (ionization energy increases). These effects tend to cancel.

▶ **SKILLBUILDER 9.7 | Ionization Energy**

Choose the element with the higher ionization energy from each pair.

(a) Mg or Sr **(b)** In or Te **(c)** C or P **(d)** F or S

▶ **FOR MORE PRACTICE** Example 9.13b; Problems 77, 78, 79, 80.

Metallic Character

As we learned in Chapter 4, metals tend to lose electrons in their chemical reactions, while nonmetals tend to gain electrons. As we move across a period in the periodic table, ionization energy increases, which means that electrons are less likely to be lost in chemical reactions. Consequently:

> As we move across a period, or row, to the right in the periodic table, **metallic character** decreases (▼ FIGURE 9.35).

As we move down a column in the periodic table, ionization energy decreases, making electrons more likely to be lost in chemical reactions. Consequently:

> As we move down a column, or family, in the periodic table, metallic character increases (see Figure 9.35).

These trends, based on the quantum-mechanical model, explain the distribution of metals and nonmetals that we were introduced to in Chapter 4. Metals are found toward the left side of the periodic table and nonmetals (with the exception of hydrogen) toward the upper right.

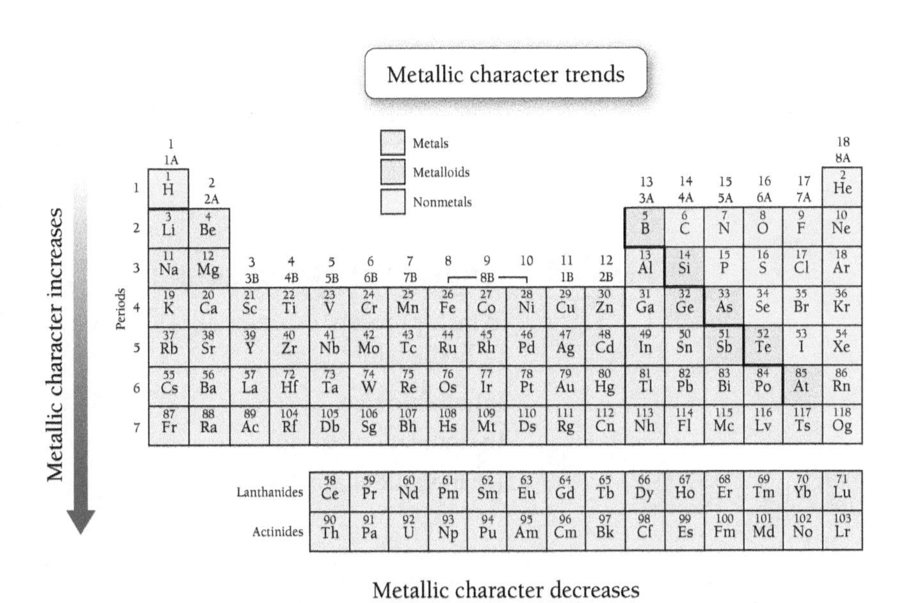

▲ FIGURE 9.35 **Periodic properties: metallic character** Metallic character decreases as we move to the right across a period and increases as we move down a column in the periodic table.

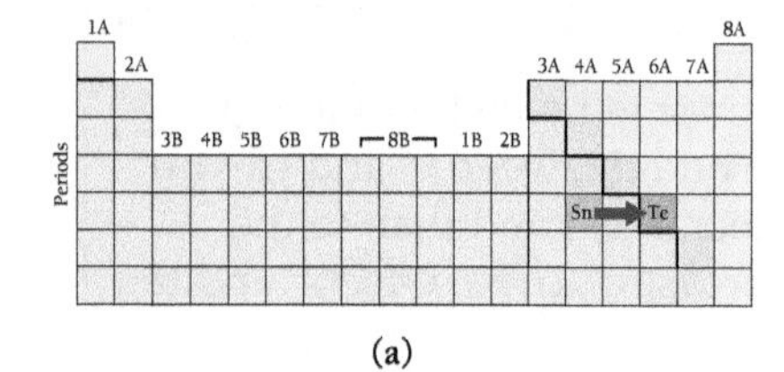

(a)

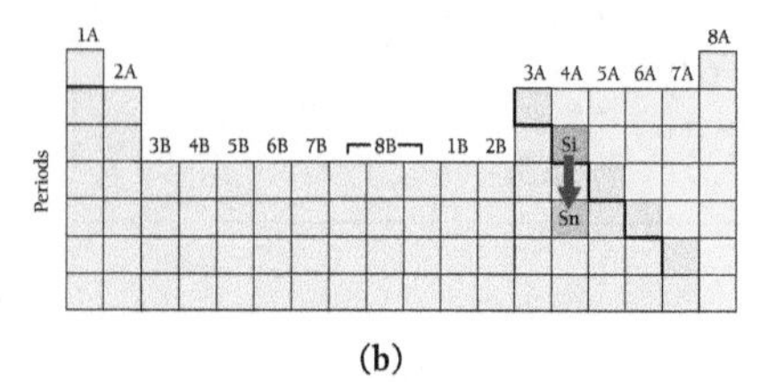

(b)

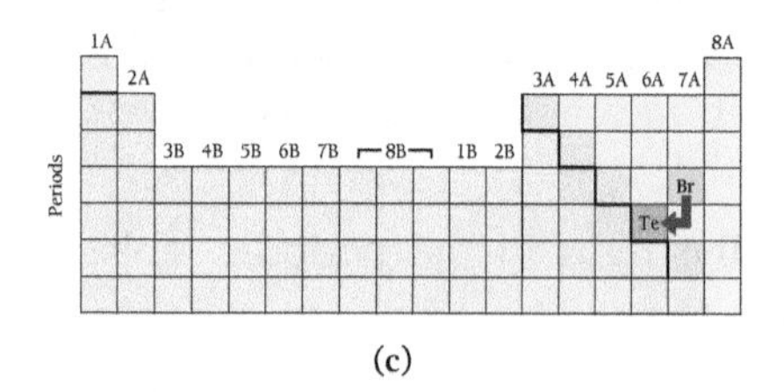

(c)

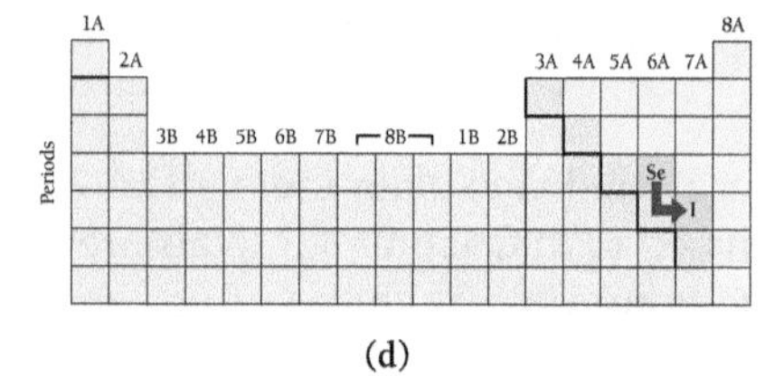

(d)

EXAMPLE 9.8 Metallic Character

Choose the more metallic element from each pair.

(a) Sn or Te
(b) Si or Sn
(c) Br or Te
(d) Se or I

SOLUTION

(a) Sn or Te
Sn is more metallic than Te because, as you trace the path between Sn and Te on the periodic table, you move to the right within the same period. Metallic character decreases as you go to the right.

(b) Si or Sn
Sn is more metallic than Si because, as you trace the path between Si and Sn on the periodic table, you move down a column. Metallic character increases as you go down a column.

(c) Br or Te
Te is more metallic than Br because, as you trace the path between Br and Te on the periodic table, you move down a column (metallic character increases) and then to the left across a period (metallic character increases). These effects add together for an overall increase.

(d) Se or I
Based on periodic properties alone, you cannot tell which is more metallic because, as you trace the path between Se and I, you go down a column (metallic character increases) and then to the right across a period (metallic character decreases). These effects tend to cancel.

▶ SKILLBUILDER 9.8 | Metallic Character

Choose the more metallic element from each pair.

(a) Ge or In
(b) Ga or Sn
(c) P or Bi
(d) B or N

▶ FOR MORE PRACTICE Example 9.13; Problems 85, 86, 87, 88.

CONCEPTUAL ✓ CHECKPOINT 9.9

Which property *increases* as you move from left to right across a row in the periodic table?

(a) atomic size
(b) ionization energy
(c) metallic character

Chapter 9 in Review

MasteringChemistry™ provides end-of-chapter exercises, feedback-enriched tutorial problems, animations, and interactive activities to encourage problem solving practice and deeper understanding of key concepts and topics.

Self-Assessment Quiz

PEARSON eText 2.0

Q1. Which set of wavelengths for light are arranged in order of increasing frequency?
(a) 250 nm; 300 nm; 350 nm
(b) 350 nm; 300 nm; 250 nm
(c) 300 nm; 350 nm; 250 nm
(d) 300 nm; 250 nm; 350 nm

Q2. Which of the listed types of electromagnetic radiation has the longest wavelength?
(a) ultraviolet (b) X-ray
(c) infrared (d) microwaves

Q3. Which electron transition in the Bohr model would produce light with the longest wavelength?
(a) $n = 2 \longrightarrow n = 1$ (b) $n = 3 \longrightarrow n = 1$
(c) $n = 4 \longrightarrow n = 1$ (d) $n = 5 \longrightarrow n = 1$

Q4. What is the electron configuration of arsenic (As)?
(a) $[Ar]4s^2 4p^3$ (b) $[Ar]4s^2 4d^{10} 4p^3$
(c) $[Ar]4s^2 3d^6 4p^3$ (d) $[Ar]4s^2 3d^{10} 4p^3$

Q5. Which orbital diagram corresponds to phosphorus (P)?

(a)

1s	2s	2p			3s	3p		
⇅	⇅	⇅	⇅	⇅	⇅	⇅	↑	

(b)

1s	2s	2p		
⇅	⇅	↑	↑	↑

(c)

1s	2s	2p			3s	3p		
⇅	⇅	⇅	⇅	⇅	⇅	↑	↑	↑

(d)

1s	2s	2p			3s	3p		
⇅	⇅	⇅	⇅	⇅	⇅	⇅	⇅	↑

Q6. How many valence electrons does tellurium (Te) have?
(a) 5 (b) 6 (c) 16 (d) 52

Q7. The element sulfur forms an ion with what charge?
(a) 2− (b) 1− (c) 1+ (d) 2+

Q8. Order the elements Sr, Ca, and Se in order of decreasing atomic size.
(a) Se > Sr > Ca (b) Ca > Se > Sr
(c) Sr > Ca > Se (d) Se > Ca >Sr

Q9. Which of the listed elements has the highest ionization energy?
(a) Sn (b) S (c) Si (d) F

Q10. Which of the listed elements is most metallic?
(a) Al (b) N (c) P (d) O

Q11. Which property decreases as you move down a column in the periodic table?
(a) atomic size
(b) ionization energy
(c) metallic character
(d) none of the above (all increase as you move down a column).

Q12. When aluminum forms an ion, it loses electrons. How many electrons does it lose, and which orbitals do the electrons come from?
(a) one electron from the 3*s* orbital
(b) two electrons: one from the 3*s* orbital and one from the 2*s* orbital
(c) three electrons: two from the 3*s* orbital and one from the 3*p* orbital
(d) five electrons from the 3*p* orbital

Answers: 1:b, 2:d, 3:a, 4:d, 5:c, 6:b, 7:a, 8:c, 9:d, 10:a, 11:b, 12:c

Chemical Principles

Light

Light is electromagnetic radiation, energy that travels through space at a constant speed of 3.0×10^8 m/s (186,000 mi/s) and exhibits both wavelike and particle-like behavior. Particles of light are called photons. The wave nature of light is characterized by its wavelength, the distance between adjacent crests in the wave. The wavelength of light is inversely proportional to both the frequency—the number of cycles that pass a stationary point in one second—and the energy of a photon. Electromagnetic radiation ranges in wavelength from 10^{-16} m (gamma rays) to 10^6 m (radio waves). In between these lie X-rays, ultraviolet light, visible light, infrared light, and microwaves.

Relevance

Light enables us to see the world. However, we see only visible light, a small sliver in the center of the electromagnetic spectrum. We use other forms of electromagnetic radiation for cancer therapy, X-ray imaging, night vision, microwave cooking, and communications. Light is also important to many chemical processes. We can learn about the electronic structure of atoms, for example, by examining their interaction with light.

The Bohr Model

The emission spectrum of hydrogen, consisting of bright lines at specific wavelengths, is explained by the Bohr model for the hydrogen atom. In this model, electrons occupy circular orbits at specific fixed distances from the nucleus. Each orbit is specified by a quantum number (n), which also specifies the orbit's energy. While an electron is in a given orbit, its energy remains constant. When an electron jumps between orbits, a quantum of energy is absorbed or emitted. Since the difference in energy between orbits is fixed, the energy emitted or absorbed is also fixed. Emitted energy is carried away in the form of a photon of specific wavelength.

The Bohr model was a first attempt to explain the bright-line spectra of atoms. While it does predict the spectrum of the hydrogen atom, it fails to predict the spectra of other atoms and was consequently replaced by the quantum-mechanical model.

The Quantum-Mechanical Model

The quantum-mechanical model for the atom describes electron orbitals, which are electron probability maps that show the relative probability of finding an electron in various places surrounding the atomic nucleus. Orbitals are specified with a number (n), called the principal quantum number, and a letter. The principal quantum number ($n = 1, 2, 3 \ldots$) specifies the principal shell, and the letter (s, p, d, or f) specifies the subshell of the orbital. In the hydrogen atom, the energy of orbitals depends only on n. In multi-electron atoms, the energy ordering is $1s$ $2s$ $2p$ $3s$ $3p$ $4s$ $3d$ $4p$ $5s$ $4d$ $5p$ $6s$.

An electron configuration indicates which orbitals are occupied for a particular atom. Orbitals are filled in order of increasing energy and obey the Pauli exclusion principle (each orbital can hold a maximum of two electrons with opposing spins) and Hund's rule (electrons occupy orbitals of identical energy singly before pairing).

The quantum-mechanical model changed the way we view nature. Before the quantum-mechanical model, electrons were viewed as small particles, much like any other particle. Electrons were expected to follow the normal laws of motion, just as a baseball does. However, the electron, with its wavelike properties, does not follow these laws. Instead, electron motion is describable only through probabilistic predictions. Quantum theory single-handedly changed the predictability of nature at its most fundamental level.

The quantum-mechanical model of the atom predicts and explains many of the chemical properties we learned about in earlier chapters.

The Periodic Table

Elements in the same column of the periodic table have similar outer electron configurations and the same number of valence electrons (electrons in the outermost principal shell), and therefore similar chemical properties. We divide the periodic table into blocks (s block, p block, d block, and f block) in which particular sublevels are filled. As we move across a period to the right in the periodic table, atomic size decreases, ionization energy increases, and metallic character decreases. As we move down a column in the periodic table, atomic size increases, ionization energy decreases, and metallic character increases.

The periodic law exists because the number of valence electrons is periodic, and valence electrons determine chemical properties. Quantum theory also predicts that atoms with eight outershell electrons (or two for helium) are particularly stable, thus explaining the inertness of the noble gases. Atoms without noble gas configurations undergo chemical reactions to attain them, explaining the reactivity of the alkali metals and the halogens as well as the tendency of several families to form ions with certain charges.

Chemical Skills

LO: Predict the relative wavelength, energy, and frequency of different types of light (Section 9.3).

- Figure 9.4 includes relative wavelengths.
- Energy per photon increases with decreasing (shorter) wavelength.
- Frequency increases with decreasing (shorter) wavelength.

Examples

EXAMPLE **9.9** **Predicting Relative Wavelength, Energy, and Frequency of Light**

Which type of light—infrared or ultraviolet—has the longer wavelength? Higher frequency? Higher energy per photon?

SOLUTION

Infrared light has the longer wavelength (see Figure 9.4). Ultraviolet light has the higher frequency and the higher energy per photon.

LO: Write electron configurations and orbital diagrams for atoms (Section 9.6).

To write electron configurations, determine the number of electrons in the atom from the element's atomic number and then follow these rules:

- Electrons occupy orbitals so as to minimize the energy of the atom; therefore, lower-energy orbitals fill before higher-energy orbitals. Orbitals fill in the order: 1*s* 2*s* 2*p* 3*s* 3*p* 4*s* 3*d* 4*p* 5*s* 4*d* 5*p* 6*s* (Figure 9.24). The *s* subshells hold up to two electrons, *p* subshells hold up to six, *d* subshells hold up to ten, and *f* subshells hold up to 14.
- Orbitals can hold no more than two electrons each. When two electrons occupy the same orbital, they must have opposing spins.
- When orbitals of identical energy are available, these are first occupied singly with parallel spins rather than in pairs.

EXAMPLE **9.10** **Writing Electron Configurations and Orbital Diagrams**

Write an electron configuration and orbital diagram (outer electrons only) for germanium.

SOLUTION

Germanium is atomic number 32; therefore, it has 32 electrons.

Electron Configuration

Ge $1s^2 2s^2 2p^6 3s^2 3p^6 4s^2 3d^{10} 4p^2$

or

Ge $[Ar]4s^2 3d^{10} 4p^2$

Orbital Diagram (Outer Electrons)

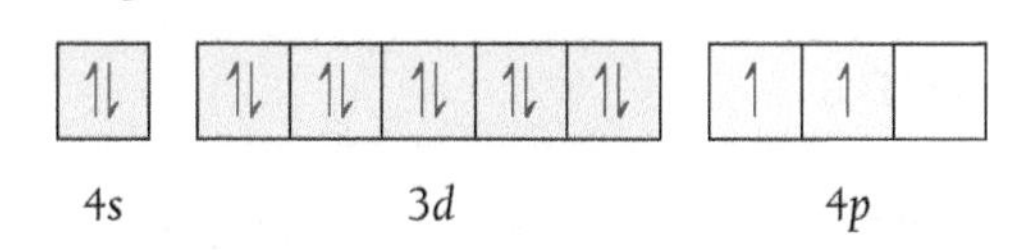

LO: Identify valence electrons and core electrons (Section 9.7).

- Valence electrons are the electrons in the outermost principal energy shell (the principal shell with the highest principal quantum number).
- Core electrons are electrons that are not in the outermost principal shell.

EXAMPLE **9.11** **Identifying Valence Electrons and Core Electrons**

Identify the valence electrons and core electrons in the electron configuration of germanium (given in Example 9.10).

SOLUTION

Ge $1s^2 2s^2 2p^6 3s^2 3p^6\ 4s^2\ 3d^{10}\ 4p^2$

28 core electrons

4 valence electrons

LO: Write electron configurations for elements based on their positions in the periodic table (Section 9.7).

- The inner electron configuration for any element is the electron configuration of the noble gas that immediately precedes that element in the periodic table. Represent the inner configuration with the symbol for the noble gas in brackets.
- The outer electrons can be determined from the element's position within a particular block (*s*, *p*, *d*, or *f*) in the periodic table. Trace the elements between the preceding noble gas and the element of interest and assign electrons to the appropriate orbitals. Figure 9.26 shows the outer electron configuration based on the position of an element in the periodic table.
- The highest principal quantum number (highest *n* value) is equal to the row number of the element in the periodic table.
- The principal quantum number (*n* value) of the outermost *d* electrons for any element containing *d* electrons is equal to the row number of the element minus 1.

EXAMPLE **9.12** **Writing an Electron Configuration for an Element Based on Its Position in the Periodic Table**

Write an electron configuration for iodine based on its position in the periodic table.

SOLUTION

The inner configuration for I is [Kr].

Begin with the [Kr] inner electron configuration. As you trace from Kr to I, add two 5*s* electrons, ten 4*d* electrons, and five 5*p* electrons. The overall configuration is:

I $[Kr]5s^2 4d^{10} 5p^5$

LO: Identify and understand periodic trends in atomic size, ionization energy, and metallic character (Section 9.9).

On the periodic table:

- Atomic size decreases as you move to the right and increases as you move down.
- Ionization energy increases as you move to the right and decreases as you move down.
- Metallic character decreases as you move to the right and increases as you move down.

EXAMPLE **9.13** **Periodic Trends: Atomic Size, Ionization Energy, and Metallic Character**

Arrange Si, In, and S in order of **(a)** increasing atomic size, **(b)** increasing ionization energy, and **(c)** increasing metallic character.

SOLUTION

(a) S, Si, In
(b) In, Si, S
(c) S, Si, In

Key Terms

atomic size [9.9]
Bohr model [9.1]
core electrons [9.7]
electromagnetic radiation [9.2]
electromagnetic spectrum [9.3]
electron configuration [9.6]
electron spin [9.6]
emission spectrum (plural, emission *spectra*) [9.4]
excited state [9.6]
frequency (ν) [9.2]
gamma ray [9.3]
ground state [9.6]
Hund's rule [9.6]
infrared light [9.3]
ionization energy [9.9]
metallic character [9.9]
microwaves [9.3]
orbital [9.5]
orbital diagram [9.6]
Pauli exclusion principle [9.6]
photon [9.2]
principal quantum number [9.6]
principal shell [9.6]
quantized [9.4]
quantum (plural, *quanta*) [9.4]
quantum-mechanical model [9.1]
quantum number [9.4]
radio waves [9.3]
subshell [9.6]
ultraviolet (UV) light [9.3]
valence electron [9.7]
visible light [9.3]
wavelength (λ) [9.2]
X-rays [9.3]

Exercises

Questions

1. When were the Bohr model and the quantum-mechanical model for the atom developed? What purpose do these models serve?
2. What is light? How fast does light travel?
3. What is white light? Colored light?
4. Explain, in terms of absorbed and reflected light, why a blue object appears blue.
5. What is the relationship between the wavelength of light and the amount of energy carried by its photons? How are wavelength and frequency of light related?
6. List some sources of gamma rays.
7. How are X-rays used?
8. Why should excess exposure to gamma rays and X-rays be avoided?
9. Why should excess exposure to ultraviolet light be avoided?
10. What objects emit infrared light? What technology exploits this?
11. Why do microwave ovens heat food but tend not to heat the dish the food is on?
12. What type of electromagnetic radiation is used in communications devices such as cellular telephones?
13. Describe the Bohr model for the hydrogen atom.
14. What is an emission spectrum? Use the Bohr model to explain why the emission spectrum of the hydrogen atom consists of distinct lines at specific wavelengths.
15. Explain the difference between a Bohr orbit and a quantum-mechanical orbital.
16. What is the difference between the ground state of an atom and an excited state of an atom?
17. Explain how the motion of an electron is different from the motion of a baseball. What is a probability map?
18. Why do quantum-mechanical orbitals have "fuzzy" boundaries?
19. List the four possible subshells in the quantum-mechanical model, the number of orbitals in each subshell, and the maximum number of electrons that can be contained in each subshell.
20. List the quantum-mechanical orbitals through 5*s*, in the correct energy order for multi-electron atoms.
21. What is the Pauli exclusion principle? Why is it important when writing electron configurations?
22. What is Hund's rule? Why is it important when writing orbital diagrams?
23. Within an electron configuration, what do symbols such as [Ne] and [Kr] represent?

24. Explain the difference between valence electrons and core electrons.

25. Identify each block in the blank periodic table.
(a) *s* block (b) *p* block
(c) *d* block (d) *f* block

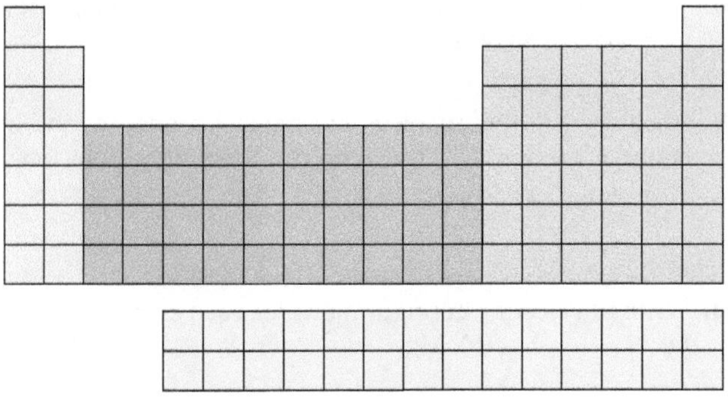

26. List some examples of the explanatory power of the quantum-mechanical model.

27. Explain why Group 1 elements tend to form 1+ ions and Group 7 elements tend to form 1− ions.

28. Explain the periodic trends in each chemical property.
(a) ionization energy
(b) atomic size
(c) metallic character

Problems

WAVELENGTH, ENERGY, AND FREQUENCY OF ELECTROMAGNETIC RADIATION

29. How long does it take light to travel:
(a) 1.0 ft (report answer in nanoseconds)
(b) 2462 mi, the distance between Los Angeles and New York (report answer in milliseconds)
(c) 4.5 billion km, the average separation between the sun and Neptune (report answer in hours and minutes)

30. How far does light travel in each time period?
(a) 1.0 s
(b) 1.0 day
(c) 1.0 yr

31. Which type of electromagnetic radiation has the longest wavelength?
(a) visible (b) ultraviolet
(c) infrared (d) X-ray

32. Which type of electromagnetic radiation has the shortest wavelength?
(a) radio waves (b) microwaves
(c) infrared (d) ultraviolet

33. List the types of electromagnetic radiation in order of increasing energy per photon.
(a) radio waves (b) microwaves
(c) infrared (d) ultraviolet

34. List the types of electromagnetic radiation in order of decreasing energy per photon.
(a) gamma rays (b) radio waves
(c) microwaves (d) visible light

35. List two types of electromagnetic radiation with frequencies higher than visible light.

36. List two types of electromagnetic radiation with frequencies lower than infrared light.

37. List these three types of radiation—infrared, X-ray, and radio waves—in order of:
(a) increasing energy per photon
(b) increasing frequency
(c) increasing wavelength

38. List these three types of electromagnetic radiation—visible, gamma rays, and microwaves—in order of:
(a) decreasing energy per photon
(b) decreasing frequency
(c) decreasing wavelength

THE BOHR MODEL

39. Bohr orbits have fixed ______ and fixed ______.

40. In the Bohr model, what happens when an electron makes a transition between orbits?

41. Two of the emission wavelengths in the hydrogen emission spectrum are 410 nm and 434 nm. One of these is due to the $n = 6$ to $n = 2$ transition, and the other is due to the $n = 5$ to $n = 2$ transition. Which wavelength corresponds to which transition?

42. Two of the emission wavelengths in the hydrogen emission spectrum are 656 nm and 486 nm. One of these is due to the $n = 4$ to $n = 2$ transition, and the other is due to the $n = 3$ to $n = 2$ transition. Which wavelength corresponds to which transition?

THE QUANTUM-MECHANICAL MODEL

43. Sketch the 1*s* and 2*p* orbitals. How do the 2*s* and 3*p* orbitals differ from the 1*s* and 2*p* orbitals?

44. Sketch the 3*d* orbitals. How do the 4*d* orbitals differ from the 3*d* orbitals?

45. Which electron is, on average, closer to the nucleus: an electron in a 2*s* orbital or an electron in a 3*s* orbital?

46. Which electron is, on average, farther from the nucleus: an electron in a 3*p* orbital or an electron in a 4*p* orbital?

47. According to the quantum-mechanical model for the hydrogen atom, which electron transition produces light with longer wavelength: 2*p* to 1*s* or 3*p* to 1*s*?

48. According to the quantum-mechanical model for the hydrogen atom, which transition produces light with longer wavelength: 3*p* to 2*s* or 4*p* to 2*s*?

ELECTRON CONFIGURATIONS

49. Write full electron configurations for each element.
(a) Sr (b) Ge (c) Li (d) Kr

50. Write full electron configurations for each element.
(a) N (b) Mg (c) Ar (d) Se

51. Write full orbital diagrams and indicate the number of unpaired electrons for each element.
(a) He (b) B (c) Li (d) N

52. Write full orbital diagrams and indicate the number of unpaired electrons for each element.
(a) F (b) C (c) Ne (d) Be

53. Write electron configurations for each element. Use the symbol of the previous noble gas in brackets to represent the core electrons.
(a) Ga (b) As (c) Rb (d) Sn

54. Write electron configurations for each element. Use the symbol of the previous noble gas in brackets to represent the core electrons.
(a) Te (b) Br (c) I (d) Cs

55. Write electron configurations for each transition metal.
(a) Zn (b) Cu (c) Zr (d) Fe

56. Write electron configurations for each transition metal.
(a) Mn (b) Ti (c) Cd (d) V

VALENCE ELECTRONS AND CORE ELECTRONS

57. Write full electron configurations and indicate the valence electrons and the core electrons for each element.
(a) Kr (b) Ge (c) Cl (d) Sr

58. Write full electron configurations and indicate the valence electrons and the core electrons for each element.
(a) Sb (b) N (c) B (d) K

59. Write orbital diagrams for the valence electrons and indicate the number of unpaired electrons for each element.
(a) Br (b) Kr (c) Na (d) In

60. Write orbital diagrams for the valence electrons and indicate the number of unpaired electrons for each element.
(a) Ne (b) I (c) Sr (d) Ge

61. How many valence electrons are in each element?
(a) O (b) S (c) Br (d) Rb

62. How many valence electrons are in each element?
(a) Ba (b) Al (c) Be (d) Se

ELECTRON CONFIGURATIONS AND THE PERIODIC TABLE

63. List the outer electron configuration for each column in the periodic table.
(a) 1A (b) 2A (c) 5A (d) 7A

64. List the outer electron configuration for each column in the periodic table.
(a) 3A (b) 4A (c) 6A (d) 8A

65. Use the periodic table to write electron configurations for each element.
(a) Al (b) Be (c) In (d) Zr

66. Use the periodic table to write electron configurations for each element.
(a) Tl (b) Co (c) Ba (d) Sb

67. Use the periodic table to write electron configurations for each element.
(a) Sr (b) Y (c) Ti (d) Te

68. Use the periodic table to write electron configurations for each element.
(a) Se (b) Sn (c) Pb (d) Cd

69. How many 2*p* electrons are in an atom of each element?
(a) C (b) N (c) F (d) P

70. How many 3*d* electrons are in an atom of each element?
(a) Fe (b) Zn (c) K (d) As

71. List the number of elements in periods 1 and 2 of the periodic table. Why does each period have a different number of elements?

72. List the number of elements in periods 3 and 4 of the periodic table. Why does each period have a different number of elements?

73. Name the element in the third period (row) of the periodic table with:
(a) three valence electrons
(b) a total of four $3p$ electrons
(c) six $3p$ electrons
(d) two $3s$ electrons and no $3p$ electrons

74. Name the element in the fourth period of the periodic table with:
(a) five valence electrons
(b) a total of four $4p$ electrons
(c) a total of three $3d$ electrons
(d) a complete outer shell

75. Use the periodic table to identify the element with each electron configuration.
(a) $[Ne]3s^2 3p^5$ (b) $[Ar]4s^2 3d^{10} 4p^1$
(c) $[Ar]4s^2 3d^6$ (d) $[Kr]5s^1$

76. Use the periodic table to identify the element with each electron configuration.
(a) $[Ne]3s^1$ (b) $[Kr]5s^2 4d^{10}$
(c) $[Xe]6s^2$ (d) $[Kr]5s^2 4d^{10} 5p^3$

PERIODIC TRENDS

77. Choose the element with the higher ionization energy from each pair.
(a) As or Bi (b) As or Br (c) S or I (d) S or Sb

78. Choose the element with the higher ionization energy from each pair.
(a) Al or In (b) Cl or Sb (c) K or Ge (d) S or Se

79. Arrange the elements in order of increasing ionization energy: Te, Pb, Cl, S, Sn.

80. Arrange the elements in order of increasing ionization energy: Ga, In, F, Si, N.

81. Choose the element with the larger atoms from each pair.
(a) Al or In (b) Si or N
(c) P or Pb (d) C or F

82. Choose the element with the larger atoms from each pair.
(a) Sn or Si (b) Br or Ga
(c) Sn or Bi (d) Se or Sn

83. Arrange these elements in order of increasing atomic size: Ca, Rb, S, Si, Ge, F.

84. Arrange these elements in order of increasing atomic size: Cs, Sb, S, Pb, Se.

85. Choose the more metallic element from each pair.
(a) Sr or Sb (b) As or Bi
(c) Cl or O (d) S or As

86. Choose the more metallic element from each pair.
(a) Sb or Pb (b) K or Ge
(c) Ge or Sb (d) As or Sn

87. Arrange these elements in order of increasing metallic character: Fr, Sb, In, S, Ba, Se.

88. Arrange these elements in order of increasing metallic character: Sr, N, Si, P, Ga, Al.

Cumulative Problems

89. What is the maximum number of electrons that can occupy the $n = 3$ quantum shell?

90. What is the maximum number of electrons that can occupy the $n = 4$ quantum shell?

91. Use the electron configurations of the alkaline earth metals to explain why they tend to form 2+ ions.

92. Use the electron configuration of oxygen to explain why it tends to form a 2− ion.

93. Write the electron configuration for each ion. What do all of the electron configurations have in common?
(a) Ca^{2+} (b) K^+ (c) S^{2-} (d) Br^-

94. Write the electron configuration for each ion. What do all of the electron configurations have in common?
(a) F^- (b) P^{3-} (c) Li^+ (d) Al^{3+}

95. Examine Figure 4.12, which shows the division of the periodic table into metals, nonmetals, and metalloids. Use what you know about electron configurations to explain these divisions.

96. Examine Figure 4.14, which shows the elements that form predictable ions. Use what you know about electron configurations to explain these trends.

97. Identify what is wrong with each electron configuration and write the correct ground-state (or lowest energy) configuration based on the number of electrons.
(a) $1s^3 2s^3 2p^9$ (b) $1s^2 2s^2 2p^6 2d^4$
(c) $1s^2 1p^5$ (d) $1s^2 2s^2 2p^8 3s^2 3p^1$

98. Identify what is wrong with each electron configuration and write the correct ground-state (or lowest energy) configuration based on the number of electrons.
(a) $1s^4 2s^4 2p^{12}$ (b) $1s^2 2s^2 2p^6 3s^2 3p^6 3d^{10}$
(c) $1s^2 2p^6 3s^2$ (d) $1s^2 2s^2 2p^6 3s^2 3p^6 4s^2 4d^{10} 4p^3$

99. Bromine is a highly reactive liquid, while krypton is an inert gas. Explain this difference based on their electron configurations.

100. Potassium is a highly reactive metal, while argon is an inert gas. Explain this difference based on their electron configurations.

101. Based on periodic trends, which one of these elements would you expect to be most easily oxidized: Ge, K, S, or N?

102. Based on periodic trends, which one of these elements would you expect to be most easily reduced: Ca, Sr, P, or Cl?

103. When an electron makes a transition from the $n = 3$ to the $n = 2$ hydrogen atom Bohr orbit, the energy difference between these two orbits (3.0×10^{-19} J) is emitted as a photon of light. The relationship between the energy of a photon and its wavelength is given by $E = hc/\lambda$, where E is the energy of the photon in J, h is Planck's constant (6.626×10^{-34} J·s), and c is the speed of light (3.00×10^{8} m/s). Find the wavelength of light emitted by hydrogen atoms when an electron makes this transition.

104. When an electron makes a transition from the $n = 4$ to the $n = 2$ hydrogen atom Bohr orbit, the energy difference between these two orbits (4.1×10^{-19} J) is emitted as a photon of light. The relationship between the energy of a photon and its wavelength is given by $E = hc/\lambda$, where E is the energy of the photon in J, h is Planck's constant (6.626×10^{-34} J·s), and c is the speed of light (3.00×10^{8} m/s). Find the wavelength of light emitted by hydrogen atoms when an electron makes this transition.

105. The distance from the sun to Earth is 1.496×10^{8} km. How long does it take light to travel from the sun to Earth?

106. The nearest star is Alpha Centauri, at a distance of 4.3 light-years from Earth. A light-year is the distance that light travels in one year (365 days). How far away, in kilometers, is Alpha Centauri from Earth?

107. The wave nature of matter was first proposed by Louis de Broglie, who suggested that the wavelength (λ) of a particle was related to its mass (m) and its velocity (ν) by the equation: $\lambda = h/m\nu$, where h is Planck's constant (6.626×10^{-34} J·s). Calculate the de Broglie wavelength of: (a) a 0.0459 kg golf ball traveling at 95 m/s; (b) an electron traveling at 3.88×10^{6} m/s. Can you explain why the wave nature of matter is significant for the electron but not for the golf ball? (*Hint:* Express mass in kilograms.)

108. The particle nature of light was first proposed by Albert Einstein, who suggested that light could be described as a stream of particles called photons. A photon of wavelength λ has an energy (E) given by the equation: $E = hc/\lambda$, where E is the energy of the photon in J, h is Planck's constant (6.626×10^{-34} J·s), and c is the speed of light (3.00×10^{8} m/s). Calculate the energy of 1 mol of photons with a wavelength of 632 nm.

109. You learned in this chapter that ionization generally increases as you move from left to right across the periodic table. However, consider the following data, which shows the ionization energies of the period 2 and 3 elements:

Group	Period 2 Elements	Ionization Energy (kJ/mol)	Period 3 Elements	Ionization Energy (kJ/mol)
1A	Li	520	Na	496
2A	Be	899	Mg	738
3A	B	801	Al	578
4A	C	1086	Si	786
5A	N	1402	P	1012
6A	O	1314	S	1000
7A	F	1681	Cl	1251
8A	Ne	2081	Ar	1521

Notice that the increase is not uniform. In fact, ionization energy actually decreases a bit in going from elements in group 2A to 3A and then again from 5A to 6A. Use what you know about electron configurations to explain why these dips in ionization energy exist.

110. When atoms lose more than one electron, the ionization energy to remove the second electron is always more than the ionization energy to remove the first. Similarly, the ionization energy to remove the third electron is more than the second and so on. However, the increase in ionization energy upon the removal of subsequent electrons is not necessarily uniform. For example, consider the first three ionization energies of magnesium:

First ionization energy	738 kJ/mol
Second ionization energy	1450 kJ/mol
Third ionization energy	7730 kJ/mol

The second ionization energy is roughly twice the first ionization energy, but then the third ionization energy is over five times the second. Use the electron configuration of magnesium to explain why this is so. Would you expect the same behavior in sodium? Why or why not?

Highlight Problems

111. Excessive exposure to sunlight increases the risk of skin cancer because some of the photons have enough energy to break chemical bonds in biological molecules. These bonds require approximately 250–800 kJ/mol of energy to break. The energy of a single photon is given by $E = hc/\lambda$, where E is the energy of the photon in J, h is Planck's constant (6.626×10^{-34} J·s), and c is the speed of light (3.00×10^{8} m/s). Determine which kinds of light contain enough energy to break chemical bonds in biological molecules by calculating the total energy in 1 mol of photons for light of each wavelength.

   (a) infrared light (1500 nm)
   (b) visible light (500 nm)
   (c) ultraviolet light (150 nm)

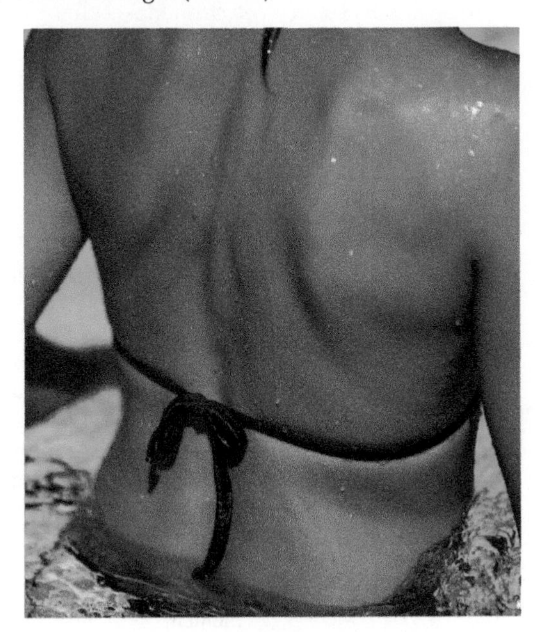

112. The quantum-mechanical model, besides revolutionizing chemistry, shook the philosophical world because of its implications regarding determinism. Determinism is the idea that the outcomes of future events are determined by preceding events. The trajectory of a baseball, for example, is deterministic; that is, its trajectory—and therefore its landing place—is determined by its position, speed, and direction of travel. Before quantum mechanics, most scientists thought that fundamental particles—such as electrons and protons—also behaved deterministically. The implication of this belief was that the entire universe must behave deterministically—the future must be determined by preceding events. Quantum mechanics challenged this reasoning because fundamental particles do not behave deterministically—their future paths are not determined by preceding events. Some scientists struggled with this idea. Einstein himself refused to believe it, stating, "God does not play dice with the universe." Explain what Einstein meant by this statement.

▲ "God does not play dice with the universe."

Questions for Group Work

Discuss these questions with the group and record your consensus answer.

113. Sketch the following orbitals (including the x, y, and z axes): $1s$, $2p_x$, $3d_{xy}$, $3d_{z^2}$.

114. Draw the best periodic table you can from memory (do not look at a table to do this). You do not need to label the elements, but you should put the correct number of elements in each block. After your group agrees that the group has done its best, spend exactly three minutes comparing your table with Figure 9.26. Make a second periodic table from memory. Is it better than your first one?

115. Play the following game to memorize the order in which orbitals fill. Go around your group and have each group member say the name of the next orbital to fill and the maximum number of electrons it can hold ("1*s* two," "2*s* two," "2*p* six,". . .). If a group member gets stuck, other members can help, referring to Figure 9.24 or 9.26 if necessary. However, when anyone gets stuck, the next player starts back at "1*s* two." Keep going until each group member can list the correct sequence up to "6*s* two."

116. Using grammatically correct sentences, describe the periodic trends for atomic size, ionization energy, and metallic character.

Data Interpretation and Analysis

117. The first graph shown here is of the first ionization energies (the energy associated with removing an electron) of the period 3 elements. The second graph shows the electron affinities (the energy associated with gaining an electron) of the period 3 elements. Refer to the graphs to answer the questions.

(a) Notice that the ionization energies are positive and the electron affinities are negative. Explain the significance of this difference.

(b) Describe the general trend in period 3 first ionization energies as you move from left to right across the periodic table. Explain why this trend occurs. (*Hint:* Consider the trend in atomic size as you move from left to right across the periodic table.)

(c) The trend in first ionization energy has two exceptions: one at Al and another at S. Write the electron configurations of Mg, Al, P, and S and refer to them to explain the exceptions.

(d) Describe the general trend in period 3 electron affinities as you move from left to right across the periodic table. Explain why this trend occurs. (*Hint:* Consider the trend in atomic size as you move from left to right across the periodic table.)

(e) The trend in electron affinities has exceptions. Write the electron configurations of Si and P and explain why the electron affinity for Si is more exothermic than that of P.

(f) Determine the overall energy change for removing one electron from Na and adding that electron to Cl. Is the exchange of the electron exothermic or endothermic?

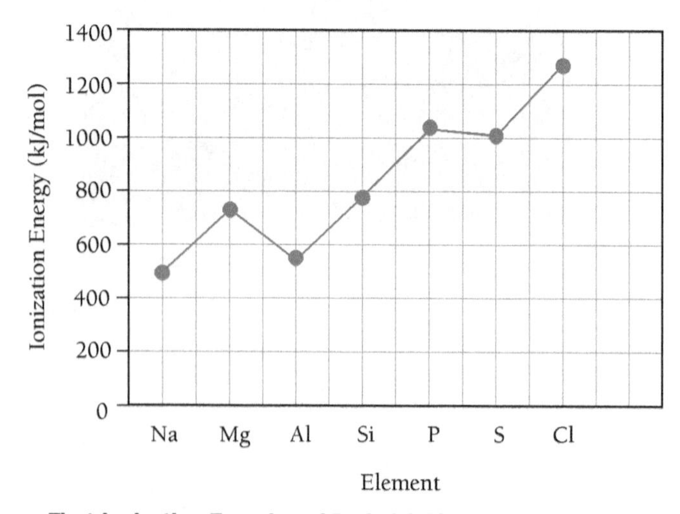

▲ **First Ionization Energies of Period 3 Elements**

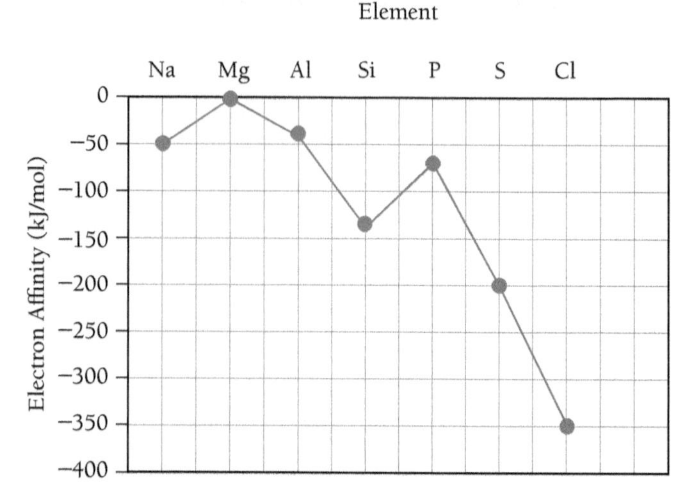

▲ **Electron Affinities of Period 3 Elements**

Answers to Skillbuilder Exercises

Skillbuilder 9.1 .... **(a)** blue, green, red
(b) red, green, blue
(c) red, green, blue

Skillbuilder 9.2

(a) Al $1s^22s^22p^63s^23p^1$ or $[Ne]3s^23p^1$
(b) Br $1s^22s^22p^63s^23p^64s^23d^{10}4p^5$ or $[Ar]4s^23d^{10}4p^5$
(c) Sr $1s^22s^22p^63s^23p^64s^23d^{10}4p^65s^2$ or $[Kr]5s^2$

Skillbuilder Plus, p. 301

Subtract one electron for each unit of positive charge. Add one electron for each unit of negative charge.

(a) Al^{3+} $1s^22s^22p^6$
(b) Cl^- $1s^22s^22p^63s^23p^6$
(c) O^{2-} $1s^22s^22p^6$

Skillbuilder 9.3

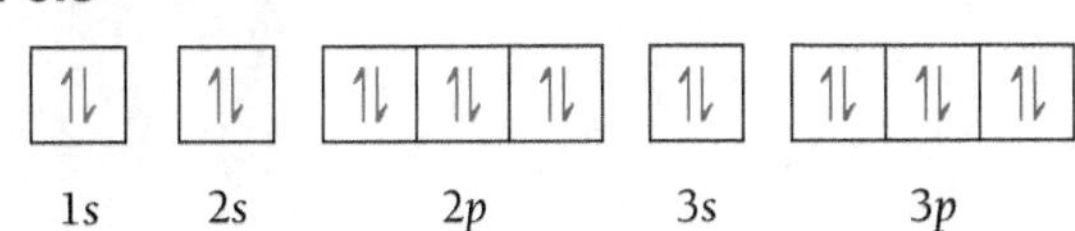

Skillbuilder 9.4

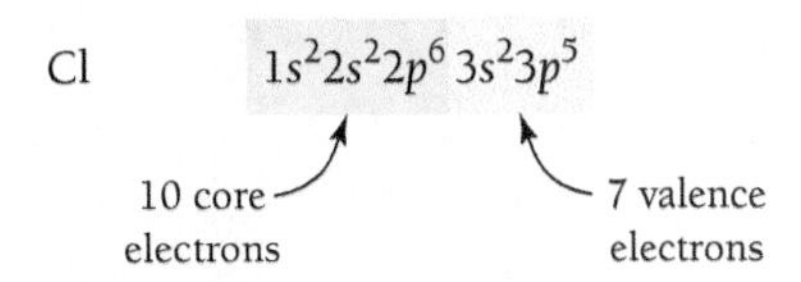

Skillbuilder 9.5 .... $[Kr]5s^24d^{10}5p^2$

Skillbuilder 9.6 .... **(a)** Pb
(b) Rb
(c) cannot determine based on periodic properties
(d) Se

Skillbuilder 9.7 .... **(a)** Mg
(b) Te
(c) cannot determine based on periodic properties
(d) F

Skillbuilder 9.8 .... **(a)** In
(b) cannot determine based on periodic properties
(c) Bi
(d) B

Answers to Conceptual Checkpoints

9.1 (a) Wavelength and frequency are inversely related. Therefore, the shortest wavelength has the highest frequency.

9.2 (b) Wavelength and energy per photon are inversely related. Since yellow light has a longer wavelength, it has less energy per photon than violet light.

9.3 (c) The higher-energy levels are more closely spaced than the lower ones, so the difference in energy between $n = 2$ and $n = 1$ is greater than the difference in energy between $n = 3$ and $n = 2$. The photon emitted when an electron falls from $n = 2$ to $n = 1$ therefore carries more energy, corresponding to radiation with a shorter wavelength and higher frequency.

9.4 (c) The $n = 3$ principal shell contains three subshells: *s*, *p*, and *d*.

9.5 (b) The *p* subshell contains three orbitals.

9.6 (d) Both have six electrons in $2p$ orbitals and six electrons in $3p$ orbitals.

9.7 (d) The outermost principal shell for K is $n = 4$, which contains only a single valence electron, $4s^1$.

9.8 (a) Calcium loses its $4s$ electron and attains a noble gas configuration (that of Ar).

9.9 (b) Ionization energy increases as you move from left to right across a row in the periodic table. Both atomic size and metallic character decrease as you move from left to right across a row.

Indinavir
O1 MK1 902 10.937 23.849 7.057
N2 MK1 902 10.193 25.953 7.094
C4 MK1 902 10.145 26.250 8.490
C5 MK1 902 9.379 27.577 8.641

10 Chemical Bonding

The fascination of a growing science lies in the work of the pioneers at the very borderland of the unknown, but to reach this frontier one must pass over well traveled roads.
—Gilbert N. Lewis (1875–1946)

CHAPTER OUTLINE

10.1 Bonding Models and AIDS Drugs

We will discuss proteins in more detail in Chapter 19.

In 1989, researchers discovered the structure of a molecule called HIV-protease. HIV-protease is a protein (a class of biological molecules) synthesized by the human immunodeficiency virus (HIV), which causes AIDS. HIV-protease is crucial to the virus's ability to replicate itself. Without HIV-protease, HIV could not spread in the human body because the virus could not copy itself, and AIDS would not develop.

With knowledge of the HIV-protease structure, drug companies set out to design a molecule that would disable the protease by attaching to the working part of the molecule (called the *active site*). To design such a molecule, researchers used **bonding theories**—models that predict how atoms bond together to form molecules—to simulate how potential drug molecules would interact with the protease molecule. By the early 1990s, these companies had developed several effective drug molecules. Since these molecules inhibit the action of HIV-protease, they are called *protease inhibitors*. In human trials, protease inhibitors in combination with other drugs decrease the viral count in HIV-infected individuals to undetectable levels. Although these drugs do not cure AIDS, HIV-infected individuals who regularly take their medication can now expect nearly normal life spans. The use of protease inhibitors has since been expanded to include treatment for Hepatitis C with similarly excellent results.

Bonding theories are central to chemistry because they predict how atoms bond together to form compounds. They predict which combinations of atoms form compounds and which combinations do not. Bonding theories predict why salt is NaCl and not $NaCl_2$ and why water is H_2O and not H_3O. Bonding theories also explain the shapes of molecules, which in turn determine many of their physical and chemical properties.

The bonding theory you will learn in this chapter is the **Lewis model**, named after G. N. Lewis (1875–1946), the American chemist who developed it. In this

◀ The gold-colored structure on the tablet screen is a representation of HIV-protease. The molecule shown in the center is Indinavir, a protease inhibitor.

model, we represent electrons as dots and draw *dot structures* or *Lewis structures* to represent molecules. These structures, which are fairly simple to draw, have tremendous predictive power. It takes just a few minutes to apply the Lewis model to determine whether a particular set of atoms will form a stable molecule and what that molecule might look like. Although modern chemists also use more advanced bonding theories to better predict molecular properties, the Lewis model remains the simplest method for making quick, everyday predictions about molecules.

10.2 Representing Valence Electrons with Dots

▶ Write Lewis structures for elements.

As we discussed in Chapter 9, valence electrons are the electrons in the outermost principal shell. Since valence electrons are most important in bonding, the Lewis model focuses on these. In the Lewis model, the valence electrons of main-group elements are represented as dots surrounding the symbol of the element. The result is a **Lewis structure**, or **dot structure**. For example, the electron configuration of O is:

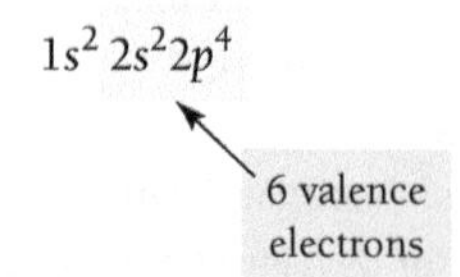

and its Lewis structure is:

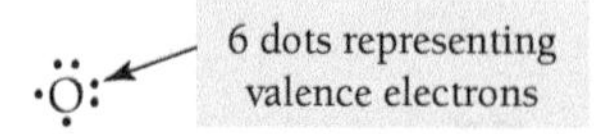

Remember, the number of valence electrons for any main-group element is equal to the group number of the element (except helium, which has two valence electrons but is in Group 8A).

Each dot represents a valence electron. We place the dots around the element's symbol with a maximum of two dots per side. Although the exact location of dots is not critical, in this book we fill in the dots singly first and then pair them (with the exception of helium, described shortly).

The Lewis structures for the period 2 elements are:

Li· ·Be· ·B· ·C· ·N: ·O: :F: :Ne:

Lewis structures allow us to easily see the number of valence electrons in an atom. Atoms with eight valence electrons—which are particularly stable—are easily identified because they have eight dots, an **octet**.

Helium is somewhat of an exception. Its electron configuration and Lewis structure are:

$1s^2$ He:

The Lewis structure of helium contains two paired dots (a **duet**). For helium, a duet represents a stable electron configuration.

In the Lewis model, a **chemical bond** involves the sharing or transfer of electrons to attain stable electron configurations for the bonding atoms. If the electrons are transferred, the bond is an **ionic bond**. If the electrons are shared, the bond is a **covalent bond**. In either case, the bonding atoms attain stable electron configurations. As we have seen, a stable configuration usually consists of eight electrons in the outermost or valence shell. This observation leads to the **octet rule**:

> In chemical bonding, atoms transfer or share electrons to obtain outer shells with eight electrons.

The octet rule generally applies to all main-group elements except hydrogen and helium. Each of these elements achieves stability when it has two electrons (a duet) in its outermost shell.

EXAMPLE **10.1** **Writing Lewis Structures for Elements**

Write the Lewis structure of phosphorus.

Since phosphorus is in Group 5A in the periodic table, it has five valence electrons. Represent these as five dots surrounding the symbol for phosphorus.

SOLUTION

·Ṗ̣:

▶ **SKILLBUILDER 10.1 | Writing Lewis Structures for Elements**

Write the Lewis structure of Mg.

▶ **FOR MORE PRACTICE** Example 10.12; Problems 25, 26.

CONCEPTUAL ✔ CHECKPOINT 10.1

Which two elements have the most similar Lewis structures?

(a) C and Si **(b)** O and P **(c)** Li and F **(d)** S and Br

10.3 Lewis Structures of Ionic Compounds: Electrons Transferred

- ▶ Write Lewis structures for ionic compounds.
- ▶ Use the Lewis model to predict the chemical formula of an ionic compound.

Recall from Chapter 5 that when metals bond with nonmetals, electrons are transferred from the metal to the nonmetal. The metal becomes a cation and the nonmetal becomes an anion. The attraction between the cation and the anion results in an ionic compound. In the Lewis model, we represent this by moving electron dots from the metal to the nonmetal. For example, the Lewis structures for potassium and chlorine are:

K· :C̣̈l:

When potassium and chlorine bond, potassium transfers its valence electron to chlorine.

K· :C̣̈l: ⟶ K^+ [:C̣̈l:]$^-$

Recall from Section 4.7 that atoms that lose electrons become positively charged and atoms that gain electrons become negatively charged.

The transfer of the electron gives chlorine an octet (shown as eight dots around chlorine) and leaves potassium with an octet in the previous principal shell, which is now the valence shell. Because the potassium lost an electron, it becomes positively charged, while the chlorine, which gained an electron, becomes negatively charged. We usually write the Lewis structure of an anion in brackets with the charge in the upper right corner (outside the brackets). The positive and negative charges attract one another, forming the compound KCl.

EXAMPLE **10.2** **Writing Ionic Lewis Structures**

Write the Lewis structure of the compound MgO.

Draw the Lewis structures of magnesium and oxygen by drawing two dots around the symbol for magnesium and six dots around the symbol for oxygen.

SOLUTION

·Mg· ·Ö̤:

In MgO, magnesium loses its two valence electrons, resulting in a 2+ charge, and oxygen gains two electrons, attaining a 2− charge and an octet.

Mg^{2+} [:Ö̤:]$^{2-}$

▶ **SKILLBUILDER 10.2 | Writing Ionic Lewis Structures**

Write the Lewis structure of the compound NaBr.

▶ **FOR MORE PRACTICE** Example 10.13; Problems 37, 38.

Recall from Section 5.4 that ionic compounds do not exist as distinct molecules, but rather as part of a large three-dimensional array (or lattice) of alternating cations and anions.

The Lewis model predicts the correct chemical formulas for ionic compounds. For the compound that forms between K and Cl, for example, the Lewis model predicts one potassium cation to every chlorine anion, KCl. As another example, consider the ionic compound formed between sodium and sulfur. The Lewis structures for sodium and sulfur are:

$$\text{Na}\cdot \quad \cdot\ddot{\underset{\cdot}{\text{S}}}:$$

Notice that sodium must lose its one valence electron to obtain an octet (in the previous principal shell), while sulfur must gain two electrons to obtain an octet. Consequently, the compound that forms between sodium and sulfur requires two sodium atoms to every one sulfur atom. The Lewis structure is:

$$\text{Na}^+ \; [:\ddot{\underset{..}{\text{S}}}:]^{2-} \; \text{Na}^+$$

The two sodium atoms each lose their single valence electron, while the sulfur atom gains two electrons and obtains an octet. The correct chemical formula is Na_2S.

EXAMPLE 10.3 Using the Lewis Model to Predict the Chemical Formula of an Ionic Compound

Use the Lewis model to predict the formula of the compound that forms between calcium and chlorine.

	SOLUTION
Draw the Lewis structures of calcium and chlorine by drawing two dots around the symbol for calcium and seven dots around the symbol for chlorine.	$\cdot\text{Ca}\cdot \quad :\ddot{\underset{\cdot}{\text{Cl}}}:$
Calcium must lose its two valence electrons (to effectively attain an octet in its previous principal shell), while chlorine needs to gain only one electron to obtain an octet. Consequently, the compound that forms between Ca and Cl has two chlorine atoms to every one calcium atom.	$[:\ddot{\underset{..}{\text{Cl}}}:]^- \; \text{Ca}^{2+} \; [:\ddot{\underset{..}{\text{Cl}}}:]^-$ The formula is therefore $CaCl_2$.

▶ **SKILLBUILDER 10.3 | Using the Lewis Model to Predict the Chemical Formula of an Ionic Compound**

Use the Lewis model to predict the formula of the compound that forms between magnesium and nitrogen.

▶ **FOR MORE PRACTICE** Example 10.14; Problems 39, 40, 41, 42.

CONCEPTUAL ✔ CHECKPOINT 10.2

Which nonmetal forms an ionic compound with aluminum that has the formula Al_2X_3 (where X represents the nonmetal)?

(a) Cl **(b)** S **(c)** N **(d)** C

10.4 Covalent Lewis Structures: Electrons Shared

▶ Write Lewis structures for covalent compounds.

Recall from Chapter 5 that when nonmetals bond with other nonmetals, a molecular compound results. Molecular compounds contain covalent bonds in which electrons are shared between atoms rather than transferred. Electrons are normally shared in pairs to form single, double, or triple bonds.

Single Bonds

In the Lewis model, we represent a single covalent bond by allowing neighboring atoms to share a pair of valence electrons to attain an octet (or duet for hydrogen). For example, hydrogen and oxygen have the Lewis structures:

$$\text{H}\cdot \quad \cdot\ddot{\underset{\cdot}{\text{O}}}:$$

In water, hydrogen and oxygen share their electrons so that each hydrogen atom has a duet and the oxygen atom has an octet.

H:Ö:H

The shared electrons—those that appear in the space between the two atoms—count toward the octets (or duets) of *both of the atoms.*

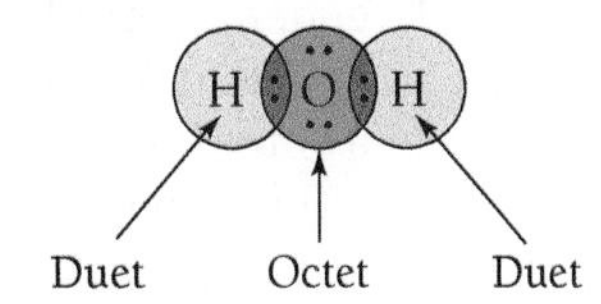

Electrons that are shared between two atoms are **bonding pair** electrons, while those that are only on one atom are **lone pair** (or nonbonding) electrons.

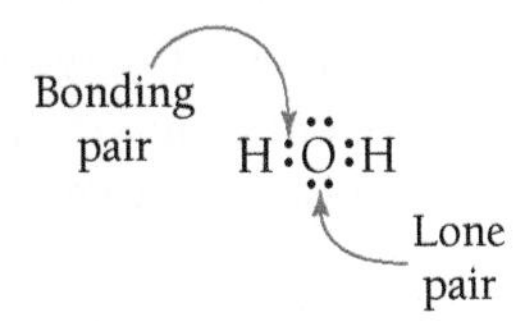

A bonding pair of electrons is often represented by a dash to emphasize that it is a chemical bond.

H—Ö—H

Remember that each dash represents a *pair* of shared electrons.

The Lewis model also explains why the halogens form diatomic molecules. Consider the Lewis structure of chlorine:

:C̤̈l·

If two Cl atoms pair, they can each attain an octet:

:C̤̈l:C̤̈l: *or* :C̤̈l—C̤̈l:

When we examine elemental chlorine, we find that it indeed exists as a diatomic molecule, just as the Lewis model predicts. The same is true for the other halogens.

Similarly, the Lewis model predicts that hydrogen, which has this Lewis structure:

H·

should exist as H_2. When two hydrogen atoms share their valence electrons, they each have a duet, a stable configuration for hydrogen.

H:H *or* H—H

Again, the Lewis model prediction is correct. In nature, elemental hydrogen exists as H_2 molecules.

Double and Triple Bonds

In the Lewis model, atoms can share more than one electron pair to attain an octet. For example, we know from Chapter 5 that oxygen exists as the diatomic molecule, O_2. The Lewis structure of an oxygen atom is:

·Ö:

If we pair two oxygen atoms and then try to write the Lewis structure, we do not have enough electrons to give each O atom an octet.

:Ö:Ö:

However, we can convert a lone pair into an additional bonding pair by moving it into the bonding region.

:Ö:Ö: ⟶ :Ö::Ö: *or* :Ö=Ö:

Each oxygen atom now has an octet because the additional bonding pair counts toward the octet of both oxygen atoms.

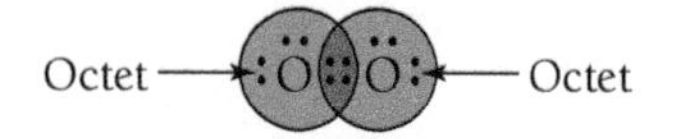

When two atoms share two electron pairs, the resulting bond is a **double bond**. In general, double bonds are shorter and stronger than single bonds. For example, the distance between oxygen nuclei in an oxygen–oxygen double bond is 121 pm. In a single bond, it is 148 pm.

1 pm = 10^{-12} m

Two atoms can also share three electron pairs. Consider the Lewis structure of N_2. Since each N atom has five valence electrons, the Lewis structure for N_2 has ten electrons. A first attempt at writing the Lewis structure looks like this:

:N̈:N̈:

As with O_2, we do not have enough electrons to satisfy the octet rule for both N atoms. However, if we convert two additional lone pairs into bonding pairs, each nitrogen atom has an octet.

:N̈:N̈: ⟶ :N:::N: *or* :N≡N:

The resulting bond is a **triple bond**. Triple bonds are even shorter and stronger than double bonds. The distance between nitrogen nuclei in a nitrogen–nitrogen triple bond is 110 pm. In a double bond, the distance is 124 pm. When we examine nitrogen in nature, we find that it exists as a diatomic molecule with a very strong short bond between the two nitrogen atoms. The bond is so strong that it is difficult to break, making N_2 a relatively unreactive molecule.

CONCEPTUAL CHECKPOINT 10.3

How many bonding electrons are in the Lewis structure of O_2?

(a) 2 (b) 4 (c) 6

10.5 Writing Lewis Structures for Covalent Compounds

▶ Write Lewis structures for covalent compounds.

To write the Lewis structure for a covalent compound, follow these steps:

When guessing at skeletal structures, we put the less metallic elements in terminal positions and the more metallic elements in central positions. Halogens, which are among the least metallic elements, are almost always terminal.

1. **Write the correct skeletal structure for the molecule.** The skeletal structure shows the relative positions of the atoms and does not include electrons, but it must have the atoms in the correct positions. For example, you *cannot* write the Lewis structure for water if you start with the hydrogen atoms next to each other and the oxygen atom at the end (H H O). In nature, oxygen is the central atom, and the hydrogen atoms are **terminal atoms** (at the ends). The correct skeletal structure is H O H.

   The only way to absolutely know the correct skeletal structure for any molecule is to examine its structure in nature. However, you can write likely skeletal structures by remembering two guidelines. First, *hydrogen atoms are always terminal*. Since hydrogen requires only a duet, it is never a central atom because central atoms must be able to form at least two bonds and hydrogen can form only one. Second, *many molecules tend to be symmetrical*, so when a molecule contains several atoms of the same type, these tend to be in terminal positions. *This symmetry guideline, however, has many exceptions.* In cases where the skeletal structure is unclear, this text provides the correct skeletal structure.

Nonterminal hydrogen atoms exist in some compounds. However, they are rare and beyond the scope of this text.

2. **Calculate the total number of electrons for the Lewis structure by summing the valence electrons of each atom in the molecule.** Remember that the number of valence electrons for any main-group element is equal to its group number in the periodic table. **If you are writing a Lewis structure for a**

polyatomic ion, you must consider the charge of the ion when calculating the total number of electrons. Add one electron for each negative charge and subtract one electron for each positive charge.

3. **Distribute the electrons among the atoms, giving octets (or duets for hydrogen) to as many atoms as possible.** Begin by placing two electrons between each pair of atoms. These are the minimal number of bonding electrons. Then distribute the remaining electrons, first to terminal atoms and then to the central atom, giving octets to as many atoms as possible.
4. **If any atoms lack an octet, form double or triple bonds as necessary to give them octets.** Do this by moving lone electron pairs from terminal atoms into the bonding region with the central atom.

A brief version of this procedure is presented in the left column. In the center and right columns, Examples 10.4 and 10.5 illustrate the procedure.

PEARSON eText 2.0 **Interactive Worked Example Video 10.4**

Writing Lewis Structures for Covalent Compounds	EXAMPLE **10.4** Write the Lewis structure for CO_2.	EXAMPLE **10.5** Write the Lewis structure for CCl_4.
1. Write the correct skeletal structure for the molecule.	**SOLUTION** Following the symmetry guideline, write: O C O	**SOLUTION** Following the symmetry guideline, write: Cl Cl C Cl Cl
2. Calculate the total number of electrons for the Lewis structure by summing the valence electrons of each atom in the molecule.	Total number of electrons for Lewis structure = $\begin{pmatrix}\text{\# valence}\\ e^-\text{ for C}\end{pmatrix} + 2\begin{pmatrix}\text{\# valence}\\ e^-\text{ for O}\end{pmatrix}$ $= 4 + 2(6)$ $= 16$	Total number of electrons for Lewis structure = $\begin{pmatrix}\text{\# valence}\\ e^-\text{ for C}\end{pmatrix} + 4\begin{pmatrix}\text{\# valence}\\ e^-\text{ for Cl}\end{pmatrix}$ $= 4 + 4(7)$ $= 32$
3. Distribute the electrons among the atoms, giving octets (or duets for hydrogen) to as many atoms as possible. Begin with the bonding electrons, proceed to lone pairs on terminal atoms, and finally go to lone pairs on the central atom.	Work with bonding electrons first. O:C:O (4 of 16 electrons used) Proceed to lone pairs on terminal atoms next. :Ö:C:Ö: (with lone pairs above and below each O) (16 of 16 electrons used)	Work with bonding electrons first. Cl Cl:C:Cl Cl (8 of 32 electrons used) Proceed to lone pairs on terminal atoms next. :Cl: :Cl:C:Cl: :Cl: (with lone pairs on each Cl) (32 of 32 electrons used)
4. If any atoms lack octets, form double or triple bonds as necessary to give them octets.	Move lone pairs from the oxygen atoms to bonding regions to form double bonds. :Ö:C:Ö: ⟶ :O::C::O:	Since all of the atoms have octets, the Lewis structure is complete. ▶ **SKILLBUILDER 10.5** | Write the Lewis structure for H_2CO.
	▶ **SKILLBUILDER 10.4** | Write the Lewis structure for CO.	▶ **FOR MORE PRACTICE** Example 10.15; Problems 47, 48, 49, 50, 51, 52.

Writing Lewis Structures for Polyatomic Ions

We write Lewis structures for polyatomic ions by following the same procedure, but we pay special attention to the charge of the ion when calculating the number of electrons for the Lewis structure. We add one electron for each negative charge and subtract one electron for each positive charge. We normally show the Lewis structure for a polyatomic ion within brackets and write the charge of the ion in the upper right corner. For example, suppose we want to write the Lewis structure for the CN^- ion. We begin by writing the skeletal structure:

CN

Next we calculate the total number of electrons for the Lewis structure by summing the number of valence electrons for each atom and adding one for the negative charge.

Total number of electrons for Lewis structure = (# valence e^- in C) + (# valence e^- in N) + 1
= 4 + 5 + 1
= 10

Add one e^- to account for 1− charge of ion.

We then place two electrons between each pair of atoms

C:N (2 of 10 electrons used)

and distribute the remaining electrons.

:C̈:N̈: (10 of 10 electrons used)

Since neither of the atoms has octets, we move two lone pairs into the bonding region to form a triple bond, giving both atoms octets. We also enclose the Lewis structure in brackets and write the charge of the ion in the upper right corner.

$[:C:::N:]^-$ *or* $[:C\equiv N:]^-$

CONCEPTUAL CHECKPOINT 10.4

How many electrons are there in the Lewis structure of OH^-?

(a) 6 (b) 7 (c) 8 (d) 9

EXAMPLE 10.6 Writing Lewis Structures for Polyatomic Ions

Write the Lewis structure for the NH_4^+ ion.	
Begin by writing the skeletal structure. Hydrogen atoms must be terminal, and following the guideline of symmetry, the nitrogen atom should be in the middle surrounded by four hydrogen atoms.	**SOLUTION** H H N H H
Calculate the total number of electrons for the Lewis structure by summing the number of valence electrons for each atom and subtracting one for the positive charge.	Total number of electrons for Lewis structure = 5 + 4 − 1 = 8 (5: # valence e^- in N; 4: 4 × (# valence e^- in H); 1: Subtract 1 e^- to account for 1+ charge of ion.)
Next, place two electrons between each pair of atoms.	H H:N̈:H (8 of 8 electrons used) H (with two dots above and below N)

Since the nitrogen atom has an octet and all of the hydrogen atoms have duets, the placement of electrons is complete. Write the entire Lewis structure in brackets indicating the charge of the ion in the upper right corner.	$\left[\begin{matrix} & H & \\ H: & \ddot{N} & :H \\ & \ddot{H} & \end{matrix}\right]^+$ *or* $\left[\begin{matrix} & H & \\ & \vert & \\ H- & N & -H \\ & \vert & \\ & H & \end{matrix}\right]^+$

▶ **SKILLBUILDER 10.6 | Writing Lewis Structures for Polyatomic Ions**

Write the Lewis structure for the ClO^- ion.

▶ **FOR MORE PRACTICE** Problems 55bcd, 56abc, 57, 58.

CONCEPTUAL CHECKPOINT 10.5

Which two species have the same number of lone electron pairs in their Lewis structures?

(a) H_2O and H_3O^+

(b) NH_3 and H_3O^+

(c) NH_3 and CH_4

(d) NH_3 and NH_4^+

Exceptions to the Octet Rule

Lewis model predictions are often correct, but exceptions exist. For example, if we try to write the Lewis structure for NO, which has 11 electrons, the best we can do is:

$$:\dot{N}::\ddot{O}: \quad or \quad :\dot{N}=\ddot{O}:$$

The nitrogen atom does not have an octet, so this is not a great Lewis structure. However, NO exists in nature. Why does the Lewis model not account for the existence of NO? As with any simple theory, the Lewis model is not sophisticated enough to be correct every time. It is impossible to write good Lewis structures for molecules with odd numbers of electrons, yet some of these molecules exist in nature. In such cases, we write the best Lewis structure that we can. Another significant exception to the octet rule is boron, which tends to form compounds with only six electrons around B, rather than eight. For example, BF_3 and BH_3—which both exist in nature—each lack an octet for B.

$$\begin{matrix} & :\ddot{F}: & \\ :\ddot{F} & :B: & \ddot{F}: \end{matrix} \qquad \begin{matrix} & H & \\ H & :\ddot{B}: & H \end{matrix}$$

A third type of exception to the octet rule is also common. A number of molecules, such as SF_6 and PCl_5, have more than eight electrons around a central atom in their Lewis structures.

$$\begin{matrix} & :\ddot{F}: & \\ :\ddot{F} & \vert & \ddot{F}: \\ & S & \\ :\ddot{F} & \vert & \ddot{F}: \\ & :\ddot{F}: & \end{matrix} \qquad \begin{matrix} & :\ddot{Cl}: & \\ :\ddot{Cl}- & P & -\ddot{Cl}: \\ :\ddot{Cl} & & \ddot{Cl}: \end{matrix}$$

We often refer to these as *expanded octets*. Expanded octets can form for period 3 elements and beyond. Beyond mentioning them, we do not cover expanded octets in this book. In spite of these exceptions, the Lewis model remains a powerful and simple way to understand chemical bonding.

10.6 Resonance: Equivalent Lewis Structures for the Same Molecule

▶ Write resonance structures.

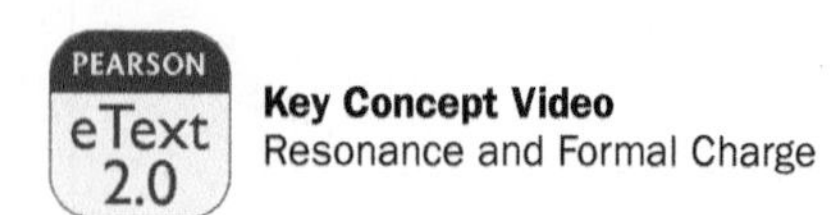

Key Concept Video
Resonance and Formal Charge

When writing Lewis structures, we may find that, for some molecules, we can write more than one good Lewis structure. For example, consider writing a Lewis structure for SO_2. We begin with the skeletal structure:

O S O

We then sum the valence electrons.

Total number of electrons for Lewis structure

$= (\#\text{ valence e}^- \text{ in S}) + 2(\#\text{ valence e}^- \text{ in O})$

$= 6 + 2(6)$

$= 18$

We next place two electrons between each pair of atoms

O:S:O (4 of 18 electrons used)

and distribute the remaining electrons, first to terminal atoms

:Ö:S:Ö: (16 of 18 electrons used)

and finally to the central atom.

:Ö:S̈:Ö: (18 of 18 electrons used)

Since the central atom lacks an octet, we move one lone pair from an oxygen atom into the bonding region to form a double bond, giving all of the atoms octets.

:Ö::S̈:Ö: *or* :Ö=S̈—Ö:

However, we could have formed the double bond with the other oxygen atom.

:Ö—S̈=Ö:

These two Lewis structures are equally correct. In cases such as this—where we can write two or more equivalent (or nearly equivalent) Lewis structures for the same molecule—we find that the molecule exists in nature as an average or intermediate between the two Lewis structures. Both of the two Lewis structures for SO_2 predict that SO_2 would contain two different kinds of bonds (one double bond and one single bond). However, when we examine SO_2 in nature, we find that both of the bonds are equivalent and intermediate in strength and length between a double bond and single bond.

We address this in the Lewis model by representing the molecule with both structures, called **resonance structures**, with a double-headed arrow between them.

:Ö=S̈—Ö: ⟷ :Ö—S̈=Ö:

The true structure of SO_2 is intermediate between these two resonance structures and is called a *resonance hybrid*. Resonance structures always have the same skeletal structure (the atoms are in the same relative positions); only the distribution of electron dots differs between them.

EXAMPLE 10.7 Writing Resonance Structures

Write the Lewis structure for the NO_3^- ion. Include resonance structures.

	SOLUTION
Begin by writing the skeletal structure. Applying the guideline of symmetry, make the three oxygen atoms terminal.	O O N O
Sum the valence electrons (adding one electron to account for the 1− charge) to determine the total number of electrons in the Lewis structure.	Total number of electrons for Lewis structure = 5 + 3(6) + 1 = 24 (5: # valence e^- in N; 3(6): 3 × (# valence e^- in O); 1: Add one e^- to account for negative charge of ion.)
Place two electrons between each pair of atoms.	O:N:O (with O above N) (6 of 24 electrons used)
Distribute the remaining electrons, first to the terminal atoms.	:Ö: above :Ö:N:Ö: (24 of 24 electrons used)
Since there are no electrons remaining to complete the octet of the central atom, form a double bond by moving a lone pair from one of the oxygen atoms into the bonding region with nitrogen. Enclose the structure in brackets and write the charge at the upper right.	[:Ö: / :Ö:N::O:]$^-$ *or* [:Ö: / :O—N=O:]$^-$
Notice that you can form the double bond with either of the other two oxygen atoms as well.	[:Ö: / :O=N—O:]$^-$ *or* [:O (double bond to N) / :O—N—O:]$^-$
Since the three Lewis structures are equally correct, write the three structures as resonance structures.	[:Ö: / :O=N—O:]$^-$ ⟷ [:O (double bond to N) / :O—N—O:]$^-$ ⟷ [:Ö: / :O—N=O:]$^-$

▶ **SKILLBUILDER 10.7 | Writing Resonance Structures**

Write the Lewis structure for the NO_2^- ion. Include resonance structures.

▶ **FOR MORE PRACTICE** Example 10.16; Problems 55, 56, 57, 58.

CONCEPTUAL CHECKPOINT 10.6

Which one of the structures that follow is NOT a resonance structure of the Lewis structure for N_2O shown here?

:N̈=N=Ö:

(a) :N≡N—Ö:

(b) :N̈—N≡O:

(c) :N̈—O≡N:

10.7 Predicting the Shapes of Molecules

▶ Predict the shapes of molecules.

We can use the Lewis model, in combination with **valence shell electron pair repulsion (VSEPR) theory**, to predict the shapes of molecules. VSEPR theory is based on the idea that **electron groups**—lone pairs, single bonds, or multiple bonds—repel each other. This repulsion between the negative charges of electron groups on the central atom determines the geometry of the molecule. For example, consider CO_2, which has the Lewis structure:

:Ö=C=Ö:

The geometry of CO_2 is determined by the repulsion between the two electron groups (the two double bonds) on the central carbon atom. These two electron groups get as far away from each other as possible, resulting in a bond angle of 180° and a **linear** geometry for CO_2.

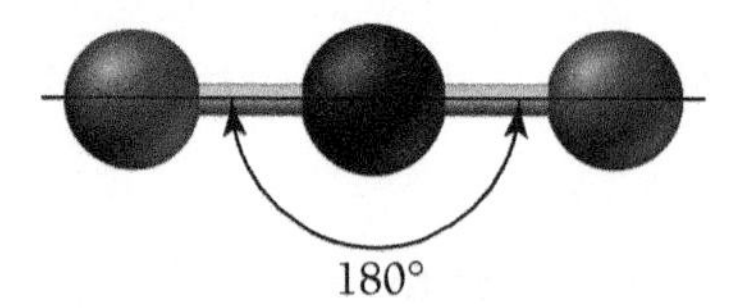

As another example, consider the molecule H_2CO. Its Lewis structure is:

```
   :O:
    ||
H — C — H
```

This molecule has three electron groups around the central atom. These three electron groups get as far away from each other as possible, resulting in a bond angle of 120° and a **trigonal planar** geometry.

The angles shown here for H_2CO are approximate. The C=O double bond contains more electron density than do C—H single bonds, resulting in a slightly greater repulsion; thus the HCH bond angle is actually 116°, and the HCO bond angles are actually 122°.

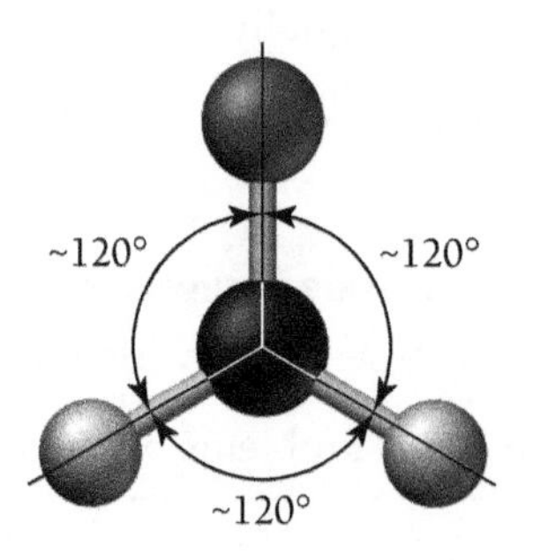

A tetrahedron is a geometric shape with four triangular faces.

If a molecule has four electron groups around the central atom, as CH_4 does, it has a **tetrahedral** geometry with bond angles of 109.5°.

CH_4 is shown here with both a ball-and-stick model (left) and a space-filling model (right). Although space-filling models more closely portray molecules, ball-and-stick models are often used to clearly illustrate molecular geometries.

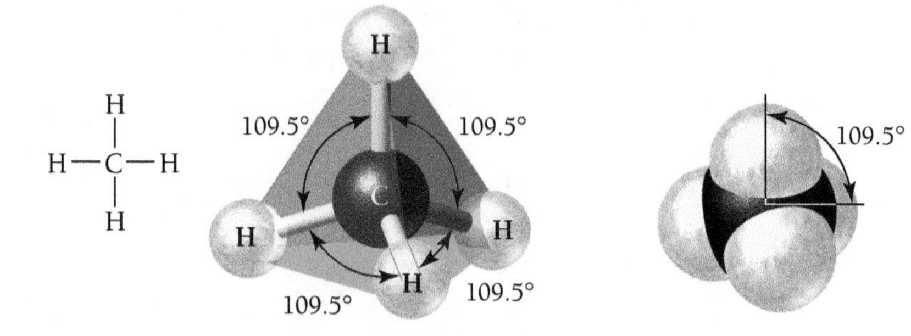

The mutual repulsion of the four electron groups causes the tetrahedral shape; the tetrahedron allows the maximum separation among the four groups. When we write the structure of CH_4 on paper, it may seem that the molecule should be square planar, with bond angles of 90°. However, in three dimensions the electron groups can get farther away from each other by forming the tetrahedral geometry.

Each of the molecules in the preceding examples has only bonding groups of electrons around the central atom. What happens in molecules with lone pairs around the central atom? These lone pairs also repel other electron groups. For example, consider the NH_3 molecule:

```
     H
     |
H—N—H
    ••
```

The four electron groups (one lone pair and three bonding pairs) get as far away from each other as possible. If we look only at the electrons, we find that the **electron geometry**—the geometrical arrangement of the electron groups—is tetrahedral.

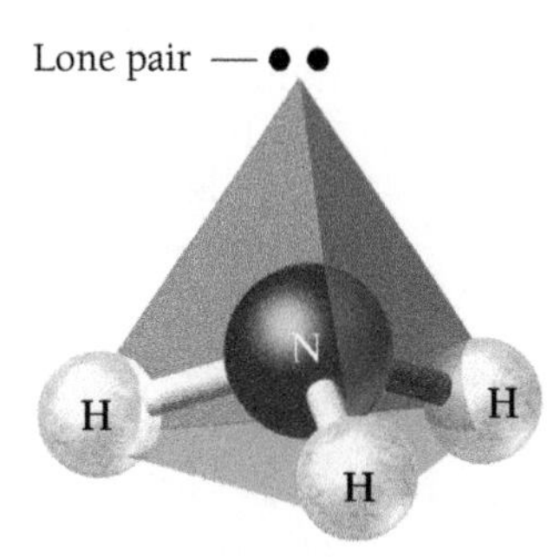

However, the **molecular geometry**—the geometrical arrangement of the atoms—is **trigonal pyramidal**.

Trigonal pyramidal structure

Notice that, although the electron geometry and the molecular geometry are different, the electron geometry is relevant to the molecular geometry. In other words, the lone pair exerts its influence on the bonding pairs.

Consider one last example, H_2O. Its Lewis structure is:

```
    ••
H—O—H
    ••
```

The bond angles in NH_3 and H_2O are actually a few degrees smaller than the ideal tetrahedral angles because lone pairs exert a slightly greater repulsion than bonding pairs.

Since it has four electron groups, its electron geometry is also tetrahedral.

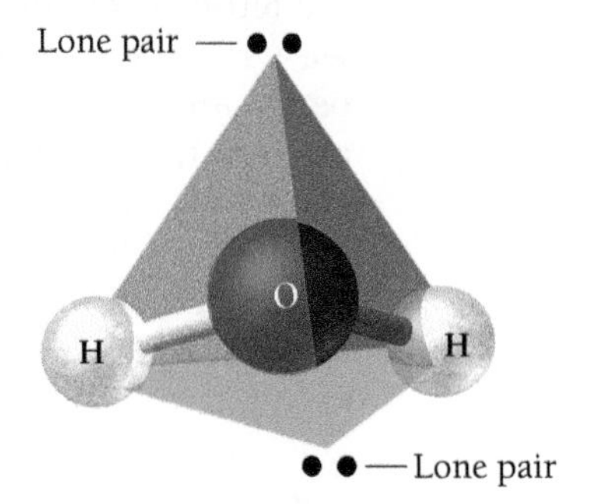

However, its molecular geometry is **bent**.

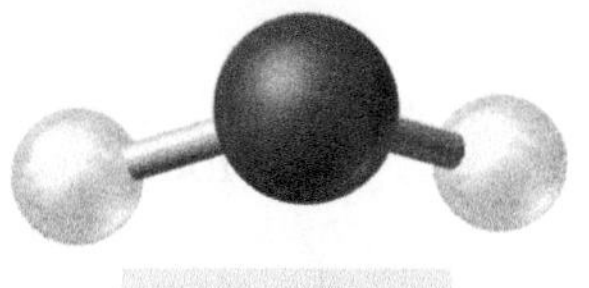

Bent structure

Table 10.1 summarizes the electron and molecular geometry of a molecule based on the total number of electron groups, the number of bonding groups, and the number of lone pairs.

To determine the geometry of any molecule, we use the procedure presented in the left column of Examples 10.8 and 10.9. As usual, the two examples of applying the steps are in the center and right columns.

TABLE 10.1 Electron and Molecular Geometries

Electron Groups*	Bonding Groups	Lone Pairs	Electron Geometry	Angle between Electron Groups**	Molecular Geometry	Example
2	2	0	linear	180°	linear	:Ö=C=Ö:
3	3	0	trigonal planar	120°	trigonal planar	H—C(=Ö:)—H
3	2	1	trigonal planar	120°	bent	:Ö=S̈—Ö:
4	4	0	tetrahedral	109.5°	tetrahedral	H—C(H)(H)—H
4	3	1	tetrahedral	109.5°	trigonal pyramidal	H—N̈(H)—H
4	2	2	tetrahedral	109.5°	bent	H—Ö—H

* Count only electron groups around the *central* atom. Each of the following is considered one electron group: a lone pair, a single bond, a double bond, and a triple bond.

** Angles listed here are idealized. Actual angles in specific molecules may vary by several degrees. For example, the bond angles in ammonia are 107° and the bond angle in water is 104.5°.

PEARSON eText 2.0 **Interactive Worked Example Video 10.8**

Predicting Geometry Using VSEPR Theory	**EXAMPLE 10.8** Predict the electron and molecular geometry of PCl_3.	**EXAMPLE 10.9** Predict the electron and molecular geometry of the $[NO_3]^-$ ion.
1. Draw a Lewis structure for the molecule.	**SOLUTION** PCl_3 has 26 electrons. :Cl: :Cl:P:Cl:	**SOLUTION** $[NO_3]^-$ has 24 electrons. [:O: :O:N::O:]$^-$
2. Determine the total number of electron groups around the central atom. Lone pairs, single bonds, double bonds, and triple bonds each count as one group.	The central atom (P) has four electron groups.	The central atom (N) has three electron groups (the double bond counts as one group).
3. Determine the number of bonding groups and the number of lone pairs around the central atom. These should sum to the result from Step 2. Bonding groups include single bonds, double bonds, and triple bonds.	:Cl: :Cl:P:Cl: Lone pair. Three of the four electron groups around P are bonding groups, and one is a lone pair.	[:O: :O:N::O:]$^-$ No lone pairs. All three of the electron groups around N are bonding groups.
4. Refer to Table 10.1 to determine the electron geometry and molecular geometry.	The electron geometry is tetrahedral (four electron groups), and the molecular geometry—the shape of the molecule—is trigonal pyramidal (four electron groups, three bonding groups, and one lone pair).	The electron geometry is trigonal planar (three electron groups), and the molecular geometry—the shape of the molecule—is trigonal planar (three electron groups, three bonding groups, and no lone pairs).
	▶ **SKILLBUILDER 10.8** | Predict the molecular geometry of ClNO (N is the central atom).	▶ **SKILLBUILDER 10.9** | Predict the molecular geometry of the SO_3^{2-} ion. ▶ **FOR MORE PRACTICE** Example 10.17; Problems 65, 66, 69, 70, 73, 74.

CONCEPTUAL CHECKPOINT 10.7

Which condition necessarily leads to a molecular geometry that is identical to the electron geometry?

(a) The presence of a double bond between the central atom and a terminal atom.

(b) The presence of two or more identical terminal atoms bonded to the central atom.

(c) The presence of one or more lone pairs on the central atom.

(d) The absence of any lone pairs on the central atom.

Representing Molecular Geometries on Paper

Because molecular geometries are three-dimensional, they are often difficult to represent on two-dimensional paper. Many chemists use this notation for bonds to indicate three-dimensional structures on two-dimensional paper:

Straight line	*Hashed lines*	*Wedge*
Bond in plane of paper	Bond projecting into the paper	Bond projecting out of the paper

The major molecular geometries used in this book are shown here using this notation:

X—A—X
Linear

Trigonal planar

Bent

Tetrahedral

Trigonal pyramidal

CHEMISTRY AND HEALTH

Fooled by Molecular Shape

Artificial sweeteners, such as aspartame (Nutrasweet™), taste sweet but have few or no calories. Why? Because taste and caloric value are entirely separate properties of foods.

The caloric value of a food depends on the amount of energy released when the food is metabolized. Sucrose (table sugar) is metabolized by oxidation to carbon dioxide and water:

$$C_{12}H_{22}O_{11} + 6\,O_2 \longrightarrow 12\,CO_2 + 11\,H_2O \quad \Delta H = -5644 \text{ kJ}$$

When your body metabolizes one mole of sucrose, it obtains 5644 kJ of energy. Some artificial sweeteners, such as saccharin, are not metabolized at all—they just pass through the body unchanged—and therefore have no caloric value. Other artificial sweeteners, such as aspartame, are metabolized but have a much lower caloric content (for a given amount of sweetness) than sucrose.

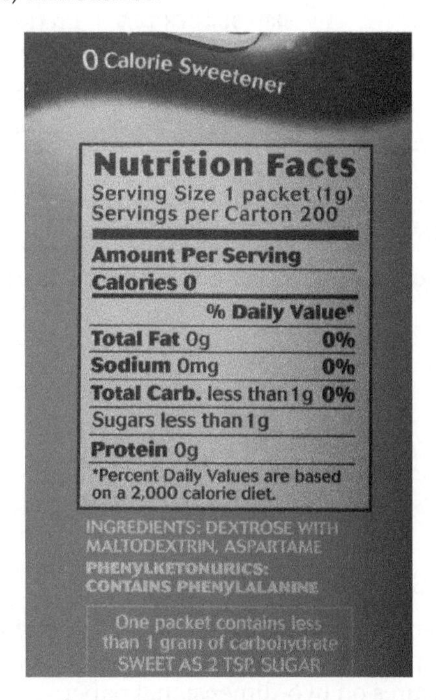

The *taste* of a food is independent of its metabolism. The sensation of taste originates in the tongue, where specialized cells called taste cells act as highly sensitive and specific molecular detectors. These cells can distinguish the sugar molecules from the thousands of different types of molecules present in a mouthful of food. The main basis for this discrimination is the molecule's *shape*.

The surface of a taste cell contains specialized protein molecules called taste receptors. Each particular *tastant*—a molecule that you can taste—fits snugly into a special pocket on the taste receptor protein called the *active site*, just as a key fits into a lock (see Section 15.12). For example, a sugar molecule fits only into the active site of the sugar receptor protein called Tlr3. When the sugar molecule (the key) enters the active site (the lock), the different subunits of the Tlr3 protein split apart. This split causes a series of events that results in transmission of a nerve signal, which reaches the brain and registers a sweet taste.

Artificial sweeteners taste sweet because they fit into the receptor pocket that normally binds sucrose. In fact, both aspartame and saccharin bind to the active site in the Tlr3 protein more strongly than sugar does! For this reason, artificial sweeteners are "sweeter than sugar." It takes 200 times as much sucrose as aspartame to trigger the same amount of nerve signal transmission from taste cells.

This type of lock-and-key fit between the active site of a protein and a particular molecule is important not only to taste but to many other biological functions as well. For example, immune response, the sense of smell, and many types of drug action all depend on shape-specific interactions between molecules and proteins. The ability of scientists to determine the shapes of key biological molecules is largely responsible for the revolution in biology that has occurred over the last 50 years.

B10.1 CAN YOU ANSWER THIS? *Proteins are long-chain molecules in which each link is an amino acid. The simplest amino acid is glycine, which has this structure:*

```
     H  :O:
     |  ||
H—N̈—C—C—Ö—H
  |  |
  H  H
```

Determine the geometry around each interior atom in the glycine structure and make a three-dimensional sketch of the molecule.

10.8 Electronegativity and Polarity: Why Oil and Water Don't Mix

▶ Determine whether a molecule is polar.

If we combine oil and water in a container, they separate into distinct regions (◀ FIGURE 10.1). Why? Something about water molecules causes them to bunch together into one region, expelling the oil molecules into a separate region. What is that something? We can begin to understand the answer by examining the Lewis structure of water.

▲ FIGURE 10.1 **Oil and water don't mix** Question: Why not?

$$\mathrm{H{-}\ddot{\underset{\cdot\cdot}{O}}{-}H}$$

The two bonds between O and H each consist of an electron pair—two electrons shared between the oxygen atom and the hydrogen atom. The oxygen and hydrogen atoms each donate one electron to this electron pair; however, like most children, they don't share them equally. The oxygen atom takes more than its fair share of the electron pair.

Electronegativity

The ability of an element to attract electrons within a covalent bond is **electronegativity.** Oxygen is more electronegative than hydrogen, which means that, on average, shared electrons are more likely to be found near the oxygen atom than near the hydrogen atom. Consider one of the two OH bonds:

The representation for depicting electron density in this figure is introduced in Section 9.6.

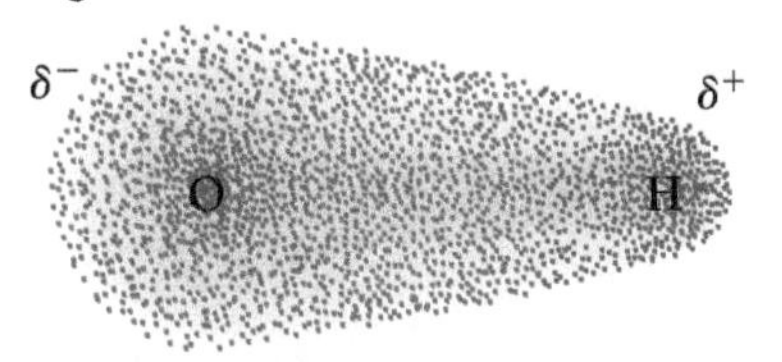

Since the electron pair is unequally shared (with oxygen getting the larger share), the oxygen atom has a partial negative charge, symbolized by $\delta-$ (delta minus). The hydrogen atom (which gets the smaller share) has a partial positive charge, symbolized by $\delta+$ (delta plus). The result of this uneven electron sharing is a **dipole moment**, a separation of charge within the bond. We call covalent bonds that have a dipole moment **polar covalent bonds**. The magnitude of the dipole moment, and therefore the degree of polarity of the bond, depend on the electronegativity difference between the two elements in the bond and the length of the bond. For a fixed bond length, the greater the electronegativity difference, the greater the dipole moment and the more polar the bond.

The value of electronegativity is assigned using a relative scale on which fluorine, the most electronegative element, has an electronegativity of 4.0. All other electronegativities are defined relative to fluorine.

▼ FIGURE 10.2 shows the relative electronegativities of the elements. Notice that electronegativity increases as we move toward the right across a period in the

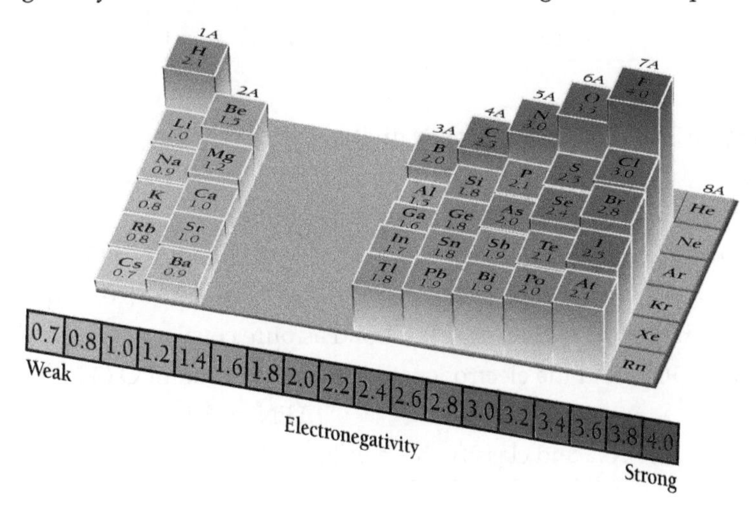

▶ FIGURE 10.2 **Electronegativity of the elements** Linus Pauling introduced the scale shown in this figure. He arbitrarily set the electronegativity of fluorine at 4.0 and calculated all other values relative to fluorine.

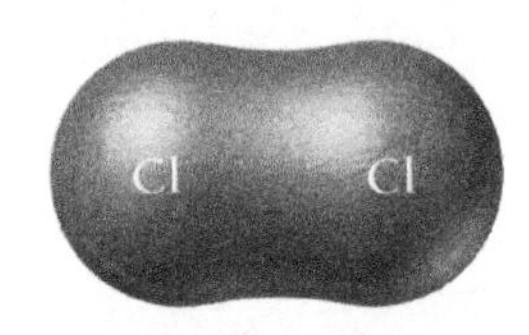

▲ FIGURE 10.3 **Pure covalent bonding** In Cl_2, the two Cl atoms share the electrons evenly. This is a pure covalent bond.

periodic table and decreases as we move down a column in the periodic table. If two elements with identical electronegativities form a covalent bond, they share the electrons equally, and there is no dipole moment. For example, the chlorine molecule, composed of two chlorine atoms (which of course have identical electronegativities), has a pure covalent bond in which electrons are evenly shared (◀ FIGURE 10.3). The bond has no dipole moment, and the molecule is **nonpolar.**

If there is a large electronegativity difference between the two elements in a bond, such as normally occurs between a metal and a nonmetal, the electron is completely transferred and the bond is ionic. For example, sodium and chlorine form an ionic bond (◀ FIGURE 10.4).

If there is an intermediate electronegativity difference between the two elements, such as between two different nonmetals, the bond is polar covalent. For example, HF forms a polar covalent bond (▼ FIGURE 10.5).

Table 10.2 and ▼ FIGURE 10.6 summarize these concepts.

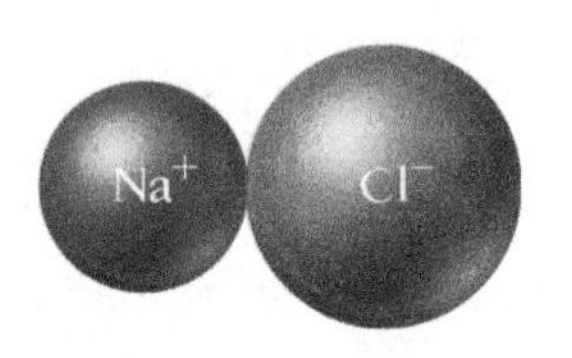

▲ FIGURE 10.4 **Ionic bonding** In NaCl, Na completely transfers an electron to Cl. This is an ionic bond.

The degree of bond polarity is a continuous function. The guidelines given here are approximate.

TABLE 10.2 The Effect of Electronegativity Difference on Bond Type

Electronegativity Difference (ΔEN)	Bond Type	Example
zero (0–0.4)	pure covalent	Cl_2
intermediate (0.4–2.0)	polar covalent	HF
large (2.0+)	ionic	NaCl

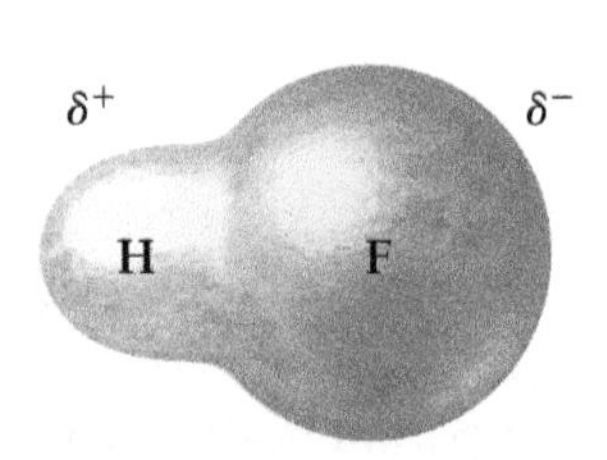

▲ FIGURE 10.5 **Polar covalent bonding** In HF, the electrons are shared, but the shared electrons are more likely to be found on F than on H. The bond is polar covalent.

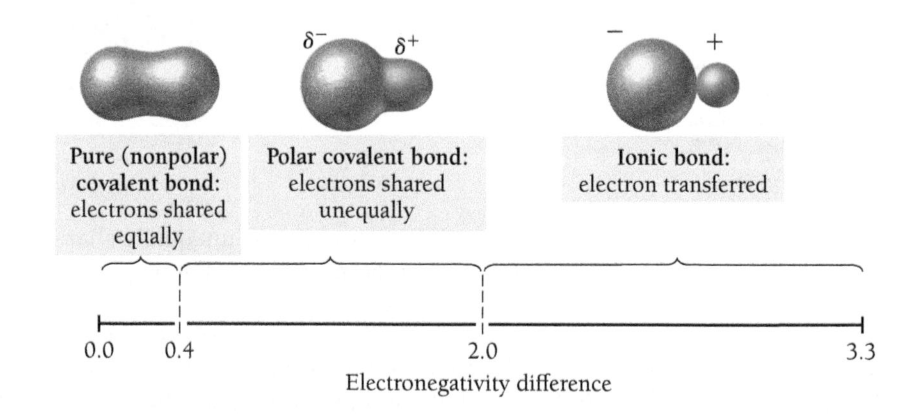

▲ FIGURE 10.6 **The continuum of bond types** The electronegativity difference between two bonded atoms determines the type of bond (pure covalent, polar covalent, or ionic).

EXAMPLE 10.10 Classifying Bonds as Pure Covalent, Polar Covalent, or Ionic

Is the bond formed between each pair of atoms pure covalent, polar covalent, or ionic?

(a) Sr and F **(b)** N and Cl **(c)** N and O

SOLUTION

(a) In Figure 10.2, find the electronegativity of Sr (1.0) and of F (4.0). The electronegativity difference (ΔEN) is:

$$\Delta EN = 4.0 - 1.0 = 3.0$$

Refer to Table 10.2 and classify this bond as ionic.

(b) In Figure 10.2, find the electronegativity of N (3.0) and of Cl (3.0). The electronegativity difference (ΔEN) is:

$$\Delta EN = 3.0 - 3.0 = 0$$

Refer to Table 10.2 and classify this bond as pure covalent.

(c) In Figure 10.2, find the electronegativity of N (3.0) and of O (3.5). The electronegativity difference (ΔEN) is:

$$\Delta EN = 3.5 - 3.0 = 0.5$$

Refer to Table 10.2 and classify this bond as polar covalent.

▶ **SKILLBUILDER 10.10 | Classifying Bonds as Pure Covalent, Polar Covalent, or Ionic**

Is the bond formed between each pair of atoms pure covalent, polar covalent, or ionic?

(a) I and I **(b)** Cs and Br **(c)** P and O

▶ **FOR MORE PRACTICE** Problems 81, 82.

CONCEPTUAL CHECKPOINT 10.8

Which bond would you expect to be more polar: the bond in HCl or the bond in HBr?

Polar Bonds and Polar Molecules

Does the presence of one or more polar bonds in a molecule always result in a polar molecule? The answer is no. A **polar molecule** is one with polar bonds that add together—they do not cancel each other—to form a net dipole moment. For diatomic molecules, we can readily tell polar molecules from nonpolar ones. If a diatomic molecule contains a polar bond, then the molecule is polar. However, for molecules with more than two atoms, it is more difficult to tell polar molecules from nonpolar ones because two or more polar bonds may cancel one another.

For example, consider carbon dioxide:

$$:\ddot{O}=C=\ddot{O}:$$

Each C=O *bond* is polar because the difference in electronegativity between oxygen and carbon is 1.0. However, since CO_2 has a linear geometry, the dipole moment of one bond completely cancels the dipole moment of the other and the *molecule* is nonpolar. We can understand this with an analogy. Imagine each polar bond to be a rope pulling on the central atom. In CO_2 we can see how the two ropes pulling in opposing directions cancel each other's effect:

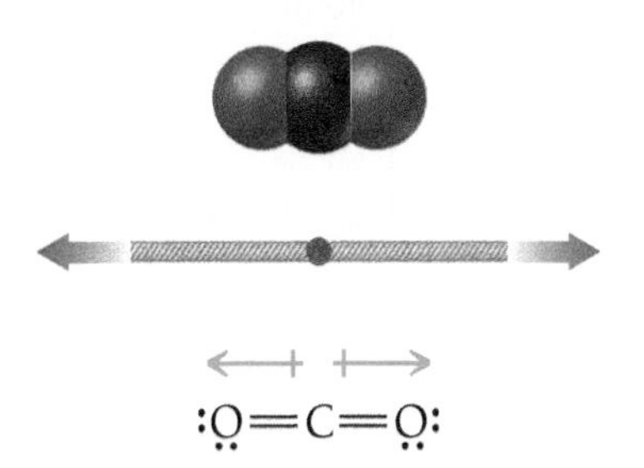

In the vector representation of a dipole moment, the vector points in the direction of the atom with the partial negative charge.

δ^+ $\longrightarrow$ δ^-

We can also represent polar bonds with arrows (or *vectors*) that point in the direction of the negative pole and have a plus sign at the positive pole (as we just saw for carbon dioxide). If the arrows (or vectors) point in exactly opposing directions as in carbon dioxide, the dipole moments cancel.

Water, on the other hand, has two dipole moments that do not cancel. If we imagine each bond as a rope pulling on oxygen, we see that, because of the angle between the bonds, the pulls of the two ropes do not cancel as shown at right.

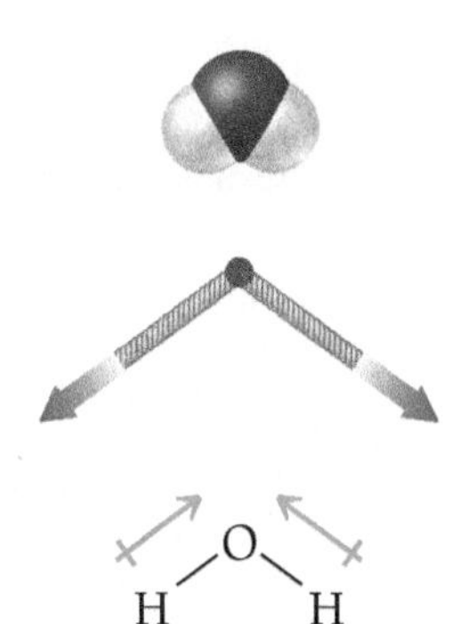

TABLE 10.3 Common Cases of Adding Dipole Moments to Determine Whether a Molecule Is Polar

Nonpolar

Two identical polar bonds pointing in opposite directions will cancel. The molecule is nonpolar.

Nonpolar

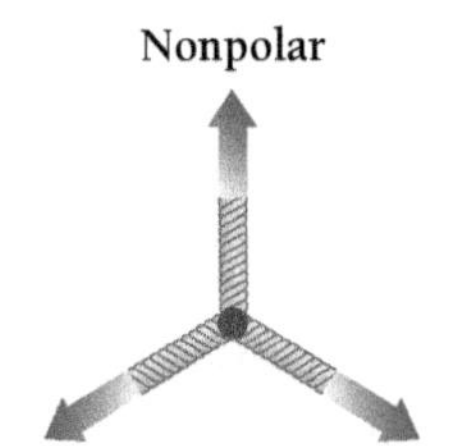

Three identical polar bonds at 120° from each other will cancel. The molecule is nonpolar.

Polar

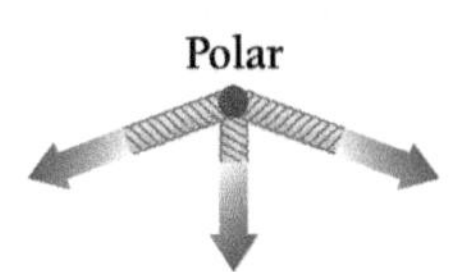

Three polar bonds in a trigonal pyramidal arrangement (109.5°) will not cancel. The molecule is polar.

Polar

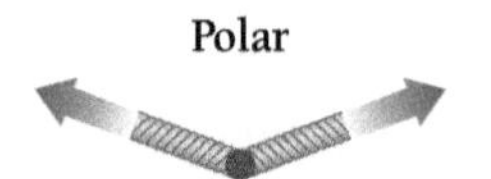

Two polar bonds with an angle of less than 180° between them will not cancel. The molecule is polar.

Nonpolar

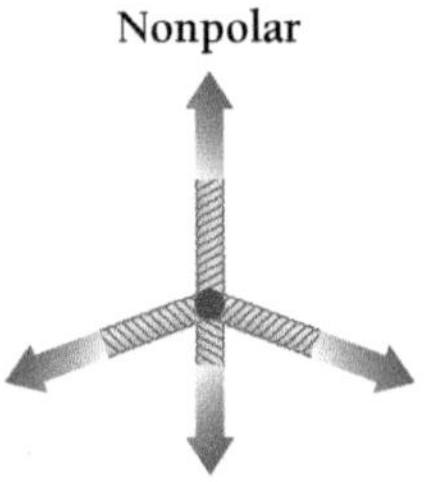

Four identical polar bonds in a tetrahedral arrangement (109.5° from each other) will cancel. The molecule is nonpolar.

Note: In all cases where the polar bonds cancel, the bonds are assumed to be identical. If one or more of the bonds are different than the other(s), the bonds will not cancel and the molecule is polar.

Consequently, water is a polar molecule. We can use symmetry as a guide to determine whether a molecule containing polar bonds is indeed polar. Highly symmetric molecules tend to be nonpolar even if they have polar bonds because the bond dipole moments (or the pulls of the ropes) tend to cancel. Asymmetric molecules that contain polar bonds tend to be polar because the bond dipole moments (or the pulls of the ropes) tend not to cancel. Table 10.3 summarizes some common cases.

In summary, to determine whether a molecule is polar:

- **Determine whether the molecule contains polar bonds.** A bond is polar if the two bonding atoms have different electronegativities. If there are no polar bonds, the molecule is nonpolar.
- **Determine whether the polar bonds add together to form a net dipole moment.** We must first use VSEPR theory to determine the geometry of the molecule. Then we visualize each bond as a rope pulling on the central atom. Is the molecule highly symmetrical? Do the pulls of the ropes cancel? If so, there is no net dipole moment and the molecule is nonpolar. If the molecule is asymmetrical and the pulls of the rope do not cancel, the molecule is polar.

Interactive Worked Example Video 10.11

EXAMPLE 10.11 Determining Whether a Molecule Is Polar

Is NH_3 polar?

	SOLUTION
Begin by drawing the Lewis structure of NH_3. Since N and H have different electronegativities, the bonds are polar.	H | H—N—H
The geometry of NH_3 is trigonal pyramidal (four electron groups, three bonding groups, one lone pair). Draw a three-dimensional picture of NH_3 and imagine each bond as a rope that is being pulled. The pulls of the ropes do not cancel and the molecule is polar.	NH_3 is polar

▶ **SKILLBUILDER 10.11 | Determining Whether a Molecule Is Polar**

Determine whether CH_4 is polar.

▶ **FOR MORE PRACTICE** Example 10.18; Problems 89, 90, 91, 92.

Polarity is important because polar molecules tend to behave differently than nonpolar molecules. Water and oil do not mix, for example, because water molecules are polar and the molecules that compose oil are generally nonpolar. Polar molecules interact strongly with other polar molecules because the positive end of one molecule is attracted to the negative end of another, just as the south pole of a

EVERYDAY CHEMISTRY

How Soap Works

Imagine eating a greasy cheeseburger without flatware or napkins. By the end of the meal, your hands are coated with grease and oil. If you try to wash them with only water, they remain greasy. However, if you add a little soap, the grease washes away. Why? As we learned previously, water molecules are polar and the molecules that compose grease and oil are nonpolar. As a result, water and grease repel each other.

The molecules that compose soap, however, have a special structure that allows them to interact strongly with both water and grease. One end of a soap molecule is polar, while the other end is nonpolar.

Soap molecule

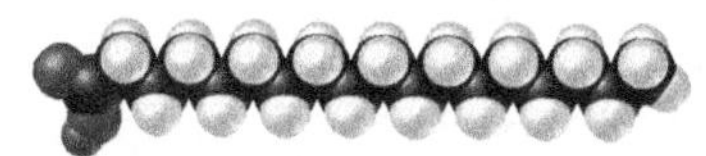

The polar head of a soap molecule strongly attracts water molecules, while the nonpolar tail strongly attracts grease and oil molecules. Soap is a sort of molecular liaison, one end interacting with water and the other end interacting with grease. Soap therefore allows water and grease to mix, removing the grease from your hands and washing it down the drain.

B10.2 CAN YOU ANSWER THIS? *Consider this detergent molecule. Which end do you think is polar? Which end is nonpolar?*

$CH_3(CH_2)_{11}OCH_2CH_2OH$

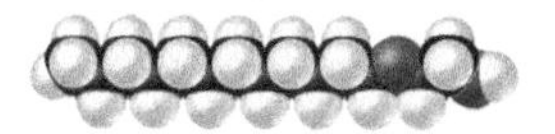

▶ FIGURE 10.7 **Dipole–dipole attraction** Just as the north pole of one magnet is attracted to the south pole of another, so the positive end of one molecule with a dipole is attracted to the negative end of another molecule with a dipole.

magnet is attracted to the north pole of another magnet (▲ FIGURE 10.7). A mixture of polar and nonpolar molecules is similar to a mixture of small magnetic and nonmagnetic particles. The magnetic particles clump together, excluding the nonmagnetic ones and separating into distinct regions (◀ FIGURE 10.8). Similarly, the polar water molecules attract one another, forming regions from which the nonpolar oil molecules are excluded (▼ FIGURE 10.9).

▲ FIGURE 10.8 **Magnetic and nonmagnetic particles** Magnetic particles (the colored marbles) attract one another, excluding nonmagnetic particles (the clear marbles). This behavior is analogous to that of polar and nonpolar molecules.

◀ FIGURE 10.9 **Polar and nonpolar molecules** A mixture of polar and nonpolar molecules, like a mixture of magnetic and nonmagnetic particles, separates into distinct regions because the polar molecules attract one another, excluding the nonpolar ones. Question: Can you think of some examples of this behavior?

Chapter 10 in Review

MasteringChemistry™ provides end-of-chapter exercises, feedback-enriched tutorial problems, animations, and interactive activities to encourage problem solving practice and deeper understanding of key concepts and topics.

Self-Assessment Quiz

Q1. Which pair of elements has the most similar Lewis structures?
(a) N and S (b) F and Ar
(c) Cl and Ar (d) O and S

Q2. What is the Lewis structure for the compound that forms between K and S?
(a) K—:S̈:—K (b) K^+ [:S̈:]$^-$
(c) K^+ [:S̈:]$^{2-}$ K^+ (d) [:S̈:]$^{2-}$ K^+ [:S̈:]$^{2-}$

Q3. Use the Lewis model to predict the correct formula for the compound that forms between K and S.
(a) KS (b) K_2S (c) KS_2 (d) K_2S_2

Q4. What is the correct Lewis structure for H_2CS?
(a) H—H—C̈—S̈:
(b) H—C—H with :S̈: single-bonded above C
(c) H=C—H with :S̈: single-bonded above C
(d) H—C—H with :S̈ double-bonded above C

Q5. How many electron dots are in the Lewis structure of NO_2^-?
(a) 17 (b) 18 (c) 19 (d) 20

Q6. Which compound has two or more resonance structures?
(a) NO_2^- (b) CO_2 (c) NH_4^+ (d) CCl_4

Q7. What is the molecular geometry of PBr_3?
(a) Bent (b) Tetrahedral
(c) Trigonal pyramidal (d) Linear

Q8. What is the molecular geometry of N_2O? (Nitrogen is the central atom.)
(a) Bent (b) Tetrahedral
(c) Trigonal pyramidal (d) Linear

Q9. Which bond is polar?
(a) A bond between C and S
(b) A bond between Br and Br
(c) A bond between C and O
(d) A bond between B and H

Q10. Which molecule is polar?
(a) SCl_2 (b) CS_2 (c) CF_4 (d) $SiCl_4$

Answers: 1:d, 2:c, 3:b, 4:d, 5:b, 6:a, 7:c, 8:d, 9:c, 10:a

Chemical Principles / Relevance

The Lewis Model

The Lewis model is a model for chemical bonding. According to the Lewis model, chemical bonds form when atoms transfer valence electrons (ionic bonding) or share valence electrons (covalent bonding) to attain noble gas electron configurations. In the Lewis model, we represent valence electrons as dots surrounding the symbol for an element. When two or more elements bond together, the dots are transferred or shared so that every atom attains eight dots (an octet), or two dots (a duet) in the case of hydrogen.

Relevance: Bonding theories predict what combinations of elements will form stable compounds, and we can use them to predict the properties of those compounds. For example, pharmaceutical companies use bonding theories when they design drug molecules to interact with a specific part of a protein molecule.

Molecular Shapes

We can predict the shapes of molecules by combining the Lewis model with valence shell electron pair repulsion (VSEPR) theory. In this model, electron groups—lone pairs, single bonds, double bonds, and triple bonds—around the central atom repel one another and determine the geometry of the molecule.

Relevance: Molecular shapes determine many of the properties of compounds. Water's bent geometry, for example, is the reason it is a liquid at room temperature instead of a gas. Its geometry is also the reason ice floats and snowflakes have hexagonal patterns.

Electronegativity and Polarity

Electronegativity refers to the relative ability of elements to attract electrons within a chemical bond. Electronegativity increases as we move to the right across a period in the periodic table and decreases as we move down a column. When two nonmetal atoms of different electronegativities form a covalent bond, the electrons in the bond are not evenly shared and the bond is polar. In diatomic molecules, a polar bond results in a polar molecule. In molecules with more than two atoms, polar bonds may cancel, forming a nonpolar molecule, or they may sum, forming a polar molecule.

Relevance: The polarity of a molecule influences many of its properties such as whether it is a solid, liquid, or gas at room temperature and whether it mixes with other compounds. Oil and water, for example, do not mix because water is polar while oil is nonpolar.

Chemical Skills

Examples

LO: Write Lewis structures for elements (Section 10.2).

The Lewis structure of any element is the symbol for the element with the valence electrons represented as dots drawn around the element. The number of valence electrons is equal to the group number of the element (for main-group elements).

EXAMPLE 10.12 Lewis Structures for Elements

Draw the Lewis structure of sulfur.

SOLUTION

Since S is in Group 6A, it has six valence electrons. Draw these as dots surrounding its symbol, S.

LO: Write Lewis structures for ionic compounds (Section 10.3).

In an ionic Lewis structure, the metal loses all of its valence electrons to the nonmetal, which attains an octet. We place the nonmetal in brackets with the charge in the upper right corner.

EXAMPLE 10.13 Writing Lewis Structures of Ionic Compounds

Write the Lewis structure for lithium bromide.

SOLUTION

Li^+ [:Br:]$^-$

LO: Use the Lewis model to predict the chemical formula of an ionic compound (Section 10.3).

To determine the chemical formula of an ionic compound, write the Lewis structures of each of the elements. Then choose the correct number of each type of atom so that the metal atom(s) lose all of their valence electrons and the nonmetal atom(s) attain an octet.

EXAMPLE 10.14 Using the Lewis Model to Predict the Chemical Formula of an Ionic Compound

Use the Lewis model to predict the formula for the compound that forms between potassium and sulfur.

SOLUTION

The Lewis structures of K and S are:

K· ·S:

Potassium must lose one electron, and sulfur must gain two. Consequently, there are two potassium atoms for every sulfur atom. The Lewis structure is:

K^+ [:S:]$^{2-}$$K^+$

The correct formula is K_2S.

LO: Write Lewis structures for covalent compounds (Sections 10.4, 10.5).

To write covalent Lewis structures, follow these steps:

1. **Write the correct skeletal structure for the molecule.** Hydrogen atoms are always terminal, halogens are usually terminal, and many molecules tend to be symmetrical.
2. **Calculate the total number of electrons for the Lewis structure by summing the valence electrons of each atom in the molecule.** Remember that the number of valence electrons for any main-group element is equal to its group number in the periodic table. For polyatomic ions, add one electron for each negative charge and subtract one electron for each positive charge.

EXAMPLE 10.15 Writing Lewis Structures for Covalent Compounds

Write the Lewis structure for CS_2.

SOLUTION

S C S

$$\begin{aligned}\text{Total } e^- &= 1 \times (\#\text{ valence } e^- \text{ in C}) \\ &\quad + 2 \times (\#\text{ valence } e^- \text{ in S}) \\ &= 4 + 2(6) \\ &= 16\end{aligned}$$

3. **Distribute the electrons among the atoms, giving octets (or duets for hydrogen) to as many atoms as possible.** Begin by placing two electrons between each pair of atoms. These are the bonding electrons. Then distribute the remaining electrons, first to terminal atoms and then to the central atom.
4. **If any atoms lack an octet, form double or triple bonds as necessary to give them octets.** Do this by moving lone electron pairs from terminal atoms into the bonding region with the central atom.

S:C:S (4 of 16 e^- used)

:S̤̈:C:S̤̈: (16 of 16 e^- used)

:S̈::C::S̈: or :S̈=C=S̈:

LO: Write resonance structures (Section 10.6).

When we can write two or more equivalent (or nearly equivalent) Lewis structures for a molecule, the true structure is an average between these. Represent this by writing all of the correct structures (called resonance structures) with double-headed arrows between them.

EXAMPLE **10.16** **Writing Resonance Structures**

Write resonance structures for SeO_2.

SOLUTION

You can write the Lewis structure for SeO_2 by following the steps for writing covalent Lewis structures. You can write two equally correct structures, so draw them both as resonance structures.

:Ö̤—S̈e=Ö: ⟷ :Ö=S̈e—Ö̤:

LO: Predict the shapes of molecules (Section 10.7).

To determine the shape of a molecule, follow these steps:

1. **Draw the Lewis structure for the molecule.**
2. **Determine the total number of electron groups around the central atom.** Lone pairs, single bonds, double bonds, and triple bonds each count as one group.
3. **Determine the number of bonding groups and the number of lone pairs around the central atom.** These should sum to the result from Step 2. Bonding groups include single bonds, double bonds, and triple bonds.
4. Refer to Table 10.1 to determine the electron geometry and molecular geometry.

EXAMPLE **10.17** **Predicting the Shapes of Molecules**

Predict the geometry of SeO_2.

SOLUTION

The Lewis structure for SeO_2 (as you determined in Example 10.16) is composed of the following two resonance structures.

:Ö̤—S̈e=Ö: ⟷ :Ö=S̈e—Ö̤:

Either of the resonance structures will give the same geometry.

Total number of electron groups = 3
Number of bonding groups = 2
Number of lone pairs = 1

Electron geometry = Trigonal planar
Molecular geometry = Bent

LO: Determine whether a molecule is polar (Section 10.8).

- **Determine whether the molecule contains polar bonds.** A bond is polar if the two bonding atoms have different electronegativities. If there are no polar bonds, the molecule is nonpolar.
- **Determine whether the polar bonds add together to form a net dipole moment.** Use VSEPR theory to determine the geometry of the molecule. Then visualize each bond as a rope pulling on the central atom. Is the molecule highly symmetrical? Do the pulls of the ropes cancel? If so, there is no net dipole moment and the molecule is nonpolar. If the molecule is asymmetrical and the pulls of the rope do not cancel, the molecule is polar.

EXAMPLE **10.18** **Determining Whether a Molecule Is Polar**

Determine whether SeO_2 is polar.

SOLUTION

Se and O are nonmetals with different electronegativities (2.4 for Se and 3.5 for O). Therefore, the Se—O bonds are polar.

As you determined in Example 10.17, the geometry of SeO_2 is bent.

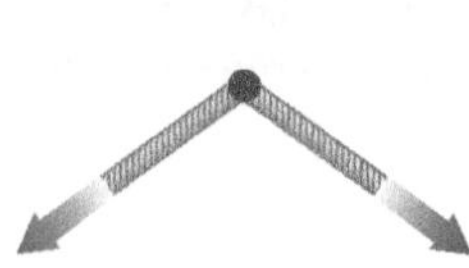

The polar bonds do not cancel but rather sum to give a net dipole moment. Therefore the molecule is polar.

Key Terms

bent [10.7]
bonding pair [10.4]
bonding theory [10.1]
chemical bond [10.2]
covalent bond [10.2]
dipole moment [10.8]
dot structure [10.2]
double bond [10.4]
duet [10.2]
electron geometry [10.7]
electron group [10.7]
electronegativity [10.8]
ionic bond [10.2]
Lewis model [10.1]
Lewis structure [10.2]
linear [10.7]
lone pair [10.4]
molecular geometry [10.7]
nonpolar [10.8]
octet [10.2]
octet rule [10.2]
polar covalent bond [10.8]
polar molecule [10.8]
resonance structures [10.6]
terminal atom [10.5]
tetrahedral [10.7]
trigonal planar [10.7]
trigonal pyramidal [10.7]
triple bond [10.4]
valence shell electron pair repulsion (VSEPR) theory [10.7]

Exercises

Questions

1. Why are bonding theories important? Cite some examples of what bonding theories can predict.
2. Write the electron configurations for Ne and Ar. How many valence electrons does each element have?
3. In the Lewis model, what is an octet? What is a duet? What is a chemical bond?
4. What is the difference between ionic bonding and covalent bonding?
5. How can the Lewis model be used to determine the formula of ionic compounds? You may explain this with an example.
6. What is the difference between lone pair and bonding pair electrons?
7. How are double and triple bonds physically different from single bonds?
8. What is the procedure for writing a covalent Lewis structure?
9. How do you determine the number of electrons that go into the Lewis structure of a molecule?
10. How do you determine the number of electrons that go into the Lewis structure of a polyatomic ion?
11. Why does the octet rule have exceptions? List some examples.
12. What are resonance structures? Why are they necessary?
13. Explain how VSEPR theory predicts the shapes of molecules.
14. If all of the electron groups around a central atom are bonding groups (that is, there are no lone pairs), what is the molecular geometry for:
    (a) two electron groups
    (b) three electron groups
    (c) four electron groups
15. Give the bond angles for each of the geometries in the preceding question.
16. What is the difference between electron geometry and molecular geometry in VSEPR theory?
17. What is electronegativity?
18. What is the most electronegative element on the periodic table?
19. What is a polar covalent bond?
20. What is a dipole moment?
21. What happens if you try to mix a polar liquid with a nonpolar one?
22. If a molecule has polar bonds, is the molecule itself polar? Why or why not?

Problems

WRITING LEWIS STRUCTURES FOR ELEMENTS

23. Write an electron configuration for each element and the corresponding Lewis structure. Indicate which electrons in the electron configuration are included in the Lewis structure.
    (a) N
    (b) C
    (c) Cl
    (d) Ar
24. Write an electron configuration for each element and the corresponding Lewis structure. Indicate which electrons in the electron configuration are included in the Lewis structure.
    (a) Li
    (b) P
    (c) F
    (d) Ne
25. Write the Lewis structure for each element.
    (a) I
    (b) S
    (c) Ge
    (d) Ca
26. Write the Lewis structure for each element.
    (a) Kr
    (b) P
    (c) B
    (d) Na

27. Write a generic Lewis structure for the halogens. Do the halogens tend to gain or lose electrons in chemical reactions? How many?

28. Write a generic Lewis structure for the alkali metals. Do the alkali metals tend to gain or lose electrons in chemical reactions? How many?

29. Write a generic Lewis structure for the alkaline earth metals. Do the alkaline earth metals tend to gain or lose electrons in chemical reactions? How many?

30. Write a generic Lewis structure for the elements in the oxygen family (Group 6A). Do the elements in the oxygen family tend to gain or lose electrons in chemical reactions? How many?

31. Write the Lewis structure for each ion.
(a) Al^{3+}
(b) Mg^{2+}
(c) Se^{2-}
(d) N^{3-}

32. Write the Lewis structure for each ion.
(a) Sr^{2+}
(b) S^{2-}
(c) Li^{+}
(d) Cl^{-}

33. Indicate the noble gas that has the same Lewis structure as each ion.
(a) Br^{-}
(b) O^{2-}
(c) Rb^{+}
(d) Ba^{2+}

34. Indicate the noble gas that has the same Lewis structure as each ion.
(a) Se^{2-}
(b) I^{-}
(c) Sr^{2+}
(d) F^{-}

LEWIS STRUCTURES FOR IONIC COMPOUNDS

35. Is each compound best represented by an ionic or a covalent Lewis structure?
(a) SF_6
(b) $MgCl_2$
(c) $BrCl$
(d) K_2S

36. Is each compound best represented by an ionic or a covalent Lewis structure?
(a) NO
(b) CO_2
(c) Rb_2O
(d) Al_2S_3

37. Write the Lewis structure for each ionic compound.
(a) NaF
(b) CaO
(c) $SrBr_2$
(d) K_2O

38. Write the Lewis structure for each ionic compound.
(a) SrO
(b) Li_2S
(c) CaI_2
(d) RbF

39. Use the Lewis model to determine the formula for the compound that forms from each pair of atoms.
(a) Ca and S
(b) Mg and Br
(c) Cs and I
(d) Ca and N

40. Use the Lewis model to determine the formula for the compound that forms from each pair of atoms.
(a) Al and S
(b) Na and S
(c) Sr and Se
(d) Ba and F

41. Draw the Lewis structure for the ionic compound that forms from Mg and each atom.
(a) F
(b) O
(c) N

42. Draw the Lewis structure for the ionic compound that forms from Al and each atom.
(a) F
(b) O
(c) N

43. Determine what is wrong with each ionic Lewis structure and write the correct structure.

(a) $[Cs:]^{+}\ [:\ddot{Cl}:]^{-}$

(b) $Ba^{+}\ [:\underset{\cdot\cdot}{\ddot{O}}:]^{-}$

(c) $Ca^{2+}\ [:\underset{\cdot\cdot}{\ddot{I}}:]^{-}$

44. Determine what is wrong with each ionic Lewis structure and write the correct structure.

(a) $[:\underset{\cdot\cdot}{\ddot{O}}:]^{2-}\ Na^{+}\ [:\underset{\cdot\cdot}{\ddot{O}}:]^{2-}$

(b) $Mg:\underset{\cdot\cdot}{\ddot{O}}:$

(c) $[Li:]^{+}\ [:\underset{\cdot\cdot}{\ddot{S}}:]^{-}$

LEWIS STRUCTURES FOR COVALENT COMPOUNDS

45. Use the Lewis model to explain why each element exists as a diatomic molecule.
(a) hydrogen
(b) iodine
(c) nitrogen
(d) oxygen

46. Use the Lewis model to explain why the compound that forms between hydrogen and sulfur has the formula H_2S. Would you expect HS to be stable? H_3S?

47. Write the Lewis structure for each molecule.
(a) PH_3
(b) SCl_2
(c) F_2
(d) HI

48. Write the Lewis structure for each molecule.
(a) CH_4
(b) NF_3
(c) OF_2
(d) H_2O

49. Write the Lewis structure for each molecule.
(a) O_2
(b) CO
(c) HONO (N is central; H bonded to one of the O atoms)
(d) SO_2

50. Write the Lewis structure for each molecule.
(a) N_2O (oxygen is terminal)
(b) SiH_4
(c) CI_4
(d) Cl_2CO (carbon is central)

51. Write the Lewis structure for each molecule.
(a) C_2H_2
(b) C_2H_4
(c) N_2H_2
(d) N_2H_4

52. Write the Lewis structure for each molecule.
(a) H_2CO (carbon is central)
(b) H_3COH (carbon and oxygen are both central)
(c) H_3COCH_3 (oxygen is between the two carbon atoms)
(d) H_2O_2

53. Determine what is wrong with each Lewis structure and write the correct structure.

(a) $:\ddot{\text{N}}=\ddot{\text{N}}:$

(b) $:\underset{\cdot\cdot}{\overset{\cdot\cdot}{\text{S}}}-\text{Si}-\underset{\cdot\cdot}{\overset{\cdot\cdot}{\text{S}}}:$

(c) $\text{H}-\text{H}-\underset{\cdot\cdot}{\overset{\cdot\cdot}{\text{O}}}:$

(d) $:\underset{\cdot\cdot}{\overset{\cdot\cdot}{\text{I}}}-\underset{|\atop :\underset{\cdot\cdot}{\text{I}}:}{\text{N}}-\underset{\cdot\cdot}{\overset{\cdot\cdot}{\text{I}}}:$

54. Determine what is wrong with each Lewis structure and write the correct structure.

(a) $\text{H}-\text{H}-\text{H}-\underset{\cdot\cdot}{\overset{\cdot\cdot}{\text{N}}}:$

(b) $:\underset{\cdot\cdot}{\overset{\cdot\cdot}{\text{Cl}}}=\text{O}=\underset{\cdot\cdot}{\overset{\cdot\cdot}{\text{Cl}}}:$

(c) $\text{H}-\overset{:\ddot{\text{O}}:\atop |}{\text{C}}-\underset{\cdot\cdot}{\overset{\cdot\cdot}{\text{O}}}-\text{H}$

(d) $\text{H}=\underset{\cdot\cdot}{\overset{\cdot\cdot}{\text{Br}}}:$

55. Write the Lewis structure for each molecule or ion. Include resonance structures if necessary.
(a) SeO_2
(b) CO_3^{2-}
(c) ClO^-
(d) ClO_2^-

56. Write the Lewis structure for each molecule or ion. Include resonance structures if necessary.
(a) ClO_3^-
(b) ClO_4^-
(c) NO_3^-
(d) SO_3

57. Write the Lewis structure for each ion. Include resonance structures if necessary.
(a) PO_4^{3-}
(b) CN^-
(c) NO_2^-
(d) SO_3^{2-}

58. Write the Lewis structure for each ion. Include resonance structures if necessary.
(a) SO_4^{2-}
(b) HSO_4^- (S is central; H is attached to one of the O atoms)
(c) NH_4^+
(d) BrO_2^- (Br is central)

59. Write the Lewis structure for each molecule. These molecules do not follow the octet rule.
(a) BCl_3
(b) NO_2
(c) BH_3

60. Write the Lewis structure for each molecule. These molecules do not follow the octet rule.
(a) BBr_3
(b) NO

PREDICTING THE SHAPES OF MOLECULES

61. Determine the number of electron groups around the central atom for each molecule.
(a) OF_2
(b) NF_3
(c) CS_2
(d) CH_4

62. Determine the number of electron groups around the central atom for each molecule.
(a) CH_2Cl_2
(b) SBr_2
(c) H_2S
(d) PCl_3

63. Determine the number of bonding groups and the number of lone pairs for each of the molecules in Problem 61. The sum of these should equal your answers to Problem 61.

64. Determine the number of bonding groups and the number of lone pairs for each of the molecules in Problem 62. The sum of these should equal your answers to Problem 62.

65. Determine the molecular geometry of each molecule.
(a) CBr_4
(b) H_2CO
(c) CS_2
(d) BH_3

66. Determine the molecular geometry of each molecule.
(a) SiO_2
(b) BF_3
(c) $CFCl_3$ (carbon is central)
(d) H_2CS (carbon is central)

67. Determine the bond angles for each molecule in Problem 65.

68. Determine the bond angles for each molecule in Problem 66.

69. Determine the electron and molecular geometries of each molecule.
(a) N_2O (oxygen is terminal)
(b) SO_2
(c) H_2S
(d) PF_3

70. Determine the electron and molecular geometries of each molecule. (*Hint:* Determine the geometry around each of the two central atoms.)
(a) C_2H_2 (skeletal structure HCCH)
(b) C_2H_4 (skeletal structure H_2CCH_2)
(c) C_2H_6 (skeletal structure H_3CCH_3)

71. Determine the bond angles for each molecule in Problem 69.

72. Determine the bond angles for each molecule in Problem 70.

73. Determine the electron and molecular geometries of each molecule. For molecules with two central atoms, indicate the geometry about each central atom.
(a) N_2
(b) N_2H_2 (skeletal structure HNNH)
(c) N_2H_4 (skeletal structure H_2NNH_2)

74. Determine the electron and molecular geometries of each molecule. For molecules with more than one central atom, indicate the geometry about each central atom.
(a) CH_3OH (skeletal structure H_3COH)
(b) H_3COCH_3 (skeletal structure H_3COCH_3)
(c) H_2O_2 (skeletal structure HOOH)

75. Determine the molecular geometry of each polyatomic ion.
(a) CO_3^{2-}
(b) ClO_2^-
(c) NO_3^-
(d) NH_4^+

76. Determine the molecular geometry of each polyatomic ion.
(a) ClO_4^-
(b) BrO_2^-
(c) NO_2^-
(d) SO_4^{2-}

ELECTRONEGATIVITY AND POLARITY

77. Refer to Figure 10.2 to determine the electronegativity of each element.
(a) Mg
(b) Si
(c) Br

78. Refer to Figure 10.2 to determine the electronegativity of each element.
(a) F
(b) C
(c) S

79. List these elements in order of decreasing electronegativity: Rb, Si, Cl, Ca, Ga.

80. List these elements in order of increasing electronegativity: Ba, N, F, Si, Cs.

81. Refer to Figure 10.2 to find the electronegativity difference between each pair of elements; then refer to Table 10.2 to classify the bonds that occur between them as pure covalent, polar covalent, or ionic.
(a) Mg and Br
(b) Cr and F
(c) Br and Br
(d) Si and O

82. Refer to Figure 10.2 to find the electronegativity difference between each pair of elements; then refer to Table 10.2 to classify the bonds that occur between them as pure covalent, polar covalent, or ionic.
(a) K and Cl
(b) N and N
(c) C and S
(d) C and Cl

83. Arrange these diatomic molecules in order of increasing bond polarity: ICl, HBr, H_2, CO.

84. Arrange these diatomic molecules in order of decreasing bond polarity: HCl, NO, F_2, HI.

85. Classify each diatomic molecule as polar or nonpolar.
(a) CO
(b) O_2
(c) F_2
(d) HBr

86. Classify each diatomic molecule as polar or nonpolar.
(a) I_2
(b) NO
(c) HCl
(d) N_2

87. For each polar molecule in Problem 85 draw the molecule and indicate the positive and negative ends of the dipole moment.

88. For each polar molecule in Problem 86 draw the molecule and indicate the positive and negative ends of the dipole moment.

89. Classify each molecule as polar or nonpolar.
(a) CS_2
(b) SO_2
(c) CH_4
(d) CH_3Cl

90. Classify each molecule as polar or nonpolar.
(a) H_2CO
(b) CH_3OH
(c) CH_2Cl_2
(d) CO_2

91. Classify each molecule as polar or nonpolar.
(a) BH_3
(b) $CHCl_3$
(c) C_2H_2
(d) NH_3

92. Classify each molecule as polar or nonpolar.
(a) N_2H_2
(b) H_2O_2
(c) CF_4
(d) NO_2

Cumulative Problems

93. Write electron configurations and Lewis structures for each element. Indicate which of the electrons in the electron configuration are shown in the Lewis structure.
(a) Ca
(b) Ga
(c) As
(d) I

94. Write electron configurations and Lewis structures for each element. Indicate which of the electrons in the electron configuration are shown in the Lewis structure.
(a) Rb
(b) Ge
(c) Kr
(d) Se

95. Determine whether each compound is ionic or covalent and write the appropriate Lewis structure.
(a) K_2S
(b) CHFO (carbon is central)
(c) MgSe
(d) PBr_3

96. Determine whether each compound is ionic or covalent and write the appropriate Lewis structure.
(a) HCN
(b) ClF
(c) MgI_2
(d) CaS

97. Write the Lewis structure for $OCCl_2$ (carbon is central) and determine whether the molecule is polar. Draw the three-dimensional structure of the molecule.

98. Write the Lewis structure for CH_3COH and determine whether the molecule is polar. Draw the three-dimensional structure of the molecule. The skeletal structure is:

```
   H  O

H  C  C  H

   H
```

99. Write the Lewis structure for acetic acid (a component of vinegar) CH_3COOH, and draw the three-dimensional sketch of the molecule. Its skeletal structure is:

```
   H  O

H  C  C  O  H

   H
```

100. Write the Lewis structure for benzene, C_6H_6, and draw a three-dimensional sketch of the molecule. The skeletal structure is the ring shown here. (*Hint:* The Lewis structure consists of two resonance structures.)

```
      H
      C
  HC     CH

  HC     CH
      C
      H
```

101. Consider the neutralization reaction.

$$HCl(aq) + NaOH(aq) \longrightarrow H_2O(l) + NaCl(aq)$$

Write the reaction showing the Lewis structures of each of the reactants and products.

102. Consider the precipitation reaction.

$$Pb(NO_3)_2(aq) + 2\,LiCl(aq) \longrightarrow PbCl_2(s) + 2\,LiNO_3(aq)$$

Write the reaction showing the Lewis structures of each of the reactants and products.

103. Consider the redox reaction.

$$2\,K(s) + Cl_2(g) \longrightarrow 2\,KCl(s)$$

Draw the Lewis structure for each reactant and product and determine which reactant was oxidized and which one was reduced.

104. Consider the redox reaction.

$$Ca(s) + Br_2(g) \longrightarrow CaBr_2(s)$$

Draw the Lewis structure for each reactant and product and determine which reactant was oxidized and which one was reduced.

105. Each compound listed contains both ionic and covalent bonds. Write the ionic Lewis structure for each one including the covalent structure for the polyatomic ion. Write resonance structures if necessary.

(a) KOH
(b) KNO_3
(c) LiIO
(d) $BaCO_3$

106. Each of the compounds listed contains both ionic and covalent bonds. Write an ionic Lewis structure for each one, including the covalent structure for the polyatomic ion. Write resonance structures if necessary.

(a) $RbIO_2$
(b) $Ca(OH)_2$
(c) NH_4Cl
(d) $Sr(CN)_2$

107. Each molecule listed contains an expanded octet (10 or 12 electrons) around the central atom. Write the Lewis structure for each molecule.

(a) PF_5
(b) SF_4
(c) SeF_4

108. Each molecule listed contains an expanded octet (10 or 12 electrons) around the central atom. Write the Lewis structure for each molecule.

(a) ClF_5
(b) SF_6
(c) IF_5

109. Formic acid is responsible for the sting you feel when stung by fire ants. By mass, formic acid is 26.10% C, 4.38% H, and 69.52% O. The molar mass of formic acid is 46.02 g/mol. Find the molecular formula of formic acid and draw its Lewis structure.

110. Diazomethane has the following composition by mass: 28.57% C, 4.80% H, and 66.64% N. The molar mass of diazomethane is 42.04 g/mol. Find the molecular formula of diazomethane and draw its Lewis structure.

111. Free radicals are molecules that contain an odd number of valence electrons and therefore contain an unpaired electron in their Lewis structure. Write the best possible Lewis structure for the free radical HOO. Does the Lewis model predict that HOO is stable? Predict its geometry.

112. Free radicals (as explained in the previous problem) are molecules that contain an odd number of valence electrons. Write the best possible Lewis structure for the free radical CH_3. Predict its geometry.

Highlight Problems

113. Some theories on aging suggest that free radicals cause a variety of diseases and aging. Free radicals (as explained in Problems 111 and 112) are molecules or ions containing an unpaired electron. As you know from the Lewis model, such molecules are not chemically stable and quickly react with other molecules. Free radicals may attack molecules within the cell, such as DNA, changing them and causing cancer or other diseases. Free radicals may also attack molecules on the surfaces of cells, making them appear foreign to the body's immune system. The immune system then attacks the cell and destroys it, weakening the body. Draw the Lewis structure for each of these free radicals, which have been implicated in theories of aging.

(a) O_2^-
(b) O^-
(c) OH
(d) CH_3OO (unpaired electron on terminal oxygen)

▲ Free radicals, molecules containing unpaired electrons (represented here as X·), may attack biological molecules such as the DNA molecule depicted here.

114. Free radicals (see Problem 113) are important in many environmentally significant reactions. For example, photochemical smog, which forms as a result of the action of sunlight on air pollutants, is formed in part by these two steps:

$$NO_2 \xrightarrow{\text{UV light}} NO + O$$

$$O + O_2 \longrightarrow O_3$$

The product of this reaction, ozone, is a pollutant in the lower atmosphere. Ozone is an eye and lung irritant and also accelerates the weathering of rubber products. Write Lewis structures for each of the reactants and products in the preceding reactions.

▲ Ozone damages rubber products.

115. Examine the formulas and space-filling models of the molecules shown here. Determine whether the structure is correct. If the structure is incorrect, sketch the correct structure.

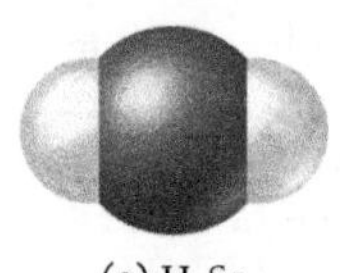

(a) H_2Se

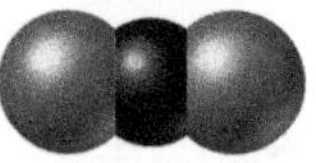

(b) CSe_2

(c) PCl_3

(d) CF_2Cl_2

Questions for Group Work

Discuss these questions with the group and record your consensus answer.

116. Draw the Lewis dot structure for the atoms Al and O. Use the Lewis model to determine the formula for the compound these atoms form.

117. Draft a list of step-by-step instructions for writing a correct Lewis dot structure for any molecule or polyatomic ion.

118. For each of the following molecules:

CS_2 NCl_3 CF_2 CH_2F_2

(a) Draw the Lewis dot structure.
(b) Determine the molecular geometry and draw it as accurately as you can.
(c) Indicate the polarity of any polar bonds with the δ symbol.
(d) Classify the molecule as polar or nonpolar.

Data Interpretation and Analysis

119. The VSEPR model is useful in predicting bond angles for many compounds. Consider the tabulated data for bond angles in related species and answer the questions.

Bond Angles in NO_2 and Associated Ions

Species	Bond Angle
NO_2	134°
NO_2^+	180°
NO_2^-	115°

(a) Draw Lewis structures for all of the species in the table.
(b) Use the Lewis structures from part a) to explain the observed bond angles in NO_2 and its associated ions.

Answers to Skillbuilder Exercises

Skillbuilder 10.1 .... $\cdot$ Mg $\cdot$

Skillbuilder 10.2 .... Na^+ $[:\ddot{\underset{..}{Br}}:]^-$

Skillbuilder 10.3 .... Mg_3N_2

Skillbuilder 10.4 .... $:C\equiv O:$

Skillbuilder 10.5 .... H—C(=Ö:)—H

Skillbuilder 10.6 .... $[:\ddot{\underset{..}{Cl}}-\ddot{\underset{..}{O}}:]^-$

Skillbuilder 10.7

$[:\ddot{O}=\ddot{N}-\ddot{\underset{..}{O}}:]^- \longleftrightarrow [:\ddot{\underset{..}{O}}-\ddot{N}=\ddot{O}:]^-$

Skillbuilder 10.8 .... bent

Skillbuilder 10.9 .... trigonal pyramidal

Skillbuilder 10.10 .... **(a)** pure covalent
(b) ionic
(c) polar covalent

Skillbuilder 10.11 .... CH_4 is nonpolar

Answers to Conceptual Checkpoints

10.1 (a) C and Si both have four dots in their Lewis structure because they are both in the same column in the periodic table.

10.2 (b) Aluminum must lose its three valence electrons to obtain an octet. Sulfur must gain two electrons to obtain an octet. Therefore, two Al atoms are required for every three S atoms.

10.3 (b) The Lewis structure of O_2 has one double bond that contains four electrons (all of them bonding electrons); therefore, the number of bonding electrons is four.

10.4 (c) The Lewis structure of OH^- has eight electrons: six from oxygen, one from hydrogen, and one from the negative charge.

10.5 (b) Both NH_3 and H_3O^+ have one lone electron pair.

10.6 (c) Resonance structures must have the same skeletal structure. Structure c differs from the other structures and is therefore not a resonance structure.

10.7 (d) If there are no lone pairs on the central atom, all of its valence electrons are involved in bonds, so the molecular geometry must be the same as the electron geometry.

10.8 The bond in H—Cl is more polar than the bond in H—Br because Cl is more electronegative than Br, so the electronegativity difference between H and Cl is greater than the difference between H and Br.

Appendix: Mathematics Review

Basic Algebra

In chemistry, you often have to solve an equation for a particular variable. For example, suppose you want to solve the following equation for V:

$$PV = nRT$$

To solve an equation for a particular variable, you must isolate that variable on one side of the equation. The rest of the variables or numbers will then be on the other side of the equation. To solve the above equation for V, divide both sides by P.

$$\frac{PV}{P} = \frac{nRT}{P}$$

$$V = \frac{nRT}{P}$$

The Ps cancel, and you are left with an expression for V. For another example, consider solving the following equation for °F:

$$°\text{C} = \frac{(°\text{F} - 32)}{1.8}$$

First, eliminate the 1.8 in the denominator of the right side by multiplying both sides by 1.8.

$$(1.8)\,°\text{C} = \frac{(°\text{F} - 32)}{\cancel{1.8}}\cancel{(1.8)}$$

$$(1.8)°\text{C} = (°\text{F} - 32)$$

Then eliminate the -32 on the right by adding 32 to both sides.

$$(1.8)°\text{C} + 32 = (°\text{F} - 32) + 32$$

$$(1.8)°\text{C} + 32 = °\text{F}$$

You are now left with an expression for °F.

In general, solve equations by following these guidelines:

- Cancel numbers or symbols in the denominator (bottom part of a fraction) by multiplying by the number or symbol to be canceled.
- Cancel numbers or symbols in the numerator (upper part of a fraction) by dividing by the number or symbol to be canceled.
- Eliminate numbers or symbols that are added by subtracting the same number or symbol.
- Eliminate numbers or symbols that are subtracted by adding the same number or symbol.
- Whether you add, subtract, multiply, or divide, **always perform the same operation for both sides of a mathematical equation**. (Otherwise, the two sides will no longer be equal.)

For a final example, solve the following equation for x:

$$\frac{67x - y + 3}{6} = 2z$$

Cancel the 6 in the denominator by multiplying both sides by 6.

$$(6)\frac{67x - y + 3}{6} = (6)2z$$

$$67x - y + 3 = 12z$$

Eliminate the +3 by subtracting 3 from both sides.

$$67x - y + 3 - 3 = 12z - 3$$

$$67x - y = 12z - 3$$

Eliminate the $-y$ by adding y to both sides.

$$67x - y + y = 12z - 3 + y$$

$$67x = 12z - 3 + y$$

Cancel the 67 by dividing both sides by 67.

$$\frac{67x}{67} = \frac{12z - 3 + y}{67}$$

$$x = \frac{12z - 3 + y}{67}$$

▶ FOR PRACTICE Using Algebra to Solve Equations

Solve each of the following for the indicated variable:

(a) $P_1V_1 = P_2V_2$; solve for V_2

(b) $\frac{V_1}{T_1} = \frac{V_2}{T_2}$; solve for T_1

(c) $PV = nRT$; solve for n

(d) K = °C + 273; solve for °C

(e) $\frac{3x + 7}{2} = y$; solve for x

(f) $\frac{32}{y + 3} = 8$; solve for y

ANSWERS

(a) $V_2 = \frac{P_1V_1}{P_2}$

(b) $T_1 = \frac{V_1T_2}{V_2}$

(c) $n = \frac{PV}{RT}$

(d) °C = K − 273

(e) $x = \frac{2y - 7}{3}$

(f) $y = 1$

Mathematical Operations with Scientific Notation

Writing numbers in scientific notation is covered in detail in Section 2.2. Briefly, a number written in scientific notation consists of a **decimal part**, a number that is usually between 1 and 10, and an **exponential part**, 10 raised to an **exponent**, n.

Each of the following numbers is written in both scientific and decimal notation:

$1.0 \times 10^5 = 100{,}000$ $\qquad$ $1.0 \times 10^{-6} = 0.000001$

$6.7 \times 10^3 = 6700$ $\qquad$ $6.7 \times 10^{-3} = 0.0067$

Multiplication and Division

To multiply numbers expressed in scientific notation, multiply the decimal parts and add the exponents.

$$(A \times 10^m)(B \times 10^n) = (A \times B) \times 10^{m+n}$$

To divide numbers expressed in scientific notation, divide the decimal parts and subtract the exponent in the denominator from the exponent in the numerator.

$$\frac{(A \times 10^m)}{(B \times 10^n)} = \left(\frac{A}{B}\right) \times 10^{m-n}$$

Consider the following example involving multiplication:

$$(3.5 \times 10^4)(1.8 \times 10^6) = (3.5 \times 1.8) \times 10^{4+6}$$

$$= 6.3 \times 10^{10}$$

Consider the following example involving division:

$$\frac{(5.6 \times 10^7)}{(1.4 \times 10^3)} = \left(\frac{5.6}{1.4}\right) \times 10^{7-3}$$

$$= 4.0 \times 10^4$$

Addition and Subtraction

To add or subtract numbers expressed in scientific notation, rewrite all the numbers so that they have the same exponent, and then add or subtract the decimal parts of the numbers. The exponents remain unchanged.

$$\begin{array}{r} A \times 10^n \\ \pm B \times 10^n \\ \hline (A \pm B) \times 10^n \end{array}$$

Notice that the numbers *must have* the same exponent.

Consider the following example involving addition:

$$\begin{array}{r} 4.82 \times 10^7 \\ +3.4 \times 10^6 \\ \hline \end{array}$$

First, express both numbers with the same exponent. In this case, rewrite the lower number and perform the addition as follows:

$$\begin{array}{r} 4.82 \times 10^7 \\ +0.34 \times 10^7 \\ \hline 5.16 \times 10^7 \end{array}$$

Consider the following example involving subtraction:

$$\begin{array}{r} 7.33 \times 10^5 \\ -1.9 \times 10^4 \\ \hline \end{array}$$

First, express both numbers with the same exponent. In this case, rewrite the lower number and perform the subtraction as follows:

$$\begin{array}{r} 7.33 \times 10^{5} \\ -0.19 \times 10^{5} \\ \hline 7.14 \times 10^{5} \end{array}$$

▶ FOR PRACTICE **Mathematical Operations with Scientific Notation**

Perform each of the following operations:

(a) $(2.1 \times 10^{7})(9.3 \times 10^{5})$

(b) $(5.58 \times 10^{12})\ (7.84 \times 10^{-8})$

(c) $\dfrac{(1.5 \times 10^{14})}{(5.9 \times 10^{8})}$

(d) $\dfrac{(2.69 \times 10^{7})}{(8.44 \times 10^{11})}$

(e) $\begin{array}{r} 1.823 \times 10^{9} \\ +1.11 \times 10^{7} \\ \hline \end{array}$

(f) $\begin{array}{r} 3.32 \times 10^{-5} \\ +3.400 \times 10^{-7} \\ \hline \end{array}$

(g) $\begin{array}{r} 6.893 \times 10^{9} \\ -2.44 \times 10^{8} \\ \hline \end{array}$

(h) $\begin{array}{r} 1.74 \times 10^{4} \\ -2.9 \times 10^{3} \\ \hline \end{array}$

ANSWERS

(a) 2.0×10^{13}

(b) 4.37×10^{5}

(c) 2.5×10^{5}

(d) 3.19×10^{-5}

(e) 1.834×10^{9}

(f) 3.35×10^{-5}

(g) 6.649×10^{9}

(h) 1.45×10^{4}

Logarithms

The logarithm (or log) of a number is the exponent to which 10 must be raised to obtain that number. For example, the log of 100 is 2 because 10 must be raised to the second power to get 100. Similarly, the log of 1000 is 3 because 10 must be raised to the third power to get 1000. The logs of several multiples of 10 are shown as follows:

$$\begin{aligned} \log 10 &= 1 \\ \log 100 &= 2 \\ \log 1000 &= 3 \\ \log 10{,}000 &= 4 \end{aligned}$$

Because $10^{0} = 1$ by definition, $\log 1 = 0$.

The log of a number smaller than 1 is negative because 10 must be raised to a negative exponent to get a number smaller than 1. For example, the log of 0.01 is -2 because 10 must be raised to the power of -2 to get 0.01. Similarly, the log of 0.001 is -3 because 10 must be raised to the power of -3 to get 0.001. The logs of several fractional numbers are as follows:

$$\begin{aligned} \log 0.1 &= -1 \\ \log 0.01 &= -2 \\ \log 0.001 &= -3 \\ \log 0.0001 &= -4 \end{aligned}$$

The logs of numbers that are not multiples of 10 can be calculated on your calculator. See your calculator manual for specific instructions.

Inverse Logarithms

The inverse logarithm or invlog function (sometimes called antilog) is exactly the opposite of the log function. For example, the log of 100 is 2 and the inverse log of 2 is 100. The log function and the invlog function undo one another.

$$\log 100 = 3$$

$$\text{invlog } 3 = 1000$$

$$\text{invlog } (\log 1000) = 1000$$

The inverse log of a number is simply 10 raised to that number.

$$\text{invlog } x = 10^x$$

$$\text{invlog } 3 = 10^3 = 1000$$

The inverse logs of numbers can be calculated on your calculator. See your calculator manual for specific instructions.

▶ FOR PRACTICE **Logarithms and Inverse Logarithms**

Perform each of the following operations:

(a) $\log 1.0 \times 10^5$

(b) $\log 59$

(c) $\log 1.0 \times 10^{-5}$

(d) $\log 0.068$

(e) invlog 7.0

(f) invlog 1.44

(g) invlog -6.0

(h) invlog -0.250

(i) invlog (log 88)

ANSWERS

(a) 5.00

(b) 1.77

(c) -5.00

(d) -1.17

(e) 1×10^7

(f) 28

(g) 1×10^{-6}

(h) 0.56

(i) 88

Answers to Odd-Numbered Exercises

NOTE: Answers in the Questions section are written as briefly as possible. Student answers may vary and still be correct.

CHAPTER 1

QUESTIONS

1. Soda fizzes due to the interactions between carbon dioxide and water under high pressure. At room temperature, carbon dioxide is a gas and water is a liquid. Through the use of pressure, the makers of soda force the carbon dioxide gas to dissolve in the water. When the can is sealed, the solution remains mixed. When the can is opened, the pressure is released and the carbon dioxide molecules escape in bubbles of gas.
3. Chemists study molecules and interactions at the molecular level to learn about and explain macroscopic events. Chemists attempt to explain why ordinary things are as they are.
5. Chemistry is the science that seeks to understand what matter does by studying what atoms and molecules do.
7. The scientific method is the way chemists investigate the chemical world. The first step consists of observing the natural world. Later observations can be combined to create a scientific law, which summarizes and predicts behavior. Theories are models that strive to explain the cause of the observed phenomenon. Theories are tested through experiment. When a theory is not well established, it is sometimes referred to as a hypothesis.
9. A law is simply a general statement that summarizes and predicts observed behavior. Theories seek to explain the causes of observed behavior.
11. To say "It is just a theory" makes it seem as if theories are easily discardable. However, many theories are very well established and are as close to truth as we get in science. Established theories are backed up with years of experimental evidence, and they are the pinnacle of scientific understanding.
13. The atomic theory states that all matter is composed of small, indestructible particles called atoms. John Dalton formulated this theory.

PROBLEMS

15. **a.** observation **b.** theory
c. law **d.** observation

17.

Mass (g)	Volume (L)	Ratio (g/L)
22.5	1.6	14
35.8	2.55	14.0
70.2	5.00	14.0
98.5	7.01	14.1

The ratio of mass to volume is constant.

19. **a.** All atoms contain a degree of chemical reactivity. The larger the size of an atom, the higher the chemical reactivity of that atom.
b. There are many correct answers. One example is: Conceivably, when the size of an atom is increased, the surface area of the atom is also increased; an atom with a greater surface area is more likely to react chemically.
25. **a.** 2.2 billion people
c. 4.8 billion people
e. 9 billion people

CHAPTER 2

QUESTIONS

1. Without units, the results are unclear and it is hard to keep track of what each separate measurement entails.
3. Often scientists work with very large or very small numbers that contain a lot of zeros. Scientific notation allows these numbers to be written more compactly, and the information is more organized.
5. Zeros count as significant digits when they are interior zeros (zeros between two numbers) and when they are trailing zeros (zeros after a decimal point). Zeros are **not** significant digits when they are leading zeros, which are zeros to the left of the first nonzero number.
7. For calculations involving only multiplication and division, the result carries the same number of significant figures as the factor with the fewest significant figures.
9. In calculations involving both multiplication/division and addition/subtraction, do the steps in parentheses first; next determine the correct number of significant figures in the intermediate answer; then do the remaining steps.
11. The basic SI unit of length is the meter. The kilogram is the SI unit of mass. Lastly, the second is the SI unit of time.
13. For measuring a Frisbee, the unit would be the meter and the prefix multiplier would be *centi-*. The final measurement would be in centimeters.
15. **a.** 2.42 cm **b.** 1.79 cm
c. 21.58 cm **d.** 21.85 cm
17. Units act as a guide in the calculation and are able to show if the calculation is off track. The units must be followed in the calculation, so that the answer is correctly written and understood.
19. A conversion factor is a quantity used to relate two separate units. They are constructed from any two quantities known to be equivalent.

21. The conversion factor is $\frac{1\text{ ft}}{12\text{ in.}}$. For a feet-to-inches conversion, the conversion factor must be inverted $\left(\frac{12\text{ in.}}{1\text{ ft}}\right)$.

23. **a.** Sort the information into the **given** information (the starting point for the problem) and the **find** information (the end point).
b. Create a solution map to get from the given information to the information you are trying to find. This will likely include conversion factors or equations.
c. Follow the solution map to solve the problem. Carry out mathematical operations and cancel units as needed.
d. Ask, does this answer make physical sense? Are the units correct? Is the number of significant figures correct?

25. The solution map for converting grams to pounds is:

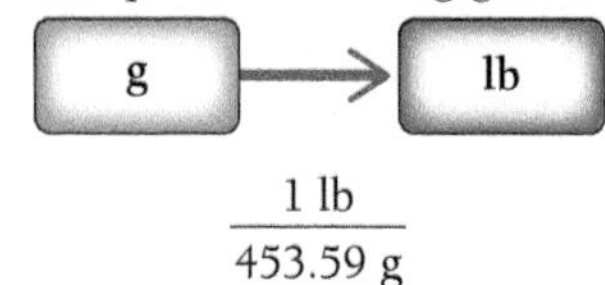

$\frac{1\text{ lb}}{453.59\text{ g}}$

27. The solution map for converting meters to feet is:

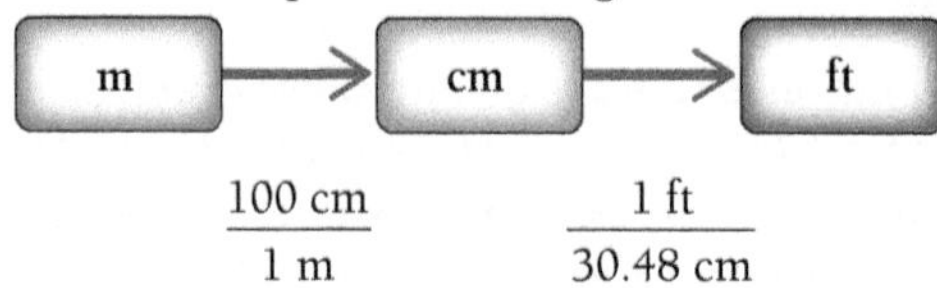

$\frac{100\text{ cm}}{1\text{ m}}$ $\frac{1\text{ ft}}{30.48\text{ cm}}$

29. The density of a substance is the ratio of its mass to its volume. Density is a fundamental property of materials and differs from one substance to another. Density can be used to relate two separate units, thus working as a conversion factor. Density is a conversion factor between mass and volume.

PROBLEMS

31. **a.** 3.8802×10^7 **b.** 1.419×10^6
c. 1.9746×10^7 **d.** 5.84×10^5

33. **a.** 7.461×10^{-11} m **b.** 1.58×10^{-5} mi
c. 6.32×10^{-7} m **d.** 1.5×10^{-5} m

35. **a.** 602,200,000,000,000,000,000,000
b. 0.00000000000000000016 C
c. 299,000,000 m/s
d. 344 m/s

37. **a.** 32,200,000
b. 0.0072
c. 118,000,000,000
d. 0.00000943

39.

2,000,000,000	2×10^9
1,211,000,000	1.211×10^9
0.000874	8.74×10^{-4}
320,000,000,000	3.2×10^{11}

41. **a.** 54.9 mL **b.** 48.7 °C
c. 46.83 °C **d.** 64 mL

43. **a.** 0.005050 **b.** 0.0000000000000060
c. 220,103 **d.** 0.00108

45. **a.** 4 **b.** 4
c. 6 **d.** 5

47. **a.** correct **b.** 3
c. 7 **d.** correct

49. **a.** 256.0 **b.** 0.0004893
c. 2.901×10^{-4} **d.** 2.231×10^6

51. **a.** 2.3 **b.** 2.4
c. 2.3 **d.** 2.4

53. **a.** 42.3 **b.** correct
c. correct **d.** 0.0456

55.

8.32466	8.325	8.3	8
84.57225	84.57	85	8×10^1
132.5512	132.6	1.3×10^2	1×10^2

57. **a.** 0.054 **b.** 0.619
c. 1.2×10^8 **d.** 6.6

59. **a.** 4.22×10^3 **b.** correct
c. 3.9969 **d.** correct

61. **a.** 110.6 **b.** 41.4
c. 183.3 **d.** 1.22

63. **a.** correct **b.** 1.0982
c. correct **d.** 3.53

65. **a.** 3.9×10^3 **b.** 632
c. 8.93×10^4 **d.** 6.34

67. **a.** 3.15×10^3 **b.** correct
c. correct **d.** correct

69. **a.** 3.55×10^3 g **b.** 8.944 m
c. 4.598×10^{-3} kg **d.** 18.7 mL

71. **a.** 0.588 L **b.** 34.1 μg
c. 10.1 ns **d.** 2.19×10^{-12} m

73. **a.** 57.2 cm **b.** 38.4 m
c. 0.754 km **d.** 61 mm

75. **a.** 15.7 in. **b.** 91.2 ft
c. 6.21 mi **d.** 8478 lb

77.

5.08×10^8 m	5.08×10^5 km	508 Mm
5.08×10^{-1} Gm	5.08×10^{-4} Tm	
2.7976×10^{10} m	2.7976×10^7 km	27,976 Mm
27976×10^1 Gm	2.7976×10^{-2} Tm	
1.77×10^{12} m	1.77×10^9 km	1.77×10^6 Mm
1.77×10^3 Gm	1.77 Tm	
1.5×10^8 m	1.5×10^5 m	1.5×10^2 Mm
0.15 Gm	1.5×10^{-4} Tm	
4.23×10^{11} m	4.23×10^8 km	4.23×10^5 Mm
423 Gm	0.423 Tm	

79. **a.** 2.255×10^7 kg **b.** 2.255×10^4 Mg
c. 2.255×10^{13} mg **d.** 2.255×10^4 metric tons

81. 1.5×10^3 g

83. 5.0×10^1 min

85. 4.7×10^3 cm^3

87. **a.** 1.0×10^6 m^2 **b.** 1.0×10^{-6} m^3
c. 1.0×10^{-9} m^3

89. **a.** 6.2×10^5 pm^3 **b.** 6.2×10^{-4} nm^3
c. 6.2×10^{-1} Å^3

91. a. 2.15×10^{-4} km^2 b. 2.15×10^4 dm^2
c. 2.15×10^6 cm^2
93. 1.49×10^6 mi^2
95. a. 2.5×10^3 km/day b. 95 ft/s
c. 29 m/s d. 1.9×10^3 yd/min
97. 3.42×10^{-3} g/lb; 0.599 g
99. 11.4 g/cm^3, lead
101. 1.26 g/cm^3
103. Yes, the density of the crown is 19.3 g/cm^3.
105. a. 4.30×10^2 g b. 3.12 L
107. a. 3.38×10^4 g (gold); 5.25×10^3 g (sand)
b. Yes, the mass of the bag of sand is different from the mass of the gold vase; thus, the weight-sensitive pedestal will sound the alarm.
109. 10.6 g/cm^3
111. $2.7 \times 10^3 \frac{\text{kg}}{\text{m}^3}$
113. 2.5×10^5 lb
115. 1.19×10^5 kg
117. 18 km/L
119. 768 mi
121. Metal A is denser than metal B.
123. 2.26 g/cm^3
125. 1.32 cm
127. 108 km; 47.2 km
129. 9.1×10^{10} g/cm^3
134. a. 8.2% c. 24.4 million cubic kilometers

CHAPTER 3

QUESTIONS

1. Matter is defined as anything that occupies space and possesses mass. It can be thought of as the physical material that makes up the universe.
3. The three states of matter are solid, liquid, and gas.
5. In a crystalline solid, the atoms/molecules are arranged in geometric patterns with repeating order. In amorphous solids, the atoms/molecules do not have long-range order.
7. The atoms/molecules in gases are not in contact with each other and are free to move relative to one another. The spacing between separate atoms/molecules is very far apart. A gas has no fixed volume or shape; rather, it assumes both the shape and the volume of the container it occupies.
9. A mixture is two or more pure substances combined in variable proportions.
11. Pure substances are those composed of only one type of atom or molecule.
13. A mixture is formed when two or more pure substances are mixed together; however, a new substance is not formed. A compound is formed when two or more elements are bonded together and form a new substance.
15. In a physical change, the composition of the substance does not change, even though its appearance might change. However, in a chemical change, the substance undergoes a change in its composition.
17. Energy is defined as the capacity to do work.
19. Kinetic energy is the energy associated with the motion of an object. Potential energy is the energy associated with the position or composition of an object.
21. Three common units for energy are joules, calories, and kilowatt-hours.
23. An endothermic reaction is one that absorbs energy from the surroundings. The products have more energy than the reactants in an endothermic reaction.
25. Heat is the transfer of thermal energy caused by a temperature difference, whereas temperature is a measure of the thermal energy of matter.
27. Heat capacity is the quantity of heat energy required to change the temperature of a given amount of the substance by 1 °C.
29. $°F = \frac{9}{5}(°C) + 32$

PROBLEMS

31. a. element
b. element
c. compound
d. compound
33. a. homogeneous
b. heterogeneous
c. homogeneous
d. homogeneous
35. a. pure substance-element
b. mixture-homogeneous
c. mixture-heterogeneous
d. mixture-heterogeneous
37. a. chemical b. physical
c. physical d. chemical
39. physical–colorless; odorless; gas at room temperature; one liter has a mass of 1.260 g under standard conditions; mixes with acetone; chemical–flammable; polymerizes to form polyethylene
41. a. chemical b. physical
c. chemical d. chemical
43. a. physical b. chemical
45. 2.10×10^2 kg
47. a. Yes b. No
49. 15.1 g of water
51. a. 2.46×10^3 J b. 4.16×10^{-3} Cal
c. 32.0 Cal d. 2.35×10^5 J
53. a. 9.0×10^7 J b. 0.249 Cal
c. 1.31×10^{-4} kWh d. 1.1×10^4 cal
55.

J	cal	Cal	kWh
225	53.8	5.38×10^{-2}	6.25×10^{-5}
3.44×10^6	8.21×10^5	8.21×10^2	9.54×10^{-1}
1.06×10^9	2.54×10^8	2.54×10^5	295
6.49×10^5	1.55×10^5	155	1.80×10^{-1}

57. 3.697×10^9 J
59. 8×10^2 kJ; 17 days

61. Exothermic.

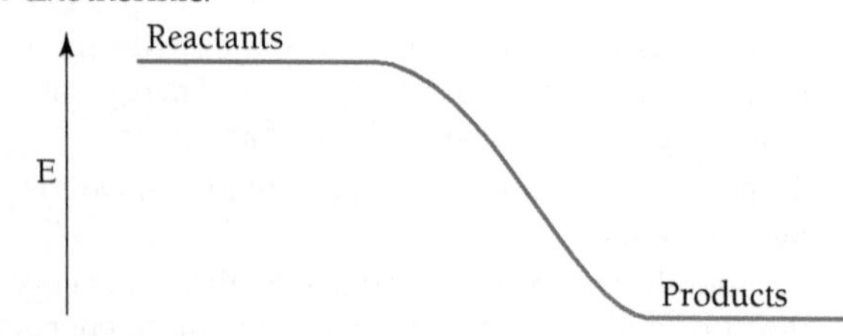

63. **a.** exothermic, $-\Delta H$ **b.** endothermic, $+\Delta H$
c. exothermic, $-\Delta H$

65. **a.** 1.00×10^2 °C **b.** -3.2×10^2 °F
c. 298 K **d.** 3.10×10^2 K

67. −62 °C, 211 K

69. 159 K, −173 °F

71. −75.5 °F

73.

0.0 K	−459.4 °F	−273.0 °C
301 K	82.5 °F	28.1 °C
282 K	47 °F	8.5 °C

75. 9.0×10^3 J

77. 8.7×10^5 J

79. 58 °C

81. 31 °C

83. 1.0×10^1 °C

85. 0.24 J/g°C; silver

87. 2.2 J/g°C

89. When warm drinks are placed into the ice, they release heat, which then melts the ice. The prechilled drinks, on the other hand, are already cold, so they do not release much heat.

91. 612 J

93. 49 °C

95. 70.2 J

97. 1.7×10^4 kJ

99. 67 °C

101. 6.0 kWh

103. 22 g of fuel

105. 78 g

107. 27.2 °C

109. 5.96×10^5 kJ; $25

111. −40°

113. **a.** pure substance **b.** pure substance
c. pure substance **d.** mixture

115. physical change

117. Small temperature changes in the ocean have a great impact on global weather because of the high heat capacity of water.

119. **a.** Sacramento is farther inland than San Francisco, so Sacramento is not as close to the ocean. The ocean water has a high heat capacity and will be able to keep San Francisco cooler in the hot days of summer. However, Sacramento is away from the ocean in a valley, so it will experience high temperatures in the summer.

b. San Francisco is located right next to the ocean, so the high heat capacity of the seawater keeps the temperature in the city from dropping. In the winter, the ocean actually helps to keep the city warmer, compared to an inland city like Sacramento.

124. **a.** Petroleum, natural gas, and coal; 80%
c. Coal and petroleum

CHAPTER 4

QUESTIONS

1. Democritus theorized that matter was ultimately composed of small, indivisible particles called atoms. Upon dividing matter, one would find tiny, indestructible atoms.

3. Rutherford's gold foil experiment involved sending positively charged alpha particles through a thin sheet of gold foil and detecting if there was any deflection of the particles. He found that most passed straight through, yet some particles showed some deflection. This result contradicts the plum-pudding model of the atom because the plum-pudding model does not explain the deflection of the alpha particles.

5.

Particle	Mass (kg)	Mass (amu)	Charge
Proton	1.67262×10^{-27}	1	+1
Neutron	1.67493×10^{-27}	1	0
Electron	0.00091×10^{-27}	0.00055	−1

7. Matter is usually charge-neutral due to protons and electrons having opposite charges. If matter were not charge-neutral, many unnatural things would occur, such as objects repelling or attracting each other.

9. A chemical symbol is a unique one- or two-letter abbreviation for an element. It is listed below the atomic number for that element on the periodic table.

11. Mendeleev noticed that many patterns were evident when elements were organized by increasing mass; from this observation he formulated the periodic law. He also organized the elements based on this law and created the basis for the periodic table being used today.

13. The periodic table is organized by listing the elements in order of increasing atomic number.

15. Nonmetals have varied properties (solid, liquid, or gas at room temperature); however, as a whole they tend to be poor conductors of heat and electricity, and they all tend to gain electrons when they undergo chemical changes. They are located toward the upper right side of the periodic table.

17. Each column within the main-group elements in the periodic table is labeled as a family or group of elements. The elements within a group usually have similar chemical properties.

19. An ion is an atom or group of atoms that has lost or gained electrons and has become charged.

21. **a.** ion charge = 1+
b. ion charge = 2+
c. ion charge = 3+
d. ion charge = 2−
e. ion charge = 1−

23. The percent natural abundance of isotopes is the relative amount of each different isotope in a naturally occurring sample of a given element.

25. Isotopes are noted in this manner: $^A_Z X$. X represents the chemical symbol, A represents the mass number, and Z represents the atomic number.

PROBLEMS

27. a. Correct.
b. False; different elements contain different types of atoms according to Dalton.
c. False; one cannot have 1.5 hydrogen atoms; combinations must be in simple, whole-number ratios.
d. Correct.

29. a. Correct.
b. False; most of the volume of the atom is empty space occupied by tiny, negatively charged electrons.
c. False; the number of negatively charged particles outside the nucleus equals the number of positively charged particles inside the nucleus.
d. False; the majority of the mass of an atom is found in the nucleus.

31. Solid matter seems to have no empty space within it because electromagnetic forces hold the atoms in a tight arrangement and the variation in density is too small to perceive with our eyes.

33. d

35. b

37. approximately 1.8×10^3 electrons

39. 5.4×10^{-4} g

41. a. 87 b. 36 c. 91 d. 32 e. 13

43. a. 18 b. 50 c. 54 d. 8 e. 81

45. a. C, 6 b. N, 7 c. Na, 11 d. K, 19 e. Cu, 29

47. a. manganese, 25 b. silver, 47 c. gold, 79 d. lead, 82 e. sulfur, 16

49.

Element Name	Element Symbol	Atomic Number
Gold	Au	79
Tin	Sn	50
Arsenic	As	33
Copper	Cu	29
Iron	Fe	26
Mercury	Hg	80

51. a. metal b. metal c. nonmetal d. nonmetal e. metalloid

53. a, d, e

55. a, b

57. c, d

59. b, e

61. a. halogen b. noble gas c. halogen d. neither e. noble gas

63. a. 6A b. 3A c. 4A d. 4A e. 5A

65. b, oxygen; it is in the same group or family.

67. b, chlorine and fluorine; they are in the same family or group.

69. d

71.

Chemical Symbol	Group Number	Group Name	Metal or Nonmetal
K	1A	Alkali Metals	Metals
Br	7A	Halogens	Nonmetal
Sr	2A	Alkaline Earth	Metal
He	8A	Noble Gas	Nonmetal
Ar	8A	Noble Gas	Nonmetal

73. a. e^- b. O^{2-} c. $2e^-$ d. Cl^-

75. a. 2− b. 3+ c. 4+ d. 1−

77. a. 11 protons, 10 electrons
b. 56 protons, 54 electrons
c. 8 protons, 10 electrons
d. 27 protons, 24 electrons

79. a. False; Ti^{2+} has 22 protons and 20 electrons.
b. True
c. False; Mg^{2+} has 12 protons and 10 electrons.
d. True

81. a. Rb^+ b. K^+ c. Al^{3+} d. O^{2-}

83. a. 3 electrons lost b. 1 electron lost c. 1 electron gained d. 2 electrons gained

85.

Symbol	Ion Commonly Formed	Number of Electrons in Ion	Number of Protons in Ion
Te	Te^{2-}	54	52
In	In^{3+}	46	49
Sr	Sr^{2+}	36	38
Mg	Mg^{2+}	10	12
Cl	Cl^-	18	17

87. a. Z = 1, A = 3 b. Z = 24, A = 52 c. Z = 20, A = 42 d. Z = 73, A = 182

89. a. $^{16}_{8}O$ b. $^{19}_{9}F$ c. $^{23}_{11}Na$ d. $^{27}_{13}Al$

91. a. $^{60}_{27}Co$ b. $^{22}_{10}Ne$ c. $^{131}_{53}I$ d. $^{244}_{94}Pu$

93. a. 11 protons, 12 neutrons
b. 88 protons, 178 neutrons
c. 82 protons, 126 neutrons
d. 7 protons, 7 neutrons

95. 6 protons, 8 neutrons, $^{14}_{6}C$

97. 85.47 amu

99. a. 49.31% b. 78.91 amu

101. 121.8 amu, Sb

103. 7.8×10^{17} electrons

105. 4.2×10^{-45} m^3; 6.2×10^{-31} m^3; 6.7×10^{-13}%

107.

Number Symbol	Number of Protons	Number of Neutrons	A (Mass Number)	Natural Abundance
Sr-84 or $^{84}_{38}Sr$	38	46	84	0.56%
Sr-86 or $^{86}_{38}Sr$	38	48	86	9.86%
Sr-87 or $^{87}_{38}Sr$	38	49	87	7.00%
Sr-88 or $^{88}_{38}Sr$	38	50	88	82.58%

Atomic mass of Sr = 87.62 amu

109.

Symbol	Z	A	Number of Protons	Number of Electrons	Number of Neutrons	Charge
Zn^{2+}	30	64	30	28	34	2+
Mn^{3+}	25	55	25	22	30	3+
P	15	31	15	15	16	0
O^{2-}	8	16	8	10	8	2−
S^{2-}	16	34	16	18	18	2−

111. 153 amu, 52.2%

113. The atomic theory and nuclear model of the atom are both theories because they attempt to provide a broader understanding and model behavior of chemical systems.

115. Atomic mass is measured as the mean value of masses of all isotopes in a sample. In the case of fluorine, only the 19.00 amu isotope is naturally occurring. In the case of chlorine, about 76% of naturally occurring atoms are 35 amu, and 24% are 37 amu.

117. 69.3% Cu-63, 30.7% Cu-65

119. **a.** Nt-304 = 72%; Nt-305 = 4%; Nt-306 = 24%

b.

120
Nt
304.5

125. **a.** There is a periodic trend in which atomic radius increases abruptly and then decreases more gradually. The trend repeats itself with each row in the periodic table.

c. They all fall in the column 8A (the rightmost column) in the periodic table. They are all noble gases.

CHAPTER 5

QUESTIONS

1. Yes; when elements combine with other elements, a compound is created. Each compound is unique and contains properties different from those of the elements that compose it.

3. The law of constant composition states that all samples of a given compound have the same proportions of their constituent elements. Joseph Proust formulated this law.

5. The more metallic element is generally listed first in a chemical formula.

7. The empirical formula gives the relative number of atoms of each element in a compound. The molecular formula gives the actual number of atoms of each element in a molecule of the compound.

9. An atomic element is one that exists in nature, with a single atom as the basic unit. A molecular element is one that exists as a diatomic molecule as the basic unit. Molecular elements include H_2, N_2, O_2, F_2, Cl_2, Br_2, and I_2.

11. The systematic name can be directly derived by looking at the compound's formula. The common name for a compound acts like a nickname and can only be learned through familiarity.

13. The block that contains the elements for Type II compounds is known as the transition metals.

15. The basic form for the names of Type II ionic compounds is to have the name of the metal cation first, followed by the charge of the metal cation (in parentheses, using Roman numerals), and finally the base name of the nonmetal anion with *-ide* attached to the end.

17. For compounds containing a polyatomic anion, the name of the cation is first, followed by the name for the polyatomic anion. Also, if the compound contains both a polyatomic cation and a polyatomic anion, one would just use the names of both polyatomic ions.

19. The form for naming molecular compounds is to have the first element preceded by a prefix to indicate the number of atoms present. This is then followed by the second element with its corresponding prefix and *-ide* placed on the end of the second element.

21. To correctly name a binary acid, one must begin the first word with *hydro-*, which is followed by the base name of the nonmetal plus *-ic* added on the end. Finally, the word *acid* follows the first word.

23. To name an acid with oxyanions ending with *-ite*, one must take the base name of the oxyanion and attach *-ous* to it; the word *acid* follows this.

PROBLEMS

25. Yes; the ratios of sodium to chlorine in both samples were equal.

27. 2.06×10^3 g

29.

	Mass N_2O	Mass N	Mass O
Sample A	2.85	1.82	1.03
Sample B	4.55	2.91	1.64
Sample C	3.74	2.39	1.35
Sample D	1.74	1.11	0.63

31. NI_3

33. **a.** Fe_3O_4 **b.** PCl_3 **c.** PCl_5 **d.** Ag_2O

35. **a.** 4 **b.** 4 **c.** 6 **d.** 4

37. **a.** magnesium, 1; chlorine, 2
b. sodium, 1; nitrogen, 1; oxygen, 3
c. calcium, 1; nitrogen, 2; oxygen, 4
d. strontium, 1; oxygen, 2; hydrogen, 2

39.

Formula	Number of $C_2H_3O_2$	Number of C Atoms	Number of H Atoms	Number of O Atoms	Number of Metal Atoms
$Mg(C_2H_3O_2)_2$	2	4	6	4	1
$NaC_2H_3O_2$	1	2	3	2	1
$Cr_2(C_2H_3O_2)_4$	4	8	12	8	2

41. **a.** CH_3 **b.** NO_2 **c.** C_2H_3O **d.** NH_3

43. **a.** molecular **b.** atomic **c.** atomic **d.** molecular

45. **a.** molecular **b.** ionic **c.** ionic **d.** molecular

47. helium ⟶ single atoms
CCl_4 ⟶ molecules
K_2SO_4 ⟶ formula units
bromine ⟶ diatomic molecules

49. **a.** formula units **b.** single atoms
c. molecules **d.** molecules

51. **a.** ionic; forms only one type of ion
b. molecular
c. molecular
d. ionic; forms more than one type of ion

53. **a.** Na_2S **b.** SrO
c. Al_2S_3 **d.** $MgCl_2$

55. **a.** $KC_2H_3O_2$ **b.** K_2CrO_4
c. K_3PO_4 **d.** KCN

57. **a.** Li_3N, Li_2O, LiF **b.** Ba_3N_2, BaO, BaF_2
c. AlN, Al_2O_3, AlF_3

59. **a.** cesium chloride **b.** strontium bromide
c. potassium oxide **d.** lithium fluoride

61. **a.** chromium(II) chloride **b.** chromium(III) chloride
c. tin(IV) oxide **d.** lead(II) iodide

63. **a.** forms more than one type of ion, chromium(III) oxide
b. forms only one type of ion, sodium iodide
c. forms only one type of ion, calcium bromide
d. forms more than one type of ion, tin(II) oxide

65. **a.** barium nitrate **b.** lead(II) acetate
c. ammonium iodide **d.** potassium chlorate
e. cobalt(II) sulfate **f.** sodium perchlorate

67. **a.** hypobromite ion **b.** bromite ion
c. bromate ion **d.** perbromate ion

69. **a.** $CuBr_2$ **b.** $AgNO_3$
c. KOH **d.** Na_2SO_4
e. $KHSO_4$ **f.** $NaHCO_3$

71. **a.** sulfur dioxide **b.** nitrogen triiodide
c. bromine pentafluoride **d.** nitrogen monoxide
e. tetranitrogen tetraselenide

73. **a.** CO **b.** S_2F_4
c. Cl_2O **d.** PF_5
e. BBr_3 **f.** P_2S_5

75. **a.** PBr_5 phosphorus pentabromide
b. P_2O_3 diphosphorus trioxide
c. SF_4 sulfur tetraflouride
d. correct

77. **a.** oxyacid, nitrous acid, nitrite
b. binary acid, hydroiodic acid
c. oxyacid, sulfuric acid, sulfate
d. oxyacid, nitric acid, nitrate

79. **a.** hypochlorous acid
b. chlorous acid
c. chloric acid
d. perchloric acid

81. **a.** H_3PO_4 **b.** HBr **c.** H_2SO_3

83. **a.** 63.02 amu **b.** 199.88 amu
c. 153.81 amu **d.** 211.64 amu

85. PBr_3, Ag_2O, PtO_2, $Al(NO_3)_3$

87. **a.** CH_4 **b.** SO_3 **c.** NO_2

89. **a.** 12 **b.** 4 **c.** 12 **d.** 7

91. **a.** 8 **b.** 12 **c.** 12

93.

Formula	Type	Name
N_2H_4	molecular	dinitrogen tetrahydride
KCl	ionic	potassium chloride
H_2CrO_4	acid	chromic acid
$Co(CN)_3$	ionic	cobalt(III) cyanide

95. **a.** calcium nitrite
b. potassium oxide
c. phosphorus trichloride
d. correct
e. potassium iodite

97. **a.** $Sn(SO_4)_2$ 310.9 amu **b.** HNO_2 47.02 amu
c. $NaHCO_3$ 84.01 amu **d.** PF_5 125.97 amu

99. **a.** platinum(IV) oxide 227.08 amu
b. dinitrogen pentoxide 108.02 amu
c. aluminum chlorate 277.33 amu
d. phosphorus pentabromide 430.47 amu

101. C_2H_4

103. 10 different isotopes can exist. 151.88 amu, 152.88 amu, 153.88 amu, 154.88 amu, 155.88 amu, 156.88 amu, 157.88 amu, 158.88 amu, 159.88 amu, and 160.88 amu.

105. **a.** molecular element **b.** atomic element
c. ionic compound **d.** molecular compound

107. **a.** $NaOCl$; $NaOH$
b. $Al(OH)_3$; $Mg(OH)_2$
c. $CaCO_3$
d. $NaHCO_3$, $Ca_3(PO_4)_2$, $NaAl(SO_4)_2$

112. **a.** 1950: 314 ppm; 2000: 371 ppm; The carbon dioxide increased by 57 ppm between 1950 and 2000.
c. 436 ppm

CHAPTER 6

QUESTIONS

1. Chemical composition lets us determine how much of a particular element is contained within a particular compound.

3. There are 6.022×10^{23} atoms in 1 mole of atoms.

5. One mole of any element has a mass equal to its atomic mass in grams.

7. **a.** 30.97 g **b.** 195.08 g
c. 12.01 g **d.** 52.00 g

9. Each element has a different atomic mass number. So, the subscripts that represent mole ratios cannot be used to represent the ratios of grams of a compound. The grams per mole of one element always differ from the grams per mole of a different element.

11. **a.** 11.19 g H ≡ 100 g H_2O
b. 53.29 g O ≡ 100 g fructose
c. 84.12 g C ≡ 100 g octane
d. 52.14 g C ≡ 100 g ethanol

13. The empirical formula gives the smallest whole-number ratio of each type of atom. The molecular formula gives the specific number of each type of atom in the molecule. The molecular formula is always a multiple of the empirical formula.

15. The empirical formula mass of a compound is the sum of the masses of all the atoms in the empirical formula.

PROBLEMS

17. 3.5×10^{24} atoms

19. **a.** 2.0×10^{24} atoms **b.** 5.8×10^{21} atoms **c.** 1.38×10^{25} atoms **d.** 1.29×10^{23} atoms

21.

Element	Moles	Number of Atoms
Ne	0.552	3.32×10^{23}
Ar	5.40	3.25×10^{24}
Xe	1.78	1.07×10^{24}
He	1.79×10^{-4}	1.08×10^{20}

23. **a.** 72.7 dozen **b.** 6.06 gross **c.** 1.74 reams **d.** 1.45×10^{-21} mol

25. 0.321 mol

27. 28.6 g

29. **a.** 2.05×10^{-2} mol **b.** 0.623 mol **c.** 0.401 mol **d.** 3.21×10^{-3} mol

31.

Element	Moles	Mass
Ne	1.11	22.5
Ar	0.117	4.67 g
Xe	7.62	1.00
He	1.44×10^{-4}	5.76×10^{-4} g

33. 8.07×10^{18} atoms

35. 8.44×10^{22} atoms

37. **a.** 1.16×10^{23} atoms **b.** 2.81×10^{23} atoms **c.** 2.46×10^{22} atoms **d.** 7.43×10^{23} atoms

39. 1.9×10^{21} atoms

41. 1.61×10^{25} atoms

43.

Element	Mass	Moles	Number of Atoms
Na	38.5 mg	1.67×10^{-3}	1.01×10^{21}
C	13.5 g	1.12	6.74×10^{23}
V	1.81×10^{-20} g	3.55×10^{-22}	214
Hg	1.44 kg	7.18	4.32×10^{24}

45. b

47. **a.** 0.654 mol **b.** 1.22 mol **c.** 96.6 mol **d.** 1.76×10^{-5} mol

49.

Compound	Mass	Moles	Molecules
H_2O	112 kg	6.22×10^{3}	3.74×10^{27}
N_2O	6.33 g	0.144	8.66×10^{22}
SO_2	156	2.44	1.47×10^{24}
CH_2Cl_2	5.46	0.0643	3.87×10^{22}

51. 6.20×10^{21} molecules

53. **a.** 1.2×10^{23} molecules **b.** 1.21×10^{24} molecules **c.** 3.5×10^{23} molecules **d.** 6.4×10^{22} molecules

55. 0.10 mg

57. \$$6.022 \times 10^{21}$ total. \$$8.5 \times 10^{11}$ Each person would be a billionaire.

59. 5.4 mol Cl

61. d, 3 mol O

63. **a.** 2.5 mol C **b.** 0.230 mol C **c.** 22.7 mol C **d.** 201 mol C

65. **a.** 2 moles H per mole of molecules; 8 H atoms present
b. 4 moles H per mole of molecules; 20 H atoms present
c. 3 moles H per mole of molecules; 9 H atoms present

67. **a.** 22.3 g **b.** 29.4 g **c.** 21.6 g **d.** 12.9 g

69. **a.** 1.4×10^{3} kg **b.** 1.4×10^{3} kg **c.** 2.1×10^{3} kg

71. 84.8% Sr

73. 36.1% Ca; 63.9% Cl

75. 10.7 g

77. 6.6 mg

79. **a.** 63.65% **b.** 46.68% **c.** 30.45% **d.** 25.94%

81. **a.** 39.99% C; 6.73% H; 53.28% O
b. 26.09% C; 4.39% H; 69.52% O
c. 60.93% C; 15.37% H; 23.69% N
d. 54.48% C; 13.74% H; 31.78% N

83. **a.** 58.50% O **b.** 42.13% O **c.** 72.71% O

85. Fe_3O_4, 72.36% Fe; Fe_2O_3, 69.94% Fe; $FeCO_3$, 48.20% Fe; magnetite

87. NO_2

89. **a.** NiI_2 **b.** $SeBr_4$ **c.** $BeSO_4$

91. C_2H_6N

93. **a.** C_3H_6O **b.** $C_5H_{10}O_2$ **c.** $C_9H_{10}O_2$

95. P_2O_3

97. NCl_3

99. C_4H_8

101. **a.** C_6Cl_6 **b.** C_2HCl_3 **c.** $C_6H_3Cl_3$

103. 2.43×10^{23} atoms

105. 2×10^{21} molecules

107.

Substance	Mass	Moles	Number of Particles
Ar	0.018 g	4.5×10^{-4}	2.7×10^{20}
NO_2	8.33×10^{-3} g	1.81×10^{-4}	1.09×10^{20}
K	22.4 mg	5.73×10^{-4}	3.45×10^{20}
C_8H_{18}	3.76 kg	32.9	1.98×10^{25}

109. **a.** CuI_2: 20.03% Cu; 79.97% I
b. $NaNO_3$: 27.05% Na; 16.48% N; 56.47% O
c. $PbSO_4$: 68.32% Pb; 10.57% S; 21.10% O
d. CaF_2: 51.33% Ca; 48.67% F

111. 1.8×10^{3} kg rock

113. 59 kg Cl

115. 1.1×10^{2} g H

117.

Formula	Molar Mass	%C (by mass)	%H (by mass)
C_2H_4	28.06	85.60%	14.40%
C_4H_{10}	58.12	82.66%	17.34%
C_4H_8	56.12	85.60%	14.40%
C_3H_8	44.09	81.71%	18.29%

119. $C_4H_6O_2$

121. $C_{10}H_{14}N_2$

123. 70.4% KBr, 29.6% KI

125. 29.6 g SO_2

127. 2.66 kg Fe

129. **a.** 1×10^{57} atoms per star

b. 1×10^{68} atoms per galaxy

c. 1×10^{79} atoms in the universe

131. $C_{16}H_{10}$

135. **a.** −20%

c. women, 4.4 L; men, 5.4 L

CHAPTER 7

QUESTIONS

1. A chemical reaction is the change of one or more substances into different substances, for example, burning wood, rusting iron, and protein synthesis.

3. The main evidence of a chemical reaction includes a color change, the formation of a solid, the formation of a gas, the emission of light, and the emission or absorption of heat.

5. **a.** gas **b.** liquid **c.** solid **d.** aqueous

7. **a.** reactants: 4 Ag, 2 O, 1 C products: 4 Ag, 2 O, 1 C balanced: yes

b. reactants: 1 Pb, 2 N, 6 O, 2 Na, 2 Cl products: 1 Pb, 2 N, 6 O, 2 Na, 2 Cl balanced: yes

c. reactants: 3 C, 8 H, 2 O products: 3 C, 8 H, 10 O balanced: no

9. If a compound dissolves in water, then it is soluble. If it does not dissolve in water, it is insoluble.

11. When ionic compounds containing polyatomic ions dissolve in water, the polyatomic ions usually dissolve as intact units.

13. The solubility rules are a set of empirical rules for ionic compounds that were deduced from observations on many compounds. The rules help us determine whether particular compounds will be soluble or insoluble.

15. The precipitate will always be insoluble; it is the solid that forms upon mixing two aqueous solutions.

17. Acid–base reactions involve an acid and a base reacting to form water and an ionic compound. An example is the reaction between hydrobromic acid and sodium hydroxide: $HBr + NaOH \longrightarrow H_2O + NaBr$

19. Gas-evolution reactions are reactions that evolve a gas. An example is the reaction between hydrochloric acid and sodium bicarbonate:
$HCl + NaHCO_3 \longrightarrow H_2O + CO_2 + NaCl$

21. Combustion reactions are a type of redox reaction and are characterized by the exothermic reaction of a substance with O_2. An example is the reaction between methane and oxygen: $CH_4 + 2\,O_2 \longrightarrow CO_2 + 2\,H_2O$

23. A synthesis reaction combines simpler substances to form more complex substances. An example is the reaction between elemental potassium and chloride: $2\,K + Cl_2 \longrightarrow 2\,KCl$. A decomposition decomposes a more complex substance into simpler substances. An example is the decomposition of water: $2\,H_2O \longrightarrow 2\,H_2 + O_2$

PROBLEMS

25. **a.** Yes; there is a color change showing a chemical reaction.

b. No; the state of the compound changes, but no chemical reaction takes place.

c. Yes; there is a formation of a solid in a previously clear solution.

d. Yes; there is a formation of a gas when the yeast is added to the solution.

27. Yes; a chemical reaction has occurred, for the presence of the bubbles is evidence for the formation of a gas.

29. Yes; a chemical reaction has occurred. We know this due to the color change of the hair.

31. **a.** Reactants: 1 Pb, 6 O, 2 Na, 2 Cl Products: 1 Pb; 6 O; 2 Na; 2 Cl; Balanced

b. Reactants: 3 C, 8 H, 2 O; Products: 3 C, 8 H, 10 O; Not balanced

33. Placing a subscript 2 after H_2O would change the compound from water to hydrogen peroxide (H_2O_2). To balance chemical reactions, one must add coefficients, not subscripts.
$2\,H_2O(l) \longrightarrow 2\,H_2(g) + O_2(g)$

35. **a.** $PbS + 2\,HCl \longrightarrow PbCl_2 + H_2S$

b. $CO + 3\,H_2 \longrightarrow CH_4 + H_2O$

c. $Fe_2O_3 + 3\,H_2 \longrightarrow 2\,Fe + 3\,H_2O$

d. $4\,NH_3 + 5\,O_2 \longrightarrow 4\,NO + 6\,H_2O$

37. **a.** $Mg(s) + 2\,CuNO_3(aq) \longrightarrow 2\,Cu(s) + Mg(NO_3)_2(aq)$

b. $2\,N_2O_5(g) \longrightarrow 4\,NO_2(g) + O_2(g)$

c. $Ca(s) + 2\,HNO_3(aq) \longrightarrow H_2(g) + Ca(NO_3)_2(aq)$

d. $2\,CH_3OH(l) + 3\,O_2(g) \longrightarrow 2\,CO_2(g) + 4\,H_2O(g)$

39. $2\,Na(s) + 2\,H_2O(l) \longrightarrow H_2(g) + 2\,NaOH(aq)$

41. $2\,SO_2(g) + O_2(g) + 2\,H_2O(l) \longrightarrow 2\,H_2SO_4(aq)$

43. $V_2O_5(s) + 2\,H_2(g) \longrightarrow V_2O_3(s) + 2\,H_2O(l)$

45. $C_{12}H_{22}O_{11}(aq) + H_2O(l) \longrightarrow 4\,CO_2(g) + 4\,C_2H_5OH(aq)$

47. **a.** $Na_2S(aq) + Cu(NO_3)_2(aq) \longrightarrow 2\,NaNO_3(aq) + CuS(s)$

b. $4\,HCl(aq) + O_2(g) \longrightarrow 2\,H_2O(l) + 2\,Cl_2(g)$

c. $2\,H_2(g) + O_2(g) \longrightarrow 2\,H_2O(l)$

d. $FeS(s) + 2\,HCl(aq) \longrightarrow FeCl_2(aq) + H_2S(g)$

49. **a.** $BaO_2(s) + H_2SO_4(aq) \longrightarrow BaSO_4(s) + H_2O_2(aq)$

b. $2\,Co(NO_3)_3(aq) + 3\,(NH_4)_2S(aq) \longrightarrow Co_2S_3(s) + 6\,NH_4NO_3(aq)$

c. $Li_2O(s) + H_2O(l) \longrightarrow 2\,LiOH(aq)$

d. $Hg_2(C_2H_3O_2)_2(aq) + 2\,KCl(aq) \longrightarrow Hg_2Cl_2(s) + 2\,KC_2H_3O_2(aq)$

51. **a.** $2\,Rb(s) + 2\,H_2O(l) \longrightarrow 2\,RbOH(aq) + H_2(g)$

b. Equation is balanced.

c. $2\,NiS(s) + 3\,O_2(g) \longrightarrow 2\,NiO(s) + 2\,SO_2(g)$

d. $3\,PbO(s) + 2\,NH_3(g) \longrightarrow 3\,Pb(s) + N_2(g) + 3\,H_2O(l)$

53. $C_6H_{12}O_6(aq) + 6\,O_2(g) \longrightarrow 6\,CO_2(g) + 6\,H_2O(l)$

55. $2\,NO(g) + 2\,CO(g) \longrightarrow N_2(g) + 2\,CO_2(g)$

57. **a.** soluble; $Na^+, C_2H_3O_2^-$ **b.** soluble; Sn^{2+}, NO_3^- **c.** insoluble **d.** soluble; Na^+, PO_4^{3-}

59. $AgCl$; $BaSO_4$; $CuCO_3$; Fe_2S_3

61.

Soluble	Insoluble
K_2S	Hg_2I_2
BaS	$Cu_3(PO_4)_2$
NH_4Cl	MgS
Na_2CO_3	$CaSO_4$
K_2SO_4	$PbSO_4$
SrS	$PbCl_2$
Li_2S	Hg_2Cl_2

63. **a.** NO REACTION
b. $K_2SO_4(aq) + BaBr_2(aq) \longrightarrow BaSO_4(s) + 2\,KBr(aq)$
c. $2\,NaCl(aq) + Hg_2(C_2H_3O_2)_2(aq) \longrightarrow Hg_2Cl_2(s) + 2\,NaC_2H_3O_2(aq)$
d. NO REACTION

65. **a.** $Na_2CO_3(aq) + Pb(NO_3)_2(aq) \longrightarrow PbCO_3(s) + 2\,NaNO_3(aq)$
b. $K_2SO_4(aq) + Pb(C_2H_3O_2)_2(aq) \longrightarrow PbSO_4(s) + 2\,KC_2H_3O_2(aq)$
c. $Cu(NO_3)_2(aq) + BaS(aq) \longrightarrow CuS(s) + Ba(NO_3)_2(aq)$
d. NO REACTION

67. **a.** correct
b. NO REACTION
c. correct
d. $Pb(NO_3)_2(aq) + 2\,LiCl(aq) \longrightarrow PbCl_2(s) + 2\,LiNO_3(aq)$

69. K^+, NO_3^-

71. **a.** $Ag^+(aq) + NO_3^-(aq) + K^+(aq) + Cl^-(aq) \longrightarrow AgCl(s) + K^+(aq) + NO_3^-(aq)$
$Ag^+(aq) + Cl^-(aq) \longrightarrow AgCl(s)$
b. $Ca^{2+}(aq) + S^{2-}(aq) + Cu^{2+}(aq) + 2\,Cl^-(aq) \longrightarrow CuS(s) + Ca^{2+}(aq) + 2\,Cl^-(aq)$
$Cu^{2+}(aq) + S^{2-}(aq) \longrightarrow CuS(s)$
c. $Na^+(aq) + OH^-(aq) + H^+(aq) + NO_3^-(aq) \longrightarrow H_2O(l) + Na^+(aq) + NO_3^-(aq)$
$H^+(aq) + OH^-(aq) \longrightarrow H_2O(l)$
d. $6\,K^+(aq) + 2\,PO_4^{3-}(aq) + 3\,Ni^{2+}(aq) + 6\,Cl^-(aq) \longrightarrow Ni_3(PO_4)_2(s) + 6\,K^+(aq) + 6\,Cl^-(aq)$
$3\,Ni^{2+}(aq) + 2\,PO_4^{3-}(aq) \longrightarrow Ni_3(PO_4)_2(s)$

73. $Hg_2^{2+}(aq) + 2\,NO_3^-(aq) + 2\,Na^+(aq) + 2\,Cl^-(aq) \longrightarrow Hg_2Cl_2(s) + 2\,Na^+(aq) + 2\,NO_3^-(aq)$
$Hg_2^{2+}(aq) + 2\,Cl^-(aq) \longrightarrow Hg_2Cl_2(s)$

75. **a.** $2\,Na^+(aq) + CO_3^{2-}(aq) + Pb^{2+}(aq) + 2\,NO_3^-(aq) \longrightarrow PbCO_3(s) + 2\,Na^+(aq) + 2\,NO_3^-(aq)$
$Pb^{2+}(aq) + CO_3^{2-}(aq) \longrightarrow PbCO_3(s)$
b. $2\,K^+(aq) + SO_4^{2-}(aq) + Pb^{2+}(aq) + 2\,CH_3CO_2^-(aq) \longrightarrow PbSO_4(s) + 2\,K^+(aq) + 2\,CH_3CO_2^-(aq)$
$Pb^{2+}(aq) + SO_4^{2-}(aq) \longrightarrow PbSO_4(s)$
c. $Cu^{2+}(aq) + 2\,NO_3^-(aq) + Ba^{2+}(aq) + S^{2-}(aq) \longrightarrow CuS(s) + Ba^{2+}(aq) + 2\,NO_3^-(aq)$
$Cu^{2+}(aq) + S^{2-}(aq) \longrightarrow CuS(s)$
d. NO REACTION

77. $HCl(aq) + KOH(aq) \longrightarrow H_2O(l) + KCl(aq)$
$H^+(aq) + OH^-(aq) \longrightarrow H_2O(l)$

79. **a.** $2\,HCl(aq) + Ba(OH)_2(aq) \longrightarrow 2\,H_2O(l) + BaCl_2(aq)$
b. $H_2SO_4(aq) + 2\,KOH(aq) \longrightarrow 2\,H_2O(l) + K_2SO_4(aq)$
c. $HClO_4(aq) + NaOH(aq) \longrightarrow H_2O(l) + NaClO_4(aq)$

81. **a.** $HBr(aq) + NaHCO_3(aq) \longrightarrow H_2O(l) + CO_2(g) + NaBr(aq)$
b. $NH_4I(aq) + KOH(aq) \longrightarrow H_2O(l) + NH_3(g) + KI(aq)$
c. $2\,HNO_3(aq) + K_2SO_3(aq) \longrightarrow H_2O(l) + SO_2(g) + 2\,KNO_3(aq)$
d. $2\,HI(aq) + Li_2S(aq) \longrightarrow H_2S(g) + 2\,LiI(aq)$

83. b and d are redox reactions; a and c are not.

85. **a.** $2\,C_2H_6(g) + 7\,O_2(g) \longrightarrow 4\,CO_2(g) + 6\,H_2O(g)$
b. $2\,Ca(s) + O_2(g) \longrightarrow 2\,CaO(s)$
c. $2\,C_3H_8O(l) + 9\,O_2(g) \longrightarrow 6\,CO_2(g) + 8\,H_2O(g)$
d. $2\,C_4H_{10}S(l) + 15\,O_2(g) \longrightarrow 8\,CO_2(g) + 10\,H_2O(g) + 2\,SO_2(g)$

87. **a.** $2\,Ag(s) + Br_2(g) \longrightarrow 2\,AgBr(s)$
b. $2\,K(s) + Br_2(g) \longrightarrow 2\,KBr(s)$
c. $2\,Al(s) + 3\,Br_2(g) \longrightarrow 2\,AlBr_3(s)$
d. $Ca(s) + Br_2(g) \longrightarrow CaBr_2(s)$

89. **a.** double displacement
b. synthesis or combination
c. single displacement
d. decomposition

91. **a.** synthesis **b.** decomposition **c.** synthesis

93. **a.** $2\,Na^+(aq) + 2\,I^-(aq) + Hg_2^+(aq) + 2\,NO_3^-(aq) \longrightarrow Hg_2I_2(s) + 2\,Na^+(aq) + 2\,NO_3^-(aq)$
b. $2\,H^+(aq) + 2\,ClO_4^-(aq) + Ba^{2+}(aq) + 2\,OH^-(aq) \longrightarrow 2\,H_2O(l) + Ba^{2+}(aq) + 2\,ClO_4^-(aq)$
$H^+(aq) + OH^-(aq) \longrightarrow H_2O(s)$
c. NO REACTION
d. $2\,H^+(aq) + 2\,Cl^-(aq) + 2\,Li^+(aq) + CO_3^{2-}(aq) \longrightarrow H_2O(l) + CO_2(g) + 2\,Li^+(aq) + 2\,Cl^-(aq)$
$2\,H^+(aq) + CO_3^{2-}(aq) \longrightarrow H_2O(l) + CO_2(g)$

95. **a.** NO REACTION
b. NO REACTION
c. $K^+(aq) + HSO_3^-(aq) + H^+(aq) + NO_3^-(aq) \longrightarrow H_2O(l) + SO_2(g) + K^+(aq) + NO_3^-(aq)$
$H^+(aq) + HSO_3^-(aq) \longrightarrow H_2O(l) + SO_2(g)$
d. $Mn^{3+}(aq) + 3\,Cl^-(aq) + 3\,K^+(aq) + PO_4^{3-}(aq) \longrightarrow MnPO_4(s) + 3\,K^+(aq) + 3\,Cl^-(aq)$
$Mn^{3+}(aq) + PO_4^{3-}(aq) \longrightarrow MnPO_4(s)$

97. **a.** acid–base; $KOH(aq) + HC_2H_3O_2(aq) \longrightarrow H_2O(l) + KC_2H_3O_2(aq)$
b. gas evolution; $2\,HBr(aq) + K_2CO_3(aq) \longrightarrow H_2O(l) + CO_2(g) + 2\,KBr(aq)$

c. synthesis; $2\ H_2(g) + O_2(g) \longrightarrow 2\ H_2O(l)$

d. precipitation; $2\ NH_4Cl(aq) + Pb(NO_3)_2(aq) \longrightarrow PbCl_2(s) + 2\ NH_4NO_3(aq)$

99. a. oxidation–reduction; single displacement

b. gas evolution; acid–base

c. gas evolution; double displacement

d. precipitation; double displacement

101. $3\ CaCl_2(aq) + 2\ Na_3PO_4(aq) \longrightarrow Ca_3(PO_4)_2(s) + 6\ NaCl(aq)$

$3\ Ca^{2+}(aq) + 6\ Cl^-(aq) + 6\ Na^+(aq) + 2\ PO_4^{3-}(aq) \longrightarrow Ca_3(PO_4)_2(s) + 6\ Na^+(aq) + 6\ Cl^-(aq)$

$3\ Ca^{2+}(aq) + 2\ PO_4^{3-}(aq) \longrightarrow Ca_3(PO_4)_2(s)$

$3\ Mg(NO_3)_2(aq) + 2\ Na_3PO_4(aq) \longrightarrow Mg_3(PO_4)_2(s) + 6\ NaNO_3(aq)$

$3\ Mg^{2+}(aq) + 6\ NO_3^-(aq) + 6\ Na^+(aq) + 2\ PO_4^{3-}(aq) \longrightarrow Mg_3(PO_4)_2(s) + 6\ Na^+(aq) + 6\ NO_3^-(aq)$

$3\ Mg^{2+}(aq) + 2\ PO_4^{3-}(aq) \longrightarrow Mg_3(PO_4)_2(s)$

103. *Correct answers may vary; representative correct answers are:

a. addition of a solution containing SO_4^{2-}; $Pb^{2+}(aq) + SO_4^{2-}(aq) \longrightarrow PbSO_4(s)$

b. addition of a solution containing SO_4^{2-}; $Ca^{2+}(aq) + SO_4^{2-}(aq) \longrightarrow CaSO_4(s)$

c. addition of a solution containing SO_4^{2-}; $Ba^{2+}(aq) + SO_4^{2-}(aq) \longrightarrow BaSO_4(s)$

d. addition of a solution containing Cl^-; $Hg_2^{2+}(aq) + 2\ Cl^-(aq) \longrightarrow Hg_2Cl_2(s)$

105. 0.168 mol Ca; 6.73 g Ca

107. 0.00128 mol NaCl, 0.0750 g NaCl

109. a. chemical b. physical

114. a. Water sample A contains Ca^{2+} and Cu^{2+}.

CHAPTER 8

QUESTIONS

1. Reaction stoichiometry is very important to chemistry. It gives us a numerical relationship between the reactants and products that allows chemists to plan and carry out chemical reactions to obtain products in the desired quantities.

For example, how much CO_2 is produced when a given amount of C_8H_{10} is burned?

How much $H_2(g)$ is produced when a given amount of water decomposes?

3. 1 mol $Cl_2 \equiv 2$ mol NaCl

5. mass A $\longrightarrow$ moles A $\longrightarrow$ moles B $\longrightarrow$ mass B (A = reactant, B = product)

7. The limiting reactant is the reactant that limits the amount of product in a chemical reaction.

9. The actual yield is the amount of product actually produced by a chemical reaction. The percent yield is the percentage of the theoretical yield that was actually attained.

11. d

13. The enthalpy of reaction is the total amount of heat generated or absorbed by a particular chemical reaction. The quantity is important because it quantifies the change in heat for the chemical reaction. It is useful for determining the necessary starting conditions and predicting the outcome of various reactions.

PROBLEMS

15. a. 2 mol C b. 1 mol C c. 3 mol C d. 1.5 mol C

17. a. 2.6 mol NO_2 b. 11.6 mol NO_2 c. 8.90×10^3 mol NO_2 d. 2.012×10^{-3} mol NO_2

19. c

21. a. 3.50 mol HCl b. 3.50 mol H_2O c. 0.875 mol Na_2O_2 d. 1.17 mol SO_3

23. a. 2.4 mol PbO(s), 2.4 mol $SO_2(g)$

b. 1.6 mol PbO(s), 1.6 mol $SO_2(g)$

c. 5.3 mol PbO(s), 5.3 mol $SO_2(g)$

d. 3.5 mol PbO(s), 3.5 mol $SO_2(g)$

25.

mol N_2H_4	mol N_2O_4	mol N_2	mol H_2O
4	2	6	8
6	3	9	12
4	2	6	8
11	5.5	16.5	22
3	1.5	4.5	6
8.26	4.13	12.4	16.5

27. $2\ C_4H_{10}(g) + 13\ O_2(g) \longrightarrow 8\ CO_2(g) + 10\ H_2O(g)$; 32 mol O_2

29. a. $Pb(s) + 2\ AgNO_3(aq) \longrightarrow Pb(NO_3)_2(aq) + 2\ Ag(s)$

b. 19 mol $AgNO_3$

c. 56.8 mol Ag

31. a. 0.157 g O_2 b. 0.500 g O_2 c. 114 g O_2 d. 2.86×10^{-4} g O_2

33. a. 4.0 g NaCl b. 4.3 g $CaCO_3$ c. 4.0 g MgO d. 3.1 g NaOH

35. a. 8.9 g Al_2O_3, 9.7 g Fe b. 3.0 g Al_2O_3, 3.3 g Fe

37.

Mass CH_4	Mass O_2	Mass CO_2	Mass H_2O
0.645 g	2.57 g	1.77 g	1.45 g
22.32 g	89.00 g	61.20 g	50.12 g
5.044 g	20.11 g	13.83 g	11.32 g
1.07 g	4.28 g	2.94 g	2.41 g
3.18 kg	12.7 kg	8.72 kg	7.14 kg
8.57×10^2 kg	3.42×10^3 kg	2.35×10^3 kg	1.92×10^3 kg

39. a. 2.3 g HCl b. 4.3 g HNO_3 c. 2.2 g H_2SO_4

41. 123 g H_2SO_4, 2.53 g H_2

43. a. 2 mol A b. 1.8 mol A c. 4 mol B d. 40 mol B

45. a. 1.5 mol C b. 3 mol C c. 3 mol C d. 96 mol C

47. a. 1 mol K b. 1.8 mol K c. 1 mol Cl_2 d. 14.6 mol K

49. a. 1.3 mol MnO_3 b. 4.8 mol MnO_3 c. 0.107 mol MnO_3 d. 27.5 mol MnO_3

51. 3 mol A, 0 mol B, 4 mol C

53. a. 2 Cl_2 b. 3 Cl_2 c. 2 Cl_2

55. a. 1.0 g F_2 b. 10.5 g Li
c. 6.79×10^3 g F_2
57. a. 1.3 g $AlCl_3$ b. 24.8 g $AlCl_3$
c. 2.17 g $AlCl_3$
59. 74.6%
61. CaO; 25.7 g $CaCO_3$; 75.5%
63. O_2; 5.07 g NiO; 95.9%
65. Pb^{2+}; 262.7 g $PbCl_2$; 96.09%
67. TiO_2: 0 g, C: 7.0 g, Ti: 5.99 g, CO: 7.00 g
69. a. exothermic, $-\Delta H$ b. endothermic, $+\Delta H$
c. exothermic, $-\Delta H$
71. a. 55 kJ b. 110 kJ c. 28 kJ d. 55 kJ
73. 4.78×10^3 kJ
75. 34.9 g C_8H_{18}
77. N_2
79. 0.152 g Ba^{2+}
81. 1.5 g HCl
83. 3.1 kg CO_2
85. 4.7 g Na_3PO_4
87. 469 g Zn
89. $2\,NH_4NO_3(s) \longrightarrow 2\,N_2(g) + O_2(g) + 4\,H_2O(l)$; 2.00×10^2 g O_2
91. salicylic acid ($C_7H_6O_3$); 2.71 g $C_9H_8O_4$; 74.1%
93. NH_3; 120 kg (CH_4N_2O); 72.9%
95. 2.4 mg $C_4H_6O_4S_2$
97. 1.0×10^3 g CO_2
99. b; the loudest explosion will occur when the ratio is 2 hydrogen to 1 oxygen, for that is the ratio that occurs in water.
101. 2.8×10^{13} kg CO_2 per year; 1.1×10^2 years
106. a. Experiments 1, 2, and 3
c. 2 A + 1 B
e. 84.8%

CHAPTER 9

QUESTIONS

1. Both the Bohr model and the quantum-mechanical model for the atom were developed in the early 1900s. These models serve to explain how electrons are arranged within the atomic structure and how the electrons affect the chemical and physical properties of each element.
3. White light contains a spectrum of wavelengths and therefore a spectrum of color. Colored light is produced by a single wavelength and is therefore a single color.
5. Energy carried per photon is greater for shorter wavelengths than for longer wavelengths. Wavelength and frequency are inversely related—the shorter the wavelength, the higher the frequency.
7. X-rays pass through many substances that block visible light and are therefore used to image bones and organs.
9. Ultraviolet light contains enough energy to damage biological molecules, and excessive exposure increases the risk of skin cancer and cataracts.
11. Microwaves can only heat things containing water; therefore the food, which contains water, becomes hot, but the plate does not.
13. The Bohr model is a representation for the atom in which electrons travel around the nucleus in circular orbits with a fixed energy at specific, fixed distances from the nucleus.
15. The Bohr orbit describes the path of an electron as an orbit or trajectory (a specified path). A quantum-mechanical orbital describes the path of an electron using a probability map.
17. The e^- has wave–particle duality, which means the path of an electron is not predictable. The motion of a baseball is predictable. A probability map shows a statistical, reproducible pattern of where the electron is located.
19. The subshells are *s* (1 orbital, which contains a maximum of 2 electrons); *p* (3 orbitals, which contain a maximum of 6 electrons); *d* (5 orbitals, which contain a maximum of 10 electrons); and *f* (7 orbitals, which contain a maximum of 14 electrons).
21. The Pauli exclusion principle states that separate orbitals may hold no more than 2 electrons, and when 2 electrons are present in a single orbital, they must have opposite spins. When writing electron configurations, the principle means that no box can have more than 2 arrows, and the arrows will point in opposite directions.
23. [Ne] represents $1s^22s^22p^6$.
[Kr] represents $1s^22s^22p^63s^23p^64s^23d^{10}4p^6$.
25.

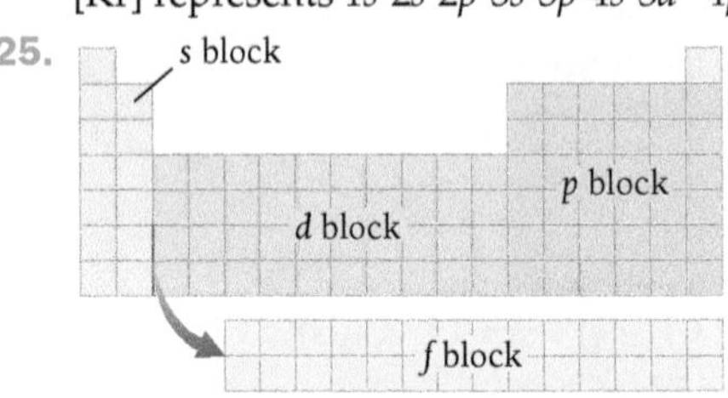

27. Group 1 elements form 1+ ions because they lose one valence electron in the outer *s* shell to obtain a noble gas configuration. Group 7 elements form 1− ions because they gain an electron to fill their outer *p* orbital to obtain a noble gas configuration.

PROBLEMS

29. a. 1.0 ns b. 13.21 ms c. 4 h 10 min
31. infrared
33. radiowaves < microwaves < infrared < ultraviolet
35. gamma, ultraviolet, or X-rays
37. a. radio waves < infrared < X-rays
b. radio waves < infrared < X-rays
c. X-rays < infrared < radio waves
39. energies, distances
41. $n = 6 \longrightarrow n = 2$: 410 nm
$n = 5 \longrightarrow n = 2$: 434 nm
43.

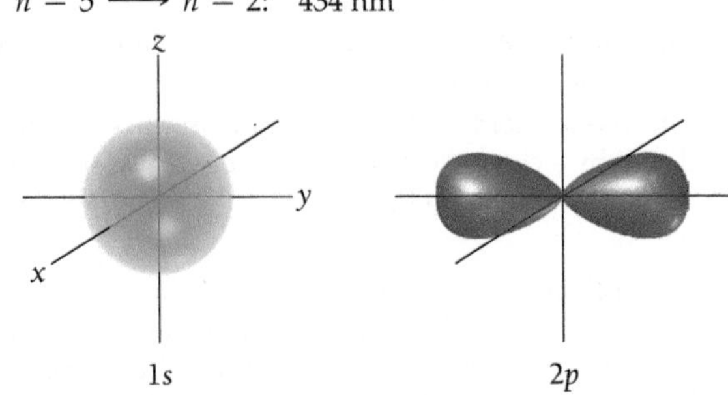

The 2*s* and 3*p* orbitals are bigger than the 1*s* and 2*p* orbitals.

45. Electron in the 2*s* orbital

47. $2p \longrightarrow 1s$

49. **a.** $1s^22p^22p^63s^23p^64s^23d^{10}4p^65s^2$
b. $1s^22s^22p^63s^23p^64s^23d^{10}4p^2$
c. $1s^22s^1$
d. $1s^22s^22p^63s^23p^64s^23d^{10}4p^6$

51. **a.** He [↑↓] [ ]
1*s* 2*s*

b. B [↑↓] [↑↓] [↑| |]
1*s* 2*s* 2*p*

c. Li [↑↓] [↑] [| |]
1*s* 2*s* 2*p*

d. N [↑↓] [↑] [↑↓|↓|↓]
1*s* 2*s* 2*p*

53. **a.** $[Ar]4s^23d^{10}4p^1$ **b.** $[Ar]4s^23d^{10}4p^3$
c. $[Kr]5s^1$ **d.** $[Kr]5s^24d^{10}5p^2$

55. **a.** $[Ar]4s^23d^{10}$ **b.** $[Ar]4s^13d^{10}$
c. $[Kr]5s^24d^2$ **d.** $[Ar]4s^23d^6$

57. Valence electrons are underlined.
a. $1s^22s^22p^63s^23p^6\underline{4s^2}3d^{10}\underline{4p^6}$
b. $1s^22s^22p^63s^23p^6\underline{4s^2}3d^{10}\underline{4p^2}$
c. $1s^22s^22p^6\underline{3s^23p^5}$
d. $1s^22s^22p^63s^23p^64s^23d^{10}4p^6\underline{5s^2}$

59. **a.** Br [↑↓] [↑↓|↑↓|↑], 1 unpaired electron
4*s* 4*p*

b. Kr [↑↓] [↑↓|↑↓|↑↓], 0 unpaired electron
4*s* 4*p*

c. Na [↑], 1 unpaired electron
3*s*

d. In [↑↓] [↑| |], 1 unpaired electron
5*s* 5*p*

61. **a.** 6 **b.** 6 **c.** 7 **d.** 1

63. **a.** ns^1 **b.** ns^2 **c.** ns^2np^3 **d.** ns^2np^5

65. **a.** $[Ne]3s^23p^1$ **b.** $[He]2s^2$
c. $[Kr]5s^24d^{10}5p^1$ **d.** $[Kr]5s^24d^2$

67. **a.** $[Kr]5s^2$ **b.** $[Kr]5s^24d^1$
c. $[Ar]4s^23d^2$ **d.** $[Kr]5s^24d^{10}5p^4$

69. **a.** 2 **b.** 3 **c.** 5 **d.** 6

71. Period 1 has two elements. Period 2 has eight elements. The number of subshells is equal to the principal quantum number. For Period 1, $n = 1$ and the *s* subshell contains only two elements. For Period 2, $n = 2$ and contains *s* and *p* subshells that have a total of 8 elements.

73. **a.** Al **b.** S
c. Ar **d.** Mg

75. **a.** Cl **b.** Ga
c. Fe **d.** Rb

77. **a.** As **b.** Br
c. cannot tell **d.** S

79. Pb < Sn < Te < S < Cl

81. **a.** In **b.** Si
c. Pb **d.** C

83. F < S < Si < Ge < Ca < Rb

85. **a.** Sr **b.** Bi
c. cannot tell **d.** As

87. S < Se < Sb < In < Ba < Fr

89. $18\ e^-$

91. Alkaline earth metals have the general electron configuration of ns^2. If they lose their two *s* electrons, they will obtain the stable electron configuration of a noble gas. This loss of electrons will give the metal a 2+ charge.

93. **a.** $1s^22s^22p^63s^23p^6$ **b.** $1s^22s^22p^63s^23p^6$
c. $1s^22s^22p^63s^23p^6$ **d.** $1s^22s^22p^63s^23p^64s^23d^{10}4p^6$
They all have noble gas electron configurations.

95. Metals tend to form positive ions because they tend to lose electrons. Elements on the left side of the periodic table have only a few extra electrons, which they will lose to gain a noble gas configuration. Metalloids tend to be elements with 3 to 5 valence electrons; they could lose or gain electrons to obtain a noble gas configuration. Nonmetals tend to gain electrons to fill their almost full valence shell, so they tend to form negative ions and are on the right side of the table.

97. **a.** Can only have 2 in the *s* shell and 6 in the *p* shell: $1s^22s^22p^63s^23p^3$.
b. There is no 2*d* subshell: $1s^22s^22p^63s^23p^2$.
c. There is no 1*p* subshell: $1s^22s^22p^3$.
d. Can only have 6 in the *p* shell: $1s^22s^22p^63s^23p^3$.

99. Bromine is highly reactive because it reacts quickly to gain an electron and obtain a stable valence shell. Krypton is a noble gas because it already has a stable valence shell.

101. K

103. 660 nm

105. 8 min, 19 sec

107. **a.** 1.5×10^{-34} m
b. 1.88×10^{-10} m
Electrons have wave–particle duality, whereas golf balls do not.

109. The ionization energy dips at column 3A because removing an electron from one of those atoms leaves the atom with a fairly stable, filled *s*-orbital as its valence shell. For the group 6A elements, special stability occurs when those elements lose an electron and achieve a half-filled *p*-orbital as their valence shell.

111. Ultraviolet light is the only one of these three types of light that contains enough energy to break chemical bonds in biological molecules.

117. **c.** Al $[Ne]3s^23p^1$ P $[Ne]3s^23p^3$ S $[Ne]3s^23p^4$
The Al exception is due to the transition from the *s* orbitals to the *p* orbital.

The S exception is due to the pairing of 2 electrons in one orbital for S (compared to P in which all of the *p* orbitals are singly occupied).

e. Si [Ne]$3s^2 3p^2$
P [Ne]$3s^2 3p^3$

The electron affinity of Si is more exothermic because in Si, the incoming electron can singly occupy a *p* orbital. However, in P, the incoming electron must pair with another electron in a *p* orbital.

CHAPTER 10

QUESTIONS

1. Bonding theories predict how atoms bond together to form molecules, and they also predict what combinations of atoms form molecules and what combinations do not. Likewise, bonding theories explain the shapes of molecules, which in turn determine many of their physical and chemical properties.
3. Atoms with 8 valence electrons are particularly stable and are said to have an octet. Atoms such as hydrogen, helium, lithium, and beryllium are exceptions to the octet rule as they achieve stability when their outermost shell contains two electrons—a duet. A chemical bond is the sharing or transfer of electrons to attain stable electron configurations among the bonding atoms.
5. The Lewis structure for potassium has 1 valence electron, whereas the Lewis structure for monatomic chlorine has 7 valence electrons. From these structures we can determine that if potassium gives up its one valence electron to chlorine, K^+ and Cl^- are formed; therefore the formula must be KCl.
7. Double and triple bonds are shorter and stronger than single bonds.
9. You determine the number of electrons that go into the Lewis structure of a molecule by summing the valence electrons of each atom in the molecule.
11. The octet rule is not sophisticated enough to be correct every time. For example, some molecules that exist in nature have an odd number of valence electrons and thus will not have octets on all their constituent atoms. Some elements tend to form compounds in nature in which they have more (sulfur) or less (boron) than 8 valence electrons.
13. VSEPR theory predicts the shape of molecules using the idea that electron groups repel each other.
15. **a.** 180° **b.** 120° **c.** 109.5°
17. Electronegativity is the ability of an element to attract electrons within a covalent bond.
19. A polar covalent bond is a covalent bond that has a dipole moment.
21. If a polar liquid and a nonpolar liquid are mixed, they will separate into distinct regions because the polar molecules will be attracted to one another and will exclude the nonpolar molecules.

PROBLEMS

23. **a.** $1s^2 2s^2 2p^3$, ·N̈: **b.** $1s^2 2s^2 2p^2$, ·Ċ· **c.** $1s^2 2s^2 2p^6 \underline{3s^2 3p^5}$, :C̈l· **d.** $1s^2 2s^2 2p^6 \underline{3s^2 3p^6}$, :Är:
25. **a.** :Ï: **b.** ·S̈: **c.** ·Ġe· **d.** ·Ca·
27. :Ẍ: Halogens tend to gain 1 electron in a chemical reaction.
29. M: Alkaline earth metals tend to lose 2 electrons in a chemical reaction.
31. **a.** Al^{3+} **b.** Mg^{2+} **c.** $[:\ddot{Se}:]^{2-}$ **d.** $[:\ddot{N}:]^{3-}$
33. **a.** Kr **b.** Ne **c.** Kr **d.** Xe
35. **a.** covalent **b.** ionic **c.** covalent **d.** ionic
37. **a.** $Na^+[:\ddot{F}:]^-$ **b.** $Ca^{2+}[:\ddot{O}:]^{2-}$ **c.** $[:\ddot{Br}:]^- Sr^{2+}[:\ddot{Br}:]^-$ **d.** $K^+[:\ddot{O}:]^{2-} K^+$
39. **a.** CaS **b.** $MgBr_2$ **c.** CsI **d.** Ca_3N_2
41. **a.** $[:\ddot{F}:]^- Mg^{2+}[:\ddot{F}:]^-$ **b.** $Mg^{2+}[:\ddot{O}:]^{2-}$ **c.** $Mg^{2+}[:\ddot{N}:]^{3-} Mg^{2+}[:\ddot{N}:]^{3-} Mg^{2+}$
43. **a.** $Cs^+[:\ddot{Cl}:]^-$ **b.** $Ba^{2+}[:\ddot{O}:]^{2-}$ **c.** $[:\ddot{I}:]^- Ca^{2+}[:\ddot{I}:]^-$
45. **a.** Hydrogen exists as a diatomic molecule because two hydrogen molecules achieve a stable duet when they share their electrons and form a single covalent bond.
b. Iodine achieves a stable octet when two atoms share electrons and form a single bond.
c. Nitrogen achieves a stable octet when two atoms share electrons and form a triple bond.
d. Oxygen achieves a stable octet when two atoms share electrons and form a double bond.
47. **a.** H—P̈—H (with H bonded below P) **b.** :C̈l—S̈—C̈l: **c.** :F̈—F̈: **d.** H—Ï:
49. **a.** Ö=Ö **b.** :C≡O: **c.** H—Ö—N̈=Ö **d.** :Ö=S̈—Ö:
51. **a.** H—C≡C—H **b.** H—C=C—H (each C bonded to an H below) **c.** H—N̈=N̈—H **d.** H—N̈—N̈—H (each N bonded to an H below)
53. **a.** :N≡N: **b.** S̈=Si=S̈ **c.** H—Ö—H **d.** :Ï—N̈—Ï: (N bonded to :Ï: below)
55. **a.** Ö=S̈e—Ö: ⟷ :Ö—S̈e=Ö
b. $[\ddot{O}=C(-\ddot{O}:)(-:\ddot{O}:)]^{2-}$ ⟷ $[:\ddot{O}-C(=\ddot{O})(-:\ddot{O}:)]^{2-}$ ⟷ $[:\ddot{O}-C(-\ddot{O}:)(=:\ddot{O}:)]^{2-}$

c. $[:\ddot{\text{Cl}}-\ddot{\text{O}}:]^-$ d. $[:\ddot{\text{O}}-\ddot{\text{Cl}}-\ddot{\text{O}}:]^-$

57. a. $\left[\begin{array}{c} :\ddot{\text{O}}: \\ | \\ :\ddot{\text{O}}-\text{P}-\ddot{\text{O}}: \\ | \\ :\ddot{\text{O}}: \end{array}\right]^{3-}$ b. $[:\text{C}\equiv\text{N}:]^-$

c. $[:\ddot{\text{O}}=\ddot{\text{N}}-\ddot{\text{O}}:]^- \longleftrightarrow [:\ddot{\text{O}}-\ddot{\text{N}}=\ddot{\text{O}}:]^-$

d. $\left[\begin{array}{c} :\ddot{\text{O}}-\ddot{\text{S}}-\ddot{\text{O}}: \\ | \\ :\ddot{\text{O}}: \end{array}\right]^{2-}$

59. a. $\begin{array}{c} :\ddot{\text{Cl}}-\text{B}-\ddot{\text{Cl}}: \\ | \\ :\ddot{\text{Cl}}: \end{array}$

b. $\ddot{\text{O}}=\dot{\text{N}}-\ddot{\text{O}}: \longleftrightarrow :\ddot{\text{O}}-\dot{\text{N}}=\ddot{\text{O}}$

c. $\begin{array}{c} \text{H}-\text{B}-\text{H} \\ | \\ \text{H} \end{array}$

61. a. 4 b. 4
c. 2 d. 4

63. a. 2 bonding groups, 2 lone pairs
b. 3 bonding groups, 1 lone pair
c. 2 bonding groups, 0 lone pair
d. 4 bonding groups, 0 lone pair

65. a. tetrahedral b. trigonal planar
c. linear d. trigonal planar

67. a. 109.5° b. 120°
c. 180° d. 120°

69. a. linear, linear
b. trigonal planar, bent
c. tetrahedral, bent
d. tetrahedral, trigonal pyramidal

71. a. 180° b. 120°
c. 109.5° d. 109.5°

73. a. linear, linear
b. trigonal planar, bent (about both nitrogen atoms)
c. tetrahedral, trigonal pyramidal (about both nitrogen atoms)

75. a. trigonal planar b. bent
c. trigonal planar d. tetrahedral

77. a. 1.2 b. 1.8 c. 2.8

79. Cl > Si > Ga > Ca > Rb

81. a. polar covalent b. ionic
c. pure covalent d. polar covalent

83. H_2 < ICl < HBr < CO

85. a. polar b. nonpolar
c. nonpolar d. polar

87. a. $(+):\text{C}\equiv\text{O}:(-)$ b. nonpolar
c. nonpolar d. $(+)\,\text{H}-\ddot{\underset{..}{\text{Br}}}:(-)$

89. a. nonpolar b. polar
c. nonpolar d. polar

91. a. nonpolar b. polar
c. nonpolar d. polar

93. a. $1s^22s^22p^63s^23p^6\underline{4s^2}$, Ca: (underlined electrons are the ones included)
b. $1s^22s^22p^63s^23p^6\underline{4s^2}3d^{10}\underline{4p^1}$, $\dot{\text{G}}\text{a}:$
c. $[\text{Ar}]\underline{4s^2}3d^{10}\underline{4p^3}$, $\cdot\dot{\text{A}}\text{s}:$
d. $[\text{Kr}]\underline{5s^2}4d^{10}\underline{5p^5}$, $:\ddot{\text{I}}:$

95. a. ionic, $\text{K}^+[:\ddot{\text{S}}:]^{2-}\text{K}^+$
b. covalent, $\begin{array}{c} \ddot{\text{O}}: \\ \| \\ \text{H}-\text{C}-\ddot{\text{F}}: \end{array}$
c. ionic, $\text{Mg}^{2+}[:\ddot{\text{Se}}:]^{2-}$
d. covalent, $\begin{array}{c} :\ddot{\text{Br}}: \\ | \\ :\ddot{\text{Br}}-\ddot{\text{P}}-\ddot{\text{Br}}: \end{array}$

97. $\begin{array}{c} \ddot{\text{O}} \\ \| \\ :\ddot{\text{Cl}}-\text{C}-\ddot{\text{Cl}}: \end{array}$ polar, $\begin{array}{c} \text{O} \\ | \\ \text{C} \\ \text{Cl} \quad \text{Cl} \end{array}$

99. $\begin{array}{c} \text{H} \quad \ddot{\text{O}}: \\ | \quad \| \\ \text{H}-\text{C}-\text{C}-\ddot{\text{O}}\text{H} \\ | \\ \text{H} \end{array}$ $\begin{array}{c} \text{H} \quad\quad \text{O} \\ | \quad\ // \\ \text{H}\cdots\text{C}-\text{C} \\ | \quad\ \backslash \\ \text{H} \quad\quad \text{OH} \end{array}$

101. $\text{H}:\ddot{\text{Cl}}: + \text{Na}^+[:\ddot{\text{O}}:\text{H}]^- \longrightarrow \text{H}:\ddot{\text{O}}:\text{H} + \text{Na}^+[:\ddot{\text{Cl}}:]^-$

103. $\text{K}\cdot$, $:\ddot{\text{Cl}}-\ddot{\text{Cl}}:$, $\text{K}^+[:\ddot{\text{Cl}}:]^-$, Cl reduced, K oxidized

105. a. $\text{K}^+[:\ddot{\text{O}}:\text{H}]^-$

b. $\text{K}^+\left[\begin{array}{c} :\ddot{\text{O}}: \\ \ddot{\text{O}}=\text{N}:\ddot{\text{O}}: \end{array}\right]^- \longleftrightarrow \text{K}^+\left[\begin{array}{c} \ddot{\text{O}} \\ \| \\ :\ddot{\text{O}}:\text{N}:\ddot{\text{O}}: \end{array}\right]^- \longleftrightarrow \text{K}^+\left[\begin{array}{c} :\ddot{\text{O}}: \\ :\ddot{\text{O}}:\text{N}=\ddot{\text{O}} \end{array}\right]^-$

c. $\text{Li}^+[:\ddot{\text{I}}:\ddot{\text{O}}:]^-$

d. $\text{Ba}^{2+}\left[\begin{array}{c} \ddot{\text{O}} \\ \| \\ \text{C} \\ :\ddot{\text{O}}\quad\ddot{\text{O}}: \end{array}\right]^{2-} \longleftrightarrow \text{Ba}^{2+}\left[\begin{array}{c} :\ddot{\text{O}}: \\ \text{C} \\ :\ddot{\text{O}}\quad\text{O}: \end{array}\right]^{2-} \longleftrightarrow \text{Ba}^{2+}\left[\begin{array}{c} :\ddot{\text{O}}: \\ \text{C} \\ :\text{O}\quad\ddot{\text{O}}: \end{array}\right]^{2-}$

107. a. PF_5: P bonded to five $:\ddot{\text{F}}:$ atoms b. SF_4: $\ddot{\text{S}}$ bonded to four $:\ddot{\text{F}}:$ atoms

c. SeF_4: $\ddot{\text{Se}}$ bonded to four $:\ddot{\text{F}}:$ atoms

109. CH_2O_2 or $\begin{array}{c} :\ddot{\text{O}} \\ \| \\ \text{H}-\text{C}-\ddot{\text{O}}-\text{H} \end{array}$

111. $\text{H}-\ddot{\text{O}}-\ddot{\text{O}}\cdot$ HOO is not stable because one oxygen atom does not have an octet. The geometry for HOO is *bent*.

113. **a.** $[:\ddot{\underset{..}{O}}-\ddot{\underset{..}{O}}\cdot]^-$ **b.** $[:\ddot{\underset{..}{O}}\cdot]^-$

c. $\cdot\ddot{\underset{..}{O}}-H$ **d.** $H-\underset{H}{\overset{H}{C}}-\ddot{\underset{..}{O}}-\ddot{\underset{..}{O}}\cdot$

115. **a.** The structure has 2 bonding electron pairs and 2 lone pairs. The Lewis structure is analogous to that of water, and the molecular geometry is bent.

b. Correct

c. The structure has 3 bonding electron pairs and 1 lone pair. The Lewis structure is analogous to that of NH_3, and the geometry is trigonal pyramidal.

d. Correct

$:\ddot{\underset{..}{O}}-\dot{N}=\ddot{O}:$

119. **a.** $[:\ddot{O}=N=\ddot{O}:]^+$

$[:\ddot{\underset{..}{O}}-\ddot{N}=\ddot{O}:]^-$

Glossary

absolute zero The coldest temperature possible. Absolute zero (0 K or –273 °C or –459 °F) is the temperature at which molecular motion stops. Lower temperatures do not exist.
acid A molecular compound that dissolves in solution to form H^+ ions. Acids have the ability to dissolve some metals and will turn litmus paper red.
acid–base reaction A reaction that forms water and typically a salt.
acidic solution A solution containing a concentration of H_3O^+ ions greater than 1.0×10^{-7} M (pH < 7).
activation energy The amount of energy that must be absorbed by reactants before a reaction can occur; an energy hump that normally exists between the reactants and products.
activity series of metals A listing of metals (and hydrogen) in order of decreasing activity, decreasing ability to oxidize, and decreasing tendency to lose electrons.
actual yield The amount of product actually produced by a chemical reaction.
addition polymer A polymer formed by addition of monomers to one another without elimination of any atoms.
addition reaction A chemical reaction in which additional atoms or functional groups add to a molecule.
alcohol An organic compound containing an —OH functional group bonded to a carbon atom and having the general formula *R*OH.
aldehyde An organic compound with the general formula *R*CHO.
alkali metals The Group 1A elements, which are highly reactive metals.
alkaline battery A dry cell employing half-reactions that use a base.
alkaline earth metals The Group 2A elements, which are fairly reactive metals.
alkaloid Organic compound that acts as a base and is typically found in plants.
alkane A hydrocarbon in which all carbon atoms are connected by single bonds. Noncyclic alkanes have the general formula C_nH_{2n+2}.
alkene A hydrocarbon that contains at least one double bond between carbon atoms. Noncyclic alkenes have the general formula C_nH_{2n}.
alkyl group In an organic molecule, any group containing only singly bonded carbon atoms and hydrogen atoms.
alkyne A hydrocarbon that contains at least one triple bond between carbon atoms. Noncyclic alkynes have the general formula C_nH_{2n-2}.
alpha particle A particle consisting of two protons and two neutrons (a helium nucleus), represented by the symbol ^4_2He.
alpha (α) radiation Radiation emitted by an unstable nucleus, consisting of alpha particles.
alpha (α)-helix The most common secondary protein structure. The amino acid chain is wrapped into a tight coil from which the side chains extend outward. The structure is maintained by hydrogen bonding interactions between NH and CO groups along the peptide backbone of the protein.
amine An organic compound that contains nitrogen and has the general formula NR_3, where *R* may be an alkyl group or a hydrogen atom.
amino acid A molecule containing an amine group, a carboxylic acid group, and an *R* group (also called a side chain). Amino acids are the building blocks of proteins.
amorphous A type of solid matter in which atoms or molecules do not have long-range order (e.g., glass and plastic).
amphoteric In Brønsted–Lowry terminology, able to act as either an acid or a base.
anion A negatively charged ion.
anode The electrode where oxidation occurs in an electrochemical cell.
aqueous solution A homogeneous mixture of a substance with water.
aromatic ring A ring of carbon atoms containing alternating single and double bonds.
Arrhenius acid A substance that produces H^+ ions in aqueous solution.
Arrhenius base A substance that produces OH^- ions in aqueous solution.
Arrhenius definitions (of acids and bases) The definitions of an acid as a substance that produces H^+ ions in aqueous solution and a base as a substance that produces OH^- ions in aqueous solution.
atmosphere (atm) The average pressure at sea level, 101,325 Pa (760 mmHg).
atom The smallest identifiable unit of an element.
atomic element An element that exists in nature with single atoms as the basic unit.
atomic mass A weighted average of the masses of each naturally occurring isotope of an element; atomic mass is the average mass of the atoms of an element.
atomic mass unit (amu) The unit commonly used to express the masses of protons, neutrons, and nuclei. 1 amu = 1.66×10^{-24} g.
atomic number (Z) The number of protons in the nucleus of an atom.
atomic size The size of an atom, which is determined by how far the outermost electrons are from the nucleus.
atomic solid A solid whose component units are individual atoms (e.g., diamond, C; iron, Fe).
atomic theory A theory stating that all matter is composed of tiny particles called atoms.
Avogadro's law A law stating that the volume (*V*) of a gas and the amount of the gas in moles (*n*) are directly proportional.
Avogadro's number The number of entities in a mole, 6.022×10^{23}.
balanced equation A chemical equation in which the numbers of each type of atom on both sides of the equation are equal.

ball-and-stick model A way to represent molecules in which an atom is represented with a ball and a bond is represented with a stick.
base A molecular compound that dissolves in solution to form OH^- ions. Bases have a slippery feel and turn litmus paper blue.
base chain The longest continuous chain of carbon atoms in an organic compound.
basic solution A solution containing a concentration of OH^- ions greater than 1.0×10^{-7} M (pH > 7).
bent The molecular geometry in which 3 atoms are not in a straight line. This geometry occurs when the central atoms contain 4 electron groups (2 bonding and 2 nonbonding) or 3 electron groups (2 bonding and 1 nonbonding).
beta (β) particle A form of radiation consisting of an energetic electron and represented by the symbol $_{-1}^{0}e$.
beta (β) radiation Energetic electrons emitted by an unstable nucleus.
beta (β)-pleated sheet A common pattern in the secondary structure of proteins. The protein chain is extended in a zigzag pattern, and the peptide backbones of adjacent strands interact with one another through hydrogen bonding to form sheets.
binary acid An acid containing only hydrogen and a nonmetal.
binary compound A compound containing only two different kinds of elements.
biochemistry The study of the chemical substances and processes that occur in living organisms.
Bohr model A model for the atom in which electrons travel around the nucleus in circular orbits at specific, fixed distances from the nucleus.
boiling point The temperature at which the vapor pressure of a liquid is equal to the pressure above it.
boiling point elevation The increase in the boiling point of a solution caused by the presence of the solute.
bonding pair Electrons that are shared between two atoms in a chemical bond.
bonding theory A model that predicts how atoms bond together to form molecules.
Boyle's law A law maintaining that the volume (*V*) of a gas and its pressure (*P*) are inversely proportional.
branched alkane An alkane composed of carbon atoms bonded in chains containing branches.
Brønsted–Lowry acid A proton (H^+ ion) donor.
Brønsted–Lowry base A proton (H^+ ion) acceptor.
buffer A solution that resists pH change by neutralizing added acid or added base.
Calorie (Cal) An energy unit equivalent to 1000 little-*c* calories.
calorie (cal) The amount of energy required to raise the temperature of 1 g of water by 1 °C.
carbohydrate A polyhydroxyl aldehyde or ketone, containing multiple —OH groups and often having the general formula $(CH_2O)_n$.
carbonyl group A carbon atom double bonded to an oxygen atom.
carboxylic acid An organic compound with the general formula RCOOH.
catalyst A substance that increases the rate of a chemical reaction but is not consumed by the reaction.
cathode The electrode where reduction occurs in an electrochemical cell.
cation A positively charged ion.
cell The smallest structural unit of living organisms that has the properties associated with life.
cell membrane The structure that bounds the cell and holds the contents of the cell together.
cellulose A common polysaccharide composed of repeating glucose units linked together.
Celsius (°C) scale A temperature scale often used by scientists. On this scale, water freezes at 0 °C and boils at 100 °C at 1 atm pressure. Room temperature is approximately 22 °C.
chain reaction A self-sustaining chemical or nuclear reaction yielding energy or products that cause further reactions of the same kind.
charge A fundamental property of protons and electrons. Charged particles experience forces such that like charges repel and unlike charges attract.
Charles's law A law stating that the volume (*V*) of a gas and its temperature (*T*) expressed in kelvins are directly proportional.
chemical bond The sharing or transfer of electrons to attain stable electron configurations among the bonding atoms.
chemical change A change in which matter changes its composition.
chemical energy The energy associated with chemical changes.
chemical formula A way to represent a compound. At a minimum, the chemical formula indicates the elements present in the compound and the relative number of atoms of each element.
chemical property A property that a substance can display only through changing its composition.
chemical reaction The process by which one or more substances transform into different substances via a chemical change. Chemical reactions often emit or absorb energy.
chemical symbol A one- or two-letter abbreviation for an element. Chemical symbols are listed directly below the atomic number in the periodic table.
chemistry The science that seeks to understand the behavior of matter by studying what atoms and molecules do.
chromosome A biological structure containing genes, located within the nucleus of a cell.
climate change Changes in the Earth's climate caused by human emission of gases (especially CO_2) into the atmosphere.
codon A sequence of three bases in a nucleic acid that codes for one amino acid.
colligative properties Physical properties of solutions that depend on the number of solute particles present but not the type of solute particles.
collision theory A theory of reaction rates stating that effective collisions between reactant molecules must take place in order for the reaction to occur.
combination reaction A chemical reaction in which simpler substances combine to form more complex substances.
combined gas law A law that combines Boyle's law and Charles's law; it is used to calculate how a property of a gas (*P*, *V*, or *T*) changes when two other properties are changed at the same time.
combustion The burning of a substance in the presence of oxygen.
combustion reaction A reaction in which a substance reacts with oxygen, emitting heat and forming one or more oxygen-containing compounds.
complementary base In DNA, a base capable of precise pairing with a specific other DNA base.
complete ionic equation A chemical equation showing all the species as they are actually present in solution.
complex carbohydrate A carbohydrate composed of many repeating saccharide units.

compound A substance composed of two or more elements in fixed, definite proportions.
compressible Able to occupy a smaller volume when subjected to increased pressure. Gases are compressible because, in the gas phase, atoms or molecules are widely separated.
concentrated solution A solution containing large amounts of solute.
condensation A physical change in which a substance is converted from its gaseous form to its liquid form.
condensation polymer A class of polymers that expel atoms, usually water, during their formation or polymerization.
condensed structural formula A shorthand way of writing a structural formula.
conjugate acid–base pair In Brønsted–Lowry terminology, two substances related to each other by the transfer of a proton.
conversion factor A factor used to convert between two separate units; a conversion factor is constructed from any two quantities known to be equivalent.
copolymers Polymers that are composed of two different kinds of monomers and result in chains composed of alternating units rather than a single repeating unit.
core electrons The electrons that are not in the outermost principal shell of an atom.
corrosion The oxidation of metals (e.g., rusting of iron).
covalent atomic solid An atomic solid, such as diamond, that is held together by covalent bonds.
covalent bond The bond that results when two nonmetals combine in a chemical reaction. In a covalent bond, the atoms share their electrons.
critical mass The mass of uranium or plutonium required for a nuclear reaction to be self-sustaining.
crystalline A type of solid matter with atoms or molecules arranged in a well-ordered, three-dimensional array with long-range, repeating order (e.g., salt and diamond).
cytoplasm In a cell, the region between the nucleus and the cell membrane.
Dalton's law of partial pressure A law stating that the sum of the partial pressures of each component in a gas mixture equals the total pressure.
daughter nuclide The nuclide product of a nuclear decay.
decanting A way to separate a mixture in which one layer is carefully poured off of another layer.
decimal part One part of a number expressed in scientific notation.
decomposition reaction A reaction in which a complex substance decomposes to form simpler substances; $AB \longrightarrow A + B$.
density (*d*) A fundamental property of materials that relates mass and volume and differs from one substance to another. The units of density are those of mass divided by volume, most commonly expressed in g/cm^3, g/mL, or g/L.
dilute solution A solution containing small amounts of solute.
dimer A molecule formed by the joining together of two smaller molecules.
dipeptide Two amino acids linked together via a peptide bond.
dipole–dipole force The interaction between two molecules having dipole–dipole moments (or between two polar molecules).
dipole moment A separation of charge within a chemical bond that produces a bond with a positive end and a negative end.
diprotic acid An acid containing two ionizable protons.
disaccharide A carbohydrate that can be decomposed into two simpler carbohydrates.
dispersion force The intermolecular force present in all molecules and atoms. Dispersion forces are caused by fluctuations in the electron distribution within molecules or atoms.
displacement reaction A reaction in which one element displaces another in a compound; $A + BC \longrightarrow AC + B$.
dissociation In aqueous solution, the process by which a solid ionic compound separates into its ions.
distillation A way to separate mixtures in which the mixture is heated to boil off the more volatile component.
disubstituted benzene A benzene in which two hydrogen atoms have been replaced by an atom or group of atoms.
DNA (deoxyribonucleic acid) Long chainlike molecules that occur in the nucleus of cells and act as blueprints for the construction of proteins.
dot structure A drawing that represents the valence electrons in atoms as dots; it shows a chemical bond as the sharing or transfer of electron dots.
double bond The bond that exists when two electron pairs are shared between two atoms. In general, double bonds are shorter and stronger than single bonds.
double displacement A reaction in which two elements or groups of elements in two different compounds exchange places to form two new compounds; $AB + CD \longrightarrow AD + CB$.
dry cell An ordinary battery (voltaic cell); it does not contain large amounts of liquid water.
duet The name for the two electrons corresponding to a stable Lewis structure in hydrogen and helium.
dynamic equilibrium In a chemical reaction, the condition in which the rate of the forward reaction equals the rate of the reverse reaction.
electrical current The flow of electric charge—for example, electrons flowing through a wire or ions through a solution.
electrical energy Energy associated with the flow of electric charge.
electrochemical cell A device that creates electrical current from a redox reaction.
electrolysis A process in which electrical current is used to drive an otherwise nonspontaneous redox reaction.
electrolyte solution A solution containing a solute that dissociates into ions.
electrolytic cell An electrochemical cell used for electrolysis.
electromagnetic radiation A type of energy that travels through space at a constant speed of 3.0×10^8 m/s (186,000 miles/s) and exhibits both wavelike and particlelike behavior. Light is a form of electromagnetic radiation.
electromagnetic spectrum A spectrum that includes all wavelengths of electromagnetic radiation.
electron A negatively charged particle that occupies most of the atom's volume but contributes almost none of its mass.
electron configuration A representation that shows the occupation of orbitals by electrons for a particular element.
electron geometry The geometrical arrangement of the electron groups in a molecule.
electron group A general term for a lone pair, single bond, or multiple bond in a molecule.
electron spin A fundamental property of all electrons that causes them to have magnetic fields associated with them. The spin of an electron can either be oriented up $\left(+\frac{1}{2}\right)$ or down $\left(-\frac{1}{2}\right)$.
electronegativity The ability of an element to attract electrons within a covalent bond.
element A substance that cannot be broken down into simpler substances.

emission spectrum (plural, emission spectra) A spectrum associated with the emission of electromagnetic radiation by elements or compounds.
empirical formula A formula for a compound that gives the smallest whole-number ratio of each type of atom.
empirical formula molar mass The sum of the molar masses of all the atoms in an empirical formula.
endothermic Describes a process that absorbs heat energy.
endothermic reaction A chemical reaction that absorbs energy from the surroundings.
energy The capacity to do work.
English system A unit system commonly used in the United States.
enthalpy The amount of thermal energy absorbed or emitted by a process under conditions of constant pressure.
enthalpy of reaction (ΔH_{rxn}) The amount of thermal energy absorbed or emitted by a chemical reaction under conditions of constant pressure.
enzyme A biological catalyst that increases the rates of biochemical reactions; enzymes are abundant in living organisms.
equilibrium constant (K_{eq}) The ratio, at equilibrium, of the concentrations of the products raised to their stoichiometric coefficients divided by the concentrations of the reactants raised to their stoichiometric coefficients.
equivalence point The point in a reaction at which the reactants are in exact stoichiometric proportions.
equivalent The stoichiometric proportions of elements and compounds in a chemical equation.
ester An organic compound with the general formula *RCOOR*.
ester linkage A type of bond with the general structure —COO—. Ester linkages join glycerol to fatty acids.
ether An organic compound with the general formula *ROR*.
evaporation A process in which molecules of a liquid, undergoing constant random motion, acquire enough energy to overcome attractions to neighbors and enter the gas phase.
excited state An unstable state for an atom or a molecule in which energy has been absorbed but not reemitted, raising an electron from the ground state into a higher energy orbital.
exothermic Describes a process that releases heat energy.
exothermic reaction A chemical reaction that releases energy to the surroundings.
experiment A procedure that attempts to measure observable predictions to test a theory or law.
exponent A number that represents the number of times a term is multiplied by itself. For example, in 2^4 the exponent is 4 and represents $2 \times 2 \times 2 \times 2$.
exponential part One part of a number expressed in scientific notation; it represents the number of places the decimal point has moved.
Fahrenheit (°F) scale The temperature scale that is most familiar in the United States; water freezes at 32 °F and boils at 212 °F at 1 atm prssure.
family (of elements) A group of elements that have similar outer electron configurations and therefore similar properties. Families occur in vertical columns in the periodic table.
family (of organic compounds) A group of organic compounds with the same functional group.
fatty acid A type of lipid consisting of a carboxylic acid with a long hydrocarbon tail.
fibrous protein Proteins with tertiary structures in which coiled amino acid chains align roughly parallel to one another, forming long, water-insoluble fibers.
film badge dosimeter Badge used to measure radiation exposure, consisting of photographic film held in a small case that is pinned to clothing.
filtration A method of separating a mixture composed of a solid and a liquid in which the mixture is poured through filter paper held in a funnel to capture the solid component.
formula mass The average mass of the molecules (or formula units) that compose a compound.
formula unit The basic unit of ionic compounds; the smallest electrically neutral collection of cations and anions that compose the compound.
fossil fuels Hydrocarbon-based fuels that originate from plant and animal life that existed on Earth in prehistoric times. The main types of fossil fuels are natural gas, petroleum, and coal.
freezing point depression The decrease in the freezing point of a solvent caused by the presence of a solute.
frequency The number of wave cycles or crests that pass through a stationary point in one second.
fuel cell A voltaic cell in which the reactants are constantly replenished.
functional group A set of atoms that characterizes a family of organic compounds.
gamma radiation High-energy, short-wavelength electromagnetic radiation emitted by an atomic nucleus.
gamma ray The shortest-wavelength, most energetic form of electromagnetic radiation. Gamma ray photons are represented by the symbol ${}^{0}_{0}\gamma$.
gas A state of matter in which atoms or molecules are widely separated and free to move relative to one another.
gas-evolution reaction A reaction that occurs in solution and forms a gas as one of the products.
Geiger-Müller counter A radioactivity detector consisting of a chamber filled with argon gas that discharges electrical signals when high-energy particles pass through it.
gene A sequence of codons within a DNA molecule that codes for a single protein. Genes vary in length from hundreds to thousands of codons.
genetic material The inheritable blueprint for making organisms.
globular protein Proteins with tertiary structures in which amino acid chains fold in on themselves, forming water-soluble globules.
glycogen A type of polysaccharide; it has a structure similar to that of starch, but the chain is highly branched.
glycolipid A biological molecule composed of a nonpolar fatty acid and hydrocarbon chain and a polar section composed of a sugar molecule such as glucose.
glycoside linkage The link between monosaccharides in a polysaccharide.
greenhouse gases Gases in the Earth's atmosphere that allow sunlight to enter the atmosphere but prevent heat from escaping.
ground state The state of an atom or molecule in which the electrons occupy the lowest possible energy orbitals available.
group (of elements) Elements that have similar outer electron configurations and therefore similar properties. Groups occur in vertical columns in the periodic table.
half-cell A compartment in which the oxidation or reduction half-reaction occurs in a galvanic or voltaic cell.
half-life The time it takes for one-half of the parent nuclides in a radioactive sample to decay to the daughter nuclides.
half-reaction Either the oxidation part or the reduction part of a redox reaction.
halogens The Group 7A elements, which are very reactive nonmetals.

heat The transfer or exchange of thermal energy caused by a temperature difference.
heat absorption One type of evidence of a chemical reaction, involving the intake of energy.
heat capacity The quantity of heat energy required to change the temperature of a given amount of a substance by 1 °C.
heat of fusion The amount of heat required to melt one mole of a solid at its melting point with no change in temperature.
heat of vaporization The amount of heat required to vaporize one mole of a liquid at its boiling point with no change in temperature.
heterogeneous mixture A mixture, such as oil and water, that has two or more regions with different compositions.
homogeneous mixture A mixture, such as salt water, that has the same composition throughout.
human genome All of the genetic material of a human being; the total DNA of a human cell.
Hund's rule A rule stating that when filling orbitals of equal energy, electrons will occupy empty orbitals singly before pairing with other electrons.
hydrocarbon A compound that contains only carbon and hydrogen atoms.
hydrogen bond A strong dipole–dipole interaction between molecules containing hydrogen directly bonded to a small, highly electronegative atom, such as N, O, or F.
hydrogenation The chemical addition of hydrogen to a compound.
hydronium ion The H_3O^+ ion. Chemists often use $H^+(aq)$ and $H_3O^+(aq)$ interchangeably to mean the same thing—a hydronium ion.
hypothesis A theory or law before it has become well established; a tentative explanation for an observation or a scientific problem that can be tested by further investigation.
hypoxia A shortage of oxygen in the tissues of the body.
ideal gas constant (R) The proportionality constant in the ideal gas law. $R = 0.0821\ \text{L} \cdot \text{atm/mol} \cdot \text{K}$
ideal gas law A law that combines the four properties of a gas—pressure (P), volume (V), temperature (T), and number of moles (n) in a single equation showing their interrelatedness: $PV = nRT$ (R = ideal gas constant).
indicator A substance that changes color with acidity level, often used to detect the end point of a titration.
infrared (IR) light The fraction of the electromagnetic spectrum between visible light and microwaves. Infrared light is invisible to the human eye.
insoluble Not soluble in water.
instantaneous (temporary) dipole A type of intermolecular force resulting from transient shifts in electron density within an atom or molecule.
intermolecular forces Attractive forces that exist between molecules.
International System (SI) The standard set of units for science measurements, based on the metric system.
ion An atom (or group of atoms) that has gained or lost one or more electrons, so that it has an electric charge.
ion–dipole force An intermolecular force that occurs between an ion and a polar molecule.
ion product constant (K_w) The product of the H_3O^+ ion concentration and the OH^- ion concentration in an aqueous solution. At room temperature, $K_w = 1.0 \times 10^{-4}$.
ionic bond The bond that results when a metal and a nonmetal combine in a chemical reaction. In an ionic bond, the metal transfers one or more electrons to the nonmetal.
ionic compound A compound formed between a metal and one or more nonmetals.
ionic solid A solid compound composed of metals and nonmetals joined by ionic bonds.
ionization energy The energy required to remove an electron from an atom in the gaseous state.
ionize Convert (an atom, molecule, or substance) into an ion or ions, typically by removing one or more electrons.
ionizing power The ability of radiation to ionize other molecules and atoms.
isomer A molecule with the same molecular formula but different structure as another molecule.
isoosmotic Describes solutions having equal osmotic pressure.
isotope One of two or more atoms with the same number of protons but different numbers of neutrons.
isotope scanning The use of radioactive isotopes to identify disease in the body.
Kelvin (K) scale The temperature scale that assigns 0 K to the coldest temperature possible, absolute zero (−273 °C or −459 °F), the temperature at which molecular motion stops. The size of the kelvin is identical to that of the Celsius degree.
ketone An organic compound with the general formula *RCOR*.
kilogram (kg) The SI standard unit of mass.
kilowatt-hour (kWh) A unit of energy equal to 3.6 million joules.
kinetic energy Energy associated with the motion of an object.
kinetic molecular theory A simple model for gases that predicts the behavior of most gases under many conditions.
law of conservation of energy A law stating that energy can be neither created nor destroyed. The total amount of energy is constant and cannot change; it can only be transferred from one object to another or converted from one form to another.
law of conservation of mass A law stating that in a chemical reaction, matter is neither created nor destroyed.
law of constant composition A law stating that all samples of a given compound have the same proportions of their constituent elements.
Le Châtelier's principle A principle stating that when a chemical system at equilibrium is disturbed, the system shifts in a direction that minimizes the disturbance.
lead-acid storage battery An automobile battery consisting of six electrochemical cells wired in series. Each cell produces 2 volts for a total of 12 volts.
Lewis model A simple model for a chemical bond in which atoms transfer or share electrons to attain a noble gas electron configuration (usually referred to as an octet).
Lewis structure A drawing that represents chemical bonds between atoms as shared or transferred electrons; the valence electrons of atoms are represented as dots.
light A form of electromagnetic radiation.
limiting reactant The reactant that determines the amount of product formed in a chemical reaction.
linear Describes the molecular geometry of a molecule containing two electron groups (two bonding groups and no lone pairs).
lipid A cellular component that is insoluble in water but soluble in nonpolar solvents.
lipid bilayer A structure formed by lipids in the cell membrane.
liquid A state of matter in which atoms or molecules are packed close to each other (about as closely as in a solid) but are free to move around and by each other.

liter (L) A unit of volume equal to 1.057 quarts.
logarithmic scale A scale involving logarithms. A logarithm entails an exponent that indicates the power to which a number is raised to produce a given number (e.g., the logarithm of 100 to the base 10 is 2).
lone pair Electrons that are only on one atom in a Lewis structure.
main-group elements Groups 1A–8A on the periodic table. These groups have properties that tend to be predictable based on their position in the periodic table.
mass A measure of the quantity of matter within an object.
mass number (A) The sum of the number of neutrons and protons in an atom.
mass percent composition (or mass percent) The percentage, by mass, of each element in a compound.
matter Anything that occupies space and has mass. Matter exists in three different states: solid, liquid, and gas.
melting point The temperature at which a solid turns into a liquid.
messenger RNA (mRNA) Long chainlike molecules that act as blueprints for the construction of proteins.
metallic atomic solid An atomic solid, such as iron, which is held together by metallic bonds that, in the simplest model, consist of positively charged ions in a sea of electrons.
metallic character The properties typical of a metal, especially the tendency to lose electrons in chemical reactions. Elements become more metallic as you move from right to left across the periodic table.
metalloids Those elements that fall along the boundary between the metals and the nonmetals in the periodic table; their properties are intermediate between those of metals and those of nonmetals.
metals Elements that tend to lose electrons in chemical reactions. They are found at the left side and in the center of the periodic table.
meter (m) The SI standard unit of length.
metric system The unit system commonly used throughout most of the world.
microwaves The part of the electromagnetic spectrum between the infrared region and the radio wave region. Microwaves are efficiently absorbed by water molecules and can therefore be used to heat water-containing substances.
millimeter of mercury (mmHg) A unit of pressure that originates from the method used to measure pressure with a barometer. Also called a *torr*.
miscibility The ability of two liquids to mix without separating into two phases, or the ability of one liquid to mix with (dissolve in) another liquid.
mixture A substance composed of two or more different types of atoms or molecules combined in variable proportions.
molality (*m*) A common unit of solution concentration, defined as the number of moles of solute per kilogram of solvent.
molar mass The mass of one mole of atoms of an element or one mole of molecules (or formula units) for a compound. An element's molar mass in grams per mole is numerically equivalent to the element's atomic mass in amu.
molar solubility The solubility of a substance in units of moles per liter (mol/L).
molar volume The volume occupied by one mole of gas. Under standard temperature and pressure conditions the molar volume of ideal gas is 22.5 L.
molarity (M) A common unit of solution concentration, defined as the number of moles of solute per liter of solution.
mole Avogadro's number (6.022×10^{23}) of particles—especially, of atoms, ions, or molecules. A mole of any element has a mass in grams that is numerically equivalent to its atomic mass in amu.
molecular compound A compound formed from two or more nonmetals. Molecular compounds have distinct molecules as their simplest identifiable units.
molecular element An element that does not normally exist in nature with single atoms as the basic unit. These elements usually exist as diatomic molecules—2 atoms of that element bonded together—as their basic units.
molecular equation A chemical equation showing the complete, neutral formulas for every compound in a reaction.
molecular formula A formula for a compound that gives the specific number of each type of atom in a molecule.
molecular geometry The geometrical arrangement of the atoms in a molecule.
molecular model A three-dimensional representation of a molecule.
molecular solid A solid whose composite units are molecules.
molecule Two or more atoms joined in a specific arrangement by chemical bonds. A molecule is the smallest identifiable unit of a molecular compound.
monomer An individual repeating unit that makes up a polymer.
monoprotic acid An acid containing only one ionizable proton.
monosaccharide A carbohydrate that cannot be decomposed into simpler carbohydrates.
monosubstituted benzene A benzene in which one of the hydrogen atoms has been replaced by another atom or group of atoms.
net ionic equation An equation that shows only the species that actually participate in a reaction.
neutral solution A solution in which the concentrations of H_3O^+ and OH^- are equal ($pH = 7$).
neutralization A chemical reaction in which an acid and a base react to form water.
neutralization reaction A reaction that takes place when an acid and a base are mixed; the $H^+(aq)$ from the acid combines with the $OH^-(aq)$ from the base to form $H_2O(l)$.
neutron A nuclear particle with no electrical charge and nearly the same mass as a proton.
nitrogen narcosis An increase in nitrogen concentration in bodily tissues and fluids that results in feelings of drunkenness.
noble gases The Group 8A elements, which are chemically unreactive.
nonbonding atomic solid An atomic solid that is held together by relatively weak dispersion forces.
nonelectrolyte solution A solution containing a solute that dissolves as molecules; therefore, the solution does not conduct electricity.
nonmetals Elements that tend to gain electrons in chemical reactions. They are found at the upper right side of the periodic table.
nonpolar Describes a molecule that does not have a dipole moment.
nonvolatile Describes a compound that does not vaporize easily.
normal alkane (or *n*-alkane) An alkane composed of carbon atoms bonded in a straight chain with no branches.
normal boiling point The boiling point of a liquid at a pressure of 1 atmosphere.

nuclear equation An equation that represents the changes that occur during radioactivity and other nuclear processes.
nuclear fission The process by which a heavy nucleus is split into nuclei of smaller masses and energy is emitted.
nuclear fusion The combination of light atomic nuclei to form heavier ones with emission of large amounts of energy.
nuclear radiation The energetic particles emitted from the nucleus of an atom when it is undergoing a nuclear process.
nuclear theory of the atom A theory stating that most of the atom's mass and all of its positive charge are contained in a small, dense nucleus. Most of the volume of the atom is empty space occupied by negatively charged electrons.
nucleic acids Biological molecules, such as deoxyribonucleic acid (DNA) and ribonucleic acid (RNA), that store and transmit genetic information.
nucleotide An individual unit of a nucleic acid. Nucleic acids are polymers of nucleotides.
nucleus (of a cell) The part of the cell that contains the genetic material.
nucleus (of an atom) The small core containing most of the atom's mass and all of its positive charge. The nucleus is made up of protons and neutrons.
observation Often the first step in the scientific method. An observation must measure or describe something about the physical world.
octet The number of electrons, 8, around atoms with stable Lewis structures.
octet rule A rule that states that an atom will give up, accept, or share electrons in order to achieve a filled outer electron shell, which usually consists of 8 electrons.
orbital The region around the nucleus of an atom where an electron is most likely to be found.
orbital diagram An electron configuration in which electrons are represented as arrows in boxes corresponding to orbitals of a particular atom.
organic chemistry The study of carbon-containing compounds and their reactions.
organic molecule A molecule whose main structural component is carbon.
osmosis The flow of solvent from a lower-concentration solution through a semipermeable membrane to a higher-concentration solution.
osmotic pressure The pressure produced on the surface of a semipermeable membrane by osmosis or the pressure required to stop osmotic flow.
oxidation The gain of oxygen, the loss of hydrogen, or the loss of electrons (the most fundamental definition).
oxidation state (or oxidation number) A number that can be used as an aid in writing formulas and balancing equations. It is computed for each element based on the number of electrons assigned to it in a scheme where the most electronegative element is assigned all of the bonding electrons.
oxidation–reduction (redox) reaction A reaction in which electrons are transferred from one substance to another.
oxidizing agent In a redox reaction, the substance being reduced. Oxidizing agents tend to gain electrons easily.
oxyacid An acid containing hydrogen, a nonmetal, and oxygen.
oxyanion An anion containing oxygen. Most polyatomic ions are oxyanions.
oxygen toxicity The result of increased oxygen concentration in bodily tissues.
parent nuclide The original nuclide in a nuclear decay.
partial pressure The pressure due to any individual component in a gas mixture.
pascal (Pa) The SI unit of pressure, defined as 1 newton per square meter.
Pauli exclusion principle A principle stating that no more than 2 electrons can occupy an orbital and that the 2 electrons must have opposite spins.
penetrating power The ability of a radioactive particle to penetrate matter.
peptide bond The bond between the amine end of one amino acid and the carboxylic acid end of another. Amino acids link together via peptide bonds to form proteins.
percent natural abundance The percentage amount of each isotope of an element in a naturally occurring sample of the element.
percent yield In a chemical reaction, the percentage of the theoretical yield that was actually attained.
periodic law A law that states that when the elements are arranged in order of increasing relative mass, certain sets of properties recur periodically.
periodic table An arrangement of the elements in which atomic number increases from left to right and elements with similar properties fall in columns called families or groups.
permanent dipole A separation of charge resulting from the unequal sharing of electrons between atoms.
pH The negative log of the concentration of H_3O^+ in a solution; the pH scale is a compact way to specify the acidity of a solution.
phenyl group The term for a benzene ring when other substituents are attached to it.
phospholipid A lipid with the same basic structure as a triglyceride, except that one of the fatty acid groups is replaced with a phosphate group.
phosphorescence The slow, long-lived emission of light that sometimes follows the absorption of light by some atoms and molecules.
photon A particle of light or a packet of light energy.
physical change A change in which matter does not change its composition, even though its appearance might change.
physical property A property that a substance displays without changing its composition.
pOH A scale used to measure basicity. $pOH = -\log[OH^-]$
polar covalent bond A covalent bond between atoms of different electronegativities. Polar covalent bonds have a dipole moment.
polar molecule A molecule with polar bonds that add together to create a net dipole moment.
polyatomic ion An ion composed of a group of atoms with an overall charge.
polymer A molecule with many similar units, called monomers, bonded together in a long chain.
polypeptide A short chain of amino acids joined by peptide bonds.
polysaccharide A long, chainlike molecule composed of many linked monosaccharide units. Polysaccharides are polymers of monosaccharides.
positron A nuclear particle that has the mass of an electron but carries a 1+ charge.
positron emission Expulsion of a positron from an unstable atomic nucleus. In positron emission, a proton is transformed into a neutron.
potential energy The energy of a body that is associated with its position or the arrangement of its parts.

pounds per square inch (psi) A unit of pressure. (1 atm = 14.7 psi)
precipitate An insoluble product formed through the reaction of two solutions containing soluble compounds.
precipitation reaction A reaction that forms a solid or precipitate when two aqueous solutions are mixed.
prefix multipliers Prefixes used by the SI system with the standard units. These multipliers change the value of the unit by powers of 10.
pressure The force exerted per unit area by gaseous molecules as they collide with the surfaces around them.
primary protein structure The sequence of amino acids in a protein's chain. Primary protein structure is maintained by the covalent peptide bonds between individual amino acids.
principal quantum number A number that indicates the shell that an electron occupies.
principal shell The shell indicated by the principal quantum number.
product A final substance produced in a chemical reaction; represented on the right side of a chemical equation.
property A characteristic we use to distinguish one substance from another.
protein A biological molecule composed of a long chain of amino acids joined by peptide bonds. In living organisms, proteins serve many varied and important functions.
proton A positively charged nuclear particle. A proton's mass is approximately 1 amu.
pure substance A substance composed of only one type of atom or molecule.
quantum (*pl.* quanta) The precise amount of energy possessed by a photon; the difference in energy between two atomic orbitals.
quantum number (n) An integer that specifies the energy of an orbital. The higher the quantum number n, the greater the distance between the electron and the nucleus and the higher its energy.
quantum-mechanical model The foundation of modern chemistry; explains how electrons exist in atoms and how they affect the chemical and physical properties of elements.
quaternary structure In a protein, the way that individual chains fit together to compose the protein. Quaternary structure is maintained by interactions between the *R* groups of amino acids on the different chains.
R group (side chain) An organic group attached to the central carbon atom of an amino acid.
radio waves The longest wavelength and least energetic form of electromagnetic radiation.
radioactive Describes a substance that emits tiny, invisible, energetic particles from the nuclei of its component atoms.
radioactivity The emission of tiny, invisible, energetic particles from the unstable nuclei of atoms. Many of these particles can penetrate matter.
radiocarbon dating A technique used to estimate the age of fossils and artifacts through the measurement of natural radioactivity of carbon atoms in the environment.
radiotherapy Treatment of disease with radiation, such as the use of gamma rays to kill rapidly dividing cancer cells.
random coil The name given to an irregular pattern of a secondary protein structure.
rate of a chemical reaction (reaction rate) The amount of reactant that changes to product in a given period of time. Also defined as the amount of a product that forms in a given period of time.
reactant An initial substance in a chemical reaction, represented on the left side of a chemical equation.
recrystallization A technique used to purify a solid; involves dissolving the solid in a solvent at high temperature, creating a saturated solution, then cooling the solution to cause the crystallization of the solid.
redox (oxidation–reduction) reaction A chemical reaction in which electrons are transferred from one reactant to another.
reducing agent In a redox reaction, the substance being oxidized. Reducing agents tend to lose electrons easily.
reduction The loss of oxygen, the gain of hydrogen, or the gain of electrons (the most fundamental definition).
rem Stands for *roentgen equivalent man;* a weighted measure of radiation exposure that accounts for the ionizing power of the different types of radiation.
resonance structures Two or more Lewis structures that are necessary to describe the bonding in a molecule or ion.
reversible reaction A reaction that is able to proceed in both the forward and reverse directions.
RNA (ribonucleic acid) Long chainlike molecules that occur throughout cells and take part in the construction of proteins.
salt An ionic compound that usually remains dissolved in a solution after an acid–base reaction has occurred.
salt bridge An inverted, U-shaped tube containing a strong electrolyte; completes the circuit in an electrochemical cell by allowing the flow of ions between the two half-cells.
saturated fat A triglyceride composed of saturated fatty acids. Saturated fat tends to be solid at room temperature.
saturated hydrocarbon A hydrocarbon that contains no double or triple bonds between the carbon atoms.
saturated solution A solution that holds the maximum amount of solute under the solution conditions. If additional solute is added to a saturated solution, it will not dissolve.
scientific law A statement that summarizes past observations and predicts future ones. Scientific laws are usually formulated from a series of related observations.
scientific method The way that scientists learn about the natural world. The scientific method involves observations, laws, hypotheses, theories, and experimentation.
scientific notation A system used to write very big or very small numbers, often containing many zeros, more compactly and precisely. A number written in scientific notation consists of a decimal part and an exponential part (10 raised to a particular exponent).
scintillation counter A device used to detect radioactivity in which energetic particles traverse a material that emits ultraviolet or visible light when excited by their passage. The light is detected and turned into an electrical signal.
second (s) The SI standard unit of time.
secondary protein structure Short-range periodic or repeating patterns often found in proteins. Secondary protein structure is maintained by interactions between amino acids that are fairly close together in the linear sequence of the protein chain or adjacent to each other on neighboring chains.
semiconductor A compound or element exhibiting intermediate electrical conductivity that can be changed and controlled.
semipermeable membrane A membrane that selectively allows some substances to pass through but not others.
SI units The most convenient system of units for science measurements, based on the metric system. The set of standard units agreed on by scientists throughout the world.

significant digits (figures) The non–place-holding digits in a reported measurement; they represent the precision of a measured quantity.
simple carbohydrate (simple sugar) A monosaccharide or disaccharide.
single-displacement reaction A reaction in which one element displaces another in a compound.
solid A state of matter in which atoms or molecules are packed close to each other in fixed locations.
solubility The amount of a compound, usually in grams, that will dissolve in a certain amount of solvent.
solubility rules A set of empirical rules used to determine whether an ionic compound is soluble.
solubility-product constant (K_{sp}) The equilibrium expression for a chemical equation that represents the dissolving of an ionic compound in solution.
soluble Dissolves in solution.
solute The minority component of a solution.
solution A homogeneous mixture of two or more substances.
solution map In this book, a solution map is a visual outline of the solution to a problem.
solvent The majority component of a solution.
space-filling model A way to represent molecules in which atoms are represented with spheres that overlap with one another.
specific heat capacity (or specific heat) The heat capacity of a substance in joules per gram degree Celsius (J/g °C).
spectator ions Ions that do not participate in a reaction; they appear unchanged on both sides of a chemical equation.
standard temperature and pressure (STP) Conditions often assumed in calculations involving gases: $T = 0$ °C (273 K) and $P = 1$ atm.
starch A common polysaccharide composed of repeating glucose units.
state of matter One of the three forms in which matter can exist: solid, liquid, and gas.
steroid A biological compound containing a 17-carbon 4-ring system.
stock solution A concentrated form in which solutions are often stored.
stoichiometry The numerical relationships among chemical quantities in a balanced chemical equation. Stoichiometry allows us to predict the amounts of products that form in a chemical reaction based on the amounts of reactants.
strong acid An acid that completely ionizes in solution.
strong base A base that completely dissociates in solution.
strong electrolyte A substance whose aqueous solutions are good conductors of electricity.
strong electrolyte solution A solution containing a solute that dissociates into ions; therefore, a solution that conducts electricity well.
structural formula A two-dimensional representation of molecules that not only shows the number and type of atoms, but also how the atoms are bonded together.
sublimation A physical change in which a substance is converted from its solid form directly into its gaseous form.
subshell In quantum mechanics, specifies the shape of the orbital and is represented by different letters (s, p, d, f).
substituent An atom or a group of atoms that has been substituted for a hydrogen atom in an organic compound.
substitution reaction A reaction in which one or more atoms are replaced by one or more different atoms.
supersaturated solution A solution holding more than the normal maximum amount of solute.
surface tension The tendency of liquids to minimize their surface area, resulting in a "skin" on the surface of the liquid.
synthesis reaction A reaction in which simpler substances combine to form more complex substances; $A + B \longrightarrow AB$.
temperature A measure of the thermal energy in a sample of matter.
temporary dipole A type of intermolecular force resulting from transient shifts in electron density within an atom or molecule.
terminal atom An atom that is located at the end of a molecule or chain.
tertiary protein structure A protein's structure that consists of the large-scale bends and folds due to interactions between the *R* groups of amino acids that are separated by large distances in the linear sequence of the protein chain.
tetrahedral The molecular geometry of a molecule containing four electron groups (four bonding groups and no lone pairs).
theoretical yield The maximum amount of product that can be made in a chemical reaction based on the amount of limiting reactant.
theory A proposed explanation for observations and laws. A theory presents a model of the way nature works and predicts behavior that extends well beyond the observations and laws from which it was formed.
thermal energy A type of kinetic energy associated with the temperature-dependent random movement of atoms and molecules.
titration A laboratory procedure used to determine the amount of a substance in solution. In a titration, a reactant in a solution of known concentration is reacted with another reactant in a solution of unknown concentration until the reaction reaches the end point.
torr A unit of pressure named after the Italian physicist Evangelista Torricelli; also called a millimeter of mercury.
transition elements The elements in columns designated with B or in columns 3–12 of the periodic table.
transition metals The elements in the middle of the periodic table whose properties tend to be less predictable based simply on their position in the periodic table. Transition metals lose electrons in their chemical reactions, but do not necessarily acquire noble gas configurations.
triglyceride A fat or oil; a tryglyceride is a triester composed of glycerol with three fatty acids attached.
trigonal planar The molecular geometry of a molecule containing three electron groups, three bonding groups, and no lone pairs.
trigonal pyramidal The molecular geometry of a molecule containing four electron groups, three bonding groups, and one lone pair.
triple bond A chemical bond consisting of three electron pairs shared between two atoms. In general, triple bonds are shorter and stronger than double bonds.
ultraviolet (UV) light The fraction of the electromagnetic spectrum between the visible region and the X-ray region. UV light is invisible to the human eye.
units Previously agreed-on quantities used to report experimental measurements. Units are vital in chemistry.
unsaturated fat (or oil) A triglyceride composed of unsaturated fatty acids. Unsaturated fats tend to be liquids at room temperature.

unsaturated hydrocarbon A hydrocarbon that contains one or more double or triple bonds between its carbon atoms.

unsaturated solution A solution holding less than the maximum possible amount of solute under the solution conditions.

valence electrons The electrons in the outermost principal shell of an atom; they are involved in chemical bonding.

valence shell electron pair repulsion (VSEPR) A theory that allows prediction of the shapes of molecules based on the idea that electrons—either as lone pairs or as bonding pairs—repel one another.

vapor pressure The partial pressure of a vapor in dynamic equilibrium with its liquid.

vaporization The phase transition between a liquid and a gas.

viscosity The resistance of a liquid to flow; manifestation of intermolecular forces.

visible light The fraction of the electromagnetic spectrum that is visible to the human eye, bounded by wavelengths of 400 nm (violet) and 780 nm (red).

vital force A mystical or supernatural power that, it was once believed, was possessed only by living organisms and allowed them to produce organic compounds.

vitalism The belief that living things contain a nonphysical "force" that allows them to synthesize organic compounds.

volatile Tending to vaporize easily.

voltage The potential difference between two electrodes; the driving force that causes electrons to flow.

voltaic (galvanic) cell An electrochemical cell that creates electrical current from a spontaneous chemical reaction.

volume A measure of space. Any unit of length, when cubed, becomes a unit of volume.

wavelength The distance between adjacent wave crests in a wave.

weak acid An acid that does not completely ionize in solution.

weak base A base that does not completely dissociate in solution.

weak electrolyte A substance whose aqueous solutions are poor conductors of electricity.

work The result of a force acting on a distance.

X-rays The portion of the electromagnetic spectrum between the ultraviolet (UV) region and the gamma-ray region.

Credits

FM Page iii: Nivaldo Tro. **Chapter 1** Page 2: GIS/Fotalia. Page 3: From What Life Means to Einstein: An Interview by George Sylvester Viereck, The Saturday Evening Post (26 October 1929) pp. 17, 110–114 & 117. Copyright © 1929 by Saturday Evening Post Society. Page 3 (left center): Cynthia Johnson/The LIFE Images Collection/Getty Images. Page 4 (center): Richard Megna/Fundamental Photographs, NYC. Page 5 (top center): Maxwell Art And Photo. Page 5 (top right): De Visu/Shutterstock. Page 5 (top left): Jaroslav74/Shutterstock. Page 6 (bottom left): The Metropolitan Museum of Art./Art Resource, NY. Page 6 (bottom right): Bettmann/Getty Images. Page 7 (top left): IBM Almaden Research Center. Page 8 (top right): SPL/Science Source. Page 10 (bottom left): Wavebreak-mediaMicro/Fotolia. **Chapter 2** Page 14: NASA. Page 15: Quoted in Arthur Koestler and J. R. Smithies, Beyond Reductionism; New York: Houghton Mifflin, 1971, p. 115. Page 16 (top left): Kim Steele/Getty Images. Page 19 (top left): Richard Megna/Fundamental Photographs. NYC. Page 19 (top right): Richard Megna/Fundamental Photographs, NYC. Page 19 (bottom left): Richard Megna/Fundamental Photographs, NYC. Page 19 (bottom right): Richard Megna/Fundamental Photographs, NYC. Page 22 (top right): NASA. Page 26 (center): Richard Megna/Fundamental Photographs, NYC. Page 27 (top left): National Institute of Standards and Technology. Page 27 (top center): AFP/Stringer/Getty Images. Page 27 (top right): National Institute of Standards and Technology. Page 27 (left center): Richard Megna/Fundamental Photographs, NYC. Page 37 (bottom left): Pearson Education, Inc. Page 39 (bottom left): Rudy Umans/Shutterstock. Page 41 (center left): Gjermund/Shutterstock. Page 41 (center right): Richard Megna/Fundamental Photographs, NYC. Page 58 (right middle): NASA. Page 58 (left bottom): NASA Goddard Space Flight Center.
Chapter 3 Page 61: As quoted in Problems of Life (1952), by Ludwig von Bertalanffy, as reported in A Dictionary of Scientific Quotations (1991) edited by Alan L. Mackay, p. 219. Page 61: National Oceanic and Atmospheric Administration (NOAA). Page 61: U.S Energy Information Administration, Annual Energy Outlook 2015. Page 61: Etienne du Preez/Shutterstock. Page 62 (center left): John A. Rizzo/Photodisc/Getty Images. Page 62 (center right): Editorial Image, LLC/Alamy Stock Photo. Page 62 (bottom left): IBM Almaden Research Center. Page 62 (bottom right): Driscoll, Youngquist, & Baldeschwieler /Caltech/Science Source. Page 63 (top left): Marek Cech/Shutterstock. Page 63 (top center): John A. Rizzo/Getty Images. Page 63 (top right): Can Balcioglu/Fotolia. Page 63 (bottom center): The Natural History Museum/Alamy Stock Photo. Page 65 (top left): Richard Megna/Fundamental Photographs, NYC. Page 65 (top center left): Richard Megna/Fundamental Photographs, NYC. Page 65 (top center right): DenisLarkin/iStock/Getty Images. Page 65 (top right): Kip Peticolas/Fundamental Photographs, NYC. Page 65 (bottom center): Sveta/Fotolia. Page 66 (left center): ShutterPNPhotography/Shutterstock. Page 66 (bottom center): Getty Images. Page 68 (top left): Michael Dalton/Fundamental Photographs, NYC. Page 68 (center right): Siede Preis/Photodisc/Getty Images. Page 69 (bottom left): Richard Megna/Fundamental Photographs, NYC. Page 69 (bottom center): Richard Megna/Fundamental Photographs, NYC. Page 70 (bottom center): Etienne du Preez/Shutterstock. Page 72 (bottom left): JLGutierrez/Getty Images. Page 80 (bottom left): Jörg Hackemann/Fotolia. Page 81 (top right): Getty Images. Page 96 (center): Nasa. Page 96 (bottom left): Peeterv/iStockphoto/Getty Images. **Chapter 4** Page 98: Pearson Education Inc. Page 99: Paraphrase of Diogenes Laertius IX, 44. Trans. R. D. Hicks (1925), Vol. 2, 453–5. Page 100 (top left): Getty Images. Page 100 (top right): Karl Maret/Corbis. Page 100 (center): Apic/Hulton Archive/Getty Images. Page 100 (bottom right): IBM Almaden Research Center. Page 103 (left center): Maxwell Art And Photo/Pearson. Page 104 (bottom center): Jeremy Woodhouse/Getty Images. Page 105 (left center): Jupiter Images/Getty Images. Page 105 (right center): John A. Rizzo/Getty Images. Page 106 (left center): Charles D. Winters/Science Source. Page 106 (bottom right): Everett Historical/Shutterstock. Page 108 (top left): Popova Olga/Fotolia. Page 109 (bottom left): Fuse/Getty Images. Page 110 (left top): Graeme Dawes/Shutterstock. Page 110 (left bottom): John A. Rizzo/Getty Images. Page 110 (center top): Richard Megna/Fundamental Photographs, NYC. Page 110 (center bottom): Sciencephotos/Alamy Stock Photo. Page 111 (top left): Richard Megna/Fundamental Photographs, NYC. Page 111 (bottom left): Richard Megna/Fundamental Photographs, NYC. Page 111 (top right): Charles D. Winters/Science Source. Page 111 (bottom right): Pearson Education, Inc. Page 115 (bottom center): TEA/Fotolia. Page 118 (bottom center): U.S. Department of Energy. **Chapter 5** Page 132: Pearson Education Inc. Page 133: Excerpt from What Mad Pursuit: A Personal View Of Scientific Discovery by Francis Harry Compton Crick. Copyright © 1988 by Basic Books. Page 133 (bottom left): Pearson Education, Inc. Page 133 (bottom center): Charles D. Winters/Science Source. Page 133 (bottom right): Diane Diederich/Getty Images. Page 134 (bottom left): Sveta/Fotolia. Page 134 (bottom right): Mike Kemp/RubberBall/Alamy Stock Photo. Page 138 (bottom left): Getty Images. Page 139 (bottom left): Harry Taylor/Dorling Kindersley, Ltd. Page 139 (bottom right): Charles D. Winters/Science Source. Page 140 (bottom right): Universal Images Group North America LLC/Alamy Stock Photo. Page 141 (top left): Paul Silverman/Fundamental Photographs, NYC. Page 148 (bottom center): Maxwell Art And Photo/Pearson. Page 148 (bottom right): Maxwell Art And Photo/Pearson. Page 164 (top left): Charles D. Winters/Science Source. Page 164 (top right): Richard Megna/Fundamental Photographs, NYC. Page 164 (center left): Richard Megna/Fundamental Photographs, NYC. Page 164 (center right): Richard Megna/Fundamental Photographs, NYC. Page 164 (bottom left): Keystone/Hulton Archive/Getty Images. Page 164 (bottom right): molekuul.be/Fotolia. Page 165 (top left): Maxwell Art And Photo/Pearson. Page 165 (top right): Maxwell Art And Photo/Pearson. Page 165 (bottom left): Maxwell Art And Photo/Pearson. Page 165 (bottom right): Maxwell Art And Photo/Pearson. **Chapter 6** Page 168: Pearson Education Inc. Page 169: Excerpt from Voices In The Labyrinth: Nature, Man, And Science by Erwin Chargaff. Copyright © 1977 by Seabury Press. Page 169 (bottom left): Richard Kittenberger/Shutterstock. Page 170 (top left): Joseph P. Sinnot/Fundamental Photographs, NYC. Page 170 (left center): Richard Megna/Fundamental Photographs, NYC. Page 170 (bottom left): Richard Megna/Fundamental Photographs, NYC. Page 171 (top left): Maxwell Art And Photo. Page 171 (bottom left): Sveta/Fotolia. Page 173 (center left): Maxwell Art And Photo/Pearson. Page 173 (center right): Maxwell Art And Photo/Pearson. Page 173 (bottom left): Richard Megna/Fundamental Photographs, NYC. Page 173 (bottom right): Richard Megna/Fundamental Photographs, NYC. Page 179 (left center): Photodisc/Photodisc/Getty Images. Page 180 (top left): GK Hart/Vikki Hart/Photodisc/Getty Images. Page 180 (top center): Photodisc/Getty Images. Page 184 (top right): Maxwell Art And Photo. Page 187 (center): Charles D. Winters/Science Source. Page 204 (top left): Stocktrek/Photodisc/Getty Images. Page 204 (top right): Jeff Maloney/Photodisc/Getty Images. Page 204 (bottom left): NASA Johnson Space Center. Page 204 (bottom right): NASA Johnson Space Center. **Chapter 7** Page 206: Pearson Education, Inc. Page 207: Excerpt from The Two Faces of Chemistry by Luciano Caglioti. Copyright © 1983 by MIT Press. Page 208 (top left): Maxwell Art And Photo. Page 208 (bottom left): Pearson Education, Inc. Page 209 (top left): Richard Megna/Fundamental Photographs, NYC. Page 209 (top right): Charles D. Winters/Science Source. Page 209 (bottom right): Pearson Education, Inc. Page 209 (bottom left): Magnascan/Getty Images. Page 210 (top left): Richard Megna/Fundamental Photographs, NYC. Page 210 (bottom left): Richard Megna/Fundamental Photographs, NYC. Page 210 (center left): Richard Megna/Fundamental Photographs, NYC. Page 210 (top right): Jeff J. Daly/Fundamental Photographs, NYC. Page 210 (bottom right): Ayazad/Shutterstock. Page 213 (center): Sami Sarkis/Getty Images. Page 217 (top center): Ryan McVay/Photodisc/Getty Images. Page 217 (top right): Ryan McVay/Photodisc/Getty Images. Page 220 (bottom left): Richard Megna/Fundamental Photographs, NYC. Page 220 (bottom right): Richard Megna/Fundamental Photographs, NYC. Page 225 (bottom left): Pearson Education, Inc. Page 226 (top left): Richard Megna/Fundamental Photographs, NYC. Page 227 (top left): Pearson Education, Inc. Page 228 (center left): Pearson Education, Inc. Page 228 (center right): Martyn F. Chillmaid/Science Source. Page 231 (bottom left): Richard Megna/Fundamental Photographs, NYC. Page 231 (bottom center): Richard Megna/Fundamental Photographs, NYC. Page 231 (bottom right): Richard Megna/Fundamental Photographs, NYC. Page 232 (top left): Richard Megna/Fundamental Photographs, NYC. Page 246 (center left): Richard Megna/Fundamental Photographs, NYC. Page 246 (center right): Richard Megna/Fundamental Photographs, NYC. Page 246 (center bottom): Richard Megna/Fundamental Photographs, NYC.
Chapter 8 Page 248: Pearson Education, Inc. Page 249: Jacob Bronowski, Science and Human Values, New York: Harper & Row, 1965, p. 10. Page 250 (bottom left): Pearson Education, Inc. Page 250 (bottom center left): Pearson Education, Inc. Page 250 (bottom center right): Pearson Education, Inc. Page 250 (bottom right):

Pearson Education, Inc. Page 251 (top center): Pearson Education, Inc. Page 251 (top right): Pearson Education, Inc. Page 251 (center left): Pearson Education, Inc. Page 251 (center right): Pearson Education, Inc. Page 257 (center left): Pearson Education, Inc. Page 257 (center): Pearson Education, Inc. Page 257 (center right): Pearson Education, Inc. Page 265 (bottom left): Richard Megna/Fundamental Photographs, NYC. Page 265 (bottom center left): Richard Megna/Fundamental Photographs, NYC. Page 265 (bottom center right): Richard Megna/Fundamental Photographs, NYC. Page 265 (bottom right): Richard Megna/Fundamental Photographs, NYC. Page 282 (center right): Nancy R. Cohen/Photodisc/Getty Images. Page 282 (center left): Nancy R. Cohen/Photodisc/Getty Images. Page 282 (bottom right): Nancy R. Cohen/Photodisc/Getty Images. Page 282 (bottom left): Nancy R. Cohen/Photodisc/Getty Images. Page 282 (top center): Katrina Leigh/Shutterstock. Page 282 (center left): Katrina Leigh/Shutterstock. Page 282 (center right): Katrina Leigh/Shutterstock. Page 282 (bottom center): Katrina Leigh/Shutterstock. Page 282 (bottom left): Katrina Leigh/Shutterstock.
Chapter 9 Page 285: Murray Becker/AP Images. Page 285: From Warner Heisenberg (author), A. J. Pomerans (translator), Physics and Beyond: Encounters and Conversations (1971), New York: Harper Collins, p. 75. Page 286 (bottom left): Photo Researchers, Inc/Alamy Stock Photo. Page 286 (bottom right): Interfoto/Alamy Stock Photo. Page 287 (top left): Martin Barraud/Getty Images. Page 287 (bottom right): Zooropa/Shutterstock. Page 289 (bottom center): Sierra Pacific Innovations Corporation. Page 289 (bottom right): Sierra Pacific Innovations Corporation. Page 290 (bottom right): Larry Mulvehill/ Science Source. Page 291 (center left): Tea/Fotolia. Page 291 (bottom center): Richard Megna/Fundamental Photographs, NYC. Page 291 (bottom right): Richard Megna/Fundamental Photographs, NYC. Page 292 (top left): Richard Megna/Fundamental Photographs, NYC. Page 295 (left center): Stringer Corbis/Reuters. Page 295 (top right): Science Source. Page 321 (bottom left): Stephane Bonnel/Fotolia. Page 321 (bottom right): Library of Congress (Photoduplication).
Chapter 10 Page 325: G. N. Lewis and M. Randall, Thermodynamics, New York: McGraw-Hill, 1923. Page 340 (bottom left): Editorial Image, LLC/Alamy Stock Photo. Page 341 (top left): Kip Peticolas/Fundamental Photographs, NYC. Page 345 (top left): Richard Megna/Fundamental Photographs, NYC. Page 345 (bottom center): Richard Megna/Fundamental Photographs, NYC. Page 355 (right center): Tom Grundy/Shutterstock. **Chapter 11** Page 358: Pearson Education Inc. Page 359: From Sir W. Ramsay, The Gases of the Atmosphere, Ch. 1, p. 10, MacMillian & Company, London, 1905. He was quoting from Robert Boyle, Memoirs for a General History of the Air: Shaw's Abridgement of Boyle's works, edition 1725, vol. iii, p. 26. Page 366 (center left): Jerry Driendl/Getty Images. Page 371 (top left): PHB.cz/Fotolia. Page 371 (bottom left): Richard Megna/Fundamental Photographs, NYC. Page 376 (bottom left): Matka_Wariatka/ Shutterstock. Page 386 (top left): Pasang Geljen Sherpa/AP Images. Page 393 (bottom right): Walter Wolfe/LatitudeStock/Alamy Stock Photo. Page 393: U.S. EPA. Page 405 (bottom left): Caspar Benson/Brand X Pictures/Getty Images. Page 406 (center left): Pearson Education, Inc. Page 407: © 2016 The University of Manchester, all rights reserved. **Chapter 12** Page 409: Roger Joseph Boscovich, Philosophiae Naturalis Theoria (1763), sec. 1. 5, translation from James Mark Child, 1922. Page 411 (bottom right): Tom Brakefield/Photodisc/Getty Images. Page 412 (center left): Anthony West/Corbis Documentary/Getty Images. Page 412 (bottom right): Richard Megna/Fundamental Photographs, NYC. Page 413 (top right): Lauri Patterson/Vetta/Getty Images. Page 414 (bottom left): Richard Megna/Fundamental Photographs, NYC. Page 414 (bottom center): Richard Megna/Fundamental Photographs, NYC. Page 414 (bottom right): Richard Megna/Fundamental Photographs, NYC. Page 415 (center): Magnascan/Getty Images. Page 419 (bottom center): Richard Megna/Fundamental Photographs, NYC. Page 422 (top left): Charles D. Winters/Science Source. Page 422 (center): Charles D. Winters/Science Source. Page 425 (top left): Richard Megna/Fundamental Photographs, NYC. Page 425 (top center): Pearson Education, Inc. Page 425 (top right): Natalie Fobes/RGB Ventures/SuperStock/Alamy Stock Photo. Page 430 (bottom left): Kiesel Und Stein/Getty Images. Page 430 (bottom center): Matejay/Getty Images. Page 430 (bottom right): Sashkin/Shutterstock. Page 431 (center left): GeoStock/ Photodisc/Getty Images. Page 431 (center right): Sashkin/Shutterstock. Page 431 (bottom left): Maciej Toporowicz, NYC/Moment/Getty Images. Page 433 (top left): Kristen Brochmann/Fundamental Photographs, NYC. Page 433 (bottom right): Bill Pugliano/Stringer/Getty Images.
Chapter 13 Page 445: John D. Barrow, New Theories of Everything: The Quest for Ultimate Explanation, Oxford, University Press, 2007, Ch. 1, p. 10. Page 446 (center right): Louise Gubb/Contributor/Corbis Historical/Getty Images. Page 449 (center left): Richard Megna/Fundamental Photographs, NYC. Page 449 (center): Richard Megna/Fundamental Photographs, NYC. Page 449 (center right): Richard Megna/Fundamental Photographs, NYC. Page 450 (left center): Kevin hill illustration/iStock/Getty Images Plus. Page 451 (top left): Pearson Education, Inc. Page 463 (left middle): Nikola Obradovic/Shutterstock. Page 464 (left center): Jane/E+/Getty Images. Page 465 (top right): Masonjar/Shutterstock. Page 468 (bottom left): Eric Audras/ONOKY/Getty Images. Page 468 (bottom right): Derek Trask Inv. Ltd./Alamy stock photo. Page 469 (top left): Pearson Education, Inc. Page 469 (top center): Pearson Education, Inc. Page 469 (top right): Pearson Education, Inc. Page 481 (bottom left): Science Source.
Chapter 14 Page 485: Inorganic Chemistry: Principles of Structure and Reactivity, New York: Harper & Row, p. 207. Page 486 (top left): Gelpi/Shutterstock. Page 486 (top right): Richard Megna/Fundamental Photographs. Page 486 (center left): Richard Megna/Fundamental Photographs. Page 487 (center left): karandaev/Fotolia. Page 488 (top left): Pearson Education, Inc. Page 488 (center left): Richard Megna/Fundamental Photographs. Page 492 (top left): Richard Megna/Fundamental Photographs. Page 492 (bottom left): Richard Megna/Fundamental Photographs. Page 494 (center left): Pearson Education, Inc. Page 494 (center right): Pearson Education, Inc. Page 496 (top left): Richard Megna/Fundamental Photographs. Page 496 (top center): Richard Megna/Fundamental Photographs. Page 496 (top right): Richard Megna/Fundamental Photographs. Page 498 (bottom left): Richard Megna/Fundamental Photographs. Page 498 (bottom right): Richard Megna/Fundamental Photographs. Page 499 (bottom left): Richard Megna/Fundamental Photographs. Page 499 (bottom right): Richard Megna/Fundamental Photographs. Page 510 (top right): Konstantin Christian/Shutterstock. Page 512 (center left): Avantgarde/Fotolia.
Chapter 15 Page 526: Pearson Education, Inc. Page 527: From Rudolf Arnheim, Entropy and Art: An Essay on Disorder and Order, p. 25, University of California Press, 1971. Page 531 (top left): Starover Sibiriak/Shutterstock. Page 548 (center right): Claude Edelmann/Science Source. Page 550 (top left): Richard Megna/Fundamental Photographs. Page 550 (top right): Richard Megna/Fundamental Photographs. **Chapter 16** Page 572: Pearson Education Inc. Page 573: From John Desmond Bernal, The Origin of Life, 1967, London: Weidenfeld & Nicholson, p. 163. Page 573 (left center): Toyota Motor Corporate Services of N. America,Inc. Page 574 (bottom left): Ron Zmiri/Shutterstock. Page 574 (bottom right): Yuratosno/Fotolia. Page 579 (top right): Sean De Burca/Photodisc/Getty Images. Page 585 (bottom right): Yellowj/Shutterstock. Page 586 (top center): Richard Megna/Fundamental Photographs. Page 586 (top right): Richard Megna/Fundamental Photographs. Page 586 (left center): Richard Megna/Fundamental Photographs. Page 586 (bottom left): Chris Cheadle/The Image Bank/Getty Images. Page 588 (left center): Richard Megna/Fundamental Photographs. Page 590 (bottom left): Wayne Barrett/Anne MacKay/All Canada Photos/Getty Images. Page 590 (bottom right): David Dent/Alamy Stock Photo. Page 594 (top left): Charles D. Winters/Science Source. Page 595 (bottom right): ImageBROKER/Alamy Stock Photo. Page 596 (top left): Dorling Kindersley Limited. Page 605 (bottom right): Tony Mcnicol/Alamy Stock. **Chapter 17** Page 609: Winston Churchill, "Fifty Years Hence," speech reprinted in Popular Mechanics, Vol. 5, No. 3, March 1932. p. 395. Page 610 (top left): Southern Illinois University/Science Source. Page 610 (bottom left): Stringer/AFP/Getty Images. Page 611 (top left): Richard Megna/Fundamental Photographs. Page 618 (center left): United States Air Force. Page 618 (center right): Hank Morgan/Science Source. Page 621 (top right): US Environmental Protection Agency. Page 623 (bottom right): Associated Press. Page 625 (top right): Emilio Segre' Visual Archives NOT IN THE PUBLIC DOMAIN. MUST LICENSE AS NEW FROM THE SOURCE. Page 625 (bottom left): Stocktrek/Photodisc/Getty Images. Page 626 (left center): H. Mark Weidman Photography/Alamy Stock Photo. Page 626 (bottom center): Associated Press. Page 627 (left center): Associated Press. Page 630 (top left): Roger Tully/The Image Bank/Getty Images. Page 620 (bottom right): Agencja Fotograficzna Caro/Alamy Stock Photo. **Chapter 18** Page 640: Pearson Education, Inc. Page 641: Epicurus, as reported by Lactantius, Divine Institutes, Book 3, Chapter 19, c. 303–311. Translated by Max Mueller, Science of Language, 1871, II, Longmans, Green; p. 81. Page 640 (top left): Alex L. Fradkin/Stockbyte/Getty Images. Page 642 (right center): Josie Elias/Stockbyte/Getty Images. Page 643 (top center): Ryan McVay/Photodisc/Getty Images. Page 643 (top right): C Squared Studios/Photodisc/Getty Images. Page 644 (bottom right): David A. Hardy/Science Source. Page 647 (bottom right): Blend Images/Alamy Stock Photo. Page 648 (top left): David Toase/Photodisc/Getty Images. Page 650 (right image 1): Ayazad/Shutterstock. Page 650 (right image 2): Jonathan Nourok/PhotoEdit. Page 650 (right image 3): David Buffington/Photodisc/Getty Images. Page 650 (right image 4): Maxwell Art And Photo. Page 650 (right image 5): Getty Images. Page 650 (right image 6): David Toase/Photodisc/Getty Images. Page 655 (left center): Melissa Brandes/Shutterstock. Page 655 (bottom left): Arthur S. Aubry/Stockbyte/Getty Images. Page 660 (top left): Fundamental Photographs. Page 664 (bottom left): StudioOneNine/Shutterstock. Page 665 (top left): Maxwell Art And Photo. Page 667 (bottom center): Spencer Jones/Photodisc/Getty Images. Page 667 (bottom right): Maxwell Art And Photo. Page 668 (bottom center): GrigoryL/Shutterstock. Page 668 (bottom right): C Squared Studios/Stockbyte/Getty Image. Page 669 (bottom left): Photodisc/Getty Images. Page 670 (top left): C Squared Studios/Photodisc/Getty Images. Page 670 (bottom right): Lawrence Lawry/Stockbyte/Getty Images. Page 672 (top left): Rubber-Ball/Alamy Stock Photo. Page 673 (top center): Maxwell Art And Photo. Page 674 (bottom right): Tro, Nivaldo Jose. **Chapter 19** Page 694: Huntstock/Getty Images. Page 695: From Scientific American, The Molecules of Life, Vol. 253, No. 4, Oct. 1985, p. 57. Page 698 (top left): Spencer Jones/Photodisc/Getty Images. Page 698 (bottom left): GeoStock/Photodisc/Getty Images. Page 712 (top left): Eye of Science/Science Source. Page 714 (center left): Tro, Nivaldo Jose. Page 714 (center right): Tro, Nivaldo Jose. Page 720 (bottom left): Science Source.

Index

C

F

G

Q

R

T

U

Fundamental Physical Constants

Atomic mass unit	1 amu = 1.660539×10^{-27} kg 1 g = 6.022142×10^{23} amu
Avogadro's number	$N_A = 6.022142 \times 10^{23}$/mol
Electron charge	$e = 1.602176 \times 10^{-19}$ C
Gas constant	$R = 8.314472$ J/(mol · K) = 0.0820582 (L · atm)/(mol · K)
Mass of electron	$m_e = 5.485799 \times 10^{-4}$ amu $= 9.109382 \times 10^{-31}$ kg
Mass of neutron	$m_n = 1.008665$ amu $= 1.674927 \times 10^{-27}$ kg
Mass of proton	$m_p = 1.007276$ amu $= 1.672622 \times 10^{-27}$ kg
Pi	$\pi = 3.1415926536$
Planck's constant	$h = 6.626069 \times 10^{-34}$ J · s
Speed of light in vacuum	$c = 2.99792458 \times 10^{8}$ m/s

Useful Geometric Formulas

Perimeter of a rectangle = $2l + 2w$

Circumference of a circle = $2\pi r$

Area of a triangle = (1/2)(base × height)

Area of a circle = πr^2

Surface area of a sphere = $4\pi r^2$

Volume of a sphere = $(4/3)\pi r^3$

Volume of a cylinder or prism = area of base × height

Important Conversion Factors

Length: SI unit = meter (m)

- 1 m = 39.37 in.
- 1 in. = 2.54 cm (exactly)
- 1 mile = 5280 ft = 1.609 km
- 1 angstrom (Å) = 10^{-10} m

Volume: SI unit = cubic meter (m^3)

- 1 L = 1000 cm^3 = 1.057 qt (U.S.)
- 1 gal (U.S.) = 4 qt = 8 pt
  = 128 fluid ounces
  = 3.785 L

Mass: SI unit = kilogram (kg)

- 1 kg = 2.205 lb
- 1 lb = 16 oz = 453.6 g
- 1 ton = 2000 lb
- 1 metric ton = 1000 kg = 1.103 tons
- 1 g = 6.022×10^{23} atomic mass units (amu)

Pressure: SI unit = pascal (Pa)

- 1 Pa = 1 N/m^2
- 1 bar = 10^5 Pa
- 1 atm = 1.01325×10^5 Pa (exactly)
  = 1.01325 bar
  = 760 mmHg
  = 760 torr (exactly)

Energy: SI unit = joule (J)

- 1 J = 1 N · m
- 1 cal = 4.184 J (exactly)
- 1 L · atm = 101.33 J

Temperature: SI unit = kelvin (K)

- K = °C + 273.15
- °C = (5/9) (°F − 32°)
- °F = (9/5) (°C) + 32°